CHEMISTRY FOR PHARMACY AND THE LIFE SCIENCES
Including Pharmacology and Biomedical Science

GARETH THOMAS
School of Pharmacy and Biomedical Science,
University of Portsmouth

PRENTICE HALL
LONDON   NEW YORK   TORONTO   SYDNEY   TOKYO   SINGAPORE
MADRID   MEXICO CITY   MUNICH

First published 1996 by
Prentice Hall Europe
Campus 400, Maylands Avenue
Hemel Hempstead
Hertfordshire, HP2 7EZ
A division of
Simon & Schuster International Group

Printed and bound in Great Britain by
T.J. International Ltd., Padstow Cornwall

Library of Congress Cataloging-in-Publication Data

Thomas, Gareth
  Chemistry for Pharmacy and the life sciences : including
pharmacology and biomedical science/ Gareth Thomas.
      p.   cm.
  Includes bibliographical references and index.
  ISBN 0-13-131699-0
  1.   Chemistry   1.   Title
   (DNLM: 1. Chemistry, Pharmaceutical.   QV 744 T456c 1996)
 QD33. T48 1996
  340' 245746-dc20
  DNLM/DLC
for Library of Congress   95-31684
                                                                    CIP

British Library Cataloguing in Publication Data

A catalogue record for this book is available from the British Library
ISBN 0-13-131699-0

---

3   4   5   00   99   98

CHEMISTRY FOR PHARMACY
AND THE LIFE SCIENCES

Contents

Contents

Foreword

As we approach the end of this century it is clear that the term 'multidisciplinary' most effectively describes the favoured approach to undergraduate teaching in the life sciences. Undergraduate courses ranging from nursing and radiography to pharmacy, dentistry and medicine all require chemistry as one of their key core subjects. However, all too often course managers are faced with the problem of finding a chemistry text of the appropriate level that contains biological or pharmaceutical information relevant to their particular course. Dr Thomas, who has considerable teaching experience in the medicinal chemistry area, has addressed this problem by providing a text that should be useful for life science undergraduates for the duration of their degree studies and beyond. The chemistry in this book is pitched at just the right level for life science students and is presented in a lively and engaging style with numerous diagrams and figures. Furthermore, the text is supported throughout by a large number of examples of either the use of a drug, or the application of a general chemical principle. Learning is consolidated through the use of an *aide memoire* section and self-test questions at the end of each chapter. In this sense the book is unique and should represent a significant contribution to life science curricula.

David E Thurston
Professor of Medicinal Chemistry
School of Pharmacy and Biomedical Science
University of Portsmouth

Preface

This text is written for the first and second years of pharmacy, biomedical sciences, pharmacology and other life science degree courses. It is also suitable as a reference text for organic chemistry and similar degree courses.

The book aims to give the reader a vocabulary that will enable him or her to understand most of the chemistry used in the life sciences literature. It revises and extends the concepts and factual information that has been covered in preliminary courses. The text is not intended to be fully comprehensive but is designed to provide the reader with sufficient background to understand much of the chemistry used in the life sciences.

Chapters 1-11 cover the chemical vocabulary used in the main areas of the life sciences and a limited review of the physical chemistry of reactions and solutions that are of importance to the pharmacist and life scientist. In addition, Chapter 11 describes analytical procedures of particular interest to pharmacists

Chapters 12-29 use the vocabulary developed in the earlier chapters to describe and discuss the chemistry of the principal classes of organic compounds.

Chapters 30-33 introduce the reader to the occurrence and uses of inorganic substances in pharmacy and the life sciences. Chapter 33 also introduces the fundamental terminology used in radiochemistry.

The complete text is supported by a further reading section, classified according to subject, at the end of the book. Cross references are normally given in brackets within the text.

The statements in this text, like all similar books, are based on generalisations so the reader should not be surprised when in practice they meet specific exceptions to these statements. During your working life you will increase your knowledge of these exceptions; it is called *gaining experience*. The text uses both the systematic and trivial nomenclature in common use in chemistry. Where appropriate the nomenclature in current use in pharmacy and the relevant life science discipline is used in preference to that used by chemists. Examples have been selected on the basis of illustrating the chemistry as well as current usage. Throughout the text the term R is used to represent any carbon hydrogen skeleton but the term Ar is restricted to aromatic species.

It is important to realise that reading this text is only the begining of learning chemistry; one must practice using the subject in order to understand chemistry and become proficient in its use. Theoretical problem solving is one of the ways you can practise using the subject, and so a selection of graded questions is given at the end of each of Chapters 2 to 33. Answers to many of these questions are given on pages 597 to 606. I cannot over-emphasise that practising by putting pen to paper is one of the secrets of coming to grips with the subject and passing examinations. Remember, if you are not making mistakes you are not trying hard enough!

Each chapter also includes an *aide memoire* entitled *What you need to know* to emphasise the points you need to be familiar with in each section of the subject. The ***bold italic script*** used in the *What you need to know* sections indicates that this fact should be memorised. Information given in normal script is useful to know but is of lesser importance.

Acknowledgements

It gives me great pleasure to thank my colleagues and many other people without whose help and encouragement this text would never have been written. I would like in particular to mention Miss L Adams, Dr E Allen, Mr J Banks, Dr V Bunker, Dr B Carpenter, Mr J Corkett, Dr C Dacke, Dr A Griffiths, Mr D Gullick, Dr P Howard, Mr A Hunt, Dr G Hall, Dr N Johnson, Miss M Lawes, Dr A Lloyd, Miss S Lobo, Mrs S Lucas, Mr R Mortlock, Dr G Otterwell, Miss J Pennock, Dr W Mclean, Dr D McLellan, Mr T Meek, Dr J Morrison, Mr S Moore, Dr M Norris, Dr A Patel, Dr C Richards, Dr O Roch, Professor D Rogers, Dr J Smart, Dr A Thompson, Dr G White, Mr S Wilson and all the others over the years whose names I have forgotten but whose advice and help have all contributed to this book.

In addition, special thanks are due to Dr R Jones, Dr A O Plunkett and Mrs P Silson who took the trouble to read and comment on the complete manuscript, Professor D Thurston for his encouragement and help in the early stages and showing me which keys to press on the computer, Richard, Matt, Penny and Simon of the computer support staff for the help they gave me in operating the mainframe computer. Finally, but not least, I would like to thank my wife for putting up with being a book widow for the past four years. Without her help and understanding I certainly could not have finished this manuscript.

1

Introduction

1.1 WHY CHEMISTRY?

The simple answer to the question 'why chemistry' is that chemistry is an essential foundation subject for the modern life sciences. We rely on chemistry to explain biological processes. It is now essential that anyone working in this field has a sound knowledge of the vocabulary, principles and applications of chemistry.

Some of the general areas of the, pharmaceutical, biomedical, biological and other life sciences in which chemistry is involved include:

(1)   The design of new drugs.
(2)   The manufacture and quality control of known drugs.
(3)   The formulation of drugs into preparations that can be administered to patients, for example, dosage forms such as injections, tablets and creams.
(4)   Understanding how drugs work at the molecular level (pharmacology).
(5)   The diagnosis and treatment of diseases.
(6)   Forensic investigations.
(7)   Research in the pharmaceutical, pharmacological, biomedical and other health sciences.

It is not possible in the space available to provide detailed examples of the use of chemistry in all these areas but an idea of its importance can be gained from a consideration of two case histories. Most of us will have met people in similar situations.

Case 1. Sue is a diabetic. Prior to 1922 she would have suffered a slow death. Today Sue leads a normal active life.

Diabetes occurs if the pancreas fails to produce sufficient insulin (Fig. 1.1) to control the concentrations of glucose and fatty acids in the patient's blood. Patients with diabetes have high glucose concentrations in their blood and urine. Chemical testing of the urine for these compounds is the first step in diagnosing the disorder. Diabetics must control the level of glucose in their blood either by following a low sugar diet and/or by injecting insulin. Sue leads an active life because she is able to carry out her own glucose checks and so regulate her diet and insulin usage.

Investigation of human insulin has revealed its structure (Fig. 1.1) and this knowledge has made it possible for insulin to be produced artificially by genetic engineering techniques.

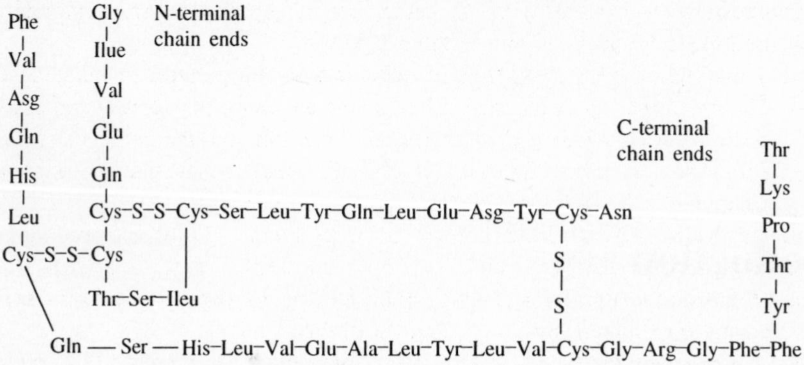

```
Phe        Gly    N-terminal
 |          |     chain ends
Val        Ilue
 |          |
Asg        Val
 |          |                                                          C-terminal
Gln        Glu                                                         chain ends    Thr
 |          |                                                                         |
His        Gln                                                                       Lys
 |          |                                                                         |
Leu        Cys–S–S–Cys–Ser–Leu–Tyr–Gln–Leu–Glu–Asg–Tyr–Cys–Asn                      Pro
 |          |                                                   |                     |
Cys–S–S–Cys                                                     S                    Thr
 |                                                                                    |
Thr–Ser–Ileu                                                   S                    Tyr
  \                                                             |                     |
   Gln — Ser —His–Leu–Val–Glu–Ala–Leu–Tyr–Leu–Val–Cys–Gly–Arg–Gly–Phe–Phe
```

Fig. 1.1. The structure of human insulin. The abbreviations, ala, gly, cys etc. refer to the structures of the amino acid residues that constitute the structure of the protein insulin. Full details of their use in representing the structure of a protein are given in 28.7.1.

Case 2. At birth John was found to have a disorder known as phenylketonuria. This, if untreated, causes irreversible brain damage leading to mental retardation. John is now twenty and perfectly healthy.

Chemical and biological investigations have shown that phenylketonuria is caused by an abnormal concentration of the amino acid phenylalanine in the blood. It is possible to test chemically for either the presence of phenylalanine in the blood of a patient or its **metabolite,** phenylpyruvic acid, in their urine. In Britain, all infants are screened routinely for phenylketonuria within a few days of birth. A small quantity of blood is taken from their heels and analysed for phenylalanine. John was diagnosed as positive when he was a young baby. His treatment was a restricted diet that kept his blood concentration of

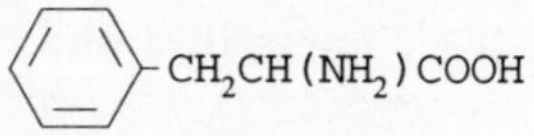

$$\text{—CH}_2\text{CH(NH}_2\text{)COOH}$$
Phenylalanine

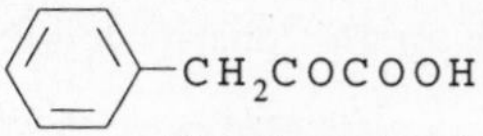

$$\text{—CH}_2\text{COCOOH}$$
Phenylpyruvic acid

phenylalanine at the correct level. Since John's parents ensured that he kept to his dietary regime his brain developed normally. Now that he has reached adulthood and his brain is fully developed John's diet has become more varied. He is now twenty and perfectly healthy. Today, there are many people in similar situations to Sue and John. They owe the quality of their lives to the doctors and scientists who studied these and other diseases. A knowledge of chemistry played a crucial role in these investigations.

This text attempts to introduce the essential vocabulary of organic and inorganic chemistry required by students and professionals in the pharmaceutical, pharmacological biomedical, biological and other life science areas. Explanations are kept as simple as possible and details to a minimum.

1.2 CHEMICAL THEORIES

Chemistry becomes much easier to understand if one has a general appreciation of the term 'theory'. Initially a theory starts out as an idea or series of ideas that are proposed to correlate and explain a large number of practical observations. At this stage it is referred to as a hypothesis and is the subject of much discussion within the scientific community. However, once a hypothesis has become established, that is, accepted by the scientific community at large, it is used to predict and communicate information.

A theory may be expressed in terms of either words and pictures or mathematical equations. Theories based on mathematical concepts are often interpreted and used in chemistry in the form of words and diagrams. For example, the mathematically based quantum theory for the structure of the atom is normally used in a pictorial form in organic chemistry (Chapter 2).

Theories are based on the statistical analysis of a large number of practical observations. Theories frequently involve the use of concepts that are not fully understood. For example, the physical nature of the electron is not fully understood but we still use electrons to explain a variety of scientific observations.

No theory is ever perfect. Its accuracy will vary from situation to situation. For example, the **valence-shell electron-pair repulsion theory** or **VSEPR** theory (the Sidgewick-Powell theory) predicts that the bond angles of methane, ammonia and water are all 109.5°. However, experimental measurements clearly show that in some instances there is a significant difference between the theoretical and practically determined values (Table 1.1). It is always important to appreciate the accuracy of a theory within the context of its use. This only comes with experience. In this text we shall attempt to indicate the limitations of the various theories presented as they arise. However, the reader should always bear in mind that theoretical deductions that do not agree with the experimental results may be due to inaccuracies in the theory and are not necessarily due to inaccurate use of the theory.

Table 1.1. A comparison of the bond angles predicted using the VSEPR theory with the values obtained by experiment for some simple molecules.

Compound	VSEPR bond angle	Practically determined bond angle
Methane	109.5°	109.5°
Ammonia	109.5°	107.0°
Water	109.5°	104.5°

Practical results can often be discussed in terms of more than one theory. However, one normally uses the theory most appropriate for the situation. For example, organic chemists consider the electron to be a negatively charged particle when discussing mechanisms of reactions but use the wave motion concept to describe molecular structures (Chapter 2).

1.3 BIOLOGICAL ACTIVITY

Until the late eighteenth and early nineteenth centuries it was thought that *supernatural vital forces* were responsible for the behaviour of living cells, and that organic molecules could only be made with the aid of these vital forces. This theory was discredited in 1828 when Friedrich Wohler discovered that it was possible to synthesise the organic molecule urea, from an inorganic compound, ammonium cyanate.

$$NH_4^+ \ OCN^- \xrightarrow{\text{heat}} NH_2CONH_2$$

Ammonium cyanate   Urea

It is now widely accepted that the processes of life itself and the biological activity of drug molecules are all based on chemical changes. The majority of these changes involve organic compounds and reactions although inorganic species also play a vital role in physiological processes. The principal factors governing these chemical changes are:

(1) The energy of the system.
(2) The functional groups present in organic molecules.
(3) The overall shapes of the chemical species (molecules, ions, free radicals) involved.

The nature and effect of each of these factors will be introduced in sections 1.4, 1.5 and 1.6, and further developed in the text as the need arises.

1.4 THE ENERGY OF THE SYSTEM

Reactions can only occur if chemical species collide. However, it has been shown that not every collision results in a reaction. This observation has been explained by postulating that a reaction occurs only if the total energy of the colliding species is above a specific minimum value. This minimum value is known as either the **free energy of activation** (ΔG^*) or the **activation energy** (E_a) of the process.

The minimum changes in the total energy of the system that occur during a reaction are shown on energy/reaction ordinate plots. The reaction ordinate shows the sequence of events in the reaction but not the actual times at which they occur. In any reaction mixture some of the reactants will rapidly obtain the required activation energy and react quickly whilst others will take longer to acquire the required activation energy and so will react more slowly. However, all the reacting species will follow the same sequence of events. Typical minimum energy pathways for simple chemical reactions are shown in Fig. 1.2 and Fig.1.4. In both cases the reactants must first gain a minimum total energy of ΔG^*, equal to the free energy of activation of the process before they can react. Once their total energy is equal to or greater than the activation energy, reaction occurs and the system loses energy to the surroundings. If the loss in energy is less than that gained to overcome the activation energy there will be an overall absorption of energy by the system (Fig. 1.2). If the energy is absorbed in the form of heat the reaction will be **endothermic** and an external observer would experience a cooling effect. If on the other hand more energy is liberated than was absorbed to overcome the activation energy, there will be an overall loss of energy from the system (Fig. 1.3). In this case the reaction will be **exothermic** if the energy is liberated as heat, and an external observer would experience an increase in temperature. Energy is usually liberated or absorbed in the form of heat in the majority of chemical reactions.

The position 'x' in both Fig. 1.2 and Fig. 1.3 is known as the **transition state** or **activated complex**. It represents the least stable arrangement of the species present at that instant in the reaction sequence. The transition state is a transitory state that two or more reacting species pass through on their way to the product(s). It does not necessarily

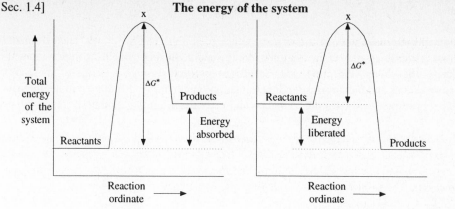

Fig. 1.2. The minimum energy pathway of a simple endothermic process.

Fig. 1.3. The minimum energy pathway of a simple exothermic process.

represent an actual intermediate that can be detected or isolated. Consider, for example the reaction of methyl bromide with sodium hydroxide (Fig. 1.4). The transition state for this reaction is given in the square brackets. It consists of the hydroxide ion moving in towards the carbon, the bromine beginning to leave, while at the same time the hydrogens are passing through a plane at right angles to the plane of the paper. Bonds that are in the process of breaking and being formed are drawn as dotted lines.

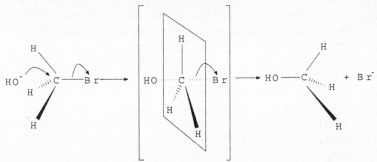

Fig. 1.4. The mechanism of the reaction of bromomethane (methyl bromide)with sodium hydroxide. Each arrow indicates the movement of two electrons during the course of the reaction. Curly arrows are useful tools used by chemists to describe the breaking and forming of bonds (See Chapter 8).

Most reactions involve the formation of intermediate species. Consider, for example, the reaction of 2-bromo-2-methylpropane [(a) below] with sodium hydroxide to form 2-methylpropan-2-ol (14.5.1 for mechanistic details).

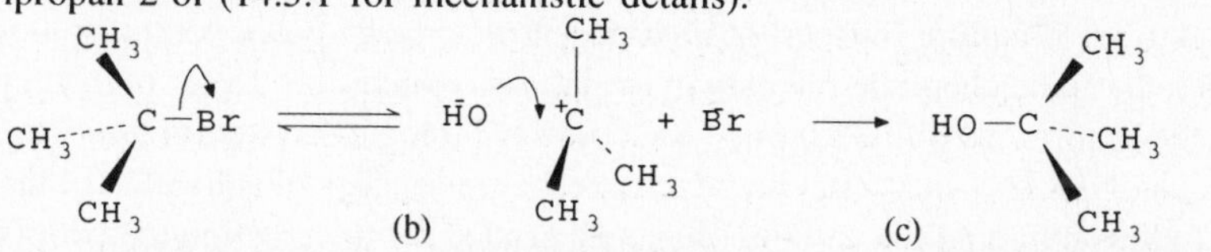

(a) 2-Bromo-2-methylpropane

(b) Metastable carbocation ion intermediate

(c) 2-Methylpropan-2-ol

The first step in this reaction is the formation of the planar carbocation intermediate (b). This intermediate is sufficiently stable to exist in the reaction mixture long enough for it to collide with a hydroxide ion and react to form the product (c).

Unstable intermediates are known generally as metastable intermediates. Their formation corresponds to a valley in the minimum energy curve, the more stable the intermediate the deeper the valley (Fig. 1.5). Most reactions proceed through a series of metastable intermediates and their corresponding transition states. In Fig. 1.6, for example, there are two transition states, x_1 for the formation and x_2 for the decomposition of the intermediate.

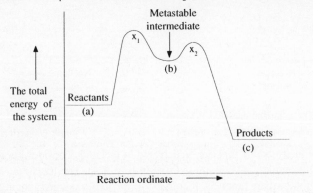

Fig.1.5. The minimum energy pathway of the reaction of 2-bromo-2-methylpropane with sodium hydroxide.

The energy needed to bring about a reaction in the chemistry laboratory is obtained by using the appropriate reaction conditions such as heat, light or pressure. In biological systems chemical reactions occur under milder conditions than those used for the same reaction in the laboratory. For example, in humans glucose is converted by a series of

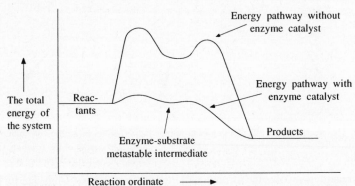

Fig. 1.6. The effect of an enzyme on the minimum energy pathway of a reaction.

chemical transformations into carbon dioxide and water at body temperature (about 37°C). However, the same chemical reaction in the laboratory requires a very high temperature, such as the heat from a Bunsen burner. Reactions in biological systems require less energy because catalysts known as **enzymes** decrease the free energy of activation of the process (Fig. 1.6) thus reducing the amount of energy needed for the reaction to occur (1.6).

1.5 FUNCTIONAL GROUPS

A functional group is part of the structure of a molecule where most of its reactions occur. The structures of functional groups often include atoms of other elements besides carbon (Table 1.2).

Table 1.2. A selection of the functional groups discussed in this book.

Functional group	Structural formula	Functional group	Structural formula
Chloro	$-Cl$	Amino	$-NH_2$
Alcohol	$-OH$	Thiol	$-SH$
Ether	$-O-$	Thioether	$-S-$
Sulphoxide	$-SO-$	Sulphone	$-SO_2$
Aldehyde	$-CHO$	Ketone	$-CO-$
Carboxylic acid	$-COOH$	Thiolic acid	$-COSH$
Ester	$-COO-$	Thioester	$-COS-$

The structures of organic compounds usually contain one or several different types of functional group (multifunctional group compounds). Compounds are classified according to the nature of these functional groups and the type of carbon-hydrogen skeleton to which they are attached. For example, compounds containing simple hydroxyl functional groups ($-OH$) are classified as either phenols or alcohols. In phenols the hydroxyl group is directly attached to a benzene ring whilst in alcohols it is bonded to a different type of skeleton (Fig. 1.7). Alcohols are further subdivided into primary, secondary and tertiary according to the nature of this carbon skeleton.

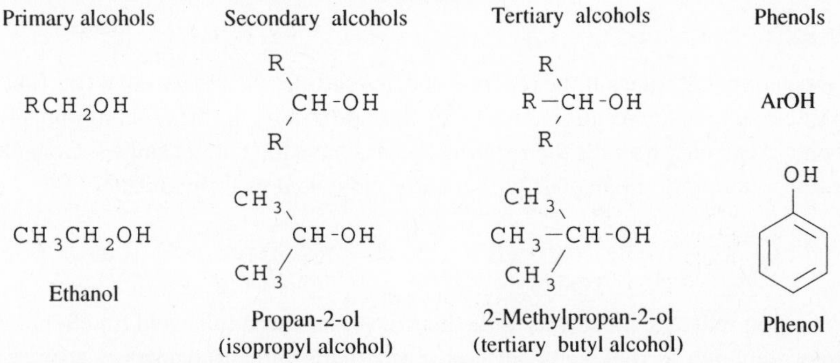

Primary alcohols

RCH_2OH

CH_3CH_2OH

Ethanol

Secondary alcohols

$$\begin{array}{c} R \\ \diagdown \\ CH-OH \\ \diagup \\ R \end{array}$$

$$\begin{array}{c} CH_3 \\ \diagdown \\ CH-OH \\ \diagup \\ CH_3 \end{array}$$

Propan-2-ol
(isopropyl alcohol)

Tertiary alcohols

$$\begin{array}{c} R \\ \diagdown \\ R-CH-OH \\ \diagup \\ R \end{array}$$

$$\begin{array}{c} CH_3 \\ \diagdown \\ CH_3-CH-OH \\ \diagup \\ CH_3 \end{array}$$

2-Methylpropan-2-ol
(tertiary butyl alcohol)

Phenols

ArOH

OH

Phenol

Fig. 1.7. Classification of compounds containing simple hydroxyl functional groups. R is currently used to represent all types of carbon hydrogen skeleton; however, Ar is only used to represent aromatic structures. It should be noted that R can represent similar or different structures.

Practical observations have shown that the same type of functional group will often react in the same manner with a particular chemical reagent irrespective of the structure of the rest of the molecule. This behaviour enables general rules concerning the nature of the chemical properties of a functional group to be formulated. These rules can be used to predict the possible course of a chemical reaction under both biological and laboratory conditions. For example, an investigation of the oxidation of the hydroxyl groups of a large number of secondary alcohols shows that in almost all cases the product is the corresponding ketone. From this we deduce that the most likely product of the oxidation of any secondary alcohol will be the appropriate ketone. For example, it would be possible to predict that the oxidation of 2-propanol should yield the ketone, propanone. In practice, this transformation has been found to occur under both laboratory and biological conditions.

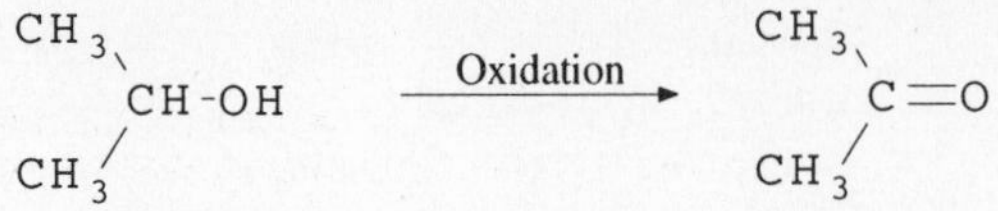

$$CH_3 \diagdown \atop CH_3 \diagup CH-OH \quad \xrightarrow{\text{Oxidation}} \quad CH_3 \diagdown \atop CH_3 \diagup C=O$$

Propan-2-ol   Propanone

Predictions of this type work well for simple compounds but are less accurate with more complex structures, especially those which contain a number of related functional groups. For example, in a large complex molecule, the shape of part of the molecule may prevent a reactant reaching a functional group and so no reaction can occur. This process is termed **steric hindrance** and is discussed in more detail in section 1.6.

It is not possible to memorise all known reactions and students may be put off chemistry by the mistaken belief that they have to. However, **it is feasible to learn the general reactions of functional groups and then use them to predict the outcome of other reactions**. It should be realised that the method is not always accurate and that predictions made for reactions in biological systems are usually less accurate than those made for reactions carried out under laboratory conditions because the involvement of enzymes can sometimes change the course of a reaction as well as reduce its activation energy. This book incorporates a *What you need to know* section at the end of each chapter. This section contains the general information that allows you to predict the theoretical outcome of many reactions.

1.6 SHAPE

Reactions can only take place if the reacting species collide. This means that the functional groups participating in the reaction must be in close proximity at 'the point of impact'. This type of contact can only occur if the reacting species have the correct shapes. Consider, for example, the reaction of an alcohol with a carboxylic acid to form an ester:

$$ROH \quad + \quad R'COOH \quad \underset{}{\overset{H^+}{\rightleftharpoons}} \quad R'COOR \quad + \quad H_2O$$

An alcohol   A carboxylic acid   An ester

This reaction will take place only if the hydroxyl and carboxylic acid functional groups can get close enough to make contact and react. If the three-dimensional shapes of the molecules to which these groups are attached prevent this contact, reaction will not occur (Fig. 1.8).

(a) No steric hindrance and
so reaction occurs

(b) Steric hindrance since
the acid and alcohol are
prevented from getting
close enough to react

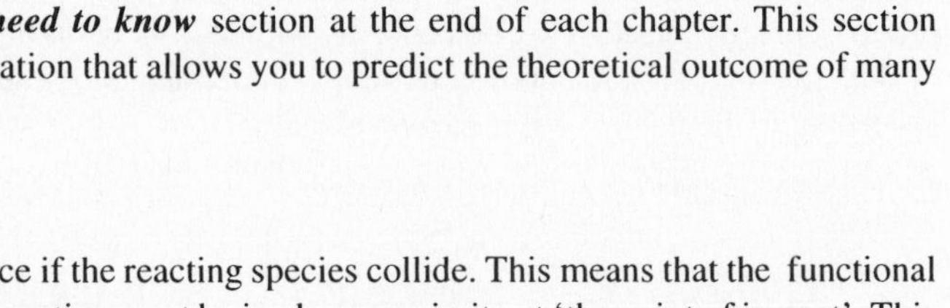

Fig.1.8. Schematic representation of the effect of the shape of a molecule on the course of a chemical reaction.

For example, the esterification of 2,4,6-trimethylbenzoic acid by an alcohol does not occur under **normal** esterification conditions because the methyl groups at positions 2 and 6 prevent the alcohol getting close enough to react with the carboxylic acid group. This is an example of a phenomenon known as **steric hindrance.**

Reaction impossible because there is insufficient space for the alcohol to approach near enough to the carbon of the carboxylic acid to react

Most biological reactions are catalysed by proteins known as enzymes. In many cases an enzyme will only catalyse a reaction in the presence of a smaller non-protein molecule generally known as a **co-enzyme** or **cofactor**. The complete enzyme/co-enzyme system is known as the **holoenzyme** whilst the enzyme itself is referred to as the **apoenzyme.** It should be noted that the names of all enzymes end in **-ase**. The reactants in enzyme catalysed reactions are called **substrates.**

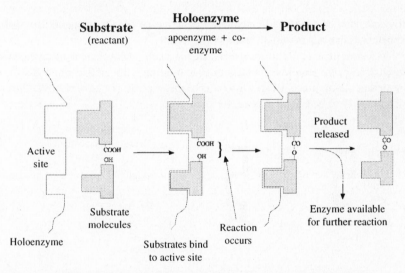

Fig. 1.9. A schematic representation of the possible course of a reaction at an active site.

It is the relative shapes of these apoenzymes, co-enzymes and substrates that determine whether a reaction will occur, provided of course that the correct functional groups are present. The substrate(s) bind to the holoenzyme which either brings the relevant functional groups into contact or weakens the appropriate bonds thus allowing reaction to occur (Fig. 1.9). The small part of the enzyme where substrate binding occurs is called the **active site**. If the substrate species do not have the correct structures and shapes to bind to the active site, they will not react. The simplest analogy, which was formulated in 1896 by Emil Fischer, is that of unlocking a door. The right key will fit the lock and open the door. The wrong key may still be inserted into the lock but will not fit it sufficiently well to unlock it.

The binding of a substrate to an active site can cause a change in the shape of an enzyme which sometimes results in the activation of a second active site on the enzyme for the same substrate (Fig. 1.10). This second site is known as an **allosteric site** and the phenomenon is called **allosteric activation**. Allosteric activation increases the efficiency of the enzyme in processing the substrate. It occurs when the concentration of the substrate rises above a certain level.

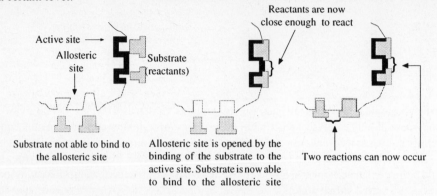

Fig. 1.10. A schematic representation of allosteric activation.

Active sites can be rendered inoperative by the use of **blocking agents (inhibitors)**. These substances can act in two different ways:

(1) The blocking agent binds to the active site in such a way that it cannot be easily dislodged. Its presence prevents the substrate binding to the active site and so stops the reaction taking place. Blocking agents of this type do not have to cover the active site completely to be effective (Fig. 1.11).

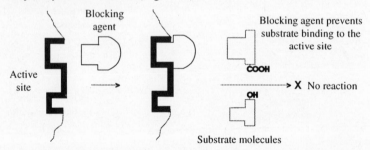

Fig. 1.11. A schematic representation of the action of a blocking agent.

(2) The blocking agent binds to an allosteric site. This disturbs the structure of the enzyme to the extent that it distorts the active site, so that the substrate is no longer able to bind to the active site and reaction becomes impossible. This type of blocking is known as allosteric inhibition (Fig. 1.12).

Enzyme substrates and blocking agents are held in position on active sites by different types of chemical bond and attractive force. The nature of these bonds and forces is discussed in Chapters 2 and 3.

Blocking agents may be used as drugs to control diseases since enzymes are intimately involved in these conditions. For example, methotrexate (Fig. 1.13) blocks the action of the enzyme dihydrofolate reductase which is the basis of its use in the treatment of certain forms of cancer and psoriasis.

Shape also governs the action of **receptors.** Receptors are enzyme-like molecules or complexes found in cells or on cell membranes which are responsible for triggering, modulating or inhibiting biological processes. Receptors are often proteinaceous in

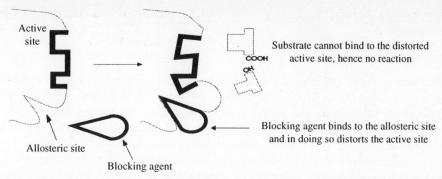

Fig. 1.12. A schematic representation of the mode of action of a blocking agent at an allosteric site.

structure but can include carbohydrates, lipids and inorganic ions in their structures. Some nucleic acids have also been implicated as receptors. When chemicals such as drugs and hormones bind to receptors they stimulate a cascade of chemical reactions or perhaps a nerve impulse which results in some physiological action. Other consequences of receptor binding include processes such as changes in cell membrane permeability, gene expression and the speed of hormone synthesis.

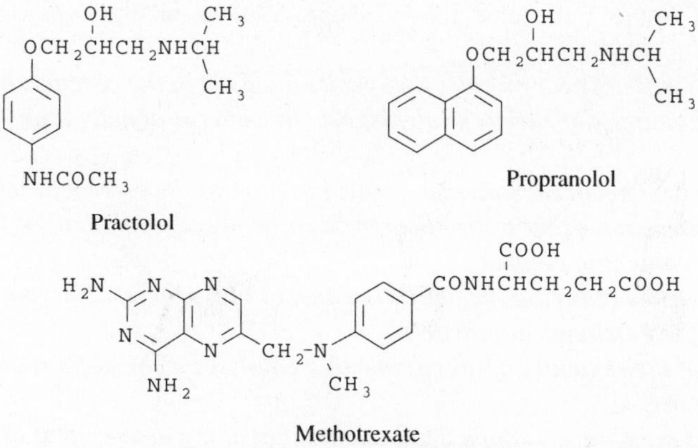

Practolol

Propranolol

Methotrexate

Fig. 1.13. The structures of some blocking agents that are used as drugs.

The action of a receptor site can be blocked by suitable blocking agents, which leads in some cases to their use as drugs. For example, practolol and propranolol (Fig. 1.13) are both β-blockers that decrease heart rate by blocking the β receptor sites for adrenaline in the sympathetic nervous system, the normal action of adrenalin at these sites being associated with an increase in heart rate. It is interesting to note that these drugs have identical side chains. This indicates that the side chain may play an important part in the action of the drug. Deductions of this type are the basis of **structure–activity relationships (SARs)** in medicinal chemistry, where attempts are made to relate similarities in the structure of drugs to their biological activity.

Blocking agents that block receptor sites are known as **antagonists** of the compound that normally binds to that site. For example, *p*-aminobenzoic acid (PABA) is an antagonist for antibacterial sulphonamides, that is, it blocks the receptor sites for the sulphonamides and so inhibits their action.

p-aminobenzoic acid (PABA) H_2N—⟨   ⟩—COOH

1.7 WHAT YOU NEED TO KNOW

The *What you need to know* sections list the points that you need to remember about the topics discussed in that chapter. Key points are printed in ***bold italics*** and should be memorised. Points printed in normal type are also useful but are not thought to be as important as those printed in bold italics. It is useful to memorise as many of the other points as possible. However, it is emphasised that learning facts alone is not sufficient to master the vocabulary of chemistry, you must also practise using those facts. To this end, questions are given at the end of all subsequent chapters and it is strongly recommended that you work through these questions before moving on to a fresh chapter. Answers are provided for selected questions at the end of the text.

(1) Theories are based on practical information and are used to communicate ideas and information.

(2) *Theories are not always accurate. An appreciation of the limitations of a theory is necessary to use it sensibly.*

(3) The main factors affecting chemical reactions are the energy of the system, types of functional groups, molecular size and shape. All these factors have to be compatible if a reaction is to occur.

(4) *A reaction will not occur unless the species involved in the reaction have a specific minimum quantity of energy known as the 'free energy of activation' or 'activation energy'.*

(5) *A reaction will not occur unless the reactants possess the correct functional groups.*

(6) *A reaction will not occur unless the reactants have shapes that allow the functional groups to come into contact.*

(7) *Catalysts act by reducing the activation energy of a process. Enzymes are proteins that catalyse biochemical reactions.*

(8) *The part of the structure of an enzyme that catalyses a chemical reaction is called the active site.*

(9) *Substrates must have a shape that matches that of the active site if they are to bind and react.*

(10) *The binding of a substrate to an active site can activate an allosteric site.*

(11) *Receptor sites are areas in or on a cell or nerve ending where the binding of a chemical species, such as a hormone or drug, can either initiate, modulate or inhibit a biological process and so can be used to treat some medical conditions.*

(12) *Substrates, for example drugs and hormones, must have the correct structure and shape if they are to activate a receptor site.*

(13) *Blocking agents are compounds that bind to an enzyme or a receptor and so prevent it functioning correctly by blocking access of the substrate.*

(14) Blocking agents can bind to either the active or allosteric sites of an enzyme to prevent its action.

(15) Blocking agents can also be used as drugs to treat some medical conditions.

2

Bonding and the shapes of molecules

2.1 INTRODUCTION

A number of theories are currently used to describe and explain the nature of the bonds
formed between the atoms and ions that form the structures of molecules and crystals. The
pictures of the structures of molecules obtained using these theories are often discussed in
terms that give the impression that they are the real structures of the compounds. This may
well be true but at the current time this has not been completely proven and so it should
always be borne in mind that the pictures given in the literature are representations of the
structure of a molecule.

All current molecular structure theories are based on the historical observation that the
noble gases (or inert gases as they were originally known) are, relatively speaking,
chemically inert. The observation that most bonded atoms are in a relatively stable situation
led to the proposal that bonded atoms try to achieve a noble gas configuration. This provides
a satisfactory explanation of the structures of many compounds. However, it did not
satisfactorily explain the structures of stable compounds where some of the atoms could
not achieve noble gas electronic configurations. For example, the boron atom of boron
trichloride (Fig. 2.1) has the bonded electronic configuration 2,6, which does not
correspond to the electronic configuration of argon 2,8, the nearest inert gas in the periodic
table.

The structures of relatively stable compounds in which some of the atoms do not have
noble gas configurations were explained by the quantum theory of the structure of the
atom. This theory suggests that a spin pair of electrons is a stable situation, and so it was
deduced that if all the electrons in a structure were spin paired, the structure would be
relatively stable even though some of its atoms did not have noble gas configurations.
Electron pairing is now the primary aim when explaining the structures of stable chemical
species. However, the reader should note that there are still chemical species whose
electronic configurations do not fit with this concept (Chapter 7).

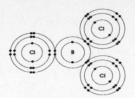

Fig. 2.1. The simple electronic structure of boron trichloride.

Stable electronic configurations are achieved *in theory* in three main ways, known as ionic, covalent and dative bonding. Once formed, covalent and dative bonds are identical (2.3) and so are represented in structures by a single line and are generally referred to as covalent bonds.

Other types of bonding have been postulated to explain situations where ionic and covalent bonding do not provide a rational explanation of the properties of a compound. The more common of these additional bonds are discussed in Chapter 3.

2.2 IONIC BONDING

Stable electronic configurations are achieved in ionic bonding by the 'donation' and 'acceptance' of electrons. This leads to the formation of cations and anions which, being oppositely charged, are held together by the electrostatic attraction of opposite charges. This attraction is known as the **ionic** or **electrostatic bond**. The attractive force of an ion occurs in all directions, in a similar manner to light radiating from an electric light bulb. Consequently, an ion can form an ionic bond in any direction.

In solid compounds the ions are arranged in a regularly repeating three-dimensional pattern called a crystal lattice. There are many different types of arrangement of the ions in these lattices. The attractive forces of the ions forming a lattice operate in all directions throughout the lattice which means that any one ion is electrostatically bonded to all of its nearest neighbours regardless of its valency. Consider, for example, the structure of a crystal of sodium chloride. The ground state electronic configurations of sodium and chlorine atoms are:

Na $1s^2, 2s^2, 2p^6, 3s^1$.   (Nearest noble gas: Ne $1s^2, 2s^2, 2p^6$)

Cl $1s^2, 2s^2, 2p^6, 3s^2, 3p_x^2, 3p_y^2, 3p_z^1$. (Nearest noble gas: Ar $1s^2, 2s^2, 2p^6, 3s^2, 3p_x^2, 3p_y^2, 3p_z^2$)

A sodium atom can acquire the stable electronic configuration of neon by 'losing' its 3s electron. This loss of one electron means that the resulting sodium ion has eleven protons and ten electrons, that is, an overall charge of +1. A chlorine atom can obtain the stable electronic configuration of argon by gaining an electron in its 3p orbital. The resulting chloride ion has eighteen electrons and seventeen protons providing a charge of $^-1$. Therefore, in sodium chloride the donation of an electron from sodium to chlorine will result in both atoms having stable electronic configurations.

Na $1s^2, 2s^2, 2p^6, 3s^1$ $\longrightarrow$ Na$^+$ $1s^2, 2s^2, 2p^6$.

one electron transferred

Cl $1s^2, 2s^2, 2p^6, 3s^2, 3p_x^2, 3p_y^2, 3p_z^1$ $\longrightarrow$ Cl$^-$ $1s^2, 2s^2, 2p^6, 3s^2, 3p_x^2, 3p_y^2, 3p_z^2$

The sodium and chloride ions are arranged in a crystal lattice of the type shown in Fig. 2.2. The non-directional nature of the ionic bond means that each ion in this structure forms electrostatic bonds with its neighbours. The ions can be thought of as being in a state of electrical balance, the electrostatic attractive forces between an ion and its neighbours, and vice versa, act like a force field holding the ions in their relative positions. It should be noted that it is not possible to specify to which chlorine atom a particular sodium atom has donated its electron. However, the lattice will be electrically neutral since the ratio of sodium ions to chlorine ions is 1:1.

Sodium ions

Chlorine ions

Fig. 2.2. The arrangement of the ions in a crystal of sodium chloride.

Simple compounds in which ionic bonding occurs are usually electrically neutral overall. Consequently, the number of positive charges must equal the number of negative charges. This may be achieved in some compounds by having unequal numbers of anions and cations. In order to achieve electrical neutrality in calcium chloride, for example, the ratio of calcium to chloride ions is 1:2. These ions are believed to be formed by the transfer of electrons from a calcium atom to two chlorine atoms (Fig. 2.3). However, calcium and chloride ions in the calcium chloride lattice will be in a different arrangement from that found in the sodium chloride crystal lattice.

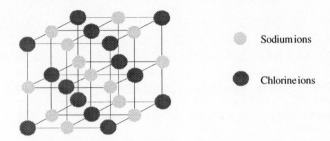

$\text{Cl } 1s^2, 2s^2, 2p^6, 3s^2, 3p_x^2, 3p_y^2, 3p_z^1 \longrightarrow \text{Cl}^- 1s^2, 2s^2, 2p^6, 3s^2, 3p_x^2, 3p_y^2, 3p_z^2$

one electron transferred

$\text{Ca } 1s^2, 2s^2, 2p^6, 3s^2, 3p^6, 4s^2 \longrightarrow \text{Ca}^{2+} 1s^2, 2s^2, 2p^6, 3s^2, 3p^6$

one electron transferred

$\text{Cl } 1s^2, 2s^2, 2p^6, 3s^2, 3p_x^2, 3p_y^2, 3p_z^1 \longrightarrow \text{Cl}^- 1s^2, 2s^2, 2p^6, 3s^2, 3p_x^2, 3p_y^2, 3p_z^2$

Fig. 2.3. A theoretical explanation of the formation of the calcium and chlorine ions in calcium chloride.

Ionic bonds occur widely in inorganic chemistry but are less common in organic chemistry where they may usually be found in the structures of salts derived from acids and amines.

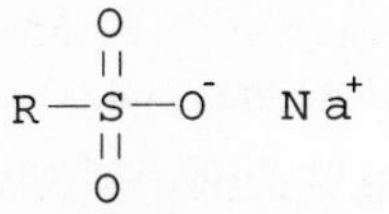

A sulphonic acid salt 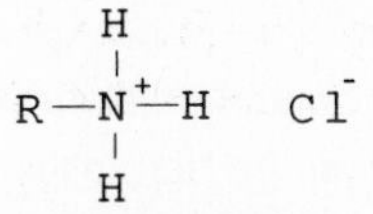   An amine salt

Some compounds that are not ionic themselves can, under the correct physical conditions, such as solution in water, dissociate into ions. The process is called **ionisation** and is reversible, the parent compound and its ions forming a dynamic equilibrium mixture (6.10). For example, carboxylic acids and amines can dissociate into ions in water.

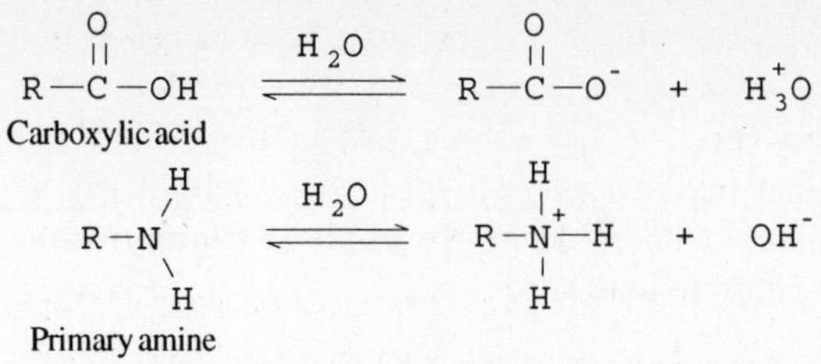

$$R-\overset{\overset{O}{\|}}{C}-OH \;\underset{\xleftarrow{\hspace{1cm}}}{\overset{H_2O}{\xrightarrow{\hspace{1cm}}}}\; R-\overset{\overset{O}{\|}}{C}-O^- \;+\; H_3^+O$$

Carboxylic acid

$$R-N\overset{\diagup H}{\diagdown H} \;\underset{\xleftarrow{\hspace{1cm}}}{\overset{H_2O}{\xrightarrow{\hspace{1cm}}}}\; R-\overset{H}{\underset{H}{\overset{|}{\underset{|}{N^+}}}}-H \;+\; OH^-$$

Primary amine

Ionic bonds play an important part in the binding of ligands to receptor sites. For example, in the dehydrogenation of lactate by the enzyme, lactate dehydrogenase, the arginine residues in the enzyme bind to the substrate and the co-enzyme by ionic bonding (Fig. 2.4). It should be observed that the structures that form the active site of the enzyme are not adjacent to each other in the structure of the enzyme. This is often a feature of active sites.

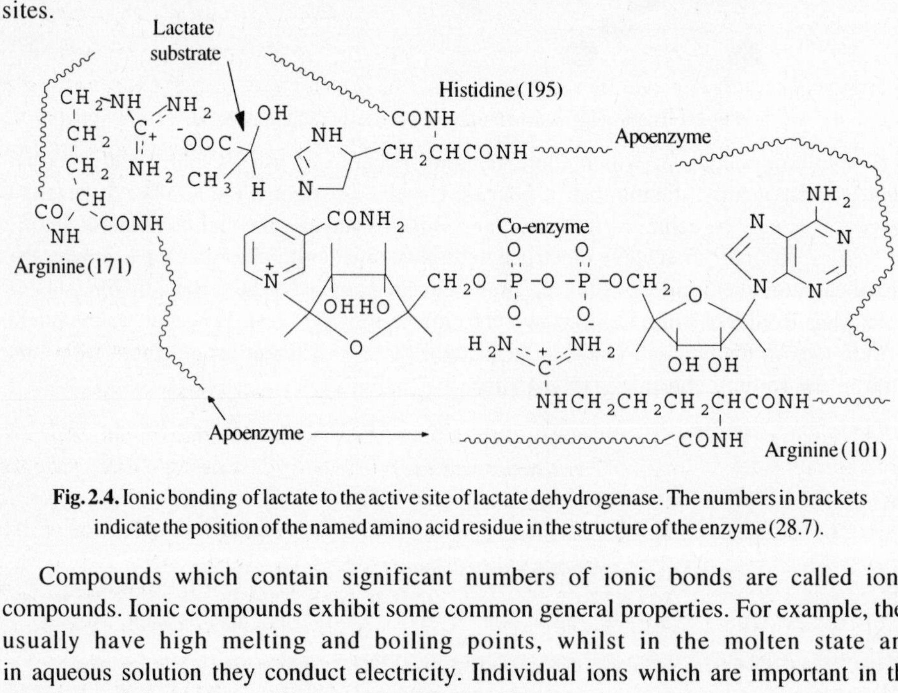

Fig. 2.4. Ionic bonding of lactate to the active site of lactate dehydrogenase. The numbers in brackets indicate the position of the named amino acid residue in the structure of the enzyme (28.7).

Compounds which contain significant numbers of ionic bonds are called ionic compounds. Ionic compounds exhibit some common general properties. For example, they usually have high melting and boiling points, whilst in the molten state and in aqueous solution they conduct electricity. Individual ions which are important in the chemistry of the body are discussed in Chapter 30.

2.3 SINGLE COVALENT AND DATIVE BONDING

2.3.1 Introduction

Covalent and dative bonding use electron sharing to achieve stable electronic configurations. Sharing occurs in the first instance by the overlap of the relevant atomic orbitals. In covalent bond formation each atom shares one of its electrons with the other atom, while in dative bond formation one of the atoms shares two of its electrons with the other atom (Fig. 2.5). Once formed, both types of bond consist of a shared pair of electrons and so have identical structures. Consequently, both covalent and dative bonds may be represented by solid lines in structural formulae although an arrow $\longrightarrow$ may be used for dative bonds, the tail of the arrow indicating the source of the pair of electrons.

Covalent bond formation: A· + ·B ⟶ A : B (or A——B)

Dative bond formation: C: + D ⟶ C : D (or C——D or C—▶D)

Fig. 2.5. The formation of covalent and dative bonds. A single dot (·) represents one electron and a double dot (: or ··) represents a pair of electrons.

The most generally satisfactory theory used to describe the *sharing of electrons* in covalent and dative bonds is the **molecular orbital (MO) theory**.

2.3.2 Molecular orbital theory

The molecular orbital theory is based on the proposal that when a bond is formed the atomic orbitals forming the bond *combine* to form a so-called **molecular orbital**. Like atomic orbitals, each molecular orbital is capable of containing one spin pair of electrons. Unlike atomic orbitals they are a property of the whole molecule not just of the atoms forming the bond. They are formed by the conversion of the atomic orbitals of all the atoms in the molecule into the relevant molecular orbitals. Consider, for example, the formation of a hydrogen molecule from its atoms shown in Fig. 2.6. The electrons forming the bond reside in the molecular orbital formed by the overlap of the 1s electrons of the two hydrogen atoms. In practice, in multi-electron-shell atoms it is usually only necessary to consider the molecular orbitals of the electrons involved in bonding (see methane, Fig. 2.9).

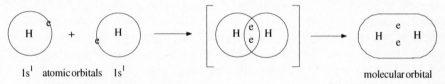

1s^1 atomic orbitals 1s^1 molecular orbital

Fig. 2.6. A pictorial representation of the molecular orbital structure of a hydrogen molecule.

At first sight the formation of one molecular orbital from two atomic orbitals raises an apparent anomaly. It implies that where there were originally two orbitals with space for four electrons, there is now only one orbital with space for two electrons, that is, atomic space has been lost. This anomaly is resolved by the mathematics of molecular orbital theory which predicts that two molecular orbitals (MOs) are formed when two atomic orbitals are combined: the **bonding** MO with a lower energy level and an **antibonding** MO with a higher energy level than the original atomic orbitals. The mathematical process for the formation of a hydrogen molecule from its constituent atoms is illustrated diagrammatically in Fig. 2.7. The bonding electrons occupy the lower energy 1s bonding MO when the bond is formed between the two hydrogen atoms because this will result in the molecule having a more stable lower energy.

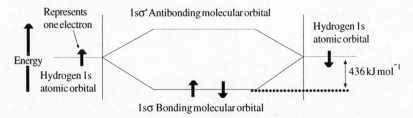

Fig. 2.7. The molecular orbital energy diagram for hydrogen. The arrows represent the positions of the electrons before and after bonding.

2.3.3 The structure of methane

The molecular orbital theory introduced in the preceding paragraphs provides a satisfactory account of the properties and shapes of many different types of molecule. The main features of the theory can be illustrated by working through the electronic structure of a molecule of methane.

The ground state electronic configuration of carbon is:

$$1s^2, 2s^2, 2p_x^1, 2p_y^1$$

Carbon has two unpaired electrons (Hund's rule) but requires four if it is to form the four C–H bonds in methane (CH_4). This problem is resolved by using an excited state of carbon instead of its ground state. The two 2s electrons are separated and one is promoted to the $2p_z$ orbital giving the carbon in methane a configuration of:

$$1s^2, 2s^1, 2p_x^1, 2p_y^1, 2p_z^1$$

In this state carbon has four unpaired electrons and is thus capable of forming four covalent bonds. The use of an excited state is only necessary when there are insufficient unpaired electrons available for bonding. The excited state used should be the lowest energy state that gives a satisfactory answer for the structure of the molecule.

The electronic structure of a molecule should give a satisfactory explanation of that molecule's properties. This raises a second difficulty in the explanation of the electronic structure of methane. All the C–H bonds in this compound appear to react chemically in the same way. Consequently, it is logical to suppose that they all have the same type of structure. However, if we simply consider the excited state configuration of carbon, the C–H bonds will have two types of structure: three bonding molecular orbitals based on an s-p atomic orbital overlap and one based on a s-s atomic orbital overlap. This implies that we should expect to have two sets of properties for the C–H bonds in methane which has never been observed in practice. The problem was resolved by Linus Pauling in 1931. He suggested that the 2s and all the 2p atomic orbitals should be converted into four 'sp^3 hybridised atomic orbitals' using the mathematical principles that underlie the structure of the atom (Fig. 2.8). Like atomic orbitals, these sp^3 hybridised atomic orbitals can accommodate a maximum of two electrons each. They are identical except that in space they are orientated towards the four corners of a regular tetrahedron.

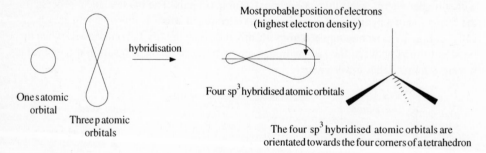

Most probable position of electrons (highest electron density)

hybridisation

One s atomic orbital

Three p atomic orbitals

Four sp^3 hybridised atomic orbitals

The four sp^3 hybridised atomic orbitals are orientated towards the four corners of a tetrahedron

Fig. 2.8. A diagrammatic representation of the formation of four sp^3 hybrid atomic orbitals from one s and three p atomic orbitals.

This hybridisation means that the four C–H bonds have identical structures, namely, four molecular orbitals based on s-sp^3 atomic overlap (Fig. 2.9).

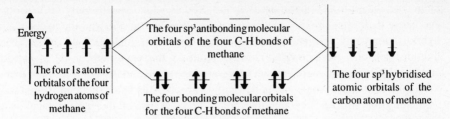

Fig. 2.9. The molecular orbital energy diagram for methane showing how the bonding and antibonding molecular orbital energy levels for the C−H bonds are formed. The arrows in the side columns represent the positions of the electrons before and in the centre column after bonding.

Consequently, in theory, the methane molecule has a tetrahedral structure with bond angles of 109°. This is in complete agreement with the experimentally determined shape of a methane molecule (Fig. 2.10).

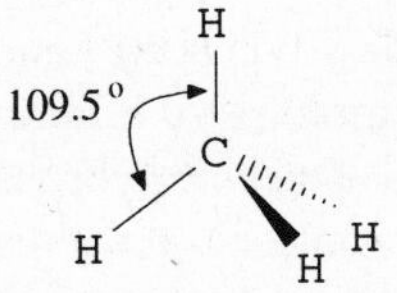

Fig. 2.10. The shape of a methane molecule.

There are many other examples where the molecular orbital theory offers a satisfactory explanation of the structure and properties of a compound and this is why the theory is so universally accepted.

2.3.4 Hybridisation

Hybridisation is a theory which correlates the experimentally observed shape and reactivity of a molecule with the concepts of molecular orbital theory. The process converts

Table 2.1. Common hybridisations.

x	Hybridisation	Orientation in space about a nucleus N	x	Hybridisation	Orientation in space about a nucleus N
2	sp	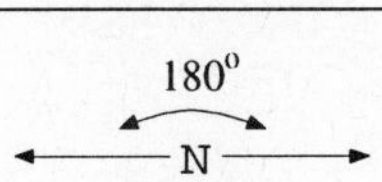 180° Linear	4	dsp^2	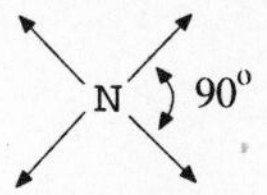 90° Square planar
3	sp^2	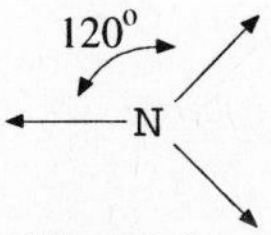 120° Trigonal planar	5	dsp^3	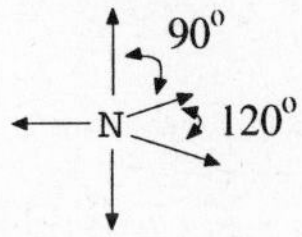 90° 120° Trigonal bipyramidal
4	sp^3	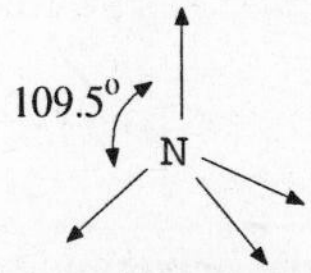 109.5° Tetrahedral	6	d^2sp^3	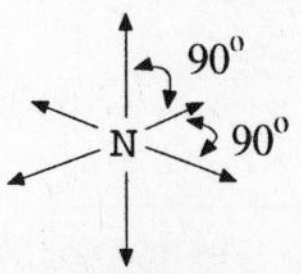 90° 90° Octahedral

atomic orbitals into new hybridised atomic orbitals which usually only differ from one another in their orientation in space. It is not necessary to understand the mathematics of hybridisation, only its general concept. However, it is important to realise that the hybridisation of x atomic orbitals always produces the same number (x) of hybridised atomic orbitals. Furthermore, it is not necessary to hybridise all the atomic orbitals in a shell to explain a structure: some can be left unhybridised. This latter situation is important in explaining the structures of multiple bonds (2.4). A list of the most commonly used hybridisations in chemistry is given in Table 2.1.

It is possible to predict the hybridisation required by the molecular orbital theory for a single covalently or datively bonded atom using the following simple method:

Count the number of bonded and lone pairs of electrons (x) in the outer shell of the atom being hybridised. Locate this value of x in Table 2.1 and read across the table at this point to obtain the hybridisation of the atom and the orientation of these hybrid orbitals in space.

If x is 4 there are two possibilities. When the compound is organic and the atom has no d orbitals available, the hybridisation is sp^3. If the compound is inorganic and the atom has suitable d orbitals available, the choice of hybridisation will depend on a knowledge of the three dimensional shape of the molecule about the atom concerned. If it is tetrahedral sp^3 will be the most suitable hybridisation. However, if it is square planar, dsp^2 is the most suitable. Some examples of the use of this method are given in Table 2.2.

Table 2.2. Examples of the deduction of the state of the hybridisation of selected atoms in some single molecules in the gas phase. The structures of some compounds will be different in the solid and gaseous phases, for example, in the solid state, phosphorus pentachloride has a crystal lattice consisting of PCl_4^+ and PCl_6^- ion.

Example	Atom whose hybridisation is required	Number and nature of bonded pairs of electrons	Lone + pairs = x ↿⇂	Hybridisation read from Table 2.1	Structure (shape)
BCl_3 (Boron trichloride)	B	3 B–Cl bonds + 0 = 3		sp^2	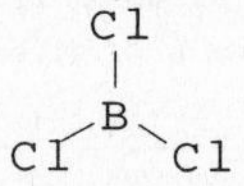 (trigonal planar)
H_2O	O	2 H–O bonds + 2 = 4		sp^3	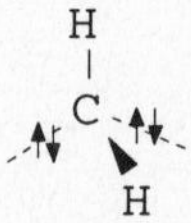 (tetrahedral)
PCl_5	P	5 P–Cl bonds + 0 = 5		dsp^3	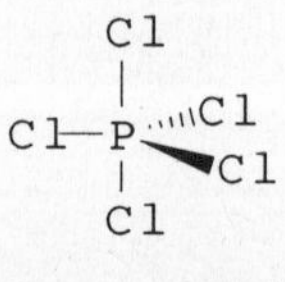 (trigonal bipyramidal)

It must be emphasised that the hybridisation selection method described in the preceding paragraphs applies only to saturated covalently and datively bonded structures.

It does not apply to unsaturated and conjugated structures. These are dealt with in sections 2.4 and 2.5.

2.3.5 Bond angles

This is the angle between a pair of bonds with an atom in common. Theoretical bond angles predicted by hybridisation are in good agreement with those found in practice (Table 2.3). Discrepancies occur between these theoretical shapes and the actual shapes determined by X-ray crystallography because the theory does not take into account the internal forces of attraction and repulsion that operate within a molecule. In a molecule of ammonia, for example, these internal forces cause a compression of its tetrahedral shape. The discrepancies between the experimentally determined and theoretically determined bond angles are usually explained by partial hybridisation. That is to say, the hybridisation is carried out so that one orbital contributes more to the hybrid orbitals than the others. This gives theoretical bond angles which are in better agreement with the experimentally determined values. However, in most cases, the differences between theoretical and experimental values are usually small enough to be ignored.

Table 2.3. A comparison of theoretical and actual bond angles. The structures in brackets are those whose bond angles are recorded. See section 2.4 for the prediction of the bond angles involving multiple bonds.

Compound	Bond angle (bond) theoretical (actual)			Compound	Bond angle (bond) theoretical (actual)		
CH_4	(H-C-H)	$109.5°$	$(109.5°)$	CH_3OCH_3	(C-O-C)	$109.5°$	$(111.3°)$
$CH_2=CH_2$	(H-C=C)	$120°$	$(121°)$	CH_3NH_2	(C-N-H)	$109.5°$	$(106°)$
Benzene	(-C=C-C-)	$120°$	$(120°)$	NH_3	(H-N-H)	$109.5°$	$(106°)$
$CH_3CH=CHCH_3$	(H-C=C-)	$120°$	$(120°)$	H_2O	(H-O-H)	$109.5°$	$(104°)$
CH_3SH	(C-S-H)	$109.5°$	$(99.4°)$				

2.3.6 Bond length

The length of a bond is defined as the distance between the two nuclei forming the bond. It is the optimum distance over which the electrons are effective in holding the nuclei together. Bond lengths are measured in nanometers (nm) but in older literature may be recorded in Angstrom units ($\overset{o}{A}$), where:

$$1 \text{ nm} = 10^{-9} \text{ m}, 1 \overset{o}{A} = 10^{-10} \text{ m, that is, } \mathbf{1 \overset{o}{A} = 0.1 \text{ nm} \text{ and } 1 \text{ nm} = 10 \overset{o}{A}}$$

Angstrom units should not now be used.

The bond length between the same two atoms will vary depending on the molecular environment of those atoms. It will be in a state of molecular vibration which will cause small changes in its length. Furthermore, the length of a particular type of bond will also vary with its molecular environment. For example, the length of the C=C bond varies from compound to compound as can be seen in Table 2.4. This is because the internal forces of attraction and repulsion will vary from molecule to molecule.

2.3.7 Bond strength

Bond strength or energy is defined as either the energy needed to break a bond or the energy liberated when a bond is formed (Fig. 2.11). Bond energy is measured in kilojoules per mole

Table 2.4. Bond lengths.

Compound	Bond length (nm) (bond)	Compound	Bond length (nm) (bond)
CH_4	0.11 (C-H)	NH_3	0.1 (N-H)
$CH_2=CH_2$	0.134 (C=C)	CH_3SH	0.181 (S-H)
Benzene	0.14 (C=C)	CH_3OCH	0.141 (C-O)
$CH_2=C=CH_2$	0.133 (C=C)	CH_3NH_2	0.147 (C-N)
$CH_3CH=CHCH_3$	0.148 (C=C)	H_2O	0.96 (H-O)

($kJ\ mol^{-1}$); the larger the value, the stronger the bond. In older literature energy is recorded in kilocalories per mole where one kilocalorie is equivalent to 4.18 kilojoules.

$$A + B = A{-}B \qquad\qquad \Delta H = {-}J\ kJ\ kJ\ mol^{-1}\ \text{(energy liberated)}$$
$$A{-}B = A + B \qquad\qquad \Delta H = {+}J\ kJ\ mol^{-1}\ \text{(energy absorbed)}$$

Fig. 2.11. Bond energy.

The strengths of bonds can be explained in terms of the degree and nature of overlap of the atomic orbitals forming the bonds. A good overlap of the regions of the orbitals where there is the maximum probability of finding the electrons results in a strong bond known as a **sigma** (σ) bond (Fig. 2.12). This type of atomic orbital overlap is sometimes referred to as **head on** overlap.

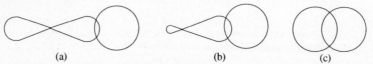

(a)   (b)   (c)

Fig. 2.12. Sigma bond formation. The diagrams represent sigma bonds formed by the overlap of (a) an s and a p orbital, (b) an sp^3 and an s orbital, and (c) two s orbitals.

Bonds may also be formed by the type of overlap shown in Fig. 2.13. This type of overlap is known as a **pi** (π) bond and the overlap itself is sometimes referred to as **sideways overlap**. Pi bonds are weaker than sigma bonds because the degree of orbital overlap is less and the overlap involves regions of the orbitals where there is a high, but not a maximum probability of finding the electrons. Pi bonds are found in unsaturated compounds (2.4).

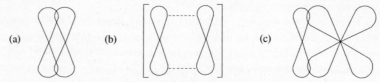

(a)   (b)   (c)

Fig. 2.13. Pi bond formation. The diagrams show (a) a p-p pi bond, (b) the way a p-p pi bond is represented in this text and (c) a p-d pi bond.

2.4 MULTIPLE COVALENT BONDING

A study of the reactions of alkenes whose structures contain a C−C double bond shows that one of the bonds of the double bond is far more reactive than the other. For example, the majority of the reactions of ethene involve only one of the C–C bonds of the double bond found in its molecule (Fig. 2.14).

$$CH_2{=}CH_2 + Br_2 \longrightarrow CH_2Br{-}CH_2Br$$

$$CH_2{=}CH_2 + HCl \longrightarrow CH_3{-}CH_2Cl$$

$$CH_2{=}CH_2 + H_2O \xrightarrow{KMnO_4} CH_2OH{-}CH_2OH$$

Fig. 2.14. Examples of the reactions of ethene. In each of these examples only one of the C–C bonds of the double bond has reacted.

This shows that one of the C–C bonds of ethene is more reactive than the other, which in turn suggests that the two C–C bonds forming the double bond of ethene have different structures. Extending these observations to other compounds containing multiple bonds, it soon becomes apparent that multiple bonds must consist of different types of bond. Molecular orbital theory explains the basic atomic structures of multiple bonds by using various combinations of sigma and pi bonds. For example, molecular orbital theory postulates that the C=C bond of ethene consists of one sigma and one pi bond. This structure is achieved through sp^2 hybridisation of each of the carbon atoms. If the p_x and the p_y orbitals are used for this hybridisation, the resulting sp^2 hybridised orbitals lie in the x/y plane. The remaining p_z orbital will lie in the z plane at right angles to the x/y plane (Fig 2.15).

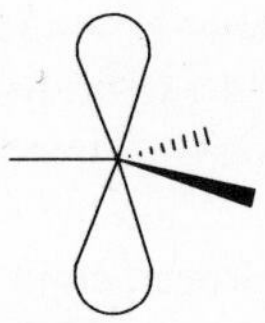

Fig. 2.15. The sp^2 hybridisation of the carbon atoms in ethene. It should be noted that the carbon atoms are in an excited state since the formation of four covalent bonds (two C–H and the two C–C bonds requires four unpaired electrons.

The stronger C–C sigma bond is formed by the overlap of one of the sp^2 hybridised orbitals of each of the two carbons and the weaker C–C pi bond by the sideways overlap of the p_z orbitals (Fig. 2.16). To complete the structure of ethene the hydrogen atoms bond to the carbons by overlapping their 1s orbitals with the remaining sp^2 hybridised orbitals of the carbon atoms.

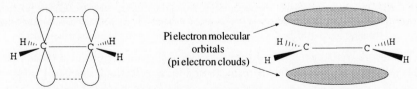

Pi electron molecular orbitals (pi electron clouds)

Fig. 2.16. The orbital and molecular orbital pictures of the structure of ethene.

The theoretical structure of ethene given in Fig 2.17 gives a satisfactory explanation of the shape and chemical reactions of that compound. It predicts a flat molecule with bond angles of 120°. Experimental observations confirm the flat shape but show that the bond angles are 121.7° for the C=C–H bond angle and 116.6° for the H–C–H bond angle. These figures indicate that the hybridisation theory is reasonably accurate.

The structure of ethene also offers a satisfactory explanation for the reactions of ethene.

It is logical to suppose that the weakest bond will break first. Therefore, reactions of the C=C bond of ethene in which only one bond takes part involve the weaker and so more reactive pi bond. The sigma bond is too strong to react except under more vigorous conditions.

The structure of the C=C of ethene forms the basis of the atomic structures of all double bonds. For example, the structure of the carbonyl group found in aldehydes and ketones shown in Fig. 2.17 is similar to that of the C=C found in ethene. However, it should be realised that many other double bonded structures are explained in terms of variations of this theme, for example, the structure of benzene (2.5).

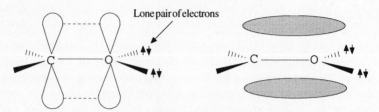

Fig. 2.17. The structure of the carbonyl group of aldehydes and ketones.

Molecular orbital theory uses two pi and one sigma bond as the basis for explaining the structures of triple bonds. The structure is based on sp hybridisation of atoms forming the bond. Consider, for example, the atomic structure of the $C \equiv C$ bond of ethyne. If the p_x orbital of the carbon atom is used for the hybridisation, the resulting sp hybrid orbitals are orientated in the x direction. The sigma C–C bond is formed by the overlap of two of these sp hybrid orbitals, one from each carbon atom while the second sp hybridised orbital of each carbon forms a sigma bond with a hydrogen atom. This leaves the p_y and p_z orbitals of each carbon atom free to form the two pi bonds (Fig. 2.18).

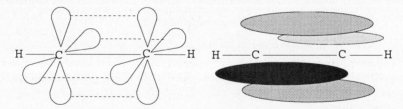

Fig. 2.18. The structure of ethyne. The carbon atoms are in an excited state, since four unpaired electrons are required to form four covalent bonds (one C–H and three C–C bonds per carbon atom). The darker shading is the part of the pi molecular orbital in the horizontal plane nearest the reader.

The theoretical structure of ethyne correlates well with the practical data. The shape and bond angles are in complete agreement with the experimental shape and values whilst the two reactive pi bonds give a satisfactory explanation of the C–C bond's reactivity. This type of structure can be used as a basis for the atomic structures of all triple bonds. For example, the nitrile group ($C \equiv N$) has the structure shown in Fig. 2.19.

2.5 CONJUGATED STRUCTURES

Structures which consist of alternating multiple and single bonds are called **conjugated structures**. They are part of the molecular structures of many naturally occurring compounds, some examples of which are given in Fig. 2.20. In each of these examples the

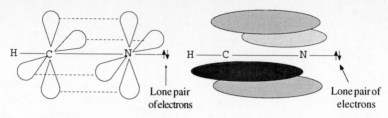

Fig. 2.19. The structure of the nitrile group.

conjugated structure is contained within the dotted box. Double bonds outside the boxes are not part of a conjugated system. They are referred to as **isolated** double bonds. Conjugated triple bond systems are rare and so will not be discussed in this text.

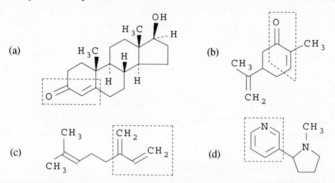

Fig. 2.20. Examples of compounds whose molecules contain conjugated structures. (a) Testosterone, the human male sex hormone. (b) Carvone, oil of spearmint. (c) Myrcene, oil of bay. (d) Nicotine.

Conjugated aliphatic structures exhibit the same general chemical properties as those shown by isolated aliphatic double bonds and in addition exhibit chemical properties that are often quite different from those of isolated double bonds. These experimental observations indicate that the structures of conjugated systems must be similar to, but different from, those of isolated double bonds. Molecular orbital theory bases its explanation of the structures of conjugated systems on those of the double and triple bonds described in section 2.4. The resulting structures provide a reasonable explanation of the chemical and physical properties of these structures. For example, the structure of 1,3-butadiene (Fig. 2.21) contains two C=C bonds in conjugation (a conjugated diene). Each of these C=C bonds has the same structure as the C=C bond of ethene (Fig. 2.16). Consequently, the C_2-C_3 sigma bond is formed by the overlap of the appropriate sp^2 hybridised orbitals and so is the same length as sigma bonds of the C=C double bonds. Therefore, purely from geometry, the sideways overlap of the p orbitals must extend along the whole length of the conjugated system (Fig. 2.22). This leads to a molecular orbital which includes all the carbon atoms. The pi electrons in this orbital are not associated with one particular carbon atom. They are said to be delocalised.

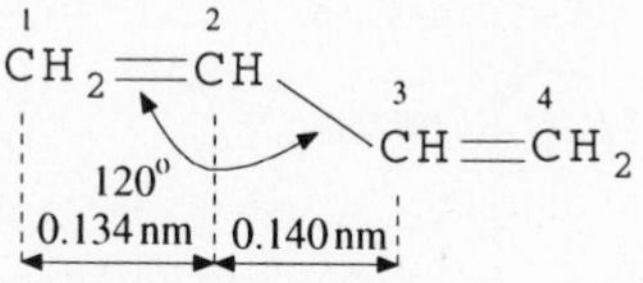

$$CH_2\!=\!\overset{2}{CH}\underset{120^0}{\diagdown}\underset{0.134\,nm}{\underset{\diagdown}{}} \overset{3}{CH}\!=\!\overset{4}{CH_2}$$

0.134 nm   0.140 nm

Fig. 2.21. Butadiene.

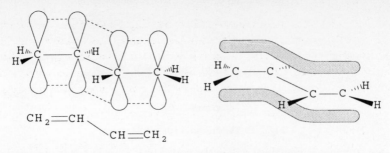

Fig. 2.22. The orbital and a pictorial representation of the molecular orbital structures of butadiene.

This picture for the structure of 1,3-butadiene implies that all the C–C bond lengths should be the same. However, experimental work shows that the C_1–C_2 and C_3–C_4 bond lengths are 0.134 nm and the C_2–C_3 bond length is 0.14 nm (Fig. 2.21). The C_2–C_3 bond is shorter than the expected value of 0.154 nm which is found for the C–C single bond in ethane but longer than that found for isolated C=C bonds. The slightly longer C_2–C_3 bond length suggests that there is slightly less interaction and hence a lower degree of attraction between the p_z orbitals of C_2 and C_3 than between the p_z orbitals of C_1 and C_2 and also C_3 and C_4. However, this explanation is not entirely satisfactory since sp^2 hybridised orbitals are shorter than sp^3 hybridised orbitals which means that the C_2–C_3, sp^2-sp^2 sigma bond of 1,3-butadiene should be shorter than the postulated sp^3-sp^3 sigma bond between the carbons of ethane, CH_3–CH_3.

In situations of this type both factors probably contribute to the structure. Consequently, a molecule of 1,3-butadiene is pictured as having a partial double bond between carbons 2 and 3, that is, the p orbitals do not overlap as much as those found in an isolated double bond (Fig. 2.23). This picture also offers a reasonable explanation of some of the chemical properties of 1,3-butadiene (12.4.3).

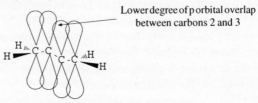

Lower degree of p orbital overlap between carbons 2 and 3

Fig. 2.23. The structure of 1,3-butadiene. The hydrogens bonded to carbons 2 and 3 are not shown.

All conjugated structures are similar to that of 1,3-butadiene. They differ in their degrees of conjugation. Benzene, for example, is completely conjugated (Fig. 2.24). All its bonds are the same length and so the p orbital overlap must be the same for every bond.

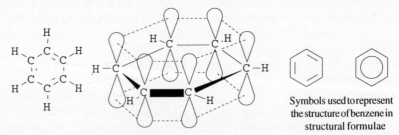

Symbols used to represent the structure of benzene in structural formulae

Fig. 2.24. The structure of benzene.

Atoms with lone pairs that are separated from a multiple bond system by a single bond in the traditional structural formula may also have a conjugated structure (Fig. 2.25). In these compounds the atom with the lone pair is normally assumed to be sp^2 hybridised with a lone pair in a p orbital that can interact with the p orbitals of the multiple bond. If the atom has more than one lone pair geometry dictates that only one of these pairs can be in conjugation.

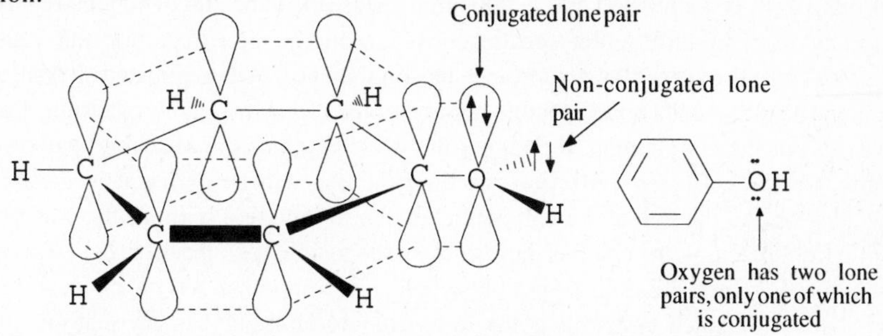

Fig. 2.25. The structure of phenol, showing the conjugation of the lone pair.

2.6 GENERAL PROPERTIES OF CONJUGATED SYSTEMS

2.6.1 Geometry

Conjugated structures are flat. Unfortunately compounds whose traditional structural formulae indicates a conjugated system are not always fully conjugated in practice. Systems where conjugation is not complete will not be flat. For example, retinol (Fig. 2.26), which plays an essential part in the chemistry of vision, is not fully conjugated. Experimental evidence shows that the ring is at an angle to the conjugated side chain which means in theory that p orbital overlap between the p orbitals of the double bond in the ring and the p orbitals of the double bond of the side chain is either very poor or does not occur.

There is no way of telling from a structural formula whether an apparently conjugated structure is in fact fully conjugated or even its degree of conjugation, that is the extent to which the electrons are delocalised. However, one usually assumes that the structure is flat in the first instance as exceptions are comparatively rare. The bond lengths will give some idea of the degree of conjugation (see butadiene, 2.5).

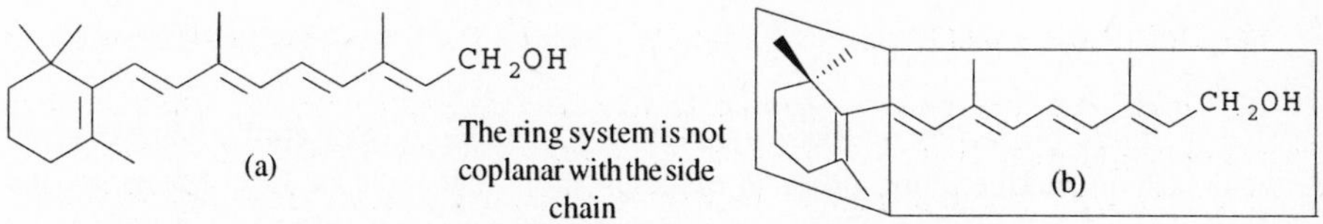

Fig. 2.26. The traditional structural formula of retinol (a) gives the impression that the structure contains a fully conjugated system, which is not correct (b).

2.6.2 Stability

Practical evidence based on heats of hydrogenation shows that conjugated structures are more stable than one would expect. It is found that the greater the conjugation the more stable the structure. Consequently, when molecules with conjugated structures react it is reasonable to expect the same conjugated structure to appear in the structure of the product. For example, in many of the reactions of benzene the benzene ring occurs in the structure of the product.

Nitrobenzene $\xleftarrow{\text{H}_2\text{SO}_4/\text{HNO}_3}$ Benzene $\xrightarrow{\text{H}_2\text{SO}_4}$ Benzene sulphonic acid

2.6.3 Spectra/colour

The wavelengths of ultra-violet and visible regions of the electromagnetic spectrum are approximately 10 to 400 and 400 to 800 nm respectively. The 200 to 400 nm region is most generally used in ultra-violet spectroscopy. Colour is the effect that the visible light reflected from the surface of a substance has on the eye. All conjugated structures absorb in either the ultra-violet and/or visible regions of the electromagnetic spectrum. The greater the extent of the conjugation, the longer the wavelength of the **absorbed radiation**. This means that the **reflected radiation** reaching the eye will be deficient in these absorbed wavelengths. As the colour we call 'white' is due to radiation with **all** the wavelengths of visible light reaching the eye this deficiency in the visible region causes the effect we know as colour. It is not possible to predict whether a compound whose molecular structure contains a conjugated system is going to be coloured but one can say that the greater the extent of the conjugation the more likely it will be coloured (Table 2.5).

Table 2.5. The variation of λ_{max} with the degree of conjugation found in molecules with the general structure: Ph(-CH=CH)$_n$-Ph. The highest wavelength of absorption of each structure is quoted.

n	λ_{max} (nm)	ε	n	λ_{max} (nm)	ε
1	309	25,000	5	403	94,000
3	358	75,000	6	420	113,000
4	384	86,000	7	435	135,000

2.7 THE THEORY OF RESONANCE

The theory of resonance, which was developed around the 1930s by Ingold, is used to describe the molecular structures of chemical compounds and species whose structures cannot be satisfactorily represented by traditional structural formulae. Consider, for example, the structure of the carboxylate ion:

$$-C \overset{\displaystyle O}{\underset{\displaystyle O^-}{}}$$

The chemical properties of this group show that both oxygen atoms behave in an identical fashion. Furthermore, physical methods show that both the C–O bonds are the same length, namely 0.127 nm. This suggests that both bonds have the same structure and not different structures as is implied by the traditional formulae normally used for their representation. The bond length of an isolated C=O in ketones is 0.122 nm whilst that for a single C–O bond in ethers is 0.142 nm. Consequently, the bond length of the C–O bonds in the carboxylate ion is closer to that of a *true* C=O bond than that of a *true* C–O bond. This suggests that both the carbon-oxygen bonds of the carboxylate ion have a high degree of double bond character which is not apparent from the traditional formula used to represent the structure of the ion.

The resonance theory uses two or more structures, known as **canonical forms**, to describe structures where traditional structural formulae are not entirely satisfactory.

Canonical forms are structures that represent the possible extremes of the structure of the substance. The use of canonical forms is indicated by a double headed arrow; $\longleftrightarrow$ linking the appropriate structures. This indicates that the actual structure of the substance or **resonance hybrid** as this structure is commonly known, lies between those used for the canonical forms. For example, the canonical forms used to represent the carboxylate ion are shown in Fig. 2.27a. These two structures imply that both oxygen atoms are in the same structural situation, that is, the pi electrons are distributed throughout the whole of the resonance structure. In the molecular orbital theory, the structure of the resonance hybrid corresponds to the delocalised structures described in the previous section.

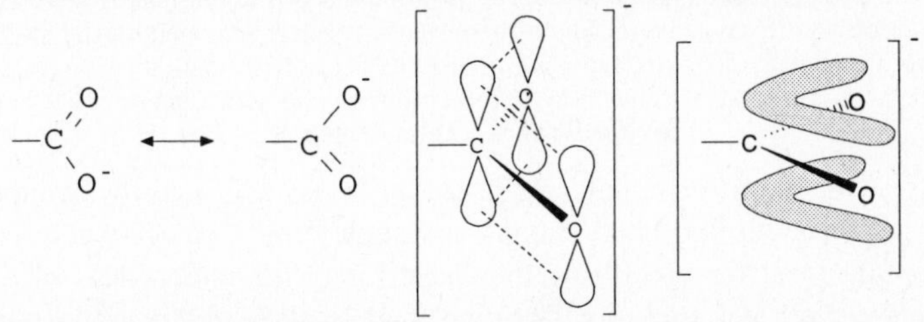

Fig. 2.27. The canonical forms and molecular structure pictures of the structure of the carboxylate ion.

The mathematics of the theory of resonance show that the energy of structures that are represented by resonance is lower than anticipated. Consequently, chemical species whose structures are represented by canonical forms are more stable, that is, less reactive, than one would expect. In general, the greater the number of canonical forms the more stable the structure. This aspect of the theory of resonance is extensively used in the theory of reaction mechanisms (see Chapters 7 and 8).

2.8 COVALENT AND DATIVE BONDS IN BIOLOGICAL SYSTEMS

Covalent and dative bonds are important contributors to the shapes of molecules and ions that are found in the living world. Substrates can bind to receptor sites using all types of covalent and dative bonding. The binding may be temporary or permanent depending on the strength of the bond. Permanent binding could have the undesirable effect of stopping a vital metabolic process. Carbon monoxide poisoning, for example, occurs because the carbon monoxide forms a strong permanent dative covalent bond with the iron of the haemoglobin in blood (Fig. 2.28). This prevents oxygen bonding to the haemoglobin which in turn stops its transport to vital areas of the body. This results in tissue anoxia and sometimes death.

Permanent binding can also be beneficial. Drugs used in the treatment of diseases often act by forming permanent bonds to receptor sites. Many anticancer drugs, for example, operate by binding permanently to receptor sites on the DNA chain. Melphalan forms covalent bonds with the bases on the two chains of DNA: it forms a permanent bridge between the chains. This stops or reduces the spread of the cancer by preventing the DNA chains from splitting and reproducing (Fig. 14.12).

2.9 MOLECULAR MODELLING

In chemistry, models are used to visualise the three-dimensional shapes of molecules and electron orbitals. The original model system used balls to represent atoms and sticks to represent bonds. This simple **ball and stick** system was used to demonstrate the relative

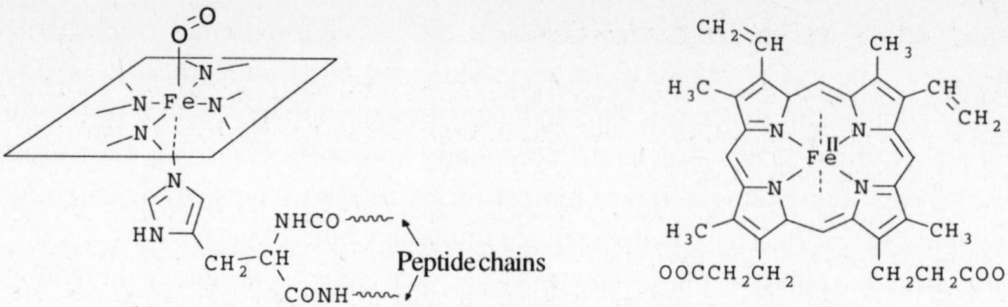

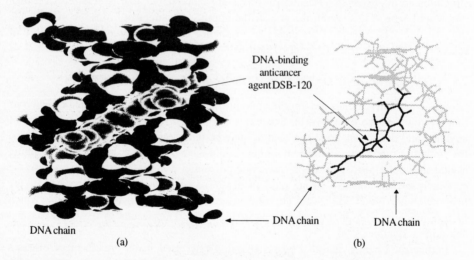

Fig. 2.28. The binding of (a) oxygen to the iron ion of haemoglobin. Carbon monoxide is bound in the same position through its oxygen atom. Figure (b) shows the structure of the porphin ring system of the haem units in haemoglobin. This structure lies in a horizontal plane at right angles to the vertical oxygen-histidine axis The iron in this structure is in the form of an Fe^{2+} ion. The structure of haemoglobin is described in more detail in Chapter 28.

arrangements of the atoms of a molecule in space. It was followed by **Dreiding** models which were machined so that bond lengths and angles could be accurately represented. These models are ideal for visualising the shape (or conformation, 4.1) of a molecule. However, neither ball and stick nor Dreiding models allow the overall volume of the molecule to be visualised. This aspect of molecular modelling was covered by the invention of Corrie, Pauling and Kellog (CPK) models. These are space fill models that show the volume of a molecule.

Fig. 2.29. The docking of the novel DNA-binding anticancer agent DSB-120 to a fragment of DNA. (a) The CPK model of DSB-120 (central diagonal section) docked in a fragment of the structure of DNA. (b) The Dreiding model of DSB-120 docked in a fragment of the structure of DNA. (Courtesy of Professor D Thurston.)

Computer programs that show the three-dimensional structure of a molecule and its volume have been developed. These programs can show either ball and stick, Dreiding or CPK models. Many of these programs allow energy minimisation. This means that bond lengths and angles can be adjusted to show the most stable shape of the molecule. In addition it is possible to visualise the fitting (**docking**) of a drug to a receptor site (Fig. 2.29). This makes it possible to either modify or design a drug to fit a particular receptor site. A further advantage is that colour prints can be made directly from the screen.

2.10 WHAT YOU NEED TO KNOW

(1) Stable chemical species are believed either to have electronic configurations which correspond to that of the nearest noble gas in the periodic table or to have all their electrons paired.

(2) The formation of bonds between atoms is explained by adjusting the electronic configurations of the atoms involved in the bond to one of the stable structures defined in (1). These stable structures are principally obtained by either the donation or the sharing of electrons to form what are known as ionic, covalent and dative bonds respectively.

(3) Ionic bonds are formed by the donation and acceptance of electrons.

(4) Ionic bonds are electrostatic forces of attraction between oppositely charged ions. These stable structures are obtained principally by either the donation or the sharing of electrons to form what are known as ionic, covalent and dative bonds respectively.

(5) Ionic bonds are important in the binding of molecules to receptor sites and to the active sites of enzymes.

(6) Covalent bonds are formed by each atom sharing a lone electron with the other atom forming the bond.

(7) Dative bonds are formed by one atom sharing a lone pair of electrons with the other atom forming the bond.

(8) Covalent and dative bonds are visualised as being formed initially by the overlap of the relevant atomic orbitals. Covalent and dative bonds have identical structures once they are formed, that is, once overlap has occurred.

(9) Molecular orbital theory postulates that when two atomic orbitals overlap to form a bond they change from two atomic orbitals into two molecular orbitals. These are known as the bonding and antibonding molecular orbitals.

(10) A bonding molecular orbital has a lower energy than its associated antibonding molecular orbital.

(11) The formation of molecular orbitals is sometimes illustrated by energy diagrams of the type:

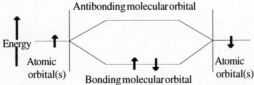

(12) The electrons forming the bond normally occupy the lower bonding molecular orbital.

(13) The overlap of the regions of two orbitals in which there is the maximum probability of finding electrons results in the formation of molecular orbitals known as sigma bonds. This type of overlap is known as end or head on overlap.

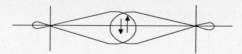

*(14) The overlap of the regions of two p orbitals in which there is not the maximum
probability of finding the electrons results in the formation of molecular orbitals
known as pi bonds. This type of overlap which usually involves p and d orbitals is
often referred to as sideways overlap.*

*(15) Hybridisation of atomic orbitals is an extension of the theory molecular structure
that explains why the geometry of atomic orbitals does not correspond to the
experimentally determined shape of the molecule. It correlates the theoretical
explanation with the practical observation.*

*(16) The hybridised state of singly bonded atoms can be predicted with a reasonable
degree of accuracy by determining the total number, N, of bonded and lone pairs
of electrons in their structures and referring to the following table.*

N	Hybridisation	N	Hybridisation	N	Hybridisation
2	sp	4	sp^3	5	dsp^3
3	sp^2	4	dsp^2	6	$d^2 sp^3$

*(17) Double bonds consist of one sigma and one pi bond. The atoms that form double
bonds are sp^2 hybridised. Double bonds have the general structure:*

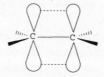 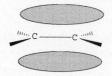

*(18) Triple bonds consist of one sigma and two pi bonds. The atoms that form triple
bonds are sp hybridised. Triple bonds have the general structure:*

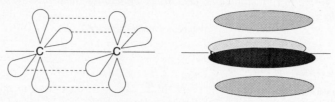

*(19) Bond length is the optimum distance between the nuclei of the atoms forming the
bond.*

(20) Bond energy is the energy needed to break a bond.

*(21) Conjugated systems consist of alternating multiple and single bonds, for
example:*

$$-CH=CH-CH=CH-CH=CH-CH=$$

(22) The atoms in a double bonded conjugated system are always sp² hybridised whilst those in a triple bonded conjugated system are always sp hybridised. These hybridisations result in conjugated systems having theoretical structures in which there is a continuous sideways overlap of the p orbitals of the atoms forming the system.

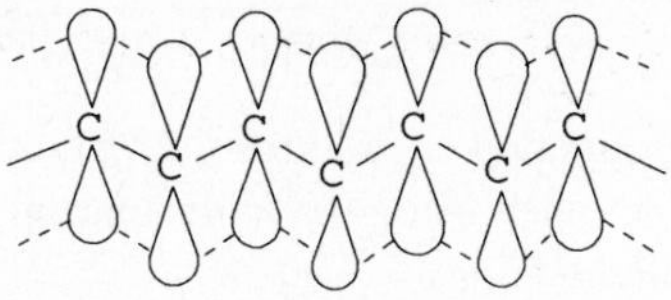

$$-CH=CH-CH=CH-CH=CH-CH=$$

(23) An atom with a lone pair that is separated by one single bond from a double bond is normally sp² hybridised and will have a structure similar to that of double bond conjugated structures.

Lone pair

$$-CH=CH-\ddot{Z}$$

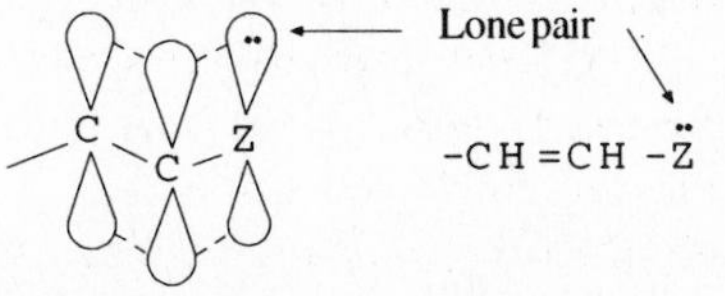

(24) The pi electrons of a conjugated system are said to be delocalised because they are able to distribute themselves over the system and are not restricted to one particular atom. The extent of this delocalisation is governed by the extent of the overlap of the p orbitals.

(25) Some compounds whose traditional structural formula show that they contain a conjugated structure do not in fact have a fully conjugated structure.

(26) Conjugated structures are flat, stable and absorb in the ultra-violet to visible regions of the electromagnetic spectrum.

(27) The theory of resonance describes the structures of unsaturated chemical species that cannot be satisfactorily explained by traditional structural formulae in terms of the canonical forms.

(28) Canonical forms are structures that represent imaginary extreme situations of the real structure of the chemical species. Canonical forms obey the traditional valency rules for molecular structure.

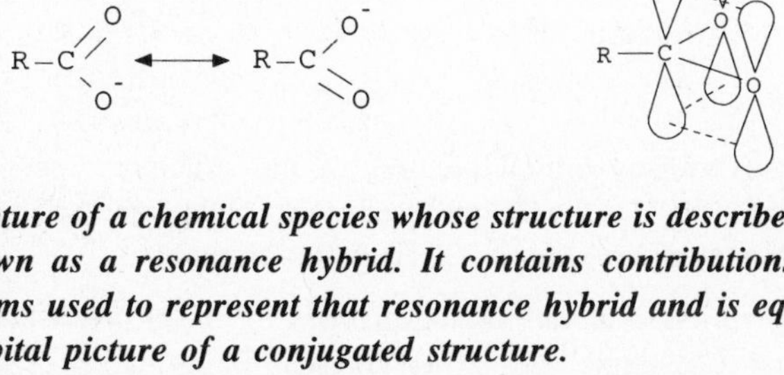

(29) The real structure of a chemical species whose structure is described by canonical forms is known as a resonance hybrid. It contains contributions from all the canonical forms used to represent that resonance hybrid and is equivalent to the molecular orbital picture of a conjugated structure.

2.11 QUESTIONS

(1) Determine the total number of electrons in the outer shell of the specified atoms in the following compounds: (a) magnesium in magnesium chloride, (b) carbon in a molecule of tetrachloromethane, and (c) oxygen in a molecule of carbon dioxide.

(2) Determine, by counting electrons and protons, the charge on the atoms (if any) in large heavy print in each of the following structures:

(a) $\overset{\bullet\bullet}{C}H_3$

(b) $CH_3CH_2CH_2\overset{\bullet}{C}HCH_2CH_2\ CH_3$

(c) $CH_3CH_2CH_2\overset{\bullet}{C}HCH_2CH_3$

(d) $HOOC\overset{\bullet\bullet}{C}HCOOH$

(e) $NC\overset{\bullet\bullet}{C}COOH$

(f) $CH_3\overset{\bullet}{C}HCH_3.$

(3) Which of the atoms in the covalently bonded structures of (a) ethanol and (b) ethylamine have to be in an excited state when using the molecular orbital theory to explain the structures of these compounds?
(Atomic numbers: C, 6; H, 1; N, 7; O, 8.)

(4) Suggest suitable hybridisations for the ringed atoms in each of the following compounds:

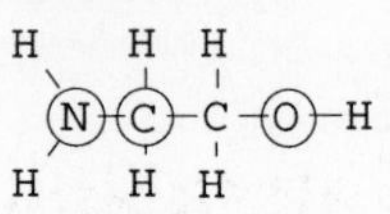

(a) 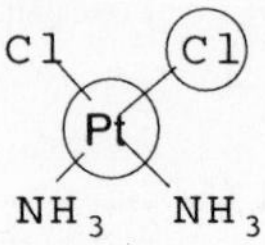   (b) 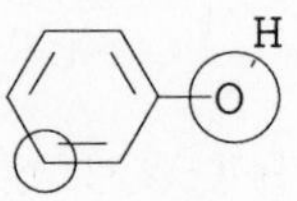   (c) 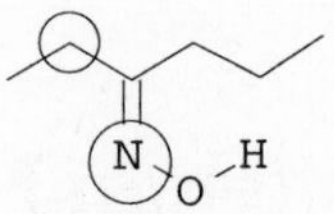   (d)

(5) Draw the structural formula of each of the compounds in question 4 and mark on the drawings the theoretical bond angles. Use these values to draw three-dimensional structures for each of these compounds.

(6) Which of the compounds (a) to (f) inclusive contain conjugated structures? Draw a line around each of the conjugated structures to show its extent.

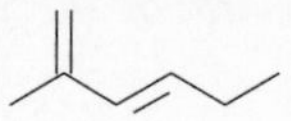

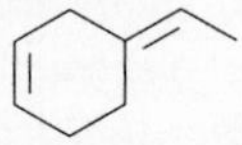

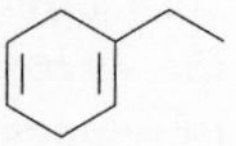

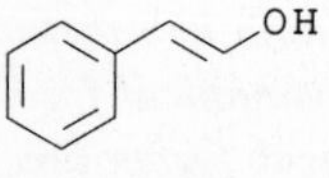

(7)

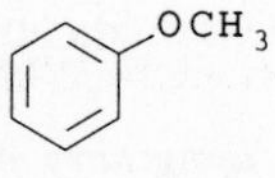

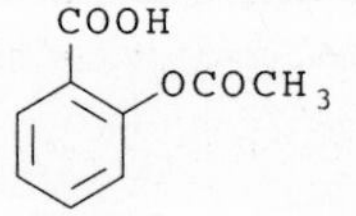

Aspirin 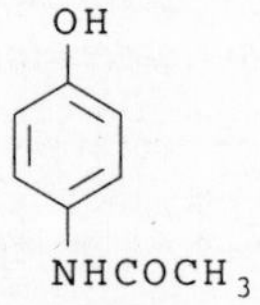   Paracetamol

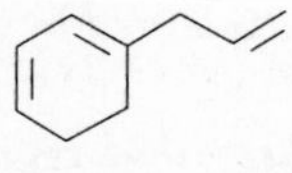

Draw the hybridised orbital pictures of the structures specified in each of the named compounds: (a) the CO group of aspirin and paracetamol, (b) the CONH group of paracetamol, and (c) the OCOCH₃ group of aspirin.

(8) Draw the hybridised atomic orbital pictures for the structures of (a) aspirin and (b) paracetamol. Use single lines — to represent the formation of sigma bonds and the appropriate orbital symbols for pi bonds.

(9) Suggest canonical forms for each of the following structures. Assume that all bonds are the same length and all oxygen atoms have identical chemical properties: (a) the nitro group and, (b) the carbonate ion.

3

Inter and intra molecular forces of attraction

3.1 DIPOLES AND DIPOLE MOMENTS

A dipole consists of two small electrical charges (q) of equal strength but opposite sign, a short distance apart (Fig. 3.1).

$$-q \xleftarrow{\quad r \quad} +q$$

Fig. 3.1. A dipole.

The strength of a dipole is recorded as its **dipole moment** which is defined as:

Dipole moment = q x r (coulomb metres)

where: r = the distance in metres between the charges
q = the strength in coulombs of the charge

Dipole moments are recorded in units known as debyes (D) where:

1 debye = 3.336 x 10^{-30} coulomb metres (Cm)

Dipoles arise in bonds because of the uneven distribution of electrons in the bonds of a molecule (8.2 and 8.3). These dipoles may give the complete molecule an overall or resultant dipole. Resultant dipoles are the average, both in charge and direction, of all the bond dipoles in a molecule. These resultant dipoles can be measured experimentally. The dipole moments of individual bonds cannot be measured experimentally, but they can be estimated by statistical analysis of the resultant dipoles of many compounds (Table 3.1).

The presence and orientation of the resultant and/or individual bond dipoles in a molecule are indicated by the symbol $\longmapsto$ (the head of the arrow points to the negative end of the dipole). The head of the arrow indicates the general area of the molecule that is rich in electrons (Table. 3.1). Bonds which have a dipole are said to be **polarised**. Compounds whose molecules have a permanent resultant dipole are referred to as **polar compounds**.

Table 3.1. Examples of dipole moments.

Compound	Structure	Resultant dipole moment (D)	Bond dipole moment (D)
Methane		0.0	C-H, 0.2
Hydrogen chloride		1.05	H-Cl, 1.05
Ammonia		1.5	N-H, 1.5
Water		1.8	O-H, 1.6
Chloroethane		2.0	C-Cl, 1.7

3.2 DIPOLE-DIPOLE BONDING

The chemical stability of substances comprised of covalently bonded molecules which have permanent dipoles is currently explained by postulating the existence of weak attractive forces between the molecules. These forces are believed to be electrostatic attractions which occur between the opposite poles of the permanent dipoles of the different molecules (Fig. 3.2).

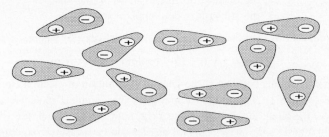

Fig. 3.2. Permanent dipole-permanent dipole interactions. Electrostatic attractions of opposites occurs between the positive and negative ends of adjacent dipoles.

The weak electrostatic attractive forces between permanent dipoles are known as **permanent dipole-permanent dipole attractions**. They are often involved in the binding of substrates to active and receptor sites.

3.3 HYDROGEN BONDING

Hydrogen bonding is a special case of a permanent dipole-permanent dipole attraction. When hydrogen is bonded to another atom (X) the single electron of hydrogen *resides* largely between the hydrogen and X as part of the bond between these atoms. This means that the nucleus of the hydrogen atom is not fully shielded by its electron cloud. As a result, the hydrogen atom of the X–H bond will have a small positive charge ($\delta+$) and the X atom a corresponding small negative charge ($\delta-$). In other words, the X–H bond possesses a dipole with the hydrogen atom being the positive end. If X has a high electronegativity the electrons forming the bond will *'live'* nearer X than the hydrogen with a subsequent increase in the strength of the X–H bonds dipole and the $\delta+$ charge of the hydrogen. For example, oxygen has a higher electronegativity than carbon and the O–H bond has a larger dipole than the C–H bond (Table 3.1).

Hydrogen bonding is the electrostatic attraction that can occur between a hydrogen atom bonded to a highly electronegative atom (the first dipole) and the lone pair of electrons of a second highly electronegative atom (Y), this second electronegative atom acting as the negative end of the second dipole (Fig. 3.3). Electrostatic attraction will only occur if the hydrogen atom and Y are close enough. Hydrogen bonds are usually formed when atoms X and Y are either fluorine, chlorine, nitrogen or oxygen.

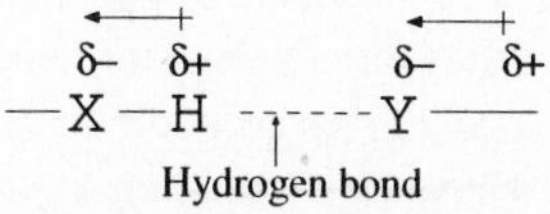

Fig. 3.3. Hydrogen bonding.

Molecules such as water, that in theory can form more than one hydrogen bond, are unlikely to form all the theoretically possible hydrogen bonds at room and physiological temperatures. However, as the temperature is lowered the number of hydrogen bonds formed will increase.

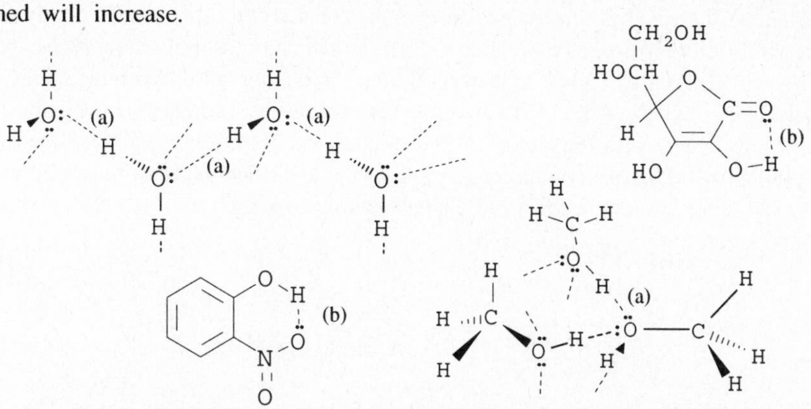

Fig. 3.4. Examples of (a) intermolecular and (b) intramolecular hydrogen bonding.

Although hydrogen bonding is a strong form of dipole-dipole attraction, the majority of hydrogen bonds are weak with bond strengths in the region of 5 to 40 kJ mol^{-1}. For example, the strength of the hydrogen bonds in water is of the order of 22 kJ mol^{-1}.

Hydrogen bonds that are formed between two separate molecules are known as **intermolecular hydrogen bonds** whilst those that are formed within the same molecule are known as **intramolecular hydrogen bonds** (Fig. 3.4). Intermolecular hydrogen bonding occurs more frequently than intramolecular hydrogen bonding. Hydrogen bonding is involved in the binding of substrates to receptor sites.

3.4 HYDROGEN BONDING AND THE SHAPES OF MOLECULES

Both inter and intramolecular hydrogen bonding have a profound influence on the structures and shapes of many naturally occurring molecules. For example, the structures of the nucleic acids, DNA (29.2) and tRNA (29.4). The structure of DNA consists of two polymer strands that are held together by intermolecular hydrogen bonding (Fig. 29.4) whilst the structures of tRNA and rRNA contain sections which are intramolecularly hydrogen bonded (Fig. 29.8 and Fig.29.9). Hydrogen bonding plays an important part in determining the shapes of the secondary, tertiary and quaternary structures of peptides and proteins (28.7).

3.5 HYDROGEN BONDING AND THE PROPERTIES OF MOLECULES

Hydrogen bonds are comparatively weak when compared with covalent and ionic bonds. However, in spite of this weakness they have a great influence on the chemical and physical properties of compounds because of the large numbers of hydrogen bonds that can be involved. They affect a variety of properties ranging from chemical activity to physical constants.

3.5.1 Chemical properties

Intermolecular and intramolecular hydrogen bonding can affect reactivity in a variety of ways, some of which are illustrated by the examples given in this section.

(1) The enolic hydroxyl group bonded to carbon-4 of ascorbic acid is sufficiently acidic to react with sodium hydrogen carbonate to form its salt and carbon dioxide. However, the enolic hydroxyl group bonded to carbon-3 does not react with this reagent as it is intramolecularly hydrogen bonded. Silver nitrate reacts in a similar way for the same reason, it only reacts with the carbon-4 hydroxyl group. Both these reactions were used in the 1980 British Pharmacopoeia monograph as the identification tests for ascorbic acid but only the latter is used in the 1993 British Pharmacopoeia.

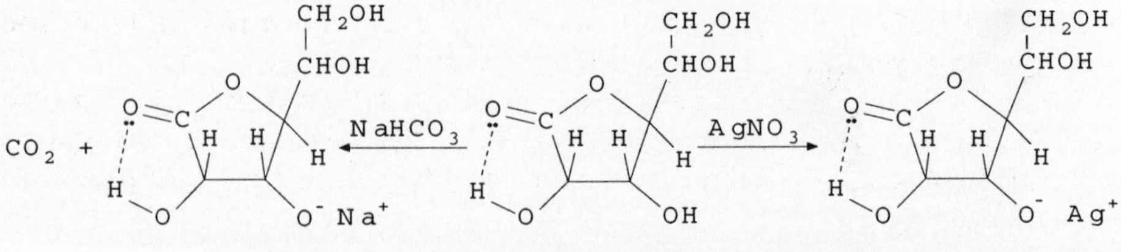

(2) A number of reactions are believed to involve hydrogen bond formation as an important step in their mechanisms. For example, experimental evidence indicates that the rapid decarboxylation of β-ketoacids is believed to be due partly to hydrogen bonding between the acid proton and the oxygen of the keto group.

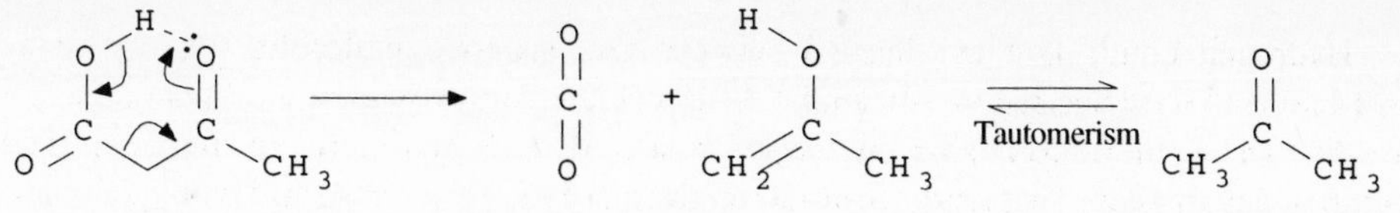

(3) The acidic strength of ortho hydroxyl substituted aromatic acids is often stronger than expected because intramolecular hydrogen bonding stabilises the anion (Fig. 3.5). This moves the position of equilibrium to the right, increasing the value of K_a and subsequently decreasing the value of the pK_a of the acid. Of course the acid itself is also intramolecularly hydrogen bonded but the negative charge of the oxygen makes the hydrogen bond of the anion stronger and so more effective.

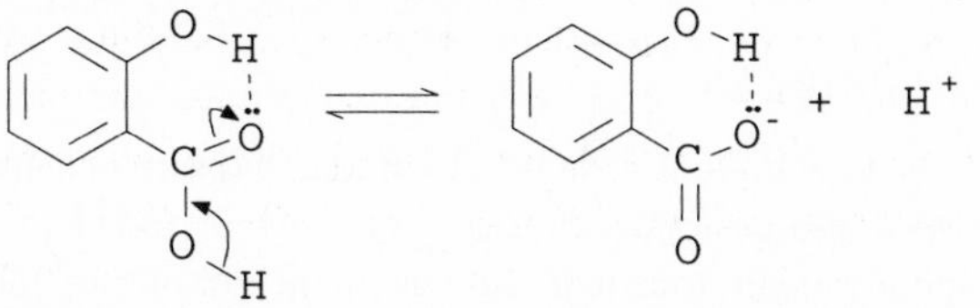

Acid	pK_a
ortho	2.98
meta	4.08
para	4.58

Fig. 3.5. Hydrogen bonding in ortho substituted aromatic acids. The ortho acid is considerably stronger than the meta and para acids as can be seen by their pK_a values.

3.5.2 Physical properties

Intermolecular hydrogen bonding has a considerable effect on physical properties that are energy related, such as, melting point, boiling point, heat of fusion and vaporisation. In each of these cases energy is required to change the state of the substance. Where intermolecular hydrogen bonding occurs extra energy is required to break the hydrogen bonds before the change of state can occur. As a result, intermolecularly hydrogen bonded compounds have higher melting points, boiling points, heats of fusion and vaporisation than similar sized and shaped compounds that are not intermolecularly hydrogen bonded. Furthermore, it follows that the greater the degree of intermolecular hydrogen bonding, the greater its effect on the physical properties of a compound. For example, the values of the physical properties of water, whose molecules are highly intermolecularly hydrogen bonded, are considerably higher than those of ammonia and hydrogen chloride whose molecules have fewer intermolecular hydrogen bonds (Table 3.2). Similarly the physical properties of hydrogen sulphide whose molecules exhibit very little intermolecular hydrogen bonding and methane that exhibits none are lower still. All these compounds are of a reasonably comparable size and shape. Their molecules are tetrahedral (in theory, O, N, S, Cl and C are all sp^3 hybridised), appropriate numbers of lone pairs, electrons and hydrogens occupying the four corners of the tetrahedra.

In general, if the physical properties of a compound are significantly different from those of similar compounds one initially considers hydrogen bonding as a possible explanation. Similarly, unexpected chemical properties can sometimes be explained by hydrogen bonding.

Hydrogen bonding plays a part in the binding of substrates to receptor sites. For example, the action of the antibiotic actinomycin D is believed to involve distortion of the DNA helix of bacteria. This distortion is believed to be caused by the antibiotic hydrogen bonding to a guanine-cytosine base pairs in the DNA of the bacteria.

Table 3.2. The physical properties of some simple compounds.

Compound	mp	bp	ΔH_{fus} kJ mol^{-1}	ΔH_{vap} kJ mol^{-1}
water	0	100	6.02	40.7
ammonia	-77	-33		19.87
hydrogen sulphide	-95	-60	2.39	18.17
hydrogen chloride	-114	-85	1.99	15.11
methane	-182	-161	0.93	8.89

3.6 LONDON DISPERSION FORCES

The electrons of a molecule are in a continuous state of movement. This movement results in the continuous appearance and disappearance of large numbers of transient dipoles. These transient dipoles can induce other transient dipoles in neighbouring molecules (Fig. 3.6). The attractive forces between these transient dipoles are known as **London dispersion forces**. They are weak electrostatic attractive forces, their strength being proportional to $1/r^6$, where r is the distance that the molecules are apart. Consequently, these attractive forces are only effective if the molecules are close enough together (within 0.4 to 0.6 nm of each other).

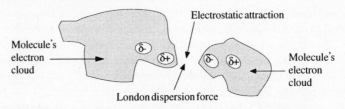

Fig. 3.6. London dispersion forces.

London dispersion forces will occur in molecules irrespective of whether they have a permanent dipole. As the size of the molecule increases so does the size of its electron cloud and the number of London dispersion forces holding the molecules together. For example, the increase in the boiling points of the halogens is due to an increase in the strengths of their London dispersion forces with the increasing size of the molecules.

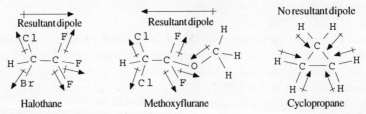

Fig. 3.7. The structures of some general anaesthetics showing their resultant and bond dipoles.

All types of dipole-dipole interaction except hydrogen bonding are often referred to as **van der Waals' forces**. They make an important contribution to the binding of substrates to receptor and active sites. For example, the anaesthetics halothane, methoxyflurane, xenon and cyclopropane are known to bind to the lipids of the neural tissue by van der Waals' forces. However, halothane and methoxyflurane are the better anaesthetics because their permanent dipoles considerably increase the strength of their van der Waals' forces (Fig. 3.7).

3.7 CHARGE TRANSFER COMPLEXES

Charge transfer complexes are formed between molecules when one molecule (the donor) is thought to donate a charge to the other (the acceptor) although the exact nature of the bonding is not fully understood. Donation normally involves pi or non-bonded electrons. Consequently, donor molecules are usually pi electron rich species, and compounds with lone pairs of electrons. Typical examples include alkenes, alcohols, thiols, electron rich benzene rings, pyrrole and some of its derivatives. Acceptor molecules are often electron-deficient compounds, such as pyridine, and halogens such as bromine and iodine. For example, a solution of iodine in cyclohexane is purple as expected but a solution in cyclohexene is brown. The colour in the latter case is thought to be due to the formation of a charge transfer complex between the iodine and the cyclohexene (Fig. 3.8).

X-ray crystallography has indicated that charge transfer complexes formed between aromatic species often have the rings face to face (Fig. 3.8).

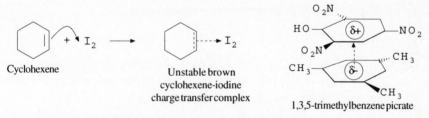

Cyclohexene

Unstable brown cyclohexene-iodine charge transfer complex

1,3,5-trimethylbenzene picrate

Fig. 3.8. Examples of charge transfer complexes.

Charge transfer complexes are believed to be frequently involved in drug-receptor interaction. Experimental evidence suggests that the antimalarial action of drugs based on 9-aminoacridine is partly due to the formation of a charge transfer complex of the drug and the purine bases of the DNA of the parasite.

3.8 WHAT YOU NEED TO KNOW

(1) Dipoles consist of two equal small charges separated by a short distance.

(2) Dipoles arise in molecules because there is an uneven distribution of the electrons within the molecule.

(3) The strength of a permanent dipole is known as a dipole moment and recorded in debyes (D).

(4) The presence of permanent dipoles in molecules explains the forces of attraction between these molecules (permanent dipole-permanent dipole attractions).

(5) Hydrogen bonding is a special case of a permanent dipole-permanent dipole attractive force.

(6) Hydrogen bonding occurs between a lone pair of electrons of a highly electronegative atom such as oxygen and nitrogen, and a hydrogen atom bonded to a highly electronegative atom such as oxygen and nitrogen.

(7) Intermolecular hydrogen bonds are formed between separate molecules.

(8) Intramolecular hydrogen bonds are formed between structures within the same molecule.

(9) Hydrogen bonding is important in determining the shapes of the molecules of many naturally occurring compounds.

(10) Hydrogen bonding can effect the reactivity of a compound.

(11) Compounds that are intermolecularly hydrogen bonded will have higher values for their energy-dependent physical properties than similar compounds that are not intermolecularly hydrogen bonded.

(12) London dispersion forces are weak attractive forces between molecules that arise because of the formation of transient dipoles in the molecules.

(13) The term van der Waals forces is commonly used for all dipole-based weak forces of attraction between molecules except hydrogen bonding.

(14) Charge transfer complexes are formed by the donation of a small electrical charge to a second molecule.

(15) Hydrogen bonding, van der Waals forces and charge transfer are all involved in the binding of substrates to active and receptor sites.

3.9 QUESTIONS

(1) Which of the following compounds have molecules which have permanent resultant dipoles? (a) Carbon dioxide, (b) hexachloroethane, (c) water, (d) phenol, (e) tetrabromomethane and (f) cis-1,2-dichloroethene.

(2) Show how intermolecular hydrogen bonding could occur between the molecules of the members of the following pairs of compounds: (a) ethanol and water, and (b) hydrogen chloride and ethanol. Indicate how intramolecular bonding could occur in (c) 2-hydroxybenzaldehyde and (d) 2-hydroxybenzoic acid.

(3) Would you expect a molecule of water to form four hydrogen bonds with other water molecules at room temperature?

(4) Suggest reasons for each of the following observations.
(a)   Ortho hydroxybenzoic acid has a pK_a of 2.2 while its meta isomer has a pK_a of 3.4.
(b)   The melting point of similar sized and shaped molecules is often very different.
(c)   Iodine and an alkene (R-CH=CH-R) form a yellow-brown complex.

(5) Describe the types of attractive force other than hydrogen bonds that one would expect to exist between molecules.

4

Conformation and configuration

4.1 CONFORMATION

The atoms of a molecule will experience and exert mutual long range repulsions (Fig. 4.1). Energy is required by the molecule to overcome this mutual repulsion and hold the atoms in their relative positions within the structure of the molecule. This means that the relative orientations that the atoms of a molecule take up in space will depend on the energy of the molecule which in turn is dependent on the energy of its environment. As the energy of the environment changes so will the relative orientations of the atoms of a molecule in space change. This change in orientation is usually visualised as the rotation of one section of the molecule about a single covalent bond relative to the other section of the molecule, no bonds being broken. It is referred to as **free rotation** although it will be seen later that this rotation is not entirely free. However, it does give molecules a degree of flexibility

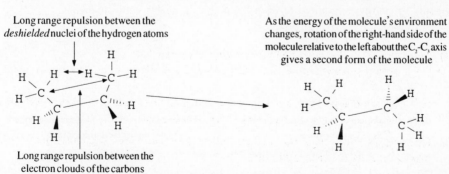

Long range repulsion between the *deshielded* nuclei of the hydrogen atoms

Long range repulsion between the electron clouds of the carbons

As the energy of the molecule's environment changes, rotation of the right-hand side of the molecule relative to the left about the C_2-C_1 axis gives a second form of the molecule

Fig. 4.1. The repulsive forces operating in a molecule and free rotation.

which explains why they are able to adjust their shapes to fit active and receptor sites. Free rotation also explains the alteration in shape of receptor and active sites which enables them to accept molecules and why the binding of substrates to a receptor site can lead to the opening of a second receptor site in the molecular structure of the receptor.

Each of the spatial orientations taken up by a molecule about a single bond is known as a **conformation.** Since there are an infinite number of conformations possible about any one single bond, the **conformation of a molecule** is its exact three-dimensional shape at any one instant.

It is not usually possible to isolate a particular conformation but chemists do distinguish between extreme conformational structures. Consider, for example, the conformations of the C−C bond in a molecule of ethane (CH_3−CH_3). Two extreme conformations of this molecule are possible: one where all the hydrogens are in alignment with each other, called the **eclipsed** conformer, and the other where they are as far out of alignment as is possible, called the **staggered** conformer (Fig. 4.2). These conformations are represented by either the **sawhorse** or **Newman projection formulae**. The sawhorse formula views the molecule from the side and uses triangular shaped bonds with dotted and solid lines to show the three-dimensional nature of the structure. Triangular bonds always have the thick end of the bond nearest the observer, dotted lines represent bonds that are behind the plane of the paper and are going away from the observer whilst solid lines are in the plane of the paper. The Newman projection formula views the structure along the axis of a bond, that is, an end on view. The atom nearest the observer is represented by a dot while the atom furthest away is represented by a circle. Bonds are drawn to either the dot or circle as is appropriate for the structure. Using this convention, the three C−H bonds of the methyl groups of ethane are at $120°$ angles to each other when viewed along the axis of the C−C bond and projected flat onto two-dimensional paper. It should be noted that the eclipsed conformer is drawn so that all the C−H bonds meet the circle representing carbon 2 at the same place (Fig. 4.2). This is to indicate that the relevant hydrogens are one behind the other when viewed along the C−C bond axis of the molecule. The C−H bonds of carbon 1 are continued to the dot that represents this atom. Molecular models show that the hydrogens are closer together in the eclipsed conformer of ethane. Consequently, as the strengths of the repulsive forces between the atoms in a molecule are stronger the closer the atoms are together this structure has the higher energy content because a larger quantity of energy is required to keep the molecule in this conformation.

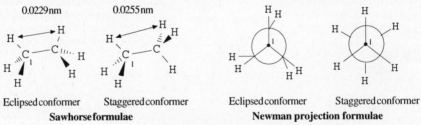

Eclipsed conformer　　　Staggered conformer　　　　Eclipsed conformer　　　Staggered conformer

Sawhorse formulae　　　　　　　　**Newman projection formulae**

Fig 4.2. The eclipsed and staggered forms of the conformers of ethane.

The eclipsed conformer of ethane can be converted into the staggered and other conformers in theory simply by rotating the right-hand side of the structural formula relative to the left (Fig. 4.3). As C_1 is rotated, its hydrogens move further away from those of C_2 and the energy of the molecule falls. Each slight movement of C_1 results in the formation of a new conformer of ethane. After rotating C_1 through $60°$ the energy of the molecule reaches a minimum because the hydrogens are at their maximum distance apart, that is, in the staggered conformation (Fig. 4.3). If the rotation is continued, the hydrogens of C_1 move nearer to those of C_2 until the energy of the molecule reaches a maximum corresponding to the eclipsed form after C_1 has been rotated $120°$. Further rotation results in the energy changes shown in Fig. 4.3. Each point on this graph represents a different conformation of the ethane molecule and so an infinite number of conformations exist between the eclipsed and staggered forms.

The energy difference between the eclipsed and staggered forms is about 12 kJ mol^{-1}. This difference acts as a barrier to free rotation about the C–C bond and is believed to be due to a slight repulsion between the hydrogens as they approach and pass each other in the eclipsed conformation.

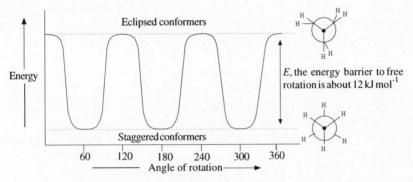

Fig. 4.3. The energy changes that occur during the rotation of one half of an ethane molecule relative to the other half.

Under normal environmental conditions the average background energy striking an ethane molecule is sufficient to keep the molecule permanently in its eclipsed state. However, the molecule is able to lose the energy it gains by processes we know little about but refer to as **relaxation processes**. This means that, in spite of the background energy, the molecule will tend to revert to the staggered conformation as the relaxation processes drain away the energy it has gained. The background radiation will, however, tend to force the molecule back into the eclipsed conformation. Since the energy striking an ethane molecule will vary with time, the molecule can be simply thought of as being in a permanent state of erratic rotation about its C–C bond.

4.2 CONFORMATION OF HYDROCARBON CHAINS

All aliphatic hydrocarbons exist as conformers. The conformations about any C–C bond in a hydrocarbon molecule can be deduced by rotating the structure on one side of the bond relative to the structure on the other side of the bond. During this rotation one can, if required, deduce the relative energies of the conformers by comparing the relative sizes and proximities of the structures on either side of the bond. Consider, for example, the

conformations of the C_2–C_3 bond in a molecule of butane (Fig. 4.4). The **fully eclipsed**
conformer in which the methyl groups are in alignment will have the highest energy. This
is because in this conformer the atoms will be closest together. As C_3 is rotated clockwise,
the methyl groups move out of alignment. The atoms of the molecule are now further apart
and the energy of the molecule will fall until it reaches a minimum at $60°$ rotation. This
conformation is called the **gauche** or **skew** conformation. Further clockwise rotation of
C_3 results in an increase in the energy of the molecule until it reaches a maximum at $120°$,
when hydrogens and methyls are eclipsed. However, the energy of this conformer will be
less than that of the eclipsed conformer because there will be less repulsion between the
smaller electron clouds of the hydrogens and the larger electron clouds of the methyl
groups. Continued rotation of C_3 results in a fall in the energy of the molecule as the
methyls move further apart until at $180°$ they are at their maximum distance apart. This
conformation is known as the **anti** conformation. It has the lowest energy of all the
conformers about the C_2–C_3 bond since the repulsion between the two electron clouds of
the methyl groups will be at its lowest. Continuation of the clockwise rotation of C_3 until
the methyl group reaches its original position gives an energy curve that is a mirror image
of the 0 to $180°$ curve.

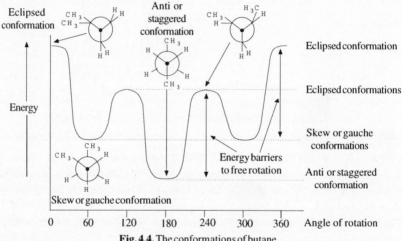

Fig. 4.4. The conformations of butane.

The barriers to the free rotation about the C_2–C_3 bond are not large enough to allow the
isolation of the skew and anti conformers of butane (Fig. 4.4) but at room temperature the
presence of both skew and anti forms have been detected by spectroscopic methods. It has
been estimated that at any one time about 60% of the molecules in a sample of butane are
in the anti form and about 40% in the skew form. The molecule apparently oscillates
between these two conformers.

At room temperature there is usually sufficient background energy to ensure that all
the possible conformers about a bond can exist. However, by far the largest percentage of
molecules will at any one time be in either the staggered or the anti forms. These
conformations are the most likely to be adopted by the C–C bonds of saturated
hydrocarbon chains (Fig. 4.5). The situation is not static and the conformations of all the

molecules will be continually changing. However, the status quo, namely the majority of molecules in the staggered or anti conformations will be maintained.

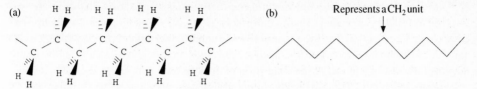

Fig. 4.5. (a) The anti (staggered) form of a hydrocarbon chain. (b) The zig-zag line used to represent saturated hydrocarbon chains. The bends in the line corresponds to CH_2 units.

4.3 CONFORMATIONS OF CYCLOHEXANE

Cyclohexane has the structural formula shown in Fig. 4.6. In theory each of the carbon atoms of the ring is sp^3 hybridised with bond angles of approximately $109.5°$. Thus it is possible to construct two model structures for cyclohexane which because of their shapes are known as the **chair** and **boat** forms. Experimental evidence shows that cyclohexane can exist in both of these conformations but the chair form is preferred at room temperature.

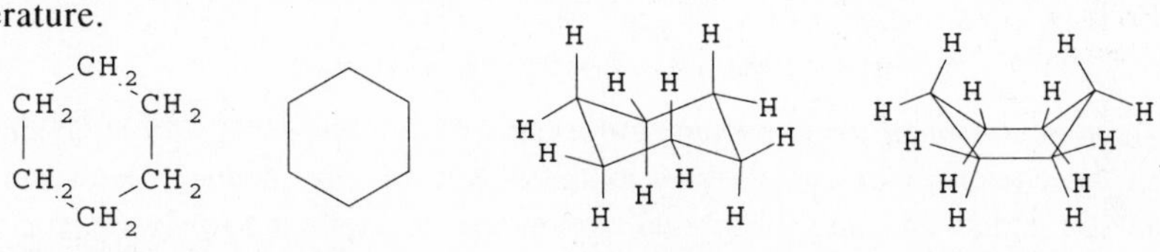

Cyclohexane   Cyclohexane   Chair form   Boat form

Fig. 4.6. The structural formulae of cyclohexane and its principal conformers.

4.3.1 The chair conformer

Spectroscopic evidence shows that this conformer is the most stable form of the molecule. The Newman projections show that all the hydrogen atoms are in the staggered conformation (Fig. 4.7). The vertical C−H bonds are called axial bonds (designated by *ax*) while the horizontal C−H bonds are referred to as equatorial bonds (designated by *eq*) because they form a 'belt' around the ring. This bond arrangement be seen by making your own

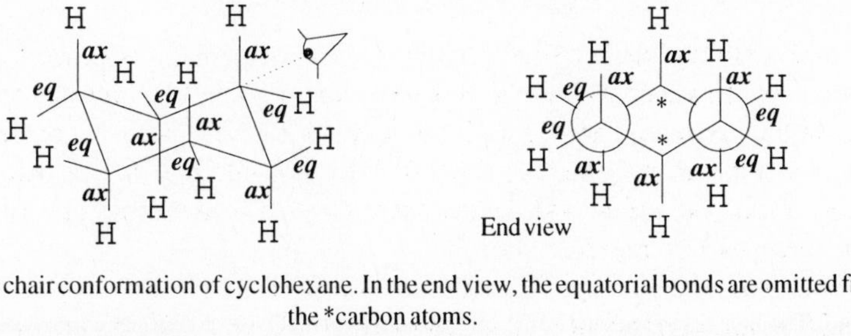

End view

Fig. 4.7. The chair conformation of cyclohexane. In the end view, the equatorial bonds are omitted from the *carbon atoms.

ball and stick models of the structures. Hydrogens occupying the axial positions are closer together than those occupying equatorial positions. This means that more energy is required to keep hydrogen atoms in axial positions than in equatorial positions. This has no significance with cyclohexane but considerable effect on the structures and chemistry of substituted cyclohexanes.

Low temperature spectroscopic and other experimental work shows that there are two interconvertible cyclohexane chair conformers (Fig. 4.8). Interconversion occurs because the energy barrier for the process is low (about 43 kJ mol^{-1}). During interconversion, the axial hydrogens move to equatorial positions and vice versa. This can be easily demonstrated using a model of the chair form of cyclohexane with say, red axial and white equatorial C–H bonds. Hold the centre four carbons stationary and move C_1 down and C_4 up. As you move the carbon atoms all the red axial C–H bonds move to equatorial positions whilst all the white equatorial C–H bonds move to axial positions. In practice this interconversion is more complex and is believed to occur via intermediate envelope and boat conformations.

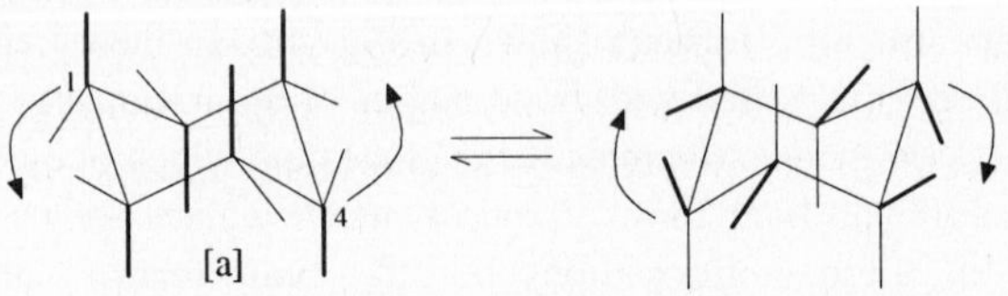

Fig. 4.8. The two chair conformers of cyclohexane. Note the axial bonds (thick lines in [a]) become equatorial and vice versa as the rings interconvert.

The interconversion of the two chair conformers of cyclohexane is very rapid and does not affect the chemistry of cyclohexane as the two forms are identical. However, it is significant for mono and polysubstituted cyclohexanes. Methylcyclohexane, for example, can exist in two conformational forms. In one the methyl group is in an axial position and in the other it is in an equatorial position (Fig. 4.9). If the Newman projection formulae along the C_1–C_2 and C_5–C_4 bond axis of these structures are compared, the methyl group is further from the axial C_3 and C_5 hydrogen atoms in equatorial methylcyclohexane (Fig. 4.9b) than it is in the axial methylcyclohexane (Fig. 4.9a). This means that the hydrogen atoms of the equatorial positioned methyl group are further away from nearest ring hydrogen than the hydrogen atoms of the axial methyl group. In other words, there is more room for the methyl group in the equatorial than in the axial position. This results in equatorial methylcyclohexane being more stable than axial methylcyclohexane.

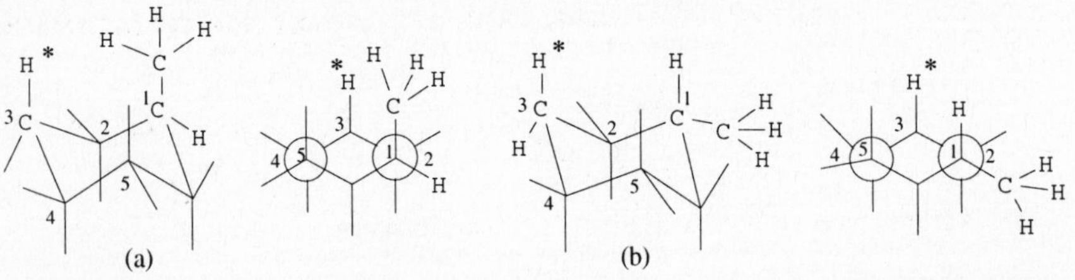

Fig. 4.9. The axial (a) and equatorial (b) chair structures of methylcyclohexane. The hydrogens marked with an asterisk are closer to the methyl group in the axial rather than the equatorial position.

Experimental work has shown that the barrier to the interconversion of the methylcyclohexane chair forms is only 7.5 kJ mol^{-1}. At any one instant, 95% of the molecules of a sample of methylcyclohexane have the methyl group in the lower energy equatorial position. As the size of the substituent group increases, the percentage of molecules with the substituent in an equatorial position also increases. With t-butyl and bigger substituents, the substituent group is found to exist almost 100% of the time in the equatorial position. The molecule is effectively locked into this lower energy structure.

The same considerations apply to all substituted cyclohexanes. Generally, the molecule will exist as its lowest energy conformer. This can be deduced from a consideration of the Newman projections for the structure. This deductive process is known as **conformational analysis**.

4.3.2 The boat and twisted conformers

The bonds of the boat conformer are designated nautically as bow sprite (**bs**) and flagpole (**fp**) as well as pseudo-axial (ax') and pseudo-equatorial (eq') (Fig. 4.10). It is more flexible than the chair form and can twist into a number of different spatial arrangements, one of which is the **twisted boat** conformer shown in Fig. 4.10. Spectroscopic measurements show that the boat conformations have significantly higher energies than the chair forms.

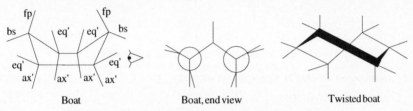

Fig. 4.10. The boat and twisted boat conformations of cyclohexane.

In practice, cyclohexane derivatives and related ring structures such as the six-membered ring of pyranose sugars (Fig. 27.4) rarely adopt the boat conformations.

4.4 CONFORMATIONS OF OTHER CYCLIC SYSTEMS

With the notable exception of cyclopropane and its derivatives, cyclic systems containing saturated atoms will not usually be flat. They are bent and twisted into conformations which give the structure the lowest energy possible. To achieve this, the atoms and substituents generally take up positions where they are as far apart in space as possible. Some examples of these structures, together with their relevant Newman projections, are given in Fig. 4.11.

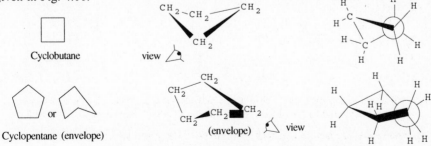

Fig. 4.11. Examples of the conformations of some simple ring systems.

The conformations of large rings are not known with any degree of certainty as they are often very flexible. This makes determination of their conformation difficult.

4.5 CONFORMATIONS OF SATURATED FUSED RING SYSTEMS

Fused ring compounds have two or more rings that share the same bond or bonds. The atoms forming the common bond are known as **bridgehead or ring junction atoms** (Fig. 4.12). Each of the rings in a saturated fused ring molecule takes up the most appropriate conformation for that ring system. Six-membered rings, for example, are nearly always found in the chair form although occasionally they may occur in the boat form. The complete structure may or may not be flexible. For example, trans-decalin has a rigid structure but cis-decalin has a flexible structure. Furthermore trans-decalin cannot be transformed into cis-decalin without breaking bonds.

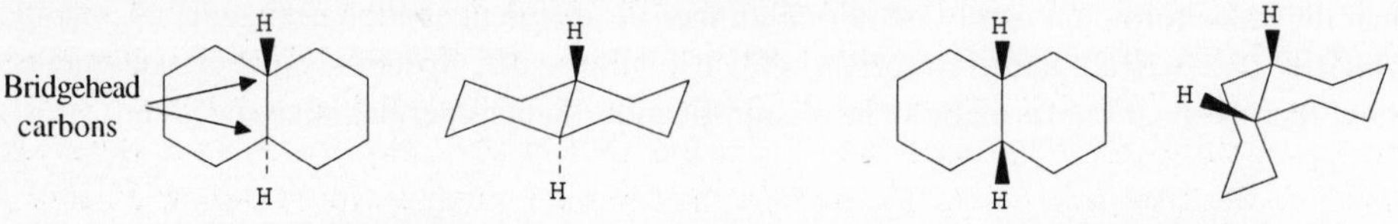

Bridgehead carbons

Trans-decalin   Cis-decalin

Fig. 4.12. Decalin, a saturated fused ring compound. The convention used for all saturated ring systems is that bonds directed above the plane of the ring system are shown as either solid or wedge shaped lines while bonds that lie below the plane of the ring system are indicated by dotted (- - -) lines.

Ring fusion can result in isomerism as can be seen in the case of decalin shown in Fig. 4.12. It is interesting to note that the 5a and 5b naturally occurring series of steroids are isomers in which the A and B rings of the steroids have been found to be fused together in the same ways as the rings of decalin (Fig. 4.13). Although ring fusion isomerism may be possible, it does not always occur in nature. The B/C and the C/D ring junctions of steroids, for example, normally occur naturally in the forms shown in Fig. 4.13.

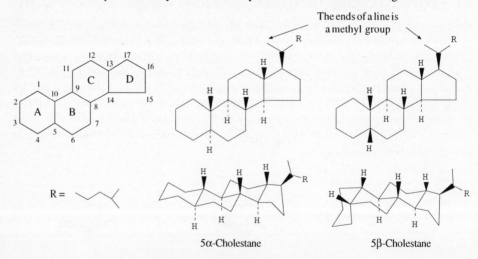

The ends of a line is a methyl group

R

5α-Cholestane   5β-Cholestane

Fig. 4.13. The 5α and 5β isomers of cholestane.

4.6 CONFIGURATION

The configuration of a molecule is the three-dimensional arrangement of the atoms forming that molecule about a fixed point in that molecule. Unlike conformations, a configuration can only be changed into another configuration by the breaking and making of bonds. It should be remembered that conformations are interconverted by rotating or twisting sections of the molecule about a bond or bonds.

Configurations are often represented on paper in two dimensions by the use of Fischer projection formulae. Their nature and the use of these formulae are best understood by the use of molecular models. Construct a tetrahedral structure like that shown in Fig. 4.14a. You can use molecular models or, if these are not available, pieces of plasticine and matchsticks are equally effective. Take the model and turn it so that two of the bonds are horizontal and pointing towards you (Fig. 4.14b). The other two bonds will now be vertical and pointing away from you. All four bonds form a cross with the central carbon atom at the point where the lines cross. This structure is now projected onto a piece of paper as a flat drawing. This flat representation of the three-dimensional structure is the Fischer projection formula of the tetrahedral structure (Fig. 4.14c). Horizontal lines represent bonds towards the observer whilst vertical lines represent bonds away from the observer. The carbon atom whose configuration is being depicted is located at the point where the lines cross.

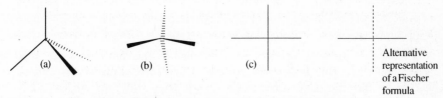

Fig. 4.14. Conversion of a tetrahedron into a Fischer formula.

Fischer formulae cannot be rotated in the plane of the paper through 90° or 270°. Rotation through these angles would break the convention because the horizontal bonds would in fact be behind the plane of the paper and not in front as required by the Fischer convention. Similarly the vertical bonds would now be in front of the plane of the paper and not behind as required by the convention. You can demonstrate this to yourself by taking your model, orientating it to yourself as shown in Fig. 4.15a and turning it through 90° (Fig. 4.15b) and 270° (Fig. 4.15d). The bonds that were further away from you will now be nearer and vice versa. At the same time one can see that rotation in the plane of the paper through 180° (Fig. 4.15c) and 360° (Fig. 4.15e) is allowed as the convention used for drawing the bonds is not broken.

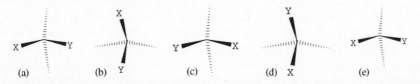

Fig. 4.15. The compromising of the Fischer convention by rotation of a Fischer projection through 360°.

An easier method of manipulating Fischer formulae is based on swapping the positions of the atoms or groups of atoms attached to the central carbon atom. Construct two identical tetrahedral structures corresponding to A and B shown in Fig. 4.16. Use structure A as a reference and manipulate the groups attached to the central carbon atom of structure B. Swap any two groups of structure B. You have now **inverted** structure B and made a mirror image of your reference structure, A. Check this fact by attempting to superimpose the two structures. This will not be possible but take care not to confuse the reference structure with the one you are manipulating. Carry out a second swap of any two structures. and again compare the result with the reference structure. This final structure will now be identical to the reference structure even though it looks quite different. Simply turn the molecule around until it can be superimposed on the reference structure.

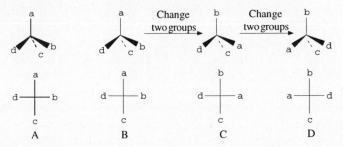

Fig. 4.16. Manipulation of a Fischer formula. Structure C is a mirror image of structure A while structure D is identical to structure A even though these structures appear to be different when drawn on paper.

It is important to realise that apparently different Fischer projections may represent the same molecule. **This is because molecules are three-dimensional and so can appear to be different structures when viewed from different angles.** A human analogy is when one initially fails to recognise a friend from behind because from this position their appearance is different. Different Fischer projections of the same molecule can be identified by taking one of the projections and attempting to convert it into the others by one of the methods previously described.

4.7 ABSOLUTE CONFIGURATION

The **absolute configuration** of a molecule is the experimentally determined configuration of a molecule. Molecules with asymmetric centres, that is, their structures have one or more centres of dissymmetry, will have more than one absolute configuration. These structures may or may not exhibit optical activity (5.2). For example, glyceraldehyde has two absolute configurations because C_2 is asymmetric (Fig. 4.17). Both these glyceraldehyde molecules are optically active. Bijvoet and his collaborators showed in 1951 that the (+) isomer had the absolute configuration A whilst the (-) isomer had the absolute configuration B. This means that the sign of optical rotation of each of these isomers can be used like a label, enabling one to identify the configuration of a particular sample of glyceraldehyde. Furthermore, this use of the sign of optical rotation can be extended to other optically active isomers once the relationship between their configurations and sign of optical rotation has been established.

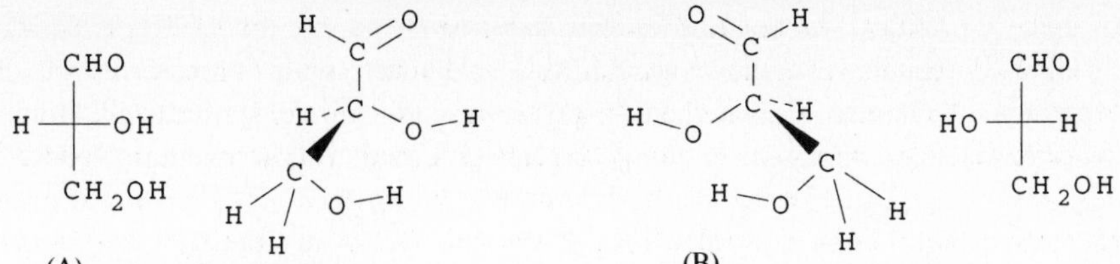

(A) (+)-Glyceraldehyde (B) (-)-Glyceraldehyde

Fig. 4.17. The absolute configurations of the isomers of glyceraldehyde.

4.8 THE NOMENCLATURE OF ABSOLUTE CONFIGURATION

Two nomenclature systems are currently used to indicate the absolute configuration of a specified point in the structure of a compound. The **D/L** system is based on the convention introduced by Fischer in 1891 whilst the **R/S** system was devised by Cahn, Ingold and Prelog in 1956. This latter system is in common use whilst the D/L system is restricted mainly to carbohydrates, amino acids and a few other natural products.

4.8.1 The D/L system

The basis of this system are the absolute configurations given to the glyceraldehyde isomers by Fischer (Fig. 4.18). Fischer assigned these configurations in a completely arbitrary fashion. Happily in 1951 Bijvoet and his collaborators showed that these arbitrary designations were correct.

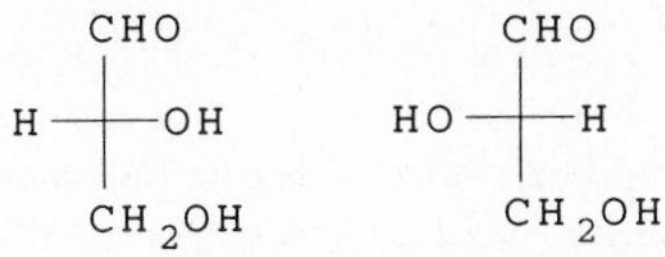

D-(+)-Glyceraldehyde L-(-)-Glyceraldehyde

Fig. 4.18. The Fischer convention representations of the absolute configurations of the isomers of glyceraldehyde.

In order to classify a particular glyceraldehyde structure as D or L, **it must be drawn**, with the aldehyde group at the top of the projection before being compared with the projections given in Fig. 4.18. However, when doing this, it should be remembered that Fischer projections represent three-dimensional structures and so they must be drawn and manipulated **on paper** in the manner described in section 4.6.

The configurational centres of other molecules can be classified as D or L by chemically relating them to the glyceraldehydes or other compounds with known absolute configurations. For example, (-)-serine is used as a standard for amino acids. These chemical relationships often use the sign of optical rotation as a label to determine precisely which isomers are chemically related. **The sign of optical rotation is not associated with a particular type of configuration.** One can have D(+), L(+), D(-) and L(-) configurations. The sign of optical rotation is used only as an indicator to link the appropriate isomers together. The reactions used for the relationship should either be of known stereospecificity, that is, their effect on the stereochemistry of the compounds must be known or they must not involve the centre whose configuration is being determined. Consider, for example, the determination of the configuration of (+)-isoserine. Practical work shows that nitrous acid reacts with (+)-isoserine to form (-)-glyceric acid. It also shows that (-)-glyceric acid can be synthesised from of (+)-glyceraldehyde by oxidation of the glyceraldehyde with yellow mercuric oxide.

$$\underset{\text{(+)-Isoserine}}{CH_2(NH_2)COOH} \xrightarrow{HNO_2} \underset{\text{(-)-Glyceric acid}}{CH_2OH.CH_2OH.COOH} \xleftarrow{HgO} \underset{\text{(+)-Glyceraldehyde}}{CH_2OH.CH_2OH.CHO}$$

These reactions do not involve the central carbon of any of the molecules in the relationship. Therefore, as D-glyceraldehyde is the glyceraldehyde isomer with the positive optical rotation, all the other isomers in the relationship must have a D configuration (Fig. 4.19). Configurations determined in this manner are called **relative configurations.** Most relative configurations have been shown to be identical to the absolute configuration of the compound and are usually referred to as absolute configurations.

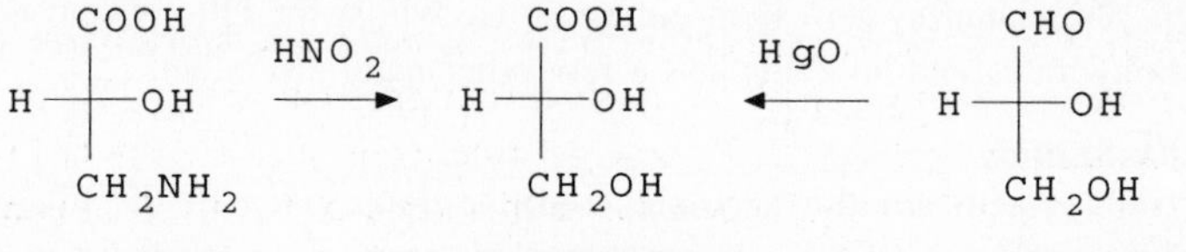

D-(+)-Isoserine D-(-)-Glyceric acid D-(+)-Glyceraldehyde

Fig. 4.19. The relationship between the configurations of (-)-glyceric acid, (+)-isoserine and the absolute configuration of D-(+)-glyceraldehyde.

There are also a number of physical methods of determining the absolute configurations of compounds. The most important of these is X-ray crystallography.

The D/L system can be ambiguous. Isomers can be designated as D or L depending on the particular series of chemical reactions used to relate them to the appropriate standard compound and the actual standard used. For example, (-)-threonine has an L configuration when related to serine but a D configuration when related to glyceraldehyde.

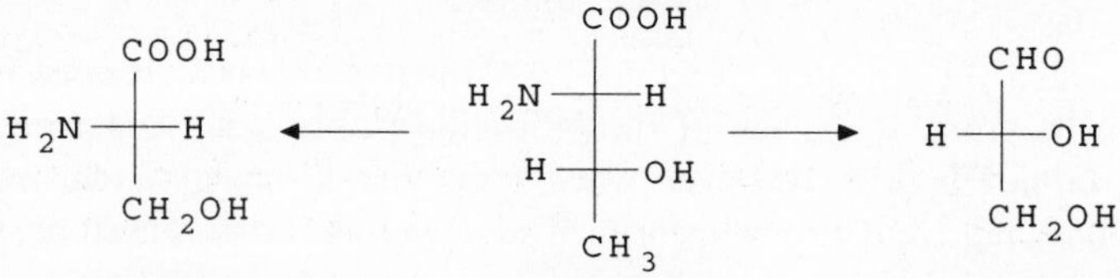

L-Serine (-)-Threonine D-Glyceraldehyde

The standard used in the determination of the configuration is sometimes indicated by the use of an appropriate subscript with D and L. For example, the subscript s is used when (-)-serine is used and g for a glyceraldehyde. Consequently (-)-threonine would be classified as being either L_s-(-)-threonine or D_g-(-)-threonine.

The use of the D/L system is restricted to one configurational centre in the molecule. In monosaccharides, for example, it is the penultimate carbon in the chain from the carbonyl group that is chemically related to the glyceraldehydes so it is this atom to which the D and L refer in the names of the monosaccharides (Fig. 4.20). The D/L system completely ignores the other configurational centres in monosaccharide molecules. The configurations of these other centres are indicated by the use of specific prefixes as part of the stem names of monosaccharides (27.2.3).

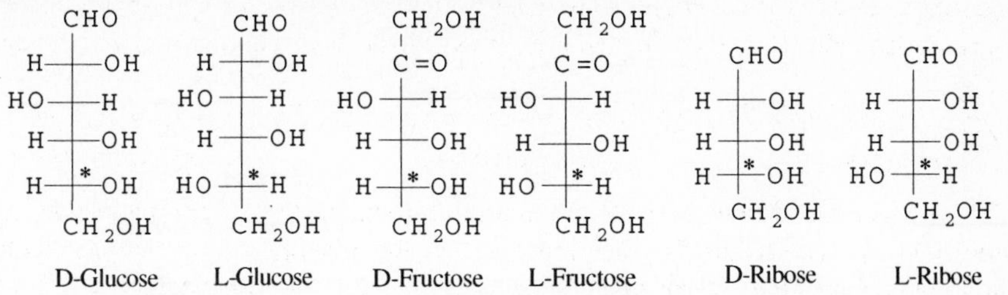

	CHO		CHO		CH$_2$OH		CH$_2$OH		CHO		CHO
H—OH		H—OH		C=O		C=O		H—OH		H—OH
HO—H		HO—H		HO—H		HO—H		H—OH		H—OH
H—OH		H—OH		H—OH		H—OH		H—*OH		HO—*H
H—*OH		HO—*H		H—*OH		HO—*H		CH$_2$OH		CH$_2$OH
CH$_2$OH		CH$_2$OH		CH$_2$OH		CH$_2$OH				
D-Glucose		L-Glucose		D-Fructose		L-Fructose		D-Ribose		L-Ribose

Fig. 4.20. The configurations of some monosaccharides. The asterisk indicates the position of the carbon atom (the penultimate atom) to which D and L refer in each of these structures.

4.8.2 The R/S system

This is an unambiguous system based on the three-dimensional structure of the compound. It can be used for any atom to which are bonded four different groups. These groups are given a priority of **a,b,c** or **d** where the order of priority is: **a > b > c > d**. Priorities are assigned using a set of rules (4.8.3). In the case of three-dimensional formulae, the atom whose configuration is required is viewed from the side furthest from the group with the **d** priority. If the sequence **a** to **b** to **c** of the remaining groups is in a clockwise direction, the atom is said to have an R configuration. If it is anticlockwise the atom is said to have an S configuration (Fig. 4.21). With Fischer projections the same sequence relationship is used but the group with the **d** priority must be at either the top or the bottom.

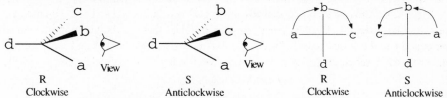

Fig. 4.21. The R and S designations for carbon atoms.

There is no predictable relationship between R, S and the sign of optical rotation of optically active compounds. R(+), R(-), S(+) and S(-) compounds are known.

4.8.3 The R/S priority rules

(1) Each atom bonded directly to the atom whose configuration is being determined is considered in turn. The atom with the highest atomic number is given the highest priority. If isotopes of the same element are present, the isotope with the highest mass number has the greater priority. In Fig. 4.22, for example, the order of priority in terms of atomic number is bromine (**a**), chlorine (**b**), carbon (**c**) and hydrogen (**d**). Since this makes the path of the sequence **a** to **b** to **c** clockwise the structure has an R configuration. It should be noted that it does not matter whether the Fischer formula has the **d** priority group at the top or bottom. However, it should be remembered that this group can only be moved

to these positions if one **interchanges two of the groups twice** (4.6) as demonstrated in Fig. 4.22.

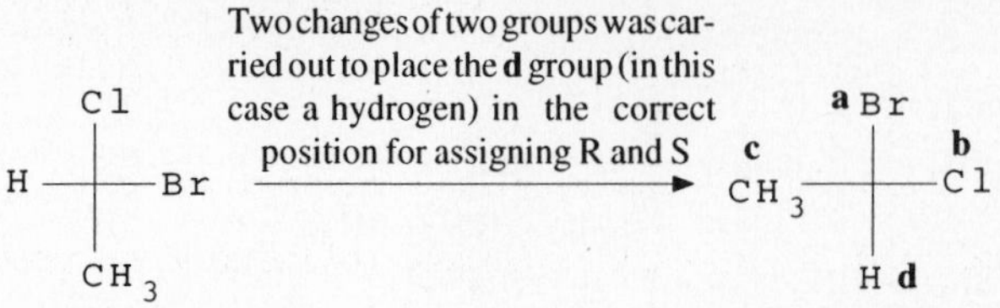

Two changes of two groups was carried out to place the **d** group (in this case a hydrogen) in the correct position for assigning R and S

$$
\begin{array}{c}
\text{Cl} \\
\text{H} \rule[0.5ex]{1.5em}{0.4pt}\!\!\rule[0.5ex]{1.5em}{0.4pt}\text{Br} \\
\text{CH}_3
\end{array}
\quad\longrightarrow\quad
\begin{array}{c}
\mathbf{a}\,\text{Br} \\
\mathbf{c}\qquad\quad\mathbf{b} \\
\text{CH}_3 \rule[0.5ex]{1.5em}{0.4pt}\!\!\rule[0.5ex]{1.5em}{0.4pt}\text{Cl} \\
\text{H}\,\mathbf{d}
\end{array}
$$

R-1-bromo-1-chloroethane.

Fig. 4.22. The determination of the configuration of 1-bromo-1-chloroethane.

(2) If two or more of the atoms being considered in rule 1 are the same, one moves along the groups in which they occur, to the first point of difference. If these groups contain substituent branches, it is necessary to examine only the branches that contain the atoms with the highest atomic numbers. Consider, for example, the configuration of C_3 of the isomer of 1-chloropentan-3-ol shown in Fig. 4.23. Applying rule one gives the hydroxyl group the **a** priority and the hydrogen atom the **d**. Both the groups marked X and Y have carbon atoms directly bonded to C_3, therefore one must apply rule 2. Moving out along the groups X and Y, the first point of difference occurs between C_1 and C_5. As the chlorine has the higher atomic number group X is given the higher **b** priority and group Y the lower **c** priority. Consequently, the sequence **a** to **b** to **c** is anticlockwise when viewed in the appropriate manner for the type of formula drawn. This means that the isomer shown in Fig. 4.23 has the S configuration.

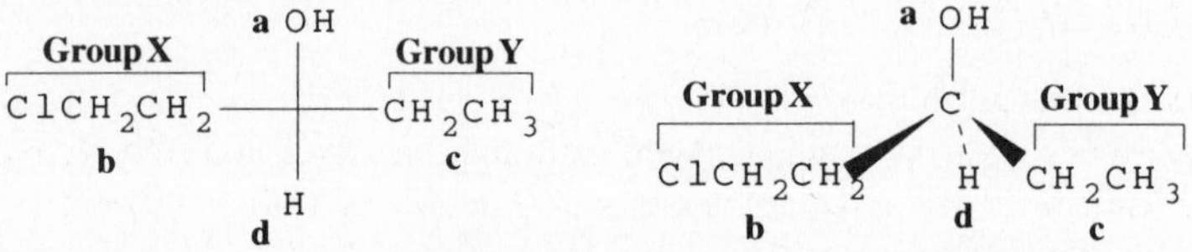

$$
\underbrace{\text{ClCH}_2\text{CH}_2}_{\mathbf{b}}\overset{\displaystyle\mathbf{a}\ \text{OH}}{\underset{\displaystyle\underset{\mathbf{d}}{\text{H}}}{\rule[0.5ex]{2em}{0.4pt}\!\!\rule[0.5ex]{2em}{0.4pt}}}\underbrace{\text{CH}_2\text{CH}_3}_{\mathbf{c}}
$$

Group X ⎰‾‾‾‾‾‾‾⎱ Group Y

Group X ⎰‾‾‾‾‾⎱ C Group Y
ClCH₂CH₂ H CH₂CH₃
b **d** **c**

Fig. 4.23. S-1-Chloropentan-3-ol.

The same principle applies to ring structures. It is only necessary to consider the arm of the ring containing the atoms with the highest atomic numbers. At the first point of difference either the structure with the atom of highest atomic number or the most substituents with higher atomic numbers has the higher priority. Consider, for example, the configuration of C_3 in the structure of the isomer of 3-chloropiperidine shown in Fig. 4.24. Applying rule 1 gives the chlorine atom the **a** and the hydrogen the **d** priority. Applying rule 2 to the arms of the ring marked X and Y, the first point of difference occurs between the N and C_5. The arm of the ring marked with an X has the higher priority since the nitrogen atom has a higher atomic number than the carbon atom at the corresponding position on the arm marked Y.

(3) Unsaturated atoms are converted into fully saturated structures by the use of **ghost or pseudo-atoms** to keep the valency of the **real atoms** correct. The valency of the pseudo-atoms is assumed to be satisfied by imaginary atoms with zero atomic numbers. This gives the pseudo-atom a lower priority than the corresponding real atom. Consider,

View from behind the plane of the paper

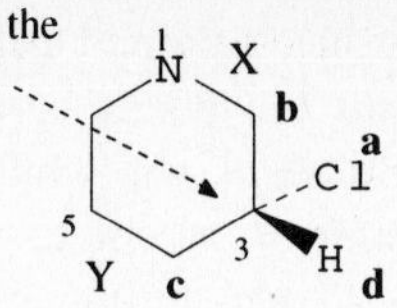

The structure must be viewed from behind the plane of the paper if it is to be viewed from the side opposite to the structure with the **d** priority. The sequence **a** to **b** to **c** is clockwise and so carbon 3 has an R configuration

Fig. 4.24. The configuration of C_3 of 3-chloropiperidine.

for example, the application of this rule to the structure of the carbonyl group. The carbon atom of the carbonyl group is bonded twice to the oxygen so the second bond is represented as a separate single bond to a ghost oxygen atom. In a similar manner, the second bond from the oxygen to the carbon is represented by a separate single bond to a ghost carbon atom (Fig. 4.25 a). This procedure may be applied to the traditional structural formula of any unsaturated structures (Fig. 4.25 b and Fig. 4.25 c). The resulting 'saturated' structures and their ghost atoms are assessed using rules 1 and 2 as appropriate.

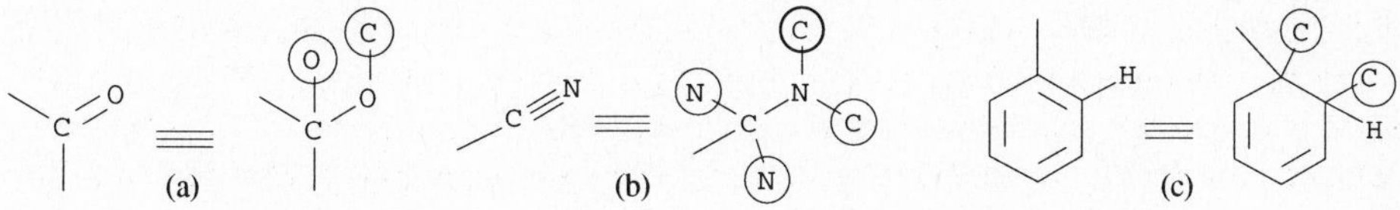

Fig. 4.25. The ghost atoms (ringed) of the carbonyl and other unsaturated structures.

(4) If isotopes of the same element are present, the isotope with the highest mass number has the highest priority. You should not confuse this with atomic number on which the rules are based. **Mass number must only be used to distinguish between the priorities of isotopes of the same element.**

The R/S system can be used to describe all the asymmetric centres in a molecule by prefixing the letter with the number of the carbon atom concerned (Fig. 4.26). It is not compatible with the D/L system (and vice versa). A compound with an R configuration will not necessarily have a D configuration, it could just as easily have an L configuration. In fact, the configurations of some compounds which can be specified as R or S cannot be specified as D or L.

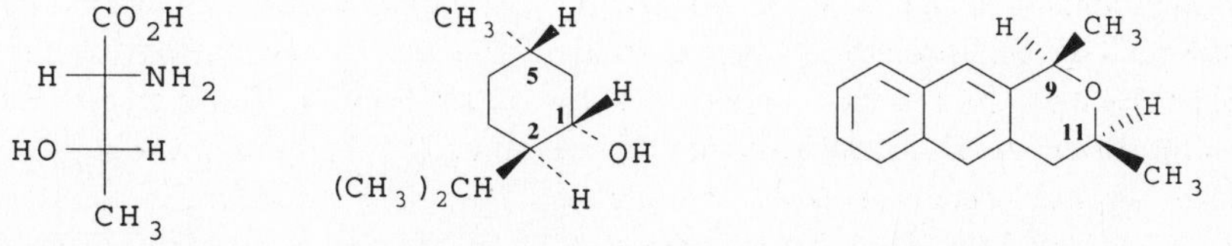

2R,3S-Threonine 5S,2R,1S-Menthol 9R,11S-Eleuthrin

Fig. 4.26. Examples of the use of the R/S system in molecules that have more than one asymmetric centre.

4.9 THREE-DIMENSIONAL FORMULAE AND FISCHER PROJECTIONS

Should you wish to convert a three-dimensional formula into a Fischer projection the simplest way is to use a model. Allocate a colour to each of the structures in the three-dimensional formula and then construct the matching tetrahedral structure. You can use either molecular models or, if they are not available, coloured balls of plasticine and matchsticks. The tetrahedral model is turned until the bonds agree with the Fischer convention, that is, horizontal bonds are pointing towards you whilst vertical bonds are pointing away from you. Draw this structure in terms of the colour codes, taking care to observe the Fischer convention. Replace the colour codes by the appropriate structures. The complete procedure is illustrated in Fig. 4.27. To convert a Fischer projection into a three-dimensional formula, once again use a model but use the process described previously in reverse.

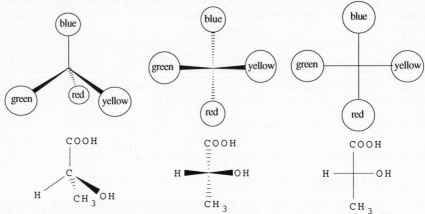

Fig. 4.27. Converting three-dimensional formulae into Fischer projections.

4.10 BIOLOGICAL SYSTEMS AND CONFIGURATION

4.10.1 Occurrence

The members of some classes of naturally occurring compounds are found to have mainly one type of configuration. For example, naturally occurring amino acids usually have L configurations while naturally occurring monosaccharides normally have D configurations (Fig. 4.28).

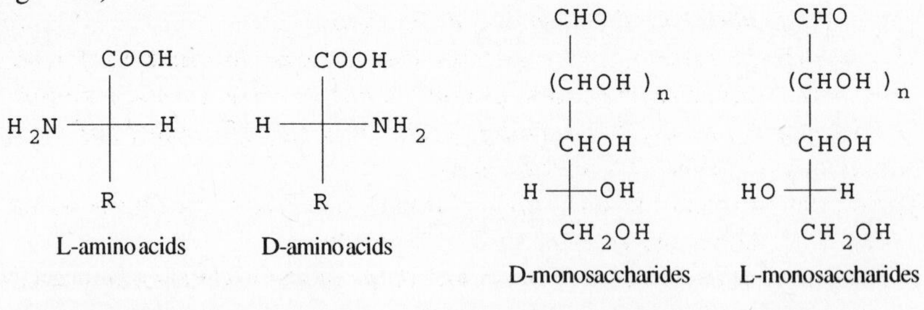

Fig. 4.28. The configurations of naturally occurring amino acids and monosaccharides.

4.10.2 Reactivity

Shape is one of the factors that governs the reactivity of compounds in biological processes (1.6). Although a configuration does not give the complete shape of a molecule it can be related to the reactivity of chemical species in biological systems. It is used to discuss and describe the stereospecific nature of reactions in these systems. Many enzymes, for example, will only catalyse reactions involving the isomer with an R configuration. They will not catalyse the reaction if the isomer has the corresponding S configurations. Other enzymes will give a product with a particular configuration as either the sole or the major product of the reaction.

4.10.3 Pharmacological activity

Different configurations of the same compound, that is, different isomers, can cause quite different physiological responses in biological systems because of the dependence of reactivity on shape. Some of the possible options are:

(1) One may cause a response, the other may be biologically inert and cause no response.
(2) Both may cause beneficial responses but one may be more efficacious than the other.
(3) Both may cause beneficial responses but they may cause different side-effects.
(4) They may have opposite effects, one having a beneficial, the other a detrimental response.

One consequence of these widely differing biological activities is that future legislation is likely to require manufacturers to use individual enantiomers (5.3) rather than racemic modifications (5.4) in clinical trials of new drugs.

4.11 WHAT YOU NEED TO KNOW

(1) Conformers are isomers formed by the free rotation of the structures on either side of all the sigma bonds in the molecule.
(2) Conformations may be represented by sawhorse or Newman projection formulae.
(3) The conformations adopted by a molecule depend on the energy of that molecule and its environment.
(4) Eclipsed conformations have the highest energy content.
(5) Anti and staggered conformations have the lowest energy content.
(6) Saturated and non-aromatic ring systems have conformations with specific names.
(7) Substituents on saturated and non-aromatic ring systems normally prefer to occupy equatorial or their equivalent positions.
(8) Six-membered saturated ring systems usually exist as chair conformers. Five-membered saturated rings are usually found to have envelope conformations.
(9) Isomers of the naturally occurring 5a and 5b series of steroids differ only in the structures of their A/B ring junctions.
(10) Free rotation (conformations) give molecules a degree of flexibility which enables them to fit active and receptor sites more easily.
(11) Configuration is the three-dimensional shape of the molecule about a fixed point in that molecule.
(12) The configuration of a molecule is indicated by the use of the D/L or R/S systems of nomenclature.
(13) The D/L and the R/S nomenclature systems are not compatible.
(14) There is no consistent relationship between the sign of optical rotation and the configuration of an isomer.

4.12 QUESTIONS

(1) Draw the Newman projections for the principal conformers of the C_1-C_2 bond of (a) propane and (b) 1,2-dichloroethane. In each case, plot energy curves for the rotation of the relevant structures about these bonds and indicate the positions of the principal conformers on these energy curves.

(2) Predict and draw the most stable conformer of each of the following compounds: (a) butylcyclohexane and (b) 1,2-dibromobutane.

(3) Convert the following structural formulae into Fischer projections.

(a)

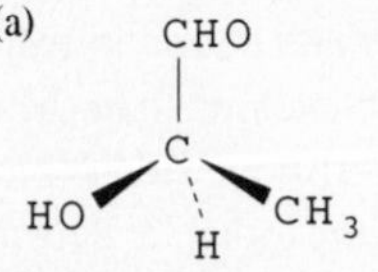

CHO

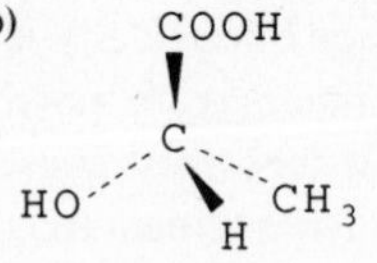

C

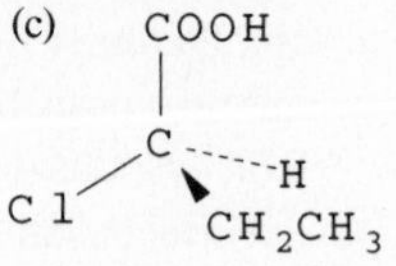

HO   H   CH₃

(b)

COOH

C

HO   H   CH₃

(c)

COOH

C

Cl   H   CH₂CH₃

(d)

H
O   CH₃

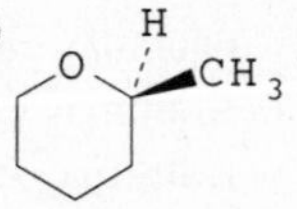

(4) Which of the following Fischer projections represent molecules with the same configurations?

(a)

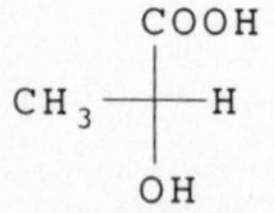

COOH
CH₃ —|— H
OH

(b)

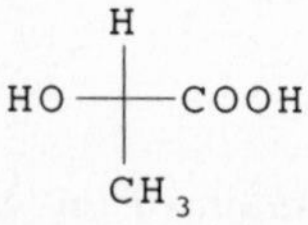

H
HO —|— COOH
CH₃

(c)

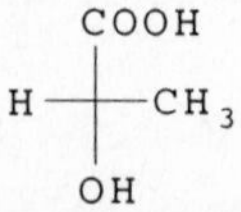

COOH
H —|— CH₃
OH

(d)

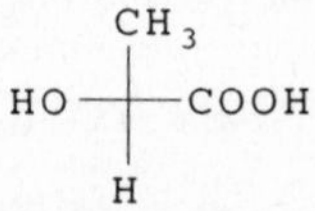

CH₃
HO —|— COOH
H

(e)

H
HOOC —|— OH
CH₃

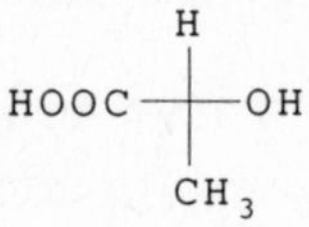

(5) Classify the following glyceraldehyde molecules as D or L.

(a)

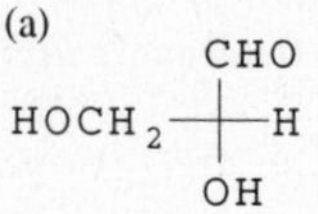

CHO
HOCH₂ —|— H
OH

(b)

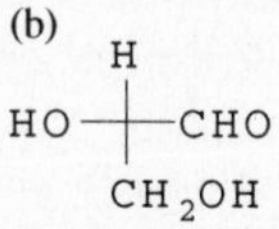

H
HO —|— CHO
CH₂OH

(c)

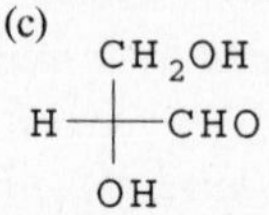

CH₂OH
H —|— CHO
OH

(d)

OH
HOCH₂ —|— H
CHO

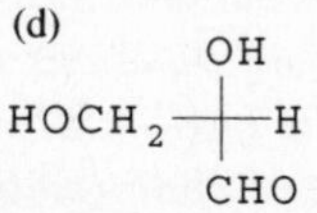

(6) Specify the priority of each of the groups bonded to the central carbon atom in each of the following molecules. Determine whether each structure has an R or S configuration.

(a)

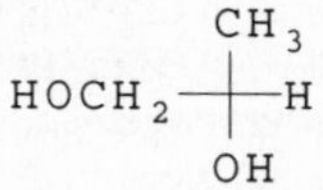

CH₃
HOCH₂ —|— H
OH

(b)

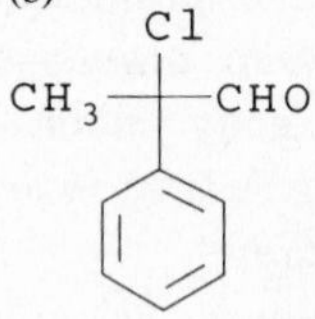

Cl
CH₃ —|— CHO

(c)

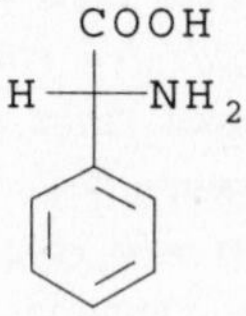

COOH
H —|— NH₂

(d)

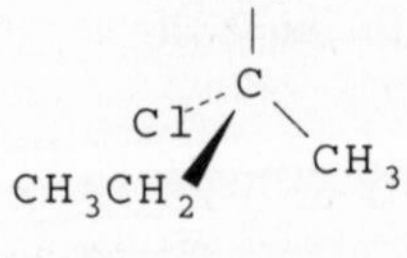

Br
C
Cl   CH₃
CH₃CH₂

(e)

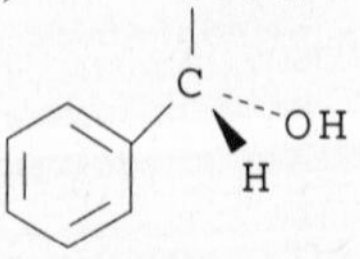

COOH
C
OH
H

(f)

H   C   N
CH₃CH₂

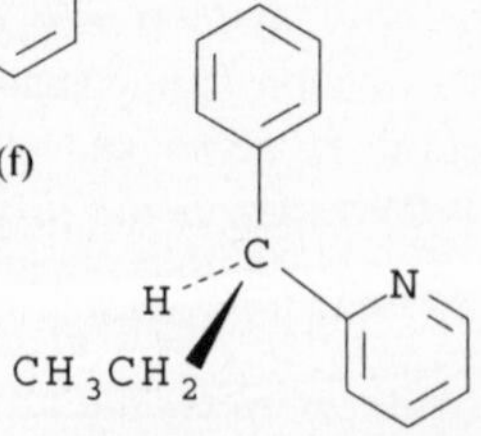

5

Stereoisomerism

5.1 INTRODUCTION

Isomers are compounds that have the same molecular formula but different configurations. This definition, and the points discussed in this chapter apply to both inorganic and organic compounds, however, most of the examples used will be organic in nature. Isomers are broadly divided into two types: structural isomerism and stereoisomerism. Although this chapter deals only with stereoisomerism it is important to appreciate the difference between stereoisomerism and structural isomerism.

5.1.1 Structural isomerism

Structural isomers have the same molecular formulae but differ in that some of the atoms are bonded together in a **different order**. This may result in their having either different functional groups, the same functional groups but in different positions on the molecule's skeleton or different molecular skeletons (Fig. 5.1).

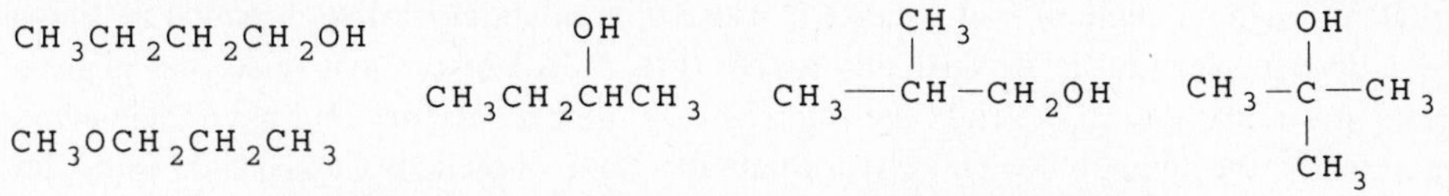

$CH_3CH_2CH_2CH_2OH$

$CH_3OCH_2CH_2CH_3$

$$CH_3CH_2\overset{\overset{\displaystyle OH}{|}}{C}HCH_3$$

$$CH_3-\overset{\overset{\displaystyle CH_3}{|}}{C}H-CH_2OH$$

$$CH_3-\overset{\overset{\displaystyle OH}{|}}{\underset{\underset{\displaystyle CH_3}{|}}{C}}-CH_3$$

Fig. 5.1. The structural isomers of butanol, molecular formula $C_4H_{10}O$.

5.1.2 Stereoisomerism

Stereoisomers are isomers with the same molecular formulae where the atoms are bonded in the **same order** but some or all of the component groups of atoms of the molecules are arranged differently in three-dimensions about the asymmetric centres in the molecule (Fig. 5.2). It is emphasised that stereoisomers differ only in the three-dimensional geometry of their structures, the order of the bonding between all the atoms in the molecules remains the same. This is in direct contrast to structural isomerism where the order of the bonding of some of the atoms is different (Fig. 5.1).

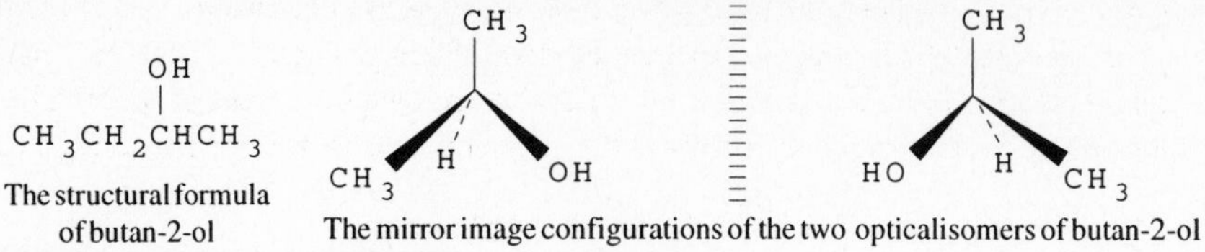

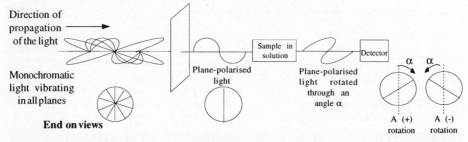

The structural formula of butan-2-ol

The mirror image configurations of the two optical isomers of butan-2-ol

Fig. 5.2. Stereoisomers of butan-2-ol.

Stereoisomers are classified into two types: optical and geometric. Optical isomers rotate the plane of plane-polarised light while geometric isomers do not rotate the plane of polarised light unless they possess a chiral centre (5.3).

5.2 OPTICAL ISOMERISM

Ordinary light consists of electromagnetic waves vibrating in all planes at right angles to the direction of propagation (Fig. 5.3). Plane-polarised light consists of electromagnetic waves (the light) vibrating in a single plane. It is produced by passing ordinary light from a suitable source through either a Nicol prism system or a diffraction grating. These devices, which are known as polarisers, act like filters. They only allow the passage of electromagnetic waves vibrating in one particular plane. The electromagnetic waves vibrating in other planes do not pass through the device (Fig. 5.3).

Fig. 5.3. The production of plane-polarised light and the effect of optical isomers on plane-polarised light.

Compounds which, when placed in a beam of plane-polarised light, cause its plane to rotate through an angle of a in either a clockwise or an anticlockwise direction as shown by a detector are said to be **optically active** (Fig. 5.3). Isomers that rotate the plane of plane-polarised light to the right are referred to as **dextrorotatory** and have their names prefixed by the symbol (+). Those that rotate the plane of plane-polarised light to the left are referred to as **levorotatory** and have their names prefixed by the symbol (-). The use of the historic symbols d and l for dextrorotatory and levorotatory respectively has been discontinued because of potential confusion with the configurational symbols D and L which cannot be related theoretically to the sign of optical rotation (4.8).

Optical activity is normally detected and measured in solution using an instrument called a polarimeter. The solvent used must not be optically active. The value of the angle of rotation (α) obtained depends on the solvent, the temperature, the wavelength of the polarised light used, the concentration of the isomer, and the path length of the polarised light through the sample. Consequently, for comparison purposes the angle of rotation is usually recorded, together with the solvent, as the **specific rotation** $[\alpha]_\lambda^T$ where the superscript and subscript indicate the conditions of temperature (T $^\circ$Kelvin) and the wavelength of the polarised light used to make the measurement respectively. For example, the specific rotation of coniine, the poison found in hemlock is quoted as:

$$[\alpha]_D^{298} = (+)16^\circ, \text{ ethanol}$$

This means that a solution of coniine in ethanol is dextrorotatory and has a specific rotation of 16° when measured at 298 K using polarised light with a wavelength corresponding to that of the D line of the sodium spectrum (589.6 nm). The superscript and subscript have no mathematical significance, they just describe the conditions under which the original measurement was made.

Specific rotations are calculated from the measured angle of rotation (α) using the expression:

$$\text{Specific rotation } [\alpha]_D^T = \frac{\text{Measured rotation } (\alpha)}{(\text{Concentration, g/cm}^3)(\text{ Path length of the polarised light, dm})}$$

5.3 MOLECULAR RECOGNITION OF OPTICAL ACTIVITY

At present there is no universally accepted theory explaining why optical rotation occurs in some molecules and not others. However, **a molecule will be optically active if it cannot be superimposed on to its own mirror image structure.** Molecules that comply with this condition are said to have a **chiral** structure.

The physical nature of non-superimposable mirror images can be readily demonstrated by constructing some simple models. Consider, for example, lactic acid (Fig. 5.4a). This compound has a tetrahedral configuration about C_2. Construct the two tetrahedra A and B as shown in Fig. 5.4b using coloured balls or their equivalent to represent the relevant groups of lactic acid. Now attempt to superimpose B onto A, that is, put the two structures together in such a way that all the atoms coincide. You will find that this is impossible to achieve with your models of A and B because they are non-superimposable mirror images of each other (Fig. 5.4c). A and B represent the structures of the two optically active isomers of lactic acid. Carbon 2 of lactic acid is referred to as an **asymmetric** or **chiral** carbon.

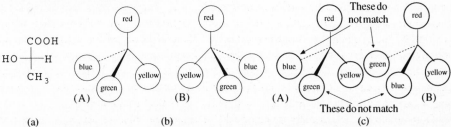

Fig. 5.4. Models of the non-superimposable mirror image structures of lactic acid.

Isomers that have structures that are mirror images of each other are known as **enantiomers**. They will rotate the plane of polarised light by equal amounts but in opposite directions. However, it is not possible to examine the two-dimensional drawing of the structural formula of an enantiomer and say whether it represents a dextrorotatory or levorotatory isomer. Enantiomers have identical chemical and physical properties except for their effect on polarised light. It is interesting to note that chiral chromatography, chiral nuclear magnetic spectroscopy (NMR), radioimmune and radioreceptor assay techniques can be used selectively to assay an enantiomer of an optically active compound.

Molecules will only be chiral, that is, have structures that are not superimposable on their mirror image structures, if they have no centre of symmetry, no plane of symmetry and no n-fold alternating axis of symmetry. *All three* of these conditions must be fulfilled if the molecule is to be optically active. However, initially one needs only to consider the first two of these conditions since molecules with an n-fold axis of symmetry are comparatively rare.

5.3.1 A centre of symmetry

A centre of symmetry is a point about which the molecule is symmetrical. Any straight line drawn through a centre of symmetry meets identical structures at equal distances along the line in opposite directions from the centre of symmetry. For example, the centre of a circle is a centre of symmetry since any straight line drawn through the centre of a circle meets the circumference at equal distances from the centre (Fig. 5.5a). The lactide in Fig. 5.5b has a centre of symmetry at x because the ring is flat, due to the two ester groups having conjugated structures (2.5). Therefore, any straight line drawn through x will meet identical structures at equal distances in opposite directions from x.

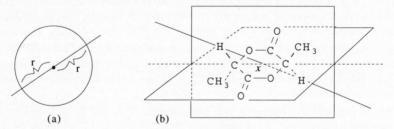

(a) (b)

Fig. 5.5. Centres of symmetry.

5.3.2 A plane of symmetry

A plane of symmetry divides a structure into two mirror image halves. For example, the isomer of tartaric acid (I) shown in Fig. 5.6 has a plane of symmetry which makes the top half of the molecule a mirror image of the bottom half. This can also be seen in structure (II) which is the Fischer projection of structure (a). You should pay particular attention to the way structure (II) is related to structure (I) since it shows a flaw in the Fischer convention. The C_2-C_3 bond is made up of two 'half bonds' both of which go behind the plane of the paper. This in effect puts a 'kink' in this bond. Kinks of this type are simply ignored when interpreting Fischer projections. Structure (III) is simply structure (I) viewed from a different angle. This can be proved by making stick models of these two

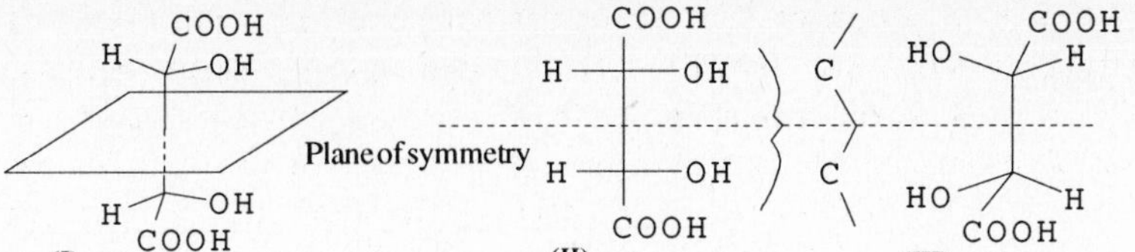

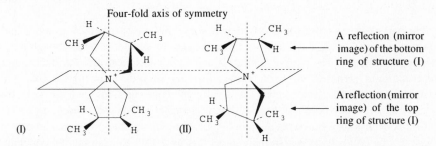

Fig. 5.6. Mesotartaric acid.

structures, turning one structure around and seeing whether it can be superimposed on the other. It is important to appreciate the three-dimensional nature of molecules and to recognise that structures like (I) and (III) while appearing to be different, are the same.

Molecules which possess a plane of symmetry have their names prefixed by the term **meso**. Meso compounds often have a centre of symmetry as well as a plane of symmetry.

5.3.3 An *n*-fold alternating axis of symmetry

An *n*-fold alternating axis of symmetry is found when it is possible to rotate the **complete** structure through $360°/n$ where *n* is a whole number, about this axis and obtain a structure that when half is reflected in a plane at right angles to the alternating axis, results in a structure that consists of reflections of the original molecule. In Fig. 5.7, for example, rotation of the complete structure (I) through $90°$ gives structure (II). The top half of this structure is a mirror image of the bottom half of the original structure (I) and the bottom half of structure (II) is a mirror image of the top half of structure (I). The compound is said to have a four-fold axis of symmetry since $360°/4 = 90°$, the angle through which the structure has been rotated.

Four-fold axis of symmetry

A reflection (mirror image) of the bottom ring of structure (I)

A reflection (mirror image) of the top ring of structure (I)

(I)   (II)

Fig. 5.7. 3,4,3',4'-Tetramethyl-1,1'-spirobipyrrolidinium p-toluenesulphonate. The rings are at $90°$ to each other. Four isomers have been prepared by McCasland and Proskow. Three are optically active but the meso compound shown is inactive because it has a four-fold alternating axis of symmetry.

5.3.4 Optical activity and asymmetric carbons

One of the commonest sources of chirality in organic compounds is the asymmetric carbon (chiral carbon). This is a carbon atom to which four different groups are bonded. Compounds with one asymmetric carbon are always optically active. Some examples of asymmetric carbon atoms are given in Fig. 5.8.

Compounds containing more than one asymmetric carbon atom fall into two categories, those in which all the asymmetric carbons have different structures and those in which they have the same structure. If all the asymmetric carbons have different structures then

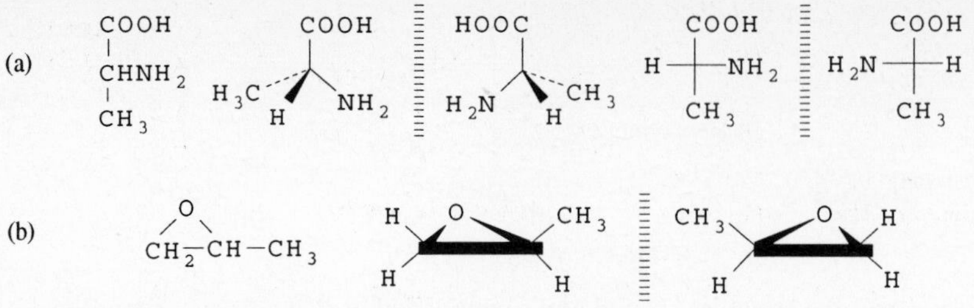

(a)

(b)

Fig. 5.8. Examples of compounds with one asymmetric carbon. (a) D and L-Alanine, and (b) propylene oxide showing the various methods of representing their structures.

the total number of optically active isomers for that compound is given by the expression:

$$\text{number of isomers} = 2^n \tag{5.1}$$

where n is the total number of asymmetric carbon atoms in the compound. The monosaccharide (sugar) threose (Fig. 5.9), for example, has two asymmetric carbon atoms (C*) and so, using the equation 5.1, has in theory, four optically active isomers. All these isomers are known although this is not always the case with predictions using equation 5.1.

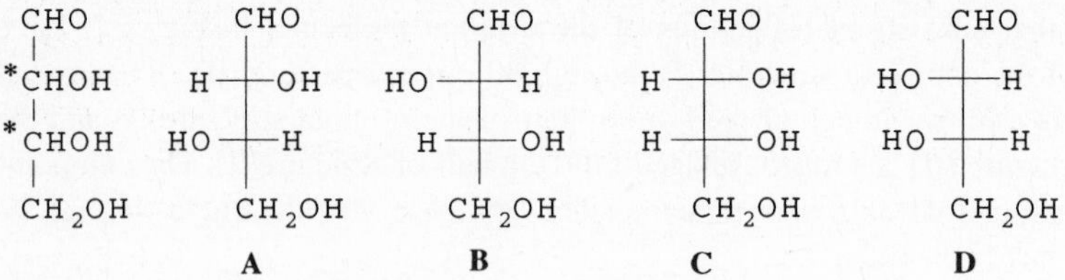

Fig. 5.9. The stereoisomers of threose.

The pair of stereoisomers A and B shown in Fig. 5.9 are enantiomers as are C and D. However, the isomers A and C are not mirror images of each other even though they are both optically active. The pairs A and D, B and C, B and D also fall into this category. Stereoisomers which do not have mirror image structures are called **diastereoisomers**. Diastereoisomers have different configurations and are not always optically active. They normally have different chemical and physical properties.

Compounds whose structures contain two or more identical asymmetric carbons may or may not be optically active. In these cases it is necessary to examine the compounds' structure for centres, planes and n-fold alternating axes of symmetry before predicting whether it is optically active. However, in most instances one has only to look for a plane of symmetry. Tartaric acid, for example, forms three stereoisomers which contain two identical asymmetric carbon atoms (Fig. 5.10). One of these isomers (A) has a plane of symmetry and is not optically active but the other two (B and C) have no planes, centres or alternating axis of symmetry and so are optically active.

5.3.5 Optically active compounds with no asymmetric carbon atoms
Both inorganic and organic compounds fall into this category (Fig. 5.11). Optically active organic compounds often have bent or twisted structures. However, the structure of any compound that is thought to be active must always be checked for centres, planes and n-fold alternating axis of symmetry.

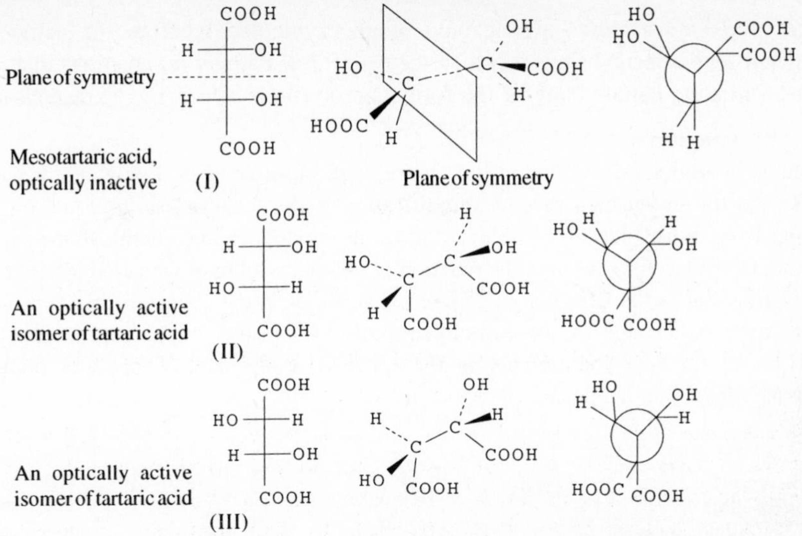

Fig. 5.10. The stereoisomers of tartaric acid. Rotation of the nearest carbon relative to the one furthest away in the Newman projections of the isomers of tartaric acid shows conclusively that unlike isomers (II) and (III) the meso isomer (I) has a plane of symmetry and so cannot be optically active. An examination of the Fischer projections of all three isomers shows that they do not have a centres of symmetry (remember that the horizontal bonds are orientated towards the observer).

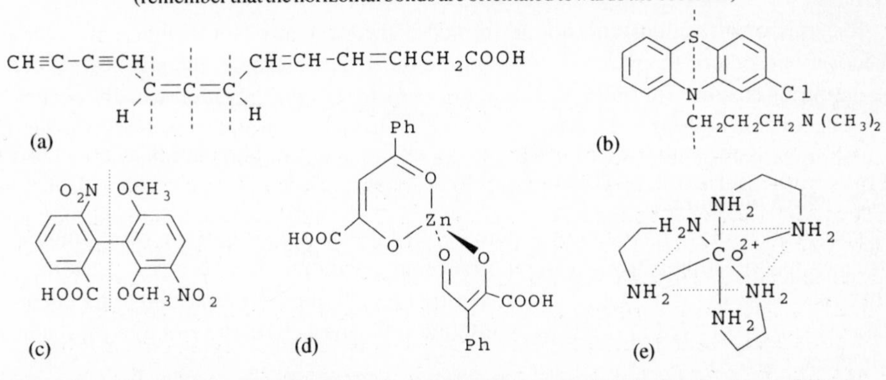

Fig. 5.11. Some examples of compounds that do not contain asymmetric carbons but are optically active. Molecules (a), (b) and (c) are either bent or twisted about the plane indicated by the dotted line. (a) Mycomycin, an antibiotic. (b) Chlorpromazine, a tranquilliser. (c) 2,6-Dimethoxy-2,2'-dinitro-6-hydroxycarbonylbiphenyl. (d) Zinc dibenzoylpruvate. (e) Triethylenediaminecobalt II chloride.

5.4 RACEMIC MODIFICATIONS

Racemic modifications are substances that contain equal amounts of a pair of enantiomers of a compound and so should exhibit no overall optical activity. For example, a racemate of lactic acid would contain 50% of the (+) and 50% of the (-) enantiomer of the acid. In practice this ideal state is seldom achieved and a sample labelled as a racemic modification will frequently contain a slight excess of one enantiomer (about 1 to 2%) and so may exhibit some optical rotation.

Racemic modifications are denoted by the symbol (±) before the name of the compound. The principal ways in which they are formed are by (a) reactions that involve either an existing chiral centre or the formation of a new centre, or (b) racemisation.

5.4.1 By reaction

Reactions involving an existing chiral centre or the formation of a new centre will only lead to the formation of a racemic modification if the reagent has an equal chance of reacting from either side of the relevant atom in the appropriate chemical species in the reaction. Consider, for example, the reaction of ethanal with hydrogen cyanide (Fig. 5.12). The cyanide ion in the first step of this reaction has an equal chance of reacting with the carbon from either side of the carbonyl group. This results in the formation of equal quantities of the two enantiomers in the form of a racemate. There are many other reactions where this happens.

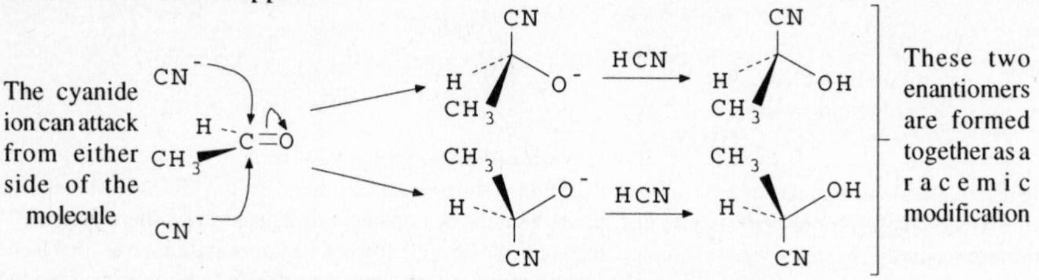

The cyanide ion can attack from either side of the molecule

These two enantiomers are formed together as a racemic modification

Fig. 5.12. The formation of (±) 2-hydroxypropanenitrile.

Racemic modifications will not be formed if there are any factors that make reaction easier on one side of the molecule than on the other. For example, the physical nature of the action of enzymes in biological systems (1.6) means that reaction usually occurs on one side of the molecule. Consequently, most enzyme-controlled reactions favour the formation of one of the enantiomers and not the racemic modification.

5.4.2   Racemisation

Racemisation is the conversion of a pure (+) or (-) enantiomer into a racemic modification. The modification is formed when 50% of an enantiomer has been converted into the other enantiomeric form (Fig. 5.13). It can be brought about by heat, light and chemical reagents and is believed to occur by both homolytic and heterolytic reaction mechanisms (Chapters 7 and 8).

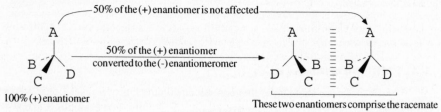

50% of the (+) enantiomer is not affected

50% of the (+) enantiomer converted to the (-) enantiomeromer

100% (+) enantiomer

These two enantiomers comprise the racemate

Fig. 5.13. A diagrammatic representation of racemisation of a pure (+) enantiomer. (-) Enantiomers also undergo this process.

A process which also involves a change in configuration of a chiral centre but not usually racemisation is **epimerisation**. **Epimers** are diastereoisomers that have more than one chiral centre but all these centres, except one, have the same configuration (Fig. 5.14a). Epimerisation is the interconversion at one chiral centre of one epimer into another, that is the interconversion of one diastereoisomer into another (Fig. 5.14b). The process normally results in the formation of an equilibrium mixture which will not necessarily contain equimolar proportions of the component epimers. This is in direct contrast to racemisation. The mutarotation of carbohydrates (27.3.2) and other compounds in suitable solvents is a spontaneous form of epimerisation.

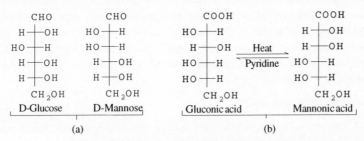

Fig. 5.14. Examples of (a) epimers and (b) epimerisation.

5.5 THE PHYSICAL PROPERTIES OF RACEMIC MODIFICATIONS

In solution, the gaseous and liquid states, most racemic modifications exhibit the same physical properties such as boiling points, refractive index, density, spectra, etc. as their enantiomers. However, in the solid state the physical properties of a racemic modification can be quite different from those of its constituent enantiomers. Solid racemates have been found to exist in three different structural forms mixtures, compounds and solid solutions. The actual form assumed by a racemic modification will depend on the temperature at which it is produced.

5.5.1 Racemic mixture

Racemic mixtures are physical mixtures of the two enantiomers. It is possible in a few cases where the crystals of the enantiomers have their own enantiomorphic shape to see the components of the mixture and manually separate them. The physical nature of racemic mixtures means that their constituent enantiomers behave as two separate phases. Consequently, racemic mixtures have a lower melting point but a higher solubility than those of their constituent enantiomers (Table 5.1).

5.5.2 Racemic compounds

Racemic compounds are generally referred to as racemates. They are formed when the two enantiomers combine to form a molecular compound. All their crystals are the same shape and contain equal amounts of each of the two enantiomers.

Racemates have physical properties that usually differ from those of the corresponding enantiomers. For example, their melting points are different (often higher) and their solubilities are usually lower than their constituent enantiomers (Table 5.1). This lower solubility means that in some cases a drug can be more effective when administered as its enantiomer rather than its racemate. The cytotoxic agent, (±)-1,2-di(2,6-dioxo-4-piperazinyl)-propane (Fig. 5.15), for example, is too insoluble to be fully effective when

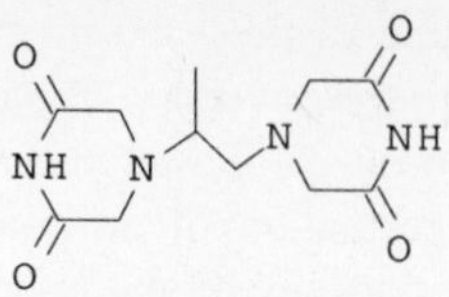

Fig. 5.15. (±)-1,2-Di(2,6-dioxo-4-piperazinyl)-propane. It should be noted that the end of a straight line is assumed to be a methyl group except in carbohydrate chemistry where it is a hydrogen atom.

administered intravenously. However, its pure enantiomers are five times more soluble and so have a much more effective therapeutic action.

As many compounds are produced initially as their racemates the solubility of racemates is also of importance in drug testing since effective drug action depends amongst other factors on the drug solubility.

5.5.3 Racemic solid solutions

Table 5.1. Some physical properties of racemates and their component enantiomers. Note the similarity of the boiling points and solubilities of the liquid enantiomers and racemic modifications.

Compound	Physical property	(+) Enantiomer	(−) Enantiomer	Racemate
Tartaric acid	Melting point ($^\circ$C)	170	170	206
	Water solubility (g/100cm^3)	139	139	125
Mandelic acid	Melting point ($^\circ$C)	133	133	121
	Water solubility (g/100cm^3)			15.8
Lactic acid	Melting point ($^\circ$C)	53	52.8	16.8
	Water solubility	Soluble	Soluble	Soluble
2-Bromobutane	Boiling point ($^\circ$C)	91	91	91
	Water solubility	Insoluble	Insoluble	Insoluble
2-Aminobutane	Boiling point ($^\circ$C)	63	63	63
	Water solubility	Miscible	Miscible	Misicible

In some instances enantiomers crystallise to form a one-phase solid solution. The physical properties of these systems resemble those of their component enantiomers. For example, the melting points and solubilities of racemic solid solutions differ only a little from those of their constituent enantiomers.

The physical nature of a racemic modifications can often be determined from the melting point-composition curve of the enantiomer-racemate system. Each type of system has a characteristic melting point-composition curve (Fig. 5.16). These curves show that the melting point of mixtures on either side of the racemic modification differs in a characteristic way from that of the racemic modification. To determine the nature of the racemic modification one simply measures the melting point of the racemic modification, adds a little of one of its enantiomers and redetermines the melting point relating any change in melting point to the shapes of the curves shown in Fig. 5.16. Solubility, X-ray crystallography and infra-red spectroscopy can also be used to determine the physical nature of a racemic modification.

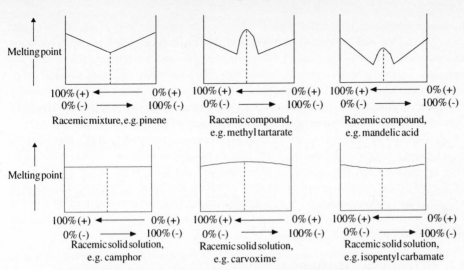

Fig 5.16. Identification of the physical nature of a racemic modification.

5.6 RESOLUTION OF RACEMIC MODIFICATIONS

The separation of a racemic modification into its component enantiomers is known as **resolution**. The main method of resolving a racemic modification into its constituent enantiomers is based on diastereoisomer formation (Fig. 5.17). The racemic modification is reacted with a **pure enantiomer** of a suitable optically active compound. This produces two diastereoisomers which can be separated because of differences in their physical properties. Separation methods include fractional distillation and crystallisation and various forms of chromatography. The separated diastereoisomers are decomposed, usually with acid or base, to yield the required enantiomers. Consider, for example, the resolution of (±)-isooctanyl hydrogen phthalate into its enantiomers. The (±)-isooctanyl

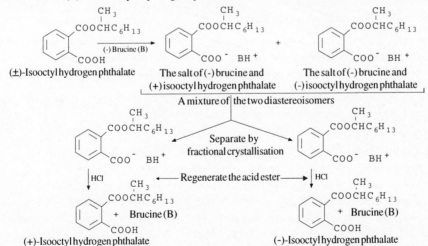

Fig. 5.17. The resolution of (±)-isooctylhydrogenphthalate into its isomers using brucine (B), a base.

hydrogen phthalate is reacted with (−)-brucine to produce a mixture of two salts (Fig. 5.17). One will be formed from (−)-isooctanyl hydrogen phthalate and (−)-brucine and the other from (+)-isooctanyl hydrogen phthalate and (−)-brucine. These salts are optically active diastereoisomers. Consequently, they will have different physical properties. In this case they have different solubilities in water and so they can be separated by fractional crystallisation. After separation, the enantiomers are regenerated from the diastereoisomers by treatment with hydrochloric acid (Fig. 5.17).

A wide variety of reactions can be used to resolve racemic modifications provided the reaction used is reversible and does not lead to racemisation. However, the enantiomer used to resolve a specific racemic modification will also depend on the chemical nature of the functional group(s) present in that modification as well as the ease of separation of the resultant diastereoisomers. For example, acidic enantiomers are used to resolve racemic modifications with basic functional groups and bases are used to resolve racemic modifications with acidic groups.

5.7 GEOMETRIC ISOMERISM

Geometric isomerism can occur in some molecules that contain structures where there is a restriction on free rotation at normal room temperatures and pressures. Consider, for example, 1,2-dibromoethene (Fig. 5.18). Free rotation about the C=C bond can only occur if the pi bond is broken. Since there is insufficient energy available to bring this about at room temperature and atmospheric pressure it is possible for this flat molecule to have the two different structures shown in Fig. 5.18. Geometric isomers with two similar structures or atoms on the same side of the molecule are traditionally known as **cis** isomers. If the structures or atoms are on the opposite sides of the molecule they are referred to as **trans** isomers. These terms are also used as prefixes to the names of geometric isomers.

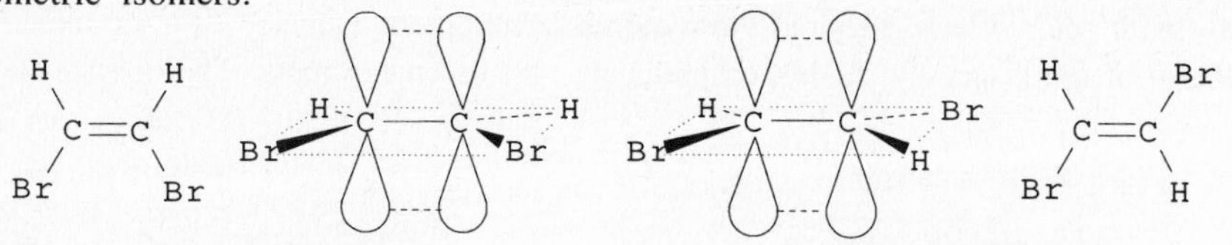

Cis-1,2-dibromoethene   Trans-1,2-dibromoethene

Fig. 5.18. Cis and trans geometric isomers.

Geometric isomerism does not occur in alkenes and similar structures if the atoms and structures bonded to one of the ends of the double bond are the same (Fig. 5.19). This is because the potential isomers are flat and so can be turned over and superimposed on each other taking into account any relevant single bond rotations.

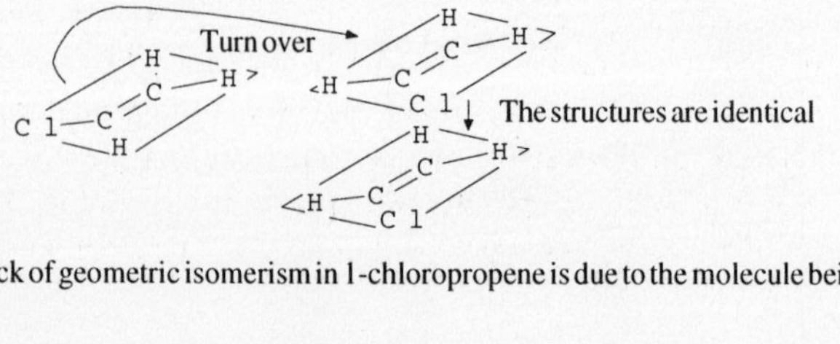

Fig. 5.19. The lack of geometric isomerism in 1-chloropropene is due to the molecule being flat.

Saturated ring compounds can also exhibit geometric isomerism. The ring has the effect of restricting free rotation at room temperature and pressure. This restriction acts in the same way as the pi bonds of alkenes and makes possible the existence of stereoisomers. Some examples of this type of geometric isomerism are given in Fig. 5.20.

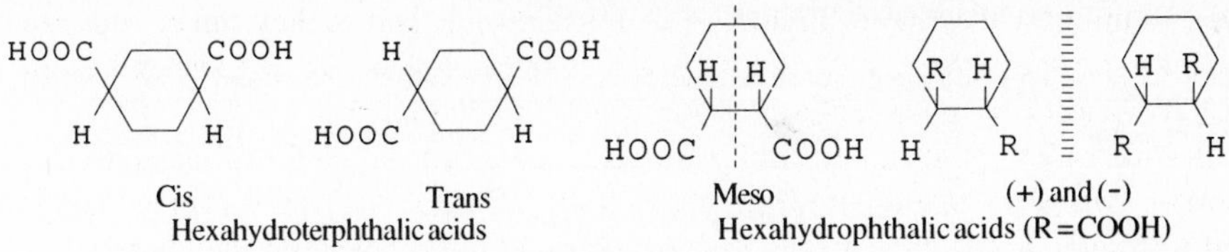

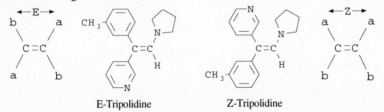

Cis   Trans   Meso   (+) and (−)
Hexahydroterphthalic acids   Hexahydrophthalic acids (R = COOH)

Fig. 5.20. The geometric isomers of some cyclic compounds.

The structures of cyclic geometric isomers are often more complex than they look because the molecule may also be chiral (Fig. 5.21). This can be further complicated from the visual point of view by the existence of ring conformations (4.3 and 4.4). **In order to appreciate the true nature of a structure it is imperative that the reader either makes or uses the relevant models.**

5.8 THE E/Z NOMENCLATURE SYSTEM

The E/Z nomenclature system is used where it is not possible to use the prefixes cis and trans to identify geometric isomers. In alkenes the structures bonded to the two carbon atoms forming the double bond are allocated an **a** or **b** priority using the same rules as the R/S system discussed in section 4.8. Each of the carbons forming the double bond is considered independently of the other when deciding the priorities of their substituent structures. If the two **a** priority substituents are **cis** to each other, the isomer is said to have a **Z** configuration. However, if they are **trans** to each other, the isomer is said to have an **E** configuration (Fig. 5.21). The same procedure and convention is used for the appropriate atoms of other geometric isomers.

E-Tripolidine   Z-Tripolidine

Fig. 5.21. The E/Z isomers of triprolidine which has antihistamine activity. The E isomer is the most active.

5.9 GENERAL PROPERTIES OF GEOMETRIC ISOMERS

Geometric isomers do not have mirror image structures, they are diastereoisomers. Their physical properties are usually significantly different (Table 5.2). This makes it comparatively easy to identify an isomer and also to separate mixtures of geometric isomers. Furthermore, in the case of cis and trans isomers the differences often show a general trend that can be useful in establishing their configuration. The melting points of trans isomers, for example, are usually higher than those of their corresponding cis isomers. However, it is not possible to say that every trans isomer will have a higher melting point than its corresponding cis isomer. The same is true of other physical properties (Table 5.2). Consequently, when assessing general trends in physical properties to establish the configuration of an isomer, it is necessary to do a statistical analysis of as many different physical properties as possible.

Table 5.2. Some physical properties of cis and trans isomers. Note the value * depends on the allotropic form.

Compound	m p °C		b p °C		Resultant dipole		Refractive index	
	Cis	Trans	Cis	Trans	Cis	Trans	Cis	Trans
1,2-Dichloroethane	-80	-50	60	48	1.85	0.00	1.4486	1.4454
1-Chloropropene			33	37	1.71	1.97	1.4054	1.4055
Diethyl butenedioate	-88	1	223	218	2.54	2.38	1.4415	1.411
Dimethyl butenedioate	-19	102	202	193			1.4416	1.4063
Cinnamic acid	68*	136	265	300				
Stilbene	1	125	141	305		0.00	1.6130	1.6264

The chemical properties of geometric isomers are different in degree or type. For example, both of the isomers of butenedioic acid (1,2-ethenedicarboxylic acid) will react with ethanol to form esters but their rates of formation as well as the esters formed are different. However, on heating only the cis isomer (maleic acid) forms a cyclic anhydride (Fig. 5.22). This is because in the trans isomer (fumaric acid) the carboxylic acid groups are too far apart to react.

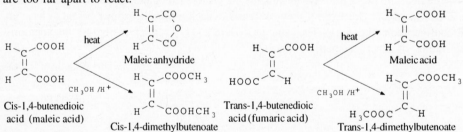

Fig. 5.22. Examples of the reactions of cis-1,4-butenedioic acid and trans-1,4-butenedioic acid.

5.10 INTERCONVERSION OF GEOMETRIC ISOMERS

The energy difference between cis and trans isomers is often small and so, provided the energy barrier is also small, it is possible to convert one isomer into the other. The size of the energy barriers to interconversion, and consequently the ease of interconversion, depends on the nature of the molecule. A simple geometric isomer may usually be converted into an equilibrium mixture of the two isomers by heating, the position of equilibrium favouring the more stable isomer under the reaction conditions. In general, trans isomers are usually more stable than the corresponding cis form of a simple molecule. With more complex molecules the situation is not so clear.

Interconversion may also be brought about by light and treatment with catalysts. For example, vitamin A_1 is deactivated by ultra-violet light (Fig. 5.23), its C_9 and C_{11} double bonds having their trans configurations changed to the cis form whilst cis-1,4-butenedioic acid can be converted into trans-1,4-butenedioic acid by heating in the presence of hydrochloric acid.

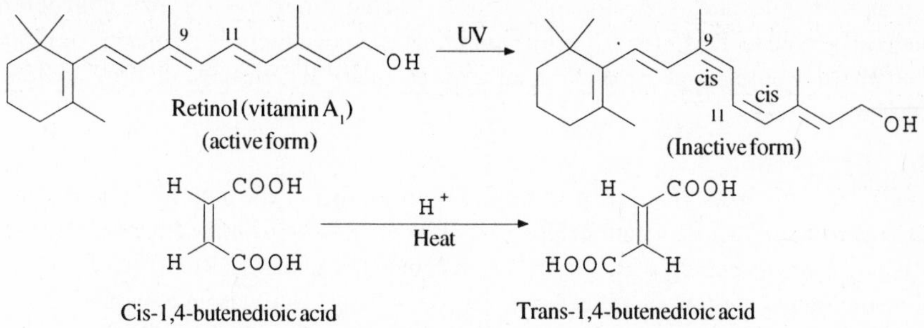

Fig. 5.23. Examples of the interconversion of cis and trans isomers.

An important biological application of the conversion of a high energy cis isomer into a lower energy trans isomer is found in the chemistry of vision (Fig. 5.24). Rhodopsin, a light-sensitive substance, is formed in the rod cells of the eye by the interaction of 11-cis-retinal with the protein opsin. Light converts rhodopsin into its trans counterpart metarhodopsin II. This results in a nerve impulse to the brain which causes the effect of sight. The metarhodopsin II is reconverted into retinal which is reused in the cycle.

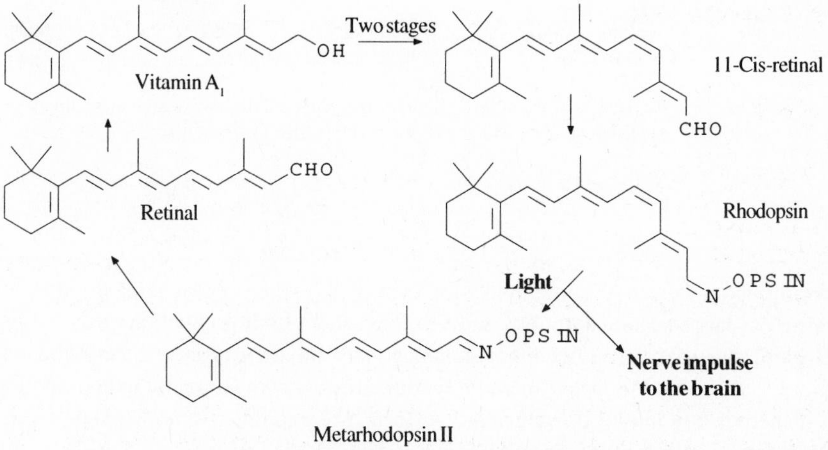

Fig. 5.24. The involvement of cis-trans interconversion in the chemistry of vision

The sensitivity of some geometric isomers to heat, light and ultra-violet radiation means that these compounds must be stored at or below room temperature in light-proof containers. Prescriptions containing these substances should be dispensed in appropriate containers.

5.11 STEREOISOMERISM AND BIOLOGICAL ACTIVITY

All geometric isomers, enantiomers and diastereoisomers are individual compounds with their own physical and chemical properties. The stereochemistry of an individual isomer will play a major part in governing its biological activity because of the physical nature of active and receptor sites (1.6). In other words, **each isomer has its own characteristic biological activity**. This stereospecific nature of stereoisomers is of major significance in the pharmaceutical, pharmacological and biomedical sciences in drug action and metabolic pathways.

5.11.1 Drug action

Currently, it is not possible to predict accurately from theory the activity of a drug *in vivo*. There are structure/action relationships (1.6) that are of some limited use and there have also been some successes with the use of computer graphics (2.9).

The biological activity of stereoisomers used as drugs can vary in a number of general ways (Table 5.3). Many drugs have more than one action. For example, the drug ketamine has anaesthetic activity and also a psychotic effect. Its anaesthetic activity resides mainly in the S isomer whilst the R isomer has both anaesthetic and pyschotic activity.

Table 5.3. A comparison of the general biological activities found in drugs that are either geometric isomers or enantiomers or diastereoisomers.

A pair of stereoisomers		Example
Enantiomer 1	Enantiomer 2	
Active	No activity	S-α-Methyldopa is an antihypertensive, the R isomer has no activity
Active	Activity is of the same type and strength	Both the isomers of the H blocker promazine have the same activity
Active	Activity is of the same type but weaker	The S isomer of the anticoagulant warfarin is more active than the R isomer
Active	Activity is different	The S isomer of ketamine is responsible for most of its anaesthetic action while the R isomer is responsible for its psychotic effects

Many optically active compounds are used as racemates because of the difficulty of carrying out large-scale resolutions and the consequent high production costs. At present, in Britain, manufacturers need not conduct tests on the separate enantiomers unless required to do so by the Department of Health. However, most manufacturers do conduct tests if there is any hint of unwanted side-effects. For example, some workers believe that the teratogenic action of the drug thalidomide (Fig. 5.25) is due to the S isomer. If this is correct then the tragic consequences of the use of this drug can be avoided by using the pure R isomer.

5.11.2 Metabolic pathways

The reactions that occur in metabolic pathways are stereospecific because of the physical nature of enzyme control (1.6). They usually produce either one or a high proportion of

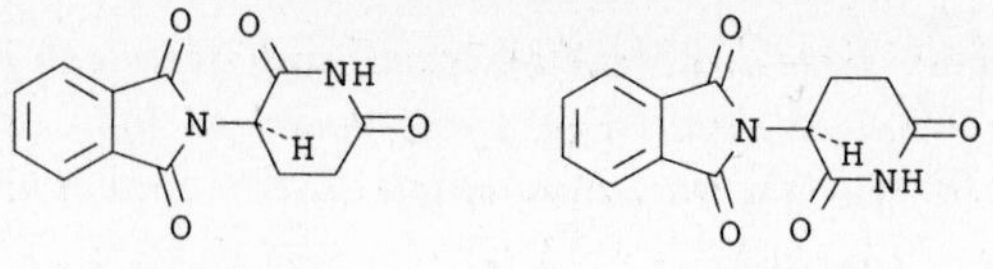

R-Thalidomide   S-Thalidomide

Fig. 5.25. R and S-Thalidomide.

one of the possible stereoisomers. For example, the hydroxylation of fumarate by the enzyme fumarase in the citric cycle yields S-malate and not a mixture of both isomers.

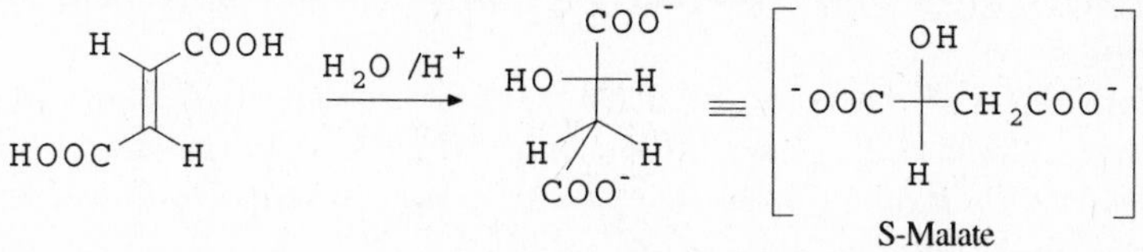

S-Malate

Practical evidence shows that stereoisomers metabolise at different rates. For example, S-verapamil is metabolised more rapidly in first pass metabolism than its corresponding R isomer.

5.12 WHAT YOU NEED TO KNOW

(1) It is essential to use models to appreciate the structures of stereoisomers.

(2) Stereoisomers have exactly the same groups of atoms in their molecules but these groups are arranged differently in space about the same fixed point(s) in their molecules.

(3) There are two general classes of stereoisomers: optical and geometric.

(4) Optical isomerism occurs in organic and inorganic molecules when it is possible to construct a non-superimposable mirror image structure.

(5) Optical isomers rotate the plane of plane-polarised light to the right (dextrorotatory) or the left (levorotatory). This phenomenon is referred to as optical activity.

(6) Optical isomers have their names prefixed by (+) if dextrorotatory and (-) if levorotatory.

(7) A compound that has no centre of symmetry, no plane of symmetry and no n-fold alternating axis of symmetry will be optically active.

(8) Meso compounds are isomers that have a plane of symmetry and so they are not optically active.

(9) Meso compounds have their names prefixed by meso.

(10) Optical activity in organic molecules is often due to the presence of an asymmetric carbon, that is a carbon to which are bonded four different groups.

(11) Enantiomers are optically active isomers whose structures are non-superimposable mirror images of each other.

(12) Enantiomers have the same chemical and physical properties except they rotate plane-polarised light equal amounts but in opposite directions.

(13) Diastereoisomers are optically active isomers whose structures are not mirror images of each other.

(14) *Diastereoisomers have different physical and chemical properties.*

(15) *The degree of optical rotation of plane-polarised light caused by an isomer is recorded as its specific rotation which is calculated using the relationship:*

$$\text{Specific rotation } [\,\alpha\,]_D^T = \frac{\text{Measured rotation } (\alpha)}{(g/cm^3)(\text{ Path length of the polarised light, dm})}$$

(16) *Racemates are optically inactive because they contain equal amounts of the two enantiomers of an optically active compound.*

(17) *Racemates often have different physical properties from their constituent enantiomers.*

(18) The separation of a racemate into its component enantiomers is known as resolution.

(19) *Geometric isomerism occurs in molecules which contain structures where there is no free rotation at room temperature, for example, non-aromatic double bond(s) and saturated ring structures.*

(20) *Geometric isomers will be optically active if they have a chiral structure.*

(21) *Cis and trans geometric isomers have the structures:*

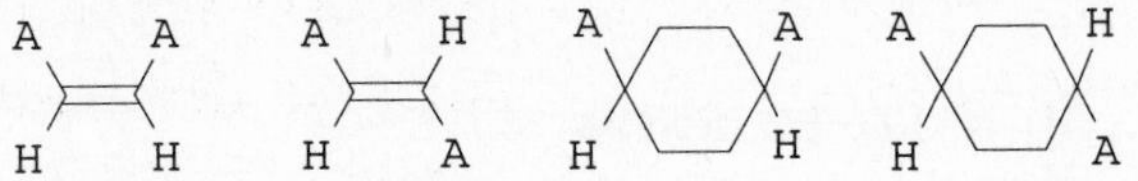

cis   trans   cis   trans

(22) *Z geometric isomers have the a priority groups in a cis configuration whilst E geometric isomers have the a priority groups in a trans configuration.*

(23) *The physical and chemical properties of geometric isomers are different.*

(24) *Heat, light, and chemicals can convert some pure geometric isomers into an equilibrium mixture of both isomers, the position of equilibrium depending on the conditions employed.*

(25) *All stereoisomers are individual compounds and have their own characteristic biological activity.*

5.13 QUESTIONS

(1) Draw all the stereoisomers of (a) $CH_2OHCHOHCH_3$ (b) $CH_3CHOHCHOHCH_3$ and (c) 1,2-dichlorocyclopentane. Indicate which of these isomers are geometric isomers, enantiomers and meso compounds. Use models to obtain your answers.

(2) State whether or not each of the following structures **could** have a centre or plane of symmetry. Use models to obtain your answers. (Me represents a methyl group).

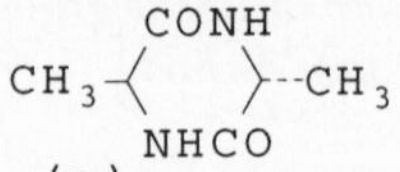

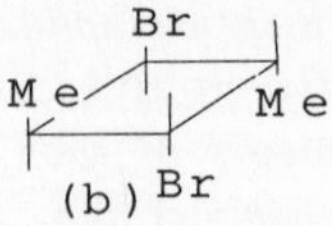

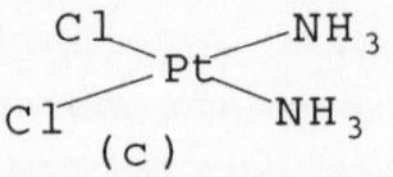

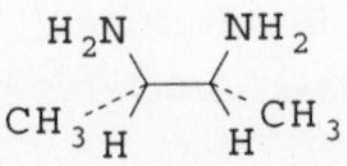

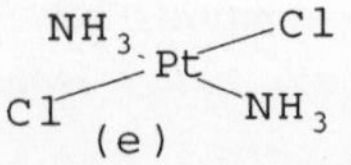

(3) Compounds (a) to (h) are drawn as Fischer projections where relevant. State which of these compounds are (a) optically active (b) enantiomers (c) meso compounds and (d) geometric isomers. Use models to obtain your answers.

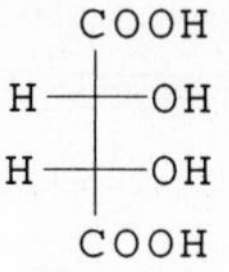

$$
\begin{array}{cccc}
\text{(a)} & \text{(b)} & \text{(c)} & \text{(d)}
\end{array}
$$

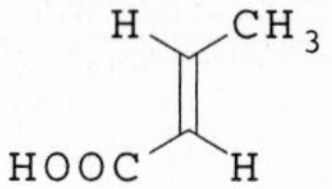

(a) Fischer: COOH / H—OH / H—OH / COOH

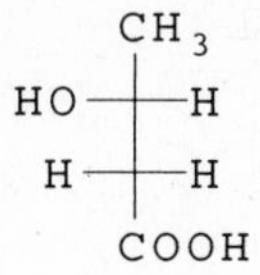

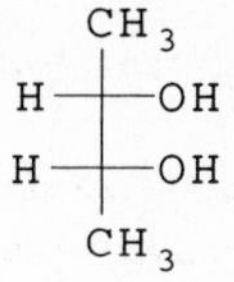

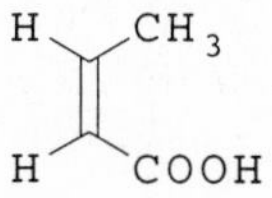

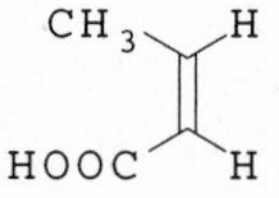

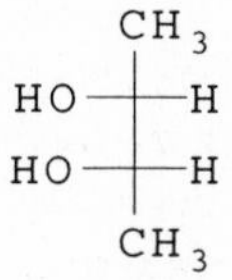

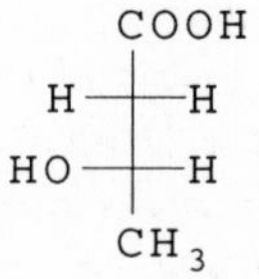

(b) H, CH$_3$ / HOOC, H (alkene)

(c) CH$_3$ / HO—H / H—H / COOH

(d) CH$_3$ / H—OH / H—OH / CH$_3$

(e) H, CH$_3$ / H, COOH (alkene)

(f) CH$_3$, H / HOOC, H (alkene)

(g) CH$_3$ / HO—H / HO—H / CH$_3$

(h) COOH / H—H / HO—H / CH$_3$

(4) Draw the structures of all the stereoisomers of each of the following compounds. State the type of stereoisomerism being exhibited by each of the compounds.

(a) $CH_3CH(OCH_3)COOH$ (b) $CH_3CH(NH_2)CH(NH_2)CH_3$

(c) $CH_3CHClCH=CH_2$ (d) $CH_3CH_2CH=CHCl$

(e) $CH_3CHClCH=CHCl$

(5) Explain how (+)-tartaric acid could be used to resolve ($\pm$) 2-aminobutane into its component enantiomers. Use equations to outline the relevant chemical reactions.

(6) The physical properties of two geometric isomers are listed in the table. Decide which of these isomers is more likely to have a cis configuration giving the reasons behind your decision.

Compound	mp °C	bp °C	Refractive index
A	-17	109	1.4017
B	-75	154	1.4762

(7) Draw all the stereoisomers of the complex with the molecular formula $Pt(NH_3)_2(OH)_2$. Predict the general relative types of pharmaceutical activity, if any, that these compounds would be expected to exhibit?

6

Chemical reactions

6.1 INTRODUCTION

All life processes are based on chemical reactions. The study of the physical properties of chemical reactions gives information concerning the mechanisms by which chemical and enzymically catalysed processes occur. The physical properties of chemical reactions that are of interest to scientists working in the medical, pharmaceutical and related fields are:

(1) The physical nature of the change: for example, is it irreversible, reversible or does it result in the formation of a dynamic equilibrium system?

(2) The rate at which the change takes place. The study of this feature of chemical reaction is known as **reaction kinetics.** The study of the absorption, distribution, metabolism and excretion together with related pharmacological, therapeutic and toxic responses is regarded as an integrated area of study known as **pharmacokinetics**.

(3) The effect of changes in particle size, temperature, pressure, concentration and catalysts on the chemical or physical changes.

(4) The spontaneity of the change. Why it should happen? These features of chemical and physical changes are discussed in the subject known as **thermodynamics**.

6.2 REACTION KINETICS

The speed at which the reactants in a chemical reaction are converted into the products is known as the **rate of the reaction**. The rate of a reaction can be defined as either the rate of change in the concentration of a reactant, or the rate of change in the concentration of a product with time (equation 6.1). Since the reactants are disappearing as the products are appearing, the rate of loss of a reactants is, by convention, designated as being negative whilst the rate of appearance of a product, positive:

$$\text{Rate} = \frac{-d[\text{reactant}]}{dt} = \frac{d[\text{product}]}{dt} \tag{6.1}$$

The rate of a chemical reaction is measured experimentally by determining the concentration of a selected substance, normally a reactant, in the reaction mixture at different times after the start of the reaction. The results are plotted as a graph of concentration against time and the rate of the reaction at a time t is given by the slope of the tangent to the resulting curve at time t (Fig. 6.1). In practice it is not always necessary to measure the actual concentration of the selected substance, instead one can measure the change in a physical property of a reactant that is directly proportional to its concentration. For example, reaction rates may be determined by following changes in optical rotation (5.2), partial pressure (6.7), radioactivity (33.4) and the intensity of absorption of either ultra-violet or visible electromagnetic radiation of a particular wavelength (10.2).

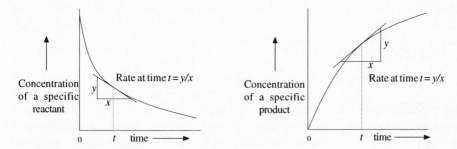

Fig. 6.1. Measurement of the rate of a reaction at a time t. The slope of the tangent at $t = 0$ is the initial rate of the reaction.

The physical conditions that affect reaction rates are particle size, temperature, pressure, reactant concentration and the presence of catalysts. Each of these conditions is discussed in this chapter.

6.3 REACTION RATE AND CONCENTRATION

Experimental work shows that at constant temperature the rate of a chemical reaction can be expressed mathematically in terms of the concentrations of its reactants. These mathematical relationships which are known as **rate equations** or **expressions,** are determined experimentally but for simple reactions often take the form:

$$\text{Rate} = k[\text{A}]^x[\text{B}]^y\text{etc.} \tag{6.2}$$

where: [A], [B], etc. are the concentrations, mol dm^{-3} of the reactants A, B, etc.

x and y are experimentally determined numerical coefficients.

k is a constant at constant temperature and is known as the rate constant.

The coefficients x and y etc. are known as the orders of the reaction with respect to the reactants A, B etc. The sum of these coefficients is known as the **overall order** of the reaction. Reactions with simple mechanisms usually have zero, first, second and occasionally third overall orders (Table 6.1). However, reactions with complex mechanisms can have fractional orders. The order of a reaction is used to classify reactions kinetically.

Table 6.1. The relationship of the overall order of a reaction to the general rate equation for reactions with simple mechanisms.

Order	Rate expression
0	Rate $= k$
1st	Rate $= k$ [A]
2rd	Rate $=$ either k [A]^2 or k [A][B]
3rd	Rate $=$ either k [A]^3 or k [A]^2[B] or k [A][B][C]

If a reaction takes place in several steps, each step will contribute to the rate expression. However, if one step is very much slower than the other steps the slow step will obviously dominate the rate expression. Consequently, rate expressions indicate the state of the reactants in the slow step of a reaction mechanism and as such are used as evidence for a proposed reaction mechanism (8.6).

6.3.1 The deduction of the rate expression of a reaction

Rate expressions are deduced from the experimental results in a number of ways. For example, the order of the reaction with respect to each reactant may be deduced from a graph of the initial rate of a reaction (Fig. 6.2) against the concentration of that reactant, the concentrations of the other reactants being kept constant. These graphs often have characteristic shapes which can be related to the order for the reactant being studied. Once the order of the reaction with respect to each of the reactants has been determined the rate expression is obtained by substituting the values of the orders in equation 6.2. For example, consider the reaction:

$$A + B \longrightarrow Products$$

Suppose that experimental investigations of this reaction show when the concentration of B is kept constant the reaction is second order with respect to A and when the concentration of A is kept constant the rate of the reaction is first order. The overall rate expression for the reaction would be: Rate $= k[A]^2[B]$.

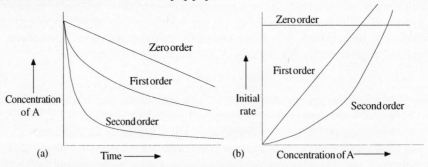

Fig. 6.2. The graphs of (a) concentration of a reactant A against time and (b) initial rate against the concentration of the reactant A for zero, first and second order reactions.

The order of a reaction (n) with respect to a particular reactant may also be determined by plotting a graph of log (initial rate) against log [A], the concentrations of all other reactants being kept constant. Allowing for experimental error, this graph should be a straight line with a slope equal to n.

6.3.2 Rate constants and their determination

Rate constants are determined experimentally by measuring the amount of the reactants that have reacted in a time t and substituting the values in a mathematical expression (**kinetic equation**) derived from the rate expression for the reaction. The form of the mathematical expression depends on the order of the reaction and the number of different reactants involved in the reaction. For example, for the first order reaction: A $\longrightarrow$ Products, the rate equation is:

$$\text{Rate} = -k[A]$$

Suppose, the initial concentration of A at a time $t = 0$ is a mol dm^{-3} and at a time $t = t$, x mol dm^{-3} of A have reacted, then:

$$\text{Rate} = -k(a\text{-}x)$$

However, the rate at which A is being converted into the products is $-dx/dt$, and so:

$$dx/dt = k(a\text{-}x)$$

that is:
$$\frac{dx}{(a\text{-}x)} = kdt \tag{6.3}$$

Integration of equation 6.3 between the limits x = 0 at time t = 0 at the start of the reaction and $x = x$ at a time $t = t$ gives:

$$\ln \frac{a}{(a\text{-}x)} = kt$$

that is:
$$\log \frac{a}{(a\text{-}x)} = \frac{kt}{2.303} \tag{6.4}$$

Therefore, a graph of log $a/(a\text{-}x)$ against t is a straight line with a slope of $k/2.303$. Thus, by measuring the value of x at different times t the value of k may be determined either graphically or by substituting suitable values of x and t in equation 6.4.

The mathematical expressions for zero and second order reactions are derived and used in a similar manner (Table 6.2). It is also possible to use these expressions to

Table 6.2. The kinetic equations for zero, first and second order reactions. The initial concentrations of A and B are a and b mol dm^{-3} respectively. The time t is normally measured in seconds (s).

Order	Rate expression	Kinetic equation	Concentration function	k (units)
Zero	Rate = k	$x = kt$	x	mol dm^{-3}s^{-1}
First	Rate = $k[A]$	$\log \dfrac{a}{a\text{-}x} = \dfrac{kt}{2.303}$	$2.303 \log \dfrac{a}{a\text{-}x}$	s^{-1}
Second	Rate=$[A]^2$ or Rate=$k[A][B]$ when $a = b$	$\dfrac{x}{a(a\text{-}x)} = kt$	$\dfrac{x}{a(a\text{-}x)}$	mol^{-1} dm^3 s^{-1}
	Rate = k[A][B] when $a \neq b$	$\log \dfrac{b(a\text{-}x)}{a(b\text{-}x)} = \dfrac{k(a\text{-}b)t}{2.303}$	$\dfrac{2.303}{(a\text{-}b)} \log \dfrac{b(a\text{-}x)}{a(b\text{-}x)}$	mol^{-1} dm^3 s^{-1}

determine the order of a reaction by plotting their concentration functions against t (Table 6.2). The plot with a straight line graph corresponds to the order of the reaction. If a reaction had a straight line plot for t against x the reaction would have a zero order with respect to the appropriate reactant and the slope of the graph would be the value of k.

The rate constant is the rate of the reaction when the concentrations of the reactants are unity. Its units depend on the overall order of the reaction (Table 6.2) while its value indicates whether a reaction is likely to be fast or slow at the specified temperature.

6.4 HALF LIFE

The half life $(t_{1/2})$ of a reactant is the time taken for the concentration of that reactant to fall to half its initial value. It can be shown mathematically that the half life of a reactant is independent of the initial concentration of that reactant for first order reactions but is a function of its concentration at the time taken as $t = 0$ for zero and second order reactions (Table 6.3).

Table 6.3. The relationships between the order, half life and concentration of the reactants at a time $t = 0$ for zero, first and second order reactions. These relationships are derived from the corresponding kinetic equations (Table 6.2) by substituting $x = a/2$ and $t = t_{1/2}$.

Order	Half life
0	$t_{1/2} = a/2k$
1	$t_{1/2} = 0.693/k$
2	$t_{1/2} = 1/ak$ (when $a = b$)

Half life can be used to determine the value of the rate constant of zero, first and second order reactions where $a = b$, by measuring the initial concentration of the reactant and determining the time taken for this concentration to halve. This is particularly useful with first order radioactive decay processes (33.5).

The **biological** or **elimination half life** of either a drug, metabolite or natural chemical component of the body is defined as the time taken for the body to reduce the concentration of the substance to half its initial value. The value of the biological half life is taken into account when deciding what dose of a drug will give the optimum therapeutic effect. Suppose for example that the metabolism of a drug follows first order kinetics. The half life of this drug will be independent of the initial concentration of the drug. Consequently, if the drug has a biological half life of three hours, once the concentration of the drug has reached a maximum in the plasma the concentration of the drug will halve every three hours. Experimental data of this type are used to help decide what dose of drug has to be administered to the patient in order to maintain an effective concentration.

6.5 PSEUDO FIRST ORDER REACTIONS

Pseudo first order reaction kinetics may be observed in second order reactions of the type:
$$A + B \longrightarrow \text{Products}$$
Consider, for example, the acid catalysed hydrolysis of methyl ethanoate in dilute aqueous solution.

$$CH_3COOCH_3 + H_2O \longrightarrow CH_3COOH + CH_3OH$$

Experimental evidence shows that under stoichiometric conditions this reaction is first order with respect to both methyl ethanoate and water. Therefore, the reaction is second order overall and its the rate equation is:

$$Rate = k_x[CH_3COOCH_3][H_2O]$$

However, if the concentration of the water is very large compared with that of the methyl ethanoate, the change in the concentration of the water as the reaction proceeds will be so small that it can be regarded as negligible, that is, the concentration of the water may be regarded as constant. Consequently, the reaction will *appear* to be first order and have an observed overall rate equation:

$$Rate = k_y[CH_3COOCH_3]$$

where k_y is the observed rate constant. Second order reactions which appear to exhibit first order kinetics are known as **pseudo first order** reactions.

Pseudo first order kinetics may be exhibited by the first step of some two-step second order biological reactions in which a slow step is coupled with a fast step that regenerates one of the original reactants (Fig. 6.3). The fast step regenerates the reactant so rapidly that its concentration in the system can be regarded as being constant. As a result, the reaction appears to be first order but in fact is second order when the effect of the concentration of the two reactants on the rate is studied separately (6.3.1) or the order is determined using the kinetic equation technique (6.3.2).

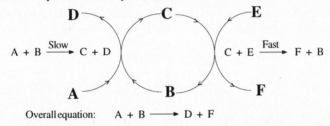

Overall equation: $A + B \longrightarrow D + F$

Fig. 6.3. A representation of a second order biological reaction that exhibits pseudo first order kinetics. The observed rate expression $= k_y[A]$ but the actual rate expression $= k_x[A][B]$.

It has already been observed that the order of a reaction is often one of the pieces of evidence on which the proposed mechanism of that reaction is based (8.6). In this context, it is essential to recognise when a reaction is exhibiting pseudo first order kinetics.

6.6 TEMPERATURE AND REACTION RATE

Experimental observation shows that an increase in the temperature of a reaction always increases the rate at which the reaction occurs irrespective of whether the reaction is endothermic or exothermic. The rates of endothermic reactions increase faster than those of exothermic reactions because endothermic reactions require heat to proceed. However, exothermic reactions need to lose heat, and so their reaction rates do not increase as quickly because the rate of loss of heat will drop as the temperature rises.

The increase in reaction rate with increase in temperature can be explained by the collision theory (1.4) of chemical reactions. In both exothermic and endothermic reactions the increase in temperature will increase the energy of the reactant species. This leads to an increase in the number of collisions between chemical species with sufficient energy to react and so to a subsequent increase in reaction rate. This increase in reaction rate must be related to an increase in the value of the rate constant for the reaction because the concentration(s) of the reactants cannot increase. The relationship between the rate constant of a reaction and the absolute temperature T at which the reaction was conducted is given by the experimentally derived **Arrhenius equation**:

$$k = Ae^{-E_a/RT}$$

(6.5)

where A is known as the Arrhenius constant, E_a is the activation energy of the reaction and R is the ideal gas constant. The values of the Arrhenius constant and E_a are not always constant. They can vary with temperature, especially when reactions are carried out in solution. However, these variations are negligible for wide changes in temperature for a large number of reactions and so it is reasonable to regard them as being constant unless the experimental evidence indicates otherwise.

The Arrhenius equation takes into account both the energetic (1.4) and steric natures (1.5) of chemical reactions. Reacting species must have both the correct energy (E_a) and must make contact at the right point if reaction is to occur. The term: $e^{-E_a/RT}$ gives the fraction of molecules with sufficient energy to react whilst the Arrhenius constant includes the probability of the reacting species being correctly orientated for reaction to occur.

The Arrhenius equation can be used to determine the activation energy of a reaction both in the laboratory and in biological systems. Taking logs to base e, equation 6.5 may be written:

$$\ln k = \ln A - E_a/RT$$

(6.6)

and so, a graph of $\ln k$ against $1/T$ is a straight line with a slope of $-E_a/R$.

6.7 PRESSURE AND REACTION RATE

Changes in the total pressure of a reaction system have little effect on the rate of a chemical reaction unless the reaction is being carried out in the gaseous state. However, an increase in the partial pressure of a gaseous reactant will increase the rate of a reaction because this increase in partial pressure increases the concentration of that reactant.

6.8 PARTICLE SIZE

The rate of a reaction in which more of the reactants are in the solid state increases with decrease in particle size. This increase in reaction rate is attributed to an increase in the surface area of the particle and hence an increase in the number of collisions between reacting species. The surface area of one mole of small particles of a compound have a larger surface area than one mole of large particles of the same compound. This explains why most reactions occur in solution. In this condition the particles are of molecular size and so at their most reactive in this respect. Decrease in particle size also increases the rate of dissolution (rate of solution) of substances. This is particularly important in the administration of poorly water soluble drugs.

The production of very small particles of the order 2 to 10 microns (1 micron = 10^{-6} metres) in pharmacy is known as micronising. Micronised particles are used in inhalation devices because larger particles are unable to reach the pulmonary regions of the lungs. Micronised particles are also used to produce tablets and pills of drugs with a low water solubility. Micronising increases the rate of dissolution of the drug with a subsequent increase in its overall absorption rate.

6.9 CATALYSTS AND INHIBITORS

Catalysts may be defined as substances that change the rate of a chemical reaction but can be recovered chemically unchanged at the end of the reaction. **Inhibitors** or negative catalysts are substances that decrease the rate of a reaction. Catalysts and inhibitors do not change the yield of the reaction, they only change the time taken to obtain that yield.

Many catalysts and inhibitors are quite specific, for example, a catalyst or inhibitor that affects one reaction will frequently have little effect on a similar reaction with a different compound. The mechanisms by which catalysts and inhibitors operate are very varied and in many cases not understood. In most cases the presence of the catalyst reduces the activation energy barrier of the reaction thereby making the reaction easier and so faster. In cases where this is not observed, it is believed that the presence of the catalyst increases the probability of correctly orientated collisions between the reactants. In other words, it increases the value of A in the Arrhenius equation (equation 6.5). This would then account for the increase in the rate of the reaction.

Catalysts can be classified as being **homogeneous** and **heterogeneous** depending on whether they dissolve in the reaction phase to form a single phase system. Homogeneous catalysts form a single phase system and usually act by forming intermediates with the reacting species.

Step 1:   A + Catalyst $\longrightarrow$ Intermediate
Step 2:   Intermediate + B $\longrightarrow$ Product + Catalyst
Overall reaction: A + B $\longrightarrow$ Products

Heterogeneous catalysts form a separate phase to the reaction system. They usually consist of finely divided solids and act by adsorbing the reactants. The adsorption correctly orientates the reactants as well as bringing them into close contact which increases the chances of a reaction occurring (1.6).

Enzymes are generally considered as being heterogeneous catalysts even though they are usually soluble in water, the phase in which many biological reactions occur. The general mechanism of their action has already been described in 1.6. Experimental observations show that the initial rates of many enzyme-catalysed reactions exhibit first order kinetics for low concentrations of a substrate provided the concentrations of all the other reactants are constant but this changes to zero order as the concentration of the substrate (S) is increased (Fig. 6.4). There is a maximum rate (V_{max}) for the enzyme-catalysed reaction which is independent of the concentration of the substrate provided the concentration of the substrate is high enough and the concentrations of all the other substrates are constant.

The relationship between the rate of an enzyme-catalysed reaction and the

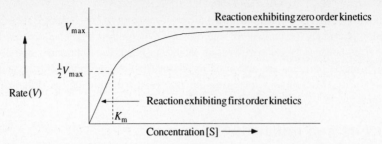

Fig. 6.4. The variation in the rate of a typical enzyme-catalysed reaction with the concentration of a substrate S, the concentrations of all other substrates being kept constant.

concentration of a substrate under a specified set of conditions when the concentrations of all other substrates are kept constant is expressed mathematically by the experimentally determined **Michaelis-Menton equation**;

$$V_o = \frac{V_{max}[S]}{K_m + [S]} \tag{6.7}$$

where K_m is a constant known as the **Michaelis constant,** V_o is the initial rate of the reaction and V_{max} is the rate of the reaction when the concentration of the substrate [S] approaches infinity. The Michaelis constant is the concentration of the substrate when the rate of the reaction is $\frac{1}{2}$ V_{max}. Rearranging equation 6.7 gives the **Lineweaver-Burk** equation:

$$\frac{1}{V_o} = \frac{1}{V_{max}} + \frac{K_m}{V_{max}[S]} \tag{6.8}$$

The values of K_m and V_{max} are characteristic properties of the enzyme and the reaction for the specified reaction conditions. They may be determined graphically by measuring the change in the initial rate V_o of the reaction with initial concentration S of the substrate. A graph of $1/V_o$ against $1/[S]$ is a straight line (equation 6.8) with a slope of K_m/V_{max}.

The values of K_m and V_{max} are used in the investigation of the mechanisms of enzyme-catalysed reactions and to determine whether a substance is a competitive, uncompetitive or non-competitive inhibitor. The values of K_m and V_{max} can also be used to determine whether an enzyme isolated from one part of an organism is **similar** to that isolated from a different part of the same organism or from a different species. Similar values for K_m and V_{max} may be regarded as good evidence for the enzymes being similar. To prove they are the same would require a full structure determination.

Inhibition of an enzyme-catalysed reaction can occur in a number of ways. The inhibitor may combine with either the apoenzyme, its coenzyme or the substrate. Substances that act as inhibitors are broadly classified as **reversible** and **irreversible** inhibitors. Reversible inhibitors react with the enzyme to form a dynamic equilibrium mixture governed by the law of mass action (6.11). As a result, a reduction in the concentration of the inhibitor present at equilibrium will reduce the extent of the inhibition. Irreversible inhibitors react with the enzyme to form a stable compound which cannot be readily dissociated to yield the enzyme.

Reversible   Enzyme + Inhibitor $\rightleftharpoons$ Enzyme-Inhibitor complex
Irreversible   Enzyme + Inhibitor $\longrightarrow$ Enzyme-Inhibitor complex

Three types of reversible inhibitor are known: competitive, non-competitive and uncompetitive.

6.9.1 Competitive inhibitors

These inhibitors (I) compete with the substrate (S) for the active site of the enzyme (E). The enzyme can only accept either the inhibitor or the substrate. Consequently, the higher the concentration of the inhibitor the lower the enzyme activity.

$$E \; + \; S \; + \; I \; \rightleftharpoons \; IE \quad \text{No further reaction, S is prevented from binding by the inhibitor}$$

Competitive inhibitors change the slope of the graph of $1/V_o$ against $1/[S]$ but not its intercept on the $1/V_o$ axis.

6.9.2 Non-competitive inhibition

Non-competitive inhibitors are believed to combine with the enzyme at an allosteric site (1.6) to form both inhibitor-enzyme (IE) and inhibitor-enzyme-substrate (IES) complexes which cannot break down to form the products.

$$
\begin{array}{ll}
\text{Either:} & E \; + \; I \; \rightleftharpoons \; IE \\
\text{Or:} & E \; + \; S \; \rightleftharpoons \; ES \;\searrow^{\;I} \\
& \qquad\qquad\qquad\qquad IES \\
& E \; + \; I \; \rightleftharpoons \; IE \;\nearrow_{\;S}
\end{array}
\left.\vphantom{\begin{array}{l}a\\a\\a\\a\\a\end{array}}\right\}
\begin{array}{l}\text{No further reaction}\\ \text{to form the products}\end{array}
$$

These inhibitors cause a change in both the slope and the intercept of the $1/V_o$ axis of the $1/V_o$ against $1/[S]$ graph.

6.9.3 Uncompetitive inhibition

These inhibitors are believed to form a complex with the enzyme-substrate complex which cannot react further to form the product.

$$E \; + \; S \; \rightleftharpoons \; ES \; + \; I \; \rightleftharpoons \; IES \quad \text{No further reaction}$$

Uncompetitive inhibitors change the intercept on the $1/V_o$ axis of the Lineweaver-Burk graph but do not change the slope of the line. Consequently, changing the concentration of the inhibitor produces a series of parallel lines for a graph of $1/V_o$ against $1/[S]$.

6.10 CHEMICAL EQUILIBRIA

All chemical reactions are believed to be thermodynamically reversible (6.15) but some are so 'biased' in one direction that for practical purposes they are regarded as irreversible. However, a number of chemical reactions form products that immediately start to react to reform the original reactants. Since the rate of the forward and backward reactions are related to the concentrations of the reactants and products respectively there comes a point where the rates of the forward and backward reactions are equal provided that the temperature remains constant:

Rate of the forward reaction = Rate of the backward reaction

At this point the reaction is said to be in a state of **dynamic equilibrium**. Therefore, at equilibrium, at constant temperature:

k [a function of the reactant's concentration] = k' [a function of the product's concentration]

that is,

$$K = \frac{k}{k'} = \frac{[\text{a function of the products concentration}]}{[\text{a function of the reactants concentration}]}$$

where, K is a constant at the temperature at which the reaction was carried out.

At equilibrium, the forward and backward reactions are still taking place but there is no overall change in the concentrations of the reactants and products. The relative concentrations of the reactants and products at equilibrium is called the **position of equilibrium** for the reaction.

Equilibria are classified into two types: **homogeneous** where all the reactants and products are in the same phase and **heterogeneous** where the reactants and products are not all in the same phase. Examples of these types of equilibrium are given in Fig. 6.5.

Homogeneous equilibria (liquid phase):

$$CH_3COOH \quad + \quad C_2H_5OH \quad \rightleftharpoons \quad CH_3COOC_2H_5 \quad + \quad H_2O$$
$$\text{(ethanol)}$$

$$CH_3CHOHCOOH \quad + H_2O \quad \rightleftharpoons \quad CH_3CHOHCOO^- \quad + H_3O^+$$
$$\text{(water)}$$

Heterogeneous equilibria (solid (s) and liquid (aq) phases):

$$K_2CO_{3(aq)} \quad + \quad BaSO_{4(s)} \quad \underset{\text{(water)}}{\rightleftharpoons} \quad K_2SO_{4(aq)} \quad + \quad BaCO_{3(s)}$$

Fig. 6.5. Examples of homogeneous and heterogeneous equilibria. The liquid phase is given in brackets.

6.11 THE LAW OF CHEMICAL EQUILIBRIUM

6.11.1 Homogeneous systems

At constant temperature, for homogeneous systems of the type;

$$a\,A \; + \; b\,B \; + \; ... \; \rightleftharpoons \; x\,X \; + \; y\,Y \; + \; ...$$

the relationship:

$$K_c = \frac{[X]^x[Y]^y\,...}{[A]^a[B]^b\,...} \tag{6.9}$$

may be written where K_c is a constant at constant temperature. K_c is known as the **equilibrium constant** for the system and equation 6.9 is known as the **law of mass action** or **the equilibrium law**. The value of K_c is only constant at concentrations exceeding about 0.01 mol dm^{-3} when the concentrations of the reactants and products are measured as their **activities**. This unit of concentration allows for the non-ideal behaviour of solutes. Activity is related to the actual concentration of the solute by the expression:

Activity = (Activity coefficient)(Concentration)

Activity coefficient values depends on the temperature and the concentration of the substance whose activity is required. The values of activity coefficients are usually deduced from colligative property measurements. In solutions with a concentration of less than about 0.01 mol dm^{-3}, the value of the activity coefficient of a solute is 1 and so for dilute solutions at constant temperature:

$$K_c = K_{eq}$$

where K_{eq} is the equilibrium constant calculated using activities in the appropriate law of mass action equation. K_{eq} is known as the **thermodynamic equilibrium constant** because its value may also be calculated from thermodynamic data. It should be noted that the use of activities as the unit of concentration is indicated by ( ) instead of [ ].

6.11.2 Heterogeneous systems

The law of mass action can also be applied at constant temperature to heterogeneous systems. However, in heterogeneous systems **the equilibrium law is only applied to either the gaseous or liquid states of these systems**. This is because experimental work has shown that the concentrations of the reactants in the other states can be regarded as being constant in the liquid or gaseous state being considered. Consequently, the law of mass action equation (equation 6.9) does not include concentration terms for these substances. Their values are incorporated into the value of the equilibrium constant. For example, consider the thermal decomposition of calcium carbonate:

$$CaCO_{3(s)} \rightleftharpoons CaO_{(s)} + CO_{2(g)}$$

As the calcium carbonate and oxide are in the solid state their concentrations in the gaseous state can be regarded as being constant and so not included in equation 6.9. Therefore, the equilibrium law expression for this reaction is:

$$K_c = [CO_2]$$

6.12 FACTORS AFFECTING THE POSITION OF EQUILIBRIUM

The factors that affect the position of equilibrium are changes in: (a) the concentration of either reactants or products, (b) the temperature, and (c) the pressure. Catalysts do not affect the position of equilibrium, they only decrease the time taken to reach equilibrium.

6.12.1 Concentration changes

The effect of changing the concentration of one of the reactants or products on the position of equilibrium of a system may be predicted by considering the law of mass action. For example, consider the equilibrium system:

$$aA + bB \rightleftharpoons cC + dD$$

At constant temperature:

$$K_c = \frac{[C]^c[D]^d}{[A]^a[B]^b} \tag{6.10}$$

Increasing the concentration of either A or B will increase the value of the denominator in equation 6.10. Therefore, if K_c is to remain constant, the concentrations of the products C and D must increase by an appropriate amount. For this to occur the reactants A and B must react to produce higher concentrations of C and D, that is the position of equilibrium of the reaction move to the right. The same argument will show that increasing the concentration of a product will move the position of equilibrium to the left. In practice it is found that these predictions are usually correct.

6.12.2 Temperature changes

Increasing the temperature increase the rates of both the forward and backward reactions of an equilibrium. This reduces the time taken to reach eqilibrium. However, increasing the temperature also increases the rate of endothermic reactions more than exothermic reactions. Therefore, the position of equilibrium will move towards the substances (reactants or products) that are produced by the endothermic process in the equilibrium. For example, an increase in the temperature at which the Haber process is conducted will reduce the yield of ammonia because the forward reaction is exothermic.

$$N_2 \; + \; 3H_2 \; \rightleftharpoons \; 2NH_3 \qquad \Delta H = \text{-}92 \, kJ$$

The relationship of the thermodynamic equilibrium constant to the temperature of the equilibrium system at constant pressure is given by the expression:

$$\log K_{eq} = \text{constant} - \frac{\Delta H}{19.14 \, T} \qquad (6.11)$$

This relationship was deduced by van't Hoff from the Arrhenius equation (equation 6.5). The enthalpy change (ΔH) for a reaction will vary with temperature but it can be regarded as being constant over small temperature ranges. Consequently equation 6.11 can be used to calculate either the value of ΔH, or the value of the thermodynamic equilibrium constant for a system at an absolute temperature T, from experimental data. Calculations can be made graphically (a plot of $\log K_{eq}$ against $1/T$ should be a straight line) or by mathematical manipulation of equation 6.11 depending on the extent of the data. Equation 6.11 can also be used to calculate standard entropy and Gibbs free energy changes (ΔG) for an equilibrium system (6.19).

6.12.3 Total pressure changes

Increasing the pressure on a gas or liquid has the effect of reducing its volume (Boyles law). Consequently, changes in the total pressure of a system have little effect on gaseous or liquid equilibrium systems at constant temperature unless the forward and backward reactions involve a volume change. In cases where there is a volume change, an increase in the total pressure of the system has been observed to favour the reaction proceeding in the direction with the reduction in volume. For example, consider the equilibrium system:

$$N_2 \; + \; 3H_2 \; \rightleftharpoons \; 2NH_3$$

The forward reaction of this system involves a reduction in the number of molecules present in the system. However, since one mole of any gas occupies a volume V at a specified temperature and pressure there must be a reduction in volume in this forward direction. Practical observations show that increasing the pressure of this equilibrium system moves the position of equilibrium to the right, that is, in the direction that produces a reduction in the volume of the system. This observation may be readily explained by the law of mass action. Suppose at equilibrium there are a moles of nitrogen, b moles of hydrogen, and c moles of ammonia in a vessel with a volume V dm^3. Then, applying the law of mass action:

$$K_c = \frac{[NH_3]^2}{[N_2][H_2]^3} = \frac{[c/V]^2}{[a/V][b/V]^3} = \frac{c^2 V^2}{ab^3} \qquad (6.12)$$

Increasing the pressure of the system will reduce the volume of the system and as a result the value of the term $c^2 V^2$ in equation 6.12. Therefore, to keep K_c constant, the value of the term ab^3 must be reduced and the value of c increased. In practical terms this means that the position of equilibrium moves to the right, that is, some of the nitrogen and hydrogen reacts to form more ammonia.

6.13 ACID AND BASE EQUILIBRIA

Two definitions of acids and bases are in common use, the Brönsted and Lowry being the more useful in the pharmaceutical and other life sciences:

(1) **The Brönsted and Lowry definition:** Acids are proton donors whilst bases are proton acceptors.
(2) **The Lewis definition:** Lewis acids are electron pair acceptors whilst Lewis bases are electron pair donors.

The strength of an acid or base is a measure of its ability to act in the manner indicated in these definitions. Strong acids will readily donate protons and accept electron pairs whilst strong bases will readily accept protons and donate electron pairs. Weak acids will not readily lose protons or accept electrons pairs whilst weak bases will not readily accept protons or donate electron pairs. Some compounds can act as either an acid or a base depending on the circumstances. These compounds are known as **amphoteric compounds**. For example, in aqueous solutions of ethanoic acid, water acts as a base, whilst in aqueous solutions of methylamine it acts as an acid. This ability of water to act as either an acid or a base is of considerable importance in chemical and biological reactions.

$$CH_3COOH \; + \; H_2O \; \rightleftharpoons \; CH_3COO^- \; + \; H_3O^+$$
Acid   Base   Conjugate base   Conjugate acid

$$CH_3NH_2 \; + \; H_2O \; \rightleftharpoons \; CH_3\overset{+}{N}H_3 \; + \; OH^-$$
Base   Acid   Conjugate acid   Conjugate base

6.13.1 Strengths of acids

The strengths of acids are recorded at a specified temperature, usually $25^{\circ}C$, as either a pH (equation 6.13) or a pK_a (equation 6.14) value (Table 6.4), the latter being the more useful. In clinical medicine the amount of acid is reported as mmoles of H^+ per litre or decimetre cubed (H^+ mmol l^{-1} or mmol dm^{-3})

$$pH = -\log [H]^+ \tag{6.13}$$

$$pK_a = -\log K_a \tag{6.14}$$

where K_a is the **dissociation constant** of the acid. The dissociation constant of an acid is defined as the equilibrium constant for its ionisation. Consider, for example, the ionisation of the acid HA:

$$HA \rightleftharpoons H^+ + A^-$$

Applying the law of mass action, at constant temperature the dissociation constant K_a is given by the expression:

$$K_a = \frac{[H^+][A^-]}{[HA]} \tag{6.15}$$

The dissociation constant and hence pK_a are valid measures of acid strength, even though they do not take into account the solvent, because K_a is directly related to the equilibrium constant K_{eq} for the acid-solvent system. This may be demonstrated by considering the dissociation of HA in water:

$$HA + H_2O \rightleftharpoons H_3O^+ + A^-$$

At constant temperature:

$$K_{eq} = \frac{[H_3O^+][A^-]}{[HA][H_2O]} \qquad (6.16)$$

However, the concentration of water (approximately 55 mol dm^{-3}) is so large compared with that of the acid, the concentration of the water can be regarded as being constant for dilute solutions of different acids. Furthermore, for HA, $[H^+] = [H_3O^+]$ and so, substituting this equality and equation 6.15 for K_a in equation 6.16:

$$K_{eq} = \frac{K_a}{[H_2O]}, \quad \text{that is:} \quad K_{eq}[H_2O] = K_a$$

Consequently, as $[H_2O]$ can be regarded as being constant, K_{eq} is directly related to K_a and therefore at the specified temperature K_a is a valid measure of an acid's strength.

Table 6.4. Examples of the pK_a values of acids and bases. These values, which are taken from various sources, were measured in different solvents but, even so, give a reasonably accurate guide to the relative strengths of these acids in water at 25°C. Strong acids have low pK_a values while strong bases have high pK_a values.

Acid	pK_a	Acid	pK_a	Base	pK_a	Base	pK_a
Benzoic acid	4.2	Methanoic acid	3.7	Ammonia	35	Isoquinoline	5.4
Boric acid	9.2	Phenol	9.9	Aminobenzene	4.6	Piperidine	11.2
Ethanoic acid	4.8	Phosphoric acid	2.1	Diethylamine	11.0	Pyridine	5.3
Glycine	2.3	Phosphorous acid	1.3	Diethylphosphine	2.7	Triethylamine	9.8
Hydrochloric acid	-7	Picric acid	0.4	Ethanolamine	9.5	Urea	1.0

6.13.2 Strengths of bases

The strengths of bases are also recorded at a specific temperature, usually 25°C, as either a pH or a pK_a value (Table 6.4). However, it should be noted that K_a refers to the dissociation constant of the conjugate acid, and not to the dissociation of the base, that is:

$$K_a = \frac{[B][H^+]}{[BH^+]} \quad \text{for the equilibrium system:} \qquad BH^+ \rightleftharpoons B + H^+$$
$$\text{Conjugate acid} \qquad \text{Base}$$

In older literature the strength of a base may be recorded as its pK_b value, where K_b is the dissociation constant, at constant temperature, for the dissociation of the base. It is given by the expression:

$$K_b = \frac{[B^+][OH^-]}{[B]} \quad \text{for the equilibrium system:} \qquad B \rightleftharpoons B^+ + OH^-$$
$$\text{Base}$$

The pK_b and pK_a values for the strength of a base are related by the equation:

$$pK_a + pK_b = pK_w = \log K_w \qquad (6.17)$$

where K_w is the ionic product of water provided all values are measured at the same temperature. At 25°C, $K_w = 10^{-14}$. It follows from equation 6.17 that strong bases with a high pK_a value will have a low pK_b value and vice versa for weak bases.

6.13.3 pH and the ionisation of acidic and basic compounds in water

Acidic and basic compounds ionise in water to form the equilibrium systems:

$$HA + H_2O \rightleftharpoons H_3O^+ + A^-$$
$$B + H_2O \rightleftharpoons BH^+ + OH^-$$

A change in the hydrogen ion concentration (pH) of a solution will change the position of equilibrium of the system. Applying the law of mass action to the relevant equilibria, a decrease in the pH (increase in H^+ concentration) of a solution will decrease the ionisation of an acid but increase the ionisation of a base. Conversely, an increase in the pH of a solution will increase the ionisation of an acid and decrease the ionisation of a base.

The affect of pH on the ionisation of acids and bases is important in a number of biological processes. For example, it is known that the uncharged form of a molecule is usually more easily transported through a membrane than the charged form of that molecule. In the very strongly acidic stomach (pH in the region of 1 to 2), the ionisation of acidic molecules will be suppressed but that of bases enhanced. Consequently, acidic compounds are more likely to be absorbed in the stomach than basic compounds. Conversely, in the alkaline intestine (pH in the region of 5 to 8), the ionisation of acidic compounds will be enhanced but that of basic compounds suppressed. As a result, basic compounds are more likely to be absorbed than acidic compounds. However, it should be realised that pH is only one of the factors that will affect the absorption of a drug and as a result these statements are not always correct. Small changes in pH will also affect the activity of proteins by suppressing or enhancing their ionisation. Similarly, changes in pH will affect reactions in which hydrogen ions are either a reactant or a catalyst.

6.14 BUFFERS

Buffers are solutions whose pH value does not change appreciably when reasonable quantities of either an acid or a base are added. It is emphasised that the pH of a buffer solution is not constant but the addition of reasonable quantities of either an acid or base will cause it to change only slightly. However, the pH of a buffer will change significantly if large quantities of an acid or base are added.

The most useful buffer mixtures are aqueous solutions of either a weak acid and its salt with a strong base, a weak base and its salt with a strong acid, or a weak acid and a weak base. Their pH stability is due to the presence of a high concentration of a weak base, usually in the form of a suitable anion, to neutralise any extra hydrogen ions and a high concentration of a weak acid which can supply hydrogen ions to neutralise any extra hydroxide ions. Each buffer solution contains a complex equilibrium system in which the ionisation equilibria of the components are in a state of mutual equilibrium. Addition of an acid or base results in a mutual change in the positions of some of the interrelated equilibria comprising the system so that very little change in the overall hydrogen ion concentration of the system occurs. Consider, for example, a buffer consisting of an aqueous solution of sodium ethanoate and ethanoic acid. The interrelated equilibria in this buffer are shown in Fig. 6.6. The $\longrightarrow$ in these equilibria indicates whether the position of equilibrium lies to the right or left. When hydrogen ions are added to this buffer solution, most are absorbed by the weakly basic CH_3COO^- ion to form more undissociated ethanoic acid molecules thereby keeping the overall hydrogen ion content of the solution almost constant. The sodium ethanoate acts as the source of ethanoate ions for this purpose. When a base is

$$CH_3COOH \;\rightleftharpoons\; CH_3COO^- \;+\; H^+$$

$$CH_3COONa \;\rightleftharpoons\; CH_3COO^- \;+\; Na^+$$

$$H_2O \;\rightleftharpoons\; H^+ \;+\; OH^-$$

Fig. 6.6. The interrelated equilibria found in an ethanoic acid-sodium ethanoate buffer.

added to the buffer solution it is converted into its salt by reaction with the hydrogen ions in the solution. These hydrogen ions are largely replaced by the ionisation of more water and ethanoic acid molecules. This keeps the pH of the buffer almost constant.

The pH of a buffer solution may be calculated using the **Henderson-Hasselbalch equation** (equation 6.18):

$$pH = pK_a + \log \frac{[\text{base component}]}{[\text{acid component}]} \qquad (6.18)$$

where the value used for pK_a is that of the acid component of the buffer mixture. For example, the acid component of an ammonia-ammonium chloride buffer is the salt because it is the ammonium ion that functions as the acid:

$$NH_4^+ \;\rightleftharpoons\; NH_3 + H^+$$

The fluids found in biological systems are usually buffers that keep the fluid at approximately the correct pH for the system, for example, at approximately pH 7.4 for blood. The proteins of these systems will have their own specialist functions but will also, because of their numerous weakly acidic and basic groups, contribute to the buffering of the system. For example, haemoglobin acts as a buffer in the transport of carbon dioxide in the venous blood supply to the lungs. Oxygen is not very soluble in water and is transported in the blood in the form of a complex with haemoglobin (Figs 32.14 and 32.15).

$$Hb + O_2 \;\rightleftharpoons\; HbO_2 \;\overset{\text{Ionisation}}{\rightleftharpoons}\; H^+ \;+\; HbO_2^-$$

Haemoglobin	Oxygenated haemoglobin	Predominant form in blood

At the site of respiration the oxygenated haemoglobin supplies oxygen to oxygen-deficient tissue and accepts carbon dioxide from carbon dioxide rich tissue. This carbon dioxide is released into the blood where it dissolves to form carbonic acid which dissociates to form hydrogen and hydrogen carbonate ions (bicarbonate ions). This ionisation is promoted by the enzyme *carbonic anhydrase* found in the erythrocytes of the blood:

$$CO_2 + H_2O \;\rightleftharpoons\; H_2CO_2 \;\overset{\textit{Carbonic anhydrase}}{\rightleftharpoons}\; H^+ + HCO_2^-$$

This process should increase the acidity of the blood, that is, decrease its pH. However, pH measurements show that the pH of venous blood is only slightly less than 7.4, the pH of oxygenated blood. Since the average human exhales between 20 and 40 moles of carbon dioxide per day, venous blood must have a very high buffering capacity. This capacity is supplied by bases which will absorb the hydrogen ions produced by the ionisation of the carbon dioxide in water. The main sources of these bases are plasma phosphates, plasma proteins and haemoglobin (Fig. 6.7) acting as a weak bases.

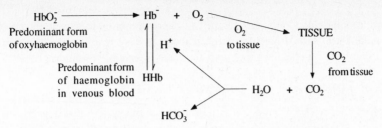

Fig. 6.7. A schematic representation of the buffering effect of haemoglobin.

6.15 THERMODYNAMICS

Thermodynamics describes the relationships between the different forms of energy and how they affect a chemical or physical change. It can be used to predict whether a change is possible and the conditions required but it does not predict the rate at which the change may occur. For example, thermodynamics predicts that glucose will react with oxygen with the release of large amounts of energy at room temperature. In practice, a mixture of oxygen and glucose is stable indefinitely at room temperature and so for thermodynamic purposes, the reaction proceeds at an infinitely slow rate. In living organisms, thermodynamics can be used to explain how energy affects the operation of the organism at the molecular level, including the conformations assumed by naturally occurring molecules, the way in which metabolic pathways are powered, the energy changes involved in the transport of substances across membranes, and the operation of muscles.

The area being studied in thermodynamics is known as the **system**. A system can be large or small, such as a complete organism or a single crystal, but all thermodynamic systems must have clearly definable boundaries. The universe outside these boundaries is known as the **surroundings**. If a reaction is carried out in a flask in the laboratory, the flask, the substances involved in the reaction and the redistribution of energy that occurs in the reaction constitute the system. The outer surface of the flask is the boundary of the system whilst the room in which the flask is being used constitutes the surroundings. Systems are said to be **open** if they are able to exchange matter or energy with the surroundings or **closed (isolated)** if the system is completely isolated from its surroundings. The state of any system is defined by its composition, temperature and pressure. Systems in which the temperature is kept constant are known as **isothermal** systems whilst those in which the temperature varies are know as **adiabatic** systems. Biological processes usually take place in closed systems at constant temperature and pressure. Most laboratory reactions and physical changes are carried out in open isothermal systems. The fundamental laws of thermodynamics which are used to explain the behaviour of these systems were derived from general observations of how the natural world behaves and from experimental evidence.

Thermodynamics treats all chemical reactions as occurring in a series of consecutive infinitesimal reversible steps, that is, **all reactions** are reversible. It regards the end of the reaction as being the point where the system reaches equilibrium. This means that reactions which go to completion have a position of equilibrium that lies well to the right-hand side of the equation for the process. Reactions which are not reversible in practice are still regarded as proceeding in a series of reversible steps. Their irreversibility is explained as being due to the *total quantity* of energy required to reverse the process once the product(s) are formed being so large that the reverse reaction becomes impractical. The use of $\rightleftharpoons$ indicates a thermodynamically reversible reaction.

6.16 THE FIRST LAW OF THERMODYNAMICS

The internal energy (U) of a system is a property of the system that depends only on its current state. It does not depend on how the system arrived at that state. Consequently, if a closed system undergoes a spontaneous change from an internal energy state Y to an internal energy state X, the change in the internal energy of the system (ΔU) is the same, no matter what route the change follows (Fig. 6.8). The properties of a system that depend only on the current state of the system and are independent of the history of the system are known as **functions of state**.

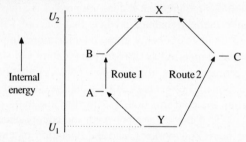

Fig. 6.8. The change in internal energy of a system when it changes from a state Y to a state X is independent of the route taken. For both routes 1 and 2 the $\Delta U = U_2 - U_1$. A,B and C are stable intermediates formed on the relevant route.

Energy can be removed from or introduced into a system. This change in the **internal energy (ΔU)** of the system can take the form of **heat (q)** absorbed or liberated by the system and the **work (w)** done on or by the system. Work, which is a form of energy, is done by a system when a force is moved through a distance in the direction of the force. It is measured by the product of the force and the distance moved by the force.

The law of conservation of energy states that energy can be neither created nor destroyed When work is done on the system this law may be expressed mathematically as

$$\Delta U = q + w \qquad (6.19)$$

Equation 6.19 is known as the **First Law of Thermodynamics.** Positive values of ΔU and q indicate a gain in energy whilst negative values indicate a loss of energy. Positive values for w show that work was done on the system, for example, the system may have contracted. A negative value for w indicates that the system has done work, for example, the system may have expanded. **By convention,** thermodynamics always considers work as being done by the system on the surroundings and so equation 6.19 becomes:

$$\Delta U = q - w \qquad (6.20)$$

In chemical and physical changes, work is usually done by the system expanding. If the change in volume of the system is ΔV then the work done by the expansion of a system at a constant pressure P is given by $-P\Delta V$. Therefore, for systems at constant pressure, equation 6.19 may be written:

$$\Delta U = q_p - P\Delta V \qquad (6.21)$$

where q_p is the heat absorbed or liberated by the system at constant pressure.

Chemical and physical changes occur in the laboratory under atmospheric pressure. This is regarded as being constant over the period required for most changes to occur. Biological processes also occur under constant pressure. The heat absorbed or liberated by chemical and biological systems when they undergo a change at constant pressure is known as the **enthalpy change** (ΔH) of the system. It is defined by the relationship:

$$\Delta H = q_p = \Delta U + P\Delta V \qquad (6.22)$$

An enthalpy change at constant pressure in which the system transfers heat to the surroundings is an **exothermic change**. Exothermic changes have a negative ΔH because energy is lost from the system to the surroundings. Conversely, an enthalpy change in which the system absorbs heat from the surroundings is an **endothermic** change and has a positive ΔH because the system has gained energy from the surroundings.

Enthalpy is a function of state, that is:

$$\Delta H = H_2 - H_1$$

where H_1 is the enthalpy of the initial state of the system and H_2 is the enthalpy of the final state of the system after the change has taken place. Consequently, enthalpy changes are dependent on the initial and final states of the system and not the route taken by the change. This is Hess's law which states that the energy change for a particular reaction is the same whether the reaction occurs in one or several steps provided all the reactions are carried out at the same constant temperature and pressure. This law enables you to calculate the energy change during a reaction from the energy changes of related reactions provided all the reactions are carried out at the same temperature and pressure. The calculation is made by treating the chemical equations as though they were algebraic equations. Whatever is done to the chemical equations must also be done to the ΔH values.

Example 6.1 *If the heats of combustion of ethanol and ethanal are -1371 kJ mol^{-1} and -1168 kJ mol^{-1} respectively, calculate the enthalpy change for the oxidisation of one mole of ethanol to ethanal.*

The equations for these oxidations are:

$$C_2H_5OH + 3O_2 \longrightarrow 2CO_2 + 3H_2O \qquad \Delta H = -1371 \text{ kJ mol}^{-1} \qquad (6.23)$$
$$CH_3CHO + 2.5\ O_2 \longrightarrow 2CO_2 + 2H_2O \qquad \Delta H = -1168 \text{ kJ mol}^{-1} \qquad (6.24)$$

Subtracting equation 6.24 from 6.23 gives:

$$C_2H_5OH - CH_3CHO + 1/2O_2 \longrightarrow H_2O \qquad \Delta H = -1371 - (-1168) \text{ kJ mol}^{-1}$$

Therefore, $\quad C_2H_5OH + 1/2O_2 \longrightarrow CH_3CHO + H_2O \quad \Delta H = -203 \text{ kJ mol}^{-1}$

Therefore, the enthalpy change for the oxidation of one mole of ethanol to ethanal is:
$$-203 \text{ kJ mol}^{-1}.$$

This type of calculation is especially useful where it is not possible to measure the enthalpy change of a reaction directly, such as in the cases of *in vivo* reactions. It may also be used to estimate the internal energy changes that occur in biological reactions where there is no appreciable change in the volume of the system. In these cases, at constant pressure $P\Delta V = 0$ and so if this is the only work done by the system, equation 6.22 becomes $\Delta H = \Delta U$. This means that the change in the internal energy during a reaction can be determined by measuring the enthalpy change for the reaction. It is not necessary to measure the enthalpy change directly: it may be determined from any reaction pathway between the same reactants and products using Hess's law.

Chemical and physical changes can also be carried out under constant volume. Under this condition $\Delta V = 0$ and so $P\Delta V = 0$. Therefore, provided that this is the only form of work done by the system, the heat absorbed by the system at constant volume (q_v) is equal to the changes in internal energy and enthalpy of the system:

$$\Delta H = \Delta U = q_v \qquad (6.25)$$

6.17 THE SECOND LAW OF THERMODYNAMICS

The second law of thermodynamics predicts whether a system will undergo a spontaneous change. It is based on the observation of spontaneous changes in the natural world. These observations indicate that spontaneous changes result in an increase in the disorder of a system and its surroundings. For example, if a bone china plate (the system) is dropped on a stone floor the plate spontaneously shatters. The relatively well ordered structure of the plate is converted into a disordered system of randomly moving small pieces of china. These possess different amounts of energy. Simultaneously the air surrounding the plate is disturbed by the explosion of the plate. However, the reverse process in which the shards of china spontaneously reform the plate has never been observed. The second law of thermodynamics summarises this type of observation. It states that **spontaneous changes occur in the direction that increases the overall disorder of the system and its surroundings.** However, it does not indicate the speed of the change and it is theoretically possible to have a spontaneous change that proceeds at such a slow rate that it does not occur in the human timescale.

The state of disorder of a system depends on the physical arrangement of the particles forming the system, their velocity, if any, and energy. It is described by a function of state known as the **entropy** (S) of the system. **The larger the entropy of a system the more disordered that system**. Since entropy can also used to describe the state of disorder of the surroundings, it follows from the second law of thermodynamics that a spontaneous change in a system is accompanied by an increase in the entropy of both the system and its surroundings, that is, for a spontaneous change to be possible:

$$\Delta S_{system} + \Delta S_{surroundings} > 0 \qquad (6.26)$$

Changes in entropy at constant temperature and pressure are defined by the expression:

$$\Delta S = q/T$$

where T is the absolute temperature at which the change occurs and q is the quantity of heat liberated or absorbed during the change. Therefore, as the heat lost by the system at constant pressure $(- \Delta H)$ is equal to the heat gained by the surroundings (q):

$$\Delta S_{surroundings} = q/T = -\Delta H_{system}/T \quad \text{J K}^{-1} \qquad (6.27)$$

Substituting equation 6.27 in equation 6.26:

$$\Delta S_{system} + (-\Delta H_{system}/T) > 0$$

that is,

$$T\Delta S_{system} - \Delta H_{system} > 0$$

rearranging,

$$\Delta H_{system} - T\Delta S_{system} < 0 \qquad (6.28)$$

Therefore, if a system is to be capable of undergoing a spontaneous change at constant temperature and pressure, equation 6.28 must have a negative value.

6.18 GIBBS FREE ENERGY

The value of the expression, $\Delta H - T\Delta S$, for changes in a system at constant pressure and temperature is known as the change in Gibbs free energy (ΔG) of the system. Since enthalpy and entropy are functions of state, it follows, since T is constant, that Gibbs free energy is also a function of state, that is:

$$\Delta G = G_2 - G_1$$

where G_1 is the Gibbs free energy of the system and G_2 is the Gibbs free energy of the system after the change. It follows from equation 6.28 that a negative value for ΔG indicates that the system could undergo a spontaneous change, that is, it can do useful work. Reactions with negative ΔG values are said to be **exergonic**. It is emphasised that a negative ΔG value does not mean that the change will occur at a measurable rate. It may occur so slowly that for all practical purposes it does not actually happen. For example, calculations show that the hydrolysis reactions of proteins and polysaccharides have negative ΔG values, that is, the proteins and polysaccharides are unstable and should spontaneously hydrolyse in biological conditions. However, these compounds are to all intents and purposes stable under biological conditions and do not spontaneously hydrolyse (at an appreciable rate) until the appropriate catalyst is present. By definition, a biological catalyst is unchanged at the end of a reaction and so cannot affect the value of ΔG. This implies that a biological catalyst can only increase the rate of a reaction not initiate it.

A zero value for ΔG is associated with the system being in a state of dynamic equilibrium. In this situation the rate of the forward change is equal to the rate of the backward change.

Positive ΔG values indicate that the system will not undergo a spontaneous change. These systems are said to be **endergonic.** It is possible to bring about a change in an endergonic system by supplying free energy from an exergonic change with a **common intermediate**. This type of action, which is known as **coupling**, forms the basis of the thermodynamic explanation of the operation of many of the steps in metabolic pathways. Consider, for example, the formation of glucose-6-phosphate, the initial step in glucose metabolism. The direct reaction of phosphate with glucose is endergonic having a ΔG value of $+13.8$ kJ mol^{-1}:

Glucose + Phosphate $\longrightarrow$ Glucose-6-phosphate + H_2O $\Delta G_1 = +13.8$ kJ mol^{-1} (6.29)

However, this reaction occurs in metabolism because it is coupled with the hydrolysis of ATP which is highly exergonic:

ATP + H_2O $\longrightarrow$ ADP + Phosphate $\Delta G_2 = -30.5$ kJ mol^{-1} (6.30)

Coupling these reactions (combining equations 6.29 and 6.30) gives a system in which the overall change in the free energy is negative, namely -16.7 kJ mol^{-1}. In other words, the coupled reactions form a system that thermodynamically can undergo a spontaneous change. In effect, the ATP supplies the energy for the reaction of the glucose with the phosphate:

Glucose + Phosphate + ATP + H_2O $\longrightarrow$ Glucose-6-phosphate + ADP + Phosphate + H_2O

Simplifying:

Glucose + ATP $\longrightarrow$ Glucose-6-phosphate + ADP $\Delta G = 13.8 - 30.5 = -16.7$ kJ mol^{-1}

The general effects of entropy and enthalpy changes on the value of ΔG and hence their influence on the spontaneity of a change are summarised in Table 6.5.

Table 6.5. The conditions governing the spontaneity of changes at constant pressure and temperature.

ΔG	ΔS	ΔG and comments
−	+	ΔG is negative at all temperatures and so system changes can be spontaneous at all temperatures
−	−	ΔG is negative at all temperatures below $T = \Delta H/\Delta S$. System changes will only be spontaneous below T
+	+	ΔG is negative at temperatures above $T = \Delta H/\Delta S$. System changes will only be spontaneous above T
+	−	ΔG is always positive. System changes are always endergonic

Gibbs free energy changes may be determined from the redox and electrode potentials of any oxidation and reduction processes taking place in the system because for a redox system:

$$\Delta G_{electrode\ process} = -nFE \tag{6.31}$$

and

$$\Delta G_{process} = \Delta G_{reduction\ process} - \Delta G_{oxidation\ process}$$

where n is the number of electrons involved in the electrode process, F is the Faraday constant and E is the electrode potential of the electrode system which is given by the appropriate Nernst equation, that is either:

$$E = E_o + \frac{2.303RT}{nF} \log [\text{Metal ion}] \qquad \text{or} \quad E = E_o + \frac{2.303RT}{nF} \log \frac{[\text{oxidised state}]}{[\text{reduced state}]}$$

Substituting the appropriate values of E in equation 6.31 it is possible to determine whether $\Delta G_{process}$ is negative and whether, as a result, the reaction is in theory spontaneous.

6.19 STANDARD GIBBS FREE ENERGIES

The change in the Gibbs free energy of a system depends on the temperature, pressure, pH and concentrations of the species present. Therefore, to compare the ΔG values of different reactions it is necessary to measure and record ΔG values under standard conditions (compare with $\Delta H^{\ominus}$ values). Standard Gibbs free energies ($\Delta G^{\ominus}$) are recorded at a temperature of 25°C (298°K), 1 atmosphere pressure and specified activities or concentrations. For example, **standard Gibbs free energies of formation** ($\Delta G^{\ominus}$) are defined as **the free energy change that occurs when one mole of a substance is formed from its constituent elements.** By convention, the standard Gibbs free energy of any pure element in its most stable state is zero. The standard free energy change for any chemical or physical change can be calculated from the relationship:

$$\Delta G^{\ominus} = \text{The sum of } \Delta G_f^{\ominus}(\text{products}) - \text{The sum of } \Delta G_f^{\ominus}(\text{reactants})$$

The change in the standard Gibbs free energy of a system is related in an isothermal system at a constant temperature T and constant pressure to the system's thermodynamic equilibrium constant by the expression:

$$\Delta G^{\ominus} = -RT \ln K_{eq} \qquad (6.32)$$

where K_{eq} is the thermodynamic equilibrium constant at T° K.

Thermodynamic data is normally recorded as standard state values. The use of standard states is indicated by use of the superscript $^{\ominus}$ with the symbol used for the thermodynamic measurement, for example, standard state values for entropy are referred to as $\Delta S^{\ominus}$ values. However, the standard state conventions used in biochemistry, biology and related subjects are sometimes different from those used in chemistry. For example, the activity of water in its standard state is defined as 1 in order to simplify free energy expressions. Furthermore, hydrogen ion activity is defined as being unity at pH 7 instead of the physical chemistry standard of pH 0. Thermodynamic values measured under these standard biochemical conditions are indicated by the use of a prime, for example, $\Delta G^{\ominus'}$

6.19.1 Gibbs free energy and temperature

The temperature at which a change in the Gibbs free energy was measured is indicated by the use of a subscript unless it was made at 298 K. For example, a measurement made at 37°C would be recorded as $\Delta G_{335} = \ldots..$kJ mol^{-1}. If the measurement was made at 298K it would be recorded as $\Delta G = \ldots$kJ mol^{-1}.

The change in Gibbs free energy with change in temperature is defined by the **Gibbs-Helmholtz** equation whose integrated form is equation 6.33. This relationship assumes that ΔH is independent of temperature, which is a valid assumption provided the change in temperature is not too large. If the values of $\Delta G^{\ominus}$, $\Delta H^{\ominus}$ are known for a system at a particular temperature, the value of the constant c may be determined and used to calculate a value for $\Delta G^{\ominus}$ at a second temperature.

$$\Delta G^{\ominus} / T = \Delta H^{\ominus}/T + c \qquad (6.33)$$

6.19.2 Gibbs free energy and chemical equilibrium

The Gibbs free energy change of a system in equilibrium at constant temperature and pressure but not at the standard concentrations or activities can be shown to be related to the standard Gibbs free energy by the expression:

$$\Delta G = \Delta G^{\ominus} + 2.303 \, RT \log K_{eq}$$

The value of K_{eq} may be determined using either activities or moles dm^{-3} depending on the concentration of the constituents of the equilibrium mixture. If hydrogen ions are involved in the reaction, a separate term for the hydrogen ion concentration may be incorporated in K_{eq}, or $\Delta G^{\ominus'}$, the biochemical standard Gibbs free energy change, may be used instead of $\Delta G^{\ominus}$ in equation 6.32. Since the concentrations of the reactants and products in most biological reactions are small, they may be measured in mol dm^{-3} and not activities (6.11). Furthermore, equilibrium constants determined using $\Delta G^{\ominus'}$ values measured under standard biological conditions are represented as K_c values.

6.19.3 Gibbs free energy changes in non-standard states

The difference (ΔG) between the Gibbs free energy for the standard and non-standard states of a compound under the same conditions of temperature and pressure can be shown to be given by equation 6.34 where (a) is the activity of the compound in its non-standard state:

$$\Delta G = RT \ln (a) \tag{6.34}$$

Equation 6.34 makes it possible to calculate the Gibbs free energy of the components of a reaction and to predict the feasibility of that reaction taking place under non-standard conditions (Table 6.5).

6.20 WHAT YOU NEED TO KNOW

(1) *The rate of a reaction is defined as:*

$$Rate = \frac{-d[reactant]}{dt} = \frac{d[product]}{dt}$$

(2) *The rate of a reaction is related to the concentrations of the reactants by experimentally determined rate expressions. These expressions often have the form:* $Rate = k[A]^x[B]^y...$

where k is a constant known as the rate constant and x and y are the orders of the reaction with respect to A and B respectively.

(3) *The sum of the indices in the rate expression for a reaction, x + y + ... is known as the overall order of the reaction.*

(4) *The order of a reactant in a rate expression may be determined experimentally by a number of methods based on determining the initial concentration of the reactant and its concentrations at various times t after the start of the reaction (Fig. 6.2a). The order of the reactant and the rate constant for the reaction can be deduced from the results either by substituting the values in the appropriate mathematical expression or graphically from a plot of the appropriate parameters (Table 6.2).*

(5) *The half life ($t_{1/2}$) of a reaction is the time taken for the concentration of a reactant to fall to half its initial value. It can be used to determine the order and rate constant for a reaction.*

(6) Biological or elimination half life is the time taken for the concentration of a drug or metabolite to fall to half its original value.

(7) *Pseudo first order reactions are second order reactions which appear to follow first order kinetics when the concentration of one of the reactants is high. Under these conditions these reactions have a first order rate expression.*

(8) Biological reactions which have a fast step in which one of the original reactants is regenerated can exhibit pseudo first order kinetics.

(9) *Increasing the temperature at which a reaction is conducted increases the rate of its reaction. The rates of endothermic reactions are increased significantly more than those of exothermic reactions.*

(10) *The relationship between temperature and the rate constant for a reaction is given by the Arrhenius equation:*
$$k = Ae^{-Ea/RT}$$

(11) Changes in the total pressure of a system have little effect on the rate of the reaction unless the reaction is being carried out in the gaseous phase.

(12) *The rate of a reaction is dependent on the sizes of the reacting particles: the smaller the size the faster the reaction.*

(13) *Catalysts increase, whilst inhibitors decrease the rate of a reaction.*

(14) Homogeneous catalysts dissolve to form a single reaction phase and act by forming intermediate species with the reactants.

(15) Heterogeneous catalysts form a separate phase to the reaction mixture and act by adsorbing the reactants. Enzymes are usually considered to be heterogeneous catalysts even though they are usually soluble in water, the phase in which most biological reactions occur.

(16) *The relationship between a catalyst and the concentration of a substrate is given by the Michaelis-Menton equation:*
$$V_o = \frac{V_{max} [S]}{K_m + [S]}$$

(17) *Inhibitors are classified as being reversible or non-reversible, competitive, uncompetitive or non-competitive inhibitors. The values of K_m and V_{max} in the Michaelis-Menton equation are used to determine whether a substance is a competitive, uncompetitive or non-competitive inhibitor.*

(18) *Competitive inhibitors compete with the substrate for the active site of the enzyme.*

(19) *Uncompetitive inhibitors form a complex with the enzyme-substrate complex which cannot react to form the product.*

(20) *Non-competitive inhibitors combine with the enzyme at an allosteric site. This either prevents the substrate from binding to the enzyme or, if this is still possible, prevents the enzyme-substrate-inhibitor complex from forming the products.*

(21) *Reactions in which the rates of the forward and backward reactions are the same are said to be in a state of dynamic equilibrium.*

(22) *Equilibrium systems are classified as homogeneous and heterogeneous systems. Homogeneous equilibrium systems are systems in which all the reactants and products occupy one homogeneous phase. Heterogeneous equilibrium systems are systems in which the reactants and products occupy more than one phase.*

(23) *The law of mass action states that for homogeneous equilibrium systems at equilibrium at constant temperature in dilute solution for the reaction:*
$$aA + bB + ... \rightleftharpoons xX + yY + ...,$$
the equilibrium constant (K_c) is given by the expression:
$$K_c = \frac{[X]^x [Y]^y ....}{[A]^a [B]^b ....}$$

(24) *In concentrated solutions the law of mass action holds provided the concentrations of the reactants are measured in activities where:*
$$Activity = Activity\ coefficient \times Concentration$$

(25) *Equilibrium constants calculated using activities are known as thermodynamic*

equilibrium constants (K_{eq}). In dilute solutions $K_c = K_{eq}$.

(26) *The activity coefficient is a measure of a solute's deviation from the ideal state in a particular solvent. It can be determined experimentally by measuring the colligative properties of the solute in that solvent.*

(27) *The law of mass action is applied to heterogeneous equilibrium systems by regarding the concentrations of substances in the solid and/or liquid phases as being constant and being incorporated into the value of the equilibrium constant.*

(28) *Increasing the concentration (or partial pressure) of a reactant will drive the position of equilibrium to the right.*

(29) *Increasing the temperature will move the position of equilibrium towards the side favoured by the endothermic process of the equilibrium.*

(30) *K_{eq} is related to temperature by the expression:*

$$log \, K_{eq} = constant - \Delta H/19.14T$$

(31) *Increasing the total pressure of an equilibrium system will move the position of equilibrium to the side whose species have the smaller volume, that is, the side which has the lowest number of molecules in the balanced equation for the system.*

(32) *Acids are proton donors whilst bases are proton acceptors. Compounds that can act as either acids or bases are known as amphoteric compounds.*

(33) *Strong acids are good proton donors whilst strong bases are good proton acceptors.*

(34) *The strengths of acids and bases are recorded as their pK_a values where:*

$$pK_a = -log \, K_a$$

$$For \; acids: \; K_a = \frac{[conjugate \; base][H^+]}{[acid]} \quad For \; bases: \; K_a = \frac{[base][H^+]}{[conjugate \; acid]}$$

(35) *Increasing the pH of a solution will increase the ionisation of an acid but suppress that of a base.*

(36) The ionisation of acids will be suppressed in the acidic stomach but increased in the alkaline intestine. Consequently, an acidic compound is more likely to be absorbed in the stomach than the intestine. Similarly, basic compounds are more likely to be absorbed in the intestine than the stomach.

(37) *Buffers are solutions whose pH does not change appreciably when reasonable quantities of either an acid or base are added. Many biological fluids contain buffers that maintain the pH of the fluid at the correct pH for that fluid.*

(38) *The pH of a buffer solution can be calculated from the Henderson-Hasselbach equation:*

$$pH = pK_a + log \frac{[base \; component]}{[acid \; component]}$$

(39) *Thermodynamics is the study of the relationships between the different forms of energy and how they affect chemical and physical changes.*

(40) *A thermodynamic system is an area with clearly defined boundaries. The region outside these boundaries is known as the surroundings. A system is defined by its composition, temperature, pressure and functions of state. A function of state is a property of a system that is dependent only on the current state of the system not on its history.*

(41) *Open systems are able to exchange matter and energy with the surroundings. Closed systems are not able to exchange matter and energy with the surroundings.*

(42) *Systems in which the temperature is kept constant are known as isothermal systems whilst those in which the temperature varies are known as adiabatic systems.*

(43) *Thermodynamics treats all changes as though they were equilibrium systems. Reactions that go to completion have a position of equilibrium that lies well to the right whilst those that are not reversible have too high an activation energy in the backward direction.*

(44) *Thermodynamically reversible reactions are indicated by the use of* $\rightleftharpoons$

(45) *The first law of thermodynamics states that the change in internal energy of the system is due to work being done on or by the system and heat being absorbed or liberated by the system.*

> *For work done on the system:* $\quad \Delta U = q + w = q + P\Delta V$
> *For work done by the system:* $\quad \Delta U = q - w = q - P\Delta V$

(46) *Chemical and biological changes occur at constant pressure. The heat liberated when a chemical or biological system undergoes a change at constant pressure is called the enthalpy change (ΔH) of the system where:*

$$\Delta H = H_2 - H_1 = q_p = \Delta U + P\Delta V$$

(47) *Enthalpy is a function of state and so ΔH is dependent only on the initial and final states of the system and not on the route taken by the change. This is the thermodynamic basis of Hess's law.*

(48) *The second law of thermodynamics states that spontaneous changes only occur in the direction that increases the overall disorder of the system and its surroundings.*

(49) *The state disorder of a system (and also its surroundings) is described by a function of state known as entropy (S). For spontaneous changes to be theoretically possible, at constant pressure:*

$$\Delta G = \Delta H_{system} - T\Delta S_{system} < 0$$

where $\Delta S = q/T$ and ΔG is known as the Gibbs free energy of the system.

(50) *Reactions which have a negative ΔG value are said to be exergonic whilst those with a positive ΔG value, endergonic. Endergonic changes can be brought about by coupling them to highly exergonic reactions to give a negative ΔG value for the coupled process. This is often found in metabolic pathways.*

(51) *Electrode and redox potentials may be used to calculate the ΔG value for a reaction and so predict whether it is possible. For an electrode process:*

$$\Delta G^{\ominus}_{process} = -nFE$$

and: $\qquad \Delta G^{\ominus}_{process} = \Delta G^{\ominus}_{reduction\ process} - \Delta G^{\ominus}_{oxidation\ process}$

(52) *The electrode potential of an electrode system can be calculated using the appropriate Nerst equation. Either:*

$$E = E_o + \frac{2.303RT}{nF} \log [metal\ ion]$$

or:
$$E = E_o + \frac{2.303RT}{nF} \log\frac{[oxidised\ state]}{[reduced\ state]}$$

(53) *Standard Gibbs free energies are* $\Delta G^{\ominus}$ *values measured at 298 K, 1 atmosphere and specified concentrations.*

(54) *The relationship of* $\Delta G^{\ominus}$ *to* K_{eq} : $\Delta G^{\ominus} = -RTlnK_{eq}$

(55) *The Gibbs-Helmholtz equation:* $\Delta G^{\ominus}/T = \Delta H^{\ominus}/T + c$

(56) *The relationship of* $\Delta G^{\ominus'}$ *to* $\Delta G^{\ominus}$: $\Delta G^{\ominus'} = \Delta G^{\ominus} + RT\ 2.303 \log pH$

(57) *The relationship of* ΔG *to* K_{eq} : $\Delta G = \Delta G^{\ominus} + RT\ 2.303 \log K_{eq}$

(58) *The relationship of* ΔG *to concentration:* $\Delta G = \Delta G^{\ominus} + RT\ ln\ (a)$

6.21 QUESTIONS

(1) What is the overall order of each of the following reactions?

(a) A $\longrightarrow$ B + C   Rate = $k[A]$

(b) A + B $\longrightarrow$ C + D   Rate = $k[A]^2[B]$

(c) A + B $\longrightarrow$ C + D   (i)   Rate = $k[A][B]$ for low concentrations of A

(ii)   Rate = $k[B]$ for high concentrations of A

(d) A + B $\xrightarrow{slow}$ C + D

D + E $\xrightarrow{fast}$ F + B   Rate = $k[A]$

(2) Investigation of a reaction between two compounds A and B gave the following data:

Initial concentration of A	Initial concentration of B	Initial rate of reaction
1 mol dm^{-3} x 10^{-3}	5 mol dm^{-3} x 10^{-3}	3 mol dm^{-3} s^{-1} x 10^{-3}
2 mol dm^{-3} x 10^{-3}	5 mol dm^{-3} x 10^{-3}	6 mol dm^{-3} s^{-1} x 10^{-3}
3 mol dm^{-3} x 10^{-3}	5 mol dm^{-3} x 10^{-3}	9 mol dm^{-3} s^{-1} x 10^{-3}
5 mol dm^{-3} x 10^{-3}	1 mol dm^{-3} x 10^{-3}	1 mol dm^{-3} s^{-1} x 10^{-3}
5 mol dm^{-3} x 10^{-3}	2 mol dm^{-3} x 10^{-3}	4 mol dm^{-3} s^{-1} x 10^{-3}
5 mol dm^{-3} x 10^{-3}	3 mol dm^{-3} x 10^{-3}	9 mol dm^{-3} s^{-1} x 10^{-3}

(a) What are the orders of the reaction with respect to (i) A and (ii) B?

(b) Write a rate expression for the reaction.

(3) The reaction: A + B $\longrightarrow$ C + D, proceeds by pseudo first order kinetics when A is present in a large excess. If the initial concentration of B was 0.01 mmol dm^{-3} determine the time taken for 0.005 mM dm^{-3} of B to react in the presence of a large excess of A if the rate constant for the reaction is 3.0 x 10^{-2} s^{-1}.

(4) Calculate the time taken for the concentration of a drug to fall to 60% of its original value if its decomposition follows first order kinetics with respect to the drug and the rate constant for the decomposition is 1.7 x 10^{-5} s^{-1}.

(5) The breakdown of a polypeptide at 25°C followed second order kinetics with respect to the polypeptide. A 20 mM solution of the polypeptide was found to be 60% degraded after 15 minutes at 25°C. Calculate (a) the rate constant at 25°C and (b) the half life of the reaction.

(6) Write K_c and K_{eq} expressions for each of the following equilibria:

(a) $2HI_{(g)} \rightleftharpoons H_{2(g)} + I_{2(g)}$

(b) $3Fe_{(s)} + 4H_2O_{(g)} \rightleftharpoons Fe_3O_{4(s)} + 4H_{2(g)}$

(c) $Fumarate^{2-}_{(aq)} + H_2O_{(l)} \rightleftharpoons Malate^{2-}_{(aq)} \; \Delta H \; +14.89 \; kJ \; mol^{-1}$

(7) State whether increasing (a) the concentration of fumarate, (b) the total pressure and (c) the temperature would increase the yield of malate in question 6.

(8) It was found that 0.02 moles of oxalic acid reacted at 25°C with 0.03 moles of methanol to form an equilibrium mixture which contained 0.01 moles of the dimethyl ester. Assuming the activity coefficients of all the substances involved in the system are unity at these concentrations, calculate the value of K_{eq} for the system at 25°C.

(9) Tris is a weak base ($K_a = 8.3 \times 10^{-9} \; mol \; dm^{-3}$) which is used as a buffer. Calculate the ratio tris/trisH$^+$ required to form a buffer solution with a pH of 9 if the equilibrium formed by tris in aqueous solution is :

$$tris + H_2O \rightleftharpoons trisH^+ \; OH^-$$

(10) Calculate the relative quantities of 0.1 M ethanoic acid and 0.2 M sodium ethanoate that have to be mixed in order to form a buffer solution with a pH of 5.6. The pK_a of ethanoic acid is 4.8.

(11) Glucose ($C_6H_{12}O_6$) is converted by a bacterium into maltose ($C_{12}H_{22}O_{11}$):

$$2 \; Glucose_{(s)} \rightleftharpoons Maltose_{(s)} + H_2O_{(l)}$$

Calculate the enthalpy change for this process given that the standard heats of combustion of solid glucose and maltose are −2816 and −5648 kJ mol^{-1} respectively.

(12) (a) Calculate the standard entropy change for the conversion of ATP to ADP at 37°C if the values of $\Delta G^{\ominus}$ and $\Delta H^{\ominus}$ at 37° are −30.96 kJ and −20.08 kJ mol^{-1}.

(b) What would be the standard entropy change for this process in a reptile with a body temperature of 10°C?

(13) The value of K_{eq} for the enzyme-catalysed decomposition of fructose-1,6-diphosphate to glyceraldehyde-3-phosphate and dihydroxyacetone phosphate is $8.91 \times 10^{-5} \; mol \; dm^{-3}$ at 25°C. Calculate the Gibbs free energy change that occurs when 0.1 moles of fructose-1,6-diphosphate are decomposed at 25°C ($R = 8.314 \; J \; K^{-1} \; mol^{-1}$).

7

Mechanisms and homolytic mechanisms

7.1 INTRODUCTION TO REACTION MECHANISMS

Reaction mechanism theory attempts to explain what happens at the electron level in chemical reactions. In recent years mechanistic theory has revolutionised the study of organic chemistry by forming a comprehensive and logical framework for classifying organic chemical reactions. It also enables chemists to predict with a reasonable degree of accuracy what is likely to happen in simple reactions. This is of considerable use in the design of new drugs and the control of disease.

Reaction mechanisms are of less value in inorganic chemistry because of the greater diversity of inorganic reactions. Consequently, this chapter is restricted to the use of mechanisms in organic chemistry.

Mechanisms are deduced using a combination of logic and practical observations (7. 5 and 8.6). They are classified into three main types: homolytic, heterolytic and pericyclic. Homolytic mechanisms are characterised by the homolytic fission of bonds. This occurs when the two electrons forming the bond return to their original atoms (Fig. 7.1). The movement of each electron is sometimes represented by the use of a fish-hook arrow, $\curvearrowright$, in the equation for the mechanism. The chemical species formed by this type of bond fission are called **free radicals** and the reactions in which they are involved, **free radical reactions.**

$$A : B \longrightarrow A\cdot + B\cdot$$

Fig. 7.1. The homolytic fission of a bond. Unpaired electrons are represented by a single dot ($\cdot$)

Simple free radicals contain one atom with an unpaired electron; however, more complex free radicals with more than one unpaired electron and/or charges are known. Simple free radicals are electrically neutral since the total number of protons in the radical equals the total number of electrons. For example, the methyl free radical formed by the homolytic fission of a C–H bond in methane (Fig. 7.2) is electrically neutral because it has nine protons and nine electrons. Similarly, the hydrogen free radical is also electrically neutral since its one electron is balanced by its single proton.

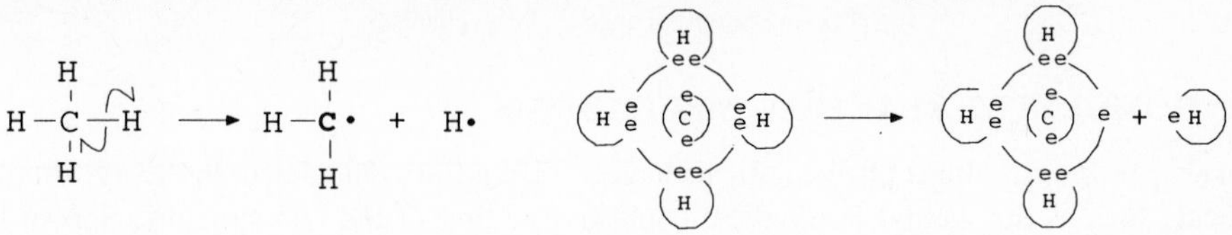

Fig. 7.2. The formation of an electrically neutral methyl free radical. Electrons are represented by a dot or e.

Heterolytic mechanisms are usually, but not always, characterised by heterolytic bond fission and the formation of ions. In heterolytic fission both the electrons of the bond move to one of the atoms forming the bond making this atom negative and leaving the other atom positively charged (Fig. 7.3). This movement is often shown by a curly arrow ($\frown$) in the equation for the mechanism.

Fig 7.3. The heterolytic fission of a bond showing (a) the movement of the electrons and (b) the charges resulting from the electron movement.

The charges arise because the number of protons in the nucleus is no longer equal to the number of electrons. Consider, for example, the homolytic fission of the C–Cl bond of 2-chloro-2-methylpropane (Fig. 7.4). After fission has occurred, the chlorine atom will be negatively charged since it now has eighteen electrons but only seventeen protons, that is, one negative electron in excess of its positive protons. The C_2 atom of the carbonium ion will be positively charged since it has six protons but only five electrons. It is important to realise that only **one** negative and **one** positive charge have been produced even though **two** electrons have been moved.

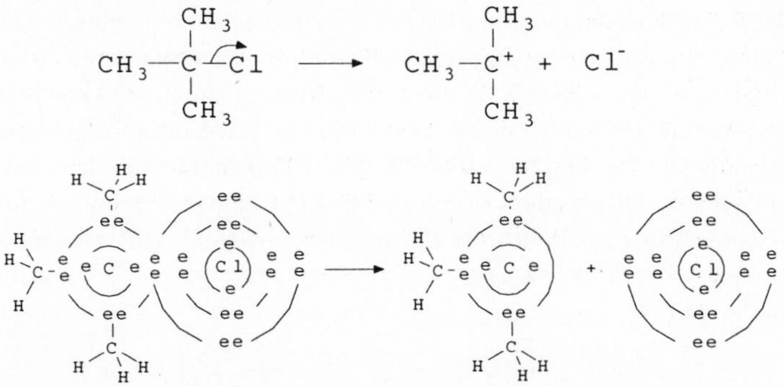

Fig. 7.4. The heterolytic fission of the C–Cl bond of 2-chloro-2-methylpropane.

Pericyclic mechanisms are a small group of mechanisms that involve no intermediate free radicals or ions. They proceed by the simultaneous rearrangement of all the electrons involved in the process via a cyclic intermediate. For example, the Diels-Alder reaction between maleic anhydride and 1,3-butadiene has a pericyclic mechanism (Fig. 7.5).

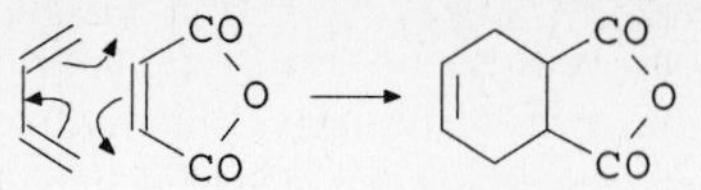

Fig. 7.5. An example of a pericyclic mechanism.

7.2 HOMOLYTIC REACTION MECHANISMS

The shapes and structures of free radicals have not been fully elucidated. In theory, simple radicals such as the methyl free radical could have either of the two structures shown in Fig. 7.6. In structure I the carbon is sp^3 hybridised with the unpaired electron located in an sp^3 hybridised orbital. This gives the radical a tetrahedral shape with bond angles near 109°. In structure II the carbon is sp^2 hybridised and the unpaired electron is in the unused p orbital. This would give the radical a flat shape. Practical evidence favours structure II.

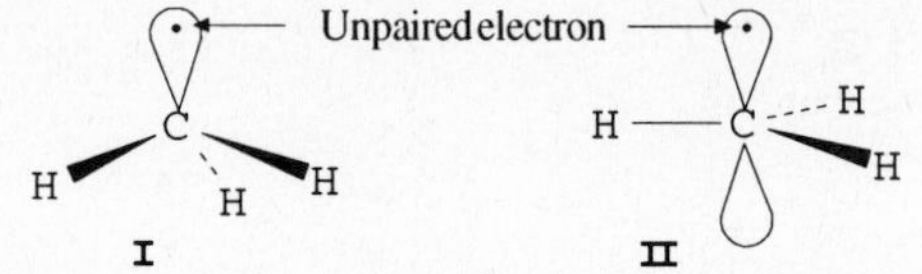

Fig. 7.6. The theoretical structures of the methyl free radical

When the hydrogens of the methyl radical are successively substituted by fluorine atoms the radical becomes more tetrahedral in shape (Fig. 7.7).

$$\cdot CF_3 \ > \ \cdot CHF_2 \ > \ \cdot CH_2F \ > \ \cdot CH_3$$
$$\underleftarrow{\text{order of increasing tetrahedral shape}}$$

Fig. 7.7. The shapes of some simple free radicals.

Radicals where the unpaired electron forms part of a conjugated structure are more stable than those in saturated structures. The stability of conjugated free radicals can be explained if the atom with the unpaired electron is sp^2 hybridised. This will give a planar structure which allows the maximum overlap of the p orbital containing the lone electron with the pi electrons and subsequent delocalisation of the unpaired electron (Fig. 7.8). This delocalisation is represented by a series of imaginary structures known as **canonical forms.** These structures indicate that the charge due to the electron is spread out over the whole of the conjugated system but tends to be more concentrated in the areas marked by the unpaired electron. This method of representing the charge distribution in a chemical species is also used in heterolytic reaction mechanisms (8.3) and resonance (2.7).

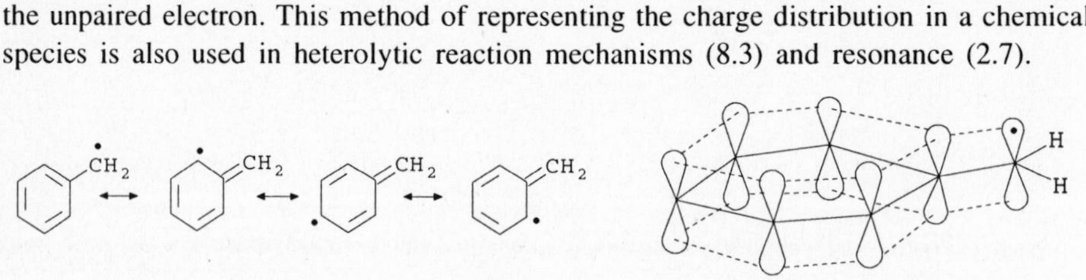

Fig. 7.8. The delocalisation of the unpaired electron in a free radical with a conjugated structure.

Conjugated free radicals can be very stable, for example, diphenylpicrylhydrazyl is a violet solid with an indefinite shelf life.

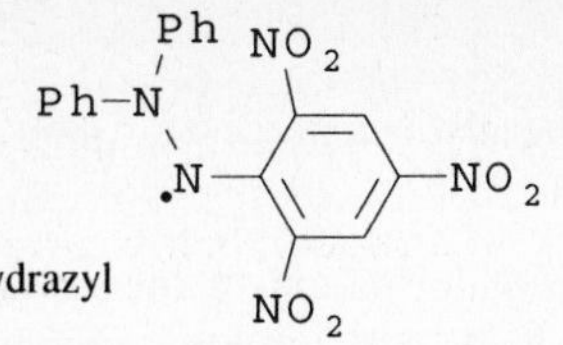

Diphenylpicrylhydrazyl

An important radical in biological systems is molecular oxygen. Practical evidence shows that the oxygen molecule is a diradical containing two unpaired electrons (Fig. 7.9). It is not a very active diradical but in biological systems its action can give rise to oxygen-derived radicals such as superoxide ($\overset{\cdot}{O}-\overset{\cdot}{O}^-$), hydroxyl ($H\overset{\cdot}{O}$) and hydroperoxy ($HO\overset{\cdot}{O}$) free radicals which are more active.

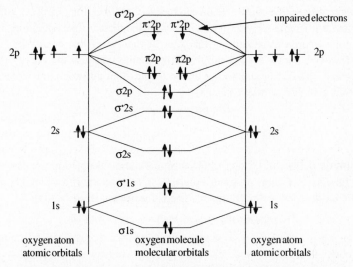

Fig. 7.9. The molecular orbital structure of oxygen. The structure of this molecule is usually written as O=O but a better representation is as the diradical $\overset{\cdot\cdot}{O}$-$\overset{\cdot\cdot}{O}$ whose molecular orbital structure is illustrated.

Radical ions where a species contains both a negative charge and an unpaired electron are known (Fig. 7.10). However, this type of radical ion is not very common.

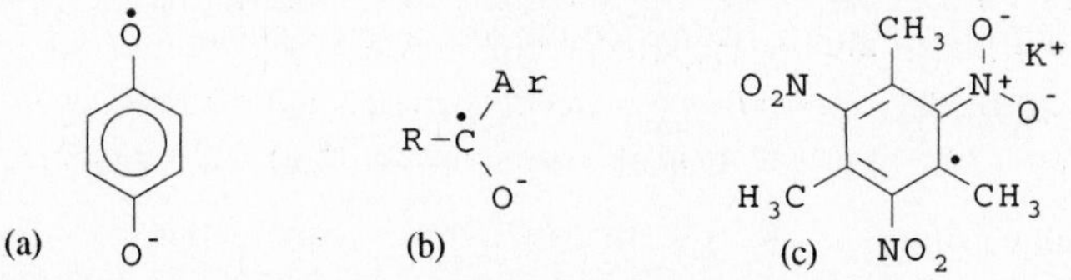

(a)   (b)   (c)

Fig. 7.10. Examples of radical ions. (a) p-Benzosemiquinone. (b) A ketyl. (c) The potassium salt of trinitromesitylene radical ion.

Most free radicals are very reactive species. This is attributed to their unpaired electron giving them an unstable electronic configuration. Free radical reactions are often very vigorous and sometimes explosive. They are usually involved in chain reactions. These are self-sustaining reactions, the production of the product being accompanied by the formation of more free radicals which keep the reaction running.

7.2.1 Chain reactions

Chain reactions consist of three distinct processes, namely **initiation**, **propagation** and **termination**.

(a) Initiation. This is the initial formation of the radicals necessary to start the chain reaction. Production of free radicals can occur by **photolysis, thermolysis** and in **redox reactions**. The radicals are more likely to be formed if the unpaired electron can be stabilised by delocalisation (Fig. 7.8). Catalysts are often employed to initiate free radical formation in the laboratory.

Photolysis occurs when a chemical reaction is brought about by irradiation with ultra-violet or visible light. Radiation with wavelengths in the region 300 to 600 nm has energy compatible with covalent bond strengths. Chlorine, for example, will form free radicals when it is irradiated with light with a wavelength less than 487.5 nm. The chlorine molecule gains energy by absorbing the light. It then loses this energy by forming free radicals. However, this is not the only way molecules can lose the energy they gain by electromagnetic irradiation and so not every species that absorbs ultra-violet and visible radiation will give rise to free radicals.

$$Cl\text{---}Cl \xrightarrow{\text{Light energy}} [Cl\text{---}Cl]^* \longleftarrow \text{Signifies a species}$$
$$[Cl\text{---}Cl]^* \xrightarrow[\text{Energy lost}]{} \cdot Cl + \cdot Cl \qquad \text{in an excited state}$$

Thermolysis is the use of heat to bring about chemical reactions. Sufficient heat can cause the homolytic fission of weak bonds (less than about 165 kJ mol^{-1}) to form free radicals. Peroxides, for example, readily form free radicals on heating

$$PhCOO\text{-}OOCPh \longrightarrow PhCOO\cdot + \cdot OOCPh$$

Thermolysis is not always a good way of producing a free radical as heating can cause further decomposition in some cases. The PhCOO· radical in the previous reaction, for example, can decompose further to Ph· and carbon dioxide.

Redox reactions are those in which one species is mutually oxidised or reduced by the other. Redox reactions that produce free radicals involve the transfer of one electron from one species to the other. Consider, for example, the reaction of Fe^{2+} salts with hydrogen peroxide to form Fenton's reagent. In this reaction the Fe^{2+} ion donates an electron to the hydrogen peroxide to form both a hydroxide ion and a free radical.

Half electronic equations:
$$Fe^{2+} \longrightarrow Fe^{3+} + e$$
$$HO\text{:}OH + e \longrightarrow OH^- + \cdot OH$$

Overall equation:
$$Fe^{2+} + HO\text{:}OH \longrightarrow Fe^{3+} + OH^- + \cdot OH$$

Molecular oxygen is converted in biosystems to a variety of free radicals. This conversion is believed to involve Fe^{2+} and other metallic ions such as Cu^+ by way of Fenton type reactions.

(b) Propagation. This is the self-perpetuating stage of the reaction where the product is formed together with more free radicals which keep the reaction going. This repeating sequence of reactions is called a chain reaction. Consider, for example, the first step in the chlorination of methane:

$$CH_4 + Cl_2 \longrightarrow CH_3Cl + HCl$$

Initiation is by photolysis which splits the chlorine to form two chlorine free radicals:

$$Cl-Cl \longrightarrow Cl\cdot + Cl\cdot$$

Propagation starts with the chlorine abstracting a hydrogen atom from the methane molecule to form hydrogen chloride and a methyl free radical:

$$H_3C-H + Cl\cdot \longrightarrow H-Cl + \cdot CH_3$$

This is followed by the methyl free radical reacting with a chlorine molecule to form the chloromethane and a chlorine free radical:

$$CH_3 + Cl-Cl \longrightarrow CH_3Cl + Cl\cdot \qquad (7.1)$$

The chlorine free radical in equation 7.1 attacks another methane molecule and starts the sequence over again to produce more chloromethane and another methyl free radical. This and subsequent methyl radicals react in the same way and so continue the chain reaction. The chlorine free radical from equation 7.1 can also react with the chloromethane to produce a chloromethane free radical which reacts with chlorine to form dichloromethane. This in turn reacts with another chlorine free radical to form trichloromethane which reacts with chlorine to form tetrachloromethane. Each of the halogen compounds specified is formed by propagation reactions that are similar to those for the formation of chloromethane. It should be noted that the sum of the equations for the processes occurring in the propagation stage of a chain reaction is equivalent to the overall equation for the reaction or that stage of the reaction reaction (Fig. 7.11).

$$CH_4 + Cl_2 \longrightarrow CH_3Cl + HCl$$
$$CH_3Cl + Cl_2 \longrightarrow CH_2Cl_2 + HCl$$
$$CH_2Cl_2 + Cl_2 \longrightarrow CHCl_3 + HCl$$
$$CHCl_3 + Cl_2 \longrightarrow CCl_4 + HCl$$

Fig. 7.11. The main sequence of reactions that occurs when methane is reacted with chlorine.

There will be many chains operating at anyone time in a chain reaction. If more than one radical is formed for each one used to produce the product, the number of chains increases rapidly and eventually an explosion can result.

(c) Termination. These reactions stop a chain continuing. They produce either the product but no other free radicals or chemical species that cannot react to form free radicals to perpetuate the chain reaction. If termination reactions stop all the chains, the chain reaction itself stops. However, it should be remembered that the chain reaction will also stop when one of the reactants is used up. Two of the most common termination reactions are:

(1)   reaction with another free radical, for example,

$$\cdot CH_3 + \cdot CH_3 \longrightarrow CH_3-CH_3$$

(2) disproportionation, for example,

$$CH_3\overset{\bullet}{C}H_2 \quad H-CH_2-\overset{\bullet}{C}H_2 \longrightarrow CH_3CH_3 + CH_2=CH_2$$

7.3 GENERAL TYPES OF HOMOLYTIC MECHANISM

Free radical reactions may be conveniently classified in the traditional manner as additions, substitutions, eliminations, oxidations, reductions and rearrangements as far as the overall reaction is concerned. However, the complexity of free radical mechanisms means that within, for example, an overall substitution reaction there may be intermediate additions, oxidations, rearrangements, etc. Furthermore, one of the most common of the propagation reactions which is not included in the traditional classification is hydrogen abstraction. This, as the name suggests, is where a radical removes a hydrogen from another species.

$$CH_4 + \cdot Cl \longrightarrow H-Cl + \cdot CH_3$$

7.3.1 Addition

Addition reactions occur when a reagent reacts with a double or triple bond in a substrate to form a single product. In the sunlight or ultra-violet catalysed reaction of chlorine with tetrachloroethene, for example, the chlorine reacts to form one product, namely, hexachloroethane. This reaction is initiated by photolysis of the chlorine. Both of these chlorine radicals can initiate a chain reaction of the type outlined in Fig. 7.12.

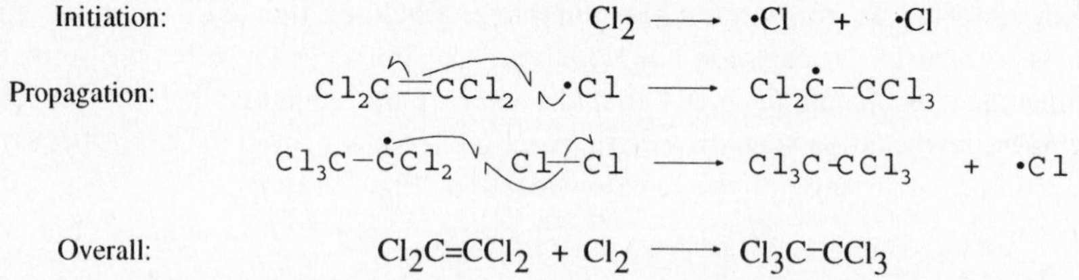

Initiation: $Cl_2 \longrightarrow \cdot Cl + \cdot Cl$

Propagation: $Cl_2C = CCl_2 \,\,\curvearrowright\, \cdot Cl \longrightarrow Cl_2\overset{\centerdot}{C} - CCl_3$

                    $Cl_3C - \overset{\centerdot}{C}Cl_2 \,\,\curvearrowright Cl \overset{\frown}{-} Cl \longrightarrow Cl_3C - CCl_3 + \cdot Cl$

Overall: $Cl_2C = CCl_2 + Cl_2 \longrightarrow Cl_3C - CCl_3$

Fig. 7.12. The mechanism of the reaction of chlorine with tetrachloroethene.

7.3.2 Substitution

Substitution reactions are those in which an atom or group is replaced by a different atom or group. In the light-catalysed reaction of alkanes with chlorine, for example, a hydrogen atom is replaced by a chlorine atom (Fig. 7.11). The mechanisms of this free radical reaction have the general form shown in Fig. 7.13.

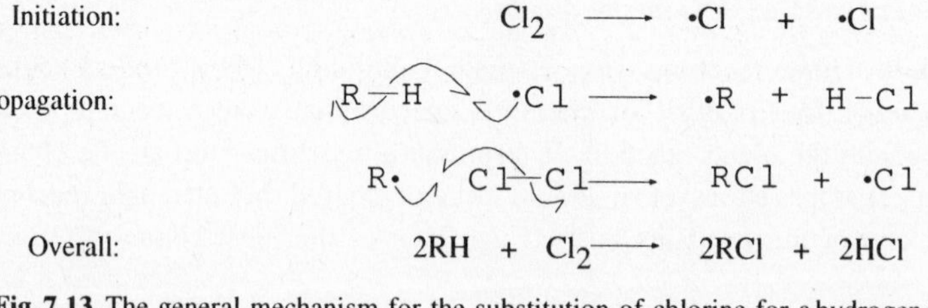

Initiation: $Cl_2 \longrightarrow \cdot Cl + \cdot Cl$

Propagation: $R \overset{\frown}{-} H \,\,\curvearrowright\, \cdot Cl \longrightarrow \cdot R + H - Cl$

                    $R \cdot \,\,\curvearrowright\, Cl \overset{\frown}{-} Cl \longrightarrow RCl + \cdot Cl$

Overall: $2RH + Cl_2 \longrightarrow 2RCl + 2HCl$

Fig. 7.13. The general mechanism for the substitution of chlorine for a hydrogen in an alkane.

7.3.3 Elimination

Elimination is in essence the reverse of addition. A molecule or stable ion is lost from a structure leaving behind a double or triple bond (Fig. 7.14). Free radical elimination mechanisms are rare. They occur mostly in the pyrolysis of poly and monohalides. The mechanisms of these reactions are believed to involve steps of the type shown in Fig. 7.14.

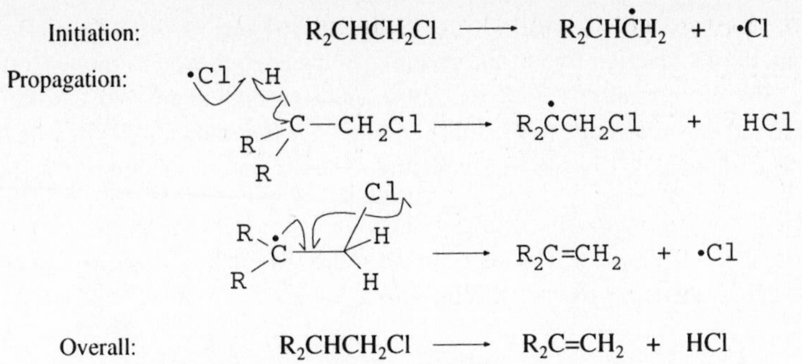

Initiation:　　　　　　$R_2CHCH_2Cl \longrightarrow R_2CH\overset{\bullet}{C}H_2 + {}^\bullet Cl$

Propagation:

$$\underset{\underset{R}{\overset{R}{|}}}{C}-CH_2Cl \longrightarrow R_2\overset{\bullet}{C}CH_2Cl + HCl$$

$$R\overset{\bullet}{\underset{R}{C}}\overset{Cl}{\underset{H}{C}}H \longrightarrow R_2C=CH_2 + {}^\bullet Cl$$

Overall:　　　　　　$R_2CHCH_2Cl \longrightarrow R_2C=CH_2 + HCl$

Fig. 7.14. The mechanism of a typical free radical elimination reaction.

7.3.4 Oxidation

Oxidation of C–H bonds by oxygen at room and body temperatures usually follows a free radical mechanism. The initiating free radical (Ra) abstracts hydrogen from the substrate to form the appropriate radical. Propagation proceeds by the addition of oxygen to form the peroxide radical. This reacts with the substrate to form the hydroperoxide radical and a substrate radical which continues the chain (Fig. 7.15).

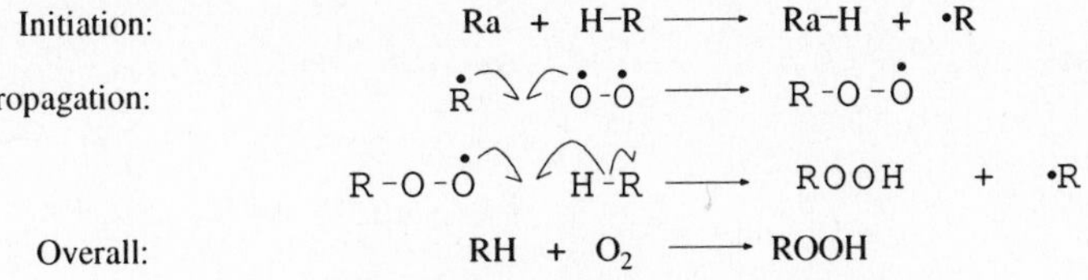

Initiation:　　　　　$Ra + H–R \longrightarrow Ra–H + {}^\bullet R$

Propagation:　　　　$\overset{\bullet}{R} \frown \overset{\bullet}{O}\text{-}\overset{\bullet}{O} \longrightarrow R\text{-}O\text{-}\overset{\bullet}{O}$

　　　　　　　　　$R\text{-}O\text{-}\overset{\bullet}{O} \frown H\text{-}R \longrightarrow ROOH + {}^\bullet R$

Overall:　　　　　　$RH + O_2 \longrightarrow ROOH$

Fig. 7.15. Free radical oxidation of an alkane.

In biological systems the hydroperoxide radical can break down to form hydroxyl and other free radicals which can have a beneficial or detrimental effect on health depending on the situation in which they are formed (7.4).

$$ROOH \longrightarrow HO^\bullet + RO^\bullet$$

7.3.5 Reduction

Free radical reductions are known. The Birch reduction of benzene by sodium in liquid ammonia is believed to proceed via free radical intermediates. The electron liberated by the ionisation of the sodium reacts to form a radical ion.

$$Na \longrightarrow Na^+ + e$$

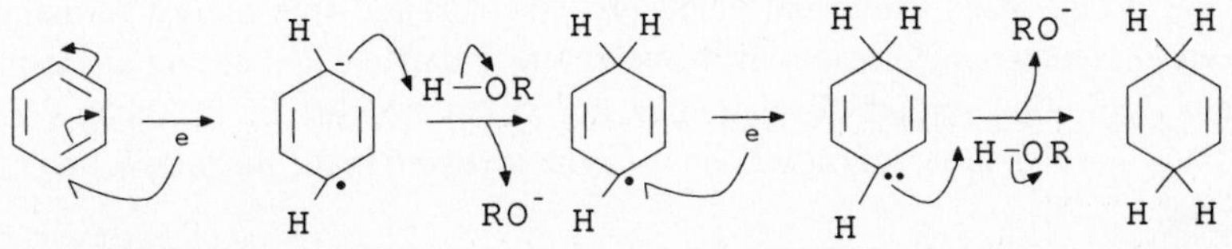

7.3.6 Rearrangement

A rearrangement occurs when one or more of the atoms in a molecule move to other positions within that molecule. The action of heat on triphenylmethyl peroxide

($Ph_3COOCPh_3$), for example, results in formation of 1,2-diphenoxy-1,1,2,2-tetraphenylethane. In this reaction two of the phenyl groups are believed to move from the carbon atoms to the oxygen atoms with the subsequent formation of two alkoxy free radicals (I). These radicals rearrange to the more stable alkyl free radicals (II) in which the unpaired electron is delocalised in the benzene ring systems before reacting to form the final product.

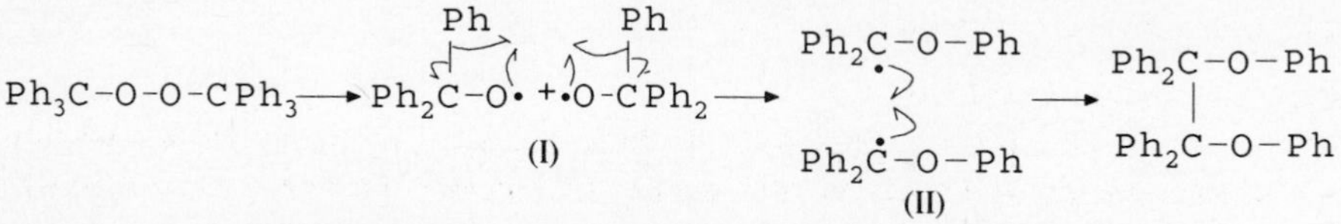

$$Ph_3C-O-O-CPh_3 \longrightarrow Ph_2C-O\bullet \ +\bullet O-CPh_2 \longrightarrow \begin{array}{c} Ph_2\overset{\bullet}{C}-O-Ph \\ \\ Ph_2\overset{\bullet}{C}-O-Ph \end{array} \longrightarrow \begin{array}{c} Ph_2C-O-Ph \\ | \\ Ph_2C-O-Ph \end{array}$$

(I)   (II)

Atoms other than carbon can be transferred in a rearrangement. The addition of hydrogen bromide to chloro alkenes can lead to rearrangement of the chlorine (Fig. 7.16). The addition is initiated by the formation of bromide free radicals in the presence of light or peroxides.

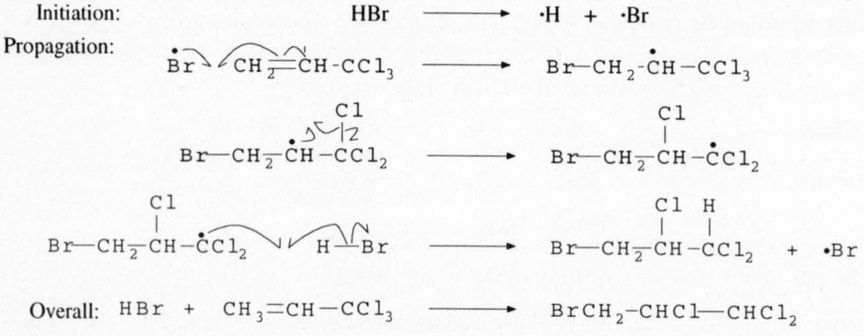

Initiation: $HBr \longrightarrow \bullet H \ + \ \bullet Br$

Propagation: $\overset{\bullet}{Br} \ \diagup CH_2 = CH-CCl_3 \longrightarrow Br-CH_2-\overset{\bullet}{C}H-CCl_3$

$$Br-CH_2-\overset{\bullet}{C}H-CCl_2 \longrightarrow Br-CH_2-CH-\overset{\bullet}{C}Cl_2$$
(with Cl above)   (with Cl above)

$$Br-CH_2-CH-\overset{\bullet}{C}Cl_2 \ \diagup H-Br \longrightarrow Br-CH_2-CH-CCl_2 \ + \ \bullet Br$$
(with Cl above)   (with Cl H above)

Overall: $HBr \ + \ CH_3 = CH-CCl_3 \longrightarrow BrCH_2-CHCl-CHCl_2$

Fig. 7.16. The rearrangement of a chlorine atom during the addition of hydrogen bromide to 3-trichloropropene.

7.4 FREE RADICALS IN BIOLOGICAL SYSTEMS

Free radical formation is an essential part of many normal metabolic processes. Radicals produced during normal metabolism are controlled and rendered inactive by a variety of mechanisms which prevent them attacking important biological macromolecules such as DNA, proteins and lipids.

The uncontrolled formation of free radicals in a biosystem can lead to extensive damage to that system. They are so reactive that they react with the nearest molecule and in doing so disrupt many biological pathways. Uncontrolled free radical formation has been linked to radiation sickness, arthritis, heart attack damage and ageing amongst other conditions. Controlled radical formation is known to be part of the body's defence mechanism against bacteria. It is also believed to be involved in the mode of action of some cancer drugs.

7.4.1 Radiation sickness

Hydroxyl free radicals are amongst the most disruptive of the free radicals that can form in the body. Luckily they are short-lived which means that they do very little damage under normal conditions. They are not usually formed from water in living cells since the O–H bonds are too strong. However, these bonds will undergo homolytic fission to form hydroxyl free radicals if sufficient energy is available. At Chernobyl the radiation levels were high enough to supply sufficient energy to bring about homolytic fission of water, and the radiation sickness suffered by the inhabitants was probably due in part to the formation of hydroxyl free radicals and their subsequent disruption of the normal biological pathways in the body.

7.4.2   Rheumatoid arthritis

Rheumatoid arthritis has been linked to the damage caused by hydroxyl and other free radicals. In the inflamed rheumatoid joint there is an increased number of phagocytic white blood cells. These cells produce superoxide anions ($\cdot O{-}O^-$) which are normally converted to hydrogen peroxide and then by various enzymes to water (Fig. 7.17). However, in the presence of Fe^{2+} the hydrogen peroxide is converted to hydroxyl free radicals which attack any available molecules especially the lipid molecules of membranes. This attack is enhanced by the presence of antioxidants (reducing agents) such as ascorbic acid which reduce the Fe^{3+} to Fe^{2+} which can be reused to produce more hydroxyl free radicals.

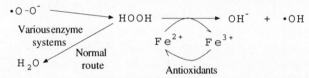

Fig. 7.17. The formation of hydroxyl free radicals from hydrogen peroxide.

7.4.3   Heart attacks

Heart damage can follow thrombolytic therapy following a heart attack. The oxygen in the restored blood supply acts as a biradical and liberates other free radicals that can damage the heart muscle. Under normal circumstances the concentration of oxygen in the body is not high enough to be a problem since the body uses antioxidants, such as vitamins C and E, to limit any free radical damage the oxygen may cause. However, when an oxygen surge occurs after a heart attack the body's defences are overwhelmed and damage can occur.

Vitamin C   Vitamin E (Tocopherol)

7.4.4   Treatment of cancer

The mode of action of the antitumour antibiotic bleomycin is believed to involve hydroxyl free radicals. It is thought that bleomycin binds to the DNA and in doing so delivers a Fe^{2+} ion to a specific site on the DNA chain. Oxygen is now able to attack this site and in a Fenton type reaction produces hydroxyl free radicals which cleave the DNA chain. This cleavage destroys the tumour. The antitumor agent streptonigrin is believed to act in a similar way but uses a Cu^+ ion instead of the Fe^{2+} ion to produce the hydroxyl radicals.

The hydroxyl free radical is short-lived and therefore its damaging effects are limited to the region in which it is generated. The presence of reducing agents such as ascorbic acid and thiols (Chapter 17) can aid the formation of hydroxyl free radicals by providing a continuous supply of reduced metal ions (Fig. 7.18). This increases the production of hydroxyl radicals and hence their effects, good or bad.

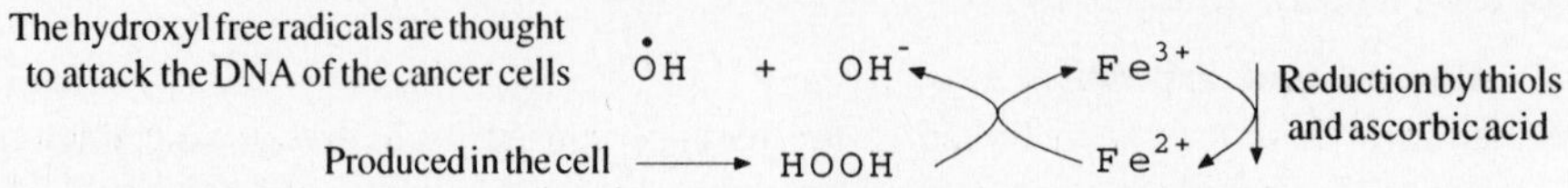

The hydroxyl free radicals are thought
to attack the DNA of the cancer cells $\dot{O}H$ + OH^- ⟶ Fe^{3+} ⟍ | Reduction by thiols
                                                                                and ascorbic acid
Produced in the cell ⟶ HOOH ⟋ Fe^{2+} ⟋ ↓

Fig. 7.18. The action of ascorbic acid and thiols in maintaining the production of hydroxyl free radicals. The reduction of Cu^{2+} to Cu^+ will also maintain the production of hydroxyl free radicals.

In cancer therapy the increased production of hydroxyl radicals could be beneficial as it would increase the extent of the attack on the cancerous DNA. It is interesting to note that there is some evidence which indicates that some cancer patients improve when large doses of ascorbic acid are included in their normal drug regime.

7.4.5 Natural bacterial defence mechanism

The production of free radicals plays a part in some of the human body's defence mechanisms. The presence of bacteria can trigger the production of free radicals which attack the bacteria. Phagocytic white cells, for example, give rise to superoxide free radicals. It is believed that these radicals probably give rise to more reactive radicals and substances such as hydrogen peroxide and hydroxyl radicals that destroy bacteria.

7.5 THE INVESTIGATION OF FREE RADICAL REACTIONS

A variety of techniques are used to detect free radicals. The most important is electron spin resonance (e.s.r.) but other methods include colour changes and chemical entrapment.

7.5.1 Colour

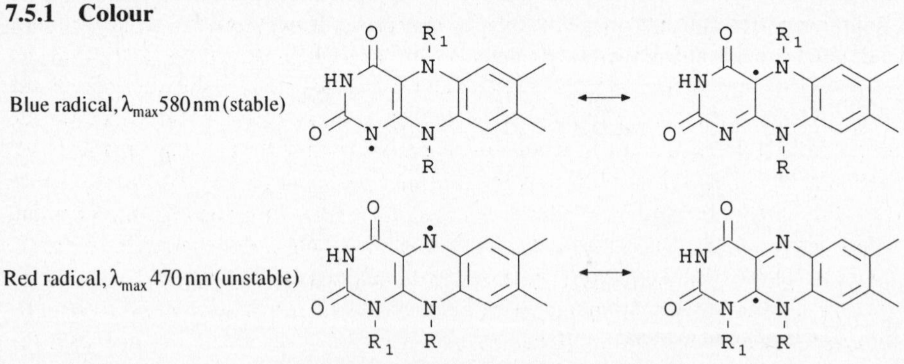

Blue radical, λ_{max} 580 nm (stable)

Red radical, λ_{max} 470 nm (unstable)

Fig. 7.19. The structures of flavin free radicals.

A number of radicals are coloured but their precursors are colourless. This fact can be used to follow free radical reactions in both chemical and biological work. For example, the isomeric flavin radicals produced by the one electron oxidation of flavohydroquinone are red and blue respectively (Fig. 7.19). Oxidation of the flavoproteins, a group of enzymes, will produce either red or blue radicals based on the flavohydroquinone radical structures This enabled workers to study the biochemistry of the flavoproteins, a group of enzymes based on flavohydroquinone, by observing the colour changes that occurred. These studies showed that one electron transfer flavoproteins form blue radicals that were essential to enzyme action. In contrast, two electron transfer enzymes form red radicals that played no part in enzyme action.

7.5.2 Electron spin resonance (e.s.r.)

Atomic silver, like the atoms of many other metals, is a free radical because it has an unpaired electron, in this case the 5s electron. Experiments by Stern and Gerlach showed that when a jet of atomised silver is passed between the poles of a permanent magnet it separates into two beams (Fig. 7.20). This and similar experiments demonstrated that free radicals were paramagnetic, that is, are attracted by a magnetic field.

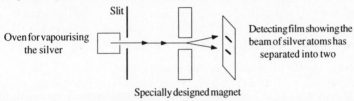

Fig. 7.20. Stern-Gerlach experiment.

Paramagnetic behaviour has been explained by considering electrons to be small bar magnets. The spin quantum numbers of a spin pair of electrons have values of $+\frac{1}{2}$ and $-\frac{1}{2}$. This is interpreted using the magnet model as two magnets spinning about their axis so that their poles are aligned in opposite directions (Fig. 7.21a). One magnet corresponds to an electron having a spin quantum number of $+\frac{1}{2}$, the other corresponds to an electron with a spin quantum number of $-\frac{1}{2}$. Using this model a spin pair of electrons has no resultant magnetic field since the magnetic field of one electron is cancelled out by that of the other electron. Consequently, atoms in which all the electrons are paired would not be deflected by a magnetic field in the Stern-Gerlach experiment. However, an unpaired electron can take up two positions in an external magnetic field. It can lie with its poles aligned with the field (parallel) or with its poles aligned in the opposite direction (antiparallel) to the magnetic field (Fig. 7.21b). It will require more energy to align the poles in a parallel than an antiparallel orientation. This means that an unpaired electron has two energy levels (Fig. 7.21c) one higher than the other. These energy levels correspond to spin quantum numbers of $+\frac{1}{2}$ and $-\frac{1}{2}$. It also means that atoms with an unpaired electron will have a small resultant magnetic field. Consequently, atoms with an unpaired electron will be attracted to one of the poles of the external magnetic field. The two beams of silver observed in the Stern-Gerlach experiment are attributed to two types of silver atom, one in which the unpaired 5s electron has a spin quantum number of $+\frac{1}{2}$ and the other in which it has a spin quantum number of $-\frac{1}{2}$.

Electron spin resonance deals with the absorption of electromagnetic radiation in the microwave region by unpaired electrons. When the unpaired electron absorbs the energy of the microwave radiation the electron moves from the lower energy level to the

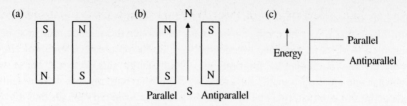

Fig. 7.21. The magnetic model of an electron.

higher (Fig. 7.22a) and in doing so produces a characteristic absorption spectrum (Fig. 7.22b). E.s.r. spectrometers usually record this absorption as a first derivative spectrum which shows the rate of change of absorption (Fig. 7.22c). The wavelength of the absorbed radiation is related to the difference in the energy levels of the electron by the expression given in Fig. 7.22a.

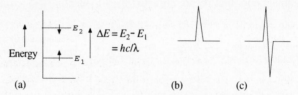

Fig. 7.22. A stylised e.s.r. absorption spectrum for a single electron and its first derivative spectrum.

The exact appearance of an e.s.r. spectrum will depend on the structural environment of the unpaired electron. Some nuclei, such as hydrogen and nitrogen, influence the nature of the absorption and hence the appearance of the spectrum. The theory of e.s.r. makes it possible to predict the effect of these atoms on the absorption, and as a consequence, interpret the spectrum of a radical in terms of the structure of that radical. Some typical e.s.r. radical spectra are shown in Fig. 7.23.

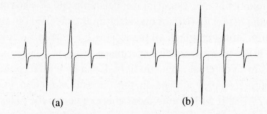

Fig. 7.23. The e.s.r. spectra of (a) the methyl and (b) the p-benzosemiquinone ion (p 113) free radicals.

E.s.r. spectra can also be used to identify a radical. The e.s.r. spectrum of the unknown radical is compared with those of known structure in a library of radical e.s.r. spectra. If the unknown's spectrum matches one of those in the library the unknown is assumed to have the same or a similar structure to the reference radical. This method will only identify radicals whose spectra are held on file but is still useful in spite of this restriction.

E.s.r. spectroscopy can detect, and hence identify, radicals at concentrations as low as 10^{-8} mol dm^{-3} provided they are stable long enough to be measured. Various techniques are used with unstable radicals to achieve the required stability. Three main methods are currently in use:

(1) **The continuous production method.** The free radical is continuously produced in the spectrometer.

(2) **The rapid freezing method.** The radical is produced in a transparent frozen solid inert matrix, for example, frozen argon. This prevents movement of the radical and hence its reaction with other species that could terminate its existence. The e.s.r. spectrum of the radical can be measured in this matrix. Its reactions can be followed by further e.s.r. spectroscopy as the solid is allowed to warm up. The superoxide radical was detected in xanthine oxidase oxidations of xanthine using this technique.

(3) **Spin trapping.** This technique is akin to derivative formation in organic analysis (11.4.1). Reactive radicals with a short life are reacted with compounds referred to as **spin traps**, or simply **traps**, to produce derivative radicals with a longer life. The reaction should be quick and produce a derivative radical with a well defined e.s.r. spectrum. The structure of the new radical, and hence the original radical, is determined from a knowledge of the reaction and the derivative radical's e.s.r. spectrum. Hydroxyl and superoxide radicals have been detected using this technique.

7.5.3 Trapping

This technique is similar in concept to derivative formation in organic analysis (11.4.1) and the spin trapping discussed in section 7.5.2. The free radical is reacted with a reagent, referred to as a scavenger, to form a non-radical product. The identity of the product is determined and the identity of the radical forming that product deduced from a consideration of the structures of the product and reagent. For example, aromatic compounds are readily hydroxylated by hydroxyl free radicals. Consequently, if a suitable aromatic compound is added to a suspected free radical reaction, the formation of the corresponding hydroxylated product(s) is taken as evidence that hydroxyl free radicals might be involved in the reaction. Phenol, for example, reacts with hydroxyl free radicals to form a mixture of 1,2-, 1,3- and the 1,4 hydroxylated products (Fig. 7.24).

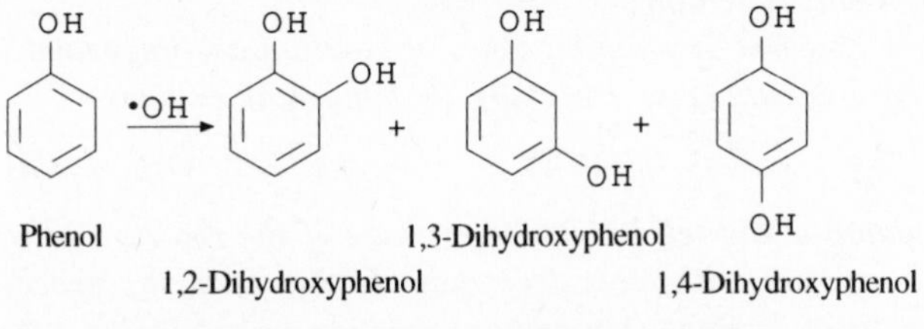

Phenol

1,2-Dihydroxyphenol

1,3-Dihydroxyphenol

1,4-Dihydroxyphenol

Fig. 7.24. The reaction of hydroxy free radicals with phenol.

Scavengers that react rapidly with radicals can be used to detect a free radical reaction. If a reaction has a free radical mechanism, the addition of a scavenger will disrupt that mechanism by reacting with the radical intermediates. This will affect the overall rate of the reaction. Consequently, if the rate of the reaction changes rapidly when a suitable scavenger is added, the mechanism of the reaction probably involves free radicals.

7.6 WHAT YOU NEED TO KNOW

(1) Reaction mechanism theory is an attempt to explain what happens at the electron level in chemical reactions.

(2) Reaction mechanisms are broadly classified into three types: heterolytic, homolytic and pericyclic.

(3) Homolytic reaction mechanisms are characterised by homolytic fission of bonds and the formation of free radicals.

(4) Simple free radicals are highly reactive electrically neutral species whose structures contain one unpaired electron.

(5) Heterolytic reaction mechanisms are characterised by heterolytic fission and the formation of charged species.

(6) Pericyclic reaction mechanisms involve no radical or ion intermediates but proceed by a mutual simultaneous rearrangement of the electrons of the reacting species.

(7) The structures and shapes of free radicals have not been fully elucidated.

(8) Free radicals in which the unpaired electron forms part of a conjugated system are more stable than radicals in which no conjugation of the unpaired electron occurs.

(9) Conjugated free radicals are easier to form than non-conjugated radicals.

(10) Molecular oxygen is a relatively stable diradical, that is, its structure has two unpaired electrons.

(11) Radical ions contain both an unpaired electron and a charge.

(12) Free radicals are reactive because they have an unstable electronic configuration. They usually react to achieve a stable electronic configuration by either combining with another free radical or abstracting an atom from another species.

(13) Free radical reaction mechanisms are usually in the form of a chain reaction.

(14) Chain reaction mechanisms contain three distinct stages: initiation, propagation and termination.

(15) Initiation is the formation of the radicals necessary to start the chain reaction. Initiation can be brought about by light (photolysis), heat (thermolysis) and certain redox reactions.

(16) Fenton's reaction and similar types of reaction are important in the formation of hydroxyl free radicals, especially in biological systems.

$$Fe^{2+} + HO:OH \longrightarrow Fe^{3+} + \cdot OH + OH^{-}$$

(17) Propagation is the self-perpetuating stage of the reaction in which the product is formed together with more free radicals that keep the process going.

(18) Termination reactions stop a chain reaction by producing substances that cannot react with any of the species present to form more free radicals to perpetuate the chain reaction.

(19) Homolytic reactions may be classified in the traditional manner as additions, substitutions, eliminations, oxidations, reductions and rearrangements.

(20) The controlled formation of free radicals is an essential part of many normal metabolic processes.

(21) The uncontrolled production of free radicals can be harmful to a biological system.

(22) The uncontrolled formation of free radicals is linked to radiation sickness, arthritis, cancer, heart damage after heart attack therapy and ageing.

(23) The uncontrolled formation of free radicals can be perpetuated by antioxidants such as ascorbic acid

(24) The principal methods of detecting free radicals are electron spin resonance (e.s.r.) and chemical trapping.

(25) The techniques used with e.s.r. are continuous production, rapid freezing and spin trapping of radicals. Radicals are identified by the characteristic shape of their esr spectrum.

(26) Chemical trapping is the reaction of the radical with a compound (scavenger) that has an affinity for radicals. Identification of the radical is made by comparing the structure of the product with that of the scavenger.

7.7   QUESTIONS

(1) Determine which of the following species and structures are free radicals:

   (a) a single sodium atom,   (b) a single calcium atom,

   (c) $CH_3\ddot{C}H_2$,   (d) $CH_3\dot{C}H_2$,

   (e) CH_3CH_3 (f) a single chlorine atom.

(2) Explain the chemical or electronic significance of each of the following arrows:

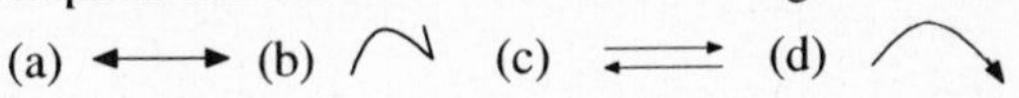

   (a) ⟷   (b) ⌒   (c) ⇌   (d) ⌒↘

(3) Determine the structure of the species resulting from each of the electron movements:

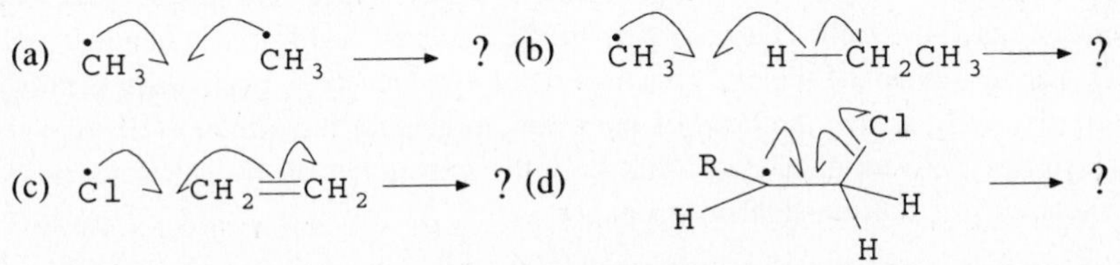

   (a) $\dot{C}H_3$ ⌒ $\dot{C}H_3$ ⟶ ? (b) $\dot{C}H_3$ ⌒ H—$\dot{C}H_2CH_3$ ⟶ ?

   (c) $\dot{C}l$ ⌒ CH_2=CH_2 ⟶ ? (d) R ⌒ $\dot{}$ ⌒ Cl

   with H, H, H ⟶ ?

(4) Suggest feasible free-radical-based mechanisms for each of the reactions.

   (a) Cu^+ + HOOH ⟶ Cu^{2+} + ·OH + OH^-

   (b) CH_2Cl_2 + Cl_2 ⟶ $CHCl_3$ + HCl

(5) Outline feasible methods for showing whether the specified situations involve a free radical.

   (a) The hydroxylation of the tranquilliser Librium.

   (b) The atmospheric oxidation of ethane.

   (c) The production of superoxide by phagocytic white blood cells.

8

Heterolytic and pericyclic reaction mechanisms

8.1 INTRODUCTION

Heterolytic mechanisms involve the formation of charged intermediates (7.1). The reaction of these intermediates is based on the principle that opposite charges attract. In other words, an electron rich (negatively charged) area of a chemical species could react with an electron deficient (positively charged) part of the same or a different chemical species (Fig. 8.1). The movement of the electrons in a heterolytic reaction mechanism is indicated by the use of a curly arrow, the head of the arrow indicating the direction of movement of *two* electrons. Convention dictates that only the movement of the electrons is shown in representations of the mechanism on paper.

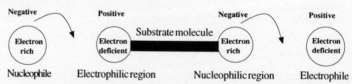

Fig. 8.1. The principle underlying heterolytic reaction mechanisms. The charges involved in mechanisms can be partial or whole. Small partial charges will often have a significant influence on the course of a reaction. Note the direction in which the arrow is drawn indicating the movement of the electrons and not the attack of the electrophile on the nucleophilic region of the substrate.

Chemical species with electron rich areas are referred to as nucleophiles whilst those with electron deficient areas, electrophiles (Fig. 8.1). Nucleophiles will react with electron deficient areas of a substrate (electrophilic regions) whilst electrophiles will react with electron rich areas of a substrate (nucleophilic regions). Some examples of nucleophiles and electrophiles are shown in Table 8.1.

The distribution of the electrons in a structure often results in parts of that structure having small positive and/or negative charges. In other words, the molecule has nucleophilic and electrophilic regions. These regions can be predicted by considering the fine nature of the covalent bonds forming the structure. For convenience, we shall treat sigma and pi bonds separately.

Table 8.1. Examples of nucleophilic and electrophilic species.

Nucleophiles				Electrophiles		
Ions	OH^- CN^- RO^-		Ion	H^+ NO_2^+ R_3C^+		
Molecules with lone pairs	$\overset{..}{N}H_3$ $H_2\overset{..}{O}$ $R\overset{..}{O}H$ $R\overset{..}{N}H_2$ $R\overset{..}{S}H$		Molecules with electron deficient centres (*)	$\overset{*}{B}F_3$ $\overset{*}{S}O_3$ $\overset{*}{I}Cl$ $\overset{*}{C}O_2$ $\overset{*}{A}lCl_3$ Br_2		
Molecules with pi electron systems						

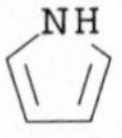

8.2 THE ELECTRON DISTRIBUTION IN SIGMA BONDS

The electrons forming a sigma bond between two different atoms will be attracted more to the atom with the highest electronegativity. However, the other atom will not completely relinquish its attraction for the electrons in the bond. As a result, the electrons will be located nearer the atom of higher electronegativity giving this atom a small negative charge with respect to the other atom. Consider the electrons of the sigma bond of hydrogen chloride. On the Pauling electronegativity scale, chlorine and hydrogen have electronegativities of 3.1 and 2.1 respectively (Fig. 8.2a). This means that chlorine has a stronger attraction for the electrons forming the H−Cl sigma bond than hydrogen. Therefore, the chlorine will acquire a very small negative charge whilst the hydrogen which is now slightly deficient in electrons will acquire a very small positive charge (Fig. 8.2b). A permanent charge distribution of this type in a sigma bond is referred to as the **inductive effect (I)**. It is represented on structural formulae by an arrowhead in the centre of the sigma bond. This arrowhead always points towards the negatively charged atom of the bond (Fig. 8.2c).

$$H \text{——} Cl \qquad H \overset{e}{\underset{e}{|}} Cl \qquad H \text{—▸} Cl$$
$$2.1 \quad 3.1 \qquad \delta+ \quad \delta- $$
$$\text{(a)} \qquad \qquad \text{(b)} \qquad \qquad \text{(c)}$$

Fig. 8.2. The inductive effect in a molecule of gaseous hydrogen chloride. (a) The electronegativities of hydrogen and chlorine. (b) A representation of electron and subsequent charge distribution in the molecule. (c) The representation of the inductive effect.

Inductive effects are permanent features of sigma bonds and therefore can be related to the physical properties of compounds in which they are found. Dipole moments, for example, are one such property which can easily be measured. Their presence or lack of it in saturated molecules gives practical support to the concept of the inductive effect.

The inductive effects of methyl chloride (Fig. 8.3), for example, predict that this molecule should have a resultant dipole moment. Measurements show that methyl chloride has a resultant dipole moment of 1.87 D. Tetrachloromethane, on the other hand, has no resultant dipole moment. This is expected since its four inductive effects (Fig. 8.3) cancel each other out.

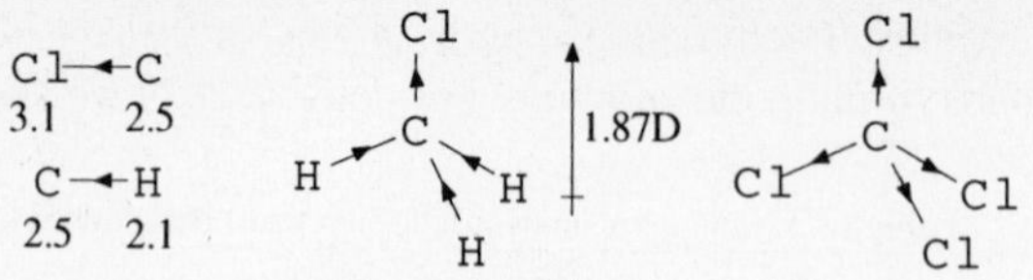

$$
\begin{array}{ll}
Cl \leftarrow C & \\
3.1 \quad 2.5 & \\
C \rightarrow H & \\
2.5 \quad 2.1 &
\end{array}
$$

Fig. 8.3. The electronegativities and inductive effects of the C–H , C–Cl bonds of chloromethane and tetrachloromethane.

Electronegative atoms and most functional groups will have inductive effects that withdraw electrons away from a carbon atom. The notable exceptions are metals and alkyl groups which have inductive effects towards a carbon atom (Table 8.2). When using the inductive effect of a structure attached to a carbon, one does not normally consider the inductive effects of sigma bonds within that structure, only the effect of the group as a whole on the structure to which it is bonded. For example, the nitro group as a whole has an electron-withdrawing inductive effect on the structure to which it is bonded. The inductive effects of the N–O bonds of this group would not normally be considered, even though they play a part in the electron-withdrawing effect of the nitro group.

Table 8.2. The inductive effects of some functional groups and structures. Inductive effects directed towards a carbon atom are referred to as **+I** effects whilst those directed from a carbon atom are referred to as **-I** effects.

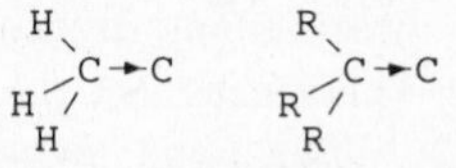

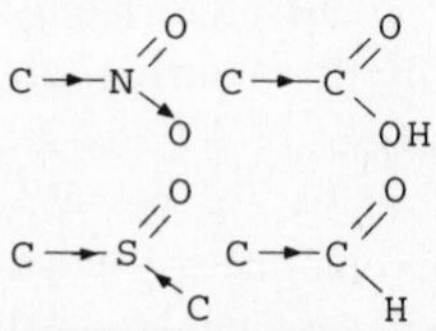

The inductive effect does not have a large sphere of influence. It usually has little effect beyond the second carbon in a chain. In chlorobutane, for example, the inductive effect of the chlorine would result in carbon 1 having a small positive charge ($\delta+$). As a result, carbon 1 will attract the electrons of the C_1–C_2 bond which results in carbon 2 having an extremely small positive charge ($\delta\delta\delta+$). This charge is too small to have any great influence on carbon 3. The transmission of the C–Cl inductive effect along the chain of sigma bonds has in effect stopped at carbon 2.

$$
\overset{\delta\delta\delta+}{CH_3}-CH_2-\overset{\delta+}{CH_2} \rightarrow \overset{\delta-}{CH_2} \blacktriangleright\blacktriangleright Cl
$$

8.3 THE ELECTRON DISTRIBUTION IN PI BONDS

Electronegativity will also affect the distribution of electrons in pi bonds. The picture is more complex than in the case of sigma bonds. In isolated pi bonds the electrons will be attracted to the more electronegative of the atoms forming the bond. As a result, the electron density of the pi bond will be greater nearer the atom of higher electronegativity and so this atom will therefore acquire a negative charge with respect to the other atom. This movement of pi electrons is referred to as the **mesomeric effect (M)**. For example, in an isolated carbonyl group, the oxygen has the higher electronegativity and so the

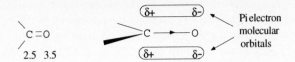

Fig. 8.4. The pi electron distribution in a carbonyl group. The inductive effect of the C–O will also make a contribution to the charges.

oxygen will become slightly negatively charged with respect to the carbon atom (Fig. 8.4).

It is not possible to predict the extent of the electron distribution in a pi bond. The distribution is currently represented on paper by the use of canonical forms (2.7). For example, the two canonical forms used for the carbonyl group are given in Fig. 8.5. The canonical form (b) can be obtained from the canonical form (a) by moving the two pi electrons to the oxygen atom. This movement is frequently represented by the use of curly arrows. The charges on structure (b) are obtained by balancing the number of electrons associated with each atom against the number of protons in the relevant nucleus.

$$\begin{array}{ccc} C\!\!=\!\!O & \longleftrightarrow & \overset{+}{C}\!-\!\overset{-}{O} \\ (a) & & (b) \end{array}$$

Fig. 8.5. The canonical forms used to represent the structure of the carbonyl group.

This picture of the structure of the carbonyl group is justified because it gives a better explanation of the chemistry of the carbonyl group than either the traditional formula Fig. 8.5a or the charged structure Fig. 8.5b.

Like the inductive effect, the mesomeric effect is a permanent feature of the molecule. Unlike the inductive effect it is transmitted by conjugated systems. This transmission is represented by a set of canonical forms that in effect indicates the overall pi electron distribution in the molecule. For example, the canonical forms that are used to represent the transmission of the mesomeric effect in benzaldehyde are given in Fig. 8.6.

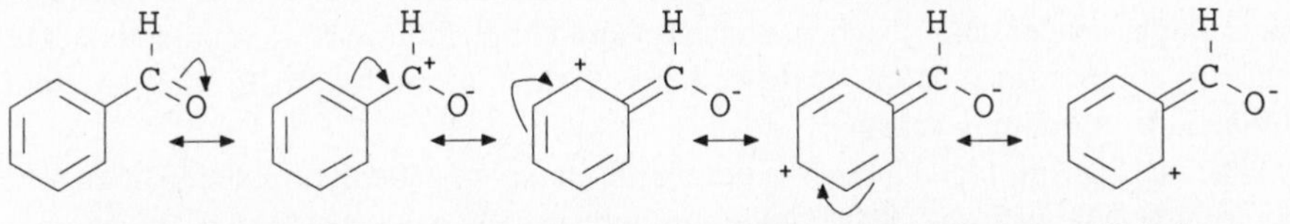

Fig. 8.6. The canonical forms used for the transmission of the mesomeric effect in benzaldehyde.

The transmission of the mesomeric effect can be quite complex depending on the nature of the conjugated system. However, the canonical forms used to represent the transmission in conjugated systems with one electron acceptor functional group follows the pattern given in Fig. 8.7 whilst the pattern for conjugated systems with one electron donor functional group is shown in Fig. 8.8. The key to deciding whether a functional group is an electron accepting or donating group is found by considering the structure of the group. **In general, the transmission of the mesomeric effect will be in the same direction as the inductive effect of the group unless the group has a lone pair in conjugation with the conjugated system, in which case the transmission of the mesomeric effect will be in the opposite direction to the inductive effect of the functional group.** The charges that arise in the canonical forms can be deduced by counting the electrons associated with an atom and balancing them against the number of protons in the nucleus of that atom (Question 2.2). It should be remembered that a curly arrow (⌒) means move two electrons and the appropriate adjustments to the structure should be made before counting the electrons. Predictions made using this method are reasonably accurate for conjugated systems with one functional group but are not always easy to interpret when there is more than one substituent.

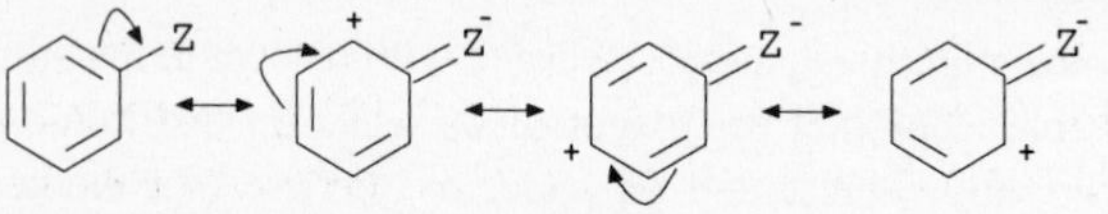

Fig. 8.7. The canonical forms of a conjugated system containing one electron acceptor functional group (Z). Z may be either negatively charged or electrically neutral in the canonical forms depending on its original structure.

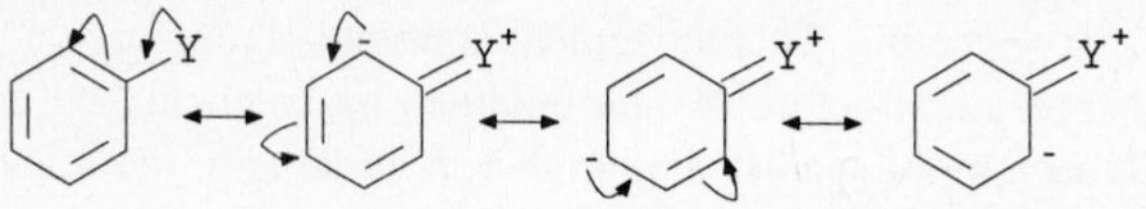

Fig. 8.8. The canonical forms of a conjugated system containing one electron donor functional group (Y). When Y has a conjugated lone pair, the conjugated lone pair is the start of the mesomeric effect pattern. Y may be positively charged or electrically neutral in the canonical forms depending on its original structure.

8.4 THE GENERAL TYPES OF HETEROLYTIC MECHANISM

Reactions that have simple heterolytic mechanisms can be broadly classified in the traditional manner as substitutions, additions, eliminations and rearrangements. These general classes of reaction have already been defined in section 7.3. For simplicity, substitutions are subdivided in this text into electrophilic, nucleophilic and nucleophilic displacements. Nucleophilic displacement reactions may be defined simply as substitution reactions which occur at an unsaturated centre. They usually proceed by a different type of mechanism from that found for nucleophilic substitutions at saturated centres. More complex reactions are usually variations and/or combinations of these types of mechanism. For example, condensation reactions consist of an addition followed by an elimination, whilst rearrangements are often followed by either a nucleophilic displacement, elimination or addition reaction.

The mechanisms found in each of the simple classes of heterolytic reaction often follow the same general pathway. Substitution reactions usually have mechanisms based on one of four mechanistic routes, namely electrophilic substitution and three nucleophilic substitution mechanisms known as S_N1, S_N2 and S_NAr2. However, in practice the precise nature of the mechanism of a reaction will depend on the structures of all the species involved, the energy pathway of the reaction and the conditions under which it is conducted.

The similarity between the mechanisms of reactions makes it possible to use the terminology of mechanisms as the basis of a method of classifying the reactivity of functional groups according to the types of mechanism they follow and the types of reagent involved in the reaction. Reversal of this procedure allows the logical prediction of the products of a reaction by classifying the reagents and logically matching this classification to the reactivity of the functional groups present in the substrate. Consider the reaction between hydrogen bromide and ethene. The structure of ethene contains a C=C functional group which acts as a nucleophile because of its pi electron structure. Its reactions can be classified as: electrophilic addition, oxidation, reduction and polymerisation. Hydrogen bromide is a weak acid that can give rise to hydrogen ions (electrophiles) and bromide ions (nucleophiles). Therefore, as the hydrogen bromide can give rise to an electrophile, the most likely reaction between the reactants is an electrophilic addition. Consequently, substituting the reactants in the general equation for electrophilic addition (Fig. 8.9) the predicted product of this reaction would be bromoethane and the mechanism would probably follow the route shown. This method of approach is reasonably successful for predicting the outcome of reactions between simple compounds.

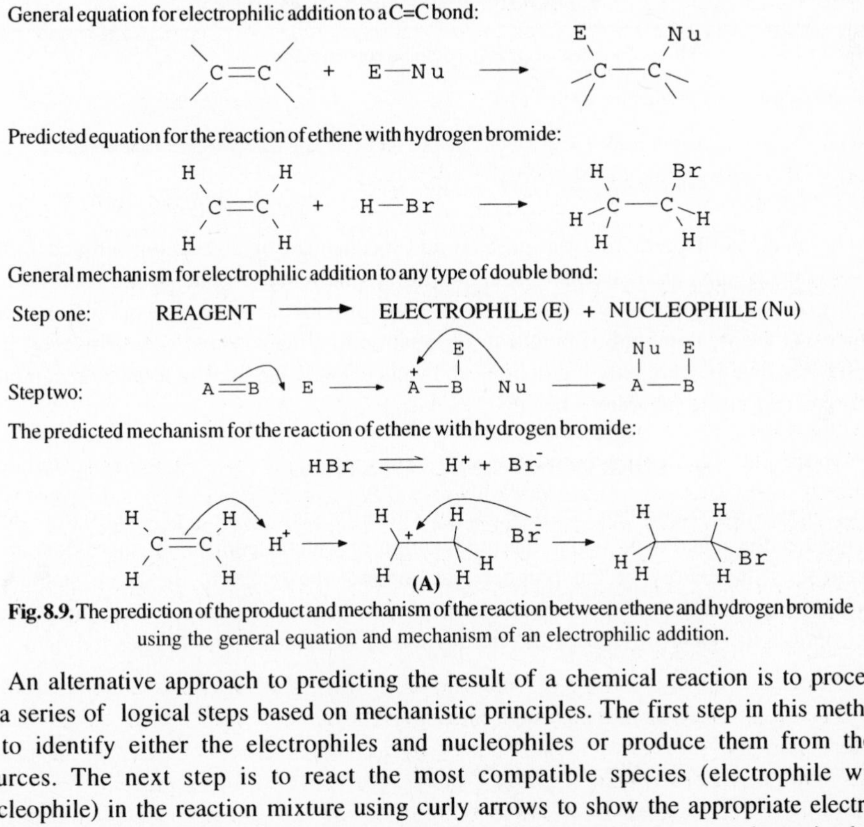

General equation for electrophilic addition to a C=C bond:

$$\underset{/}{\overset{\backslash}{C}} = \underset{\backslash}{\overset{/}{C}} \quad + \quad E - Nu \quad \longrightarrow \quad \overset{E \qquad Nu}{\underset{/}{\overset{\backslash}{C}} - \underset{\backslash}{\overset{/}{C}}}$$

Predicted equation for the reaction of ethene with hydrogen bromide:

$$\underset{H}{\overset{H}{\backslash}} C = C \underset{H}{\overset{H}{/}} \quad + \quad H - Br \quad \longrightarrow \quad \overset{H \qquad Br}{H - C - C - H}$$

General mechanism for electrophilic addition to any type of double bond:

Step one: REAGENT ⟶ ELECTROPHILE (E) + NUCLEOPHILE (Nu)

Step two: $A = B \quad \overset{E}{\searrow} \quad E \longrightarrow \underset{+}{\overset{E}{A - B}} \quad Nu \longrightarrow \underset{A - B}{\overset{Nu \ E}{|\ \ |}}$

The predicted mechanism for the reaction of ethene with hydrogen bromide:

$$HBr \ \rightleftharpoons \ H^+ + Br^-$$

Fig. 8.9. The prediction of the product and mechanism of the reaction between ethene and hydrogen bromide using the general equation and mechanism of an electrophilic addition.

An alternative approach to predicting the result of a chemical reaction is to proceed in a series of logical steps based on mechanistic principles. The first step in this method is to identify either the electrophiles and nucleophiles or produce them from their sources. The next step is to react the most compatible species (electrophile with nucleophile) in the reaction mixture using curly arrows to show the appropriate electron movements. The charges on the new species formed by this process are determined by counting electrons and balancing them against the protons in the atoms concerned (Question 2.2). This is followed by the next logical reaction or electron movement and so on until a stable product is obtained. For example, in the reaction between ethene and hydrogen bromide, the first step is to form the hydrogen ion electrophile and the bromide ion nucleophile by the ionisation of the hydrogen bromide. Since ethene is a nucleophile because of its pi electrons it will react with the hydrogen ion to form the intermediate the carbocation intermediate (A in Fig. 8.9). This intermediate is an electrophile and so will now react with the nucleophilic bromide ion to form a stable product, bromoethane. This procedure is illustrated in the lower part of Fig. 8.9.

A list of generalisations that have been found to be useful when making predictions using either of the preceding methods is given in Table 8.3. In cases where a reaction can form more than one product, a mixture of all the possible products is often obtained when it is carried out in the laboratory. However, the general mechanism sometimes indicates which substance will be the major product. Further information can be obtained if it is known whether the reaction is under thermodynamic or kinetic control (8.5).

Table 8.3. Useful generalisations for predicting the products of chemical reactions.

Reagent	The type of reaction in which it can participate
Nucleophiles	Substitution (displacement), addition, condensation
Electrophiles	Substitution, addition
Acids	Salt formation with bases; catalysts for elimination, displacement and addition
Bases	Salt formation with acids; catalysts for elimination, displacement and addition

The methods of predicting the product and mechanism of a reaction outlined in the previous paragraphs are reasonably accurate for simple reactions and so are useful when starting to study organic chemistry since they provide a route for classifying and condensing the vast amount of practical information to a manageable size. However, it is emphasised that they are only predictions and what actually happens in a reaction can only be discovered at the laboratory bench.

8.5 KINETIC OR THERMODYNAMIC CONTROL

In reactions where more than one product is formed the yield of each product will depend on whether the reaction is under thermodynamic or kinetic control. In thermodynamic control it is the energy of the product that controls the reaction: the more stable the product the greater the yield. In kinetic control it is the rate of production of the product that controls the yield: the faster the production of the product the greater its yield.

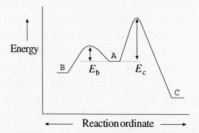

Fig. 8.10. The energy profile of a hypothetical reaction under either kinetic or thermodynamic control.

Consider the hypothetical reaction: A $\longrightarrow$ B + C. Suppose that the activation energy (E_b) for the production of B is less than the activation energy (E_c) for the production of C (Fig. 8.10). This means that the rate at which B is produced will be faster that the rate at which C is produced. Therefore if both reactions are irreversible the yield of B will be greater than that of C even though B has a higher energy than C. Consequently, the reaction is under kinetic control. If the reactions are reversible, in the early stages well before equilibrium is established the reaction will still be kinetically controlled and B will still have the larger yield. However, as the reactions approach equilibrium the rate of the conversion of B back to A increases faster than the rate of the conversion of C back to A. Consequently, the yield of C is likely to be higher at equilibrium and the reaction is said to be under thermodynamic control.

8.6 METHODS OF INVESTIGATING HETEROLYTIC MECHANISMS

The mechanism of a reaction is logically deduced from practical data gathered from experimental sources. These deductions can never be proved outright but when supported by sufficient experimental information they are accepted as being correct. Experimental evidence is obtained from a variety of sources, some of which are the same as those already described in section 7.5. Other sources of data are listed below.

8.6.1 The structure of the products
The structure of the products is one of the most important pieces of experimental evidence. A mechanism must account for all the products formed in the reaction including the by-products. It should also give an explanation of the relative yields of the products.

8.6.2 Reaction kinetics
Reaction kinetics is a major source of information. The rate expression for the reaction gives information concerning the nature of the slow step(s) in the mechanism. However, interpretation can be difficult. For example, the solvent may be involved in the mechanism but the rate expression does not contain a term for the solvent because the relatively small changes in solvent concentration do not appear to affect the rate. As a result, it is important that all the experimental data are taken into account when interpreting kinetic information.

8.6.3 Identification of intermediates
Intermediates can sometimes be isolated by stopping the reaction after a short space of time. Alternatively, intermediates have been isolated and identified by slowing down a reaction by using milder conditions. For example, the isolation of the intermediates $RCONHBr$, $RCONBr^-$ and $RNCO$ in the study of the Hofmann reaction for the conversion of amides to amines, helped establish the mechanism of this reaction (Fig. 8.11).

Spectroscopy, such as infra-red (10.4), n.m.r. (10.6) and e.s.r. (7.5) can be used to

$$RCONH_2 + BrO^- \xrightarrow{NaOH} RNH_2 + HCO_3^-$$

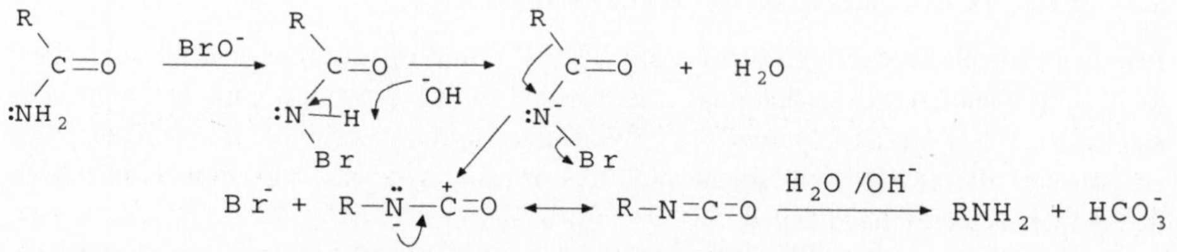

Fig. 8.11. The mechanism of the Hofmann reaction.

detect and identify intermediates. The disappearance and appearance of absorption bands can indicate the presence of intermediates. For example, the nitronium ion (NO_2^+) was detected in aromatic nitrations with concentrated sulphuric and nitric acid mixtures by Raman spectroscopy.

Intermediates may be isolated by trapping (7.5). For example, evidence concerning the nature of the intermediates in the addition of bromine to ethene was obtained trapping with chloride ions (Fig. 8.12). When the reaction is carried out in the presence of chloride ions 1-chloro-2-bromoethane (I) was isolated. The formation of this product is thought to be due to the attack of the chloride ion on the bromonium ion intermediate (II) whose existence has been indicated by n.m.r. spectroscopy.

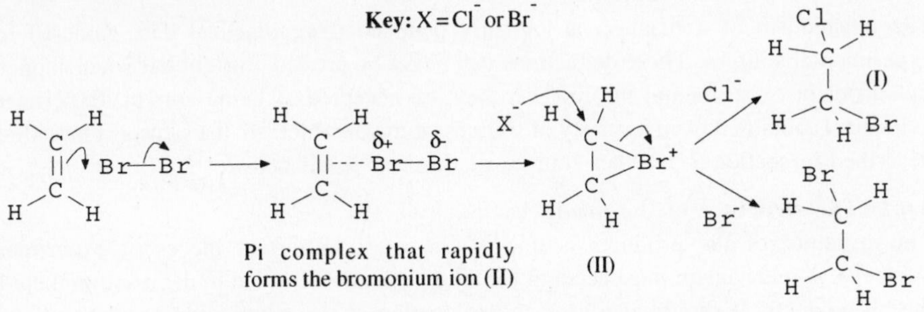

Key: $X = Cl^-$ or Br^-

Pi complex that rapidly
forms the bromonium ion (II)

(II)

Fig. 8.12. The mechanism of the addition of bromine to ethene.

8.6.4 Isotopic labelling

Both radioactive and non-radioactive isotopes have been used to label a particular atom in a structure. Radioactive isotopes are detected by means of a suitable counter (33.8) while mass spectrometers and ^{13}C n.m.r are used for non-radioactive isotopes. Deuterium can be detected by the absence of the corresponding proton signal in p.m.r. spectroscopy (10.6) The path followed by the isotope is found by analysis of the products and any intermediates that can be isolated from the reaction mixture. For example, hydrolysis of esters RCOOR' with $H_2^{18}O$ gave $RCO^{18}OH$ and R'OH (Fig. 8.13). This indicates that the mechanism of this reaction must involve C-OR cleavage (acyl oxygen fission).

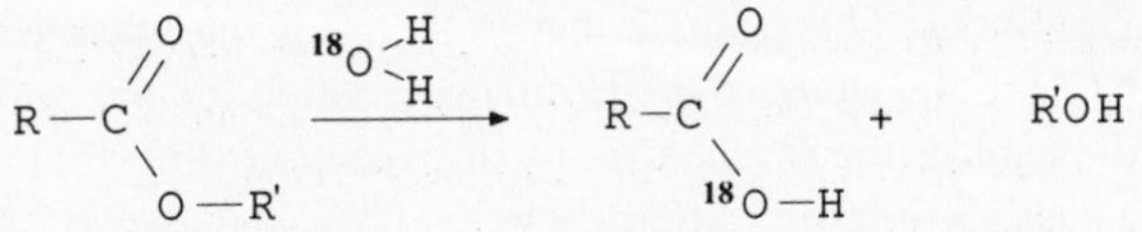

Fig. 8.13. The hydrolysis of an ester using ^{18}O labelled water.

8.7 PERICYCLIC REACTION MECHANISMS

Pericyclic (or electrocyclic) reactions are a small group of reactions that do not appear to be either homolytic or heterolytic in nature. Unlike reactions with heterogeneous mechanisms they are largely unaffected by changes in the polarity of the solvent. They are also not affected by free radical inhibitors or other catalysts. All attempts to trap or detect intermediates have failed.

Pericyclic reactions are thought to occur through the concerted movement of all the electrons involved (Fig. 8.14). The arrows used in these mechanisms could equally well be in the anticlockwise direction. They are simply included to emphasise the concerted nature of the movement of the electrons. The currently accepted explanation of this type of electron movement is that it occurs by a rearrangement of the relevant molecular orbitals. A detailed explanation of this rearrangement is beyond the scope of this book.

Reaction	Example

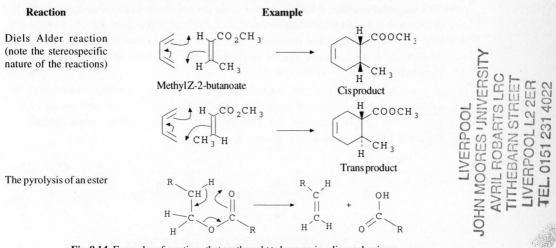

Diels Alder reaction (note the stereospecific nature of the reactions)

Methyl Z-2-butanoate

Cis product

Trans product

The pyrolysis of an ester

Fig. 8.14. Examples of reactions that are thought to have pericyclic mechanisms.

Reactions that proceed by pericyclic mechanisms are affected by heat and light. The effects are normally quite specific. If a pericyclic reaction is affected by light it is unlikely to be affected by heat. For example, the Diels Alder reaction (Fig. 8.14) is often initiated by heat but seldom by irradiation with light.

8.8 WHAT YOU NEED TO KNOW

(1) Heterolytic mechanisms are based on the heterolytic fission of bonds and the attraction of oppositely charged species. They are a combination of experimental observation and logic.

(2) Chemical species with an electron deficient centre are called electrophiles.

(3) Chemical species with an electron rich centre are called nucleophiles.

(4) Nucleophiles react with electrophiles and vice versa.

(5) The inductive effect (I) represents the electron distribution in sigma bonds. It is shown by an arrow head on the bond which points towards the most electronegative and so most electron rich end of the bond, that is: ——▶

(6) The inductive effect is a permanent feature of these bonds but is not transmitted significantly beyond the second carbon in a saturated chain.

(7) Structures containing more than one atom can be classified as having either a +I or -I effect depending on whether they have an inductive effect towards or away from a carbon atom. Alkyl groups and metals have a +I effect whilst most other groups have a -I effect.

(8) In isolated pi bonds the pi electron density is greatest nearest the atom with the

higher electronegativity. This electron distribution is shown by the appropriate canonical forms. The formation of the canonical forms can be predicted by the use of the mesomeric effect (M).

(9) The mesomeric effect is depicted on the canonical forms by a curly arrow (⌢ᷢ) showing the direction and position of movement of two electrons.

(10) The mesomeric effect operates in the same direction as adjacent inductive effects, that is, towards the most electronegative end of the pi bond except when there is a lone pair in conjugation with an unsaturated system. The mesomeric effects of conjugated lone pairs are drawn in the opposite direction to that of an adjacent inductive effect.

(11) The mesomeric effect is transmitted through a conjugated system. In conjugated systems with one functional group the transmission is usually drawn in one of two basic forms depending on whether it involves an electron-withdrawing or electron-donating functional group Z.

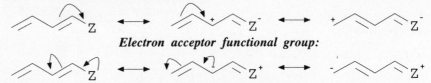

Electron acceptor functional group:

Electron donor functional group:

(12) Reactions with simple heterolytic mechanisms may be classified as substitutions (including nucleophilic displacements), additions, eliminations and rearrangements.

(13) The reactions found within a class will often have similar mechanistic routes.

(14) Complex reactions and their mechanisms are often variations and/or combinations of these simple types of reaction.

(15) The principles of reaction mechanisms can be used to predict the products of simple chemical reactions with a reasonable degree of accuracy.

(16) Reactions are believed to be either kinetically or thermodynamically controlled.

(17) Kinetic control is based on the rate of production of the product: the faster the rate the greater the yield of product.

(18) Thermodynamic control is based on the energy of the product: the product with the lowest energy will be formed in the greatest yield.

(19) Methods of investigating heterolytic reaction mechanisms include: determination of the structure of the product, determination of the rate expression, identification of intermediates and isotopic labelling.

(20) Pericyclic reaction mechanisms involve a concerted movement of all the electrons involved. They do not involve charged or radical intermediates.

8.9 QUESTIONS

(1) Classify each of the following species as either electrophiles or nucleophiles: (a) ethanol, (b) dimethylamine, (c) boron trichloride, (d) water, (e) bromine and (f) hydride ions.

(2) Explain the structural or electronic significance of each of the following the arrows.

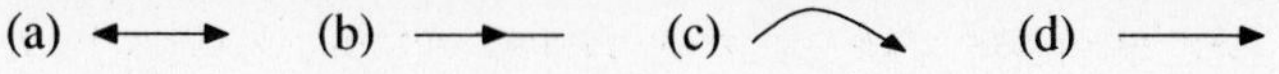

   (a) ⟷   (b) ⟶   (c) ⌒↘   (d) ⟶

(3) Predict the electron distribution in each of the following molecules. Show all inductive and mesomeric effects that you use to make the prediction.

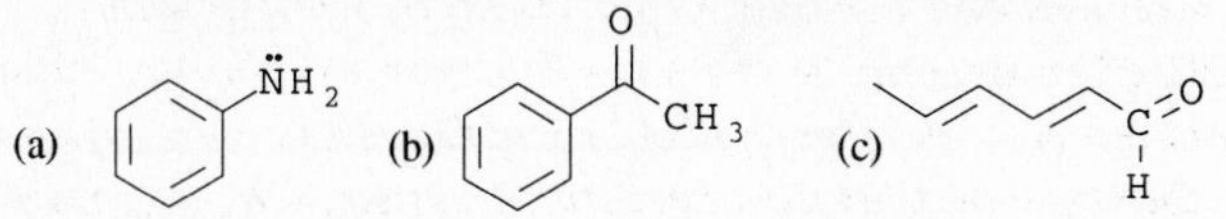

   (a)   (b)   (c)

(4) Complete each of the specified electron movements.

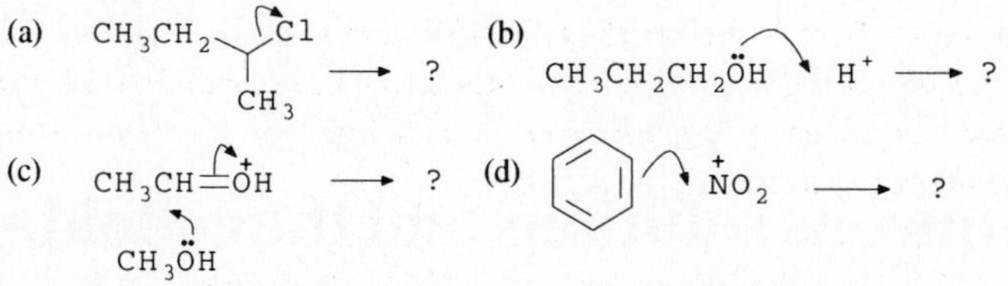

   (a) CH_3CH_2—Cl   (b)

                    | $CH_3CH_2CH_2\ddot{O}H$ H^+ ⟶ ?
                    CH_3 ⟶ ?

   (c) CH_3CH=$\overset{+}{\ddot{O}}H$ ⟶ ?   (d) $\overset{+}{N}O_2$ ⟶   ?

         $CH_3\ddot{O}H$

(5) Suggest feasible mechanisms for each of the reactions;

   (a) $CH_3CH=CH_2 + Br_2 \longrightarrow CH_3CHBrCH_2Br$

   (b)   ⬡ + CH_3^+ $[AlCl_4]^-$ $\xrightarrow{\text{two steps}}$ ⬡CH_3 + $AlCl_3$ + HCl

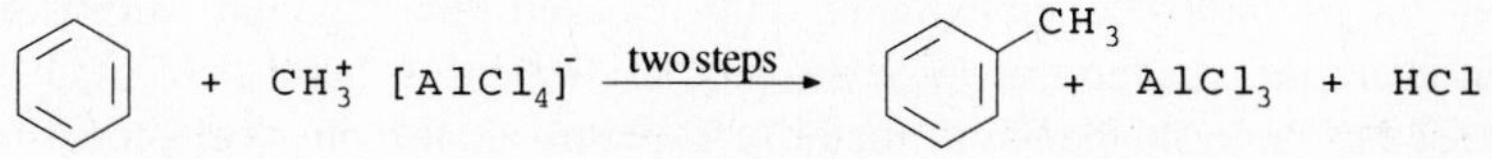

   (c)

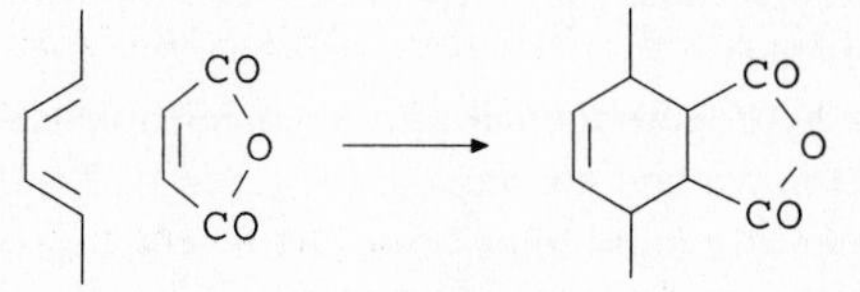

(6) Suggest and describe a feasible method of investigating reaction 5a.

9

Water, aqueous solutions and the colloidal state

9.1 THE STRUCTURE OF BULK SAMPLES OF WATER

In spite of its biological importance and extensive investigation, surprisingly little is known about the structure of large samples of water. A number of models have been proposed but none of them satisfactorily explains all the physical properties of water. However, they all postulate that hydrogen bonds are formed between the water molecules and that ice-like structures are involved.

In ice, the water molecules are hydrogen-bonded into a three-dimensional crystal lattice. Various allotropic forms are known and are named, Ice I, Ice II, Ice III, etc. They occur under varying conditions of pressure and temperature. Ice I which is formed at 0° C and 1 atmosphere pressure is the common form of ice. The basis of its structure is shown in Fig. 9.1a. It should be noted that this structure is very open and contains large empty spaces. However, the actual structure is not as perfect as Fig. 9.1a would suggest. Various types of lattice defect can be found, for example, the D and L defects shown in Fig. 9.1b. These defects are probably due to incorrect alignment of the molecules as the water freezes.

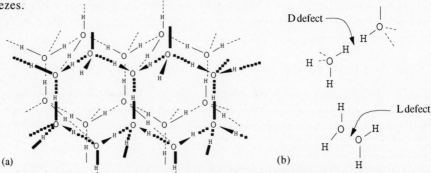

(a) (b)

Fig. 9.1. (a) A fragment of the structure of ice I. (b) D and L defects in the structure of ice.

The **flickering cluster** model of water is probably the most popular of the current theories for the structure of water. This model postulates that hydrogen-bonded clusters of water molecules with ice-like structures are suspended in a medium of non-hydrogen-bonded water molecules (Fig. 9.2). These clusters are temporary structures and have only a very short life-span: almost as soon as they are formed they disintegrate. However, as some clusters are disintegrating others are being formed at different points in the liquid. The clusters flicker in and out of existence giving the structure a very dynamic nature.

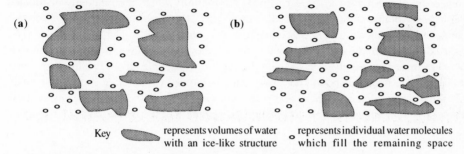

Key represents volumes of water with an ice-like structure   represents individual water molecules which fill the remaining space

Fig. 9.2. The flickering cluster model for the structure of water. (**a**) The situation at a time t. (**b**) The situation at a time t_1 a fraction of a second later.

9.2 SOLUTIONS

A solution is a dispersion of solute particles in a solvent . The sizes of the solvent and solute particles that form solutions are usually of the order of 0.1 to 1 nm. The solution has a uniform appearance because the particles of the solute are so small they cannot be detected by the naked eye. The nature of the interaction between the solute and solvent particles is not fully understood but it results in the solute particles being prevented from congregating (**coagulating**) into particles large enough to be seen. Any structural features of the solute that results in forces of attraction between the solute and the solvent molecules will enhance the solubility of the solute, that is, its dispersion in the solvent.

The bonding of solvent molecules to a solute is called **solvation**, the solute being said to be **solvated**. When water is the solvent the binding is referred to as **hydration** and the solute is said to be **hydrated**. The bonding is usually weak and of a temporary nature but in some cases the bonding is strong enough for a solute to form stable compounds that can be isolated from solution as crystalline solids. Sodium carbonate, for example, crystallises as its decahydrate ($Na_2CO_3 \cdot 10H_2O$) from water.

9.3 THE SOLUBILITY OF NON-POLAR SUBSTANCES IN WATER

Non-polar compounds are compounds whose structures do not have dipole moments (3.1) However, many molecules with a small dipole moment are also referred to as non-polar. It is not possible to define rigidly how small a dipole moment must be before the compound is classified as non-polar. However, non-polar substances are not usually very soluble in water.

Prediction of the water solubility of non-polar compounds is largely a matter of experience. In organic compounds, the larger the ratio of the non-polar carbon-hydrogen skeleton of a molecule to the polar structures in the organic molecule, the lower its water solubility. However, water solubility will probably improve if the non-polar compound contains structures which can either hydrogen bond (3.3) to the water molecules or form ions by interacting with the water.

It is believed that when some non-polar solute particles dissolve, they occupy the spaces in the ice like structures found in water. Consequently, non-polar molecules are surrounded in solution by protective structured layers of water molecules. Non-polar solutes are also believed to form **clathrate compounds** which are randomly distributed throughout volume of the water. Clathrates consist of a cage-like structure of hydrogen bonded water molecules in which is trapped the solute particle. The size of these water molecule cages and the size of the solute particles they can contain depends on the number of water molecules involved in forming the cage. For example, about twenty or so water molecules can hydrogen bond into a cage with an internal cavity large enough to contain small molecules such as methane and propanone (Fig. 9.3). If the clathrate is particularly stable it is sometimes possible to obtain a crystalline hydrate of the non-polar species.

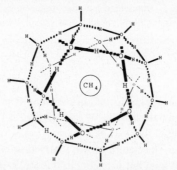

Fig. 9.3. An example of a cagework of water molecules showing the internal cavity in which a small molecule the size of a methane molecule could be trapped.

In aqueous solution, non-polar solutes can form particles held together by van der Waals forces (about 45% of the bonding) and a group of attractive forces collectively is known as **hydrophobic bonding**. Hydrophobic bonding occurs when the non-polar molecules are forced into contact with each other by the structured nature of the surrounding water molecules (Fig. 9.4). Its exact nature is not known but it does result in the partial removal of the non-polar species from the aqueous solution. This reduces the surface area of the solute molecules in contact with the water which leads to a reduction in the energy of the system.

Hydrophobic bonding occurs
between the molecules

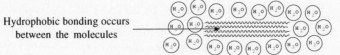

Fig. 9.4. Hydrophobic bonding.

9.4 THE SOLUBILITY OF POLAR MOLECULES AND IONS IN WATER

Polar molecules have permanent dipole moments which are significantly larger than those found in non-polar molecules. They are more soluble in water than non-polar molecules.

This solubility is believed to be due to their immobilising the water molecules in their vicinity by weak dipole-dipole attractive forces to form relatively stable hydrates (Fig. 9.5). Hydrogen bond formation between the solute molecules and water will increase the solubility of the solute. In general, the greater the number of hydrogen bonds formed between the solute and water the greater the solute's water solubility. The solubility of alcohols, phenols, carbohydrates, ethers and many other polar organic compounds in water is due largely to hydrogen bonding between these substances and water.

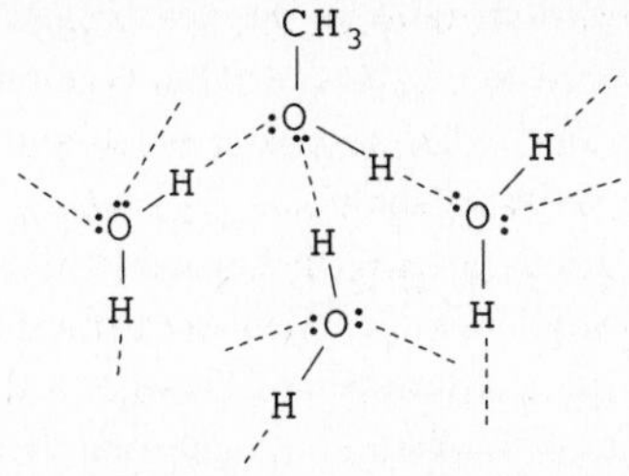

Fig. 9.5. Hydrogen bonding between methanol and water. At room temperature it is unlikely that all the hydrogen bonds shown would be formed. This is only likely to happen at very low temperatures.

Both positive and negative ions are very effective in immobilising water molecules. The ions electrostatically attract the oppositely charged ends of the dipoles of the water molecules to form ion-dipole bonds (Fig. 9.6). This attraction extends beyond the first layer of water molecules, and ions can be encased by several layers of water molecules. These layers of water molecules prevent the ions reforming the original compound and effectively keep the compound dispersed and in solution. Compounds that can either ionise or can be converted into a charged form are more likely to be water soluble.

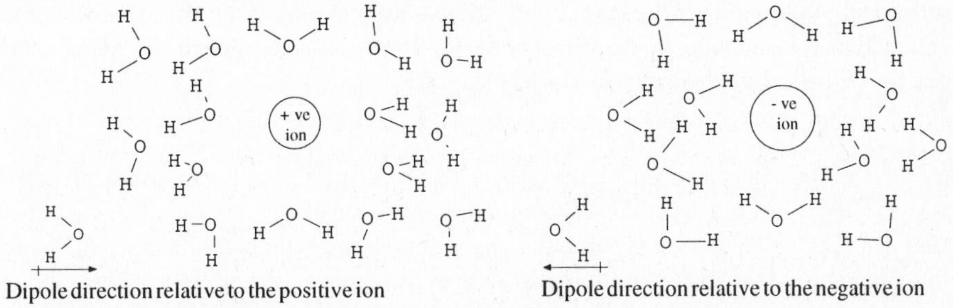

Dipole direction relative to the positive ion   Dipole direction relative to the negative ion

Fig. 9.6. A representation of the hydration of a cation and an anion. An electrostatic attraction exists between the water molecules and the ion. The lone pairs of the oxygen atoms are omitted for clarity.

Hydrogen and hydroxyl ions are hydrated in dilute aqueous solution. Practical evidence indicates that, at the very least, they are tetrahydrates ($H_9O_4^+$) in which the water molecules are held together by hydrogen bonds. In spite of this, hydrogen ions are referred to as though they are monohydrates (H_3O^+) in aqueous solution.

The conversion of a compound to the appropriate ion is often used in medical work to improve the water solubility of the compound since it does not usually affect its biological activity. The naturally occurring preservative benzoic acid which is almost insoluble in cold water is frequently used in the form of its water soluble sodium salt.

9.5 SOLUBILITY

The solubity of substances, whatever their physical state, can be recorded in a number of ways:

(1) Percentage weight for weight (%w/w). This is the number of grams of solute per hundred grams of solvent.
(2) Grams per hundred cubic centimetres of solution. Although this is not strictly a percentage it is commonly referred to as the percentage weight for volume (%w/v).
(3) The percentage volume for volume (%v/v). This is sometimes used for solutions of gases and liquids in water and other solvents. It is the number of volumes of solute in one hundred volumes of the solution.
(4) Moles per decimetre cubed (litre) of solution. This is currently called the concentration or molarity (mol dm^{-3} or M) of the solution. Since volume varies with temperature, the molarity of a solution will change with change in temperature.
(5) Moles per thousand grams of solvent. This is known as the molality (mol kg^{-1} or m) of the solution. Unlike molarity, molality does not vary with temperature.

Very dilute solutions are often recorded using micromoles (μmol), millimoles (mmol) or milligrams (mg) instead of grams and moles in the preceding definitions. With very dilute solutions, parts per million (ppm) or parts per billion (ppb) are frequently used.

The solubility of most solids and liquids increases with temperature. However, the solubility of calcium hydroxide and sodium sulphate in water decreases with increase in temperature.

The solubility of all gases in any solvent decreases with increase in temperature but increases with increase in pressure. At constant temperature the solubility is directly proportional to the pressure or, in mixtures, the partial pressure of the gas. This statement is called Henry's law after William Henry (1775-1836) who discovered the relationship. It can be expressed mathematically for any gas as:

$$C_g = k\,P_g \tag{9.1}$$

where C_g is the concentration of the gas in the solvent, P_g is the pressure or partial pressure of a gas if it is a constituent of a mixture of gases and k is a constant characteristic of that gas. The value of k for a particular gas will not vary with pressure but will change with temperature. It is important to realise that in mixtures of gases, Henry's law applies separately to each component of the mixture, and not to the mixture as a whole. It can be used to calculate the change in the solubility of a gaseous component of the mixture with variations in that gas's partial pressure.

The pressure or partial pressure of a gas in biological fluids is usually measured in kiloPascals (kPa). For example, the carbon dioxide in arterial blood is normally present at a pressure of about 4.3 kPa.

The solubility of gases is important in human and animal respiration. The venous blood returning to the lungs will have a relatively low PO_2 and a high PCO_2. As gases will always diffuse from regions of high concentration to areas of low concentration, the relatively high concentration of oxygen in the air in the lungs will diffuse into the blood. Conversely, the relatively high pressure of carbon dioxide in venous blood means that this carbon dioxide will diffuse from the blood into the air in the lungs and be exhaled.

The partial pressures of oxygen and carbon dioxide in the blood taken in conjunction with other measurements give valuable information concerning the acid-base and respiratory state of a patient.

9.6 SPARINGLY SOLUBLE SALTS AND COMPOUNDS

The saturated solutions of sparingly soluble salts and compounds that ionise in water constitute heterogeneous equilibrium systems (6.11.2) provided solid salt or compound is present and the temperature is constant. In this situation the law of mass action constant for the equilibrium is known as the solubility product ($K_{(sp)}$) of the salt. Consider, for example, a saturated aqueous solution of sparingly soluble silver chloride in contact with solid silver chloride:

$$AgCl_{(s)} \rightleftharpoons Ag^+_{(aq)} + Cl^-_{(aq)}$$

Applying the law of mass action for **dilute** heterogeneous systems at constant temperature:

$$K_{(sp)} = [Ag^+][Cl^-] \quad (= 1.7 \times 10^{-10} \text{ mol}^2 \text{ dm}^{-6})$$

where the right-hand side of this equation is known as the ionic product.

The solubility product is a measure of the limit of solubility of a salt in water. If the experimentally determined value of the ionic product in the $K_{(sp)}$ expression for a salt exceeds the value of $K_{(sp)}$ the solubility of the salt is exceeded and a precipitate will form or the solid will not completely dissolve. The addition of a solution containing ion(s) in common with those of the sparingly soluble salt will reduce the solubility of that salt because the solubility product expression does not differentiate between the sources of the ions. For example, the solubility product expression for a saturated solution of silver chloride in contact with solid silver chloride in the presence of hydrochloric acid will also include the concentration of chloride ions from the hydrochloric acid:

$$K_{(sp)} = [Ag^+][Cl^-_{AgCl} + Cl^-_{HCl}]$$

Since $K_{(sp)}$ is a constant at constant temperature, the contribution of Cl^-_{HCl} to the value of the ionic product means that the contributions from Ag^+ and Cl^-_{AgCl} are reduced, that is, the solubility of the silver chloride is reduced in the hydrochloric acid solution. This reduction can only be brought about by the suppression of the ionisation of the silver chloride.

Suppression of ionisation of a compound by the addition of a common ion is known as the **common ion** effect. It is a factor in the control of the transfer of substances across membranes in some regions of the body. For example, an important factor in the absorption of aspirin in the stomach is the supression of the ionisation of the carboxylic acid group of aspirin by the hydrochloric acid produced by the stomach lining.

9.7 THE PHYSICAL PROPERTIES OF WATER

9.7.1 Specific heat and heat capacity

Specific heat (Jg^{-1}) is the heat required to raise the temperature of one gram of a substance by $1°C$. Water has a high specific heat by comparison to other common substances. This means that it requires a considerable amount of energy to cause a temperature change of any significance in water. Consequently, the water in biological systems, such as the human body, will protect those systems from changes in temperature which would disrupt their normal modes of operation. If, for example, water had a low specific heat then very small amounts of energy would cause rapid and significant increases in the temperature of aqueous body fluids which could result in the denaturation of proteins (28.9) with a consequent disruption of their function. Rapid changes in temperature would also cause rapid fluctuations in reaction rates with the result that many metabolic processes would proceed in an erratic manner.

Specific heat should not be confused with heat capacity. Heat capacity is the heat absorbed or liberated when the temperature of the whole sample rises or falls by $1°C$ while specific heat is the heat required to raise the temperature of one gram of a substance by $1°C$. The average human adult body contains about 60% water and its heat capacity is of the order of 20×10^4 $J/°C$. Absorption of this amount of energy will only raise an adult's body temperature by $1°C$ but it would raise the equivalent mass of iron by about $10°C$. This demonstrates the effectiveness of water in protecting our bodies since a $10°C$ rise in body temperature would be lethal.

9.7.2 Heats of vaporisation and fusion

The heats of vaporisation and fusion of water are both high compared with those of other liquids. This high heat of vaporisation minimises water loss by evaporation, that is, it protects the system from dehydration because a large amount of energy is required to evaporate one gram of water. Conversely, vaporisation will significantly cool the system as a large amount of energy is removed when one gram of water evaporates. About 40% of the human body's total water intake and production is lost by evaporation from the skin and lungs. The heat this evaporation requires is an essential part of the body's temperature control system.

The high heat of fusion protects living tissue against freezing because a large amount of heat would have to be lost from the biological system before freezing could occur.

9.7.3 Dielectric constant

The dielectric constant of water is 78.54 at $25°C$. This is a high value for a liquid and is believed to be the reason why water is such a good solvent for ionic substances. It is known that the force F necessary to separate two unlike charges in solution is proportional to the reciprocal of the dielectric constant D of the solvent, that is:

$$F \propto 1/D \qquad\qquad (9.2)$$

It follows, therefore, that the higher the dielectric constant of a solvent the lower the force necessary to bring about the ionisation of an ionic compound in that solvent.

The high dielectric constant of water means that many ionic compounds will dissolve easily in water because the force (energy) required to separate the ions is low and available. The separated ions can then pass into solution by interacting with the water molecules in the manner described in section 9.4. When ionic compounds are insoluble or sparingly soluble in water, insufficient energy is available to separate the ions.

9.7.4 Viscosity

Water has a high viscosity because of its highly hydrogen bonded nature. This contributes to its suitability as a transport medium in biological systems because the flow rates of liquids with a high viscosity can be more easily controlled. Relatively small ions, such as

F^-, Li^+, Na^+, H^+, Mg^{2+} and OH^- increase water's viscosity and so reduce its flow rate in biological systems. Larger monovalent ions, such as K^+, NH_4^+, Br^- and ClO_4^- tend to decrease its viscosity thereby increasing its flow rate in biological systems.

9.7.5 Surface tension

The polar nature of water molecules means that there are quite strong forces of attraction between adjacent water molecules (Chapter 3). Consider a sample of water in a container (Fig. 9.7). A molecule in the interior will experience attractive forces from the surrounding molecules. These forces will tend to cancel each other out. However, the situation at the surface is quite different. A molecule on the surface will only experience attractive forces towards the interior of the water sample because of the non-polar nature of the molecules that constitute air. The outcome is an attractive force pulling the molecule towards the interior of the sample. Surface tension is the force exerted by the surface of the water which just balances this resultant force towards the interior. Water has a high surface tension compared with other common liquids.

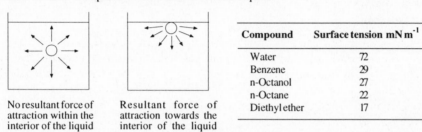

No resultant force of attraction within the interior of the liquid

Resultant force of attraction towards the interior of the liquid

Compound	Surface tension mN m^{-1}
Water	72
Benzene	29
n-Octanol	27
n-Octane	22
Diethyl ether	17

Fig. 9.7. The surface tension of a liquid is the force balancing the resultant force of attraction towards the interior of the liquid.

The high value of the surface tension of water indicates that there are strong forces of attraction between water molecules. Because of their solubility there must be strong forces of attraction between water and other polar molecules but only weak forces of attraction between water and non-polar molecules. This means water will *wet* polar surfaces but will not *wet* non-polar surfaces. The flow of water in the capillaries found in the membranes of plants and animals is due to this wetting of their polar surfaces by the water. The presence of lipid molecules in membranes is important in the control of this flow because lipid molecules are not wetted by water.

Surfactants, or **surface-active agents** as they are also known, are compounds that reduce the surface tension of water. This enables water to wet surfaces that are normally water repellent. Aqueous solutions of surfactants will often dissolve substances that are normally insoluble in water and emulsify substances that do not normally mix with water. These features make surfactants extremely useful in the formulation of water insoluble drugs.

The structures of surfactants (Table 9.1) can be divided into **hydrophilic** (water-liking) and **hydrophobic** (water-hating) regions. The polar hydrophilic section of the molecule is soluble in water while the non-polar hydrophobic section of the molecule is insoluble in water but is soluble in non-polar solvents and can bind to non-polar substances. The polar hydrophilic end of a surfactant has such a strong affinity for water (9.4) that it can pull a non-polar substance into solution in water. The nature and degree of solubility achieved by this action varies from total solubility through partial solubility to colloidal solutions (9.8).

Surfactants are classified as **anionic**, **cationic**, **ampholytic** and **non-ionic** according to the nature of their hydrophilic structures. Ampholytic surfactants are electrically neutral dipolar ions having both negatively and positively charged atoms in their structure. They are also known as zwitterions. Surfactants are also classified in terms of their use, for example, detergent, emulsifier, dispersant and wetting agent. Anionic surfactants are the most widely used as they are cheap and effective.

Table 9.3. Examples of surfactants.

Compound	Structural formula
	Hydrophobic end ┤ Hydrophilic end
Cationic surfactants	
Sodium stearate	$CH_3(CH_2)_{16}$ ┤ $COO^-\ Na^+$
Anionic surfactant	
Dodecylpyridinium chloride	$C_{12}H_{25}$ ┤ $C_5H_5N^+\ Cl^-$
Dodecylamine hydrochloride	$CH_3(CH_2)_{11}$ ┤ $NH_3^+\ Cl^-$
Ampholytic surfactants	
Dodecyl betaine	$C_{12}H_{25}N(CH_3)_2CH_2$ ┤ COO^-
Non-ionic surfactants	
Heptaoxyethylene monohexyldecyl ether	$CH_3(CH_2)_{15}$ ┤ $(OCH_2CH_2)_7OH$
Polyoxyethylene sorbitan monolaurate	The polyoxyethylene ethers of the lauric acid esters of sorbitan

Surfactants are used as detergents. Because of its high surface tension, water on its own is not very effective in removing the greases and oils that bind bacteria and dirt to the skin and other surfaces. The presence of a surfactant decreases the surface tension of water and increases its wetting power, greatly improving its cleansing effect. Cationic surfactants, such as dodecylpyridinium chloride, are especially useful as detergents since they also act as germicides, being effective against a wide range of gram negative and positive bacteria. The antibacterial cream Cetrimide BP contains a mixture of cationic surfactants, the main one being tetradecyltrimethylammonium chloride.

Naturally occurring surfactants are important in some bodily functions. Bile salts, for example, are surfactants that are produced in the liver and pass into the intestine where they play an essential part in the digestion of lipids and the absorption of the relatively non-polar vitamins A and D (Fig. 9.8). The surfactants that are produced in the membranes of the alveoli of the lungs prevent water accumulating in the lungs which would collapse the alveoli and make breathing difficult.

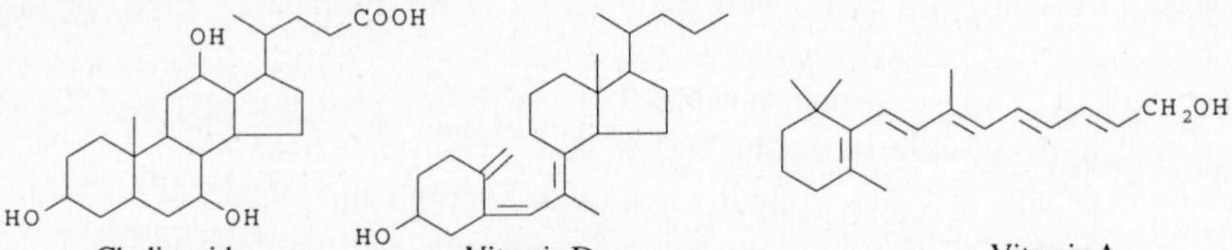

Cholic acid   Vitamin D_3 Vitamin A

Fig. 9.8. Examples of some naturally occurring surfactants

A large number of water insoluble drugs are formulated with surfactants to make them water soluble. Non-ionic surfactants such as polysorbates are frequently used to prepare aqueous solutions of steroids that are not very water soluble for use in opthalmic preparations where clarity of the solution is very important.

9.8 OSMOSIS

Osmosis is the process whereby the solvent molecules of a solution diffuse through a semi-permeable membrane known as an osmotic membrane. Theoretically, ideal osmotic membranes have pores which are large enough to allow the passage of the solvent molecules, but small enough to prevent the passage of larger solute particles (Fig. 9.9a). In practice the pores of many semi-permeable membranes are also large enough to allow the passage of small solute particles (9.11). The diffusion of solvent molecules in osmosis will occur in both directions but there is always a net diffusion of solvent from a region of low solute particle concentration to one of high solute particle concentration. This has the effect of equalising the concentration of the solute particles on both sides of the membrane.

$$\text{Low concentration of solute particles} \xrightleftharpoons[\text{Net solvent flow} \longrightarrow]{\text{Solvent diffusion}} \text{High concentration of solute particles}$$

Discussions concerning osmosis are normally confined to aqueous solutions. Osmosis can be demonstrated with the apparatus shown in Fig. 9.9b A solution of sucrose in the inverted thistle funnel is separated from the pure water in the beaker by an osmotic membrane. At the start of the demonstration the sugar solution in the thistle funnel must be at the same level as the solution in the outer container. As time passes, the solution in the tube rises as there is a net diffusion of water into the sugar solution. When the sugar solution stops rising, the system is in equilibrium and the rate of flow of water into the sugar solution equals the rate of flow of water out of the sugar solution. The hydrostatic pressure (h) exerted by the column of liquid has reduced the rate of diffusion of water into the sugar solution until it equals the rate of diffusion out of the sugar solution.

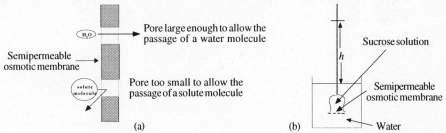

Fig 9.9. (a) A cross-section through a semi-permeable membrane showing how pore size governs the passage of a species through the membrane. (b) A simple experiment to demonstrate osmosis.

The pressure that must be exerted to just prevent the net diffusion of solvent (osmosis) is known as the osmotic pressure of the solution. Osmotic pressure is normally measured in kiloPascals (kPa) where $1\ Pa = 1\ Nm^{-2}$. However, in biomedicine the osmotic pressure of a solution is usually expressed as its **osmotic potential** which is recorded as its **osmolality**.

Osmotic pressure is a **colligative property** of water, that is, a property whose value depends on the number of solute particles present rather than their nature. For example, consider the osmotic pressures of equal volumes of 0.01 M aqueous solutions of sodium chloride and glucose. Sodium chloride is fully ionised in aqueous solution while glucose is not ionised (Fig. 9.10). As a result, there are twice as many solute particles in the 0.01 M sodium chloride solution as there are in the 0.01 M glucose solution. Consequently, the osmotic pressure of the 0.01 M sodium chloride solution would be double that of the 0.01 M glucose solution if they were each separated from pure water by *ideal* osmotic membranes. In practice, because most semi-permeable membranes are *non-ideal* and allow the passage of some solute particles, the osmotic pressure of the sodium chloride solution will be a little less than double that of the glucose solution. The actual figure depends on the membrane used.

$$NaCl \longrightarrow Na^+ + Cl^-$$

0.01 mol   0.01 mol   0.01 mol

(0.02 mol of particles in solution)

$$C_6H_{12}O_6 \longrightarrow$$ No ionisation and so a total of 0.01 moles of particles in solution

Glucose

0.01 mol

Fig. 9.10. A solution containing 0.01 mole of NaCl has twice the number of particles in solution than a solution containing 0.01 mole of glucose.

The osmotic pressure (π) of very dilute solutions is related to the concentration of the solute particles by the expression:

$$\pi = cRT \qquad (9.3)$$

where c is the concentration of the solute, $T\,°$K is the temperature of the solutions and R is the ideal gas constant (8.314 J K^{-1} mol^{-1}). Concentration may be expressed as either the temperature-dependent molarity (M or mol dm^{-3}) or more commonly in biomedicine and biochemistry as the temperature-independent molality (m or mol kg^{-1}).

The practically determined values of the osmotic pressure of a solution and the values calculated using equation 9.3 do not correspond for compounds that associate or dissociate in solution. To overcome these discrepencies Van't Hoff introduced a correction factor, i (equation 9.4) into equation 9.3.

$$i = \frac{\text{Observed osmotic pressure}}{\text{Osmotic pressure expected if no association or dissociation occurs}} \qquad (9.4)$$

As a result, equation 9.3 becomes: $$\pi = icRT \qquad (9.5)$$

Experimental evidence shows that for strong electrolytes in very dilute aqueous solution, the value of i equals the total number of moles of ions (v) formed per mole of the compound. For weak electrolytes, the degree of ionisation is so small that $i \sim 1$.

In biochemistry and biomedicine osmotic pressure is recorded in terms of the **osmolality** of the solution where osmolality is the molality (mol kg^{-1}) of all the particles in the solution that are responsible for the osmotic pressure of that solution. In the case of ideal membranes: osmolality = im, and for dilute solutions of strong electrolytes where dissociation is complete: $i = v$. For example, a 0.01 m sodium chloride solution would with an ideal membrane have an osmolality of 0.02 mol kg^{-1} (20 mmol kg^{-1}) since $v = 2$. However, for non-electrolytes and very weak electrolytes, osmolality = m since $i = 1$.

In the case of non-ideal membranes: osmolality = ϕmi, where:

$$\phi = \frac{\text{Observed osmotic pressure or other colligative property}}{\text{Calculated osmotic pressure or other colligative property assuming complete dissociation into ions}}$$

Consequently, with non-ideal membranes: osmolality = $\phi m\nu$ for dilute solutions of strong electrolytes and ϕm for non-electrolytes and weak electrolytes. In this case osmolality must be calculated from a colligative property.

The osmolality of a solution containing a mixture of solutes is the sum of the osmolalities of the individual solutes. For example, under ideal conditions, the osmolality of a mixture containing 0.1 mol of sodium chloride and 0.1 mol of glucose per 1000 g of water is 0.3 mol kg^{-1}. Human plasma usually has an osmolality of 285-295 mmol kg^{-1}.

Osmolality is determined using an osmometer which measures either the depression of freezing point or the vapour pressure of the solution. These are also directly related to the molality of the solution. Osmometers are usually calibrated using solutions of sodium chloride, allowance being made for the value of ϕ which varies with the molality (e.g. 0.944 at 0.1 mol kg^{-1} and 0.910 at 0.9 mol kg^{-1}).

Solutions with equal osmotic pressures are said to be **iso-osmotic.** Therefore, if two **iso-osmotic** aqueous solutions at the same temperature are separated by an ideal semipermeable membrane, both solutions will contain equal numbers of solute particles (solutions A and B in Fig. 9.11) and **there will be no net diffusion of water molecules**. However, if the membrane and solutions are not ideal, as is usually the case, but the two solutions are still iso-osmotic, then they are referred to as being **isotonic** with respect to that membrane. Isotonic solutions have the same osmotic pressures but do not contain the same number of solute particles. This occurs because the particles that contribute to the magnitude of the osmotic pressure are those that **do not pass** through the membrane. Solute particles that pass through a semi-permeable membrane do not contribute to the osmotic pressure of the solution. Such particles act as though they were solvent molecules because they can diffuse in both directions through the membrane. Consider, for example, a solution C containing 0.1 moles of urea and 0.1 moles of glucose separated from a solution D containing 0.1 moles of glucose by a semi-permeable membrane that will allow the diffusion of the urea molecules but not the glucose molecules (Fig. 9.11). Initially, because it contains a higher number of solute particles, solution C will exert a greater osmotic pressure than solution D. Solution C is initially **hyperosmotic** with respect to D. However, the urea molecules will pass through the membrane and eventually equilibrium will be reached where there will be equal numbers of urea molecules passing through the membrane in both directions. At this point solutions C and D will exert the same osmotic pressure, that is be iso-osmotic. Since solution C **initially** contained twice as many solute particles as solution D the two solutions are said to be isotonic with respect to the membrane.

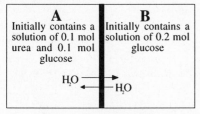 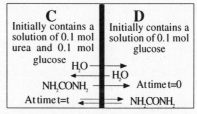

Ideal semipermeable membrane:
only water can diffuse through

Non-ideal semi-permeable membrane:
both water and urea can diffuse through

Fig. 9.11. Iso-osmotic and isotonic solutions.

An isotonic solution cannot exert its full osmotic pressure because some of the particles that would have contributed to the osmotic pressure of the solution are able to pass through the semi-permeable membrane. The **tonicity** of such a solution is the effective concentration of the solute, that is, the concentration of solute that is responsible for the osmotic pressure of the solution. It is quoted in terms of either the experimentally determined osmotic pressure or, more usually in biomedicine, as the **osmolality** of the solution.

The cell walls and membranes of many of the cells found in animals and plants can act as osmotic membranes. Therefore, it is important for the health of the cell that the fluid surrounding it is approximately isotonic with its contents, especially if the cell has a non-rigid membrane. Consider, for example, the effect that changes in the concentration of the surrounding medium would have on red blood cells whose cell membranes act as osmotic membranes. If the fluid surrounding the cell has a higher tonicity (hypertonic solution) and therefore a higher osmotic pressure than the fluid inside the cell, water will diffuse out of the cell and the dehydrated cell will shrink (Fig. 9.12). This contraction in shape is sometimes referred to as **crenation**. In the reverse situation where the red blood cell is surrounded by a fluid which has a lower tonicity (hypotonic solution) and as a result a lower osmotic pressure than the fluid inside the cell, water passes into the cell and the cell expands and changes from a biconcave disc to a sphere. If sufficient water flows into the cell it will burst (**haemolysis for red blood cells, cytolysis for all types of cell**) releasing haemoglobin into the plasma (Fig. 9.12). In view of this behaviour it is important that the solutions used for injection of drugs into the blood stream are isotonic with the blood. Important exceptions are the hypertonic solutions that are injected into large veins for total parental nutrition (TPN). In contrast, most rigid cell walls are strong enough to contain large increases in internal pressure due to osmosis, unless there is a weak point in their structure. These weak points can be caused by the action of enzymes such as lysozyme which partially dissolve the membrane, or drugs such as penicillin and cycloserine which can interfere with the formation of bacterial cell walls. Cells with weak areas in their cell walls could burst if placed in a hypotonic solution with an osmotic pressure lower than the cell contents.

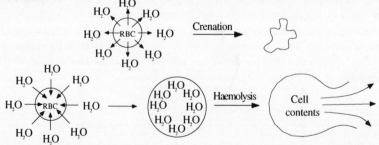

Fig. 9 12. A diagrammatic representation of crenation and haemolysis of red blood cells (RBC).

9.9 COLLOIDAL DISPERSIONS

The varied nature of colloids makes an exact definition of the colloidal state difficult. A simple but adequate definition is that colloidal dispersions are homogeneous mixtures of microscopic particles of substance which range in size from 1 to 1000 nm dispersed in a dispersing medium. In biological systems this medium is usually water. The physical nature of colloidal systems ranges from dusty air through semi-opaque liquids which are referred to as sols, to semi-solids such as whipped cream. The structure of the dispersed colloidal particles ranges from aggregates of molecules (**micelles**), to large single molecules such as proteins. These particles are in a state of constant erratic movement known as Brownian movement after its discoverer Robert Brown (1773-1858). Blood is part colloidal, part solution and part suspension.

Micelles are formed at certain concentrations by solute molecules whose structures contain non-polar and polar structural sections. At low concentrations it is believed that the molecules congregate together to form spherical micelles, the non-polar sections of the molecules forming the interior of the sphere whilst their polar sections form its surface (Fig. 9.13). Hydrophobic bonding is believed to be mainly responsible for holding the molecules together in the interior of the micelle. As the concentration increases, the spherical micelles are thought to change to a cylinder and then to a liquid crystalline form which is based on a laminar structure.

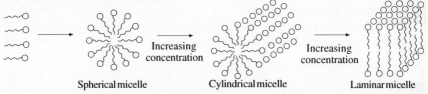

Spherical micelle   Cylindrical micelle   Laminar micelle

Fig. 9.13. Micelle formation and its variation with concentration.

Early in the digestion and absorption process of fats the lipids form micelles which are slowly broken down as the fat is metabolised. These micelles can be seen, by means of a microscope, in the plasma after a meal.

A number of pharmaceutical preparations use micelle formation as a delivery system for drugs with a poor water solubility. The drug occupies the core of the micelle which in effect keeps the drug in suspension in an aqueous medium. For example, the antiseptic disinfectant chloroxylenol is solubilised in water by incorporation in micelles formed from oleic acid and caster oil.

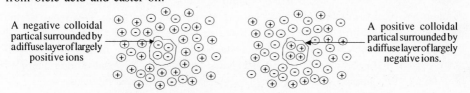

A negative colloidal partical surrounded by a diffuse layer of largely positive ions

A positive colloidal partical surrounded by a diffuse layer of largely negative ions.

Fig. 9.14. The charged double layer structure of colloidal particles.

The particles in a colloid are either all positively charged or all negatively charged. These charges are either due to ionisation of the molecules forming the particle or the adsorption of specific ions on its surface (Fig. 9.14). This layer of fixed charges is balanced, in the dispersed phase by a diffuse layer consisting mainly of ions of the opposite charge. However, the random movement of ions in solution means that this diffuse layer will contain some ions with the same charge as the colloidal particle. The charged particles of a colloid will repel each other and it is this mutual repulsion that stabilises the system: it opposes the attractive forces that would normally hold the particles together.

9.10 SOME PHYSICAL PROPERTIES OF THE COLLOIDAL STATE

9.10.1 The Tyndall effect

When either a narrow beam of light or a beam of laser light is passed through a sol or a gaseous colloid the path of the beam can be seen because the colloidal particles are large enough to scatter the light (Fig. 10.15). This phenomenon is called the Tyndall effect after John Tyndall (1820-1893), its discoverer.

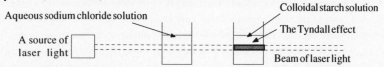

Fig. 9.15. The Tyndall effect. The beam of laser light can be seen as it passes through the colloidal starch. The laser source provides a very narrow beam of light which is easy to see when it has been scattered.

9.10.2 Osmotic pressure

The osmotic pressures of colloids are usually very small. This distinguishes them from true solutions which have comparatively large osmotic pressures.

9.10.3 Electrophoresis

When a potential difference is applied to a colloid all the colloidal particles move towards the electrode with the opposite charge to that of the particles (Fig 10.17). This movement is called electrophoresis. It is used in the form of zone electrophoresis to analyse colloidal biological materials such as proteins in serum (28.4) and identifying nucleic acid fragments (29.5). The technique normally uses an almost inert homogeneous gel as a support for the sample being analysed. The sample is injected into the gel at a specific point. When the potential is applied the component compounds separate because they move at different speeds through the gel (Fig. 9.16a). Positive ions move to the cathode while negative ions move to the anode. The speed of movement of the ions depends on their molecular geometry and size. In general, the larger the molecular mass of the ion the slower the ion moves. Gels usually contain substances that will combine with the components of the mixture to form charged particles of one type. Treatment of the mixture prior to electrophoresis can also result in the formation of charged particles of one type. This means that on electrophoresis all the components of the mixture will usually move in the same direction. The separated components are located and identified by a range of techniques including specific spot tests for known compounds.

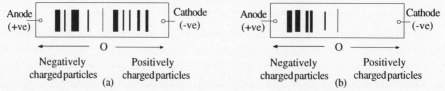

Fig. 9.16. (a) A simulation of an electrophoresis separation. O is the point of injection of the sample. (b) A simulation of the separation of a mixture of proteins using a gel containing sodium dodecyl sulphate.

9.10.4 Precipitation by electrolytes

The addition of small amounts of electrolyte to aqueous colloids will depend on the nature of the colloid (section 9.10). Addition of electrolyte has little affect on the stability of lyophilic colloids (**hydrophilic** colloids). Coagulation will only occur when the electrolyte is added in sufficient concentration. In contrast, only small amounts of electrolyte are needed to precipitate aqueous lyophobic colloids (**hydrophobic** colloids). This precipitation technique is referred to as **salting out**. It is used to isolate many biological substances. The proteins found in blood plasma, for example, may be precipitated by the addition of ammonium sulphate to the plasma. Different concentrations of ammonium sulphate will 'salt-out' different proteins.

9.11 CLASSIFICATION OF COLLOIDS

Colloidal systems are classified for convenience as either **lyophilic** or **lyophobic**. The classification is based on an appraisal of their properties (Table 9.2). Many colloidal systems have properties which lie between those that generally characterise lyophilic and lyophobic colloids.

Table 9.2. Properties characteristic of lyophilic and lyophobic colloids.

Lyophilic colloids	Lyophobic colloids
Concentrations can be high	Dilute solutions only
High viscosity	Low viscosity
Reversible, can be reconstituted after Coagulation	Irreversible, cannot be reconstituted after Coagulation
Poor Tyndall effects	Good Tyndall effects
Stable to the addition of electrolytes	Coagulated by the addition of electrolytes

9.11.1 Lyophilic colloids

These are thermodynamically stable colloids whose stability is due to the strong forces of attraction between the colloidal particles and the dispersing medium. In other words the colloid particles are stabilised by solvation. Lyophilic colloids differ from true solutions only in the size of the solute particles. Many biologically important molecules, such as nucleic acids, proteins and polysaccharides, form lyophilic sols. Gels, such as agar, gelatin and pectin, are lyophilic colloids.

9.11.2 Lyophobic colloids

These colloids have little affinity for the dispersing medium. The particles are charged and it is this charge that stabilises the colloid. The charged particles repel each other thereby maintaining the dispersed state of the colloid. Aerosols, emulsions and foams are common forms of lyophobic colloids.

9.12 DIALYSIS

Solute particles always diffuse from regions of high concentration to those of low concentration. For example, when a drop of concentrated copper sulphate solution is carefully added to a beaker of water the blue colour spreads slowly through the solution. Dialysis utilises the diffusion of solutes from high to low concentration through a semi-permeable membrane to separate small molecules and ions from sols. The pores in the membranes are small enough to retain colloidal particles but large enough to allow the diffusion of small solute molecules and ions. However, they are not as small as those that are found in osmotic membranes (9.8).

A simple dialysis apparatus is shown in Fig. 9.17. The sol is placed in the dialysis tube which is made from a suitable membrane such as Cellophane or Visking. Small molecules and ions diffuse from the region of high concentration in the sol through the membrane into the surrounding solvent where their concentration is lower. The colloid particles are too large to pass through the membrane so they remain inside the dialysis tube. The solvent must be replaced from time to time in order to maintain a good concentration gradient across the membrane otherwise diffusion will stop when the concentration of solute in the solvent equals that in the dialysis tube. Stirring the solvent improves the rate of separation.

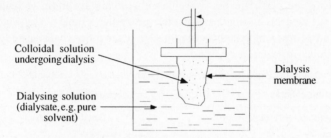

Fig. 9.17. A simple experiment to demonstrate dialysis.

Dialysis is used to stabilise and to purify colloidal solutions. Artificial kidneys (dialysis machines) use the same basic principle (Fig. 9.18). Blood is part colloid, part suspension and part solution. When the kidneys stop working efficiently there is a build up of waste products in the patient's blood. Dialysis, using an artificial kidney, is one way of dealing with this problem. In dialysis machines a dialysis tube is wound on a drum and rotated in a bath through which flows a suitable fluid referred to as the dialysate. As the blood flows through the tube, unwanted small molecules that are present in the blood in too high a concentration diffuse through the membrane into the dialysate. The colloidal and suspended particles such as cells remain in the blood as they cannot pass through the membrane. Diffusion can also take place in the opposite direction substances in which the blood is deficient can be made to diffuse from the dialysate into the patient's bloodstream by increasing their concentration in the dialysate. Careful formulation of the dialysate can also prevent an overall change in the concentration of substances in the blood that are present in their normal concentrations.

Cell membranes are semi-permeable. Dialysis is one of the ways by which ions and molecules are transported across some of these membranes.

9.13 PARTITION

Partition is the distribution of a solute between two immiscible liquids. Experimental evidence shows that at constant temperature a solute x will distribute itself between two immiscible liquids A and B according to the partition law which may be expressed as either: $K = \dfrac{[\,x\,]\text{ in solvent A}}{[\,x\,]\text{ in solvent B}}$ or $K' = \dfrac{[\,x\,]\text{ in solvent B}}{[\,x\,]\text{ in solvent A}}$

where K and K' are constants at constant temperature known as partition coefficients. Since $K = 1/K'$ it is important to quote the relationship of the solvents with numerical values of a partition coefficient. If the solute x associates or dissociates in one of the solvents, the partition law applies to the species that is present in both solvents. The concentrations of the species present in a solvent will be related by the equilibrium law (6.11).

The relative solubilities of a substance in water and lipids is an important factor in the simple diffusion of ions and molecules through a membrane (26.11.3).

9.14 PHARMACEUTICAL WATERS

The water used in pharmaceutical preparations must agree with the specifications laid down for that preparation in the relevant literature. British Pharmacopoeia preparations, for example, must be prepared (Table 9.3) using the type of water specified in the relevant British Pharmacopoeia monograph.

Specially prepared aqueous solutions are often used as the basis of pharmaceutical preparations. Chloroform water, for example, has a sweet taste and is currently used as a preservative in potassium citrate mixture (used to control the symptoms of cystitis) and Benylin (a cough syrup) amongst many other pharmaceutical preparations. However, the use of chloroform water in pharmaceutical products is currently under review as chloroform is known to be carcinogenic.

Table 9.3. Definitions of some of the types of water used in pharmaceutical preparations.

Type	Preparation
Water	This is potable water drawn direct from the public supply
Purified water	Water prepared by distillation or ion exchange or other means from suitable potable water
Distilled water	Purified water of the British Pharmacopoeia prepared by distillation
Water for injection	This is sterilised, distilled water free from pyrogens
Aromatic waters	These are saturated aqueous solutions of volatile oils and other aromatic substances. They are used mainly as flavourings

9.15 WHAT YOU NEED TO KNOW

(1) Several theories for the structure of bulk samples of water have been proposed. The flickering cluster model is the most universally accepted.

(2) *The flickering cluster model postulates that in a bulk sample of water, at any instant in time, regions of the water have an ice-like structure (the cluster) while the rest of the volume of water is filled with a random arrangement of individual water molecules. The clusters are constantly disintegrating and being reformed.*

(3) A solution is a dispersion of solute particles of the order of 0.1 to 1 nm in size in a solvent.

(4) *Non-polar substances are not very soluble in water.*

(5) Non-polar solute particles can dissolve by forming clathrates with the solvent.

(6) Non-polar solute particles can consist of groups of molecules held together by van der Waals' forces and hydrophobic bonding.

(7) The nature of hydrophobic bonding is not understood but it occurs between the non-polar sections of some molecules.

(8) Polar molecules, that is, molecules with a medium to large permanent dipole, are usually more soluble in water than non-polar solutes.

(9) Hydrogen bonding between a substance and water will usually improve the solubility of that substance in water.

(10) Ions form ion dipole bonds with water molecules.

(11) Ionic compounds are usually water soluble, the extent depending on the compound.

(12) The conversion of a compound to an ionic derivative will usually improve water solubility.

(13) Solubility is recorded as:
   Grams per 100g of solution (%w/w).
   Grams per 100cm^3 of solution (%w/v).
   The volume of a solute in a 100 volumes of solvent (%v/v).
   Moles per litre of solution (M or mol dm^{-3}).
   Moles per 1000g of solvent (m or mol kg^{-1}).

(14) The solubility of most solids and liquids increases with temperature.

(15) The solubility of gases decreases with increase in temperature.

(16) The solubility of a gas is proportional to the pressure of the gas.

(17) The solubility product expression for a salt is obtained by applying the law of mass action to a heterogeneous system comprising a saturated solution of the salt in contact with solid salt.

(18) The solubility product of a salt is a measure of the limit of its solubility in water.

(19) Water has a high specific heat, heat of vaporisation, heat of fusion, dielectric constant, viscosity and surface tension.

(20) Surfactants are molecules that reduce the surface tension of water.

(21) The structures of surfactants have polar, water-liking (hydrophilic) and non-polar, water-hating (hydrophobic) regions.

(22) Aqueous solutions of surfactants will dissolve or emulsify solutes that are normally insoluble in water.

(23) Osmosis is the diffusion of a solvent through a semi-permeable membrane from a region of high concentration to one of low concentration.

(24) The osmotic pressure of a solution is the externally applied pressure that would just stop the flow of solvent through the membrane. Osmotic pressure can be measured in kPa (1 Pa = 1 Nm^{-2}) but is more commonly measured in biological systems in terms of the solutions osmolality (mmol kg^{-1}).

(25) Osmotic pressure is a colligative property of a solution. Its value is proportional to the number of discrete particles in solution.

(26) Osmotic semi-permeable membranes are not ideal and will allow the diffusion of some chemical species through the membrane.

(27) The osmotic pressure of a solution is given by the equation: $\pi = icRT$.

(28) For ideal solutions and ideal osmotic semi-permeable membranes $i = 1$.

(29) Solutions that have the same osmotic pressure are said to be iso-osmotic.

(30) Iso-osmotic solutions that are separated by an ideal semi-permeable membrane will contain equal numbers of particles.

(31) Solutions that have equal osmotic pressures but are separated by a non-ideal semi-permeable membrane are said to be isotonic with respect to that membrane.

(32) Isotonic solutions do not exert their full osmotic pressures because some of the solute particles have diffused through the membrane.

(33) The effective concentration of an isotonic solution, that is, the concentration that is responsible for its osmotic pressure, is known as its tonicity.

(34) The tonicity of a solution is recorded as osmolality (mmol kg^{-1}).

(35) Osmolality = im (ideal membrane) = ϕmi (non-ideal membrane).

(36) If the aqueous fluid surrounding a cell has a higher tonicity than the aqueous fluid inside the cell, there will be a net diffusion of water out of the cell, and the cell will crenate

(37) If the aqueous fluid surrounding a cell has a lower tonicity than the aqueous fluid inside the cell, there will be a net diffusion of water into the cell, and the cell will increase in size and could burst (cyctolysis).

(38) Colloids are dispersions of particles of about 1 to 1000 nm in size, dispersed in a solvent.

(39) The particles of a colloid are either all positively or all negatively charged. These charged particles are surrounded by a layer of ions that consists mainly of ions of the opposite charge to that of the colloidal particles.

(40) The structures of colloidal particles range from large single molecules to micelles.

(41) Micelles can be used as a delivery system for drugs with poor water solubility.

(42) Colloids exhibit the Tyndall effect, have low osmotic pressures and may be separated by electrophoresis.

(43) Colloids are generally classified according to their physical properties as either lyophilic or lyophobic colloids.

(44) Lyophilic colloids are stabilised by the strong forces of attraction between the colloidal particles and the dispersing medium.

(45) Many biologically important molecules form lyophilic colloids.

(46) Lyophobic colloids are stabilised by the charges on the particles. They do not exhibit strong forces of attraction for the surrounding medium.

(47) Dialysis is the diffusion of small solute molecules through the pores in a semi-permeable membrane from regions of high concentration to a region of low concentration.

(48) Dialysis can be used to purify colloids and remove unwanted waste compounds from blood.

(49) Dialysis is one of the ways that substances are transported across cell membranes.

(50) At constant temperature a solute will distribute itself between the two immisicible solvents according to the partition law (9.13).

(51) The partition of solutes between water and lipids is an important factor in the transfer of ions and molecules through membranes by simple diffusion.

(52) The water used in pharmaceutical preparations must agree with the specifications laid down in the appropriate literature.

9.16 QUESTIONS

(1) Outline the flickering cluster model for the structures of bulk samples of water. What features in the structure of a solute molecule will tend to improve water solubility and what will reduce water solubility?

(2) Predict the water solubility of each of the following molecules as excellent, good, poor and almost insoluble. Give a reason for your prediction.

(a)

OH
|
$CH_3CHCOOH$

(b)

OH

(c)

H_2N——NHCOCH$_3$

(3) Calculate the volume of 36% w/w hydrochloric acid that would be required to prepare 250 cm^{-3} of a 2 M solution of the acid. The density of the 36% w/w hydrochloric acid solution is 1.2 g cm^{-3}.

(4) Suggest a way of improving the water solubility of (a) amphetamine $(PhCH_2CH(NH_2)CH_3$ and (b) aspirin (structure, p. 34) that would not change their pharmaceutical activity.

(5)

A   HO——
                    OH
                    |
                    —NH—CH$_3$
         HO

Compound A is the central nervous system stimulant adrenalin. Predict, giving reasons for your predictions, the relative solubilities of A in (a) water, (b) an aqueous solution with a pH below 7, and (c) an aqueous solution with a pH above 7. Give your answer in general terms, not specific solubilities.

(6) The partial pressure of oxygen in a gaseous mixture being used as a general anaesthetic is 16.625 kPa. If at this pressure the concentration in the patient's blood is 0.275 g dm^{-3}, what would be the partial pressure of the oxygen in the gaseous mixture if the concentration of oxygen in the patients blood needed to be increased to 0.300 g dm^{-3}? Assume that the general anaesthetic obeys Henry's law.

(7) What are the characteristic features of the structure of a surfactant molecule? Outline the general mode of action of a surfactant.

(8) Explain the essential difference between osmosis and dialysis.

(9) Define the meaning of the term osmolality. Calculate the osmolality of a solution that contains 0.1 mol of sodium carbonate, 0.1 mol of urea and 0.2 mol of glucose per kilogram of water assuming ideal conditions.

(10) Explain the meaning of the terms isotonic and tonicity.

(11) Explain what is meant by the terms: (a) colloidal system, (b) micelle and (c) lyophobic colloid.

10

Spectroscopy

10.1 INTRODUCTION

Spectroscopy is widely used in structure determination and the qualitative and quantitative analysis of substances. All forms of spectroscopy involve the interaction of a sample with a form of energy. It is the way in which this energy is assimilated that gives information concerning the identity and quantity of the chemical species present in the sample. This chapter discusses only the most widely used forms of spectroscopy, namely, ultra-violet, visible, fluorescence, infra-red, atomic absorption and nuclear magnetic resonance spectroscopy. It also discusses mass spectrometry which is not a form of spectroscopy but is usually treated as such.

10.2 ULTRA-VIOLET AND VISIBLE SPECTRA

Ultra-violet (UV) and visible radiation are believed to cause an increase in the energy of some of the bonding and lone pairs of electrons in a molecule with their resultant promotion to either an empty non-bonding or antibonding molecular orbital. Useful ultra-violet and visible absorption spectra are produced by the absorption of electromagnetic radiation with wavelengths lying in the 200 to 400 nm (UV) and 400 to 800 nm (Visible) regions of the electromagnetic spectrum. Below 200 nm ultra-violet spectra are difficult to measure and relatively uninformative. The energy of the radiation absorbed by the molecule is often enough to promote chemical reactions. Consequently, samples should not be left in a spectrometer for any length of time.

10.2.1 Spectra
Spectra are normally measured in solution. Most of the common non-aromatic solvents can be used provided they do not absorb at a wavelength that corresponds to those in the

absorption spectrum of the substance under test. One of the most useful solvents is 95% ethanol in which measurements can be taken down to 210 nm. Absolute ethanol is not suitable since it contains some residual benzene which is used in its production. Other solvents that are employed are listed in Table 10.1.

Table 10.1. Solvents commonly used in ultra-violet and visible spectroscopy. The wavelength quoted is the minimum value for accurate measurements using a 1 cm cell.

Solvent	Wavelength (nm)	Solvent	Wavelength (nm)
Methyl nitrile (Methyl cyanide)	190	Ethanol	204
Water	191	Diethyl ether	215
Cyclohexane	195	Dichloromethane	220
Hexane	201	Trichloromethane (Chloroform)	237
Methanol	203	Tetrachloromethane (Carbon tetrachloride)	257

The intensity of the absorption depends on the intensity of the incident radiation (I_o), the concentration of the sample (c) and the distance (l cm) the radiation passes through the solution of the sample. This distance is known as the path length. The relationship between these variables is expressed by the Beer-Lambert law (equation 10.1) where I is the intensity of the transmitted radiation and k is known as the absorption coefficient. It should be noted that there are many instances where this law is not obeyed.

$$\text{Absorbance } (A) = \log_{10} \frac{I_o}{I} = kcl \tag{10.1}$$

The extinction coefficient k is a constant at a specified wavelength and is characteristic of the substance being examined. It is recorded as the molar absorption coefficient (formerly extinction coefficient) when the concentration is measured in mol dm^{-3}. The molar absorption coefficient (e) is the absorbance, at a specified wavelength, of a 1 M solution in a 1 cm cell. This value is determined by calculation using the Beer-Lambert law. The units for the molar absorption are seldom quoted in the literature but are 1000 cm^2 mol^{-1}. The wavelength and temperature at which the spectrum was recorded, together with solvent used for the determination must be quoted with the molar absorption coefficient as its value can vary with these parameters.

The British Pharmacopoeia also records extinction coefficients as specific absorbance values A (1%, 1 cm), that is, the absorbance of a 1% solution in a 1 cm cell. Specific absorbance is the value of k when the concentration is measured as a % w/v (g/100 cm^3).

Spectra are recorded for general use as plots of either, absorbance (formerly optical density), percentage transmittance, molar absorption coefficient or log molar absorption extinction against wavelength which is normally measured in nanometres (Fig. 10.1). Transmittance (T) is defined as:

$$T = I/I_o$$

where, $A = \log_{10}(1/T)$. For convenience, spectra are often quoted simply as the molar absorption coefficient(s) for the wavelength(s) corresponding to the maximum absorption(s) (λ_{max}).

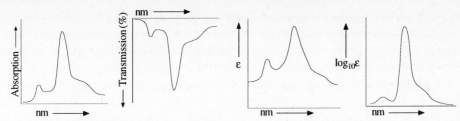

Fig. 10.1. Examples of the different ways of recording ultra-violet and visible spectra.

Structures that are capable of absorbing ultra-violet and visible radiation are known as **chromophores**. Both saturated and unsaturated structures can absorb ultra-violet radiation. Saturated structures tend to absorb at lower and therefore less useful wavelengths than many unsaturated structures. The exact wavelength at which absorption occurs will depend on the nature of the structure and its molecular environment. For example, the C=C of ethene has a broad absorption band with a maximum at 160 nm. Conjugation of this bond with other C=C bonds increases the wavelength of the maximum absorption until it reaches the visible region at 400 nm (Table 10.2). Similar variations are observed with substituted polyethenes.

Table 10.2. The effect of increasing conjugation on the maximum wavelength of absorption.

	H-(CH=CH)$_n$-H (trans) l_{max}	CH$_3$-(CH=CH)$_n$-CH$_3$(trans) l_{max}	Ph-(CH=CH)$_n$-Ph (trans) l_{max}
n=1	160	190	
n=3	268	275	358
n=6	364	380	420
n=7		401	435
n=8	410	411	

The introduction of other structures (groups) into a molecule can either increase or decrease the **maximum wavelength of absorption**. For example, the introduction of NH_2 and OH groups to an aromatic system can increase the conjugation (8.3) and so increase the wavelength of maximum absorption. On the other hand, the protonation of an amino group can reduce conjugation and hence the maximum wavelength of absorption by forming the amine salt. Structures that increase the wavelength of absorption, that is cause a **bathochromic or red shift** are called **auxochromes**. Reduction in the wavelength of absorption is known as a **hypsochromic or blue shift**. In both cases the introduction of the substituent structure changes the original chromophore to a new chromophore.

The introduction of a group into a structure can also change the **intensity of the absorption**. Decreases in absorption intensity are referred to as **hypochromic** effects, whilst increases in intensity are known as **hyperchromic** effects.

The main use of ultra-violet and visible spectroscopy is in quantitative analysis. Many plasma constituents, drugs and other substances are assayed by methods that are based on the measurement of the absorption of a solution of the substance at a specified wavelength in the ultra-violet and visible regions. The calculation of the concentration is made either by substituting the appropriate quantities in the equation for the Beer-Lambert law (equation 10.1) if the compound obeys the law or by plotting a calibration curve of the concentrations of solutions of known strength against their absorbance at a specified wavelength and reading the concentration of the unknown solution from the graph after measuring its absorbance (Fig. 10.2). The latter can be carried out automatically by the spectrometer, which can also include an internal standardisation routine. Even if the compound does not strictly obey the Beer-Lambert law the concentration of a solution can still be obtained using a calibration curve provided a sufficient number of points are plotted.

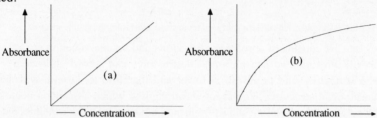

Fig. 10.2. Typical calibration curves. The Beer-Lambert law is obeyed in (a) but not in (b).

Ultra-violet and visible spectra are also used to identify chromophores in qualitative analysis. Identification is carried out by comparing the spectrum of the unknown compound to those of known chromophores by consulting suitable source books, such as *Organic Electronic Spectral Data* (published by Wiley). If, within experimental error, the spectrum of the unknown compound matches that of a chromophore in the source book, it is taken as evidence that the chromophore is found in the structure of the unknown compound. The process is similar to that used to identify a person from their fingerprints. The procedure follows no set rules and it is largely a matter of experience. In some cases a number of empirical rules have been devised which enable calculation of a theoretical value for the λ_{max} for the absorption of a chromophore based on a particular type of structure. For example, a set of rules for calculating the λ_{max} values for the absorptions of substituted dienes and trienes based on work by Woodward and Fieser has been published by Scott in *Interpretation of the Ultra-violet Spectra of Natural products,* (published by Pergamon Press, Oxford). Again identification is made by matching this calculated value to the observed value for λ_{max}. Empirical rules of this type are of limited use as they do not take into account any special factors that may operate within the molecule.

A more accurate and useful method of identifying chromophores is to use the first derivative or higher order derivative spectrum. Experimentally, absorbance is related to the wavelength of the absorption by the zero order derivative equation:

$$A = f(\lambda) \tag{10.2}$$

where f is a function.

The first, second, etc. order derivatives are obtained by the appropriate differentiation of equation 10.2. For example, the first and second order derivative expressions would be:

$$dA/d\lambda = f'(\lambda) \text{ and } dA^2/d\lambda^2 = f''(\lambda) \text{ respectively.}$$

The first, second and third order spectra of a Gaussian peak are shown in Fig. 10.3.

Derivative spectrometry magnifies the fine structure of the spectral curves which results in more accurate comparisons and identifications. The spectrometer automatically calculates the first or higher order derivatives of the absorbance of the spectrum and plots these against the wavelength of the absorption. The first derivative often reveals

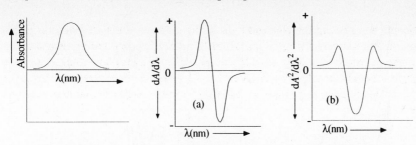

Fig. 10.3. The (a) first and (b) second order derivative spectra of a Gaussian peak.

hidden peaks (Fig. 10.4) since at λ_{max} for each peak, $dA/d\lambda = 0$. Second and other order derivative spectra are more complex and because of this are potentially more useful in identifying a chromophore by direct comparison of the spectrum of the unknown compound with the spectra of known compounds. This technique is known as **fingerprinting**.

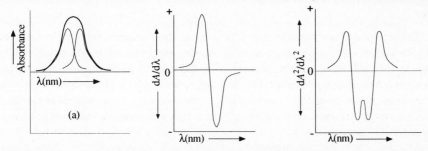

Fig. 10.4. The fine structure of an absorption band shown by plotting the first and second derivative spectra. The spectrum shown in (a) is the resultant plot for two absorption bands. This and its first order derivative spectrum are not significantly different from the corresponding plots for one absorption band shown in Fig. 10.3. However, its second derivative spectrum is significantly different.

The use of derivative spectroscopy in qualitative analysis has widespread application in clinical, forensic and the biomedical fields, for example the detection of xenobiotics (foreign compounds) in tissue samples (Fig. 10.5). However, it should also be realised that it is not usually possible to identify a chromophore from only its ultra-violet or visible spectrum. Information from other sources, such as chemical tests and other forms of spectroscopy, is normally required.

Ultra-violet radiation in the 240 to 280 nm wavelength region is used as a bactericide. It is used to sterilise air and thin layers of water but because of poor penetrative power it is not used to sterilise medical equipment or pharmaceutical preparations.

10.3 FLUORESCENCE AND PHOSPHORESCENCE

The absorption of ultra-violet or visible radiation by a molecule results in that molecule gaining energy. Fluorescence and phosphorescence are the two routes by which *some* molecules lose this energy by emitting radiation in the visible region (Fig. 10.6).

Fluorescence **(fluorimetry)** and phosphorescence **(phosphorimetry)** are used mainly for quantitative work However, phosphorimetry has few applications in the biomedical field and so this section is primarily concerned with fluorimetry.

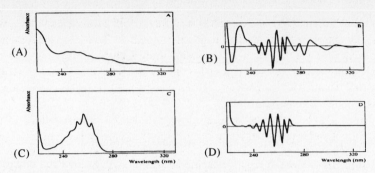

Fig. 10.5. The detection of amphetamine in homogenised liver extract. The presence of amphetamine cannot be proved by comparing the zero order spectra of the homogenised liver extract (A) with that of amphetamine (C). Comparison of their first order derivative spectra (B and D) however, indicates the presence of amphetamine in the homogenised liver. (R Gill *et al., J. Foren.Sci.Soc.,* 1982; 22(2): 165-171.)

Compounds that exhibit fluorescence and phosphorescence have two characteristic spectra: the absorption spectrum that results from the excitation of the electrons of a molecule and the emission spectrum that results when the molecule exhibits either fluorescence or phosphorescence. The emission spectrum of a compound is similar to its absorption spectrum except that it occurs at a longer wavelength. The wavelength at which the maximum emission occurs is known as λ_{em}. The absorption wavelength that produces λ_{em}, is known as the excitation wavelength λ_{exc}. The wavelength at which λ_{em} occurs is independent of λ_{exc}.

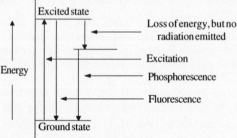

Fig. 10.6. The electronic transitions involved in fluorescence and phosphorescence

The intensity of the fluorescent radiation is related to the concentration of the compound exhibiting the fluorescence. It can be shown that in dilute solutions, the intensity of the fluorescent radiation (F) of a compound is related to the concentration of that compound (c) by the expression:

$$F = 2.31\, I_0\, \phi\, \varepsilon\, l\, c \qquad (10.3)$$

where I_0 is the intensity of the exciting radiation, ϕ is the quantum yield of the fluorescence process, ε is the molar absorption coefficient of the compound and l is the path length of the radiation through the sample. The quantum yield (ϕ) is defined as the ratio of emitted quanta to absorbed quanta and can have values between 0 and 1. It is a measure of the efficiency of the fluorescence process of the compound and as such is a characteristic property of the compound. However, the value of ϕ and hence that of F does vary with temperature, pH and solvent. By using excitation radiation with a wavelength of λ_{exc} which will give a maximum value for ε, and adjusting the values of I_o, and l to give maximum fluorescence it is possible to use high resolution fluorimeters to detect and measure concentrations as low a 10^{-9} g cm^{-1} (1ng cm^{-1}). It should be noted that it is not possible to use equation 10.3 to calculate the concentration of a compound since this expression does not take into account machine error.

Fluorimetry is used in a number of ways to determine the concentrations of compounds in solution. In all cases the efficiency of the fluorimeter must be taken into account when determining the concentration of the compound. The method most commonly used is to plot a calibration curve of the intensity of the fluorescence (F) at a specified wavelength (usually λ_{em}) for solutions of known strength against their concentrations (c) for an excitation radiation of λ_{exc}. The concentration of the unknown solution is read from the graph after measuring the intensity of its fluorescence with the same fluorimeter and settings that were used for the calibration curve measurements (Fig. 10.7a.). This automatically allows for instrument error.

The calibration curves are usually linear at low concentrations but show deviations at higher concentrations (Fig. 10.7b). These deviations are due to **quenching** effects. Quenching is the general term used to describe a diverse group of processes by which the energy level of the molecule is modified so that the molecule is no longer able to return to its ground state by emitting energy as fluorescence. For example, quenching can occur by energy transfer due to collisions between molecules of the same compound, the compound and the molecules of different solutes, or the compound and the solvent. Quenching may also be brought about by the addition of chemical species, such as alkali metal halides, to the compound.

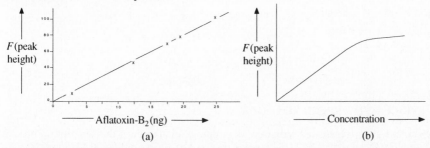

(a)   (b)

Fig. 10.7. (a) The fluorescence intensity (F)/concentration calibration curve for Aflatoxin-B2, a toxic fungal metabolite produced by some fungi that grow on crops such as cereals and peanuts. (Courtesy of Dr Olive Roch, University of Portsmouth.) (b) A calibration curve with a negative deviation.

It is not possible to predict whether a compound will exhibit fluorescence. However, compounds are likely to exhibit fluorescence if their structure contains a rigid planar conjugated double bond system. In general, the larger the conjugated system the greater the fluorescence. The intensity of the fluorescence of a conjugated system is either increased or decreased by the presence of substituents and their position. In general, electron donors, such as methoxy, hydroxy and amino, tend to improve fluorescence, whilst electron-withdrawing groups, such as nitro, cyano and bromo, reduce fluorescence. Aromatic heterocyclic systems which have electron donor hetero atoms, such as nitrogen, oxygen and sulphur, usually exhibit fluorescence. However, it should be noted that the pH of the solution will affect the possibility and intensity of the fluorescence especially if the molecule contains a polarisable group. For example, conversion of a phenol to the corresponding phenolate increases the possibility and intensity of fluorescence but conversion of an aromatic amine to its salt reduces the possibility and intensity of fluorescence.

Compounds that do not exhibit fluorescence may still be assayed using fluorimetry by quantitatively converting them into compounds that exhibit fluorescence. For example, the diphenylhydantoin content of plasma can be determined by oxidation of the diphenylhydantoin *in situ* to benzophenone which exhibits fluorescence. The benzophenone is isolated from the reaction mixture and the fluorimetric measurements are made in

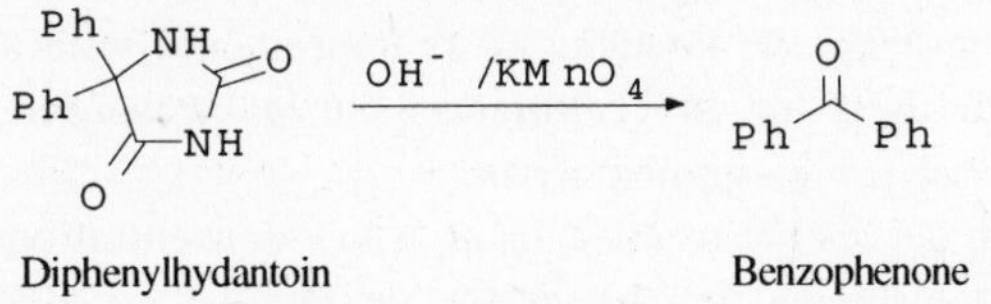

Diphenylhydantoin   Benzophenone

sulphuric acid at λ_{exc} 355 nm and λ_{em} 485 nm. Some organic compounds and inorganic ions may be assayed fluorometrically by either reaction or chelation with a compound that exhibits fluorescence. For example, dansyl chloride is used for primary amines and benzoin has been used in the assay of Zn^{2+}. This method of assay suffers from the disadvantage that it is usually necessary to remove any excess reagent before fluorimetric assay can be attempted.

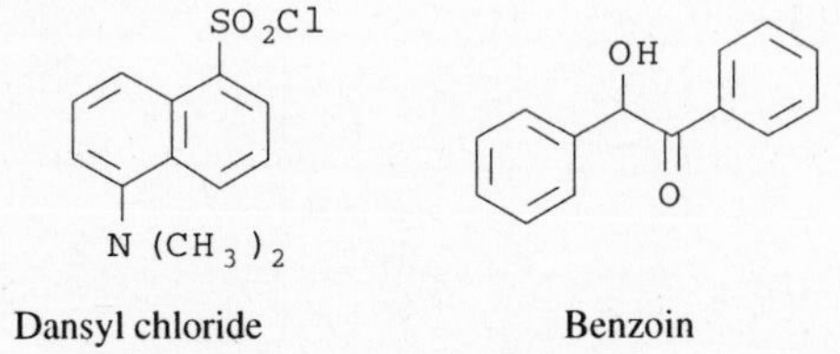

Dansyl chloride   Benzoin

10.4 INFRA-RED SPECTROSCOPY

All substances absorb infra-red radiation. A simple explanation of this absorption process can be obtained by picturing the molecules forming the substance as solid balls representing the atoms, linked by springs representing the bonds. This structure will be in a state of perpetual wobbly motion, this movement taking the form of the bonds stretching, contracting, bending, twisting, etc. (Fig 10.8). The energy of the absorbed radiation is dissipated within the molecule by increasing the intensity of this molecular movement. It is possible to relate the wavelength of the absorption to a specific bond.

10.4.1 Spectrum

The infra-red spectrum of a compound is usually presented as a plot of transmittance against wavenumber, the reciprocal of wavelength. Absorptions are recorded as downward facing peaks (Fig. 10.9). The spectrum is normally taken using either a dilute solution of the compound in a suitable non-aqueous solvent or a solid solution in potassium bromide

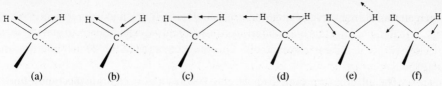

(a)	(b)	(c)	(d)	(e)	(f)

Fig 10.8. The vibrational modes of the methylene group. The arrows show the relative directions of the vibrations of the bonds. (a) Symmetrical stretching, the bonds move in and out towards the carbon in unison. (b) Asymmetrical stretching, the bonds move in the same direction at the same time. (c) Bending, the bonds move in the plane of the paper towards and away at the same time. (d) Rocking, the bonds move in the same direction in the plane of the paper at the same time. (e) Twisting, the bonds move in opposite directions out of the plane of the paper at the same time. (f) Wagging, the bonds move in the same direction out of the plane of the paper at the same time.

or a nujol mull. Spectra may also be taken in the form of liquid films and the vapour state. These different sampling techniques can affect the appearance of the spectrum of a particular compound (Fig. 10.9). Spectra taken in non-polar solvents, such as tetrachloromethane and alcohol free trichloromethane are preferred since fewer intermolecular forces are found in these solutions. As a result, the resolution of the spectrum will be better, that is, it will have sharper and better defined peaks. Intermolecular forces such as hydrogen bonding, tends to broaden the absorption peaks and in some cases the result is a broad absorption band rather than a narrow peak in the spectrum.

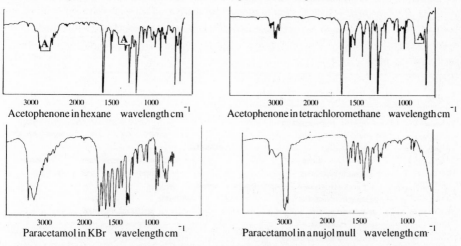

Fig. 10.9. The infra-red spectra of acetophenone and paracetamol recorded using different solvents. Note the differences between the spectra of each compound and also the 'dead pan peaks' (A) where the interaction of the absorptions of the solvent and compound results in no signal reaching the recorder's pen. Note also the peaks due to the nujol at 2850-2950 (strong), 1460 and 1370 cm^{-1} (weak).

Many functional groups absorb at characteristic wavelengths in the infra-red region of the electromagnetic spectrum. The positions of these absorptions show little variation with change in molecular environment and so can be used in reverse to identify the presence of a functional group in a molecule. The initial interpretation of a spectrum is made using correlation tables of the form shown in Table 10.3. These tables are of a general nature and are not likely to have been compiled under the same conditions that an investigator would use to run a spectrum in the laboratory. A functional group may give rise to a peak in the spectrum that is at a significantly different wavenumber from that recorded for that structure in the correlation table. This must be borne in mind when interpreting spectra.

It is not feasible to interpret all the peaks in a spectrum from tables of this type but with practice it is possible to pick out the key ones and relate them to functional groups within the molecule. Carbonyl groups, for example, exhibit strong absorptions in the 1600 to 1780 cm^{-1} region. If a compound has a strong absorption in this region its structure could contain a carbonyl group. Further examination of the spectrum may enable one to speculate further on the exact nature of this carbonyl group. Aldehydes, for example, have a C–H stretching absorption at about 2700 to 2900 cm^{-1}; ketones and esters do not absorb in this region whilst acids and amides have broad O–H stretching absorption bands in the 2700 to 3600 cm^{-1} region. Many functional groups can be detected by this method but deductions of this nature should be backed up by other evidence such as chemical tests and other forms of spectroscopy.

A more detailed interpretation of a spectrum can be obtained by consulting specialised tables of absorptions for the particular type of structure being studied.

Table 10.3. An example of an infra-red correlation chart. The letters s, m and w indicate that the absorptions are usually either strong, medium or weak. Groups of letters indicate that the peaks may have different strengths. Note m' refers to peaks due to conjugated C=C bonds only.

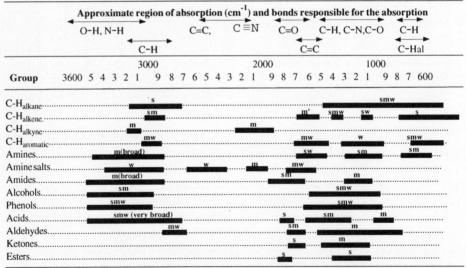

In practice it is difficult to correlate absorptions with functional group structures in the region below 1500 cm^{-1}. This area, because of its complexity, is known as the **fingerprint region**. It is particularly useful for identifying an unknown compound by comparing its spectrum with the spectra of known compounds recorded under the same conditions. If the spectrum of the unknown matches that of a known compound it is probably the same compound. This procedure is commonly referred to as *fingerprinting*. Deductions made by infra-red fingerprinting should be supported by additional evidence.

10.4.2 Uses

The main use of infra-red spectroscopy is in qualitative analysis. Fingerprinting is used extensively as an identification test in the British Pharmacopoeia and similar publications. It should be appreciated that, when used for this purpose, an identical match is seldom possible, and most books of this nature will state that only a reasonable agreement as far as the wavelengths and relative intensities of absorption are concerned is required for a positive result.

Infra-red spectroscopy is used to follow reactions by observing the disappearance and appearance of relevant peaks, for example, the synthesis of ethyl benzoate from benzonitrile (Fig. 10.10). However, samples must be isolated from the reaction for accurate analysis.

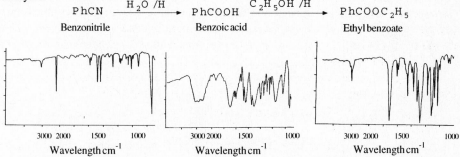

$$PhCN \xrightarrow{H_2O/H} PhCOOH \xrightarrow{C_2H_5OH/H} PhCOOC_2H_5$$

Benzonitrile   Benzoic acid   Ethyl benzoate

Fig. 10.10. Following the course of the synthesis of ethyl benzoate by infra-red spectroscopy. In the first step the sharp peak at 2228 cm^{-1} due to the nitrile disappears and a broad peak due to the absorption of the O-H bond of the carboxylic acid appears. In step two the broad carboxylic acid peak is replaced by a narrow peak due to aromatic C-H bond absorption whilst the peak at 1682 cm^{-1} due to the carbonyl group of the carboxylic acid is replaced by a narrow peak at 1718 cm^{-1}, due to the carbonyl group of the ester. It should be noted that although the spectra in this figure are run in different solvent systems it is still possible to use them to follow the synthesis. However, due allowances must be made if different solvents are used when following a synthesis.

Infra-red spectroscopy is seldom used for quantitative analysis as it is less accurate than other analytical methods. However, it does have one advantage over ultra-violet and visible spectroscopy. The large number of well defined absorption peaks means that it is possible to assay the individual components of a mixture provided a peak that is due only to the substance being assayed (analyte) can be found. For example, automated quantitative infra-red spectroscopy has been used to determine the concentrations of contaminants, such as carbon monoxide, chloroform and methanol in air. However, it can be difficult to locate peaks due solely to the analytes in an infra-red of a mixture.

10.5 NUCLEAR MAGNETIC RESONANCE SPECTROSCOPY

Nuclear magnetic resonance spectroscopy (n.m.r.) is concerned with the absorption of electromagnetic radiation in the radio frequency range by the nuclei of some isotopes. These isotopes behave as though they are spinning charged particles and generate a magnetic field along the axis about which they are spinning. As a result, these nuclei can be pictured as being like tiny spinning bar magnets which, in the absence of a strong external magnetic field, are randomly orientated in space. It should be noted that the nucleus is pictured as having a precessional movement, that is, the bar magnet (nucleus) spins in such a way that its ends trace circles in space (Fig. 10.11). When a strong external magnetic field is applied to these spinning magnets, the magnetic fields interact and the spinning magnets assume different orientations in the external field. Certain nuclei, notably ^{1}H and ^{13}C, can only take up two orientations in this field, a low energy parallel orientation in which the magnetic field of the nucleus is aligned in the same direction as that of the external field and an antiparallel high energy orientation where the nuclear magnetic field is opposite to that of the external field. Under normal conditions there are slightly more nuclei in the lower energy parallel orientation than the higher energy antiparallel orientation.

A number of other isotopes, such as ^{14}N, ^{19}F and ^{31}P also exhibit nuclear magnetic resonance in a strong magnetic field. ^{31}P Nuclear magnetic resonance spectroscopy is the basis of the CAT scanners used as a diagnostic tool in hospitals.

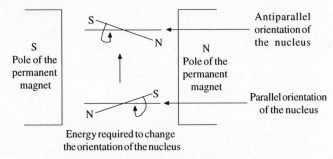

Fig.10.11. The orientation of nuclei in a magnetic field.

Absorption of electromagnetic radiation in the radiofrequency region by ^{1}H, ^{13}C and other suitable nuclei when placed in a strong magnetic field can cause the nuclei in the lower energy parallel state to spin-flip to the higher energy antiparallel state. When this occurs the nucleus is said to be in **resonance** and the absorbed radiation is commonly referred to as a **signal**. Nuclei do not stay in the higher energy state but dissipate their energy through so-called **relaxation processes**, the exact nature of which is not under-stood. If relaxation did not occur, all the nuclei in a sample would eventually be promoted to the higher energy state and there would be none left to absorb the radiofrequency energy (Fig. 10.12). In other words, no absorption signal would be observed. However, this does not happen unless the sample is irradiated with radiation of such high intensity that all the nuclei are forced to remain in the higher energy state. In this situation the sample is said to be **saturated**. Nuclear magnetic resonance spectrometers have to be adjusted so that the intensity of the radiofrequency radiation being used does not cause this to happen.

Resonance can be brought about in two ways: either the external field can be kept constant and the radiofrequency varied, or the radiofrequency is kept constant and the external magnetic field varied. The former is known as a frequency sweep whilst the latter is known as a field sweep. In practice most instruments use a field sweep as it is easier to achieve a homogeneous magnetic field.

The amount of energy needed to cause nuclei to resonate depends on the external magnetic field, the isotope and its molecular environment. For example, with an external

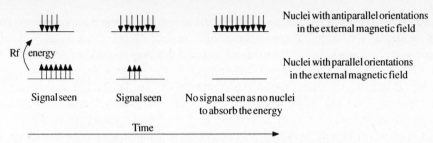

Nuclei with antiparallel orientations in the external magnetic field

Nuclei with parallel orientations in the external magnetic field

Signal seen   Signal seen   No signal seen as no nuclei to absorb the energy

Time

Fig. 10.12. A representation of saturation in nuclear magnet resonance.

magnetic field strength of 14,000 gauss, a radiofrequency of the order of 60 megahertz (MHz) is required to cause protons (^1H nuclei) to resonate whilst a radiofrequency of the order of 25.14 MHz is necessary to bring ^{13}C nuclei into resonance. Other field strengths can be used, some instruments use field strengths up to 140,000 gauss. These machines are very sensitive but require very high radiofrequencies to cause resonance. For example, a machine with a field strength of 234,900 gauss uses a radiofrequency of the order of 100 MHz to bring ^1H nuclei to resonance.

Since all the nuclei of an isotope are identical, one would expect all the ^1H nuclei to absorb at the same frequency. Similarly, all the ^{13}C nuclei would be expected to absorb at the same frequency but a different one from the ^1H nuclei. However, this is not the case because the actual frequency of the absorption will depend on the electrons in the structure which are also influenced by the external magnetic field (H_o). This field is believed to cause the electrons to circulate the nucleus in a plane perpendicular to H_o (Fig. 10.13). This produces a small magnetic field in the opposite direction to H_o which reduces the effect of the external magnetic field experienced by the nucleus. The nucleus is said to be **shielded**. As a result, the magnetic field H_E actually experienced by a particular nucleus in the compound will be the resultant of the external magnetic field and the small opposing local fields due to the electron clouds of its neighbouring nuclei. As each nucleus in a molecule is in a slightly different electron environment it will experience a slightly different degree of shielding relative to the other nuclei in the molecule. Therefore, each nucleus will require a slightly different radiofrequency to cause it to resonate.

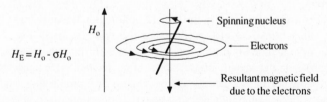

H_o

$$H_E = H_o - \sigma H_o$$

Spinning nucleus

Electrons

Resultant magnetic field due to the electrons

Fig. 10.13. The magnetic field of the electrons reduces the strength of the applied magnetic field H_E acting on the nucleus.

The resonance frequencies of ^1H and ^{13}C atoms are measured relative to a reference point, usually tetramethylsilane (TMS). Tetramethylsilane is used as an internal standard for both ^1H and ^{13}C spectroscopy because it has a single sharp proton absorption signal that occurs above the signals of most organic molecules. For convenience, spectra are recorded as plots of signal intensity against chemical shift where the chemical shift of

TMS is arbitrarily set at zero. Chemical shift (δ) is defined by equation 10. 4. It has no units and is normally recorded as parts per million (ppm). Chemical shifts in ^{13}C n.m.r. spectroscopy are much larger than those found in ^1H n.m.r. spectroscopy.

$$\delta = \frac{\text{Difference between the frequency of the signal and that of TMS (Hz)}}{\text{Spectrometer operating frequency (MHz)}} \qquad (10.4)$$

The chemical shifts of a spectrum recorded on a 60 MHz spectrometer will have the same values when the same spectrum is recorded on a 100 MHz machine even though the absorption occurs at a different radiofrequency for each instrument. Absorption signals that occur to the left of the TMS signal on the spectrum are referred to as being downfield of TMS whilst those that occur to the right are referred to as being upfield.

Chemical shift values are not only affected by the nature of the structure but are also influenced by hydrogen bonding, temperature and the solvent. These factors are discussed as they arise in the text.

10.6 PROTON MAGNETIC RESONANCE SPECTROSCOPY

Proton magnetic resonance (^1H n.m.r. or p.m.r.) spectra are obtained using a solution of the sample containing a little TMS, in a solvent that does not absorb in the radiofrequency region being studied. Tetrachloromethane (CCl_4) and deuterated solvents, such as deuterated trichloromethane ($CDCl_3$), methanol (CD_3OD) and propanone (CD_3COCD_3) are commonly used. These deuterated solvents all contain a small amount of the corresponding protonated compound because of incomplete deuteration during their manufacture. Deuterotrichloromethane ($CDCl_3$), for example, will contain a little trichloromethane ($CHCl_3$). Since p.m.r. spectra are additive, the signals of these impurities will occur in the spectrum and it is important that they are recognised and taken into account when interpreting a spectrum. The nature of the solvent will also affect the value of the chemical shift of a signal. Changing from tetrachloromethane to deutrotrichloromethane has little effect on the chemical shift, but changing to more polar solvents can cause a significant change in its value. These changes are sometimes used to help identify signals.

10.6.1 Spectra

Most proton chemical shifts occur in the 0 to 12 ppm region. Identification is made in the first instance using broad correlation charts of the type shown in Fig. 10.14. This is followed by a more detailed interpretation using tables for the type of structure believed to be responsible for the signal. These tables often include an empirical formula that enables the investigator to estimate the chemical shift of a proton in a model compound and compare it with the value obtained in practice. If the comparison gives a good match it is taken as evidence that the signal on the spectrum is due to a similar type of structure.

Protons that are in the same chemical and therefore the same magnetic environment will have identical chemical shifts. They are known as equivalent protons. For example, the three protons of the methyl group of bromoethane will be equivalent and have the same chemical shift (Fig. 10.15). Similarly, the two protons of the methylene group will also be equivalent because there is unrestricted free rotation about the C−C bond. However, it is not safe to assume that protons attached to the same atom are always equivalent.

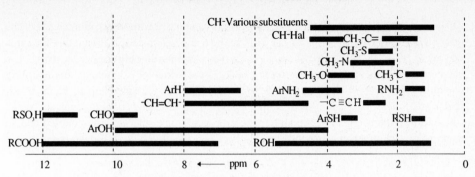

Fig. 10.14. A p.m.r. general correlation chart.

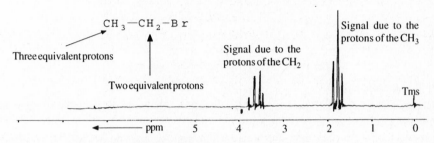

Fig. 10.15. Equivalent protons in bromoethane.

10.6.2 Spin-spin coupling

The signals seen on a p.m.r. spectrum vary from a single peak to groups of peaks. The division of a signal into a group of peaks occurs because the magnetic fields of adjacent protons influence the magnetic field strength at which a proton comes into resonance. Consider, for example, the p.m.r. spectrum of a sample of dichloroethanal. For convenience, the protons of this compound will be referred to as H_A and H_B respectively.

Dichloroethanal

The strength of the magnetic field (H_R) required to bring the proton H_A to resonance in a spectrometer is given by equation 10.5 where (H_o) is the strength of the external magnetic field and (H_l) is the strength of the local magnetic fields of adjacent nuclei.

$$H_R = H_o - H_l \qquad\qquad (10.5)$$

The only other nuclei in dichloroethanal that have local magnetic fields that will affect the H_A protons is the proton H_B. This proton will have one of two local magnetic fields (H_x and H_y) depending whether they are in a high energy or low energy orientation (Fig 10.12). Consequently, there will be two values for H_o, (H_o^1 and H_o^2) where H_A protons can be brought into resonance, namely:

$$H_R = H_o^1 - H_x$$

and

$$H_R = H_o^2 - H_y$$

As a result, the spectrum shows a pair of peaks (a doublet) for the absorption signal of the proton H_A. Similarly, the signal of H_B will also be a doublet since it is affected in the same way by H_A (Fig. 10.16). This behaviour is known as **spin-spin coupling** or **splitting**. The value of the chemical shift separating the peaks in each doublet is known as the **coupling constant J**. It is a constant characteristic of protons that are spin-spin coupled. In the p.m.r. spectrum of dichloroethanal, for example, H_A will have the same coupling constant as H_B. This enables one to pick out the signals of hydrogen atoms that are spin-coupled and so adjacent to each other in a structure, which is of considerable help in the interpretation of a spectrum.

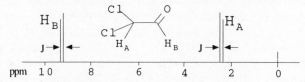

Fig. 10.16. The p.m.r. spectrum of dichloroethanal.

It should be noted that;

(1) Chemically equivalent protons do not couple with each other even if they are bonded to different carbon atoms.

(2) Protons that are further than two single 'bond lengths' apart do not usually couple.

(3) Protons that are spin-coupled with each other have the same J values.

More complex splitting patterns are observed when more than two protons are involved in the coupling. In theory, the number of peaks occurring in a signal will be $n + 1$, where n is the number of equivalent protons whilst their relative intensities are predicted by Pascal's triangle (Fig. 10.17). Consider, for example, the p.m.r. spectrum of bromoethane (CH_3CH_2Br). Unrestricted free rotation about the C–C bond means that the three protons of the methyl group are equivalent and the two protons of the methylene group are equivalent. Therefore, *in theory*, the methyl protons with their two equivalent neighbouring protons will have a signal that is a triplet (2+1) with the intensities of the peaks in the ratio 1:2:1. On the other hand, the methylene group has three equivalent neighbours and so its signal will be predicted to be a quartet (3+1) with the peaks having relative intensities of 1:3:3:1. This agrees reasonably well with the spectrum of bromoethane which also shows that the signal for the methyl's protons is upfield from that of the methylene (Fig. 10.15). This type of prediction is reasonably accurate for simple molecules but less accurate for more complex molecules because the three-dimensional nature of these molecules sometimes makes it difficult to identify all the nuclei that can affect a signal.

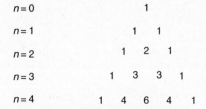

Fig. 10.17. Pascal's triangle.

10.6.3 Signal intensity

The area under a signal in a p.m.r. spectrum is proportional to the number of equivalent protons responsible for that signal. Consequently, electronic integration of the area under each signal in a p.m.r. spectrum enables one to determine the ratio of the numbers of equivalent protons responsible for each signal. This can be of considerable help in interpreting a p.m.r. spectrum.

Electronic integrators present the data as either a numerical printout above each signal or a stepped line running across the spectrum where the relative height of each step is proportional to the area under the corresponding signal (Fig. 10.18). This latter method is not as accurate as the numerical printout as it is difficult to decide where a step starts and ends. The integral measurements are reduced to their simplest ratio by dividing through by the lowest figure. This gives the ratio of the numbers of equivalent protons responsible for each signal and, if necessary, it is converted to a set of whole numbers. Consider, for example, the p.m.r. spectrum of ethanol (Fig. 10.18). Both methods of recording integrals are illustrated on this spectrum. Dividing through by the lowest figure in each case it can be seen that the nearest whole number ratio of the signals in both instances is 1:2:3 which corresponds with the ratio of equivalent protons in the structure of ethanol. It must be emphasised that these figures represent the ratio of the groups of equivalent protons in the molecule and not the actual numbers of protons bonded to each atom in the compound. This ratio may only be converted to the actual numbers if other information such as the molecular formula of the compound is known.

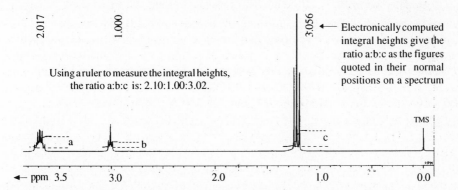

Fig 10.18. Calculation of the ratio of the equivalent protons in ethanol from its integrated p.m.r. spectrum.

10.6.4 Deuterium exchange

Deuterium does not absorb radiofrequency radiation in the same region as protons. However, it will undergo rapid exchange reactions with some acidic protons such as those in hydroxy and amino groups. The mixing of D_2O with a sample (known as a D_2O shake) causes either a reduction in the intensity or the complete removal of the signals in a spectrum due to exchangeable protons. This loss will be accompanied by the appearance of a weak signal at 4.8 ppm produced by the formation of HOD provided the HOD is soluble in the solvent used. The changes in a spectrum caused by a D_2O shake enables one to identify the signals of groups that contain exchangeable protons.

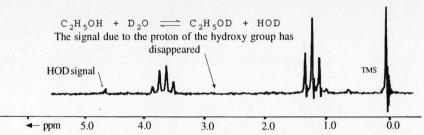

$$C_2H_5OH + D_2O \rightleftharpoons C_2H_5OD + HOD$$

The signal due to the proton of the hydroxy group has disappeared

HOD signal

TMS

$\longleftarrow$ ppm   5.0   4.0   3.0   2.0   1.0   0.0

Fig 10.19. The p.m.r. spectrum of ethanol after treatment with D_2O. Compare this spectrum with Fig. 10.18.

10.6.5 Uses

P.m.r. spectroscopy is mainly used either to identify or to determine the structures of compounds. This is normally carried out by interpreting the spectrum but it can also be done by comparing the unknown to standard reference spectra (cf. fingerprinting in IR identifications). P.m.r. is also used for quantitative analysis since:

$$\text{Concentration} \propto \frac{\text{Area under a signal}}{\text{Number of protons causing that signal}}$$

Assays are carried out using a mixture of the sample A and a known concentration of a pure standard B. The areas under suitable signals for both A and B are determined and the concentration of A calculated from:

$$\frac{\text{Concentration of A}}{\text{Concentration of B}} = \frac{\text{(Ia)(Number of protons causing the signal used for B)}}{\text{(Ib)(Number of protons causing the signal used for A)}}$$

Where Ia and Ib are the integrals of the signals used for A and B respectively.

The accuracy of assays based on p.m.r. are largely dependent on the accuracy of the integrator used with the spectrometer. However, as p.m.r .spectra are additive it is possible to use this technique to determine the concentration of a component of a mixture.

10.7 ^{13}C NUCLEAR MAGNETIC RESONANCE SPECTROSCOPY

The origins and interpretation of ^{13}C nuclear magnetic resonance spectra are similar to proton magnetic resonance spectra. However, they are usually simpler and therefore easier to interpret. This is due to two main factors: (1) the low natural abundance of the isotope and (2) the mode of operation of the spectrometer. Both these factors simplify the spectrum by preventing the splitting of signals by spin-spin coupling between nuclei.

10.7.1 Abundance

The natural abundance of ^{13}C isotopes is 1.1%, that is, roughly one carbon atom in every hundred is a ^{13}C isotope. This means that it is highly unlikely that adjacent carbon atoms in a molecule will be ^{13}C isotopes. Since ^{12}C nuclei do not absorb radiofrequency radiation when placed in a strong magnetic field, most of the carbon atoms in a molecule will be incapable of spin-spin coupling with any ^{13}C nuclei present. However, there will be sufficient ^{13}C isotopes in *all* the molecules found in a sample to obtain an *average absorption spectrum* of all the molecules in that sample. This average spectrum is in effect that of a theoretically impossible molecule whose carbon atoms are all non-coupled ^{13}C isotopes.

The rarity of ^{13}C–^{13}C spin-spin coupling means that the intensity of any signals originating from such couplings is usually too weak to be of use. However, an increasing amount of mechanistic and biological work is being carried out using ^{13}C enriched compounds. In these compounds, ^{13}C–^{13}C spin coupling is significant and splitting of the signals useful.

10.7.2 Mode of operation of the spectrometer

^{13}C–^{1}H spin-spin coupling is usually prevented by operating the spectrometer in the **proton-decoupled mode**. In this mode all the protons in the compound are saturated by the use of a broad band of radiofrequencies covering the region in which protons come into resonance. As a result, there is only one local field for each proton and so no coupling occurs. This process is known as broad-band decoupling. It is the most common way of operating the spectrometer and results in a spectrum in which each carbon appears as a single sharp peak. Each signal represents one or a group of magnetically equivalent carbon atoms. Consider, for example, the ^{13}C proton-decoupled spectra of butan-2-one and acetophenone shown in Fig. 10.20. The spectrum of butan-2-one has four signals corresponding to the four carbon atoms in the molecule. However, the spectrum of acetophenone has only five signals even though this molecule has eight carbons. This is because the molecule contains a symmetrical structure in which the ring carbon atoms 2 and 6 are in identical molecular environments and so have signals with the same chemical shifts. Similarly, carbons 3 and 5 are chemically and magnetically equivalent and also have signals with the same chemical shift value. Coincidentally, in this instance both pairs of atoms have identical chemical shifts which results in only one signal for all four atoms. However, the area under each signal is not proportional to the number of carbon atoms responsible for that signal when spectra are run in the proton-decoupled mode.

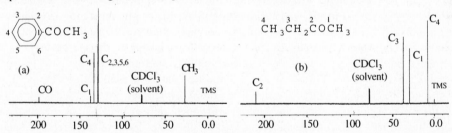

Fig. 10.20. The proton-decoupled ^{13}C spectra of (a) acetophenone and (b) butan-2-one. Note the signal from the carbon of the solvent, deuterochloroform, which is commonly used for running ^{13}C n.m.r. spectra.

10.7.3 Interpretation

Chemical shifts are initially identified from general correlation tables of the type shown in Fig. 10.21 (cf. p.m.r.). One should always bear in mind that the structure might contain equivalent carbons when making these deductions. These initial deductions are expanded and/or confirmed by consulting more accurate tables concerned only with the appropriate structures. These specialised tables are often associated with empirical formulae that enable one to calculate a theoretical value for the chemical shift for the relevant ^{13}C atom. A good correlation between the calculated value and the observed value is taken as evidence for the molecule containing the structure used in the calculation.

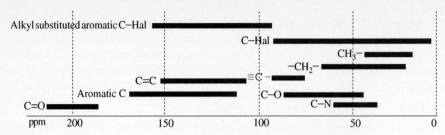

Fig 10.21. A general chemical shift correlation chart for ^{13}C n.m.r.

^{13}C spectra can be integrated in the **gated-decoupled mode**. In this mode of operation the area under a signal is now proportional to the number of equivalent carbons responsible for that signal. It is recorded in the same way as p.m.r. integrations, namely, either as a stepped line or number related to each signal. Calculation of the relative numbers of equivalent carbon atoms are also made in the same way (Fig. 10.19). However, integration is of limited use since the gated-decoupled mode is two to three times less sensitive than the proton-decoupled mode and so calculations are correspondingly less accurate.

10.7.4 Use

^{13}C n.m.r. is used in the same way as p.m.r. to identify compounds (10.6.5). However, at present, ^{13}C n.m.r. is not used for quantitative analysis.

10.8 MASS SPECTROMETRY

Mass spectrometry uses an instrument called a mass spectrometer to determine relative molecular masses and to investigate the structures of molecules. There are several different types of instrument available but they all operate by vaporising the sample in an ion chamber. The molecules of the sample are normally converted into a mixture of neutral and positively charged particles in this chamber. The number of charges on the ions is normally one, but very occasionally multicharged ions are formed. All the particles pass along a curved tube and through a strong magnetic field. This field deflects the positive ions round the curve and onto a detector. The neutral particles are not deflected and are lost on the walls of the tube. Not only does the instrument deflect the ions but it sorts them according to their mass/charge ratio (m/z) and plots a bar graph of their relative abundances (vertical) against m/z (horizontal) (Fig.10.22). This bar graph is known as the **mass spectrum** of the sample. Since the charge on most ions is 1, this spectrum is in effect a plot of the mass of the ions against their abundance.

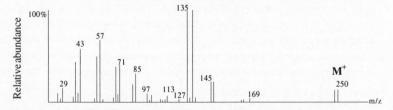

Fig. 10.22. The EI mass spectrum of bromododecane.

Two methods, known as the **electron impact** (EI) and **chemical ionisation** (CI) mass spectrometry, are used to decompose the sample. In EI mass spectrometry the vapour of the sample is bombarded with a stream of high energy electrons. The energy transferred from these electrons to the molecules of the sample causes the molecules of the sample to fragment into a mixture of neutral and positively charged particles (Fig. 10.23).

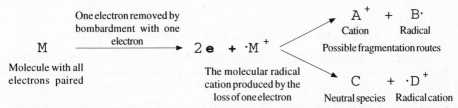

Fig. 10.23. A schematic representation of the formation of neutral and positively charged ions by EI in a mass spectrometer.

In CI instruments, a gas, such as methane, ammonia or 2-methylpropane (isobutane), is introduced into the spectrometer and ionised by bombardment with high energy electrons. The ions produced by these gases react to form very reactive acidic species. These acidic species react with the sample vapour to form the mixture of ions that is analysed by the instrument (Fig. 10.24.). It should be noted that in CI mass spectrometry, the extent of fragmentation is often reduced, especially when using ammonia, which makes CI mass spectroscopy an excellent method for determining relative molecular masses.

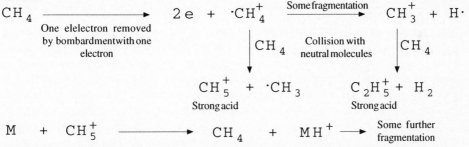

Fig. 10.24. A schematic representation of the formation of neutral fragments and ions by CI in mass spectrometry.

10.8.1 The spectrum

The most abundant ion in a mass spectrum is arbitrarily given an abundance of 100% and the abundance of the other ions is measured relative to this standard. It is not always the heaviest ion in the spectrum. The fragmentation patterns of EI and CI mass spectra are usually significantly different, as is exemplified by the EI and CI spectra of bromododecane shown in Figs 10.22 and 10.25 respectively.

The ion formed by the loss of one electron from the molecule in EI spectroscopy is known as the molecular ion. It is *usually* the most abundant peak on the far right-hand side of the spectrum and its mass is the same as the relative molecular mass of the compound. However, it should be noted that *some* EI spectra of compounds with relative molecular masses of less than 300 do not show a molecular ion peak. Furthermore, identification can

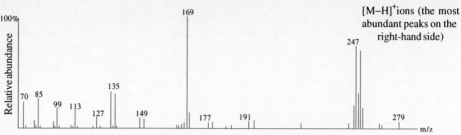

Fig. 10.25. The CI mass spectrum of bromododecane.

be complicated by the presence of isotopes in some of the molecular ions collected and counted by the instrument. For example, bromododecane has two molecular ions with relative molecular mass (RMM) values of 248 and 250, marked as M^+ in Fig. 10.22. These peaks are due to the presence of the bromine isotopes ^{79}Br and ^{81}Br in the molecules of the sample of bromododecane. It should be noted that not all peaks due to the presence of isotopes are small. Their size will depend on the relative abundance of the isotope in the compound.

In CI spectra the peak in the corresponding position to the *molecular ion* peak of EI spectra is that due to the MH^+ ion whose mass is one unit higher than the RMM of the compound. In rarer instances it may be due to the presence of the $[M-H]^+$ ion whose RMM is one unit less than the RMM of the molecular ion (Fig. 10.25). One of these **molecular ions** always occurs even in compounds with relative molecular masses of less than 300 and so CI spectra are very useful for determining the relative molecular masses of these compounds.

10.8.2 Interpretation

Mass spectra can be difficult to interpret because of the complexity of the fragmentation. Generally, one identifies the peak due to the M^+ or MH^+ ion and uses this as a reference point. The mass differences between this peak and the other peaks are determined and the *likely* nature of the fragments lost from the reference peak deduced from tables of the type shown in Table 10.4. One can also work the other way around and consult tables which indicate the most likely composition of a fragment. The peaks on either side of a main peak may be formed because of the presence of isotopes in a fragment. These peaks must be taken into account when making any deductions, and initially it is best to consider only the most abundant peaks. Unfortunately, the great diversity of fragments produced in a mass spectrum means that these deductions are of limited value on their own but, taken in conjunction with other experimental evidence, can be of considerable use in either identifying or determining the structure of a compound.

10.8.3 Uses

The main uses of mass spectrometry are relative molecular mass and structure determination. It is always used in conjunction with information from other sources.

Mass spectrometers are often used in conjunction with gas chromatography (GC) and high pressure liquid (HPLC) chromatography to provide structural information about the components of mixtures as they elute from the column. They are also used to a lesser extent as detectors in gas chromatography.

Table 10.4. The possible structures of some of the fragments found in a mass spectrum.

Mass loss from M	Structure possibly responsible for the mass loss	Mass loss from M	Structure possibly responsible for the mass loss
1	$H^{\cdot}$	29	C_2H_5
15	CH_3 or CH_3^{+}	30	$CH_2{=}\overset{+}{N}H_2$
16	O or NH_2	31	CH_3O or $CH_2{=}\overset{+}{O}H$
17	OH or NH_3	43	$CH_3\overset{+}{C}O$ or $\overset{+}{C}_3H_7$
18	H_2O or $H_2\overset{+}{O}$		

10.9 ATOMIC EMISSION AND ABSORPTION SPECTROSCOPY

10.9.1 Atomic emission spectra

When the atoms of an element are raised to an excited state by the absorption of energy they can lose energy to return to their ground energy state by emitting electromagnetic radiation in the visible and/or ultra-violet regions. The wavelengths of the emitted radiation are characteristic of the element and their intensity is related to the concentration of the element.

Atomic emission spectra are used for both the quantitative and qualitative analysis of elements in a biological sample. The technique is known generally as emission spectrometry and the type of emission spectroscopy used to determine sodium, potassium and lithium, is known as **flame photometry** because the energy used to excite the atoms of these elements is supplied by a flame. Flame photometers are modified visible and ultra-violet spectrophotometers (Fig. 10.26). The sample is atomised and carried into a flame by a stream of air or oxygen which is required to burn the fuel. The flame from the burning fuel supplies thermal energy that excites some of the atoms in the sample. The electromagnetic radiation produced when these excited atoms lose energy to regain their ground state is analysed by the spectrometer. Careful control of the fuel, oxidant gas and sample spray supplies ensures reproducible results.

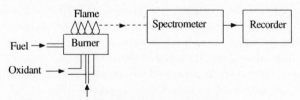

Nebuliser, the sample is injected as a spray or aerosol

Fig. 10.26. The principles of a flame photometer.

Qualitative analyses are carried out by comparing the wavelengths of the emission with those recorded in tables whilst quantitative analyses are performed using the calibration curve method used in other forms of spectroscopy (10.2 and 10.3). In this type of analysis the calibration curve is a plot of instrument response rather than intensity against element concentration. Flame photometry is used mainly for the determination of the alkali metals and alkaline earths. Its use in biomedical science is being superseded by the use of ion-selective electrodes.

10.9.1 Atomic absorption spectrometry

In atomic emission spectroscopy only a small percentage (0.01 to 1.0%) of the atoms of an element are excited. However, those atoms of the element that are not excited can absorb radiation with a specific wavelength from an external source. The degree of this absorption is proportional to the concentration of the element in the sample. Graphite furnaces are often used instead of flames to produce the atoms from their parent substances: they are more sensitive and solid samples can be used.

Atomic absorption spectroscopy is used only for quantitative determinations. which are carried out using an atomic absorption spectrometer (Fig. 10.27). The instrument is calibrated using standard solutions and the concentration of the sample determined either by computation or the calibration curve method used for other forms of spectroscopy (10.2 and 10.3).

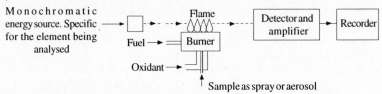

Fig. 10.27. The principles of an atomic absorption photometer.

10.10 WHAT YOU NEED TO KNOW

(1) Spectroscopy is a study of how molecules behave when they absorb electromagnetic radiation.

(2) *The absorption of electromagnetic radiation of a specific wavelength by a substance in the ultra-violet, visible and infra-red regions is related to the concentration of the substance in solution by the empirical Beer-Lambert Law;*

$$Absorbance\ (A) = log_{10} \frac{I_0}{I} = kcl$$

(3) *Ultra-violet and visible spectra are quoted as either the molar extinction coefficient e or the specific absorbance A (1%,1cm) for the wavelength corresponding to the maximum absorption.*

(4) *Structures that absorb in the ultra-violet and visible regions are known as chromophores. They can be saturated or unsaturated structures.*

(5) The greater the degree of conjugation in a structure, the higher the wavelength corresponding to the maximum absorption in the ultra-violet and visible regions.

(6) First and higher order derivative visible and ultra-violet spectra are of more use than the absorption spectrum in identifying compounds by the *fingerprinting* method.

(7) *Molecules which are raised to an excited state by the absorption of energy but lose this energy by emitting light exhibit either fluorescence or phosphorescence.*

(8) *Fluorescence occurs when the molecule goes from the excited state to the ground state by a single step or series of steps each of which results in the emission of electromagnetic radiation.*

(9) Phosphorescence occurs when the molecule goes from the excited state to the ground state by a series of steps, some of which do not emit electromagnetic radiation.

(10) Molecules exhibiting fluorescence and phosphorescence have both an absorption and an emission spectrum. The wavelengths of maximum absorption and emission are known as λ excitation (λ_{exc}) and λ emission (λ_{em}).

(11) The intensity of the fluorescent radiation is related to the concentration of the substance responsible for the emission. Unknown concentrations may be determined by plotting a calibration curve of intensity of fluorescence for standard solutions of a compound and reading the concentration of test solutions of the compound from the calibration curve after measuring the intensities of their emissions under the same conditions.

(12) Fluorescence is of wider application than phosphorescence which usually requires low temperatures.

(13) Conjugated structures are more likely to exhibit fluorescence than non-conjugated structures.

(14) Molecules absorb electromagnetic radiation in the infra-red region as bond vibrations.

(15) Infra-red spectra are recorded as plots of percentage transmission or absorbance against wavenumber (reciprocal wavelength, cm^{-1}).

(16) Infra-red spectroscopy is used to identify the functional groups in a molecule by comparison of the wavelengths of absorption with the values given in correlation charts and tables.

(17) Infra-red spectra can be used to confirm the identity of a compound by directly comparing the spectrum of the compound with that of an authenticated sample.

(18) The nuclei of some isotopes can take up more than one orientation in a strong magnetic field. Nuclear magnetic resonance occurs when a nucleus in a strong magnetic field changes from an orientation of low energy to one of high energy by the absorption of electromagnetic radiation in the radiofrequency region.

(19) Absorption of energy (resonance) will only occur if the number of the nuclei in the low energy orientation is higher than that in the higher energy orientation.

(20) Saturation occurs when the number of nuclei in the lower energy orientation is lower than the number in the higher energy orientation. When saturation occurs, no nuclear magnetic resonance spectrum will be observed.

(21) Nuclear magnetic resonance spectra are recorded as plots of signal intensity against chemical shift.

(22) Proton magnetic resonance spectroscopy is concerned with the nuclear magnetic resonance of ^{1}H.

(23) Protons in the same magnetic environments (usually the same chemical environment) have the same chemical shifts.

(24) The identity of the protons causing a signal can be determined by consulting correlation charts and tables.

(25) Spin-spin coupling occurs when signals are split into groups of characteristic peaks by the influence of adjacent protons in the molecule.

(26) The area under a signal is proportional to the number of equivalent protons responsible for that signal. It is used to determine the relative numbers of equivalent protons in a molecule and as the basis of assay procedures.

(27) ^{13}C nuclear magnetic resonance spectroscopy is concerned with the nuclear magnetic resonance of ^{13}C isotopes.

(28) *Each signal on a ^{13}C nuclear magnetic spectrum corresponds to the resonance of a carbon atom or a group of equivalent carbon atoms. The environment of the atoms responsible for a signal can be identified by reference to correlation charts and tables.*

(29) ^{13}C nuclear magnetic spectra are used to identify compounds by determining their structures.

(30) *Mass spectrometry is based on the conversion of a substance into a mixture of different neutral and positively charged particles.*

(31) Mass spectra are classified as either electron impact (EI) or chemical ionisation (CI) spectra depending on the way the ions are produced in their ionisation chambers.

(32) *Mass spectrometers normally sort the mixture of ions according to their mass/charge ratio (m/z) and display the result as a bar graph of abundance against m/z. This graph is known as the mass spectrum of a substance.*

(33) *The most abundant peak on a mass spectrum is arbitrarily assigned an abundance of 100%.*

(34) *The most abundant peak on the far right-hand side of a mass spectrum is the M$^+$ ion in EI spectra and MH$^+$ ion in CI spectra. Not all EI spectra have an M$^+$ peak.*

(35) *Peaks are identified by the use of mass loss and peak-mass correlation tables.*

(36) The presence of isotopes in a substance gives rise to peaks adjacent to a main peak in a mass spectrum. This must be allowed for when interpreting a mass spectrum.

(37) *Atoms in an excited state can emit radiation in the visible and ultra-violet regions when they lose energy and return to their ground state.*

(38) Thermal energy is commonly used to excite the atoms and the subsequent emission spectrum is known as an atomic emission spectrum. The instruments used to measure the spectra are known as flame photometers.

(39) *The intensity of the emitted radiation is related to the concentration of the element emitting the radiation. Quantitative analyses are carried out using either calibration curve methods or by calibrating the instrument.*

(40) Flame photometry is used for the determination of the alkali metals and alkaline earths in biological fluids.

(41) *Atomic absorption spectra are produced when the atoms of an element absorb electromagnetic radiation from an external light source in the presence of a thermal energy source.*

(42) *The intensity of the absorption is related to the concentration of the element absorbing the radiation. Quantitative analyses are carried out using either calibration curve methods or by calibrating the instrument.*

10.11 QUESTIONS

(1) Explain the meaning of the terms (a) chromophore, (b) auxochrome, (c) hyperchromic effect and (d) bathochromic shift.

(2) The molar extinction coefficient of $NADH + H^+$ is 6.3×10^3 at 340 nm. What would be the concentration of a solution of $NADH + H^+$ with an absorbance of 1.17 in a 1 cm cell?

(3) Explain the difference between fluorescence and phosphorescence. Determine the concentration of warfarin in a urine sample if the sample has a fluorescence intensity of 40.1 intensity units. The fluorimeter when calibrated under the same operating conditions with standard warfarin solutions gave the following results:

   Fluorescence intensity: 10.1   20.5   30.3   40.7   50.8   intensity units
   Warfarin concentration: 10.0   22.5   30.0   43.0   53.0   ng cm^{-3}

(4) Explain, giving details of the absorptions that would be observed, how infra-red spectroscopy could be used to follow the oxidation of ethanol to ethanal and its subsequent oxidation to ethanoic acid.

(5) Deduce, from their ^1H n.m.r. spectra, possible molecular structures for the ketones A and B.

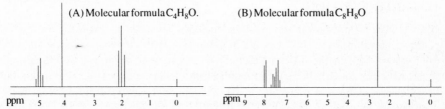

(A) Molecular formula C_4H_8O.   (B) Molecular formula C_8H_8O

(6) Deduce molecular structures for compounds D and E from their ^{13}C n.m.r. spectra and molecular formulae. Compound D is aromatic whilst compound E is an aliphatic acid.

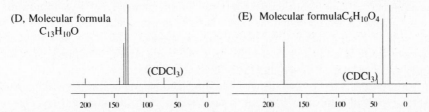

(D, Molecular formula $C_{13}H_{10}O$ (E) Molecular formula $C_6H_{10}O_4$

(CDCl$_3$)   (CDCl$_3$)

(7) Differentiate between EI and CI mass spectra. Explain why the EI mass spectrum of a compound F shows two molecular ion lines. Suggest possible structures for the fragments a, b and c in the EI mass spectrum of the hydrocarbon A.

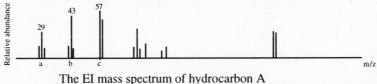

The EI mass spectrum of hydrocarbon A

11

Drug identification and analysis in pharmacy

11.1 INTRODUCTION

Official drug identification and analysis methods are used to maintain the quality of pharmaceutical products once they are in production. The chemical, physical and biological methods used in the official literature will have been approved by the appropriate statutory bodies. In Britain the Ministry of Health is the principal decision maker but the Royal Pharmaceutical Society also plays an important part in deciding what tests and methods are employed.

Drugs are evaluated by comparing the results of specific experiments with a set of standard specifications. Both methods and specifications will often be detailed in the appropriate official publication in a form known generally as a **monograph**. These monographs list the permitted limits of tolerance for the quality and purity of the product as well as general information concerning the product. They take into account all the known methods of manufacture, the most likely impurities, conditions of storage, the probable shelf life of the product and the dosage used.

The chemical test procedures given in official monographs can be broadly divided into three types: identification tests, limit tests and assays. Identification tests are used to establish the identity of the compound or preparation. Limit tests are used to check whether the level of a specified impurity is below the maximum limit allowed, whilst assays are used to determine the quantity of a specified compound in the substance. However, the results obtained from any type of test must be interpreted within the limitations given in the source publication. For example, identification tests in the British Pharmacopoeia only distinguish between the compounds and preparations listed in that publication. They do not distinguish listed substances from unlisted substances which give the same test results as the listed substances. The limitations governing tests and assays are normally given in the introduction and appendices of a publication.

Substances are said to comply with the specification if *all* the tests described in the publication give results that are in agreement with those given in the official publication. In many cases it is a matter of personal judgement as to whether a substance passes a test. Consequently, it is important to realise that safety is the overriding consideration and if only one test is suspect the substance fails.

11.2 SOURCES OF IMPURITY

Impurities in a product are due to either the manufacturing process or poor storage leading to degradation. It is not possible to remove all impurities from a substance during manufacture, the important thing is to keep their concentration to an acceptable level. Degradation on storage is largely due to atmospheric oxidation and hydrolysis by atmospheric moisture.

11.2.1 From raw materials

Compounds, reagents and solvents used in the manufacturing process can contain impurities which are transferred to the product during the course of the manufacturing process. Water, for example, can act as a source of contamination by Na^+, Ca^{2+}, Mg^{2+}, CO_3^{2-}, Cl^- and SO_4^{2-} ions.

11.2.2 Unreacted starting materials

Unreacted starting materials, both substrate and reagent, are always a possible impurity in a compound or preparation. Paracetamol, for example, is often contaminated with 4-aminophenol from which it is prepared.

$$H_2N-\!\!\!\langle \rangle\!\!\!-OH \xrightarrow{CH_3COOCOCH_3} CH_3CONH-\!\!\!\langle \rangle\!\!\!-OH$$

4-Aminophenol   Paracetamol

11.2.3 Solvents

It is often difficult to remove the last traces of a solvent from a preparation. For example, amoxapine can contain traces of the ethyl ethanoate used as a solvent in its manufacture. Compounds are often contaminated with water when it is used either in their preparation or purification.

11.2.4 Machinery

Corrosion and wear and tear on the machinery used to produce a product give rise to metal and plastic contaminants in products. For example, the high speeds at which modern tableting machines operate means that very small particles of metal are broken off the moving parts of the machine. To cope with this, many tableting machines have built-in metal detectors which can detect these very small particles of metal. Contaminated tablets are then diverted from the main tablet flow and reprocessed.

11.2.5 Atmospheric contaminants

These can be either solid, such as dust, aluminium oxide, silica and soot, or gaseous such as sulphur dioxide, carbon dioxide, water and hydrogen sulphide. Many of these gaseous substances can react with a product and cause its slow deterioration. This may be minimised by correct storage in tightly sealed containers.

11.2.6 Human errors

Human errors range from the use of incorrect strengths of solutions, wrong materials, incorrect solvents to cross-contamination due to poor cleaning of equipment.

11.3 SAMPLING

The accuracy of all analytical work depends on the accuracy of the sampling step. If the sampling procedure does not produce a sample that is representative of the whole of the material under test then the analytical results will be worthless as they will not represent the true situation in that material. Consequently, it is of paramount importance in all analytical work to obtain a truly representative sample of the substance under test by following the correct sampling procedures.

Sampling procedures vary depending on the nature of the sample. For example, the British Pharmacopoeia recommends that a minimum of twenty tablets should be taken and it specifies that these tablets should be taken from widely different parts of the container. Liquid samples are comparatively easy to obtain although all liquids must be thoroughly shaken before they are sampled as heavier components have a tendency to settle to towards the bottom of the container. Sampling of solids often involves crushing, grinding, sieving, mixing and further sampling by successively dividing the sample until it is small enough to use. This treatment can cause contamination through wear and tear of equipment and a change in the water content of the sample. This change can be very important if the water content of the original sample is required. In this case it is advisable to determine the water content of a sample immediately the material arrives.

Water occurs in samples in two ways: as part of the structure of the substance in the form of water of crystallisation and as moisture absorbed from the atmosphere. In both cases the concentration of water can vary with ambient temperature and humidity. This means that identical quantitative analyses carried out on the same sample on different days could give different results if the temperature and humidity on those days were different. Consequently, quantitative results are frequently expressed in terms of the dry mass of the sample in order to obtain reproducible results that can be compared with those from the same or different samples of the same material.

The dry mass of a sample is usually obtained either by drying to constant mass and using the dried sample or by determining the water content and making a correction for it in any subsequent calculations (Example 11.1). Drying to constant mass is normally achieved by drying the sample in an oven. The sample is cooled in a controlled environment such as a desiccator and weighed. This procedure is repeated until the sample's mass remains constant within the desired limits of accuracy.

It is important to realise that variation in analytical data from different sources can be caused by differences in sampling techniques. In order to reduce these errors it is important to use the same sampling method as other workers. This is especially important when comparing the results of an experiment with the standard results given in an official publication. Most official documents will indicate the sampling procedure required and this must be followed if the results are to be considered valid. If no procedure is given it is imperative that a representative sample is taken. If possible, bulk samples should be thoroughly mixed before taking the analytical sample.

Example 11.1

A substance contains 83.63% w/w of glycine. If 0.9517 g of the substance was found to weigh 0.9388 g after drying to constant mass, calculate the % w/w of glycine in the dry sample.

$$Mass\ of\ glycine\ in\ 0.9517\ g\ =\ \frac{83.63 \times 0.9517}{100} = 0.7959\ g$$

But the same mass of glycine will be in the dried sample, therefore:

$$\%\ w/w\ of\ glycine\ in\ the\ dried\ sample\ =\ \frac{0.7959 \times 100}{0.9388} = 84.77\ \%$$

11.4 IDENTIFICATION TESTS

Identification tests, as the name implies, are used to identify a compound *within the limits specified in the official publication* being used. They are based on either the chemical reactions or physical chemistry of the compound. The number of tests listed in an official monograph for a compound is the minimum number that can be used to distinguish between that compound and all the other compounds listed in the publication. As a result, they cannot be used on their own to identify that compound in the wider world of chemistry because other compounds could exist which would give the same results. In this context, identification can only be made by carrying out a full inorganic or organic analysis which takes into account all the other compounds that could have positive results for the identification tests.

11.4.1 Drug identification tests based on chemical reactions

The reagents used must react only with the drug and not the other substances present in the material. Identification tests can be based on chemical reactions in which either *an observable visible change occurs* or a solid compound known as a **derivative** is formed. The former makes use of reactions involving colour changes, gas evolution, precipitate and emulsion formation while in the latter case, the solid is purified and identified by measuring a physical property such as its melting point (Fig. 11.1). Any reaction may be used to form a derivative which can be made the basis of an identification test provided the reaction yields a product that is easily purified and has a sharp, reproducible melting point or other accurately measurable physical constant.

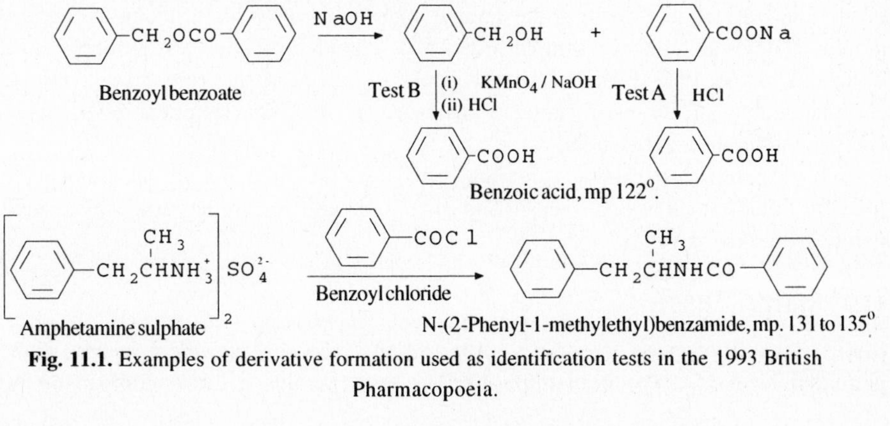

Fig. 11.1. Examples of derivative formation used as identification tests in the 1993 British Pharmacopoeia.

The chemistry of the reactions used for identification tests may or may not be known.

11.4.2 Identification tests based on physical chemistry

Identification tests based on physical chemistry involve either the measurement of a physical property such as melting point, refractive index and viscosity or the use of a physical technique such as chromatography and spectroscopic methods. Melting points are limited in their use as variations can arise when the equipment used for their measurement is different from that used to determine the values given in the literature. This problem can be avoided by simultaneously measuring the melting points of the sample under test, an authentic sample and a 50-50 mixture of the test and authentic samples. A positive identification is made if all three have melting points that agree to within 2°.

The use of infra-red spectra is normally restricted to a comparison of the spectrum of the test sample with that of an authentic sample. A positive identification is made if they have the same absorptions at the same wavenumbers. The intensities of the absorptions do not have to be the same, however, the relative intensities of the peaks of the test sample spectrum should be of the same order as those on the authentic samples spectrum. The fingerprint region is particularly useful in this type of identification test.

Ultra-violet and visible spectra based identification tests usually take the form of measuring either the percentage absorbance or the transmission at a specified wavelength and calculating either the absorbance for a solution of a specified strength, molar absorbance or $A(1\%,1cm)$ at that wavelength (10.2.1). If the calculated value does not correspond, within experimental error, to that given in the literature, the sample fails the test. For example, the 1993 British Pharmacopoeia identification test B for the preservative benzyl hydroxybenzoate requires a 0.001%w/v ethanolic solution of the drug to have an absorbance of about 0.76 in a 1cm cell at a λ_{max} of 260 nm.

Benzyl 4-hydroxybenzoate

Bisacodyl

Comparative thin layer chromatography is the basis of some identification tests. A positive identification is made if the sample under test travels as far along the chromatographic plate as an authentic sample (Fig. 11.3). The identification test D for the laxative bisacodyl in the 1993 British Pharmacopoeia is based on the comparative thin layer chromatography of the test compound and bisacodyl EPCRS (European Pharmacopoeia chemical reference substance).

Weight per cm^3, refractive index and viscosity measurements are used as specifications for liquids, such as cod liver oil and the liquid paraffins, in the same way as melting points are used for solids.

11.5 LIMIT TESTS

Limit tests are designed to show whether a substance contains more or less than the permitted level of a specified impurity. The tests involve a direct comparison of an experimental result with either a standard experimental result or a literature value. It is important to realise that in all limit tests **the reactions and/or physical measurements involve the impurities and not the drug**. Consequently, any chemical reaction, analytical technique or physical measurement can be made the basis of a limit test provided it does not involve the drug and gives results that are consistent, reproducible and an accurate measure of the concentration of the impurity.

11.5.1 Nessler tube methods

These methods involve comparing the sample under test with a standard sample containing the maximum allowed concentration of the impurity (the maximum permitted level of impurity) under examination. Comparison is made using a pair of optically matched flat bottomed Nessler tubes. One of the tubes is used for the sample under test and the other for the standard. Each tube is treated separately and the results of a test are assessed by simultaneously viewing both the test and standard tubes against a uniform background. They may be viewed either horizontally from the side or vertically from above and against a suitable background (Fig. 11.2).

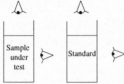

Fig. 11.2. The Nessler tube method of carrying out a limit test. Note the alternative viewing angles.

All the tests are based on converting the impurity under examination in both the test and standard sample tubes into a derivative. The reactions are carried out in such a way that this derivative is usually formed as either a coloured solution, emulsion or colloidal suspension, the intensity of the colour or density of the suspension being proportional to the concentration of the original impurity. Therefore, it follows that if the colour intensity or density of the derivative formed from the test sample is greater than that formed from

Table 11.1. Examples of reactions used as the basis of Nessler tube limit tests. The chemistry of some of the reactions that are used as the bases of limit tests for some organic compounds is not known.

Impurity	Basic chemistry of the test	Nature of the product
Pb^{2+}	$Pb^{2+} \xrightarrow{Na_2S} PbS$	Brown colloidal precipitate
Fe^{2+} and Fe^{3+}	$Fe^{3+} \xrightarrow{HSCH_2COOH} Fe^{2+}$	
	$Fe^{2+} \xrightarrow{HSCH_2COOH}$ (Fe complex)	Pink-purple solution
Cl^-	$Cl^- \xrightarrow{AgNO_3} AgCl$	White colloidal precipitate
SO_4^{2-}	$SO_4^{2-} \xrightarrow{BaCl_2} BaSO_4$	White colloidal precipitate
Phenols	Phenols $\xrightarrow{FeCl_3}$ Coloured complexes	Various coloured solutions

the standard, the test sample must contain more than the permitted level of that impurity and so fails the test. This technique is used to evaluate a wide variety of the inorganic and organic impurities found in drugs (Table 11.1).

11.5.2   Volumetric methods

In these limit tests the impurity is titrated with a suitable reagent and the titration value compared with that given in the literature. A correction must be made for any differences between the molarity and mass of sample used in the test and those used to calculate the value in the literature (Example 11.2). The acidity of benzyl hydroxybenzoate, for example, can be determined by titration with sodium hydroxide using methyl red as indicator.

Example 11.2

A 0.2031 g sample of benzyl hydroxybenzoate was dissolved in 10 cm³ of 50% ethanol previously neutralised to methyl red. If 1.07 cm³ of a 0.113 M solution of sodium hydroxide was required to change the colour of the solution, determine whether the sample meets the official specification for the acidity of benzyl hydroxybenzoate. The official requirement for the acidity of benzyl hydroxybenzoate is that a 0.2 g sample of the drug should require not more than 0.1 cm³ of 0.1 M sodium hydroxide to cause a change in colour.

The first step is to calculate the volume V of 0.1 M sodium hydroxide equivalent to 0.47 cm³ of 0.113 M sodium hydroxide. Since the number of moles of sodium hydroxide in the 0.113 M NaOH solution must be the same as the number of moles in the 0.1 M NaOH solution:

$$\frac{0.113 \times 1.07}{1000} = \frac{0.1 \times V}{1000}$$

that is: $V = 1.21 \ cm^3 \ of \ 0.1 \ M \ NaOH$

However, this is the volume of 0.1 M sodium hydroxide solution required by the 0.2031 g sample and so by simple proportion the volume of 0.1 M sodium hydroxide solution required by a 0.2 g sample as stated in the official specification would be:

$$\frac{1.21 \times 0.2}{0.2031} = 1.19 \ cm^3$$

As this corrected volume is above the limit quoted this sample of benzyl hydroxybenzoate does not meet to the required specification and should therefore be rejected.

11.5.3   Chromatographic and electrophoresis methods

All forms of chromatographic separation can be used as the basis of limit tests. In most limit tests based on chromatographic methods the results using the sample under test are compared with those from a standard sample. For example, limit tests utilising comparative thin layer chromatography are based on a comparison of the spots due to the impurities in the test sample with either the spots due to standard solutions of the impurities (Fig. 11.3a) or the drug spot on a chromatogram of a standard solution (Fig. 11.3b). Consider, in the latter case, the limit test for related substances in the monograph for the antibiotic chloramphenicol in the 1993 British Pharmacopoeia. A solution of the sample of chloramphenicol (T) under test, a solution of an European Pharmacopoeia Chemical Reference Substance (EPCRS) sample of chloramphenicol (A) and a dilute solution of solution A (solution B) are chromatographed on the same plate. Sample A is used to locate the spot due to the drug (the primary spot) in the other two samples. The drug spot from solution B is used as the reference standard for the impurities (the secondary spots) in the test sample T. The chloramphenicol sample T passes the test if the secondary spots on its chromatogram are less intense when viewed under a UV light at 254 nm than that of the primary spot in sample B. In other thin layer chromatography based limit tests, the spots may be formed by spraying with a reagent.

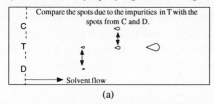
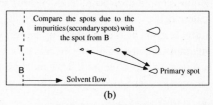

(a)   (b)

Fig. 11.3. Thin layer chromatography as the basis of a limit test. (a) The comparison of impurity spots in a sample T with those from standard solutions of the impurities (C and D). (b) The comparison of the impurity spots in a sample T with that of the drug spot from a standard solution of the drug.

Limit tests utilising paper chromatography and electrophoresis may also be based on a comparison of 'spots' from test and standard samples. The spots may be visualised either by spraying with a reagent or by the use of UV light.

Limit tests based on the different forms of column chromatography usually rely on a comparison of peak sizes. This comparison may be made between the peak due to the impurity and either a compound that has been added as an internal reference standard or peaks due to other compounds in the sample. For example, the 1993 British Pharmacopoeia limit test for neomycin C in the antibiotic neomycin sulphate is based on separating the components of a sample of the drug on an anion-exchange resin. Neamine is eluted first followed by neomycin C and then neomycin B. The fractions are analysed by treating with ninhydrin (28.5.4) and measuring the absorbance of the solution. A graph of absorbance against fraction number is plotted and the areas under the peaks due to neomycin B and C calculated. The area of the peak corresponding to neomycin C should be 3 to 15% of the sum of the areas of the peaks corresponding to neomycin B and C.

11.5.4   Spectroscopic methods

These limit tests usually involve the ultra-violet and visible regions of the electromagnetic spectrum. The absorbance or transmission of a suitable solution of a sample of the drug is measured at a wavelength where the absorption is solely due to the impurity. Either it may be compared with a value specified in the literature or the value of the molar absorbance, or specific absorbance may be calculated and compared with that recorded in the literature. In some procedures a derivative of the impurity is formed and it is the absorption of this derivative that is measured.

11.5.5   pH measurement

pH measurement is used as a limit test for acidity and alkalinity. The pH of a solution of the drug is measured and the result compared with the specification given in the literature. For example, the 1993 British Pharmacopoeia specifies that the pH of a 5% w/v solution of the antitubercular drug isoniazid must have a pH value in the range 6 to 8.

11.6 ASSAY METHODS

The techniques commonly used in the evaluation of drugs are: volumetric, gravimetric, spectroscopic and chromatographic analysis. The results obtained by these methods are usually recorded in the literature as either the %w/w, %w/v or %v/v (9.5). Specifications given in the literature are quoted in terms of a specific molecular formula. For example, the %w/w of aspirin in a sample of aspirin powder would be quoted as the %w/w of $C_9H_8O_4$ in the sample of aspirin. This ensures that the results of experimental determinations are calculated to the same standard as the specification and so can be compared with the specification. It is important to appreciate that a figure obtained for a sample using one analytical method will not necessarily be the same as that obtained for the same sample using a different method. This must be taken into consideration when comparing experimentally determined figures with those quoted in the literature.

11.6.1 Volumetric methods

All volumetric methods are based on determining the volume of a solution of known concentration (**standard solution**) of a reagent that exactly reacts with the species (analyte) being analysed. The point in the reaction where this happens is known as its **equivalence point**. Equivalence points cannot be determined directly by experiment but can be indicated by observing related chemical and physical changes. These changes occur at what is known as the **end point** of the reaction. End points do not usually correspond exactly to the equivalence point but are sufficiently accurate to be used in calculations as though they were the equivalence point.

Reagents that are used as reference compounds in volumetric analysis are known as **primary standards**. They are stable compounds that have been purified to a very high degree. However, it is not essential for a compound to be a primary standard for it to be used as the reagent in a volumetric analysis. It is necessary only for the concentration of its solution to be accurately known at the time of its use. This may involve a separate assay (**standardisation**) of the reagent before it can be used.

11.6.2 Simple volumetric titrations

Simple volumetric titration may be used for both inorganic and organic substances. An accurately weighed sample of the analyte is dissolved in a suitable solvent. The solution either is used directly or may be diluted and a specified volume taken for use in the assay. The reagent is added to the analyte solution from a burette until the end point of the reaction is reached, the procedure being known as a **titration.** This end point may be detected by either a change in the colour of a previously added chemical known as an **indicator** or a measurable change in a physical property.

The indicator used and its mode of operation will depend on the nature of the reaction used in the assay. In titrations that involve the reaction of an acid with a base, for example, the indicators are compounds that change colour when the pH of the reaction mixture rapidly changes. The indicators used in ethylenediaminetetraacetic acid (EDTA), metal ion titrations are dyes that form coloured complexes with the metal ions being titrated. At the end point of the titration the decomposition of these complexes is complete and this causes a change in the colour of the reaction mixture (32.7.2).

Physical methods of determining the end point include: potentiometric, measuring changes in the potential between two electrodes placed in the test solution, conductometric, measuring changes in the conductivity of the test solution, and amperometric, measuring changes in the current flowing through the test solution during a titration. The techniques are known as potentiometric, conductometric and amperometric titrations respectively. These methods are temperature-dependent and so the temperature should be kept as constant as possible during the titration.

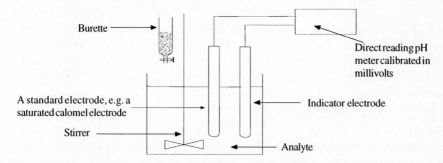

Fig 11.4. The arrangement of the equipment for a potentiometric titration.

Potentiometric titrations are based on measuring the change in the e.m.f. (E) of a cell consisting of a standard electrode, the analyte and an indicator electrode (Fig. 11.4). Any electrode can be used as the standard electrode provided its potential remains constant during the titration. However, the type of the indicator electrode will depend on the nature of the titration. For example, an inert metal such as platinum is normally used for redox titrations whilst a glass electrode is used for acid-base titrations. The end point of a titration is found by plotting either the e.m.f. of the cell, dE/dV or d^2E/dV^2 against volume V of the titrant (Fig. 11.5). In the first two cases the end point (P) corresponds to a rapid change in the gradient whilst in the third case it corresponds to where the curve intercepts with the x axis. The differential plots give a more accurate value for the end point than the linear plot.

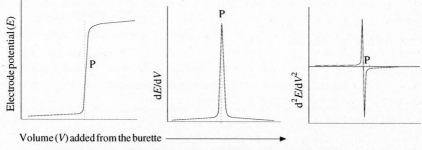

Fig. 11.5. Methods of detecting the end point in a potentiometric titration.

In conductometric titrations the conductance (c) of the reaction mixture is measured and plotted against the volume (V) of the titrant. As the titration progresses the conductance will change in a regular manner as one ion is replaced another. As a result, the plot of c against V will be a straight line until the equivalence point is reached. At the equivalence point the gradient of this line will change as the replacement is complete (Fig. 11.6). Further additions of the reagent will result in a second straight line with a different gradient. The graph is usually curved at the equivalence point and so the equivalence point position (P) is estimated by extrapolating the two straight lines.

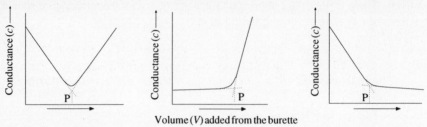

Volume (*V*) added from the burette

Fig. 11.6. Typical plots of conductance against the volume of the titrant. The plots for amperometric titrations are similar except conductance is replaced by current (mA).

Amperometric titrations are carried out in a cell consisting of either the analyte and a microelectrode, usually a rotating platinum electrode, coupled to a depolarised reference electrode (Fig. 11.7) or a pair of identical solid state microelectrodes immersed in the analyte. A constant e.m.f. is applied to the electrodes and the current flowing in the cell during the titration is determined. This current is proportional to the volume of the titrating agent added and so a plot of current against volume is a straight line. The end point occurs where the gradient of this straight line changes. Its position can be found by extrapolation of the straight lines on either side of this change. The plots obtained are similar to those shown in Fig. 11.6 except conductance is replaced by current (milliamps, μA).

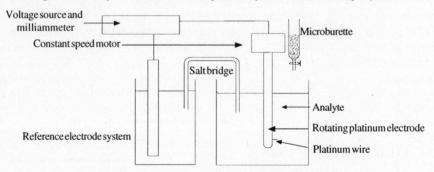

Fig. 11.7. The typical cell arrangement for an amperometric titration.

Amperometric titrations are the basis of sulphonamide assays and the Karl Fischer method for determining the amount of water in drugs and food.

The calculations for all types of simple volumetric titration are based on the balanced equation for the reaction because for the reaction of A with B:

$$aA + bB \longrightarrow \text{Products}$$

$$\frac{a}{b} = \frac{\text{Number of moles of compound A used in the reaction}}{\text{Number of moles of compound B used in the reaction}} \qquad (11.1)$$

where *a* and *b* are the coefficients in the balanced equation for the reaction.

The procedure is to convert all quantities into moles, carry out the required computation and then reconvert the molar quantities to the required units (Example 11.3). Some useful formulae are:

$$\text{Moles} = \frac{\text{Mass in grams of a compound}}{\text{Relative molecular mass of the compound}} = \frac{\text{Molarity x Volume}}{1000}$$

Example 11.3

2.809 g of benzoic acid was dissolved in 20 cm^3 of neutral 96% ethanol. The solution was titrated with 1.003 M sodium hydroxide using phenol red as indicator. If 22.09 cm^3 of the 0.1003 M sodium hydroxide was required, calculate the % w/w of $C_7H_6O_2$ in the sample. (RMM, $C_7H_6O_2$ = 122.1).

The equation for the reaction is:

$$PhCOOH + NaOH \longrightarrow PhCOONa + H_2O \qquad (11.2)$$

Using the appropriate definitions :

$$\text{Moles of NaOH} = \frac{22.09 \text{ x } 1.003}{1000} = 0.02216 \text{ moles}$$

and,
$$\text{Moles of } C_7H_6O_2 = g/121.1$$

where g is the grams of $C_7H_6O_2$ in the benzoic acid sample. But in equation 11.2, a and b = 1 and so substituting in equation 11.1:

$$\frac{1}{1} = \frac{\text{Number of moles of benzoic acid}}{\text{Number of moles of sodium hydroxide}} = \frac{g}{(0.002216)(121.1)}$$

Therefore:
$$g = 0.02216 \text{ x } 121.1 = 2.684 \text{ g}$$

and the % w/w of $C_7H_6O_2$ in the benzoic acid sample
$$= \frac{2.684 \text{ x } 100}{2.809} = 95.55 \text{ %}$$

British Pharmacopoeia (BP) assay methods usually include an expression called the **ml equivalent** which can also be used to calculate the answer for an assay. The ml equivalent is the mass of the analyte that would react exactly with 1 ml (as far as BP assays are concerned, 1ml is the same as 1 cm^3) of the specified reagent. It is used by converting the volume of the reagent measured in the experiment to an equivalent volume of a reagent solution with the molarity specified in the ml equivalent. This corrected volume is multiplied by the ml equivalent to give the mass of the analyte in the sample.

The compounds used in an assay sometimes contain impurities that can react with the titrating reagent. Consider, for example, the assay of benzoic acid by titration with sodium hydroxide described in Example 11.3. The titration is carried out in aqueous ethanol since benzoic acid is not very soluble in cold water. However, ethanol is often contaminated with a small quantity of ethanoic acid which is formed by atmospheric oxidation of the ethanol during storage. This ethanoic acid would react with the sodium hydroxide and give an inaccurate high titration reading. To prevent this happening the ethanol is neutralised prior to use, hence the use of the term *neutralised ethanol* in the description of the assay in Example 11.3. However, it is not always convenient to remove an impurity in this way. A more common way of resolving this problem is to carry out a **blank** determination.

$$C_2H_5OH \xrightarrow[\text{oxygen}]{\text{Atmospheric}} CH_3COOH \xrightarrow[\text{In the titration}]{NaOH} CH_3COONa$$

| Ethanol | Ethanoic acid | Sodium ethanoate |

A blank determination is carried out using the same method as the assay but without using the sample. As a result, the titre obtained in the blank is due solely to the impurities in the various solvents and reagents used in the titration. Since the blank and assay have been carried out under identical conditions, the true volume of the reagent that reacts with the analyte is the difference between the volumes required for the assay and blank titrations that is:

True volume = Assay volume − Blank volume

11.6.3 Back titrations

This technique is used when it is not possible to use the simple titration method. The sample containing the analyte A is reacted with a **known volume** of a standard solution of a reagent B which contains an excess of the reagent B. Once the reaction is completed the quantity of B in excess is found by titration with a standard solution C. The quantity of B reacting with A is found by subtracting the experimentally determined excess of B from the amount of B used at the start of the assay. The required value of A can then be calculated using a similar procedure to that used for simple titrations (Example 11.3). The procedure is referred to as a back titration since one works back from the titration in a series of steps to obtain the required answer. A typical assay procedure and its associated calculation are illustrated by Example 11.4.

Example 11.4

0.8755 g of a sample of chalk that is to be the basis of an antacid preparation was reacted with 50 cm³ of a 0.4878 M solution of hydrochloric acid. The reaction mixture was boiled for one minute to remove any dissolved carbon dioxide and cooled. The cooled solution was titrated with 0.5113 M sodium hydroxide using methyl orange as indicator. It required 31.57 cm³ of the sodium hydroxide for neutralisation. Calculate the % w/w of CaCO₃ in the sample of chalk,. The RMM value of CaCO₃ is 100.

The chemistry underlying this procedure is:

$$CaCO_3 + 2HCl = CaCl_2 + CO_2 + H_2O + (HCl_{(excess)}) \qquad (11.3)$$
$$HCl_{(excess)} + NaOH = NaCl + H_2O \qquad (11.4)$$

The first step is to determine the number of moles of hydrochloric acid (m) in excess. The number of moles of sodium hydroxide used in the titration is:

$$\frac{31.57 \times 0.5113}{1000} = 0.01614 \text{ moles}$$

Therefore, substituting the values for a and b from equation 11.4 in the appropriate form of equation 11.1 for the titration of sodium hydroxide with the excess hydrochloric acid:

$$\frac{1}{1} = \frac{\text{Moles of excess hydrochloric acid}}{\text{Moles of sodium hydroxide}} = \frac{m}{0.01614}$$

and so:

$$m = 0.01614 \text{ moles}$$

The number of moles of hydrochloric acid (n) used to react with the sample of chalk is:

$$\frac{50 \times 0.4787}{1000} = 0.02394 \text{ moles}$$

but the number of moles of hydrochloric acid that react with the CaCO₃ in the chalk is the difference between the moles used and the moles in excess n - m, that is:

$$0.02394 - 0.01614 = 0.0078 \; moles$$

Therefore, substituting the values of a and b from equation 11.3 in the appropriate form of equation 11.1 for the reaction of the calcium carbonate with the hydrochloric acid:

$$\frac{Moles \; of \; calcium \; carbonate}{Moles \; of \; hydrochloric \; acid \; reacting \; with \; the \; calcium \; carbonate} = \frac{1}{2} = \frac{x}{0.0078}$$

therefore, the mass of $CaCO_3$ in 0.8755g of the sample = 0.0039 x RMM of $CaCO_3$,

that is: *0.0039 x 100 = 0.39 g*

and so the % w/w is:

$$\frac{0.39 \; x \; 100}{0.8755} = 44.59 \; \%$$

Blank determinations are also used in back titrations to allow for impurities and other sources of error. Their use also simplifies the calculation of the result. Consider, for example, the 1993 British Pharmacopoeia back titration assay procedure for aspirin tablets. In this procedure an accurately weighed sample of powdered aspirin tablets is heated under reflux in a glass volumetric flask for about ten minutes with a known volume of standard sodium hydroxide. The excess sodium hydroxide is titrated with a standard solution of hydrochloric acid using phenol red as indicator (Fig. 11.8).

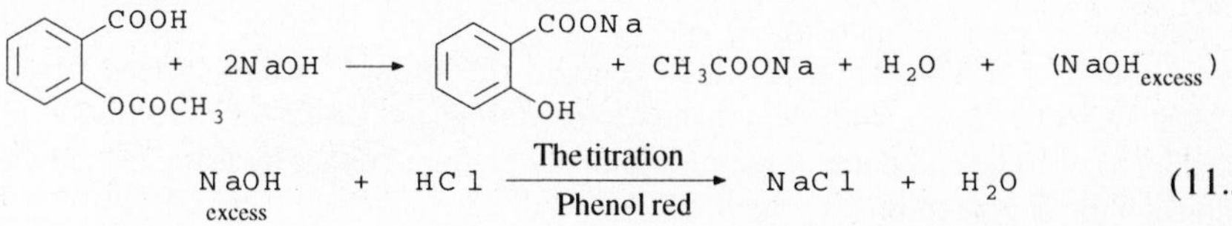

$$NaOH_{excess} + HCl \xrightarrow[Phenol\;red]{The\;titration} NaCl + H_2O \qquad (11.5)$$

Fig. 11.8. The chemistry of the assay of aspirin tablets by back titration with hydrochloric acid.

A blank determination is necessary in this assay because the hot sodium hydroxide can extract alkali from the glass. This extraction will vary depending on the nature of the glass and how many times the flask has been used. Consequently it is not possible to completely eliminate this type of error by the use of a blank, it can only be minimised. In the blank determination, the volume of hydrochloric acid (b cm^3) used in the titration will be equivalent to the volume of sodium hydroxide (t cm^3) used initially plus any alkali extracted from the glass. Similarly, in the assay, the volume of hydrochloric acid used in the titration will be equivalent to the excess sodium hydroxide and also any alkali extracted from the glass. Therefore, for the blank and the assay:

b cm^3 of HCl is equivalent to the NaOH at the start + OH$^-$ from the glass

t cm^3 of HCl is equivalent to the NaOH in excess + OH$^-$ from the glass

The two quantities of alkali (OH$^-$) from the glass will be slightly different but the overall accuracy of this assay is such that in this instance they can be assumed to be the same, within experimental error. Therefore:

$$(b - t) \; cm^3 \; of \; HCl \; is \; equivalent \; to \; (NaOH_{start} - NaOH_{excess})$$

That is $(b - t)$ cm^3 of HCl is equivalent to the NaOH that reacts with the aspirin.

However, from the balanced equation for the reaction of aspirin with sodium hydroxide:

$$\frac{1}{2} = \frac{Moles \; of \; aspirin}{Moles \; of \; sodium \; hydroxide}$$

but from equation 11.5:

1 mole of NaOH is equivalent to 1 mole of HCl

Therefore, the quantity of aspirin in the sample may be calculated from the expression;

$$\frac{1}{2} = \frac{\text{Moles of aspirin}}{\text{Moles of hydrochloric acid}}$$

It should be noted that the molarity of the sodium hydroxide is not used in the calculation and so need not be accurately known provided the same sodium hydroxide solution is used for both the blank and assay.

11.6.4 Non-aqueous titrations

Acids and bases that are not very soluble in water can sometimes be titrated in non-aqueous media. The most common solvents used for this purpose are anhydrous ethanoic acid (glacial acetic acid), ethanol and 1,2-diaminoethane (ethylenediamine). Ethanoic acid is a good solvent for the titration of weak bases, 1,2-diaminoethane for weak acids while ethanol can be used for both weak acids and bases. End points can often be determined using chemical indicators but it is more usual to use potentiometric methods.

The part the solvent plays in the titration depends on the compound being titrated. Consider, for example the titration of adrenaline by perchloric acid using anhydrous ethanoic acid as solvent (Fig. 11.9). The adrenaline reacts with the ethanoic acid to form a solution of its corresponding ethanoate. This solution is titrated with a solution of perchloric acid in ethanoic acid which contains the strongly acidic species $CH_3COOH_2^+$ The $CH_3COOH_2^+$ reacts with the ethanoate ion during the course of the titration to form ethanoic acid. The end point of the titration may be found either by using crystal violet as indicator or a potentiometric method.

In the burette: $HClO_4 + CH_3COOH \longrightarrow CH_3COOH_2^+ + ClO_4^-$

Initially in the flask:

$HO-\langle\rangle-CH(OH)CH_2NHCH_3 + CH_3COOH \longrightarrow HO-\langle\rangle-CH(OH)CH_2\overset{+}{N}H_2CH_3 + CH_3COO^-$

During the titration: $CH_3COOH_2^+ + CH_3COO^- \longrightarrow 2CH_3COOH$

Overall reaction:

$HO-\langle\rangle-CH(OH)CH_2NHCH_3 + HClO_4 \longrightarrow HO-\langle\rangle-CH(OH)CH_2\overset{+}{N}H_2CH_3 + ClO_4^-$

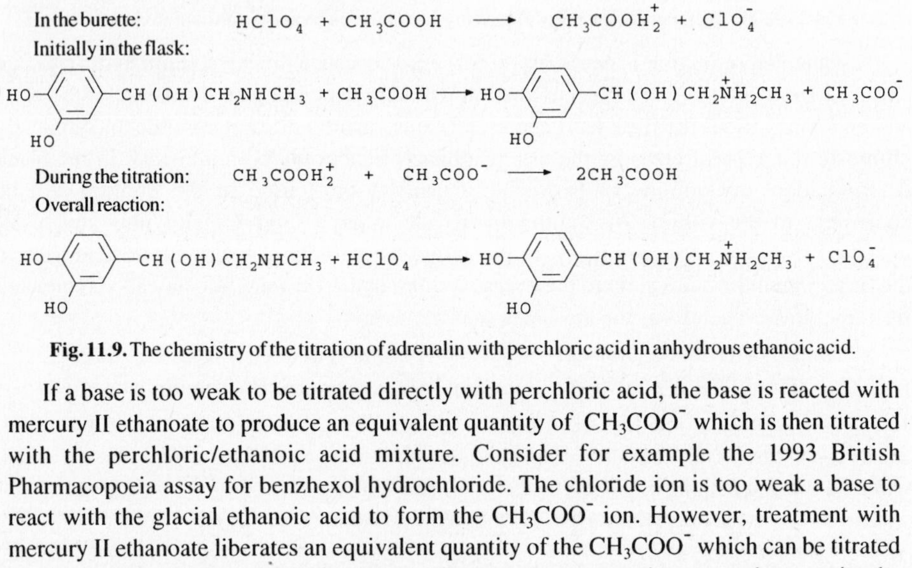

Fig. 11.9. The chemistry of the titration of adrenalin with perchloric acid in anhydrous ethanoic acid.

If a base is too weak to be titrated directly with perchloric acid, the base is reacted with mercury II ethanoate to produce an equivalent quantity of CH_3COO^- which is then titrated with the perchloric/ethanoic acid mixture. Consider for example the 1993 British Pharmacopoeia assay for benzhexol hydrochloride. The chloride ion is too weak a base to react with the glacial ethanoic acid to form the CH_3COO^- ion. However, treatment with mercury II ethanoate liberates an equivalent quantity of the CH_3COO^- which can be titrated by the perchloric/ethanoic acid mixture. The reactions occurring at each stage in the titration are similar to those shown in Fig. 11.9.

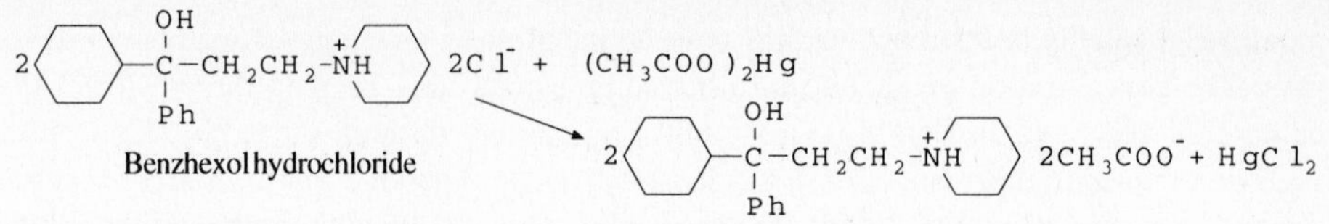

$$2 \left\langle \underset{Ph}{\overset{OH}{\underset{|}{\overset{|}{C}}}} -CH_2CH_2 -\overset{+}{N}H \right\rangle 2C\overset{-}{l} + (CH_3COO)_2Hg$$

Benzhexol hydrochloride

$$\longrightarrow 2 \left\langle \underset{Ph}{\overset{OH}{\underset{|}{\overset{|}{C}}}} -CH_2CH_2 -\overset{+}{N}H \right\rangle 2CH_3CO\overset{-}{O} + HgCl_2$$

Solutions of perchloric acid in anhydrous ethanoic acid are used to titrate many drugs that are weak bases. They are also used to titrate inorganic and organic salts.

11.6.5 Gravimetric methods
Gravimetric analysis is based on weighing to constant mass. The analyte is completely converted to a pure compound with a known molecular formula. This product is weighed to constant mass and this mass, together with a knowledge of the chemistry of the conversion is used to calculate the required answer (Example 11.5). The method is very accurate but suffers from the disadvantage that it is often very time consuming. It is normally used to assay inorganic species.

Example 11.5
50 cm³ of an aqueous solution was treated with 25 cm³ of a 2 M solution of barium chloride. The solution was filtered and the white precipitate of barium sulphate washed with dilute sulphuric acid before being dried to constant mass. If the mass of the precipitate is 0.4461 g calculate the % w/v of sulphate ions in the aqueous solution. The relative ion mass of sulphate ions is 96.1 and the relative atomic mass of barium sulphate is 233.4.

The calculation is based on the assumption that all the sulphate ions in the aqueous solution react with the barium chloride according to the balanced equation:

$$SO_4^{2-} + BaCl_2 = BaSO_4 + 2Cl^-$$

The fraction of sulphate ions in a barium sulphate molecule is 96.1/233.4. Therefore, the mass of sulphate in 0.4461 g of barium sulphate is:

$$\frac{0.4461 \times 96.1}{233.4} = 0.1837 \, grams$$

As all the sulphate ions in the aqueous solution have reacted to form the barium sulphate, this must also be the mass of the sulphate ions in that solution. Therefore, the % w/v of sulphate in the solution is:

$$\frac{0.1837 \times 100}{50} = 0.37\% \, w/v$$

11.6.6 Spectroscopic methods
These methods are discussed in Chapter 10.

11.6.7 Chromatographic methods
Gas-liquid (GL) and high pressure liquid chromatography (HPLC) are normally the preferred chromatographic methods in quantitative analysis. In both cases the simplest way of carrying out an analysis is to plot a calibration curve of either peak height or area against concentration for a series of standard analyte solutions. Peak heights are measured perpendicularly from the top of the peak to a straight line that links the base lines on either side of the peak (Fig. 11.10a). They are used when the peaks are sharp. Peak area can be

measured manually or the machine may have an integrator which automatically records the area. The concentration of analyte in other samples can now be read directly from this calibration curve which should be a straight line passing through the origin. Concentrations determined in this manner have a consistency error of about 2 to 5%. A more accurate method is to use an internal standard (if possible). This can be any compound that elutes from the column well clear of the analyte. The same concentration of standard is added to each of the standard analyte solutions and the peak heights (areas) of both the standard and analyte are measured. A graph of the ratio of these peak heights (areas) for each standard/ analyte solution against the concentration of the analyte is plotted and the concentrations of unknown analyte solutions read from this calibration curve (Fig. 11.10b). Consistency error of 0.5 to 1% can be achieved using this method.

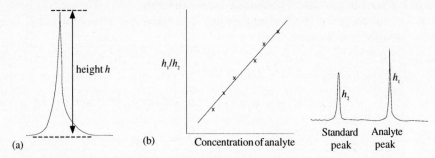

Fig. 11. 10. (a) Measurement of peak height in GL and HPLC. (b) Example of a calibration curve using an internal standard.

11.7 WHAT YOU NEED TO KNOW

(1) The quality of pharmaceutical products is governed by various official bodies who detail the specifications required in the monographs found in official publications.

(2) *Identification tests in an official publication can only be used to distinguish between substances found in that publication.*

(3) *A limit test is used to check whether a substance contains more or less than the permitted amount of a specified impurity.*

(4) *Assays are used to determine the percentage of a specified component in a sample.*

(5) The main sources of impurities in a substance are: (a) degradation often due to hydrolysis by atmospheric moisture and atmospheric oxidation; (b) impurities in the raw materials; (c) unreacted starting material; (d) solvent contamination; (e) machinery wear and tear; (f) atmospheric pollution; (g) human error.

(6) *Incorrect sampling procedures can invalidate the analytical results. The sampling procedure laid down for a substance must be followed if consistent results are to be obtained.*

(7) Quantitative results are sometimes expressed in terms of the dry mass of the sample because the moisture content of some samples can vary quite considerably.

(8) *Reactions used for identification tests must either involve an observable change, such as a gas evolved, change in colour or smell and precipitate*

formation or form a solid compound (derivative) which has a sharp reproducible
  melting point.

(9) The measurement of physical properties, such as, melting point, infra-red spectrum, the absorption of a solution of a specific strength in a specified cell at a designated wavelength in the UV or visible regions, refractive index, specific gravity and n.m.r. spectra are used as identification tests.

(10) Comparative thin layer chromatography can be used to identify compounds.

(11) Limit tests involve the comparison of a test sample result with either the result from a standard or the result quoted in the literature. A test sample fails if it exceeds the standard result or the literature result.

(12) Limit tests are commonly based on:
  (a) Colour comparison (e.g. Nessler tube methods).
  (b) UV and visible spectroscopy (measure absorbance or transmission at a specific wavelength).
  (c) Chromatography: TLC, paper (compare intensities of spots), all forms of column (compare relative size of peaks).
  (d) Electrophoresis

(13) Assays are used to assess the quantity of a substance in a sample. They are commonly based on:
  (a) Volumetric analysis.
  (b) Gravimetric analysis.
  (c) Spectroscopic analysis
  (d) GL and HPLC chromatography.

(14) Primary standards are compounds that are pure and stable enough to be used as reference compounds in assays.

(15) In simple volumetric titrations the analyte reacts directly with the standard reagent used for the titration. The end point of the titration may be found by chemical or physical means.

(16) Indicators are compounds that change colour at the end point of a reaction.

(17) Physical methods of determining end points are based on measuring changes in:
  (a) The potential between two electrodes in the titration mixture during the course of the titration (potentiometric titrations).
  (b) The conductivity of the titration mixture during the course of the titration (conductometric titrations).
  (c) The current flowing through the titration mixture during the course of the titration (amperometric titrations).

(18) Calculations for simple volumetric titrations are based on the molar ratio obtained from the balanced equation for the reaction (see equation 11.1).

(19) Back titration is a technique that is used when it is not possible to assay an analyte by means of the simple titration method. In back titrations a known quantity of a reagent, which contains an excess of that reagent, is reacted with the analyte. The excess reagent is determined by titration. The quantity of the reagent reacting with the analyte is the difference between this excess and the original quantity of reagent used. Calculation of analyte concentration is then made in the usual way.

(20) Blank determinations are used to allow for the presence of impurities that can react with the titrating reagent. They are also used to allow for other parameters that affect the accuracy of an assay. They can be used in all types of assay.

(21) Non-aqueous titrations are usually carried out in anhydrous ethanoic acid (for bases), 1,2-diaminoethane (for acids) and ethanol (for both acids and bases). End points are usually found by potentiometric means but indicators can be used.

(22) Gravimetric analysis is based on converting the analyte to a substance of known chemical constitution (a derivative) which is weighed to constant weight. The concentration of the analyte is calculated from a knowledge of the chemical constitution of this derivative and the assumption that all the analyte is converted into this derivative.

(23) Spectroscopic assay methods are based on the principles described in Chapter 10.

(24) GL and HPLC are the main chromatographic methods used in assays. In both cases either peak heights or areas of standard analyte samples are used to produce a calibration curve which is used to make the determination.

11.8   QUESTIONS

(1)   A sample of chloroquine phosphate weighing 1.4326 g lost 190 mg when dried to constant weight. Calculate the % w/w of water in the drug.

(2)   What features are required if a chemical reaction is to be used as (a) an identification test, (b) the basis of a limit test and (c) the basis of an assay?

(3)   The drug benzyl benzoate was prepared using the route:

$$PhCH_2OH \quad + \quad PhCOOH \underset{}{\overset{HCl}{\rightleftharpoons}} PhCH_2OCOPh \quad + \quad H_2O$$

(a)   Devise two feasible identification tests for benzyl benzoate. One of these tests must be based on physical methods.

(b)   List *all* the possible substances for which limit tests may be required in the monograph for benzyl benzoate. Describe feasible limit tests for the substances you have listed.

(c)   Outline, by means of equations and notes, a possible chemically based assay procedure for benzyl benzoate.

(4)   Outline, by means of equations, chemical reactions that could be made the basis of two identification tests for (a) aspirin, (b) paracetamol and (c) amphetamine. The structural formulae of these compounds can be found by consulting the index.

(5)   A 25 cm³ solution of glycine is treated with methanal. The resulting reaction mixture required 25.7 cm³ of 0.1 M sodium hydroxide for neutralisation. A blank determination required 0.4 cm³ of the same molarity sodium hydroxide solution. Calculate the concentration of glycine in the original solution. See 28.5.2 for the chemistry of this assay.

(6)   A 2.1373 g sample of chalk was reacted with 50.0 cm³ of 0.961 M hydrochloric acid. The excess acid required 8.7 cm³ of 1.07 M sodium hydroxide solution for neutralisation. Calculate the %w/w of $CaCO_3$ in the sample of chalk.

(7) A 0.5354 g sample of aspirin powder was reacted with 30 cm^3 of a hot 0.5012 M sodium hydroxide solution. The resulting mixture required 20.2 cm^3 of a 0.4956 M hydrochloric acid solution for neutralisation. A blank determination was carried out, the reaction mixture from this experiment requiring 31.6 cm^3 of the same molarity hydrochloric acid for neutralisation. Calculate the %w/w of $C_9H_8O_4$ in the powder.

(8)

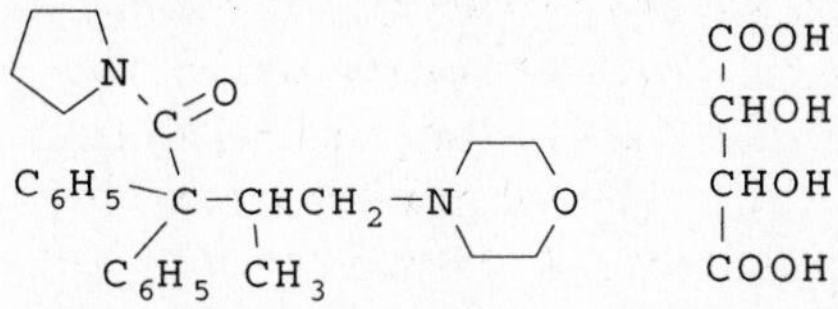

A solution of dextromoramide tartrate (A) in glacial ethanoic acid was titrated with 0.0491M perchloric acid. If 30 cm^3 of the drug solution required 25.7 cm^3 of the perchloric acid for complete reaction, calculate:

(a) the ml equivalent of the 0.05 M perchloric acid in this assay. (A reacts in a 1:1 ratio with the acid.)

(b) the %w/v of $C_{25}H_{32}N_2O_2 \cdot C_4H_6O_6$ in the sample of A. (RMM of $C_{25}H_{32}N_2O_2 \cdot C_4H_6O_6$ is 542.6.)

(9)

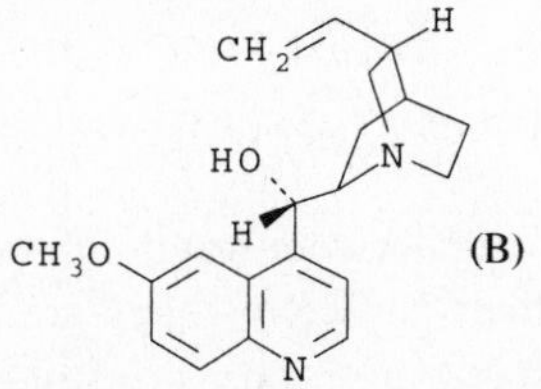

A 0.2964 g sample of quinine hydrochloride (B) lost 0.0103 g on drying. If 0.3386 g of the same sample reacted exactly with 14.52 cm^3 of 0.1187 M perchloric acid in the presence of mercury II ethanoate ,calculate the %w/w of $C_{20}H_{24}N_2O_2$ in (a) the original sample of B and (b) the dried sample of B. (RMM of $C_{20}H_{24}N_2O_2$ is 396.6, 1 ml of 0.1 M perchloric acid is equivalent to 18.04 mg of B.)

(10) A solution containing 10.2 mg of a sample of testosterone ($C_{19}H_{28}O_2$) in 100 cm^3 of absolute ethanol was prepared and 5 cm^3 of this solution was diluted to 50 cm^3 with absolute ethanol. The dilute solution was found to have an absorbance of 0.555 at 240 nm in a 1 cm cell. Calculate the % purity of the original sample of testoserone. (A 1%,1 cm for testosterone in absolute ethanol is 560.)

(11) Analysis, by HPLC, of a series of urea / alanine solutions containing 12.5 mg/dm^{-3} of alanine gave the results shown. Analysis of a urine solution after the addition of 12.5 mg/dm^{-3} of alanine gave a 3.70 cm peak that was identified as urea. Determine the concentration of urea in the sample if the average height of the alanine peaks in each of the standard solutions was 6.30 cm.

Concentration of urea / alanine solution:

µg/dm^{-3} of urea	2.50	5.00	6.70	8.40
Urea peak height (cm)	1.20	2.45	3.25	4.10

12

Hydrocarbons

12.1 INTRODUCTION

Hydrocarbon is the name given to organic compounds whose structures contain only carbon and hydrogen. On the basis of their structures, they are broadly classified as aromatic and aliphatic. Aromatic hydrocarbons are conjugated ring systems that have $(4n + 2)$ pi electrons where n is an integer (12.6). Aliphatic hydrocarbons consist of chain structures and ring systems that do not have a conjugated system containing $(4n + 2)$ pi electrons. Both aliphatic and aromatic hydrocarbons are further subdivided into different classes according to the nature of structures of their carbon skeleton (12.1). These carbon skeletons form the basis of the structures of all other organic molecules.

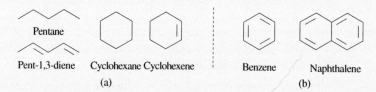

Pentane		
Pent-1,3-diene	Cyclohexane Cyclohexene	Benzene	Naphthalene
	(a)		(b)

Fig 12.1. Examples of some of the different classes of organic hydrocarbons. (a) Aliphatic. (b) Aromatic.

The C–H bond found in all types of hydrocarbon is not usually as reactive as the structures that are traditionally thought of as functional groups. Consequently, the C–H bond is not classified as a functional group. However, the reactivity of all C–H bonds depends on the nature of the carbon skeleton to which they are attached. For example, the presence of strong electron acceptor group(s) bonded to an adjacent carbon will increase the scope of the C–H bonds reactivity.

12.2 ALKANES AND CYCLOALKANES

The alkanes were originally known as the paraffins. Their carbon skeletons consist of either branched or unbranched chains of sigma-bonded carbon atoms (Fig. 12.2). The unbranched chains are often referred to as **straight chains**. However, the chains are not straight since each C–C bond in the structure will exist in the appropriate conformation for the prevailing physical conditions and molecular environment (4.2). It should be noted that the shorthand system for representing alkane carbon chains is normally drawn in the staggered conformation, each angle in the chain representing a CH_2 or **methylene** structure whilst the end of a line represents a CH_3 or **methyl** group.

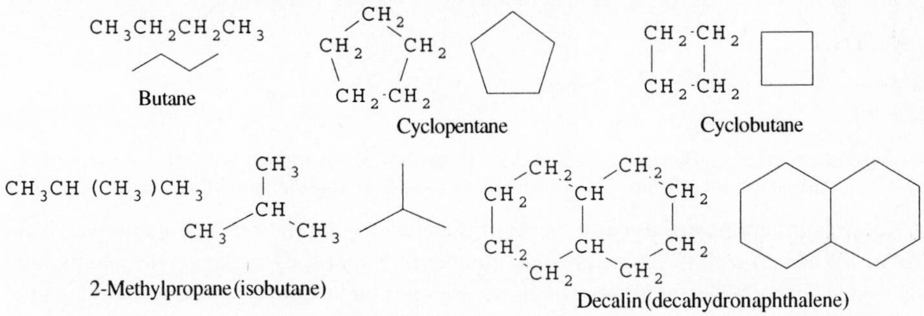

$CH_3CH_2CH_2CH_3$

Butane

Cyclopentane

Cyclobutane

$CH_3CH(CH_3)CH_3$

2-Methylpropane (isobutane)

Decalin (decahydronaphthalene)

Fig 12.2. Examples of alkanes and cycloalkanes. The line structures given with each of the structural formulae are often used to represent these structures either as such or when they form parts of the structures of more complex molecules. The end of a line is always a CH_3 group. The prefix **iso-** is sometimes used to indicate a terminal dimethyl structure.

Simple cycloalkanes are the cyclic analogues of the straight chain alkanes. Their rings may have straight or branched chain alkane residues as substituents. The simplest of the cycloalkanes, cyclopropane has a flat ring, cyclopentane has a puckered ring and spectroscopic evidence suggests that cyclobutane is slightly bent. All the others exist in the form of the most appropriate conformation(s) for their environment (4.3, 4.4 and 4.5). Cycloalkanes, such as decalin (Fig. 12.2), consisting of two or more rings joined by common bond(s) are said to possess a **fused ring system**. The rings will exist in the appropriate conformations for the system.

12.2.1 Nomenclature

Table 12.1. The names and structures of the first ten straight chain alkanes.

N°· of C atoms	Structure	Name
1	CH_4	Methane
2	CH_3CH_3	Ethane
3	$CH_3CH_2CH_3$	Propane
4	$CH_3CH_2CH_2CH_3$	Butane
5	$CH_3CH_2CH_2CH_2CH_3$	Pentane
6	$CH_3CH_2CH_2CH_2CH_2CH_3$	Hexane
7	$CH_3CH_2CH_2CH_2CH_2CH_2CH_3$	Heptane
8	$CH_3CH_2CH_2CH_2CH_2CH_2CH_2CH_3$	Octane
9	$CH_3CH_2CH_2CH_2CH_2CH_2CH_2CH_2CH_3$	Nonane
10	$CH_3CH_2CH_2CH_2CH_2CH_2CH_2CH_2CH_2CH_3$	Decane

The names of the simplest 'straight chain' alkanes are listed in Table 12.1. The names of compounds containing 1, 2, 3 and 4 carbon atoms are traditional and have no logical basis. The names of straight chain alkanes with 5 or more carbon atoms consist of a stem derived from the Greek number corresponding to the number of carbon atoms in the chain combined with the suffix **-ane**. This suffix is used to indicate that a carbon-hydrogen skeleton is fully saturated, that is, its structure does not contain any pi bonds.

The names of branched chain alkanes are based on the name of the longest straight chain structure in the molecule. The carbon atoms of this chain are numbered from one end of the chain so that the numbers of the atoms to which the side chains are attached are kept as low as possible (Fig.12.3). These numbers are known as **locants**.

Fig. 12.3. Numbering of branched alkane chains. The numbering shown in structure (a) is correct whilst that shown for structure (b) is incorrect since 4 and 5 are a higher locants than 2 and 3.

Side chains are treated as substituents of the main chain. Their names are derived from that of the alkane with the corresponding number of carbons, by replacing the ending -ane with **-yl**. These substituent alkane structures are referred to as **alkyl groups**. For example, a side chain having the structure CH_3CH_2— would be called an ethyl group. The positions of side chains in a structure is indicated by prefixing their names by the number of the carbon to which they are attached (Fig. 12.4). This number and substituent name are used as prefixes to the name of the longest straight chain in order to obtain the name of the complete structure. Should there be more than one side chain of the same type, the alkyl name of that side chain is prefixed by di, tri, etc. and the appropriate numbers. In all names it is customary to separate numbers from letters by a dash (-) and number from number by a comma (,). When drawing structures abbreviations consisting of the first letters of the name of the alkyl group are often used to represent the structure of an alkyl group. For example, Me and Et are used for methyl group and ethyl groups respectively.

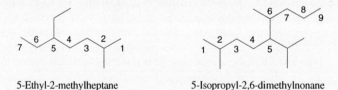

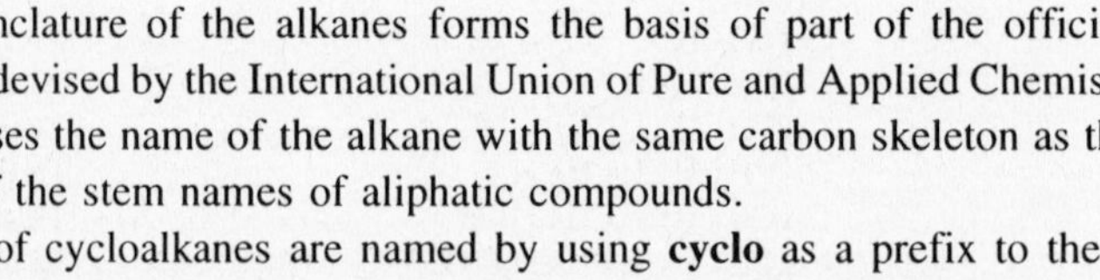

5-Ethyl-2-methylheptane   5-Isopropyl-2,6-dimethylnonane

Fig. 12.4. Examples of the nomenclature of branched chain alkanes.

The nomenclature of the alkanes forms the basis of part of the official systematic nomenclature devised by the International Union of Pure and Applied Chemistry (IUPAC). This system uses the name of the alkane with the same carbon skeleton as the compound as the basis of the stem names of aliphatic compounds.

The rings of cycloalkanes are named by using **cyclo** as a prefix to the name of the straight chain alkane with the same number of carbon atoms as the ring (Fig. 12.5). Substituted cycloalkanes are named in the same way as branched chain alkanes.

(a) 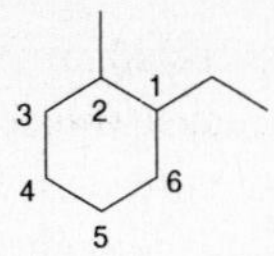 1-Ethyl-2-methylcyclohexane   (b) 1,2,3-Trimethylcyclopentane   (c) 1-Methyl-2-(2-Methylbutyl)cyclohexane

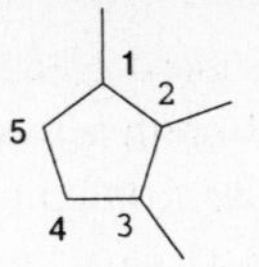

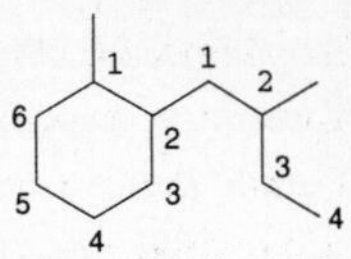

Fig 12.5. Examples of the nomenclature of cycloalkanes. The locant 1 is omitted in the name when there is no possibility of ambiguity (example a). The ending -yl for the substituent indicates that it is carbon 1 of the side chain that is attached to the cyclohexane ring (example c). It should be noted that a bracket indicates that the structure inside the bracket is a complete substituent with its own number system (example c).

12.2.2 Physical properties

Alkanes and cycloalkanes are insoluble in water because their molecules do not contain any structures that are capable of forming hydrogen or ion-dipole bonds with water molecules. The only forces of attraction between the alkane and water molecules are van der Waal's forces which in this situation are very weak. In general, organic compounds with a low number of functional groups, that can either hydrogen bond or form ion-dipole bonds with water, and a large carbon-hydrogen skeleton are not very soluble in water. However, the solubility of alkanes in lipids and other alkanes is good because there are strong van der Waal's forces of attraction between these molecules. Consequently, in biological systems alkanes will readily move from aqueous media, such as plasma, into non-aqueous media, such as lipids. For example, alkanes such as hexane that are used as solvents in many household products are readily absorbed via the lungs into the blood stream. They rapidly pass from the blood stream into the nervous system and brain because they are highly soluble in lipids. Low concentrations in these organs can cause hallucinations whereas high concentrations result in loss of consciousness and in extreme cases, death. The absorption is reversible and cyclopropane is used, mixed with oxygen, as a general anaesthetic.

12.2.3 Chemical properties

Alkanes and cycloalkanes are not very reactive. Their chemical properties with the exception of cyclopropane and cyclobutane, are generally very similar. These compounds, because of their small rings, are more reactive than the alkanes and cycloalkanes with five or more carbons in their rings. The reactions of both cyclopropane and cyclobutane normally involve the opening of the ring. For example, cyclopropane reacts with bromine, hydrogen bromide and sulphuric acid to form 'straight chain' compounds. This last reaction is the basis of an assay procedure for cyclopropane.

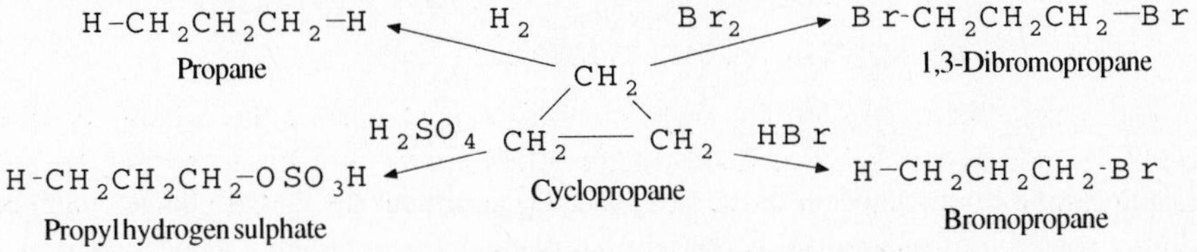

The reactivity of cyclopropane is currently explained in terms of poor orbital overlap and conformational repulsions. Consider, for example, the flat ring of cyclopropane. In order to obtain such a structure the sp^3 orbitals of the carbons must overlap in the manner

shown in Fig. 12.6. This overlap is not as good as the normal C–C bond orbital overlap and so will result in a weaker and therefore more reactive C–C bond. The unusual reactivity of cyclobutane can be explained in a similar fashion.

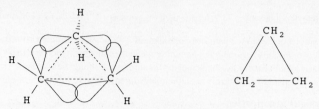

Fig 12.6. The orbital structure of cyclopropane showing the bent bonds and poor sp^3 orbital overlap. The carbons in this structure are all in the eclipsed state.

Alkanes and cycloalkanes with the exceptions of cyclopropane and cyclobutane, exhibit a limited number of **substitution** and **oxidation** reactions under both laboratory and biological conditions. The laboratory reactions are often difficult to carry out and so the alkanes and the larger cycloalkanes are regarded as being almost chemically inert. This general lack of reactivity is also exhibited by alkane and cycloalkane residues in the structures of other compounds.

Substitution is usually carried out under laboratory conditions that are far removed from those found in the body. Mixtures of isomers are frequently obtained. However, the ease of substitution is usually in the order:

tertiary (methine CH) > secondary (methylene CH$_2$) > primary (methyl CH$_3$)

For example, chlorine reacts with 2-methylpropane to form a mixture of 2-chloro-2-methylpropane and 2-methylchloropropane in the ratio of 2:1 respectively.

$$CH_3 \xrightarrow{Cl_2} \quad + \quad$$

2-Methylpropane   2-Chloro-2-methylpropane   2-Methylchloropropane

Laboratory oxidation of alkanes is rare and often difficult. Methyl groups are often oxidised to carboxylic acids whilst methylene and methine groups are usually either hydroxylated or converted to ketones. The rate of oxidation appears to be in the order:

methine (CH) > methylene (CH$_2$) > methyl (CH$_3$)

$$\xrightarrow{KMnO_4}$$

2-Methylpropane   2-Methylpropan-2-ol

Biological oxidations occur in the body usually at or near the end of alkane chains and yield the same types of product as the laboratory oxidations. Oxidations occurring in the liver involve mixed-function oxidases with nicotinamide adenine dinucleotide acting as coenzyme for the process. Tracer experiments with $^{18}O_2$ and $H_2^{18}O$ have shown that these reactions require molecular oxygen and not water as the source of oxygen.

Metabolism of drugs can occur by hydroxylation of the alkane residues in their structures, for example, pentobarbitone is metabolised by hydroxylation of its butyl side chain.

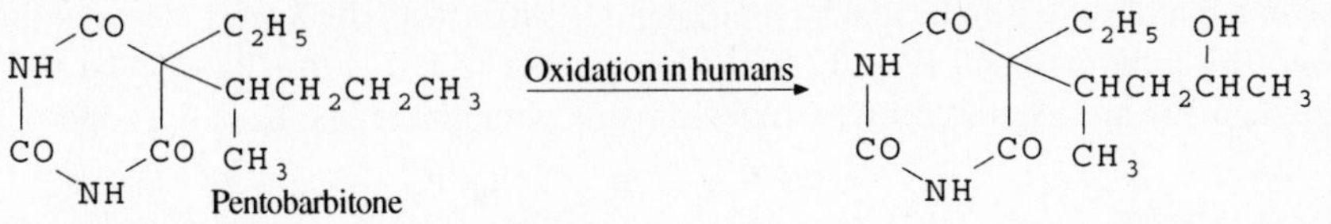

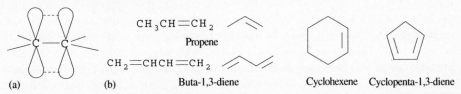

12.2.4 Pharmaceutical and other uses

Alkanes and cycloalkanes are used mainly as solvents in industry and many household products. Their vapours irritate the respiratory system and in high concentrations have a narcotic action and cause death. This has given rise to the modern phenomenon known as glue sniffing which is fast becoming a considerable social problem.

Liquid paraffin is used as a laxative. It is not absorbed in its passage through the body and rapidly passes into the bowel where it acts as a lubricant for faecal material. However, its use is not recommended since it reduces the absorption of lipid-soluble vitamins from food. It can also cause lipid pneumonia if sufficient enters the lungs.

Yellow soft paraffin is a semi-solid translucent mixture of hydrocarbons produced from petroleum residues. Its composition varies but its density is always about 0.82 to 0.85 g cm^{-3} at room temperature. It is used as a lubricant and a moisture-resistant base for ointments and creams. *Vaseline* is a homogeneous form of yellow soft paraffin. Soft white paraffin is bleached yellow soft paraffin. It is used for the same purposes as yellow soft paraffin. Hard paraffin is a mixture of solid paraffins with a density of about 0.9 g cm^{-3}.

Cyclopropane is used in conjunction with oxygen as a general anaesthetic. It is readily absorbed via the lungs into the bloodstream but its low water solubility and high lipid solubility results in it being rapidly transferred from the blood stream to the nervous system where it causes anaesthesia.

12.3 ALKENES AND CYCLOALKENES

Alkenes are straight chain hydrocarbons whose structures contain the planar alkene (C=C) functional group. The structure of this bond is discussed in 2.4 and also outlined in Fig. 12.7. They will exhibit geometric isomerism (5.7) at room temperature provided the individual carbons of the C=C bond are unsymmetrically substituted.

Cycloalkenes are the cyclic analogues of the alkenes (Fig. 12.7). They do not exhibit geometric isomerism because the ring structure restricts them to one configuration. However, the rings will exist in the most appropriate conformation (4.4) for the prevailing molecular and atmospheric conditions.

Fig 12.7. (a) The orbital structure of a C=C bond. (b) Some examples of simple alkenes and cycloalkenes.

12.3.1   Nomenclature

Alkenes and cycloalkenes are named by replacing the ending **-ane** of the name of either the corresponding alkane or cycloalkane with the suffix **-ene**. The position of the C=C bond in the molecule is indicated by prefixing this suffix with the lowest locant of the two carbons forming the bond (Fig. 12.8). If this number is 1 it is usually omitted from the name. Isomers are identified by prefixing a name with **cis**, **trans**, **E** or **Z** as appropriate.

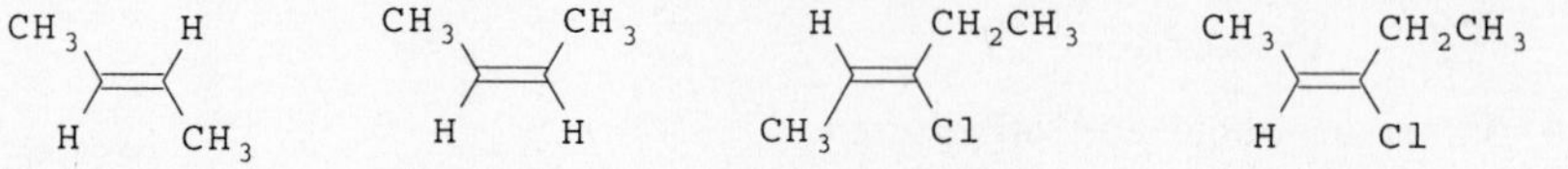

Trans-but-2-ene   Cis-but-2-ene   Z-3-chloropent-2-ene   E-3-chloropent-2-ene

Fig 12.8. Examples of the nomenclature of alkenes and cycloalkenes.

Alkene substituents have the ending **-ene** changed to **-enyl**. For example, a propene substituent becomes propenyl or prop-2-enyl depending on the point of attachment of the substituent to the parent structure.

Propenyl   Prop-2-enyl

12.3.2   Physical properties

(a) Physical constants. The values of the physical constants of geometric isomers are usually different (5.9). This make it possible to identify a geometric isomer by measuring an appropriate physical constant and comparing its value with those in the literature.

(b) Solubility. The water solubility of alkenes and cycloalkenes is poor because the C=C bond is not capable of forming either hydrogen or ion-dipole bonds with water. Furthermore, the van der Waals' forces between C=C bonds and water are also very weak. However, lipid solubility is good as the van der Waals' forces between the lipid molecules and the alkene are strong.

12.3.3   Chemical properties

The main reactions of the C=C bonds of alkenes and cycloalkenes fall into the general categories **electrophilic addition, free radical addition, oxidation, reduction** and **polymerisation**. It is important to realise that these types of C=C bond can react in this general manner when they occur in more complex structures. However, practical observations do not always agree with the theoretical predictions based on this premise.

(a) Electrophilic addition. Many compounds, such as, the halogens and halogen acids, that can act either as electrophiles or give rise to electrophilic species, add to the C=C of alkenes and cycloalkenes (Fig. 12.9). These reactions are classified as electrophilic additions because the rate governing step of their mechanisms is believed to be the reaction of the pi electrons with the electrophile (attraction of opposites). This results in the formation of a metastable intermediate whose structure depends on the nature of the electrophile. However, it is normally thought of as being a positively charged ion, usually a carbocation. This intermediate reacts rapidly with any nucleophiles present to form products which will be mainly in the staggered (anti) conformation. For example, in the case of the halogen acids the intermediate is believed to be a carbocation which reacts mainly with the halide ion of the acid (Fig. 12.10). This carbocation can also react with

General equation

$$\text{C}=\text{C} \quad + \quad \text{A}-\text{B} \quad \longrightarrow \quad \underset{\overset{|}{\text{C}}-\overset{|}{\text{C}}}{\overset{\text{A}\ \ \text{B}}{}}$$

General examples

$$\text{C}=\text{C} \quad + \quad \text{Br}-\text{Br} \quad \longrightarrow \quad \underset{}{\overset{\text{Br Br}}{-\text{C}-\text{C}-}}$$

$$\text{C}=\text{C} \quad + \quad \text{H}-\text{Hal} \quad \longrightarrow \quad \underset{}{\overset{\text{H Hal}}{-\text{C}-\text{C}-}}$$

$$\text{C}=\text{C} \quad + \quad \text{H}-\text{OSO}_3\text{H} \quad \longrightarrow \quad \underset{}{\overset{\text{H OSO}_3\text{H}}{-\text{C}-\text{C}-}}$$

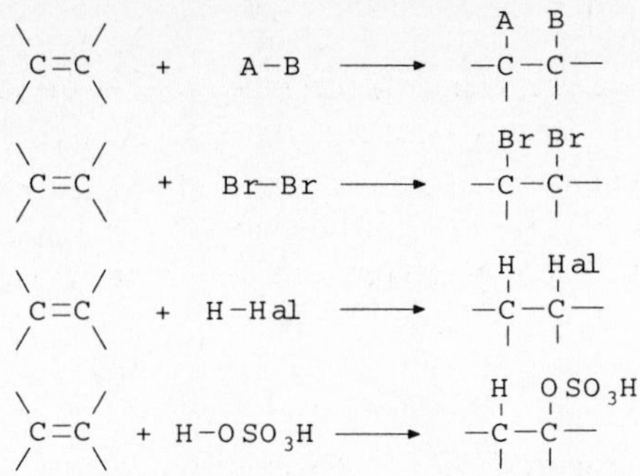

Fig. 12.9. General examples of the electrophilic addition reactions of alkenes. The electrophilic region (A) of the reagent molecule is on the left of the single bond.

any nucleophilic solvents, such as ethanol, used for the reaction which gives rise to some of the by-products found in these electrophilic addition reactions.

$$\text{CH}_3\text{CH}=\text{CHCH}_3 \xrightarrow{\text{H}^+} \text{CH}_3\overset{+}{\text{CHCHCH}}_3 \xrightarrow{\text{Cl}^-} \text{CH}_3\overset{\text{H Cl}}{\underset{}{\text{CHCHCH}}}_3$$

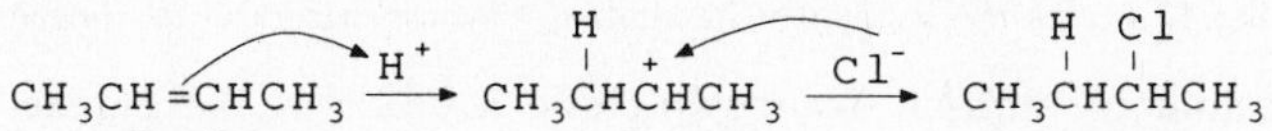

But-2-ene   2-Chlorobutane

Fig. 12.10. The mechanism of the electrophilic addition of hydrogen chloride to but-2-ene.

In reactions where there is the possibility of the formation of more than one intermediate positive ion the reaction usually appears to proceed by forming the most stable of these ions (8.5). The most stable intermediate has the lowest energy transition state (TS) and so will be formed in preference to less stable intermediates with higher energy transition states. It will also remain in the reaction mixture long enough to collide with a suitable reactant and form the product. Less stable intermediates would probably decompose back to the starting material before a suitable collision could occur. For example, the addition of hydrogen chloride to propene could in theory produce the two intermediate carbocations (a) and (b) (Fig. 12.11). The positive charges of both these carbocations could be delocalised, that is, they are effectively reduced by being spread over the adjacent C–C sigma bonds. However, the degree of delocalisation will be greater in intermediate (a) since it has two inductive effects to (b)'s one. Consequently, in theory, (a) will be more stable than (b). In practice the reaction results in the exclusive formation of 2-chloropropane and so is assumed to proceed entirely through intermediate (a). The addition of the nucleophile, in electrophilic addition, to the most substituted carbon of an alkene is known as **Markownikov addition**.

Nucleophiles, such as, water and alcohols will also undergo reactions with alkenes under acid conditions. These reactions are believed to proceed by an electrophilic addition mechanism, a proton from the acid acting as the initial electrophile. For example, the industrial and pharmaceutical solvent propan-2-ol is produced by the reaction of water with propene:

$$\text{CH}_3\text{CH}=\text{CH}_2 \xrightarrow{\text{H}^+/\text{H}_2\text{O}} \text{CH}_2\text{CH(OH)CH}_3$$

Propene   Propan-2-ol (isopropanol, IPA)

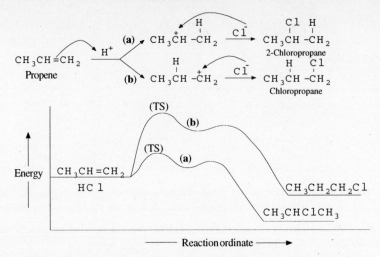

Fig. 12.11. The mechanism for the addition of hydrogen chloride to propene.

Decolourisation of bromine by electrophilic addition is the classic test for $C=C$ bonds. Bromination can also be used to assay alkenes and cycloalkenes. Bromine solutions are not used in this assay since bromine is too volatile to give an accurate result. Instead a solution of potassium bromide and potassium bromate is used in the presence of an acid to generate the bromine (15.11). The assay is a back titration, the excess bromine being treated with potassium iodide and the iodine liberated titrated with sodium thiosulphate. A blank determination is carried out to allow for any loss of bromine or iodine vapours.

$$\underset{/}{\overset{\backslash}{C}}=\underset{\backslash}{\overset{/}{C}} + Br_2 \longrightarrow \underset{|}{\overset{Br}{-C}}-\underset{|}{\overset{Br}{C-}} + Br_{2\,(Excess)}$$

$$Br_{2(Excess)} + 2KI \longrightarrow I_2 + 2KBr \qquad I_2 + 2NaS_2O_3 \longrightarrow 2NaI + Na_2S_4O_6$$

Halogen derivatives, such as iodine monochloride and iodine bromide can be used instead of bromine. These methods are usually satisfactory for compounds that contain only isolated $C=C$ bonds. Iodine monochloride is also used to determine the iodine number of fats and oils (26.10). This determination involves the electrophilic addition of the iodine monochloride to $C=C$ bonds in the fat. The reaction with iodine monochloride is also the basis of a limit test for unsaturated hydrocarbons in the general anaesthetic cyclopropane.

(b) Free radical addition. Addition to $C=C$ bonds can occur by a free radical mechanism in the presence of sunlight and peroxides. In cases where there is a possibility of more than one product the orientation is opposite to that found in electrophilic addition and so the reaction is referred to as **Anti-Markownikov** addition. The free radical addition of hydrogen bromide to propene, for example, yields bromopropane whilst electrophilic addition would result in the formation of 2-bromopropane.

$$\underset{\substack{| \ | \\ \text{2-Bromopropane}}}{\overset{Br\ H}{CH_3CH-CH}} \xleftarrow[\text{Electrophilic addition}]{HBr} CH_3CH=CH_2 \xrightarrow[\text{Free radical addition}]{HBr} \underset{\substack{| \ | \\ \text{Bromopropane}}}{\overset{H\ Br}{CH_3CH-CH_2}}$$

Markownikov addition   Anti-Markownikov addition

On storage many alkenes react slowly with the oxygen of the air to form a little of the corresponding peroxide. These peroxide impurities can initiate anti-Markownikov addition to the alkene when an addition reaction is carried out.

(c) Biological addition. Addition to C=C bonds occurs in a number of biological processes. These reactions are often reversible. For example, hydratases control the addition of water (**hydration**) to fumarate and citrate in the citric cycle of living cells.

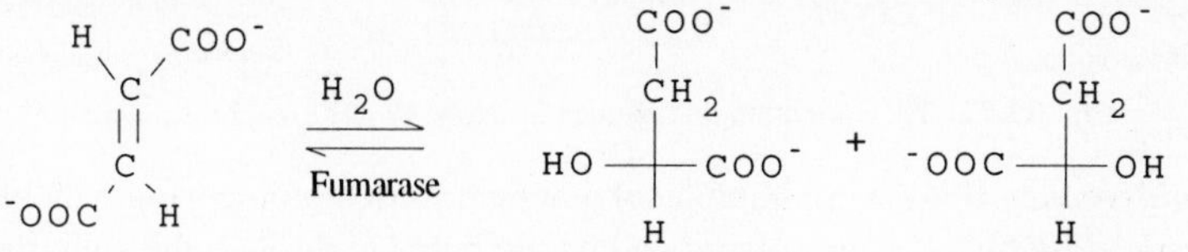

S-malate (80%)   R-malate (20%)

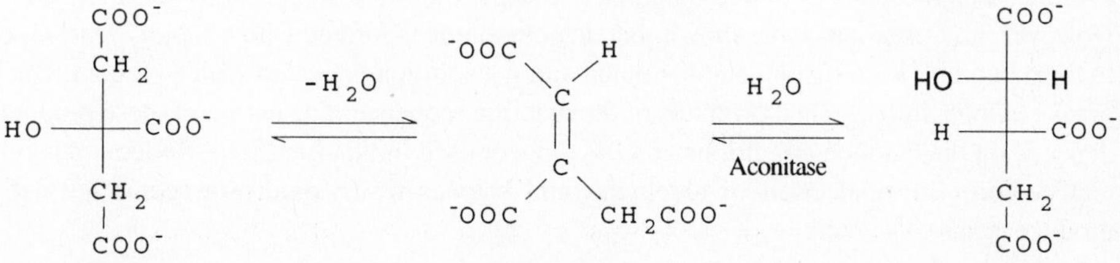

Citrate   cis-Aconite   Isocitrate

In biological systems the C=C bond is often found in the 2-3 position relative to a COZ group where Z can be a variety of structures such as $-OH$, $-OR$, $-SH$ and $-SR$. This structure polarises the C=C and influences the orientation of the addition especially in acid conditions. For example, the conversion of $\alpha\beta$-unsaturated fatty acids to β-hydroxy fatty acids in fatty acid metabolism.

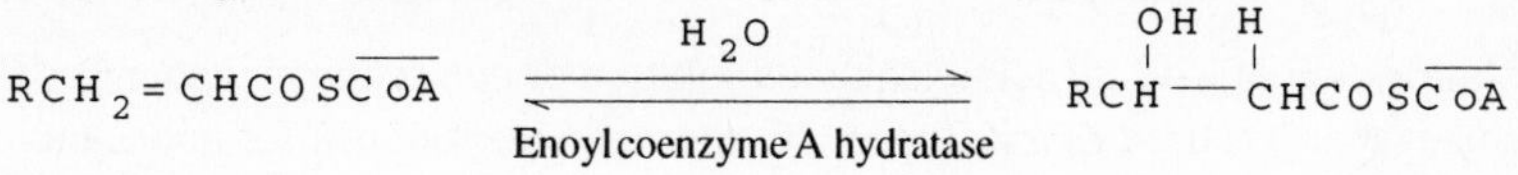

Key

$\overline{CoA}$ = coenzyme A

$Z = \overline{SCoA}$

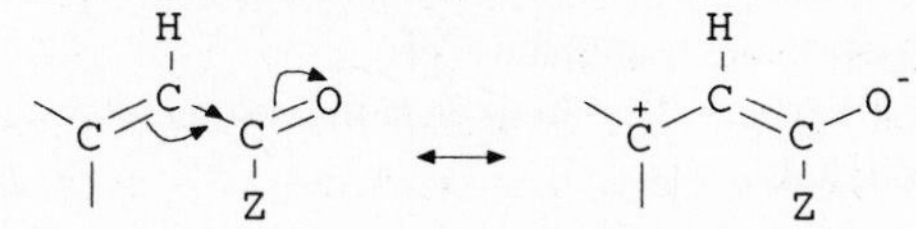

12.3.4 Oxidation

Laboratory oxidation of alkenes and cycloalkenes is relatively easy. The reactions fall into two general categories: those in which only the pi bond is broken and those where both the pi and sigma bonds are broken. Oxidation usually yields either epoxides (oxirans), peroxides, carbonyl compounds, carboxylic acids or diols depending on the oxidising agent and the conditions employed.

Oxidation with ozone is used in structure determination to locate the position of C=C bonds in alkenes. Depending on the number of C=C bonds in its structure, the alkene is converted into two or more compounds by a sequence of reactions based on oxidative cleavage of the C=C bond. The procedure is referred to as **ozonolysis**. The products of ozonolysis are identified and the structure of the original alkene deduced from these products.

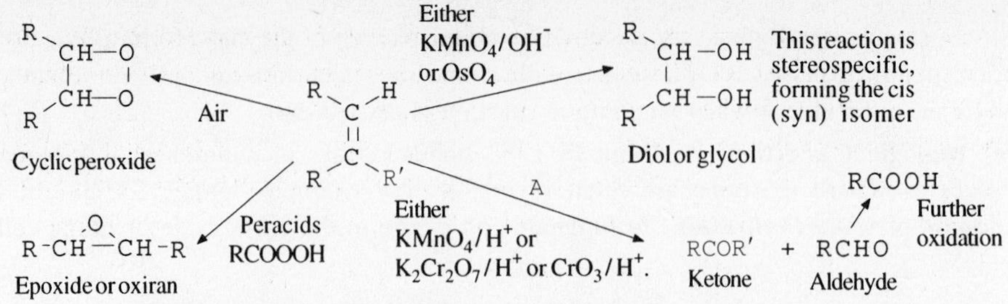

Fig. 12.12. The oxidation of alkenes. Note, R′=H except for reaction A

Ozonolysis is carried out in a suitable solvent such as trichloromethane or dichloromethane. The alkene is dissolved in the solvent and ozone bubbled through the solution to form the ozonide. The mechanism of the reaction is complex and the subject of much discussion. However, investigations have shown that a molozonide is formed which rapidly rearranges to the ozonide. This ozonide is not isolated and the solution is treated with a reagent which breaks it down into the final products of the reaction sequence. The nature of these products depends on the reaction conditions and the reagent used in this final step. Reducing agents tend to form either alcohols or aldehydes and ketones whilst oxidising agents normally produce acids.

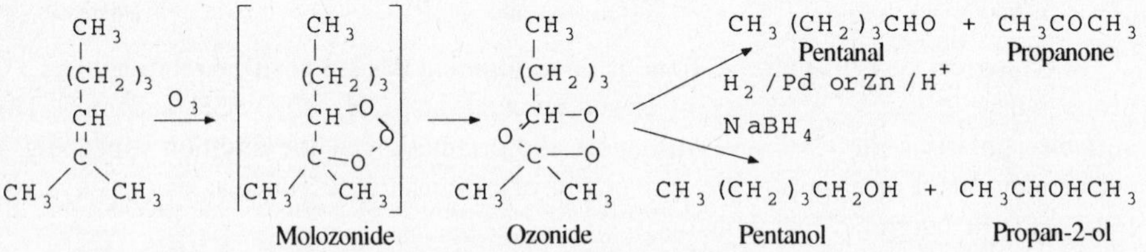

Decolourisation of a dilute acidic aqueous solution of potassium manganate VII (potassium permanganate) is used as a test for C=C and as an identity test for undecanoic acid in the 1993 British Pharmacopoeia. The test is not specific since many other functional groups are oxidised by potassium manganate VII.

Oxidation of alkenes and cycloalkenes in biological systems produces mainly epoxides, peroxides and alcohols.

12.3.5 Reduction

The C=C bonds of alkenes and cycloalkenes are reduced by hydrogen in the presence of a suitable metal catalyst to the corresponding saturated derivative. Reduction may be carried out at room temperature and pressure using either palladium or platinum as catalysts but less reactive metals require higher temperatures and pressures. Reaction occurs on the surface of the catalyst and with the exception of Raney nickel is essentially a **cis** addition of hydrogen to the C=C. The use of Raney nickel often results in a trans addition. The reaction is known generally as **hydrogenation**.

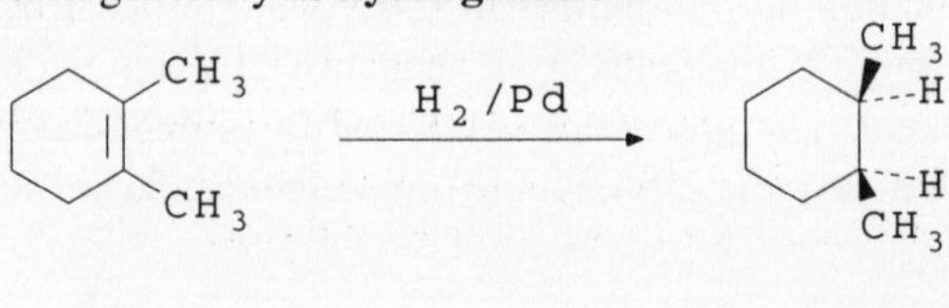

In humans and other mammals, reduction takes place in the liver and is controlled by a group of enzymes collectively known as reductases. The enzyme delivers what is effectively a hydride ion (H ⁻) directly to one end of the C=C bond whilst a proton donor or the surrounding buffer supplies H^+ to the other end. For example, in the reduction of crotonyl-ACP to butyryl-ACP in the biosynthesis of fatty acids the flavine mononucleotide (FMN) acts as the source of what is in effect a hydride ion:

$$CH_3CH=CHCOSCoA + NADPH + H^+ \xrightleftharpoons{FMN} CH_3CH_2CH_2COSCoA + NADP^+$$

Hydrogenation of C=C bonds is used to change the physical properties of fats and oils. The process is known as **fat hardening**. It can also be used to estimate the number of C=C double bonds in an alkene. The volume of hydrogen absorbed by a known mass of the alkene is measured and converted to moles using Avogadro's relationship, **1 mol of a gas occupies 22.4 dm³ at S.T.P**. Since 1 mole of C=C bonds requires 1 mole of hydrogen for complete reduction, the number of C=C can be estimated from the relationship:

$$\text{Number of C=C bonds} = \frac{\text{Moles of hydrogen absorbed}}{\text{Moles of alkene used}}$$

12.3.6   Polymerisation

This is the name given to the process by which the molecules of either the same or two or more different compounds react with each other to form large molecules with large relative molecular masses. The products of these reactions are known as **polymers** and the starting compounds **monomers**. Polymers are used extensively in everyday life and also have a number of specific uses in medicine.

Alkenes can be polymerised under a variety of conditions by both electrophilic and free radical addition mechanisms. In both cases a chain reaction occurs, the product of one reaction initiating the next reaction.

Monomer   Polymer

Polymerisation occurs in some metabolic processes (27.12) but many of the reactions yield only small polymers consisting of a few monomer residues.

12.4   POLYENES

Polyenes are **aliphatic** hydrocarbons that contain more than one C=C bond in their structures (Fig. 12.13). These bonds may or may not be conjugated. Bonds that are not conjugated are referred to as **isolated** C=C bonds. Aromatic hydrocarbons (12.7) which have conjugated double bond structures are not classified as polyenes because they have properties that are significantly different from those of aliphatic polyenes.

12.4.1   Nomenclature

Aliphatic polyenes are named in the same way as alkenes and cycloalkenes. The number of C=C bonds in the molecule is indicated by the addition of the prefixes **di, tri, tetra**, etc to the ending **-ene**. Locants are dropped if there is no possibility of ambiguity (Fig. 12.13). Trivial names are normally used for more complex polyenes.

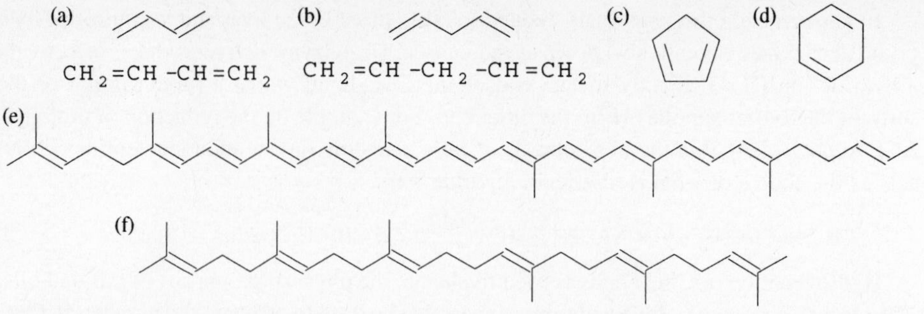

(a)

$CH_2=CH-CH=CH_2$

(b)

$CH_2=CH-CH_2-CH=CH_2$

(c)

(d)

(e)

(f)

Fig. 12.13. Examples of the nomenclature of polyenes whose structures contain conjugated and non-conjugated double bonds. (a) 1,3-Butadiene (conjugated). (b) 1,4-Pentadiene (isolated double bonds). (c) Cyclopentadiene (conjugated, no locants as no possibility of ambiguity). (d) 1,4-Cyclohexadiene (isolated double bonds). (e) Lycopene, a red pigment found in tomatoes (the eleven centre double bonds are conjugated but the end double bonds are isolated). (f) Squalene, the biological precursor of steroids and triterpenes (12.6). All the double bonds are isolated.

12.4.2 Physical properties

Polyenes are insoluble in water but soluble in lipids and non-polar solvents (compare with simple alkenes and cycloalkenes). Both types of aliphatic polyene will absorb in the ultra-violet and visible regions of the electromagnetic spectrum. The spectra of conjugated polyenes can be utilised as the basis of spectroscopic assay procedures for compounds containing these structures.

12.4.3 Chemical properties

The C=C bonds in these compounds exhibit the same general properties as those in the simple alkenes and cycloalkenes. For example, they undergo electrophilic and free radical addition, oxidation, reduction and polymerisation. In theory, if sufficient reagent is used, all the C=C bonds could react but in the laboratory a mixture of products is often obtained.

Conjugated dienes will often form 1,4 addition products with electrophiles. For example, 1,3-butadiene reacts with hydrogen bromide in molar proportions to form a mixture of 1-bromobut-2-ene and 3-bromobut-1-ene.

$$CH_2=CH-CH=CH_2 \xrightarrow{HBr} \overset{H}{C}H_2-CH=CH-\overset{Br}{C}H_2 + \overset{H}{C}H_2-\overset{Br}{C}H-CH=CH_2$$

1,3-Butadiene   1-Bromobut-2-ene   3-Bromobut-1-ene

These 1,4 addition products are believed to arise because the intermediate allylic carbocation is stabilised by delocalisation. This means that the nucleophilic attack can now take place at either carbon 2 or 4 (Fig. 12.14). The proportion of the attack that takes place at a particular carbon depends on the reaction conditions. For example, at -60^0 the yield of 3-bromobut-1-ene is about 25%, but at higher temperatures it rises to about 75%.

Free radical additions to conjugated dienes also yield 1,4 addition products because of the formation of an allylic intermediate free radical which is stabilised by the delocalisation of its unpaired electron.

An important 1,4 addition with a pericyclic mechanism is the Diels-Alder reaction. Conjugated dienes react with isolated alkene C=C bonds to form cyclic derivatives often referred to generally as **adducts**. For simple alkenes the yields are poor, but are normally

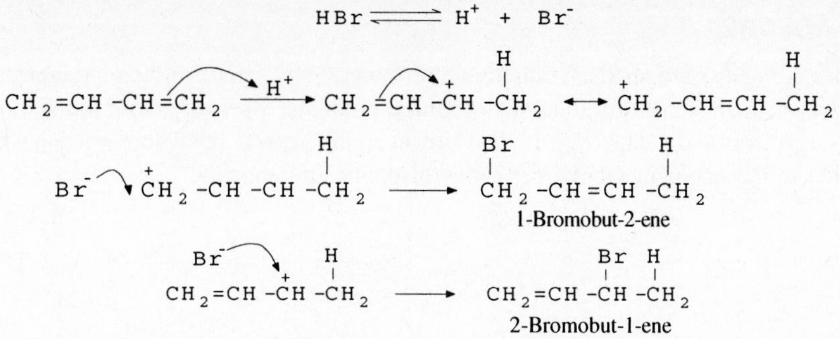

$$HBr \rightleftharpoons H^+ + Br^-$$

$$CH_2=CH-CH=CH_2 \xrightarrow{H^+} CH_2=CH-\overset{+}{C}H-CH_2 \longleftrightarrow \overset{+}{C}H_2-CH=CH-CH_2$$

$$Br^- + \overset{+}{C}H_2-CH-CH-CH_2 \longrightarrow CH_2-CH=CH-CH_2$$
1-Bromobut-2-ene

$$CH_2=CH-\overset{+}{C}H-CH_2 \longrightarrow CH_2=CH-CH-CH_2$$
2-Bromobut-1-ene

Fig. 12.14. An explanation of the electrophilic addition to 1,3-butadiene.

good to excellent if the C=C bond has electron acceptor substituents. However, the reaction is stereospecific: only dienes in which the two C=C bonds have a cis geometry (cistoid) are able to undergo a Diels-Alder reaction (Fig. 12.15).

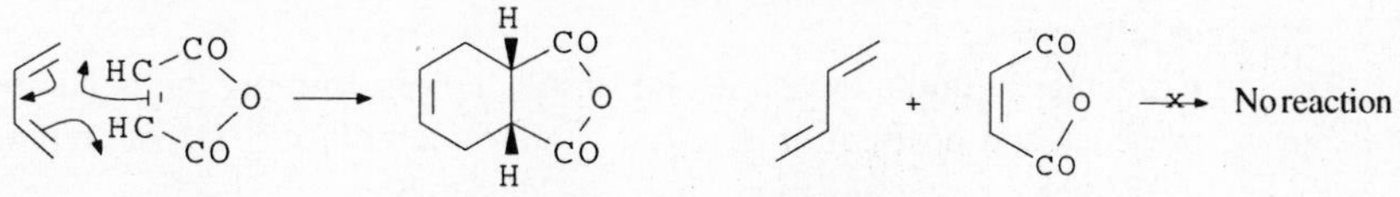

Cis-diene   Maleic   Diels-Alder   Trans-diene
(cistoid)   anhydride   adduct   (transoid)

Fig. 12.15. The stereospecific nature of the Diels-Alder reaction.

The use of cyclic dienes can lead to geometric isomer formation. For example, the reaction between cyclopentadiene and maleic anhydride produces two compounds that are known as the **endo** and **exo** adduct respectively.

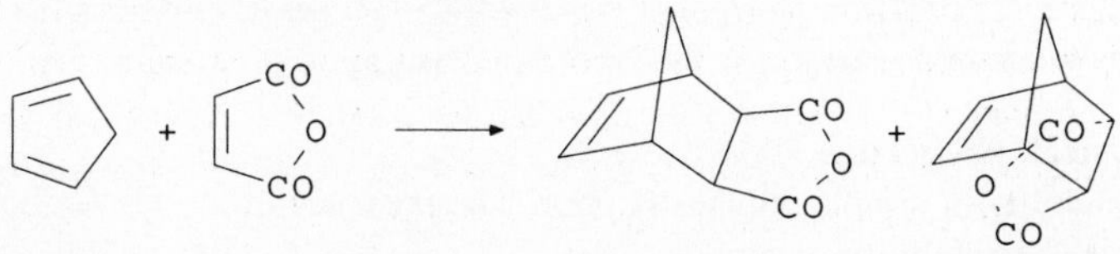

1,3-Cyclopentadiene   Maleic anhydride   Exo-adduct   Endo-adduct

The Diels-Alder reaction is used to manufacture a range of pharmaceutically active compounds, such as the insecticides aldrin and dieldrin. Aldrin and dieldrin are very toxic to man and their use should be strictly controlled.

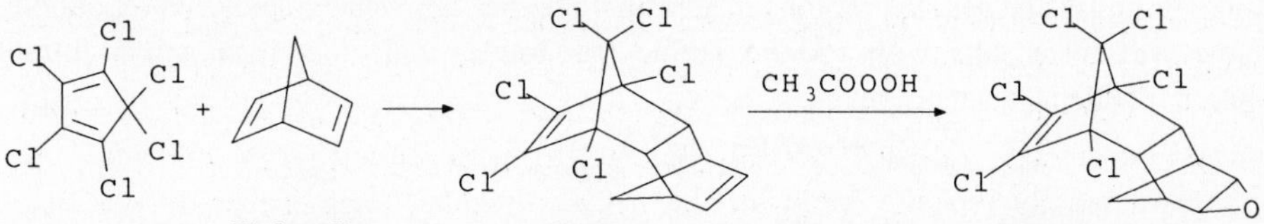

$$CH_3COOOH$$

Norboradiene   Aldrin   Dieldrin

Many polyenes undergo free radical oxidations, polymerisations and rearrangements that are initiated by light. Consequently, compounds such as retinol (vitamin A_1), dienoestrol (oestrogen) and calciferol (antirachitic) which contain a polyene structure should be stored in the absence of light and in some cases under nitrogen.

12.5 ALKYNES

Alkynes are hydrocarbons that contain one or more $C\equiv C$ bonds in their structures. This structure occurs in some drugs and many natural products, especially those found in fungi, micro-organisms and plants (Fig. 12.16). However, the $C\equiv C$ bond does not appear to be involved in the major metabolic processes of plants and animals.

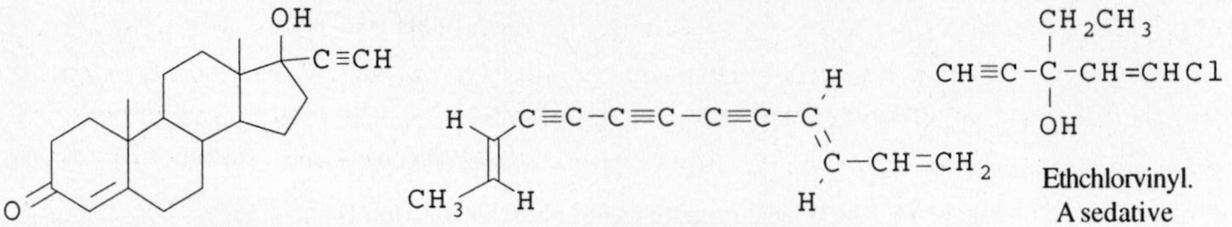

Ethisterone. A progestogen   1,3E,11E-Tridecatrien-5,7,9-triyne.   Ethchlorvinyl.
                                   A nematicide found in plants.   A sedative

Fig. 12.16. Some examples of drugs and natural products whose structures contain C≡C bonds.

12.5.1 Nomenclature

The suffix **-yne** together with the lowest locant of the carbons forming the bond are used to indicate the presence and position of a $C\equiv C$ bond in a molecule. **Diyne, triyne**, etc. are used when the molecule contains more than one $C\equiv C$ bond.

$CH\equiv CH$ $CH_3C\equiv CH$ $CH_3C\equiv CCH_3$ $CH_3C\equiv C-C\equiv CCH_3$

Ethyne (acetylene)   Propyne   But-2-yne   Hex-2,4-diyne

Alkyne substituents are named by using the ending **-ynyl**. For example, a propyne residue becomes either propynyl if the $C\equiv C$ is at position 1 of the substituent or prop-2-ynyl if it is at position 2 of the substituent.

$$\overset{3}{C}H_3-\overset{2}{C}\equiv\overset{1}{C}-\qquad\qquad\overset{3}{C}H\equiv\overset{2}{C}-\overset{1}{C}H_2-$$

Propynyl   Prop-2-ynyl

12.5.2 Physical properties

The water solubility of simple alkynes is poor. The presence of a $C\equiv C$ bond does not enhance the water solubility of a compound.

12.5.3 Chemical properties

The main reactions of the $C\equiv C$ bond of alkynes are **electrophilic addition, oxidation** and **reduction**. In these respects its reactivity resembles that of the C=C bond of alkenes. Unlike terminal alkenes, the protons of terminal alkynes are weakly acidic. For example, ethyne will form salts with sodium amide (sodamide) and ammonical silver nitrate solution (Tollens reagent).

$$CH\equiv CH \xrightarrow{\ Na/NH_3\ } CH\equiv C^-\ Na^+\ +\ NH_3$$

$$CH\equiv CH \xrightarrow{\ Ag(NH_3)_2^+\ } Ag-C\equiv C-Ag\ +\ 2H^+$$

(a) Electrophilic addition. Electrophiles and also some nucleophiles in the presence of an acid catalyst can react with the $C\equiv C$ bond of alkynes. The addition appears to occur in consecutive steps, the orientation following the Markownikov rule at both stages.

$$R-C\equiv CH \xrightarrow{HCl} R-\underset{\underset{H}{|}}{\overset{\overset{Cl}{|}}{C}}=CH \xrightarrow{HCl} R-\underset{\underset{Cl}{|}\;\underset{H}{|}}{\overset{\overset{Cl}{|}\;\overset{H}{|}}{C}}-CH$$

Surprisingly, the $C\equiv C$ bond of alkynes is less reactive than the C=C bonds of alkenes. Their reactions often require more drastic conditions than those required by the corresponding alkene. This is probably due to the formation of vinylic carbocations as intermediates. Experimental evidence shows that these ions are less stable than tertiary and secondary carbocations produced in alkene additions (12.3). Consequently, some of them will revert to the alkyne before they have had time to collide with a nucleophile and form the product. As a result, it will take longer for a significant number of alkyne molecules to react.

$$R-C\equiv CH \xrightarrow{H^+} R-\overset{+}{C}=CH \xrightarrow{Cl^-} R-\underset{\underset{Cl}{|}\;\underset{H}{|}}{\overset{\overset{Cl}{|}\;\overset{H}{|}}{C}}=CH$$

Vinyl carbocation

(b) Oxidation. The $C\equiv C$ bonds of alkynes are less reactive than the C=C bonds of alkenes. Strong oxidising agents cleave the $C\equiv C$ bond and form either a mixture of acids or one acid and carbon dioxide if a terminal alkyne is present.

$$R-C\equiv C-R' \xrightarrow{\text{Oxidation}} RCOOH \;+\; R'COOH$$

$$R-C\equiv C-H \xrightarrow{\text{Oxidation}} RCOOH \;+\; CO_2$$

In structure determination this reaction is used to locate the positions of $C\equiv C$ bonds in the molecule by identifying the products of the oxidation. For example, the structure of tariric acid, a fatty acid isolated from the seeds of *Picramnia tariri* was established by oxidising the compound with aqueous potassium permanganate. The oxidation gave adipic acid which enabled researchers to identify the position of the $C\equiv C$ bond in the molecule

$$CH_3(CH_2)_{10}C\equiv C(CH_2)_4COOH \xrightarrow{KMnO_4} HOOC(CH_2)_4COOH + CH_3(CH_2)_{10}COOH$$

Tariric acid   Adipic acid   Decanoic acid (lauric acid)

(c) Reduction. Catalytic hydrogenation of alkynes is relatively easy. The reduction can stopped at the alkene stage by the use of a suitable catalyst, such as Lindlars' catalyst which is finely divided palladium that has been partly deactivated. Reduction using this catalyst yields the cis alkene. Trans derivatives can be obtained using sodium in liquid ammonia as reducing agent.

$$CH_3(CH_2)_3C\equiv C(CH_2)_3CH_3$$
Dec-5-yne

Lindlars catalyst / H_2 →

$$\underset{CH_3(CH_2)_3}{\overset{H}{\diagdown}}C=C\underset{(CH_2)_3CH_3}{\overset{H}{\diagup}}$$
Cis-dec-5-ene

Na / NH_3 →

$$\underset{CH_3(CH_2)_3}{\overset{H}{\diagdown}}C=C\underset{H}{\overset{(CH_2)_3CH_3}{\diagup}}$$
Trans-dec-5-ene

(d) Acidic nature of terminal alkyne groups. Terminal alkynes are weak acids. They react with strong bases to form salts known as acetylides. These ions can act as a

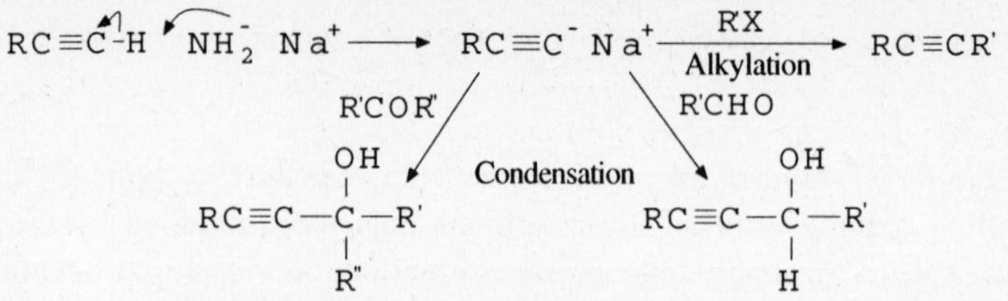

$$RC\equiv\overset{\curvearrowleft}{C}-H \quad \overset{\curvearrowright}{NH_2^-} \; Na^+ \longrightarrow \quad RC\equiv C^- \; Na^+ \xrightarrow[\text{Alkylation}]{RX} \quad RC\equiv CR'$$

R'COR'

$$\underset{\underset{R''}{|}}{\overset{\overset{OH}{|}}{RC\equiv C-C-R'}}$$

Condensation

R'CHO

$$\underset{\underset{H}{|}}{\overset{\overset{OH}{|}}{RC\equiv C-C-R'}}$$

nucleophile in nucleophilic substitution and addition reactions. Nucleophilic addition is used to manufacture a number of sedatives, such as methylpentynol (*Oblivon*) and Ethchlorvynol.

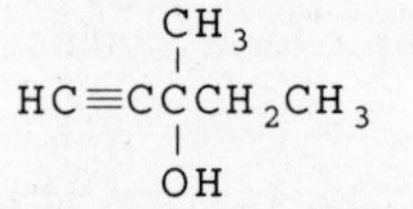

$$\underset{OH}{\overset{CH_3}{HC\equiv CCCH_2CH_3}}$$

Methylpentynol (*Oblivon*)

$$\underset{OH}{\overset{CH=CHCl}{HC\equiv CCCH_2CH_3}}$$

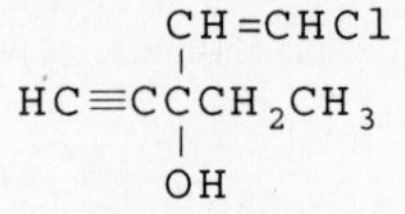

Ethchlorvynol

12.6 AROMATIC HYDROCARBONS

Aromatic hydrocarbons (**arenes**) are **flat conjugated cyclic polyenes**. They have physical and chemical properties that are distinctly different from other conjugated cyclic polyenes. Theoretical calculations made in 1931 by the German physicist Erich Huckel showed that an aromatic compound should have a fully conjugated cyclic structure in which the number of pi electrons in the structure should equal $4n + 2$ where n is a whole number. This prediction has now been substantiated by experimental observations and the Huckel rule is used as the criterion for aromaticity. For example, benzene has six pi electrons (two per C=C bond, Fig. 12.17). Substituting this figure in the Huckel rule, $n=1$ and so benzene is an aromatic compound. However, cyclooctatetraene has eight pi electrons. Substituting this figure in the Huckel rule gives a value for n which is not a whole number and so cyclooctatetraene is not an aromatic compound. Experimental work shows that cyclooctatetraene does not have a flat ring and so cannot be a fully conjugated structure even though its traditional structural formula shows a conjugated system. Cyclooctatetraene does not exhibit aromatic chemical properties but its C=C bonds behave like those of simple alkenes, readily reacting with bromine, hydrogen chloride and aqueous potassium permanganate.

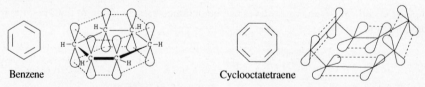

Benzene   Cyclooctatetraene

Fig 12.17. Benzene and cyclooctatetraene. The ring of cyclooctatetraene is not planar and so the sets of p orbitals forming the pi bonds lie in different planes which means that the continuous overlap of p orbitals which is a feature of aromatic and conjugated polyene systems does not occur in cyclooctatetraene.

12.6.1 Benzene

Benzene is a colourless, toxic sweet-smelling volatile liquid which can be obtained from coal tar. Prolonged exposure to either the liquid or the vapour causes leukopenia (reduction of the white blood cell count) in humans. Benzene is the simplest of the aromatic hydrocarbons. Its properties are fairly typical of aromatic hydrocarbons and can be used to predict the properties of other aromatic compounds. However, it should be remembered that these predictions will not always be accurate.

(a) Physical properties. Benzene is insoluble in water but soluble in non-polar solvents and lipids. It has a characteristic ultra-violet spectrum which can be used to detect the presence of a benzenoid group in the molecular structure of a compound.

(b) Chemical properties. Benzene is chemically stable being resistant to oxidation, reduction and addition. Free radical addition occurs with both water and the halogens. Hydroxyl radicals formed from water convert benzene into phenol and biphenyl. This reaction could occur when aqueous suspensions or solutions of drugs containing benzene rings are stored, sterilised by or exposed to ionising radiation. Gamma benzenehexachloride, a powerful insecticide used in horticulture, is one of a mixture of six isomers of 1,2,3,4,5,6-hexachlorocyclohexane produced by the free radical addition of chlorine to benzene.

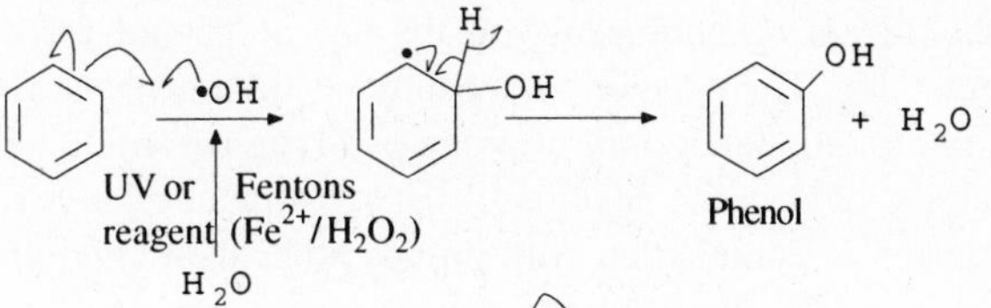

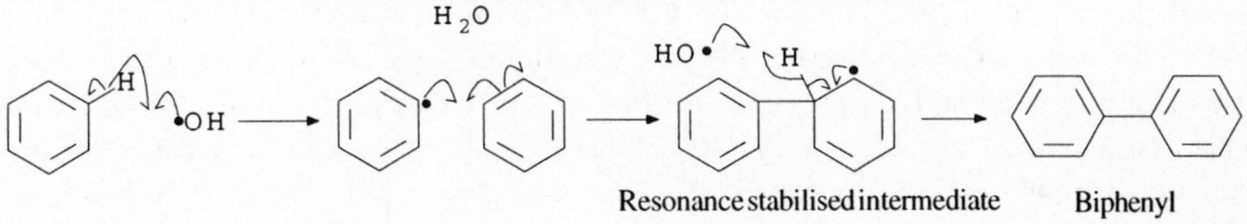

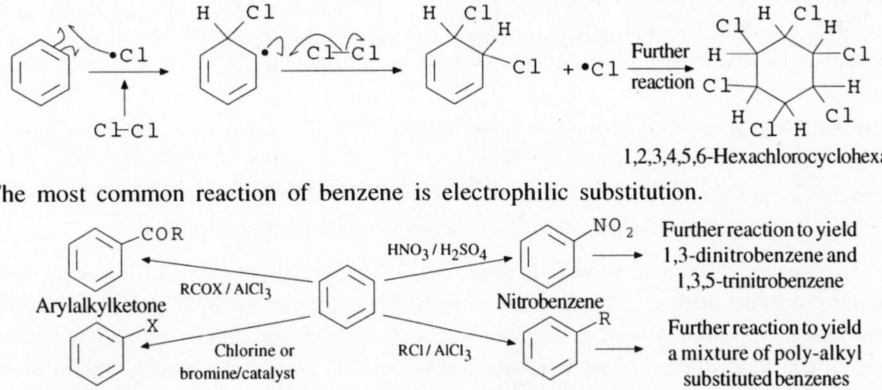

The most common reaction of benzene is electrophilic substitution.

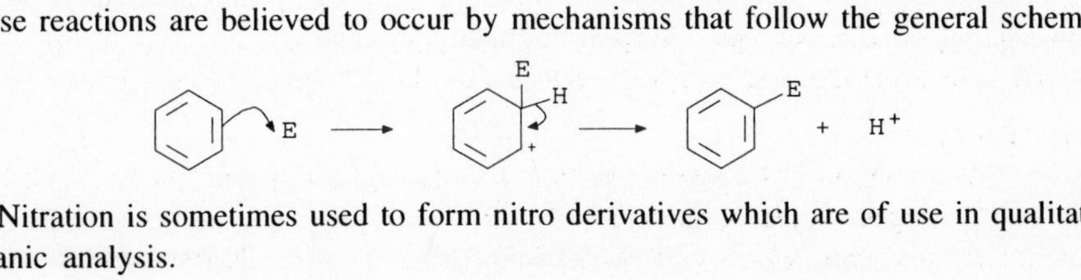

These reactions are believed to occur by mechanisms that follow the general scheme:

Nitration is sometimes used to form nitro derivatives which are of use in qualitative organic analysis.

(c) Biological systems. Benzene is not very reactive in biological systems. It is metabolised mainly in the liver by hydroxylation to phenol and 1,2-dihydroxybenzene (catechol). The process is catalysed by oxidases and is believed to involve an epoxide intermediate. Hydroxylation often occurs in the metabolism of drugs whose structures contain benzene rings.

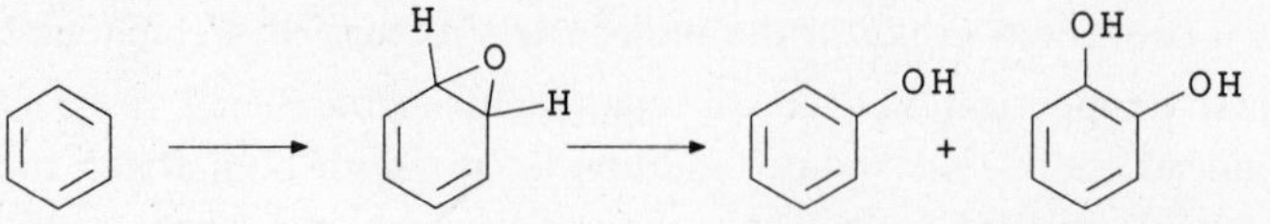

Phenol   1,2-Dihydroxybenzene
(catechol)

12.6.2 Alkylbenzenes

These are benzene derivatives that have alkyl group substituents (Fig. 12.18). They are named either as substituents of benzene or with benzene as the substituent depending which is the most convenient. In the latter case **phenyl** (Ph) is used as the radical name for benzene. It should not be confused with the use of phenol for hydroxy derivatives of benzene (Chapter 15). The position of substituents on a benzene ring are indicated either by locants or the traditional system of prefixes, **ortho** (*o*), **meta** (*m*) and **para** (*p*). These traditional prefixes refer to the position of the substituent relative to the alkyl group. The system is also used in conjunction with groups other than alkyl groups.

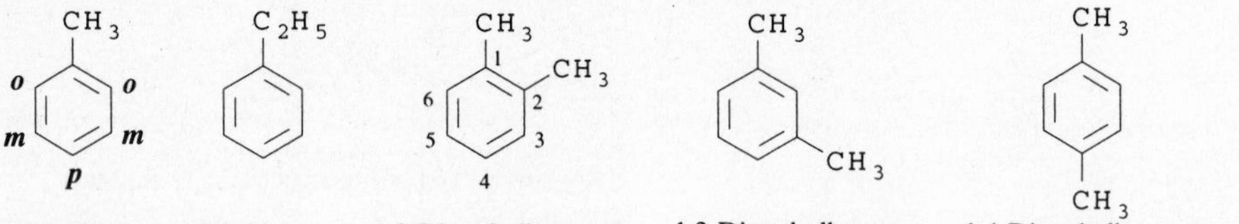

Methylbenzene Ethylbenzene 1,2-Dimethylbenzene   1,3-Dimethylbenzene   1,4-Dimethylbenzene
(toluene)   (*ortho*-xylene, bp 144°C) (*meta*-xylene, bp 139°C) (*para*-xylene, bp 138°C)

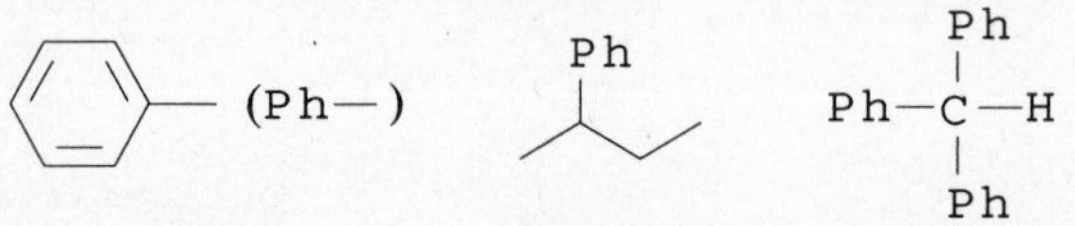

Phenyl   2-Phenylbutane   Triphenylmethane

Fig. 12.18. Examples of the names and structures of alkylbenzenes.

The general chemical, biological and physical properties of the alkylbenzenes are a mixture of those of their constituent structures. For example, methylbenzene will exhibit benzene-like properties such as electrophilic substitution and alkane properties such as free radical substitution. However, substituents will influence the precise nature of the benzene ring's properties and vice versa. For example, mononitration of methylbenzene, an electrophilic substitution, results in the formation of mainly *o*-nitromethylbenzene and *p*-nitromethylbenzene but very little *m*-nitromethylbenzene.

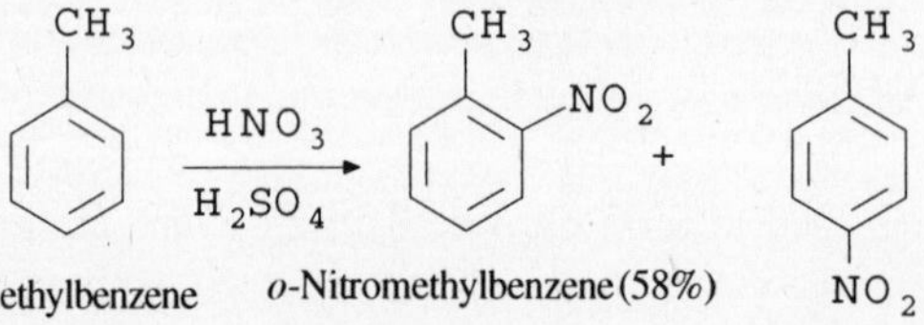

Methylbenzene   *o*-Nitromethylbenzene (58%)   NO$_2$ *p*-Nitromethylbenzene (37%)

This substitution pattern may be *simply* explained in terms of the relative stabilities of the intermediate carbocations. In each case the inductive effect of the methyl group will stabilise the carbocation by reducing its positive charge. However, the inductive effect is only effective over short distances and so will have its greatest effect at position 1. Since the charges shown on the canonical forms describe the electron distribution in the structure of the resonance hybrid, the *ortho* and *para* carbocations which have canonical forms with a positive charge at position 1 will experience the greatest stabilising effect (Fig. 12.19) because these carbocations will be less positively charged than the corresponding *meta*-carbocation. The positive charge in the case of the *meta*-carbocation is too far away from the methyl group for the inductive effect to be effective. It should be noted that in many cases the charge distribution of the substituent can also affect the stability of the intermediate carbocation. Other explanations involve a concept known as hyperconjugation which is beyond the scope of this text.

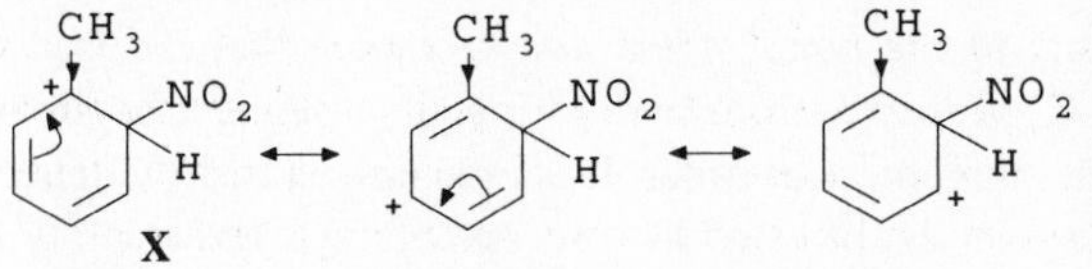

<table>
<tr>
<td>

CH_3 ... NO_2 ... H (**X**) ⟷ CH_3 ... NO_2 ... H ⟷ CH_3 ... NO_2 ... H

</td>
<td>

The inductive effect of the methyl group directly reduces the positive charge of the canonical form **X** for this intermediate carbocation. Consequently, the methyl will be effective in stabilising this intermediate carbocation

</td>
</tr>
<tr>
<td>

CH_3 ... NO_2 ... H ⟷ CH_3 ... NO_2 ... H ⟷ CH_3 ... NO_2 ... H

</td>
<td>

The inductive effect of the methyl group does not directly reduce the positive charge of any of the canonical forms for this intermediate carbocation. This means that the methyl group will not be so effective in stabilising this intermediate carbocation

</td>
</tr>
<tr>
<td>

CH_3 ... H NO_2 ⟷ CH_3 (**Y**) ... H NO_2 ⟷ CH_3 ... H NO_2

</td>
<td>

The inductive effect of the methyl group directly reduces the positive charge of the canonical form **Y** for this intermediate carbocation. Consequently, the methyl will be effective in stabilising this intermediate carbocation

</td>
</tr>
</table>

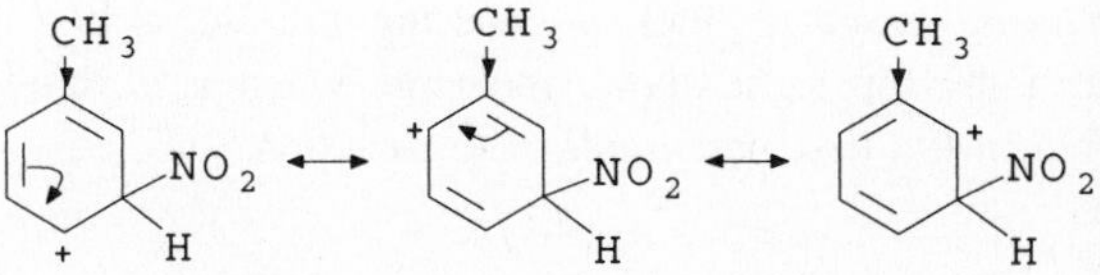

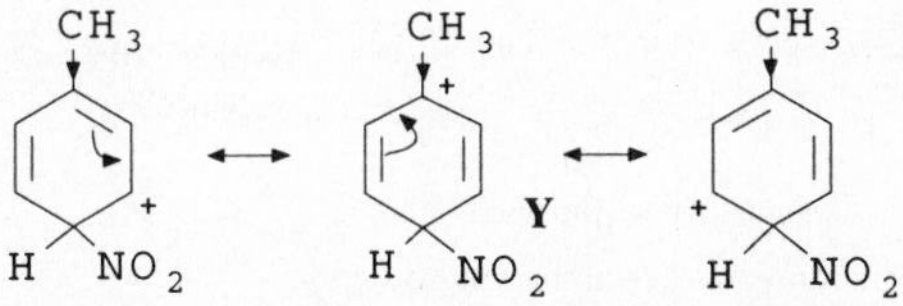

Fig. 12.19. The relative stabilities of the carbocations formed in the mononitration of methylbenzene. Relative stability is assessed by comparing the numbers and stabilities of the canonical forms. The number of forms for each intermediate carbocation is the same so in this case one has to consider their relative stabilities.

12.7 FUSED RING AROMATIC HYDROCARBONS

Fused ring aromatic hydrocarbons are polycyclic aromatic compounds in which the rings are joined by a common bond (Fig. 12.20). They normally obey the Huckel rule.

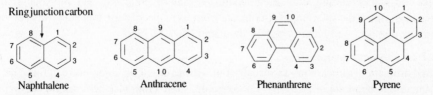

Ring junction carbon

Naphthalene   Anthracene   Phenanthrene   Pyrene

Fig. 12.20. Some examples of fused aromatic hydrocarbons. Pyrene does not obey the Huckel rule.

12.7.1 Nomenclature

The simple ring systems have trivial names (Fig. 12.20). By convention, their structures are normally drawn with the maximum number of rings in a horizontal line and, as far as possible, the remainder on the top right-hand side of the structure. Their carbons are numbered clockwise starting from the right-hand atom to the furthest top right ring junction carbon. Ring junction carbons are not included in the number system because any reaction involving these atoms will completely change the nature of the ring. Anthracene is an exception to this numbering system.

Complex ring systems are named as derivatives of the largest ring system (the parent system) having a trivial name. Substituent ring systems have the endings of their names changed to **o** which is dropped before a vowel. For example, benzene would become benzo- whilst naphthalene, naphtho. These names would be changed to benz and naphth respectively if the parent ring system were anthracene. The position of attachment is indicated by the insertion of a letter in square brackets between the substituent and parent names (Fig. 12.21). These letters correspond to *the bond of the parent system* that is fused to the substituent system. Bond C_1–C_2 of the parent is always **a** and subsequent bonds of the parent system become **b**, **c**, **d**, etc. in numerical order. It is emphasised that the letters are allocated to the parent ring system on the basis of its own number system which will not be the same as the overall number system of the complete molecule. This is obtained as before by numbering the *complete ring structure* clockwise starting with the carbon to the right of the ring junction nearest the top right of the molecule when it is drawn according to the convention specified in the first paragraph of this section.

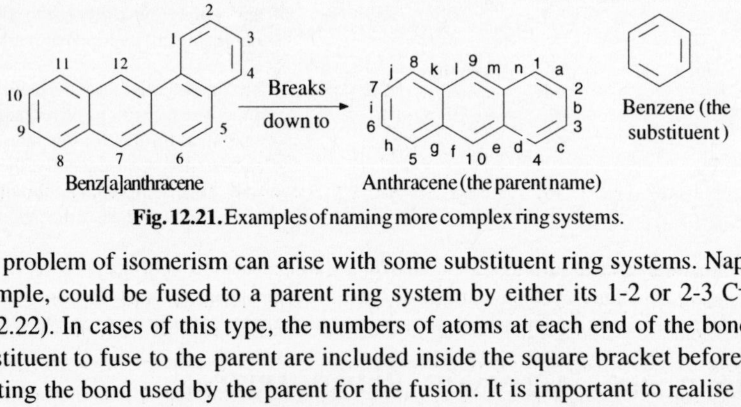

Fig. 12.21. Examples of naming more complex ring systems.

The problem of isomerism can arise with some substituent ring systems. Naphthalene, for example, could be fused to a parent ring system by either its 1-2 or 2-3 C–C bonds (Fig. 12.22). In cases of this type, the numbers of atoms at each end of the bond used by the substituent to fuse to the parent are included inside the square bracket before the letter designating the bond used by the parent for the fusion. It is important to realise that these numbers belong to the substituent's own numbering system and are not related to the number system used for the complete molecule. In this connection it is often helpful to make flat numbered and lettered cardboard cut-outs of the parent and substituent structures to work out the structure of a molecule from its name because it is not always obvious how the numbers and letters arose in the name (Fig. 12.22).

12.7.2 Physical properties

Fused aromatic hydrocarbons are usually flat. However, steric hindrance can force part of a structure to bend out of planarity. For example, steric hindrance between the methyl and benzene ring D of 7,12-dimethylbenz[a]anthracene is the cause of ring D being at an

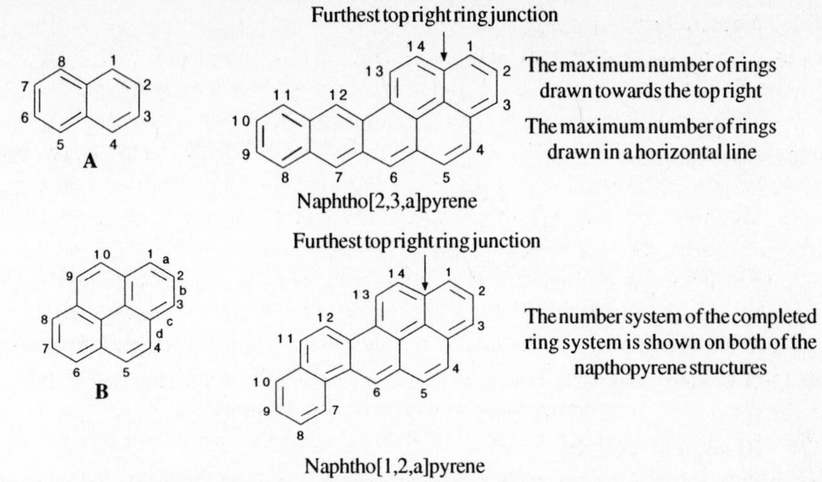

Furthest top right ring junction

The maximum number of rings drawn towards the top right

The maximum number of rings drawn in a horizontal line

Naphtho[2,3,a]pyrene

Furthest top right ring junction

The number system of the completed ring system is shown on both of the napthopyrene structures

Naphtho[1,2,a]pyrene

Fig. 12.22. Naming naphtho isomers of pyrene. To understand how the names of the systems were obtained, make flat cardboard cutouts of shapes **A** and **B**. Letter and number the shapes as shown and manipulate them so that the appropriate numbers and letters in the square brackets used in the name of the compound line up.

angle of 20^0 to the rest of the molecule. It is possible to predict whether steric hindrance is going to be important in a structure by examining the molecular formula and making models or computer simulations of the structure (2.9).

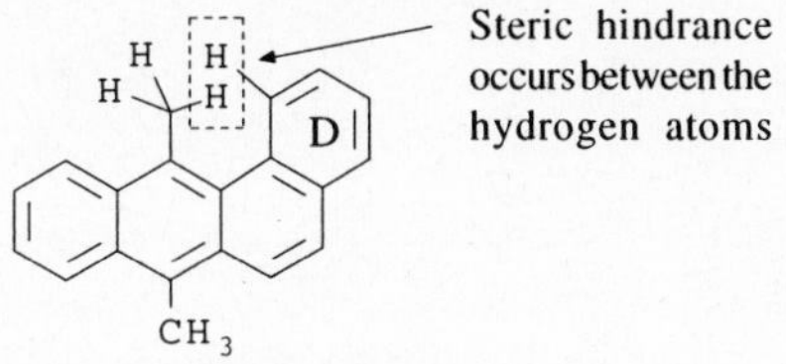

Steric hindrance occurs between the hydrogen atoms

Fused aromatic hydrocarbons are not soluble in water but are soluble in non-polar solvents and lipids.

12.7.3 Chemical properties

All fused aromatic ring systems exhibit similar general chemical and physical properties to those of benzene. The greater parts of their structures will be resistant to oxidation, reduction and addition but can be substituted by electrophiles, the position of substitution being impossible to predict accurately using mechanistic theory. However, many fused aromatic hydrocarbons have C=C bonds in their conjugated structures which behave like typical alkenes and polyenes. In other words, these bonds behave as though some of the pi electrons of the conjugated system were concentrated in these bonds rather than spread out over the whole of the ring system. This phenomenon is called **bond fixation**. The bonds in which it occurs are often the most reactive in the molecule. They exhibit electrophilic addition, oxidation and reduction like typical alkene and polyene C=C bonds (Fig. 12.23). Unfortunately, it is not possible to predict where bond fixation will occur by examining the structural formula of a fused aromatic ring system. It can only be memorised as it arises, and discovered by investigation of the system's chemical and physical properties.

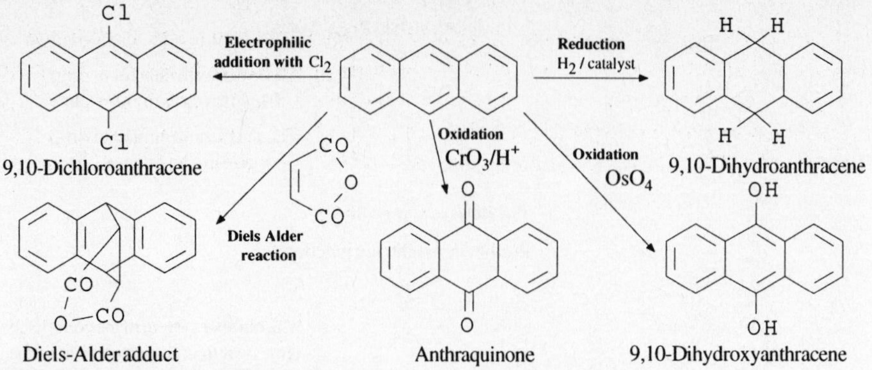

Fig. 12.23. Bond fixation in anthracene. Bond fixation occurs between the 9 and 10 positions. This region of the structure of anthracene acts like a typical diene.

12.7.4 Biological activity

Many fused aromatic hydrocarbons are carcinogenic. Their mode of action is still the subject of discussion but often appears to be related to the reactivity of the regions in which bond fixation occurs. For example, benz[a]pyrene which occurs in cigarette smoke and soot is metabolised in the body to a metabolite which is believed to bind to DNA and so disrupt normal metabolism.

12.8 NON-FUSED RING AROMATIC HYDROCARBONS

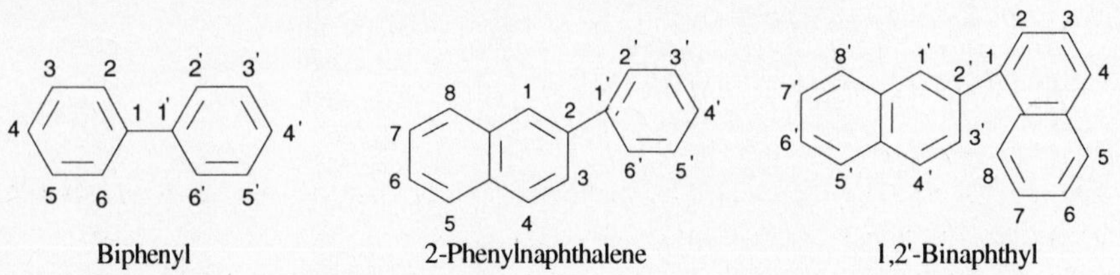

Fig. 12.24. Examples of the structures and nomenclature of non-fused aromatic hydrocarbons.

These are aromatic hydrocarbons that consist of several ring systems linked in the manner shown in Fig. 12.24. Each ring system is numbered in the appropriate manner for that structure but it is customary to use primes (') when it is necessary to distinguish between two of the number systems in the hydrocarbon. **It is emphasised that primes are used only if there is a possibility of ambiguity.** Symmetrical compounds are named by prefixing the radical name of the ring system by bi, tri, etc. and the numbers of the atoms involved in the links between the ring systems when they are not joined 1 to 1'. The names of unsymmetrical compounds use the radical name(s) of the smallest ring system(s) as prefixes to that of the largest system. The positions of these substituents are located in the usual way by the appropriate number of the parent system.

12.8.1 Physical properties

The physical properties of non-fused aromatic hydrocarbons are similar to those of the aromatic compounds discussed in the previous section. However, it is interesting to note that derivatives of these hydrocarbons will exhibit optical activity provided the ortho positions to each ring are substituted by structures large enough to cause steric hinderance between the two ring systems. For example, as a result of steric hindrance 2,2'-dinitro-6,6'-dihydroxycarbonylbiphenyl has a twisted molecular structure and has been resolved into two enantiomers with specific rotations of $[\alpha]_D^{25}$ +225°. In practice, compounds with one large or two medium to large ortho substituents on each ring system may be optically active:

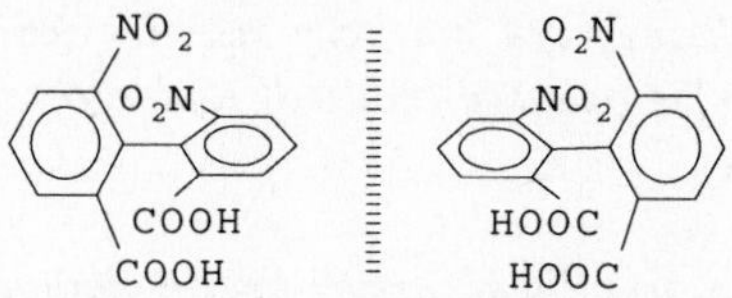

Optical activity is believed to occur because the presence of bulky groups in the molecule prevents the free rotation of one ring relative to the other. As a result, the compound will have two non-superimposable mirror image structures and so will exhibit optical activity (5.3). This theory is well supported by practical evidence.

12.8.2 Chemical properties

In general, the compounds exhibit a mixture of the chemical properties of their constituent aromatic ring systems. However, because their overall shapes are different from those of their constituent ring systems they will not necessarily behave in the same fashion as these structures in biological systems. Furthermore, it should be noted that electrophilic substitution of the benzene rings of any type of aromatic system is not very common in biological processes.

12.9 WHAT YOU NEED TO KNOW

12.9.1 Nomenclature

(1) The parent name of a compound indicates the number of carbon atoms and their arrangement in its main carbon skeleton.

(2) Prefixes and suffixes to the parent name are used to indicate the presence of other structures in the molecule. The prefixes and suffixes used for hydrocarbons are:

Structure	Suffix	Suffix used when a substituent
Alkane, cycloalkane	-ane	-yl
Aromatic ring system	-ene	-yl
Alkene	-ene	-enyl
Alkyne	-yne	-ynyl

(3) The position of substituent structures on the parent are indicated by numbers known as locants.

(4) The [letter] in the name of a fused polynuclear aromatic compound indicates the bond to which substituent ring system is fused.

(5) The [numbers] in the name of a fused polynuclear aromatic compound indicates the bond of the substituent ring system that is fused to the parent ring system. The

numbers refer to the number system of the substituent ring system.

(6) Primes (') are only used in a name if there is a possibility of ambiguity.

12.9.2 Alkanes and cycloalkanes

(1) Alkanes and cycloalkanes are insoluble in water but soluble in non-polar solvents and lipids.

(2) *Alkanes and cycloalkanes are not very reactive (exceptions cyclopropane and cyclobutane) but can be oxidised and will undergo substitution, usually by free radical mechanisms.*

$$R-H \ + \ X-Y \longrightarrow R-X \ + \ H-Y \qquad \text{(X is often the same as Y)}$$

(3) Metabolism of alkanes usually occurs by hydroxylation of the carbon of the chain.

12.9.3 Alkenes, cycloalkenes and polyenes

(1) Alkenes and cycloalkenes are hydrocarbons with C=C bonds in their structures.

(2) Molecular orbital theory pictures the C=C bond as consisting of one sigma and one pi bond. The pi bond is the more reactive bond because it has a lower degree of orbital overlap.

(3) Alkenes and cycloalkenes are insoluble in water but are soluble in non-polar solvents and lipids.

(4) The main reactions of alkenes and cycloalkenes are: electrophilic addition, free radical addition, oxidation, reduction and polymerisation.

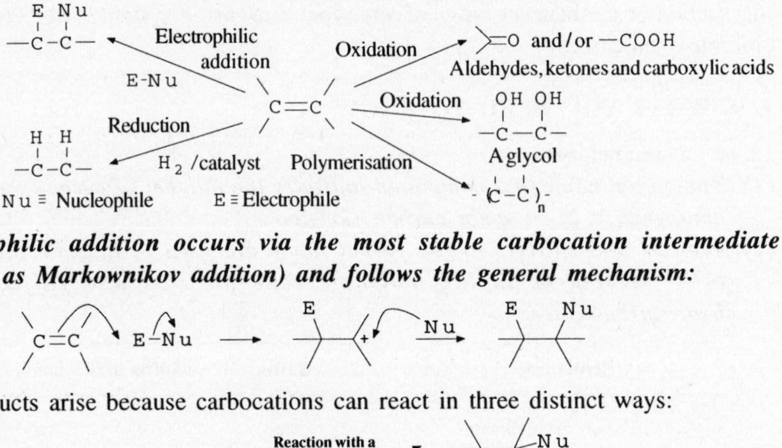

(5) Electrophilic addition occurs via the most stable carbocation intermediate (known as Markownikov addition) and follows the general mechanism:

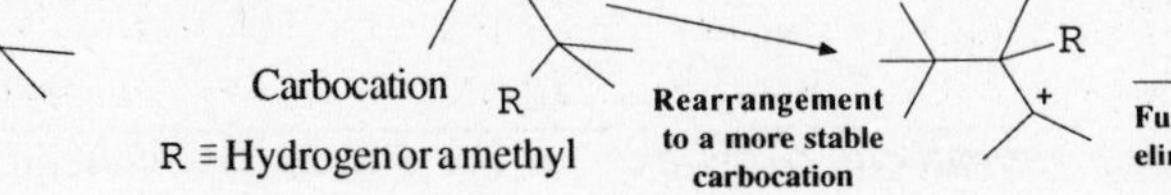

(6) By-products arise because carbocations can react in three distinct ways:

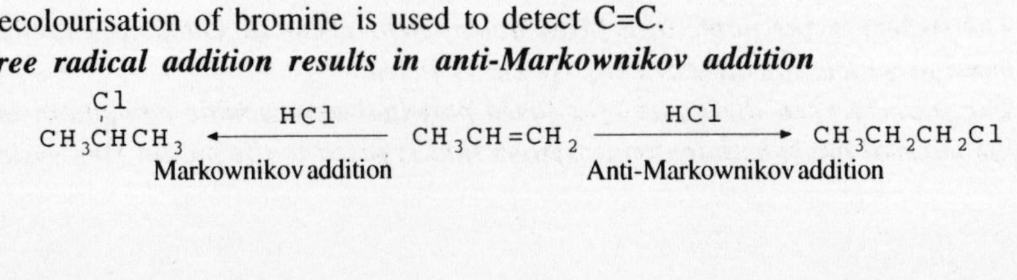

(7) Decolourisation of bromine is used to detect C=C.

(8) Free radical addition results in anti-Markownikov addition

(9) Hydration is the addition of water to an alkene in the presence of an acid catalyst.

(10) Oxidation of alkenes may involve either the pi bond or both the pi and sigma bonds of the C=C bond. The products of an oxidation depend on the reagent and the reaction conditions. It usually results in the formation of epoxides, peroxides, diols, carboxylic acids, aldehydes and ketones. Peroxideformation can occur on storage.

(11) Oxidations in biological systems usually produce epoxides, peroxides and alcohols.

(12) The C=C bonds can be reduced by hydrogen in the presence of a metal as catalyst. The reaction is referred to as hydrogenation.

(13) Biological reduction of C=C bonds is brought about by the addition of H^+ from a proton donor such as $NADH^+$ and the transfer of what is in effect H^- from a reductase.

(14) Alkenes can polymerise.

(15) Polyenes are hydrocarbons with more than one C=C. These C=C bonds may be either isolated or conjugated.

(16) Polyenes exhibit the same general properties as simple alkenes and cycloalkenes, namely, electrophilic and free radical addition, oxidation, reduction and polymerisation.

(17) Conjugated dienes will form 1,4-addition products with electrophiles and other reagents.

(18) The Diels-Alder is a 1,4-addition reaction with a pericyclic mechanism.

X = or ≠ X' = electron donor
X and X' can also be a hydrogen atom
Y = or ≠ Y' = electron acceptor

(20) Polyenes drugs should be stored in airtight lightproof containers as they often undergo free radical oxidations, polymerisations and rearrangements initiated by light.

12.9.4 Alkynes

(1) Alkynes are hydrocarbons that contain one or a number of C≡C bonds in their structures. These bonds do not appear to be important in human metabolism.

(2) The general types of reaction exhibited by all alkynes are: electrophilic addition, oxidation and reduction.

(3) Terminal alkynes are acidic, reacting with strong bases to form acetylides.

(4) Acetylides can act as nucleophiles in nucleophilic addition and substitution reactions.

(5) Drugs containing C≡C bonds are prone to light-catalysed atmospheric oxidation and should be stored in lightproof containers.

12.9.5 Aromatic hydrocarbons

(1) Most aromatic hydrocarbons have flat conjugated ring structures. Steric hindrance may distort this planarity.

(2) Aromatic hydrocarbons obey the Huckel rule, viz. the number of pi electrons in the conjugated system = 4n + 2 where n is 1, 2, 3, 4, ...

(3) All aromatic hydrocarbons are insoluble in water but are soluble in non-polar solvents and lipids.

(4) Simple aromatic hydrocarbons are resistant to oxidation, reduction and addition.

(5) The most common reaction of simple aromatic hydrocarbons is electrophilic substitution. It occurs by mechanisms based on the general scheme:

(6) Electrophilic substitution of benzene rings does not often occur in biological processes.

(7) Hydroxylation of the benzene ring is the main metabolic route for all types of aromatic hydrocarbon.

(8) Fused polynuclear aromatic hydrocarbons have regions where the molecule exhibits alkene like properties, namely, addition, oxidation, reduction, etc. These are the most reactive areas in the molecule.

(9) The chemical properties of the greater part of a fused polynuclear aromatic hydrocarbon structure resemble those of benzene, that is, undergoes electrophilic substitution but is resistant to oxidation, reduction and addition.

(10) Many fused polynuclear aromatic hydrocarbons are carcinogenic.

(11) Non-fused aromatic hydrocarbons may exhibit optical activity if the compounds which have either two large or two medium to large ortho substituents.

12.10 QUESTIONS

(1) Draw structural formulae for each of the following compounds: (a) 2,3,4,6-tetra-methyldecane, (b) 3-prop-2-enylcyclohexene, (c) cis-1,4-diethylcyclohexane, (d) 1,4-dimethylcyclopentene, (e) Z-2-chloro-3-methylpent-2-ene, (f) 3-methyl-3E,5Z-hepta-1,3,5-triene, (g) 2-phenyl-2-cyclohexylpropane, (h) 9,10-dimethyl-benz[a]anthracene.

(2)

Determine whether compound (A) is an E or a Z isomer. What general chemical reactions would A be expected to exhibit? Show, by means of equations, the products that would be expected when A reacts with (a) a mild oxidising agent, (b) a strong oxidising agent, (c) water in the presence of an acid catalyst, (d) hydrogen bromide and (e) dilute sulphuric acid. Suggest feasible reaction mechanisms for reactions (c), (d) and (e).

(3) Predict the main products of the reaction of 1 mole of pent-1,4-diene with 1 mole of each of the following reagents (a) bromine, (b) hydrogen chloride and (c) hydrogen.

(4)

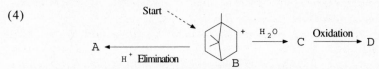

The scheme shown is a fragment of the metabolic pathway for the formation of some natural products. Deduce the possible structures of A, C and D. Suggest a feasible electronic mechanism for the conversion of B to C.

(5) Explain the meaning of the term 'conjugated system' as applied to the structure of an organic molecule. Draw a 'ring' around the conjugated system (if any) in each of the following compounds.

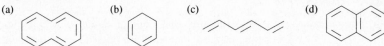

(6) Determine which of the following compounds might exhibit aromatic properties. Describe the general, physical and chemical properties that each of these compounds might exhibit.

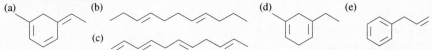

(7) The C_9-C_{10} bond of phenanthrene exhibits bond fixation. Predict, by means of equations, how phenanthrene would react with each of the following reagents: (a) bromine, (b) a solution of potassium permanganate in dilute hydrochloric acid and (c) hydrogen in the presence of palladium at room temperature. Nitration of phenanthrene using a mixture of concentrated sulphuric and nitric acids yields 5-nitrophenanthrene among other products. Suggest a feasible electronic mechanism for the formation of this compound.

13

Heterocyclic systems

13.1 INTRODUCTION

Heterocyclic compounds have ring structures that contain atoms of other elements besides carbon (Fig. 13.1). The atoms of these elements are referred to as **heteroatoms**, the commonest being nitrogen, oxygen and sulphur. This chapter will be restricted to heterocyclic compounds containing these elements.

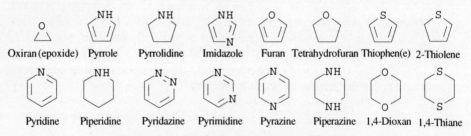

Oxiran (epoxide) Pyrrole Pyrrolidine Imidazole Furan Tetrahydrofuran Thiophen(e) 2-Thiolene

Pyridine Piperidine Pyridazine Pyrimidine Pyrazine Piperazine 1,4-Dioxan 1,4-Thiane

Fig 13.1. Examples of some simple heterocyclic systems.

Heterocyclic systems may be broadly classified as alicyclic and aromatic. They are found as parts of the structures of many important naturally occurring substances, for example, carbohydrates (Chapter 27), nucleic acids (Chapter 29), alkaloids (Chapter 24) and haemoglobin (Chapter 28).

13.2 NOMENCLATURE

Both trivial and systematic names are in use for mono- and multicyclic systems (Fig. 13.1)

The larger multicyclic systems are more likely to have an allowed trivial name rather than a systematic name. However, with a few exceptions, the numbering is systematic and follows the appropriate rules for that particular system.

13.2.1 Monocyclic systems

The systematic names of monocyclic systems consist of two parts, a prefix and a suffix. The prefix or 'a' term, as it is sometimes called, indicates the nature of the heteroatom(s) in the structure, the ending **-a** being dropped before a vowel (Table 13.1). Furthermore, when following -in or -ine, phospha becomes phosphor, arsa, arsen, and stiba, antimon. Di, tri, etc. and the appropriate locants are used when the ring contains two or more heteroatoms of the same type. If the ring contains heteroatoms of different types, the prefixes are arranged in descending periodic table group order, the prefix with the highest group number first. For example, oxa (oxygen) would come before aza (nitrogen) or phospha (phosphorus). Elements higher up a group are placed before those lower down that group. For example, oxa would come before thia (sulphur).

Table 13.1. Some of the prefixes used in the systematic naming of monocyclic heterocyclic systems. Prefixes are in order of seniority, the most senior being top left and the order of seniority decreasing down a column.

Element	Prefix	Element	Prefix	Element	Prefix
Oxygen	oxa	Nitrogen	aza	Silicon	sila
Sulphur	thia	Phosphorus	phospha	Tin	stanna
Selenium	selena	Arsenic	arsa	Lead	plumba
Tellurium	tellura	Antimony	stilba	Mercury	mercura

The suffix describes the size and the maximum degree of unsaturation of the ring. However, in the case of unsaturated four- and five-membered rings with one double bond a different prefix is used if there can be more than one double bond (Table 13.2). Ring systems that contain less than the maximum number of double bonds are prefixed by dihydro, tetrahydro, etc. It should be noted that in all cases nitrogen heterocyclics often use different suffixes from those used for other heteroatoms (Fig. 13.2).

Table 13.2. Some of the suffixes used in the systematic naming of monocyclic heterocyclic systems. *Use the prefix perhydro to the name of the corresponding fully unsaturated compound.

Number of atoms in the ring.	Hetero atom	Saturated rings	Unsaturated rings Number of double bonds in the ring		
			(1)	(2)	(3)
3	O and S	irane	irene		
	N	iridine	irine		
4	O and S	etan	eten	et	
	N	etidine	etine	ete	
5	O and S	olan	olen	ole	
	N	olidine	oline	ole	
6	O and S	an			in
	N	*			ine

The numbering of monocyclic systems containing one heteroatom starts with the heteroatom and proceeds in the direction that gives the lowest locants to any other substituents (Fig. 13.2). In monocyclic systems with more than one heteroatom, number 1 is usually allocated to the heteroatom that is highest in the table of heteroatom prefixes (Table 13.2) and numbering proceeds in the direction that gives the other heteroatoms the lowest locants irrespective of their position in the table and other substituents present. These locants are cited in the same order as the heteroatom prefixes in the name of the system (Table 13.1). It should be noted that the **locant-H** is used to denote the position of a saturated atom in unsaturated compounds. Furthermore, the prefixes **di-** and **tetrahydro-**, together with the appropriate locants if required, are used to name certain of the reduced and partly reduced derivatives of some unsaturated ring systems.

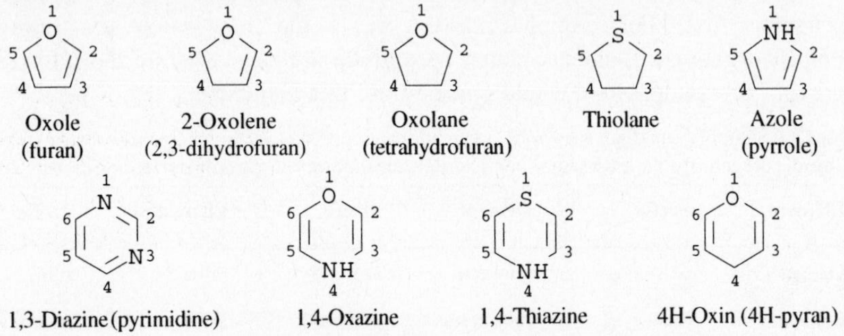

| Oxole | 2-Oxolene | Oxolane | Thiolane | Azole |
| (furan) | (2,3-dihydrofuran) | (tetrahydrofuran) | | (pyrrole) |

1,3-Diazine (pyrimidine)   1,4-Oxazine   1,4-Thiazine   4H-Oxin (4H-pyran)

Fig. 13.2. Examples of the systematic nomenclature used for monocyclic ring systems. The commonly used trivial name is given in brackets where applicable.

13.2.2   Heterocyclic systems as substituents

Substituent structures are often referred to as radicals. The names of monovalent heterocyclic radicals are derived from the name of the corresponding heterocyclic system by replacing the ending of that name with **-yl** (Fig. 13.3). The atom of the substituent that is attached to the parent structure is indicated by its number in the substituents' number system whereas its position on the parent structure is shown by the parent's locant. This may require the use of brackets around the name of the radical and its number, however, by convention the use of brackets is always kept to a minimum. If the

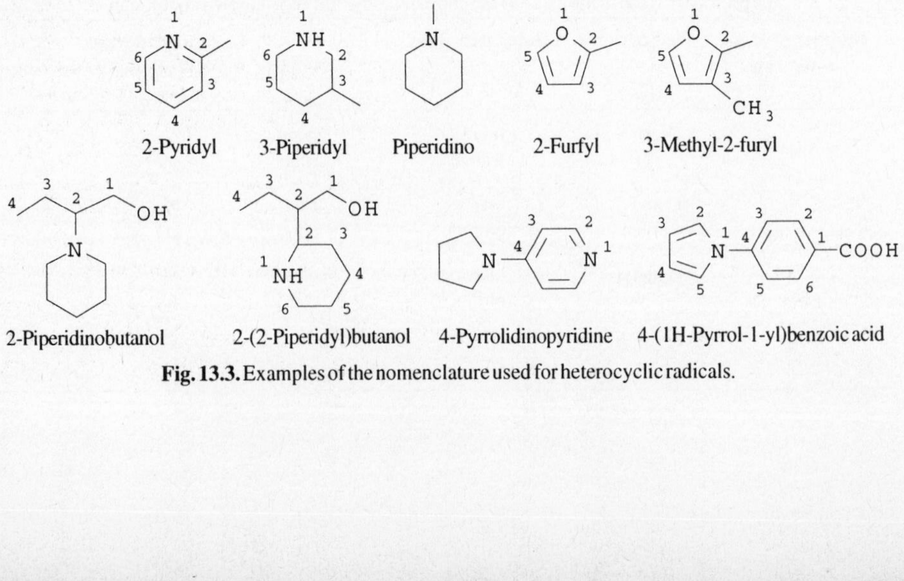

2-Pyridyl   3-Piperidyl   Piperidino   2-Furfyl   3-Methyl-2-furyl

2-Piperidinobutanol   2-(2-Piperidyl)butanol   4-Pyrrolidinopyridine   4-(1H-Pyrrol-1-yl)benzoic acid

Fig. 13.3. Examples of the nomenclature used for heterocyclic radicals.

heteroatom is the point of attachment to the parent, the ending **-o** is sometimes used instead of **-yl**.

13.2.3 Multicyclic systems

These use a mixture of allowed trivial and systematic nomenclature. Trivial names are used for some simple and most complex systems. The systematic system is broadly based on those used for the corresponding hydrocarbon systems. Consider, for example, the nomenclature of fused polynuclear aromatic systems. The biggest system with a systematic or allowed trivial name is selected as the stem name for the structure. However, the parent ring structure should also be selected, in the order of preference:

(1) A nitrogen heterocycle.
(2) If no nitrogen is present, the component containing the heteroatom highest in the periodic table.
(3) The structure with the largest number of rings.

The ending **-o** (dropped before a vowel) is used for fused substituent ring systems. Lower case letters and numbers in square brackets are used in the same way as in fused polynuclear aromatic systems to indicate the point of fusion of the substituent ring to the parent (12.7). However, the complete structure is numbered so that the numbers of the heteroatoms are kept as low as possible (Fig.13.4). This is different from that used for fused polynuclear hydrocarbons.

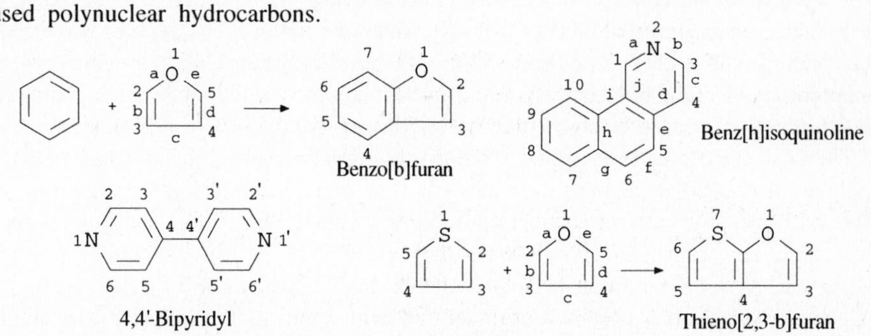

Benzo[b]furan

Benz[h]isoquinoline

4,4'-Bipyridyl

Thieno[2,3-b]furan

Fig. 13.4. Examples of the systematic nomenclature used for polynuclear heterocyclic systems. Attention is drawn to the way the numbering of the constituent rings is written prior to fusion with the complete system. This is to keep the letters used as near the beginning of the alphabet as possible. The numbers on the completed structures are the locants used for the complete fused ring system (12.7).

Non-fused aromatic systems are named in a similar fashion to the appropriate non-fused aromatic hydrocarbon system (12.8). In these systems the number system is again based on keeping the numbers of the heteroatoms as low as possible (Fig. 13.4).

13.3 ALICYCLIC HETEROCYCLIC SYSTEMS

13.3.1 Structure

The minimum energy conformations assumed by both saturated and unsaturated monocyclic aliphatic heterocyclic systems are similar to those of their carbocyclic analogues (Fig. 13.5). The same conformations are found in more complex structures, for example, the chair form of the 1,3,5-triazane ring is repeated in the structure of urotropine (hexamethylenetetramine). Alicyclic heterocyclic ring systems may also exhibit geometric and optical isomerism (Chapter 5).

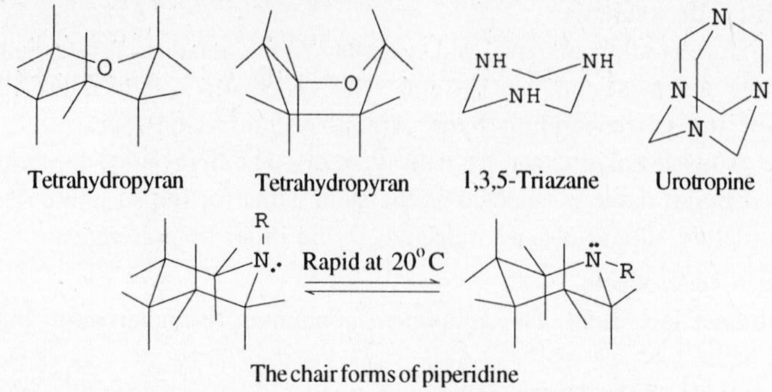

| Tetrahydropyran | Tetrahydropyran | 1,3,5-Triazane | Urotropine |

The chair forms of piperidine

Fig. 13.5. The conformations of some aliphatic heterocyclic systems. It has been estimated that the energy of the chair forms of piperidine and tetrahydropyran are 20 kJmol^{-1} less than their boat forms. Piperidine has two chair forms which undergo rapid interconversion at room temperature (compare with cyclohexane, 4.3.1).

13.3.2 General physical properties

In general the physical properties of aliphatic heterocyclic systems are similar to those of the corresponding carbocyclic systems. However, they are modified to some extent by the presence of the heteroatom. For example, saturated heterocyclic systems are usually more water soluble than the corresponding carbocyclic systems. Furthermore, systems containing oxygen and nitrogen will tend to be more water soluble than those containing other heteroatoms because these atoms can hydrogen bond to the water molecules.

The presence of one heteroatom in a ring system may result in the structure having a dipole because of polarisation of the bonds adjacent to that heteroatom. These dipoles will affect both the physical and chemical properties of the system. Increased dipole-dipole interactions are probably responsible for the tendency of the boiling points of simple heterocyclic systems with one heteroatom to be higher than those of their carbocyclic analogues. Systems that contain more than one heteroatom will be affected in a similar fashion provided the molecule has a resultant dipole moment.

13.3.3 General chemical properties

The chemical properties of heterocyclic systems depend on the size of the ring, the heteroatom and the structure of the carbon skeleton. Three- and four- membered rings are more reactive than five- and larger membered ring systems. This increased reactivity is due to weak bonds which can be theoretically explained in terms of a combination of poor orbital overlap (12.2.3) and the polarisation of the bonds by the heteroatom. For example, in oxiranes (epoxides) the oxygen polarises the C–O bonds sufficiently to allow attack of nucleophiles on the carbon atoms.

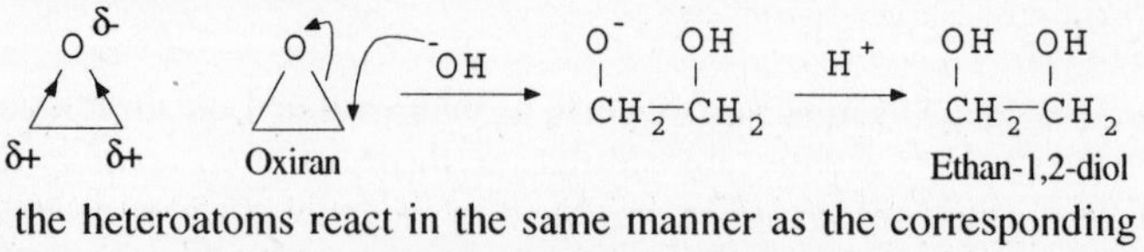

In general the heteroatoms react in the same manner as the corresponding functional group taking into account any restrictions imposed by the ring system. For instance, tetrahydrofuran is relatively inert like any other ether (Chapter 17), thiolane behaves like a sulphide (Chapter 17) and piperidine behaves as a typical secondary amine (Chapter 24).

Saturated carbon skeletons are relatively inert but unsaturated aliphatic systems react in a similar manner to alkenes and polyenes. For example, the C=C bond 2,3-dihydropyran reacts like a typical alkene readily undergoing oxidation, reduction and addition.

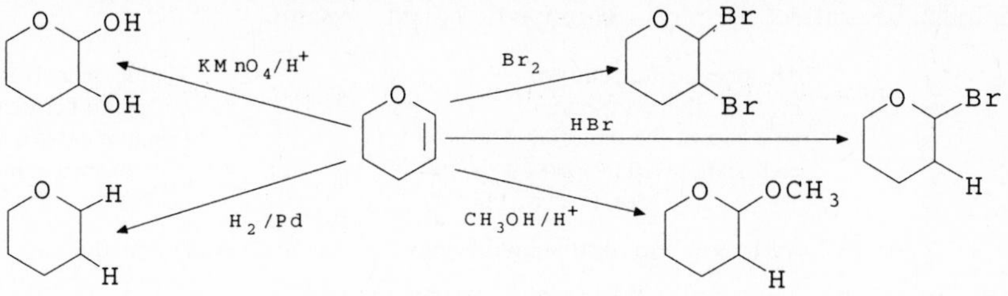

13.4 MONOCYCLIC HETEROAROMATIC SYSTEMS

Monocyclic heteroaromatic systems obey the Huckel rule (12.6). Their pi electron systems include a lone pair of electrons supplied by the heteroatom. The dipole moments of monocyclic heteroaromatic systems with one heteroatom show that this type of heterocyclic can be subdivided into structures in which the heteroatom acts as an electron donor and those in which it acts an electron acceptor (Figs 13.6 and 13.9). In the former case the carbons of the ring are pi electron rich (pi excessive compounds) whilst in the latter they are pi electron deficient (pi deficient compounds). This variation in pi electron density accounts for many of the differences in their chemical properties.

More complex systems may have both types of heteroatom in their structures (Fig. 13.6). This may make it more difficult to explain or predict their chemical and physical properties.

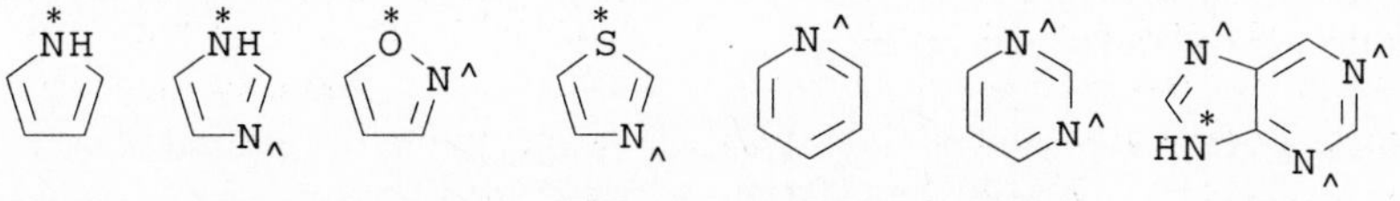

Pyrrole Imidazole Isoxazole Thiazole Pyridine Pyrimidine Purine

Fig. 13.6. Examples of heterocyclic systems with electron donor (*) and acceptor (^) heteroatoms.

The presence of a chromophore in both types of aromatic heterocyclic system means that UV and visible spectroscopy can be used as the basis of identification tests, assay and limit test procedures for compounds containing these systems. For example, the 1993 British Pharmacopoeia identification test A for the hypnotic/sedative and anticonvulsant chlormethiazole (5-(2-chloroethyl)-4-methylthiazole) states that a 0.004% solution of this compound in 1 M hydrochloric acid has a maximum absorption at 257 nm with an absorbance of about 1.1 in a 1 cm cell.

13.5 MONOCYCLIC HETEROAROMATIC PI EXCESSIVE SYSTEMS

These are systems in which the lone pair of the heteroatom is part of the conjugated pi electron system (Fig. 13.7). As a result, the lone pair is not readily available for reaction and so the heteroatom does not usually exhibit the same general chemical properties as the corresponding functional group. For example, the secondary amine group ($^-$NH$^-$) of pyrrole is almost non-basic (24.5) because the lone pair of the nitrogen atom are part of the delocalised pi electron system unlike in pyrrolidine where the lone pair is not part of a delocalised pi electron system. This participation of the lone pair in the structure of a system can also affect the physical properties of the system.

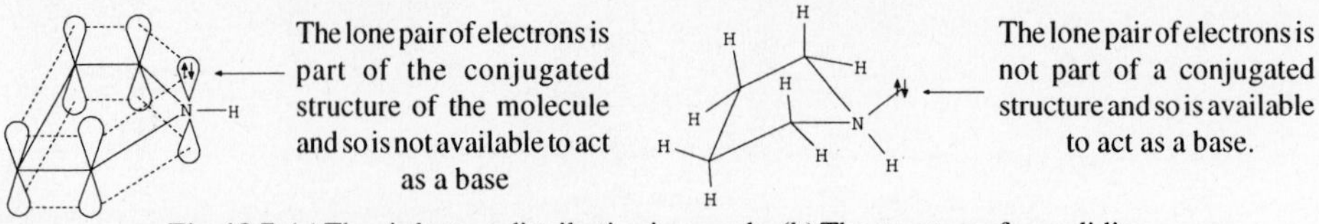

The lone pair of electrons is part of the conjugated structure of the molecule and so is not available to act as a base

The lone pair of electrons is not part of a conjugated structure and so is available to act as a base.

Fig. 13.7. (a) The pi electron distribution in pyrrole. (b) The structure of pyrrolidine

13.5.1　Physical　properties

The flat shape of the rings and their electron-rich nature makes it possible for these systems to form charge transfer complexes with suitable electron-deficient compounds. This is believed partly to account for the binding of drugs, whose molecules contain these structures, to active and receptor sites. It should be noted that the presence of dipoles in heterocyclic structures could also have an affect on the binding of substances to active and receptor sites.

Water solubility tends to be low in monocyclic systems even in the case of heterocyclics that contain heteroatoms, such as oxygen, which contain more than one lone pair.

13.5.2　Chemical　properties

Pi excessive heterocyclics behave similarly to aromatic hydrocarbons in that they prefer electrophilic substitution to addition. Substitution is relatively easy and usually occurs at carbons 2 and 5 (Fig.13.8) or carbons 3 and 4 if carbons 2 and 5 are unavailable. However, both pyrrole and furan are sensitive to acid conditions. This can result in polymerisation and in the case of furan, ring opening.

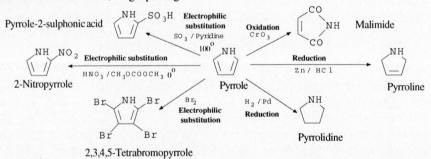

Fig. 13.8. Examples of the reactions of pyrrole. (Bromination of furan in dioxan yields 2-bromofuran while bromination of thiophen produces 2,5-dibromothiophen. Sulphonation of both furan and thiophen forms the corresponding 2-substituted product whilst nitration of furan yields 2-nitrofuran. Both furan and thiophen are reduced by zinc and hydrochloric acid but the reaction produces the 2,3 unsaturated ring compound. Oxidation of furan, thiophen and pyrrole yields a variety of products depending on the oxidising agent and the conditions.)

Biological examples of electrophilic substitution of pi electron rich heterocyclic systems are rare. One of the few examples is found in the biogenesis of the porphin ring of haem. The formation of this ring which forms part of the structure of haemoglobin, is believed to involve the electrophilic substitution of a pyrrole ring (Fig. 13.9).

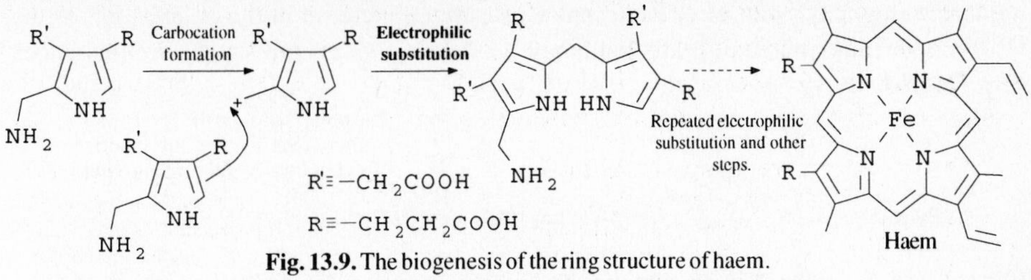

R'≡−CH₂COOH   NH₂

R≡−CH₂CH₂COOH

Fig. 13.9. The biogenesis of the ring structure of haem.

Monocyclic pi excessive heterocyclic systems are relatively easy to oxidise and reduce (Fig. 13.8). This is in direct contrast to benzene and other aromatic hydrocarbons.

Thiophen behaves as a typical aromatic compound. Furan, and to a limited extent pyrrole, exhibit some properties that are typical of alkenes. For example, both furan and pyrrole polymerise in acid solution and oxidise on exposure to air. Furan also acts as a diene and undergoes the Diels-Alder reaction. The reaction can give rise to stereoisomers.

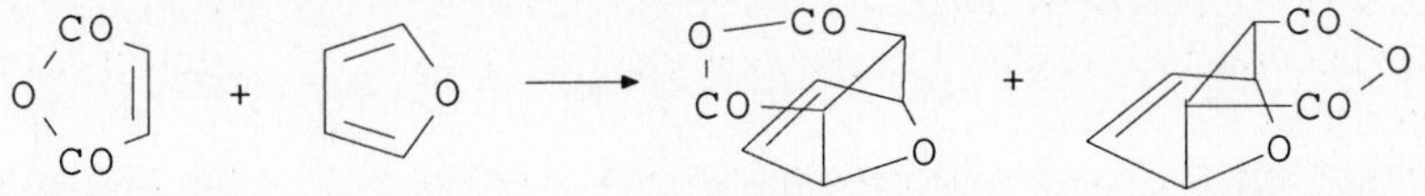

Maleic anhydride   Furan   Endo Diels-Alder adduct   Exo Diels-Alder adduct

These differences in the chemical properties of pyrrole, thiophen, and furan can be accounted for by the differences in the electronegativities of their heteroatoms, namely oxygen 3.5, nitrogen 3.0 and sulphur 2.5. The high electronegativity of oxygen reduces the degree of lone pair delocalisation because the oxygen **holds its electrons** more tightly than nitrogen and sulphur. This causes a degree of bond fixation (12.7.3) in the structure which gives rise to its alkene properties. Although nitrogen has a high electronegativity, it is lower than that of oxygen and so is less effective in causing bond fixation, hence pyrrole has few alkene-like properties. The electronegativity of sulphur has almost no affect on the delocalisation of the lone pair, consequently, thiophen has almost no alkene-like properties.

Pyrrole and furan are extremely weak bases. Pyrrole will only from salts under strongly acidic conditions. The cation is not aromatic but appears to react like a conjugated diene.

The N−H bond of pyrrole is weakly acidic (pK_a about 15). For example, it reacts with dry potassium hydroxide to form the potassium salt which under anhydrous conditions reacts with alkyl halides to form N-alkyl derivatives of pyrrole. This acidity is partly due to the stabilisation of the anion by delocalisation of its negative charge in the ring. The pyrrole anion is still aromatic. Thiophen is almost non-basic.

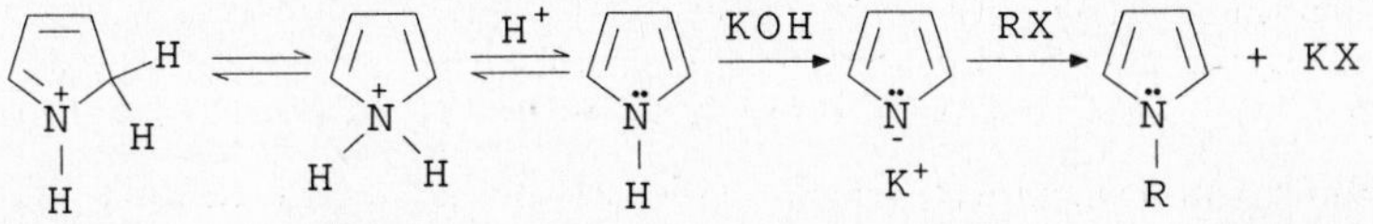

Pyrrole derivatives with electron acceptor substituents are usually more acidic whilst

those with electron donor substituents are generally less acidic than pyrrole. For instance, imidazole is a stronger acid than pyrrole. The double bonded nitrogen in the imidazole ring acts as an electron acceptor causing an increase in the polarisation of the N−H bond thus increasing its tendency to break, under the right circumstances (Fig. 13.10).

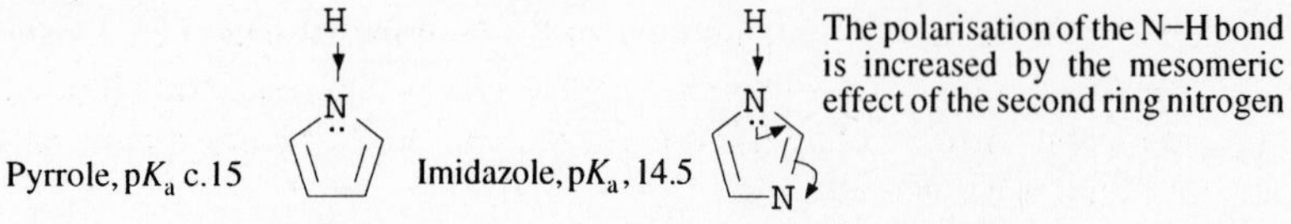

Pyrrole, pK_a c.15   Imidazole, pK_a, 14.5

The polarisation of the N−H bond is increased by the mesomeric effect of the second ring nitrogen

Fig 13.10. The relative acidic strengths of pyrrole and imidazole.

The increased acidic nature of the N−H bond of many pyrrole derivatives with electron acceptor substituents on the pyrrole ring results in them forming stable salts with organic bases. These salts are often more water soluble than the parent compound which can be useful in the formulation of drugs. For example, the xanthine bronchodilator theophylline is usually administered as aqueous solutions of either its choline or 1,2-diaminoethane salts.

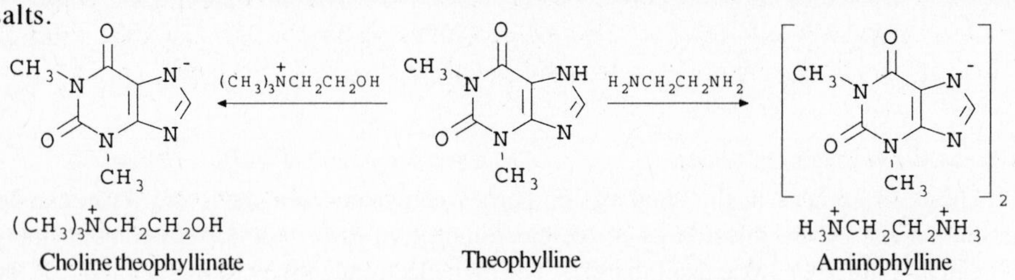

$(CH_3)_3\overset{+}{N}CH_2CH_2OH$

Choline theophyllinate

Theophylline

$H_3\overset{+}{N}CH_2CH_2\overset{+}{N}H_3$

Aminophylline

13.6 MONOCYCLIC HETEROAROMATIC PI DEFICIENT SYSTEMS

In pi deficient systems, such as pyridine, the nitrogen's lone pair is not part of the conjugated aromatic pi electron system but occupies an sp^2 hybrid orbital. Consequently, the nitrogen atom acts as an electron acceptor reducing the electron density at C_2, C_4 and C_6 (Fig. 13.11). This accounts for the relative ease of nucleophilic substitution of pyridine at these positions and the great difficulty of bringing about electrophilic substitution at the C_3 and C_5 positions.

In heterocyclics where there is more than one heteroatom in the ring, the pi electron distribution will be the resultant of the characteristic effects exerted by each heteroatom.

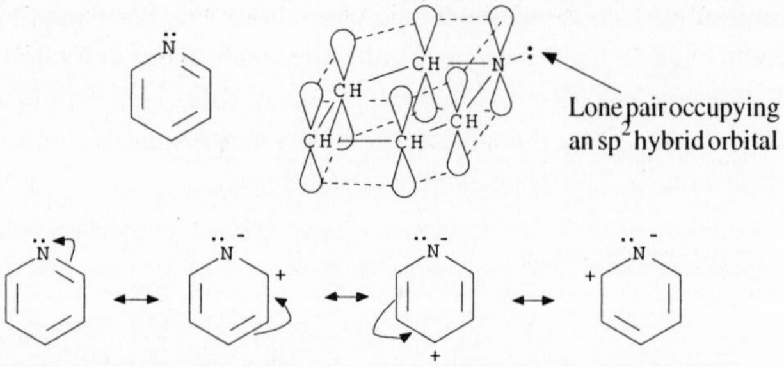

Lone pair occupying an sp^2 hybrid orbital

Fig. 13.11. The structure and pi electron distribution of pyridine.

In this context it should be noted that many heterocyclic compounds will have both electron acceptor and donor heteroatoms in the same ring (Fig. 13.6). Consequently, it is often difficult to predict this resultant distribution and the affect it will have on the chemical and physical properties of the system.

Many substituted pi electron deficient aromatic heterocyclic species exhibit tautomerism, that is, they exist as equilibrium mixtures of the appropriate tautomers (isomers). For example, 2-hydroxypyridine and 4-hydroxypyridine exist as an equilibrium mixture of the hydroxypyridine and corresponding pyridone. The same type of tautomerism occurs with uracil, a base found in the nucleic acids.

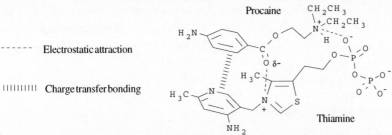

2-Hydroxypyridine   Uracil

The existence of tautomeric forms has a considerable affect on the basic strengths of pi deficient aromatic heterocyclic species. For example, in aqueous solution 2-hydroxypyridine, 4-hydroxypyridine and uracil exist almost exclusively in the amide form and as a result are very weak bases (25.5).

13.6.1   Physical properties

The lone pairs of the heteroatoms of pi deficient aromatic heterocyclic systems are available to form hydrogen bonds as they are not involved in the delocalised pi electron structure of the ring (Fig. 13.11). This may have an effect on the physical properties of the compound. For example, hydrogen bonding between the nitrogen atom and water molecules is believed to be responsible for the slight water solubility of pyridine. This is in complete contrast to the water insolubility of its hydrocarbon analogue benzene. However, substituents do not always have the expected effect on water solubility: hydroxy and carboxylic acid substituents, for example, reduce water solubility. This is the opposite effect to what is normally observed. The reason for this behaviour is not clear. However, the water solubility of this type of heterocyclic system appears to be an important factor in the widespread biological activity of compounds whose molecular structures are based on these systems.

The flat nature of the heterocyclic ring system may result in charge transfer complex formation that contributes to the stabilisation of drug-active/receptor site complexes. For

Fig. 13.12. The binding of procaine to thiamine pyrophosphate.

example, charge transfer is believed to contribute to the stability of the local anaesthetic-thiamine pyrophosphate complexes which are thought to be involved in the diffusion of local anaesthetics through membranes (Fig. 13.12).

13.6.2　Chemical properties

Nitrogen heterocyclics constitute the main class of pi electron deficient aromatic heterocyclics and so this section will be restricted to these systems. The nitrogen acts as a **base**, a nucleophile and can be oxidised whilst the carbon ring can undergo **electrophilic substitution, oxidation** and **reduction**. These general properties are very similar to those exhibited by other types of aromatic system (Chapter 12). However, the nitrogen atoms in these structures are strong enough electron acceptors to allow **nucleophilic substitution** of the carbon ring. This should be contrasted with the behaviour of the aromatic hydrocarbon systems discussed in Chapter 12 and the pi electron rich heterocyclic systems described earlier in this chapter.

(a) Basic nature. Pyridine and other pi electron deficient nitrogen aromatic heterocyclic systems can function as bases because a lone pair of the hetero atom is not delocalised in the pi electron structure. However, they are not such strong bases as their corresponding aliphatic heterocyclic analogues. For example, piperidine (pK_a, 11.2) is a much stronger base than pyridine (pK_a, 5.3). The presence of more than one electron acceptor heteroatom and/or electron acceptor substituents usually results in weaker bases because electron acceptors reduce the availability of the lone pair and the relative stability of the conjugate acid. On the other hand, the presence of electron donor groups generally increases basic strength by increasing the availability of the lone pair and stabilising the conjugate acid (24.5 for a fuller explanation).

A strongly basic nitrogen appears to be an important factor in the biological activity of many pi electron deficient heterocyclics. For example, increased basicity, due to the presence of electron donor methoxy groups, is believed to be responsible for the increased level of serum-protein binding of the long acting sulphonamides, sulphamethoxydiazine, and sulphamethoxypyridazine.

Sulphamethoxydiazine pK_a 7.0　　　　　　　　Sulphamethoxypyridazine pK_a 7.2

Pi electron deficient nitrogen aromatic heterocyclic readily form salts with all types of acid and quaternary salts with alkyl halides. For example, pyridine forms pyridinium chloride with hydrochloric acid, a picrate with picric acid and quaternary salts with methyl

Pyridinium picrate　　　　Picric acid　　　　　Pyridinium chloride　　　N-acetylpyridinium chloride iodide and cetyl chloride. Picrates and quaternary methyl salts are useful as derivatives of pyridine and similar heterocyclic compounds in organic analysis. Cetylpyridinium chloride acts as a detergent (9.6.5). It also exhibits antifungal and antibacterial activity and so is used as a surface active antiseptic.

The positive charge of a quaternary salt enhances the electron withdrawal effect of the nitrogen and makes nucleophilic attack of the ring easier. For example, alcohol dehydrogenation in the liver involves the attack of a transient hydride ion formed from a hydrogen of the alcohol. The process is catalysed by alcohol dehydrogenase which requires nicotinamide adenine dinucleotide (NAD^+) as co-enzyme.

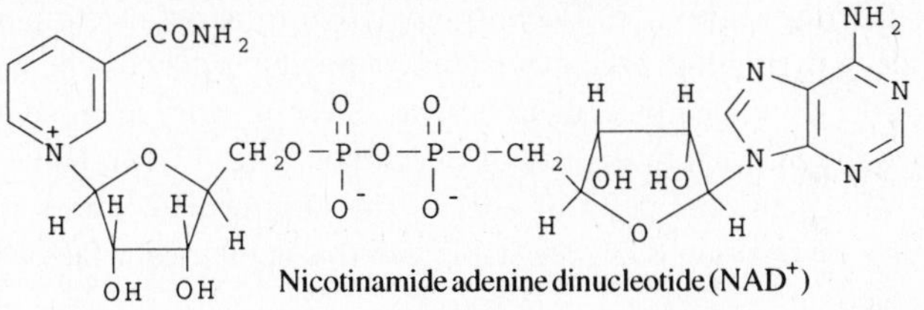

Nicotinamide adenine dinucleotide (NAD^+)

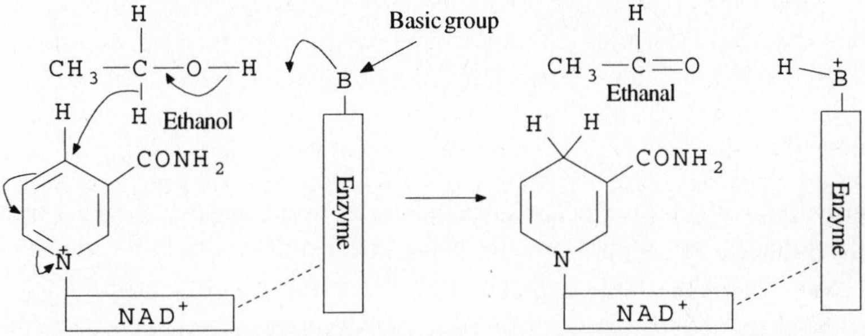

(b) Electrophilic substitution. The reduction in the pi electron density of the carbon atoms of the ring by the electron acceptor effect of the heteroatom in monoheteroatom systems makes electrophilic substitution of these systems much more difficult to carry out in the laboratory than the corresponding electrophilic substitutions of benzene. For example, electrophilic substitution of pyridine occurs at the electron deficient positions 3 and 5 (Fig. 13.13). It requires a far higher temperature and longer reaction time than those needed for the corresponding reactions of benzene (Fig. 13.13). Yields are usually poor especially in acidic conditions when the formation of pyridinium salts further reduces the pi electron density of the ring carbons. However, the ease of electrophilic substitution is improved by the presence of electron donor substituents such as, alkyl, amino and hydroxyl groups. In these compounds, substitution usually occurs ortho and para to the substituent; however, position 4 of the ring does not tend to react unless positions 2 and 6 are blocked.

Electrophilic substitution does not occur in unsubstituted pi electron deficient heterocyclic systems such as pyrimidine and pyrazine that possess more than one electron acceptor heteroatom. The combined electron-accepting effect of the heteroatoms is strong enough to prevent reaction. However, compounds with strong electron donor substituents such as amino and pyridone groups will react. These electron donors counter the electron acceptor effects of the heteroatoms and the resultant pi electron density

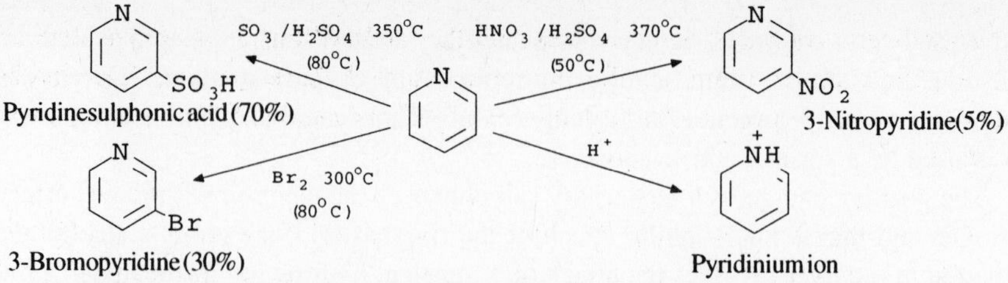

Fig.13 13. Examples of electrophilic substitution of pyridine. The temperatures in the brackets are those necessary for the corresponding electrophilic substitution of benzene.

makes some of the ring carbons sufficiently attractive to allow electrophilic attack. For example, cytosine, a pyrimidine base that occurs in nucleic acids, can be easily nitrated at position 5 where the combined mesomeric effects of the primary amino and the N_1 of the pyridone enhance the pi electron density of the carbon (Fig. 13.14). However, this is not the only factor influencing the point of attack. The intermediate carbocation formed by electrophilic attack on position 5 is very stable and this is the major factor in deciding the course of the reaction:

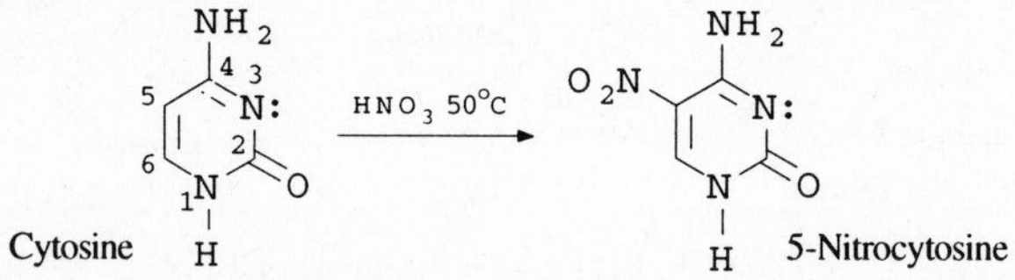

Pyrimidines with electron donor substituents can also behave in a similar fashion with other electrophiles, substitution usually being ortho and/or para to the substituent.

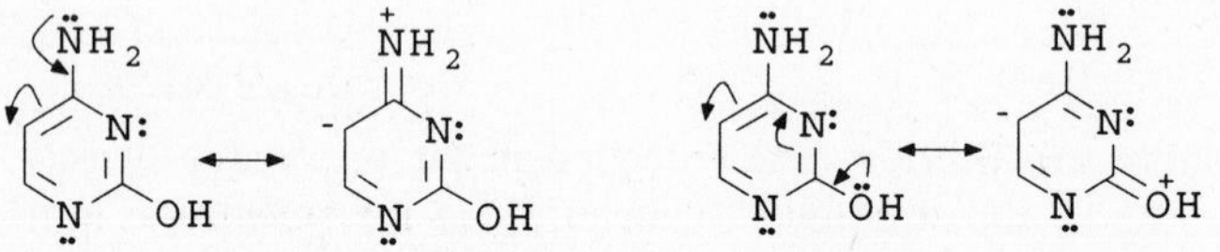

Fig. 13.14. The pi electron distribution in cytosine is best shown using the tautomeric hydroxy form of the pyridone group. The mesomeric effects of the primary amine and pyridone groups reinforce each other at the 5 position making this position very electron rich. It should be noted that these mesomeric effects will also increase the electron density at positions 1 and 3.

(c) Nucleophilic substitution. The electron withdrawing affect of the heteroatom reduces the pi electron density to the point where nucleophilic attack can occur. For example, the electron deficient carbons 2, 4 and 6 (Fig 13.11) of pyridine are attacked by strong nucleophiles, positions 2 and 6 being preferred.

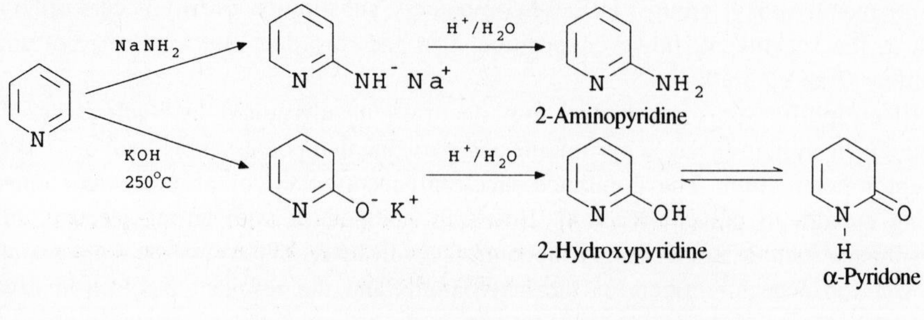

These reactions are believed to occur by the nucleophilic substitution mechanisms of the type illustrated in Fig. 13.15 for the formation of 2-aminopyridine. It should be noted that the intermediate anion (**A**) is stabilised by delocalisation of the negative charge over the ring.

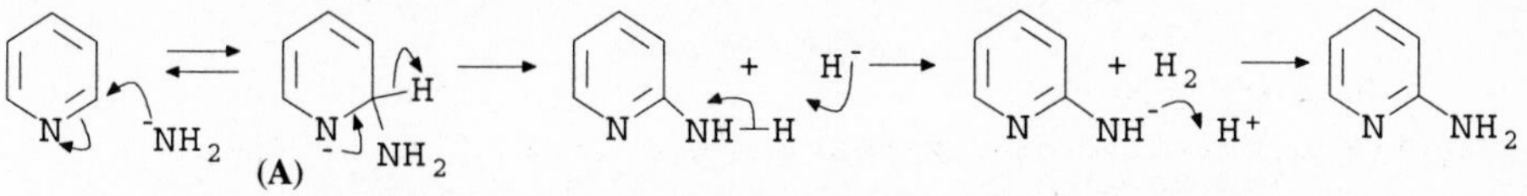

Fig. 13.15. The proposed mechanism for the nucleophilic substitution of pyridine.

The conditions required for these reactions are somewhat extreme and would not occur in biological systems. However, the nucleophilic substitution of halo substituents in the 2, 4 and 6 positions is comparatively easy.

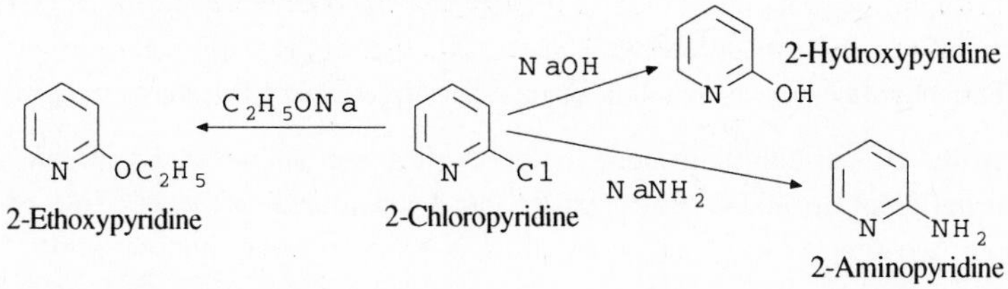

2-Aminopyridine is used in the synthesis of a number of sulphonamide drugs, for example, sulphapyridine which is used to treat dermatitis herpetiformis and the antihistamine, pyribenzamine.

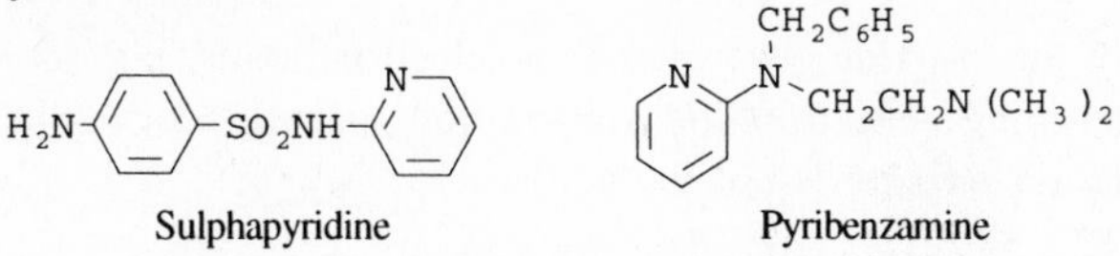

The presence of more than one electron acceptor heteroatom in the ring also substantially increases the ease of nucleophilic attack. For example, pyrimidine and a number of its 2 and 4 substituted derivatives such as 4-chloropyrimidine, are substituted by nucleophiles since the electron acceptor effects of the two heteroatoms complement each other (Figs 13.16 and 13.17). Similarly, the electron acceptor effects of the heteroatoms of pyridazine, pyrazine and 1,3,5-triazine make all the carbon atoms susceptible to nucleophilic attack. As would be expected 1,3,5-triazine is the most reactive,

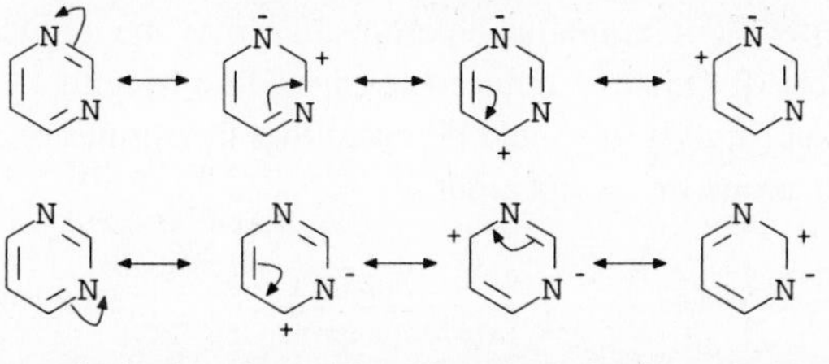

Fig. 13.16. The pi electron distribution of pyrimidine. The mesomeric effects of the two nitrogens are shown separately; however, they will both contribute to the resonance hybrid and so will complement each other at positions 2, 4 and 6.

reacting rapidly with water in the presence of an acid to form ammonium methanoate as the only isolatable product (Fig. 13.17).

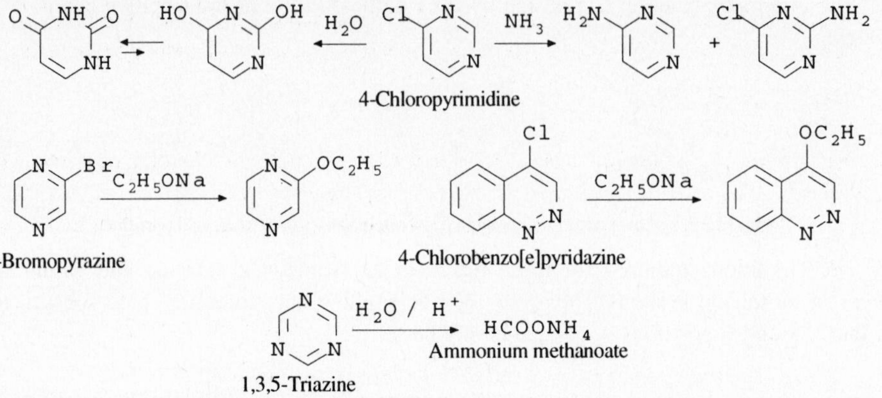

4-Chloropyrimidine

2-Bromopyrazine

4-Chlorobenzo[e]pyridazine

1,3,5-Triazine

Fig 13.17. Examples of the nucleophilic substitution reactions of pi electron deficient heterocyclic ring systems.

(d) Oxidation. Pi deficient aromatic heterocyclics are stable to oxidation. However, nitrogen pi deficient aromatic heterocyclics can be oxidised to the N-oxide by peracids. For example, pyridine is oxidised to pyridine-N-oxide by perethanoic acid:

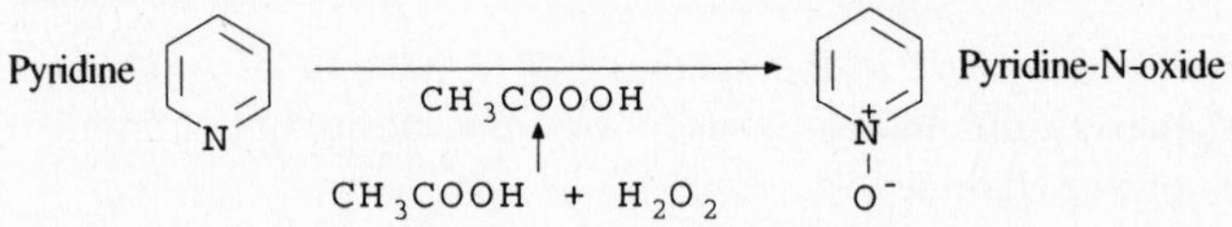

Pyridine   Pyridine-N-oxide

$$CH_3COOH$$

$$CH_3COOH + H_2O_2$$

The oxygen of the N-oxide group acts as an electron donor (Fig. 13.18) which results in the ring undergoing electrophilic substitution, usually occurs in the 4 position, provided the reaction mixture is not too acidic.

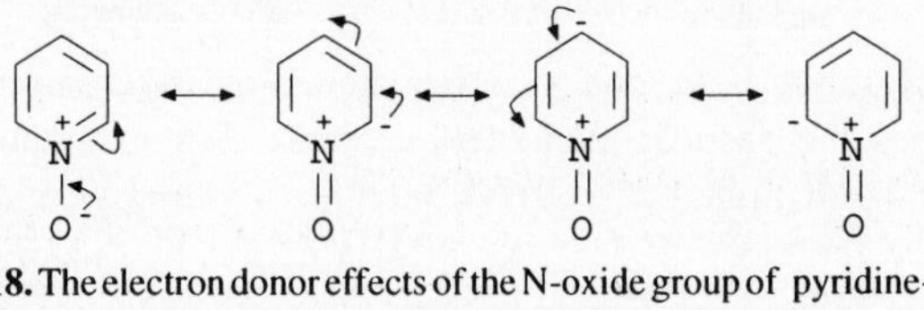

Fig. 13.18. The electron donor effects of the N-oxide group of pyridine-N-oxide.

Metabolic oxidation of pi deficient aromatic heterocyclics yields hydroxy compounds. This is similar to the metabolism of aromatic hydrocarbons (12.6.1c).

(e) Reduction. Pi deficient aromatic heterocyclics may be reduced by a number of reagents. For example, pyridine is reduced to piperidine by either sodium in ethanol or hydrogen and a nickel catalyst. It should be noted that the former reagent does not reduce benzene and similar aromatic hydrocarbons.

Pyridine   Various   Piperidine
                       reducing agents

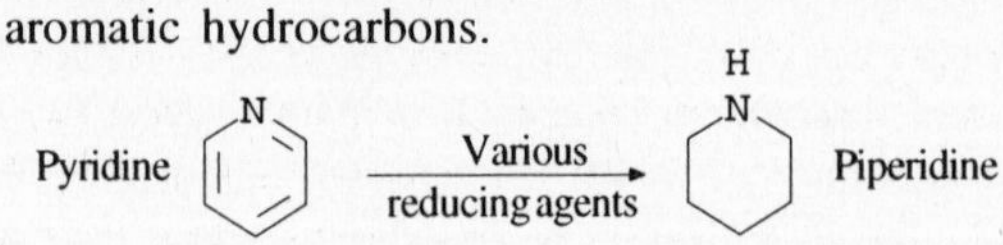

Partial hydrogenation of pyridine has not been achieved although derivatives of these systems are known. For example, NADH is a 1,4-dihydropyridine derivative (13.6.2a).

13.7 FUSED AND NON-FUSED AROMATIC HETEROCYCLIC SYSTEMS

The ring structures of these systems can consist of either all heterocyclic or a mixture of aromatic carbocyclic and heterocyclic rings. They are analogues of the fused and non-fused aromatic hydrocarbons discussed in 12.8 and 12.9 and occur widely in the structures of many natural products and drugs (Fig. 13.19).

13.7.1 Nomenclature

This uses a mixture of allowed trivial and systematic nomenclature. Systematic names are based on the rules used for fused and non-fused polynuclear aromatic hydrocarbons (12.8 and 12.9). However, trivial names are used for some simple and most complex heterocyclic ring systems (Fig. 13.19).

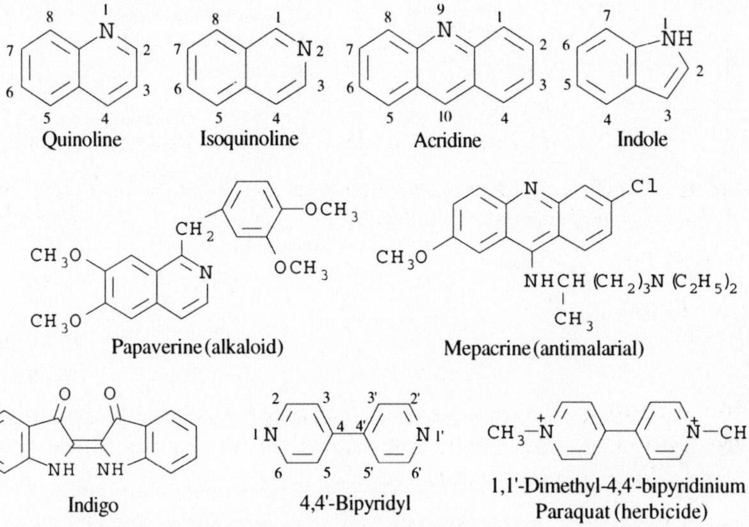

Fig. 13.19. Examples of fused and non-fused aromatic heterocyclic systems.

13.7.2 General chemical and physical properties

The general properties of a system are based on those of the individual ring structures forming that system. These properties are modified by the electronic effects of the other rings and any bond fixation (12.8) that may occur in the complete system. Consequently, it is possible to predict a system's general properties by considering those of each of the constituent ring systems. This method does not take into account the possibility and effects of bond fixation and other peculiarities of the complete system. These must be taken into consideration when predicting and interpreting the reactions of the system. A useful guide is that if bond fixation or other structural phenomena occur in the corresponding hydrocarbon analogue, it is likely to be present in the same part of the heterocyclic system, provided no heteroatoms are present. For example, consider the prediction of the general properties of quinoline (Fig. 13.19). One would predict that the benzene ring would be substituted by electrophiles but be resistant to oxidation and reduction whilst the pyridine ring would act as a base, undergo nucleophilic substitution and reduction but be resistant to oxidation and electrophilic substitution since electrophilic substitution of the benzene ring is easier. Furthermore, by comparison with its aromatic hydrocarbon analogue naphthalene (Fig. 12.22), one would expect position 3 to be relatively unreactive.

In practice, quinoline is a weak base (pK_a 4.9). It usually undergoes electrophilic substitution at positions 5 and 8 of the benzene ring. Nucleophilic substitution occurs mainly at positions 2 and 4 of the pyridine ring (Fig.13.20). Reduction of the pyridine ring is comparatively easy but reduction of the benzene ring is comparatively difficult. Oxidation of both rings is difficult.

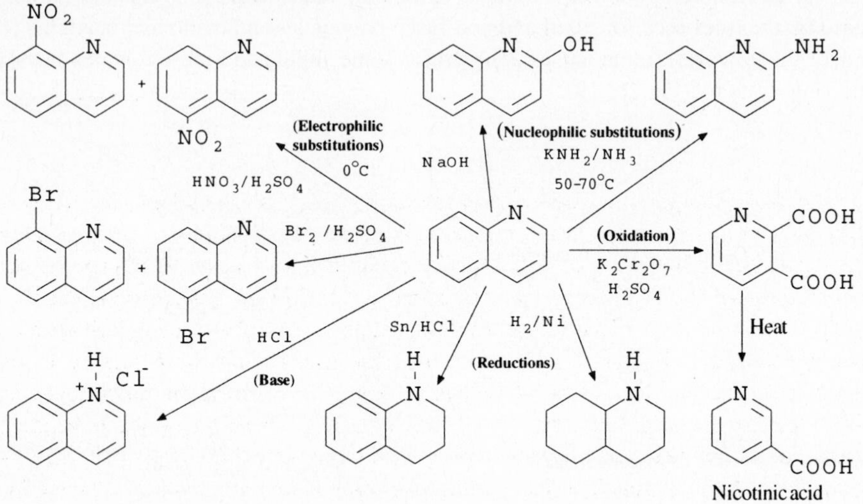

Fig. 13.20. Examples of the general types (in brackets) of reaction of quinoline.

13.7.3 Tautomerism

Fused aromatic heterocyclic compounds whose structures contain a pi deficient ring may exhibit tautomerism. For example, spectroscopy shows that the pyrimidine residue of uric acid exists mainly in the amide form. The formation of the enolic tautomer accounts for the solubility of uric acid in aqueous alkaline solutions. The hydroxy group of the enol forms a resonance-stabilised water soluble salt with the alkali. Similar behaviour is observed in theobromine, caffeine and similar compounds in aqueous alkali solutions.

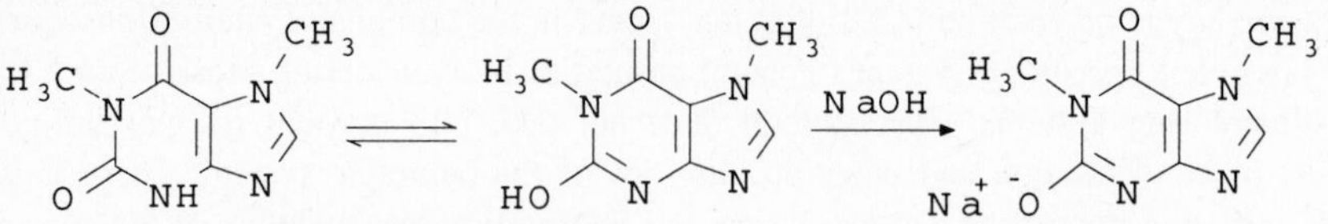

Theobromine, a stimulant found in tea and coffee

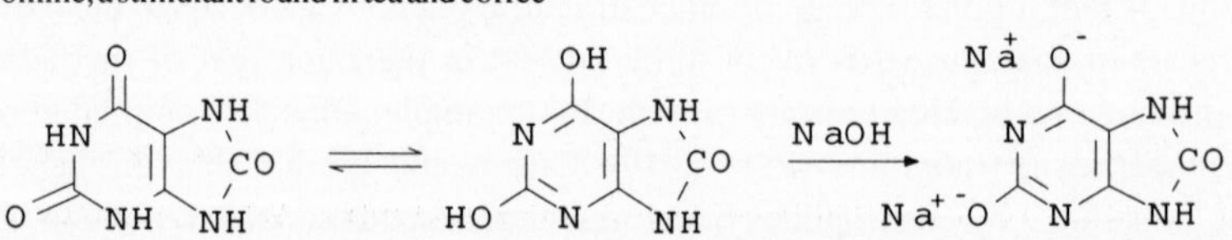

Uric acid. The deposition of sodium urate crystals triggers inflammation of the joints. Uric acid is the final nitrogen metabolite of birds and reptiles. Their urine is a 'paste' of crystals with a little water.

13.7.4 Basic nature

Compounds whose ring structures contain a nitrogen will be basic if the nitrogen is in a similar structural situation as the nitrogen of pyridine. The strength of the base is usually increased by the presence of electron donor substituents and electron donor ring heteroatoms. For, example, 4-aminoquinoline (pK_a 9.2) is a stronger base than quinoline (pK_a 4.9). whilst the antibacterial aminacrine (9-aminoacridine) (pK_a 10.0) is a stronger base than acridine (pK_a 5.6). In the amino derivatives, the electron donor effect of the primary amino group increases the electron density of the ring, nitrogen making it more electron rich and therefore more basic.

4-Aminoquinoline   9-Aminoacridine

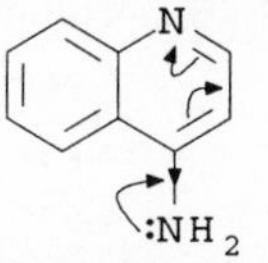

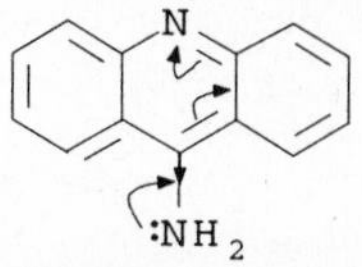

The basic strength of compounds is particularly important in biological systems since the degree of ionisation is a factor, not only in the compounds ability to cross membranes but also in any pharmacological activity. Neutral organic molecules are transported across membranes more readily than molecules that have charged structures. However, the charged form of the molecule is often responsible for its pharmacological action. For example, unionised organic base molecules are transported more readily across a membrane than their cations but these cations are believed to be the pharmacologically active form of the base. The antibacterial activity of aminacrine is believed to be due to it being almost completely ionised in aqueous solution at pH 7.4 (physiological pH). Other examples of the importance of basic strength in drug action are given in Chapter 24.

13.7.5 Acidic nature

The NH bonds of residues with pyrrole type structures are weakly acidic, the anion being stabilised by delocalisation of the negative charge in the ring. Electron-attracting ring substituents and heteroatoms can increase the acidic strength of this bond by withdrawing electrons from the vicinity of the nitrogen (Fig. 13.13). This has the effect of increasing the polarisation of the N−H bond which makes it easier for the molecule to lose the hydrogen as a hydrogen ion. For example, the secondary amine in the five-membered ring of theophylline acts as an acid and is responsible for the compound forming stable water soluble salts with choline and other organic bases. The negative charge of the anion of choline theophyllate is stabilised by resonance.

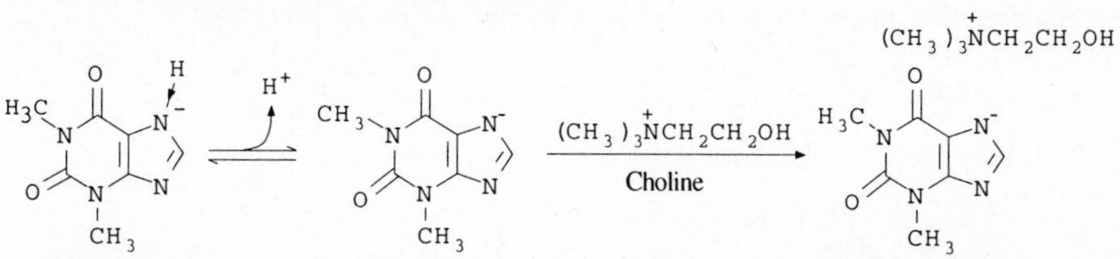

Theophylline   Choline Theophyllinate

Many purines, such as, mercaptopurine, and azathioprine whose structures contain electron acceptor substituents and heteroatoms are sufficiently acidic to form soluble metal salts in aqueous alkali solution.

Mercaptopurine

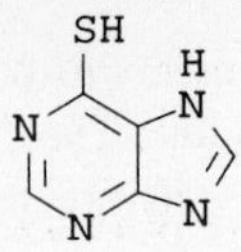

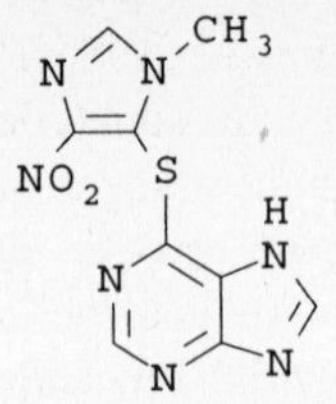

Azathioprine

Porphyrins, a class of compound based on the porphin ring system form very stable metallic salts. These salts include haem which forms part of the structure of haemoglobin and its magnesium analogue that is found in the structure of chlorophyll, the compound that is responsible for the green colour of plants. Both these compounds play an essential part in the life processes (32.8.2).

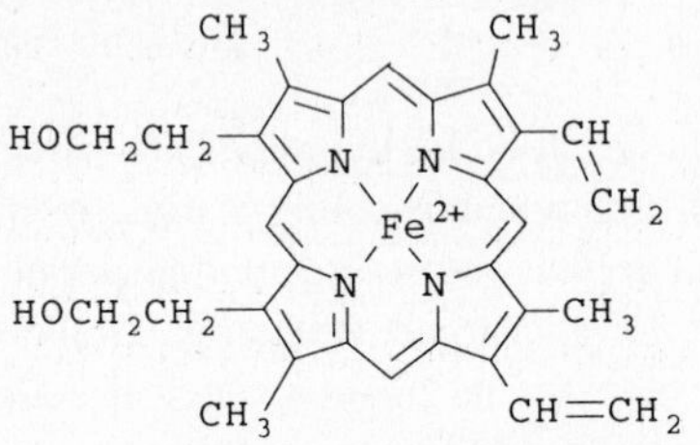

Haem as it occurs in haemoglobin

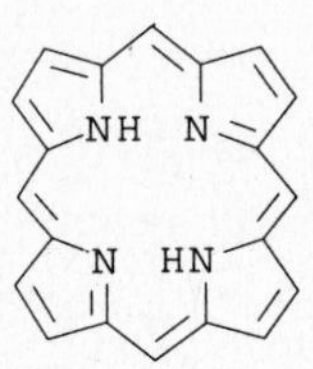

The phorphin ring system

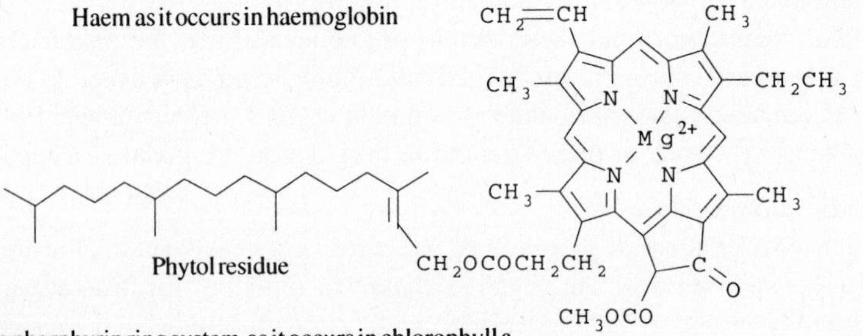

Phytol residue

The phorphyrin ring system as it occurs in chlorophyll a

13.7.6   Electrophilic and nucleophilic substitution

Electrophilic substitution will occur in the aromatic hydrocarbon residues and in pi excessive rings of these heterocyclic systems. In general, the heterocyclic rings are more reactive than any benzene residues in the molecule, as is the case, in the electrophilic substitutions of indole. The pyrrole ring of indole is readily substituted in the 3 position by electrophiles, the reaction with sulphur trioxide being an exception.

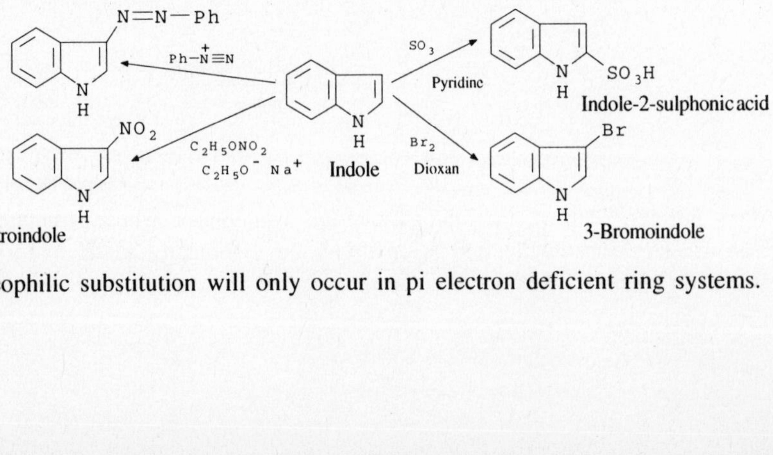

3-Nitroindole

Indole

Indole-2-sulphonic acid

3-Bromoindole

Nucleophilic substitution will only occur in pi electron deficient ring systems.

13.7.7 Oxidation and reduction

The different types of ring usually exhibit their characteristic reactivity towards oxidation and reduction (Table 13.3).

Table 13.3. A comparison of the ease of oxidation and reduction of aromatic ring systems.

Ring system	Oxidation	Reduction
Benzene	Difficult but easier than for pyridine	Difficult
Pyridine	Difficult	Difficult but easier than for benzene
Pyrrole, furan, thiophen	Fair, easier than for benzene	Fair, easier than for pyridine

Oxidative metabolism of both benzene and heterocyclic rings normally results in hydroxylation. For example, xanthine is formed by the hydroxylation of the purine hypoxanthine and is then converted into uric acid by further hydroxylation. Both transformations are catalysed by the enzyme xanthine oxidase. This enzyme is inhibited by allopurinol which itself is hydroxylated in the process. Allopurinol is used to treat gout. Its effect is to inhibit the formation of uric acid and increase the rate at which hypoxanthine and xanthine are removed from the kidney.

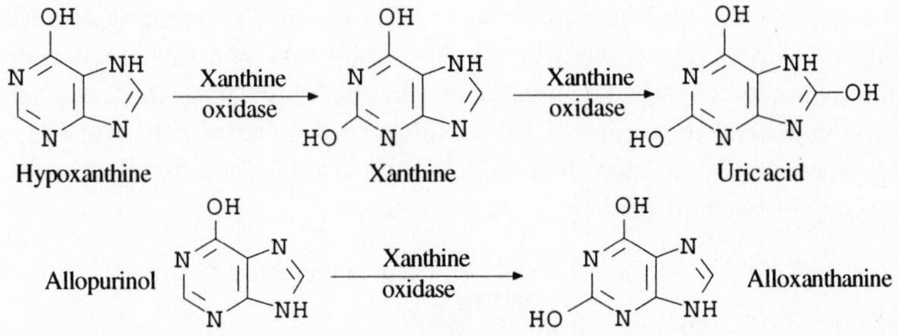

Reduction of aromatic and heterocyclic rings generally follows the characteristic routes for each type of ring. Pi electron rich heterocyclic rings can usually be reduced under mild conditions whilst aromatic hydrocarbon and pi electron deficient heterocyclic rings require more vigorous conditions.

The formation of a bright red dye by the reduction of colourless 2,3,5-triphenyl-1,2,3,4-tetrazolium chloride is used as a test for reducing agents such as ascorbic acid, in biological systems. This reagent should be stored in lightproof containers.

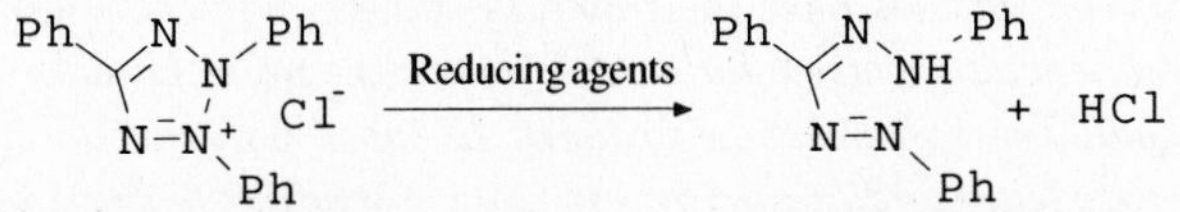

2,3,5-Triphenyl-1,2,3,4-tetrazolium chloride (colourless)   1,3,5-Triphenylformazan (red)

13.7.8 Bond fixation

Nitrogen pi deficient systems can exhibit a degree of bond fixation. It can occur in quaternary pyridine compounds and heterocyclics with more than one nitrogen atom in the ring and results in these compounds undergoing addition reactions with weak acids. For example, hydroxylamine and methoxyamine add to the 5,6 double bond of cytosine, methylcytosine and hydroxymethylcytosine residues in DNA. The addition is accompanied by the conversion of the aromatic amino group to an oxime. This type of addition, followed by elimination at the 5 and 6 positions and the conversion of the remaining oxime group to a carbonyl, is believed to be an important factor in the mutagenic effects of these compounds in the nucleic acids.

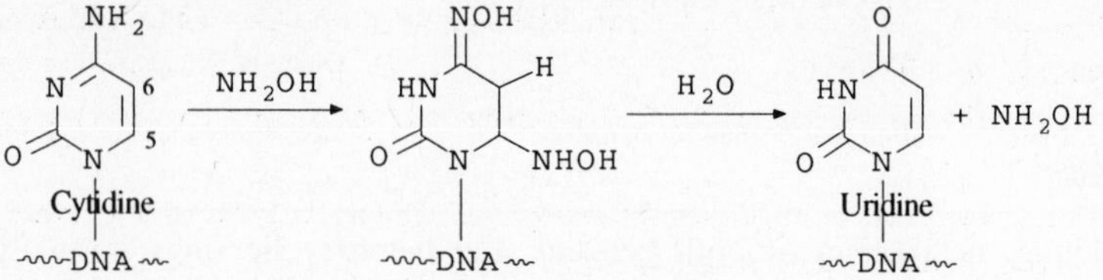

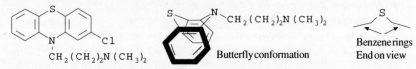

13.8   OTHER MULTICYCLIC HETEROCYCLIC SYSTEMS

This covers a wide range of compounds whose structures and shapes can be explained using the molecular orbital theory outlined in chapters 2, 3, 4 and 5. Their physical and chemical properties may be predicted and/or explained by analogy with those of their component ring systems and functional groups. For example, the antipsychotic (major tranquilliser) chlorpromazine has a folded structure (Fig. 13.21) because the central ring has a boat conformation. The presence of the aryl chloride makes this molecule chiral and so it has two enantiomers. Its general chemical and physical properties can be predicted by considering the general properties of its constituent functional groups and ring systems. This enable one to predict the most likely possibilities in a specific situation. However, these predictions should be confirmed by a literature search.

Fig. 13.21. The structure and general shape of chlorpromazine.

13.9 WHAT YOU NEED TO KNOW

(1)   Heterocyclic structures are ring structures that contain other atoms besides carbon. These atoms are known as heteroatoms.

(2)   Nomenclature may be systematic or allowed trivial.

(3)   Monocyclic systems have their own systematic nomenclature system.

(4)   The nomenclature used for multicyclic systems is broadly based on that of the corresponding hydrocarbon systems. In fused polynuclear heterocyclics it uses a lower case letter and/or numbers in square brackets to show where substituent rings are fused to the parent.

(5)   The number systems used in all heterocyclic systems are based on keeping the numbers of the heteroatoms as low as possible.

(6)   Many heterocyclic ring systems have dipoles. The resulting dipole moments form

part of the practical evidence underlying the mechanistic explanations of some of the physical and chemical properties of heterocyclic systems and how they differ from those of their carbocyclic analogues.

(7) Aliphatic heterocyclic molecules usually exist as conformers which usually resemble those of their carbocyclic analogues. Six-membered saturated rings can exist as both chair and boat forms, the chair being most common.

(8) Aliphatic heterocyclics are usually more water soluble than their carbocyclic counterparts. This is due to hydrogen bonding between the heteroatom and the water molecules.

(9) Three- and four- membered aliphatic rings are more reactive than five and larger membered rings.

(10) The heteroatoms of aliphatic heterocyclic systems behave in the same manner as the corresponding functional group.

(11) Monocyclic aromatic systems are flat. They may have more than one heteroatom in their structures and are classified as either pi excessive or deficient.

(12) Ultra-violet and visible spectra are frequently used as the basis of assays, identification, and limit tests for all types of aromatic heterocyclic system.

(13) In pi excessive systems the heteroatom acts as an electron donor.

(14) The flat electron rich structure of pi excessive systems makes it possible to form charge transfer complexes with suitable electron deficient structures of receptor and active sites.

(15) Water solubility of pi electron rich structures is low.

(16) Pi electron rich aromatic sulphur and nitrogen heterocyclic systems exhibit typical aromatic chemical properties. These are:

    (a) Electrophilic substitution (easier that of benzene).

    (b) Oxidation (easier than oxidation of benzene).

    (c) Reduction (easier than reduction of benzene).

(17) Pi excessive oxygen heterocyclics exhibit alkene as well as aromatic chemical properties such as polymerisation, addition and the Diels-Alder reaction.

(18) The N⁻H bond of nitrogen aromatic pi electron rich compounds is acidic and can form salts which are often more water soluble than the parent compound.

(19) In pi deficient aromatic heterocyclic systems the heteroatom acts as an electron acceptor. Nitrogen heterocyclics are the main class of aromatic pi electron deficient heterocyclics.

(20) Some aromatic nitrogen pi deficient heterocyclic compounds exhibit tautomerism. In some hydroxyl compounds, such as hydroxypyridines and nucleic acid bases, the amido form predominates.

(21) The water solubility of pi deficient heterocyclic systems is better than that of the corresponding carbocyclic hydrocarbon because of hydrogen bonding between the heteroatom and the water molecules.

(22) Hydrogen bonding by the heteroatom assists compounds with heterocyclic residues to bind to active and receptor sites.

(23) The general chemical reactions of pi deficient nitrogen aromatic heterocyclic systems are:

(a) *As a base.*

(b) *Electrophilic substitution (very much more difficult than electrophilic substitution of benzene).*

(c) *Nucleophilic substitution (does not occur with benzene).*

(d) *Oxidation (difficult).*

(e) *Reduction (easier than reduction of benzene).*

(24) Aromatic heterocyclic systems are often metabolised by hydroxylation of the ring (compare with aromatic hydrocarbons).

(25) The general physical and chemical properties of multicyclic compounds are the result of the properties of their functional groups and constituent ring systems.

(26) *The general chemical and physical properties of hydrocarbon, N, S and O heterocyclic constituent ring systems are listed in Table 13.4.*

Table 13.4. A summary of the general properties of hydrocarbon and heterocyclic ring systems.
* Indicates that this applies to nitrogen heterocyclics only, other types of heterocyclic system give negative results. Tautomerism occurs with some hydroxynitrogen heterocyclics.

Reaction/ property	Aliphatic ring systems	Aromatic ring systems	Aliphatic heterocyclic systems	Pi electron rich aromatic systems	Pi electron deficient heterocyclic systems
Shape	Various conformations	Flat	Various conformations	Flat	Flat
Tautomerism	No	No	No	No	Yes*
Water soluble	No	No	Yes, poor	No	Yes, poor
Hydrogen bonding	No	No	Yes, not S heterocyclic atoms	No	Yes
UV/visible	Only if unsaturated	Yes	Only if unsaturated	Yes	Yes
Base	No	No	Yes*	No or very weak	Yes*
Electrophilic substitution	No	Yes	No	Yes, easier than benzene	Yes, more difficult than in benzene
Nucleophilic substitution	No	No	No	No	Yes
Oxidation	Yes, usually difficult	Difficult	Yes especially S heterocyclic systems	Yes	No
Reduction	No	Difficult	Yes if the ring is unsaturated	Yes	Yes

3.10 QUESTIONS

(1) Draw structural formulae for each of the following: (a) 2,3-dimethyloxole, (b) 3-ethyl-4H-pyran, (c) 2,5-dimethyl-1,3-diazine, (d) 2-(2-pyridyl)naphthalene, (e) 3-piperidinomethylbenzene, (f) 2-methylfurano[2,3-b]furan, and (g) 2,2'-dimethyl-4,4'-bipyridyl.

(2) Comment on the expected solubilities of quinoline and azathioprine in water, dilute aqueous hydrochloric acid, and dilute aqueous sodium hydroxide. Suggest reasons for your predictions, using chemical equations where necessary.

(3) List the general chemical properties (if any) you would expect each of the following ring systems to exhibit: (a) piperidine, (b) isoquinoline, (c) tetrahydrofuran, (d) benzo[b]furan, and (e) 2,3-dihydrofuran.

(4) Suggest *feasible* electronic mechanisms for: (a) the hydroxylation of quinoline by hydroxide ions to form 2-hydroxyquinoline and (b) the sulphonation of pyridine with a mixture of sulphur trioxide and concentrated sulphuric acid to yield pyridine-3-sulphonic acid.

(5) Outline, by means of equations, chemical reactions that could be made the basis of identification tests and assays for each of the following compounds. Mention any relevant physical measurements that would be required. (a) Pyridine, (b) Pyrrole. (c) 2,3,5-Triphenyl-1,2,3,4-tetrazolium chloride.

14

Simple halogen functional groups

14.1 INTRODUCTION

Simple halogen functional groups consist of a halogen atom or atoms directly bonded to a carbon atom of the carbon-hydrogen skeleton of the molecule. This type of structure is found in the alkyl, allyl, alkenyl (vinyl) and aryl halides. These general names are also used to classify halogen functional groups when they are found in more complex compounds.

Alkyl halides are halogen functional groups in which the halogen is bound to a saturated carbon atom (Fig. 14.1). These functional groups are classified as either primary (p or 1°), secondary (s or 2°) and tertiary (t or 3°) structures depending on the degree of substitution of the carbon to which the halogen is attached. Compounds containing these functional groups are often referred to as alkyl halides.

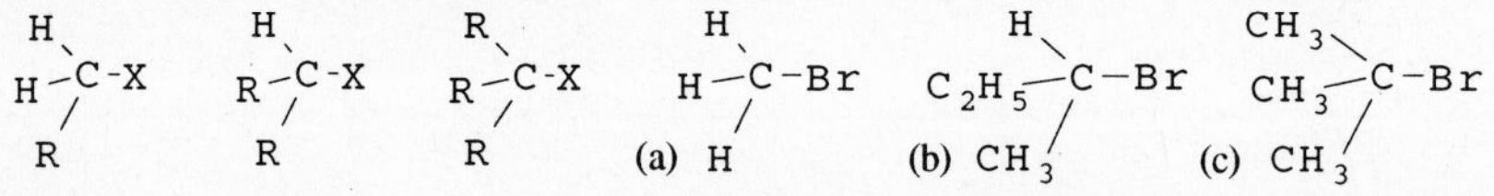

| Primary | Secondary | Tertiary | Bromomethane | 2-Bromobutane | 2-Bromo-2-methylpropane |
| alkyl halides | alkyl halides | alkyl halides | (methyl bromide) | (isobutyl bromide) | (tertiarybutyl bromide) |

Fig. 14.1. The general formulae (X is any halogen) of primary, secondary and tertiary alkyl halides. The R groups may be the same or different. Some examples of (a) primary, (b) secondary and (c) tertiary alkyl halides.

Allyl halides are alkyl halides in which the halogen is bonded to a saturated carbon atom directly bonded to an alkene C=C bond whilst in alkenyl halides, the halogen is bound to one of the carbon atoms of an alkene C=C bond (Fig. 14.2). The halogen atom of an aryl halide is bound to a carbon atom that forms part of the ring of an aromatic system (Fig. 14.2). Halogens which are separated from an **aromatic ring system** by one or more CH_2 groups are **alkyl,** not aryl halides.

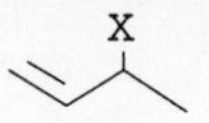

CH$_2$=CH—CH$_2$—Br

CH$_2$=CH—Cl

| Allyl halide | 3-Bromopropene, an allyl halide | Alkenyl or vinyl halides | Chloroethene(vinyl chloride), an allenyl or vinyl halide |

| Aryl halide | Chlorobenzene, an aryl halide | 2-Chloropyridine, an aryl halide | Benzyl chloride, an *alkyl* halide |

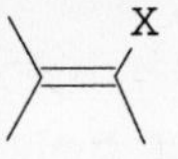

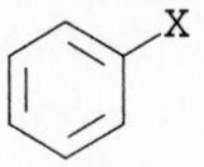

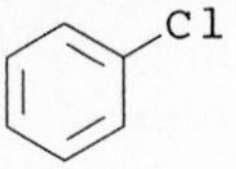

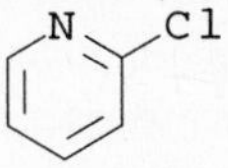

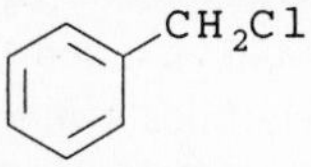

Fig. 14.2. The general structures and examples of allenyl, allyl and aromatic halides. The difference between structures of benzyl chloride (an alkyl halide) and the other compounds should be noted.

Naturally occurring compounds containing halogen functional groups come largely from marine sources (Fig. 14.3). They do not often occur naturally in land-based biological materials. However, simple halogen functional groups are found in many synthetic pesticides, anaesthetics and other drugs (Fig. 14.3).

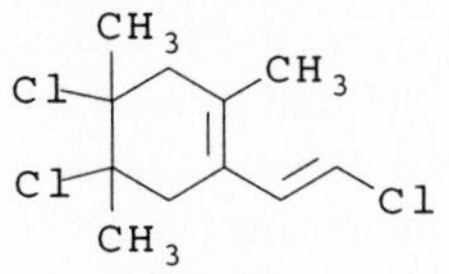

| Plocamene B (from the alga, *Plocamium violaceum*) | Drosophillin (from the basidiomycete, *Drosophilia subatrata*) | Bronopol (preservative and mild antiseptic) | Chlormethathiozole (hypnotic, sedative and anticonvulsant) |

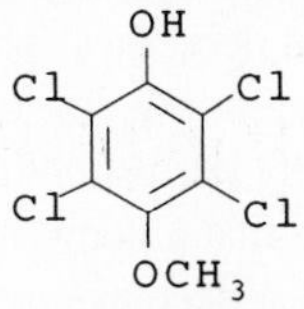

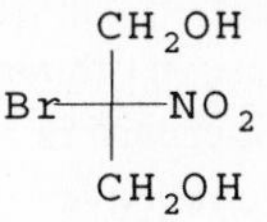

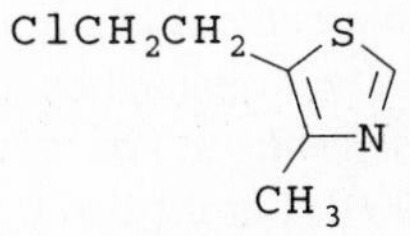

Fig. 14.3. Examples of naturally occurring and pharmaceutically active halogen compounds.

14.2 NOMENCLATURE

The prefix **halo** and the suffix **halide** are used with a number indicating the point of attachment (Figs 14.4 and 14.5). The presence of more than one halogen of the same type is indicated by the use of the prefix **di**, **tri**, etc. before the name of the halogen together with the appropriate numbers for their points of attachment.

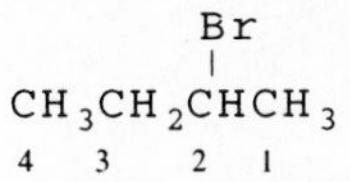

| 2-Bromobutane | 1,2-Dichloroethene | 2,3-Dichloropyridine | 3-Bromomethylbenzene |

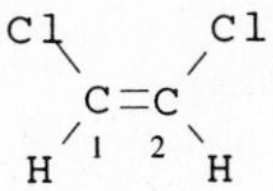

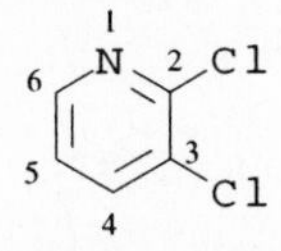

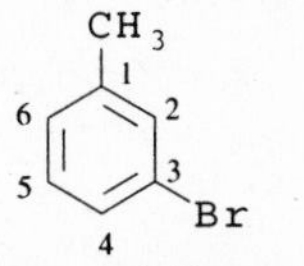

Fig. 14.4. Examples of the nomenclature of simple halogen compounds.

14.3 PHYSICAL PROPERTIES

Alkyl, vinyl and aryl halide functional groups do not improve the water solubility of compounds. However, they can increase a compound's lipid solubility.

C–Cl, C–Br and C–I bonds have relatively strong absorptions in the 800 to 500 cm^{-1} region of the infra-red whilst C–F bonds exhibit a relatively strong absorption in the 1400 to 1000 cm^{-1} region. All these absorptions are characteristic enough to be used as evidence for the existence of a halogen group in organic analysis.

14.4 GENERAL   CHEMICAL   PROPERTIES

The most common reactions of simple halogen functional groups are **nucleophilic substitution** and in the case of the alkyl halides, **base-catalysed elimination** (Fig. 14.5). It should be noted that the simple halogen functional groups do not normally function as nucleophiles even though they possess three lone pairs of electrons. Nucleophilic substitution of the alkyl and allyl halides is relatively easy but nucleophilic substitution of vinyl and aryl halides is usually quite difficult.

Nucleophilic Substitution: $RX + Nu \longrightarrow RNu + X^-$

Base Elimination: $RCH_2CH_2X + B \longrightarrow RCH=CH_2 + BH^+ \, X^-$

Fig. 14.5. General equations for the nucleophilic substitution and elimination reactions involving halides. X represents a halogen atom, Nu a nucleophile and B a suitable base.

In both these types of reaction the reactivity of the halogen increases down the periodic table as the strength of the carbon-halogen bond decreases. This decrease in bond strength can be explained by a decrease in orbital overlap as the size of the halogen increases. It is interesting to note that this **increase** in reactivity corresponds to a **decrease** in the dipole moment of the bond. One would have expected that a decrease in the polarity of the bond would result in decreased reactivity because of a decrease in the polarisation of the bond (8.2 and 8.3.). However, in this instance the increasing weakness of the bond overrides the effect of the polarity. This demonstrates the fact that our current mechanistic theory does not take into account all the parameters that affect a reaction and so one must expect some inaccuracies when using the theory of mechanisms to predict and explain practical work.

Simple halogen functional groups are resistant to oxidation but free radical oxidations can occur. Sunlight will often catalyse these reactions. Trichloromethane (chloroform), for example, is slowly oxidised by air in sunlight to phosgene and other products. The substances produced are toxic, and so small amounts of ethanol are added to commercially produced chloroform to prevent their formation. Most pharmaceutical preparations involving the use of trichloromethane do not require the removal of the ethanol.

$$CHCl_3 \; + \; O_2 \longrightarrow COCl_2 \; + \; \text{other products}$$
Phosgene
$$C_2H_5OH + COCl_2 \longrightarrow C_2H_5OCOOC_2H_5$$
Diethyl carbonate

The slow oxidation of alkyl halides in air means that compounds containing this functional group should be stored in airtight lightproof containers in a cool place.

14.5 NUCLEOPHILIC SUBSTITUTION OF ALKYL AND ALLYL HALIDES

The halogen atoms of alkyl and allyl halides can be substituted by many different nucleophiles, for example, by hydroxide (OH^-), alkoxide (RO^-), cyanide (CN^-), and sulphydryl (SH^-) from the appropriate alkali metal salts as well as hydride ions (H^-) from lithium aluminium hydride. The equations for these reactions can be obtained by substituting for Nu in the general reaction given in Fig. 14.5.

Many of these nucleophilic substitutions have been found to exhibit first or second order kinetics with their rate expressions having the general forms:

$$\text{Rate} = [\text{alkyl halide}] \tag{14.1}$$

$$\text{Rate} = [\text{alkyl halide}][\text{nucleophile}] \tag{14.2}$$

It was deduced from these general rate equations that the nucleophilic substitutions of alkyl halides proceeded by two different mechanisms which are now referred to as S_N1 (Substitution Nucleophilic monomolecular) and S_N2 (Substitution Nucleophilic bimolecular) respectively. In general, it now appears that nucleophilic substitution proceeds either via one of these mechanisms or a mechanism broadly based on one of them or a mixture of both these mechanisms.

14.5.1   Monomolecular nucleophilic substitution S_N1

Reactions that proceed by this type of mechanism will have rate expressions of the form given in equation 14.1. This suggests that the slow step of the mechanism involves only the halide and not the nucleophile. Consequently, it was deduced that the mechanism was a two-stage process. In the first stage the halide slowly ionised to form a relatively stable carbocation, the second stage being the fast attack of the nucleophile on this carbocation. Consider, for example, the reaction of 2-chloro-2-methylpropane with sodium hydroxide (Fig. 14.6). The C–Cl bond of the 2-chloro-2-methylpropane is already polarised by the difference in electronegativities of the carbon and the chlorine atoms (8.2). Under the reaction conditions this polarisation moves to completion and the compound slowly ionises to form the carbocation. The hydroxide ion (the nucleophile) immediately reacts with this ion to form the product. Other nucleophilic reagents are believed to react with 2-chloro-2-methylpropane in a similar manner but the logical sequence of the mechanism may be modified slightly by the nature of the reagent. Water, for example, reacts by a similar mechanism except it appears to involve an extra step where the hydroxonium ion formed by the reaction of water with the carbocation loses a proton to form the product (Fig. 14.6). **Formation of the next species in the mechanism by the loss of a proton is a logical step included in many heterolytic mechanisms.**

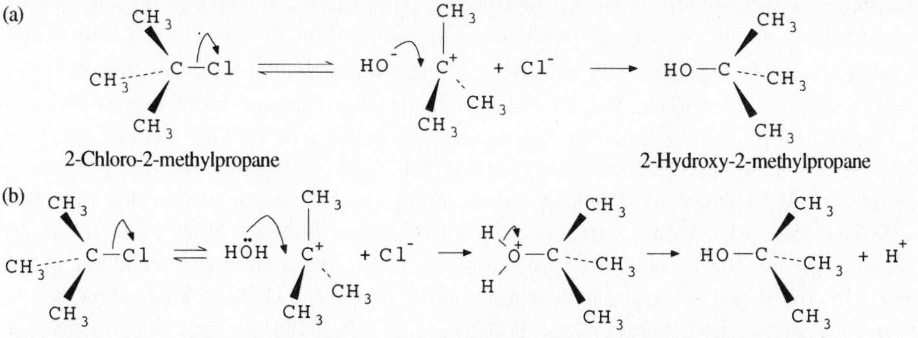

Fig. 14.6. Nucleophilic substitution of 2-chloro-2-methylpropane by (a) sodium hydroxide and (b) water.

The carbocation formed in an S_N1 mechanism is planar (sp^2 hybridisation) and so it can be attacked from either side by the nucleophile. Consequently, S_N1 nucleophilic substitutions involving chiral centres should result in the formation of a racemate. In the laboratory, very few S_N1 substitutions result in complete racemisation. They usually produce a mixture of the racemate and a product with a configuration opposite to that of the original halide (Fig. 14.7). It is thought that as the leaving group diffuses away from the carbocation it prevents the nucleophile attacking from that side. As a result, during this diffusion period, the nucleophile can attack only from the opposite side of the carbocation which results in the inversion of the molecules configuration. Once the leaving group has diffused away from the carbocation the nucleophile can attack from both sides. This leads to racemate formation (Fig. 14.7). The degree of racemisation has been found to depend on the structure of the halide and the solvent.

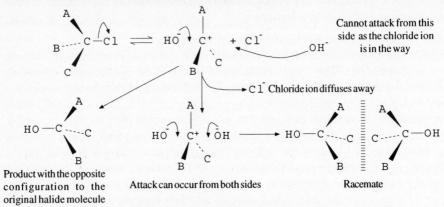

Fig. 14.7. The stereochemistry of S_N1 mechanisms.

The planar nature of the carbocation also means that there is little or no steric hindrance of the attacking nucleophile by the structures attached to the positive carbon atom.

14.5.2 Factors affecting S_N1 mechanisms

Two main factors appear to play a part in deciding whether a compound undergoes a nucleophilic substitution by an S_N1 mechanism. They are the stability of the carbocation intermediate and the polarity of the solvent. The nature of the nucleophile has little effect.

(a) **Carbocation stability**. Compounds that can form stable carbocations are likely to react using this mechanism. Benzyl and allyl halides, for example, form carbocations that are stabilised by delocalisation of the positive charge (Fig. 14.8). This form of stabilisation results in relatively stable carbocations. The carbocations formed from primary, secondary and tertiary alkyl halides are stabilised to a lesser degree by the inductive effects of their alkyl groups reducing the positive charge of the ion (Fig. 14.8). It follows that the greater the number of these inductive effects the greater the reduction in the positive charge and hence the greater the stability of the carbocation. This means that in terms of carbocation stability, the probability of alkyl halides reacting via an S_N1 mechanism is usually in the order:

benzyl = allyl > tertiary > secondary > primary

(b) **Solvent**. The rates of S_N1 reactions are usually much faster in polar than non-polar solvents. This increase is believed to be due to the stabilisation of the carbocation intermediate by solvation. The nature of this stabilisation is not fully understood but is

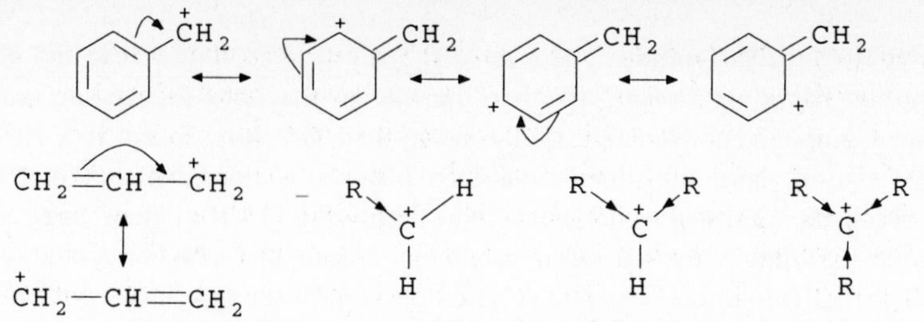

Fig. 14.8. The stabilisation of the intermediate carbocations.

believed to take the form of ion dipole bonds between the carbocation and the solvent molecules (Fig. 9.6). Solvation also lowers the energy of the transition state (activation energy) of the system thus making the reaction easier and so faster.

In some cases the solvent instantaneously acts as both the nucleophile and the reaction medium. For example, water can act as both a nucleophile and a solvent.

14.5.3   Bimolecular nucleophilic substitution S_N2

Reactions utilising this type of mechanism have rate expressions of the type given in equation 14.2. This expression suggested that the slow step of the mechanism involves both the nucleophile and the halide. Logic and practical evidence led workers to deduce that the nucleophile attacks from the opposite side to the halide and in effect forces the halide out. The process is believed to involve a planar transition state where the nucleophile is not fully bonded to the carbon and the carbon–halogen bond is not completely broken (Fig. 14.9).

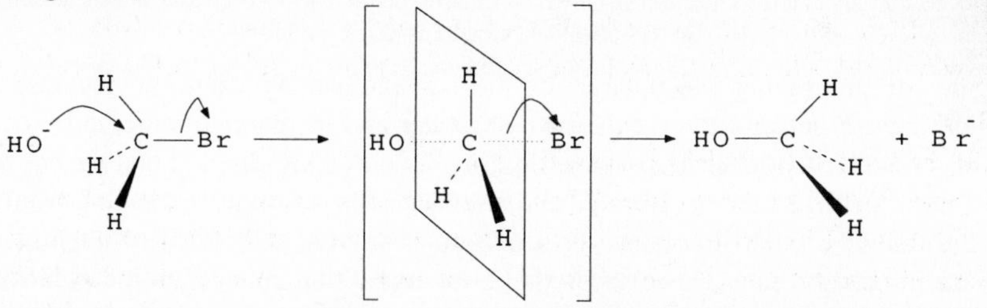

Fig. 14.9. The S_N2 mechanism for the reaction of methyl bromide with sodium hydroxide to form methanol.

The stereochemical nature of the S_N2 mechanism means that the product molecule is formed with its configuration opposite to that of the original reactant molecule (Fig. 14.9). This process which is similar to an umbrella being blown inside out, is referred to as **inversion** of the structure. It is only significant when the nucleophilic attack involves a chiral centre. Consider, for example, the reaction of 2R-bromobutane with hydroxide ions via an S_N2 mechanism. This would result in the formation of 2S-hydroxybutane.

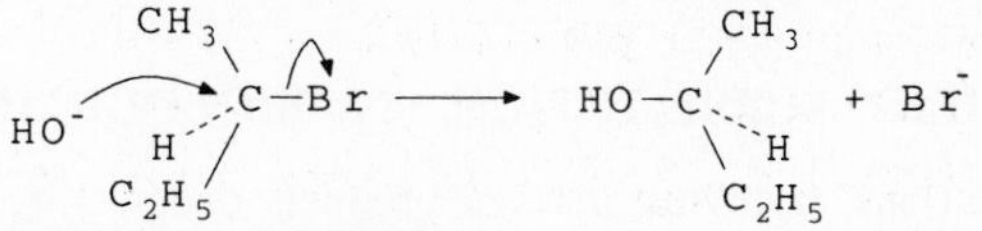

14.5.4   Factors affecting S_N2 mechanisms

Reactions proceeding by S_N2 mechanisms are affected by the structure of the halide, solvent and the nature of the nucleophile.

(a) The structure of the halide. The size of the substituents attached to the positively charged carbon has a significant effect on the rate of reactions proceeding via an S_N2 mechanism. Compounds with large bulky substituent structures react at a slower rate because the size of these substituents makes it physically more difficult for the nucleophile to reach the positively charged carbon atom (Fig. 14.10). Their large size also results in overcrowding at the transition state which means that a higher activation energy is required for its formation. The rate of reaction is considerably faster with molecules with smaller substituents. This is because the approach of the nucleophile is not so sterically hindered and the transition state has a lower energy as it is less crowded (Fig. 14.10). Consequently, S_N2 mechanisms are often favoured by primary alkyl halides.

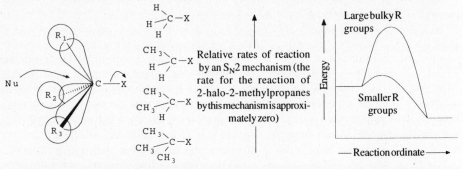

Relative rates of reaction by an S_N2 mechanism (the rate for the reaction of 2-halo-2-methylpropanes by this mechanism is approximately zero)

Fig. 14.10. Steric hindrance in an S_N2 mechanism. Kinetic studies carried out on substitution reactions proceeding only by an S_N2 mechanism have shown that the rate of reaction decreases with increase in size of the substituent structures, R_1, R_2 and R_3.

(b) Solvent. Increasing the polarity of the solvent usually causes a decrease in S_N2 reaction rates. Polar protic solvents, such as, water and methanol often greatly reduce the rate of an S_N2 reaction. In these cases hydrogen and/or ion-dipole bonding between the nucleophile stabilises the nucleophile and so reduces its reactivity. Changing from a polar protic to either a polar non-protic or a non-polar solvent will often result in a marked increase in reaction rate. However, it should be noted that changes in the polarity of the solvent can change the reaction mechanism.

(c) The nucleophile. The effect of the chemical nature of nucleophiles, that is, their **nucleophilicity**, on an S_N2 mechanism is difficult to correlate and quantify. However, their nucleophilicity appears to increase with increasing basicity of the nucleophile's conjugate acid provided one compares nucleophiles having the same attacking atom (Table 14.1). This correlation is not so accurate when one compares nucleophiles with different attacking atoms. However, nucleophilicity does appear to increase down a group in the periodic table.

14.6 NUCLEOPHILIC SUBSTITUTION IN BIOLOGICAL SYSTEMS

Nucleophilic substitutions involving *natural* products occur widely in metabolic processes. They normally involve functional groups other than halides. However, these nucleophilic substitutions follow similar mechanisms and are governed in a similar way by the same factors as the corresponding halide substitutions.

Table 14.1. The relative reaction rates of different oxygen nucleophiles with bromomethane.

Nucleophile	Conjugate acid	pK_a	Relative reaction rate
$CH_3CH_2O^-$	CH_3CH_2OH	16	
OH^-	H_2O	15.7	Increasing
PhO^-	$PhOH$	10	reaction rate and
CH_3COO^-	CH_3COOH	4.8	nucleophilicity
H_2O^-	H_3O^+	-1.7	

Nucleophilic substitution of halides in biological systems usually involves molecules that have been produced synthetically and introduced into the body artificially or by accident. The toxicity of many of these alkyl halides is believed to be due to the nucleophilic substitution of the halide by the amino and thiol groups of vital enzymes (Fig. 14.11). The resulting transfer of an alkyl group to the enzyme interferes with its action probably by blocking the active site of the enzyme. It is interesting to note that detoxification of alkyl halides also involves this type of reaction. The thiol group of glutathione reacts with the halide and then, by a series of steps, the S-alkylated glutathione is converted into the corresponding S-alkylated mercapturic acid and excreted (Fig. 14.11).

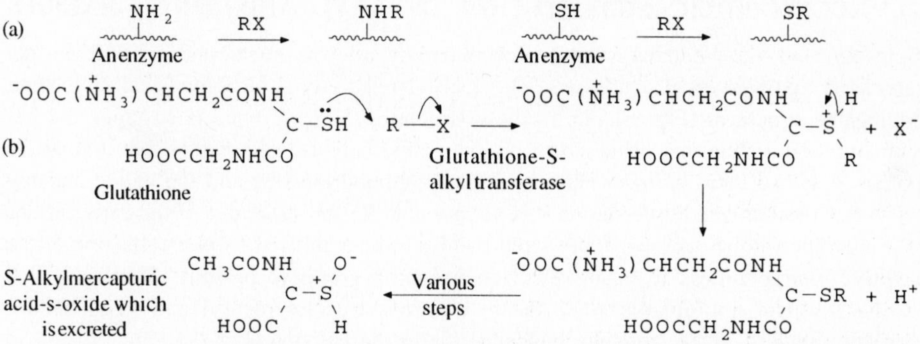

Fig. 14.11. (a) General examples of the nucleophilic substitution reaction of enzymes with alkylhalides. (b) Nucleophilic substitution in the detoxification of alkyl halides.

Nucleophilic substitution is often modified by a neighbouring nucleophilic group. Mustard gas, whose structure contains two primary alkyl halides, for example, is believed to alkylate proteins by an S_N2 mechanism that incorporates the formation of a cyclic intermediate sulphonium ion (Fig. 14.12). The action of the related nitrogen mustards is also believed to be due to similar type of reaction. For example, the anticancer drug melphalan is thought to act by preventing the replication of malignant DNA. A molecule of melphalan is believed to bond covalently to both strands thereby forming a bridge between the strands. This holds the strands together which prevents the double helix unwinding and the subsequent formation of new daughter strands (29.3). Unfortunately melphalan, like many similar anticancer drugs, also prevents the reproduction of healthy DNA. Consequently, great care has to be taken in its administration. It should be noted that the development of the nitrogen mustards is an example of drugs being developed by synthesising compounds with similar structures to an active compound, in this case, mustard gas.

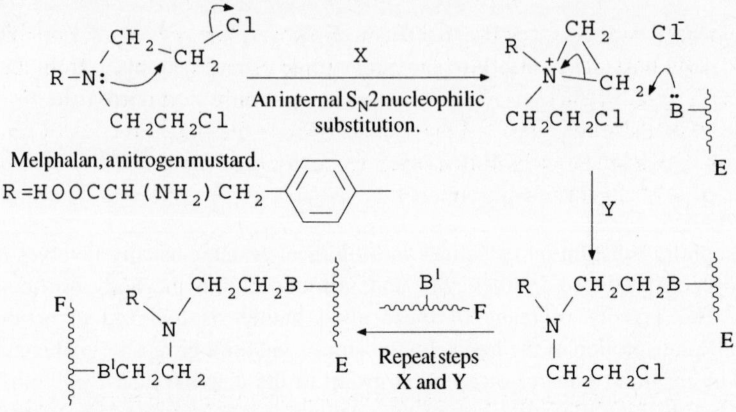

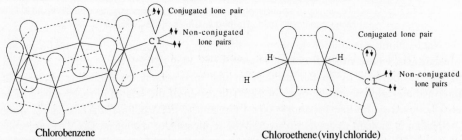

Melphalan, a nitrogen mustard.

$R = HOOCCH(NH_2)CH_2 -$ ⟨benzene ring⟩ $-$

Fig. 14.12. The mode of action of the nitrogen mustard melphalan. E and F are the two strands of a DNA double helix. Melphalan bonds to the bases B and B^1 of these strands permanently holding the two strands together thereby preventing them unwinding and reproducing.

14.7 NUCLEOPHILIC SUBSTITUTION OF ARYL AND VINYL HALIDES

Both aryl and vinyl halides exhibit little reactivity towards nucleophiles under normal laboratory and biological conditions. This lack of reactivity is attributed mainly to their conjugated structures (Fig. 14.13). In these structures the C−X bond is stronger and has a smaller dipole than that of the corresponding alkyl halides. This means that it is not so polarised and so there is lower chance of the C−X bond ionising and the halide reacting by an S_N1 mechanism. Furthermore, it is not possible for the structures of the carbocations that could be formed from aryl and vinyl halides to be stabilised by delocalisation of the positive charge unless a strong electron acceptor group is present. Similarly, S_N2 mechanisms are not suitable since the pi electron structure of aryl and vinyl halides sterically hinders attack from the opposite side to the halogen (Fig. 14.13).

Chlorobenzene   Chloroethene (vinyl chloride)

Fig. 14.13. The pi electron clouds of chlorobenzene and chloroethene will repel nucleophiles and so prevent the nucleophiles attacking carbon 1 from the opposite side to the chlorine.

Nucleophilic substitution of aryl halides can occur when strong electron acceptor groups are present, either as substituents or as heteroatoms. For example, nitro, cyano, some carbonyl groups and the ring nitrogen of pi deficient systems. Kinetic studies show that the rate expressions for these reactions have the general form shown in equation 14.3.

$$\text{Rate} = k[\text{aryl halide}][\text{nucleophile}] \tag{14.3}$$

This would suggest an S_N2 type of mechanism. However, this type of nucleophilic reaction is not sterically possible as attack of the nucleophile cannot take place from the opposite side to the halogen. Reaction of the nucleophile can only occur from the side. Further investigation of these reactions led workers to deduce that they proceed by an addition-elimination mechanism. 4-Nitrobenzene, for example, reacts with aqueous sodium hydroxide at 130° to form 4-nitrophenol.

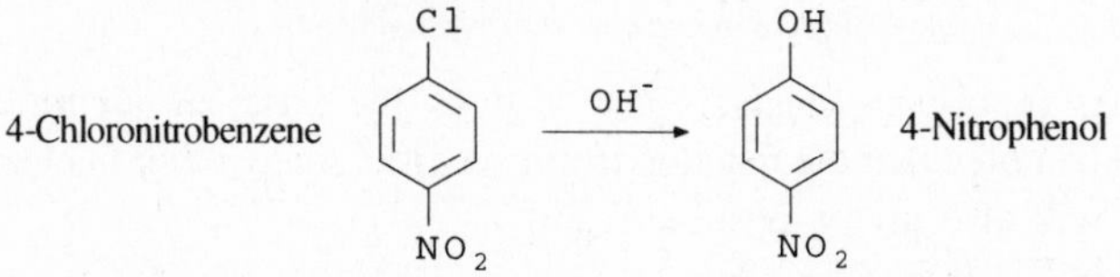

This reaction is believed to occur because the combined electron-withdrawing effects of the chlorine and the nitro groups make carbon 1 sufficiently electron deficient to allow nucleophilic attack.

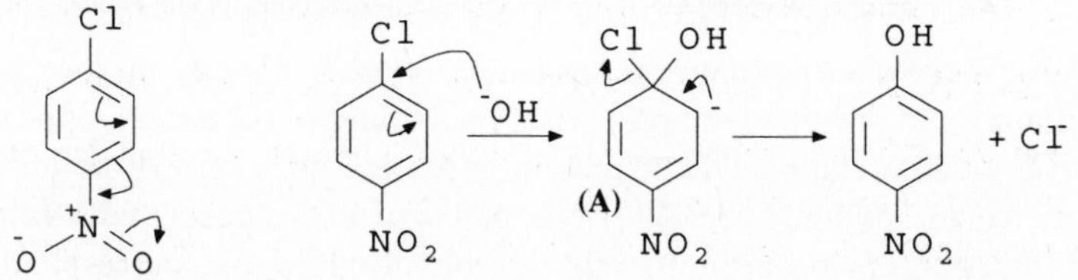

Experimental evidence suggests that the nucleophile forms an intermediate (A), known as the Meisenheimer complex. This intermediate is stabilised by the electron-withdrawing effect of the nitro group.

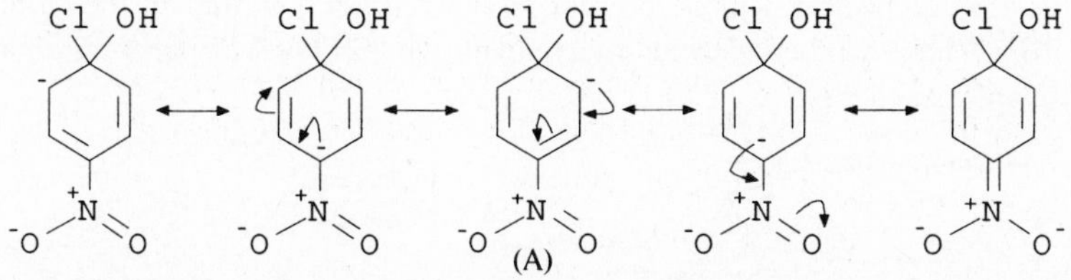

The second step of the mechanism is the elimination of the halogen to reform the stable aromatic system. Mechanisms of this type are referred to as $S_N2(aromatic)$ or S_N2Ar.

Electron acceptors in the ortho position will also have the same effect but substituents in the meta position are far less effective because they are unable directly to stabilise the negative charge of the Meisenheimer complex (Fig.14.14). They are in the wrong position for their mesomeric effects to be effective.

The very reactive 2,4-dinitrochlorobenzene and its fluorine analogue are used in qualitative and quantitative analysis of amines and related compounds.

(a) Qualitative analysis. Both 2,4-dinitrochlorobenzene and 2,4-dinitroflurobenzene react with primary amines to form bright yellow derivatives. The reaction is used as a test

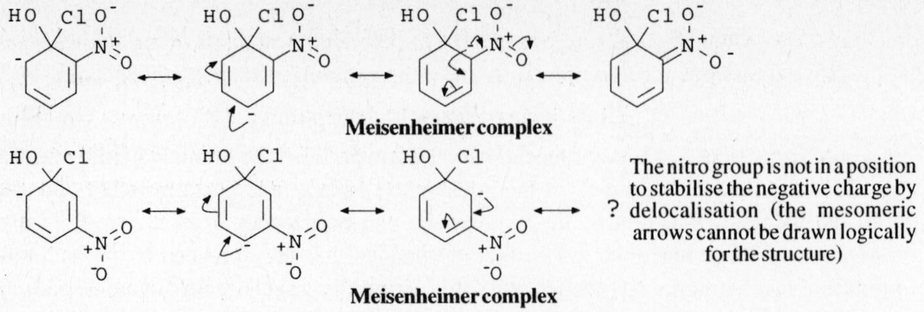

Meisenheimer complex

Meisenheimer complex

The nitro group is not in a position to stabilise the negative charge by
? delocalisation (the mesomeric arrows cannot be drawn logically for the structure)

Fig. 14.15. The relative stabilities of the Meisenheimer intermediates in the reactions of 2-nitrochlorobenzene and 3-nitrochlorobenzene with hydroxide ions.

for primary amines in organic analysis and also as the basis of identification tests. However it should be noted that the reaction is not specific: many other nucleophiles, such as thiols (16.6.1), will also give a positive result.

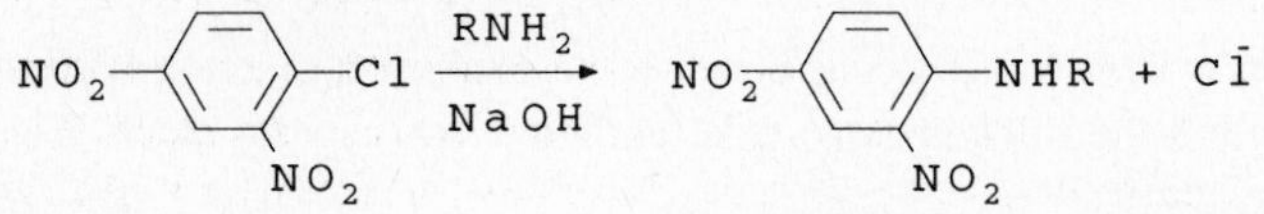

2,4-Dinitrochlorobenzene   N-(2,4-Dinitrophenyl)alkylamine (bright yellow)

2,4-Dinitrofluorobenzene (Sanger's reagent) is used to identify the N-terminal amino acids and amino acid residues with primary amine groups in the structures of peptides and proteins (Fig. 14.15). The peptide or protein is reacted with the 2,4-dinitroflurobenzene in the presence of sodium hydroxide. The amino acid residue(s) with primary amine groups in the peptide or protein will be converted to the corresponding yellow 2,4-dinitrophenyl derivatives (DNPs). This yellow product is purified by crystallisation or chromatography before being completely hydrolysed by heating with 5 M hydrochloric acid for several hours. The amino acid residues which had primary amino groups in their original peptide or protein will be present in the reaction mixture in the form of their yellow DNPs. These can be identified by comparative thin layer or paper chromatography against suitable standards.

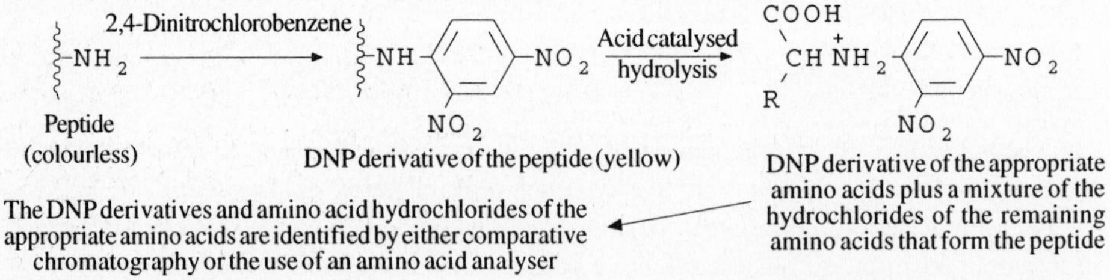

Peptide (colourless)

DNP derivative of the peptide (yellow)

DNP derivative of the appropriate amino acids plus a mixture of the hydrochlorides of the remaining amino acids that form the peptide

The DNP derivatives and amino acid hydrochlorides of the appropriate amino acids are identified by either comparative chromatography or the use of an amino acid analyser

The information from experiments of this type can be used to identify a peptide or protein, distinguish between two small peptides that are isomers and play a part in determining the structure of a peptide or protein (28.12.3).

(b) Quantitative analysis. DNP derivatives have been used as the basis of colorimetric assays (10.2) of amino acids, peptides and proteins. They have also been used as derivatives for HPLC, for example, in the analysis of the antibiotic *Neomycin*.

14.8 ELIMINATION

Reaction of a base with either an alkyl or a vinyl halide can result in the elimination of HX and formation of either an alkene or an alkyne. Almost any base is suitable, for example, sodium hydroxide, pyridine and quinoline will all bring about eliminations under suitable conditions. In some cases simply warming the halide with the solvent is sufficient to bring about elimination.

$$\underset{\text{Alkyl halide}}{\overset{H}{\underset{X}{\overset{\beta\;\;\alpha}{C-C}}}} \quad\xrightarrow{\;B\;}\quad \underset{\text{Alkene}}{C=C} \;+\; HB \;+\; X^- \qquad \underset{\text{Vinyl halide}}{\overset{H}{\underset{X}{C=C}}} \quad\xrightarrow{\;B\;}\quad \underset{\text{Alkyne}}{-C\equiv C-} \;+\; HB \;+\; X^-$$

Most reactions involve halogen and hydrogen atoms on adjacent carbon atoms. This type of elimination is referred to as a 1,2- or β elimination in order to distinguish it from other types of elimination where the atoms or groups being eliminated may be more widely separated. With alkyl halides, β elimination can result in the formation of either a mixture of structural and geometric isomers of the product or one specific structural or geometric isomer. Usually the most stable alkene is the predominant product. It is useful to note that the most stable alkene is frequently the most substituted alkene and that eliminations that yield the most stable alkene are sometimes referred to as **Saytzeff eliminations**.

$$\underset{\text{2-Bromobutane}}{CH_3CH_2CHBrCH_3} \xrightarrow{\text{Base}} \underset{\text{2-Butene (80\%)}}{CH_3CH=CHCH_3} \;+\; \underset{\text{Butene (20\%)}}{CH_2CH_2CH=CH_2}$$

Meso-1,2-dibromo-1,2-diphenylethane   E-1-bromodiphenylethane (100%)

Elimination is usually accompanied by nucleophilic substitution since bases can also act as nucleophiles. The relative yields of the substitution and the elimination reactions will depend on the reaction conditions, the base and the nature of the halide.

$$\underset{\text{2-Chloro-2-methylpropane}}{(CH_3)_3CCl} \xrightarrow[65^\circ C]{H_2O\,/C_2H_5OH} \underset{\text{2-Methylpropan-2-ol (64\%)}}{(CH_3)_3COH} \;+\; \underset{\text{2-Methylpropene (36\%)}}{(CH_3)_2C=CH_2}$$

Practical evidence suggests that elimination reactions can follow a variety of mechanisms. The mechanisms that represent the extremes of the most common possible electron movements are known as E1 and E2 respectively. Many eliminations appear to follow these mechanisms but in other cases they represent the extreme limits of the electron movements.

14.8.1 The E1 mechanism

Reactions proceeding by this mechanism exhibit first order kinetics with rate expressions of the form:

$$\text{Rate} = k[\text{alkyl halide}] \tag{14.3}$$

This means that the slow step of the mechanism is independent of the concentration of the base. Consequently, the E1 mechanism is believed to have the general form shown in Fig. 14.16 with the initial step being the slow dissociation of the halide (cf. S_N1 mechanism). This is followed by the fast attack of the base on the appropriate proton.

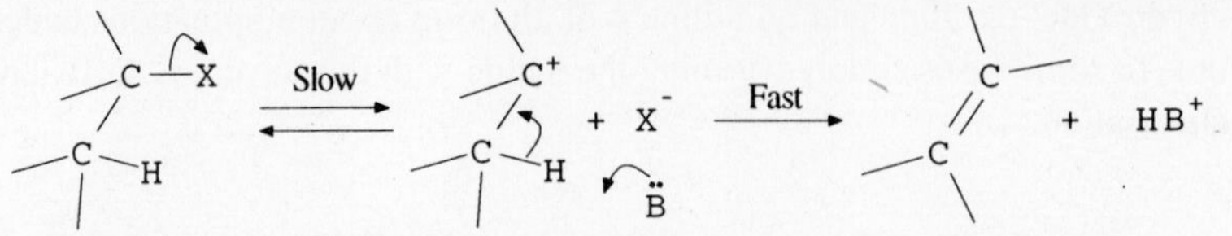

Fig. 14.16. The E1 reaction mechanism. The R groups may or may not be the same.

It is not always necessary to add a base to a reaction following an E1 mechanism because in many cases the solvent can act as the base. For example, simply warming a mixture of 2-chloro-2-methylpropane in 80% aqueous ethanol results in the formation of 2-methylpropene in about 36% yield. However, the accompanying nucleophilic substitution gives a higher yield of the substituted product under these conditions.

E1 mechanisms are favoured by the same factors that favour S_N1 mechanisms, namely polar solvents and halides that form stable carbocations. Consequently, the possibility of an alkyl halide elimination reaction proceeding by an E1 mechanism in a polar solvent is in the order:

$$tertiary > secondary > primary$$

The possibility of a primary alkyl halides undergoing elimination via an E1 mechanism is extremely low.

14.8.2 The E2 mechanism

Reactions with this mechanism exhibit second order kinetics. Their rate expressions take the form:

$$Rate = k \text{ [alkyl halide][base]} \tag{14.4}$$

which indicates that both the base and the halide are involved in the slow step of the mechanism. This and other practical evidence led to the concerted mechanism shown in general terms in Fig. 14.17 (cf. S_N2). It is the most common mechanism for all types of elimination, not just those involving halogen atoms.

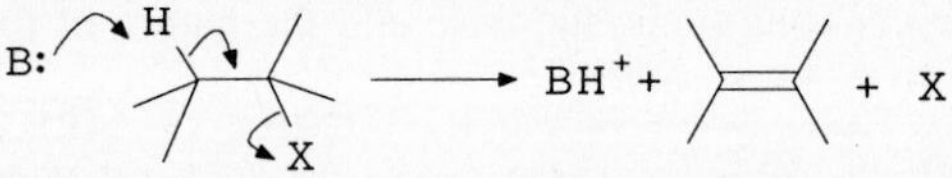

Fig. 14.17. A general outline of the E2 elimination mechanism.

The main factors that affect reactions using an E2 mechanism are the strength of the base, the relative positions of the atoms being eliminated and the solvent. It is not surprising that as basic strength increases so the rate of reaction increases. The stereochemical picture is more complicated. The hydrogen can be on the same side as (**syn**) or on the opposite side (**anti**) to the halogen (Fig. 14.18). In practice, alkyl halides show a marked preference for anti elimination provided their structure is suitable. This is because syn elimination requires the molecule to be in the less energetically favourable eclipsed conformation. Consequently, the rates of elimination reactions proceeding by an anti route are usually significantly faster than those proceeding by the syn route. This often results in a predominance of one of the stereoisomers of the alkene.

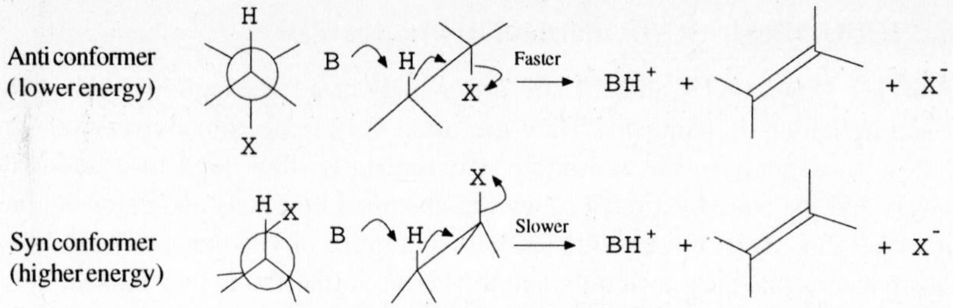

Fig. 14.18. Syn and anti elimination of HX.

E2 eliminations are more likely to occur in non-polar solvents. Changing to a more polar solvent can cause in a change from an E2 to an E1 mechanism.

14.9   METABOLISM

Alkyl and activated aryl halides, such as 2,4-dinitrochlorobenzene, are metabolised in mammals by nucleophilic substitution of the halogen by the peptide glutathione (Fig. 14.11). The thiol group of the glutathione acts as the nucleophile and the reaction is catalysed by the enzyme, glutathione-S-alkyl transferase. The product of the substitution, an S-alkylglutathione, is converted by a series of steps to the S-alkylmercapturic acid. Some of this acid is excreted and the rest converted to the S-oxide before being excreted.

Vinyl and unactivated aryl halides such as the aryl chloro groups of Dicophane (DDT) and the vinylic chloro groups of trichloroethene and Aldrin, are not metabolised by this route because of the unreactive nature of aryl and allenyl halides. However, DDT is metabolised by elimination of HCl to dichlorodiphenyldichloroethylene (DDE) by insects that are resistant to the drug. Generally, vinyl and unactivated aryl halides are metabolised, where possible, by reactions of the other functional groups present in their structures. Aldrin, for example, is metabolised to Dieldrin (Fig. 14.19) by epoxidation of its alkene group.

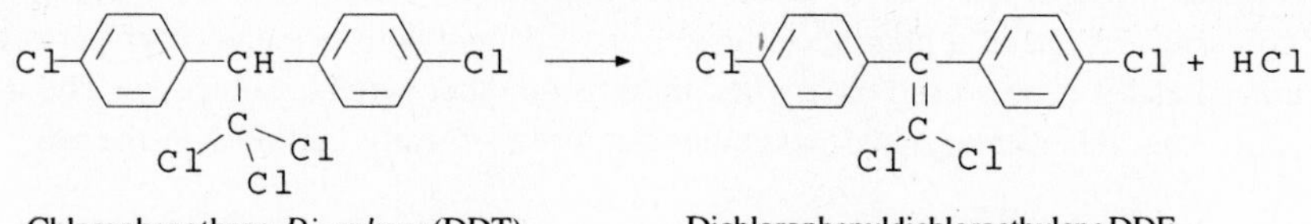

Chlorophenothane, *Dicophane* (DDT)   Dichlorophenyldichloroethylene DDE

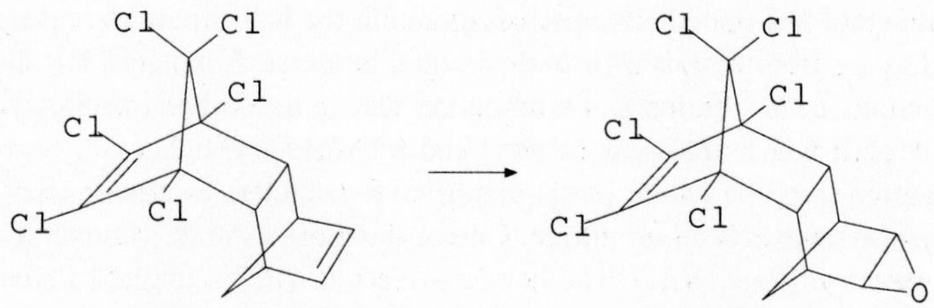

Aldrin   Dieldrin

Fig. 14.19. The metabolism of DDT and Aldrin.

14.10 AGRICULTURAL AND INDUSTRIAL HALIDES

Organochlorides (Figs 14.19 and 14.20) are widely used throughout the world as industrial and agricultural chemicals. They are often very toxic, some are carcinogenic. Most are slow to degrade in the biosphere. Consequently, they tend to accumulate in streams, rivers and the soil. As a result, they are absorbed by plants and enter the human food chain either directly or through the animals that graze on the plants. Their slow rate of metabolism means that they accumulate in the body. Furthermore, their metabolites can also accumulate in the body. The biological effects of these accumulations are not clear but current investigations do give some cause for concern. Dieldrin and Aldrin, for example, are known to produce liver tumours in mice.

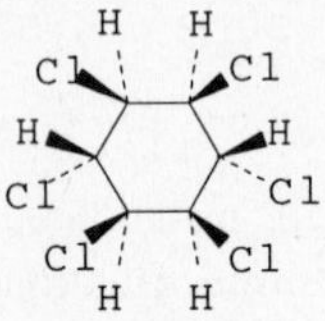

Lindane (1α,2α,3β,4α,5α,6β-hexachlorocyclohexane) used as a garden insecticide

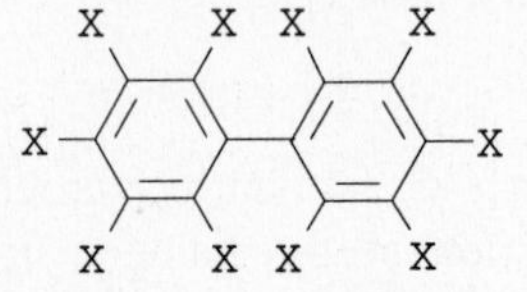

PCBs, X can be either chlorine or hydrogen. PBBs, X can be either bromine or hydrogen

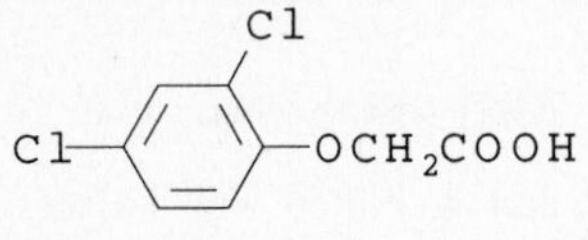

2,4-D, (2,4-dichlorophenoxyacetic acid), a herbicide. Several herbicides based on phenoxyacetic acid are in use

Fig.14.20. The structural formulae of some agricultural and industrial organohalides.

Organochloride pesticides, such as DDT are a particular environmental hazard because of their widespread use. So far no human deaths have been attributed to the use of DDT but worldwide, many birds and animals die each year because they have fed on plants contaminated by DDT and other pesticides. DDT and its metabolites have been found in appreciable quantities in mothers' milk and maternal blood. What effect this has on babies and foetuses is not known. However, there is no evidence to date of it being carcinogenic to humans. The WHO recommends that DDT be still available but used with discretion.

The general use of DDT as an agricultural insecticide is banned in Britain and many other countries. Its use as a pharmaceutical to treat people with lice and other parasites is discouraged and it is now used only when there is no other suitable substitute. The use of DDT and many other chlorinated pesticides has been severely curtailed in the US.

14.11 IDENTIFICATION AND ASSAYS

All identification and assay methods involve liberating the halogen as its halide ion. The reactions used range from fusion with molten sodium, which is suitable for all types of halide, to nucleophilic substitution and elimination reactions involving sodium hydroxide and sodium ethoxide which are used for alkyl and activated aryl halides.

In identification tests the halide ion is identified by standard inorganic tests while in assays it is often titrated with silver nitrate. One of the most accurate volumetric methods is the Volhard method (Fig. 14.21). The halide is reacted with an excess of silver nitrate in the presence of nitric acid. The excess silver nitrate in the reaction mixture is back titrated with potassium thiocyanate using ferric ions as an indicator. At the end point reddish-brown iron III thiocyanate is produced. This indicator change is very sensitive.

Titration reactions: $X^- + AgNO_{3(excess)} \longrightarrow AgX + NO_3^- + AgNO_{3(excess)}$

$KSCN + AgNO_3 \longrightarrow KNO_3 + AgSCN$

Indicator reaction: $Fe^{3+} + 3KSCN \qquad Fe(SCN)_3 + 3K^+$

Fig. 14.21. The chemistry of the Volhard titration.

Volhard's method works well for bromide and iodide but nitrobenzene must be added when titrating chloride. This is to prevent the silver chloride produced in the titration from reacting with the thiocyanate and giving an artificially high titration reading.

Halide ions can also be estimated gravimetrically, colorimetrically and by ampometric titration using ion selective electrodes.

14.12 WHAT YOU NEED TO KNOW

(1) The general structural formulae of the simple halogen compounds are:

R–X   A r–X

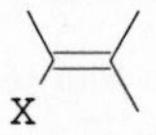

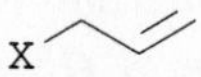

Alkyl halide   Aryl halide   Allenyl halide (Vinyl halide)   Allyl halide

(2) Alkyl halides are far more reactive than alkenyl and aryl halides.

(3) Alkyl halides undergo slow atmospheric oxidation and so must be stored in lightproof and airtight containers.

(4) The main reactions of alkyl and allyl halides are nucleophilic substitution and elimination.

Substitution: $RX + Nu \longrightarrow RNu + X^-$

Elimination:

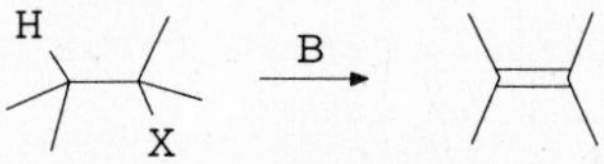

$+ BH^+ X^-$

(5) Aryl and alkenyl halides do not usually undergo nucleophilic substitution.

(6) Aryl halides will undergo nucleophilic substitution of the halogen provided at least one strong electron acceptor substituent is present on the ring.

(7) The main mechanisms found in nucleophilic substitution reactions are known as S_N1, S_N2 and $S_N(Ar)2$ mechanisms respectively.

(8) S_N1 mechanisms are based on the sequence;

$$RX \longrightarrow R^+ + X^-$$
$$R^+ + Nu \longrightarrow RNu$$

(9) S_N1 mechanisms are favoured by reactions carried out in polar solvents using compounds that can form stable carbocations. They are not sensitive to steric hindrance.

(10) S_N1 mechanisms involving a chiral centre usually result in the formation of a mixture of the racemate and an enantiomer with the opposite configuration to that of the original molecule.

(11) S_N2 mechanisms are based on the sequence:

$$Nu + RX \longrightarrow Nu\cdots R\cdots X \longrightarrow NuR + X^-$$

(12) S_N2 mechanisms are favoured by primary alkyl halides, nucleophiles with high basicities, polar non-protic solvents and non-polar solvents.

(13) S_N2 mechanisms involving a chiral centre produce product molecules whose configuration has been inverted (opposite to that of the original halide).

(14) Nucleophilic substitution of aryl halides under normal conditions only occurs if there are strong electron acceptor substituents present on the ring system.

(15) S_N2Ar mechanisms are based on the general sequence:

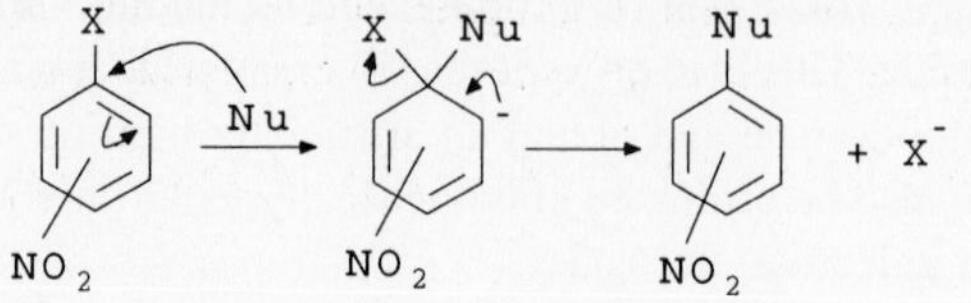

(16) 2,4-Dinitrochlorobenzene and 2,4-dinitrofluorobenzene are used to identify amines and locate the positions of primary amino groups in peptides and proteins.

(17) Elimination of HX by a base (or simply the solvent) tends to yield the most stable product (Saytzeff elimination).

(18) Elimination of HX by a base has the general form:

$$-CH_2-CHX- + B \longrightarrow -CH=CH- + BH^+$$

(19) The main mechanisms for elimination reactions are known as E1 and E2.

(20) E1 mechanisms have the general form:

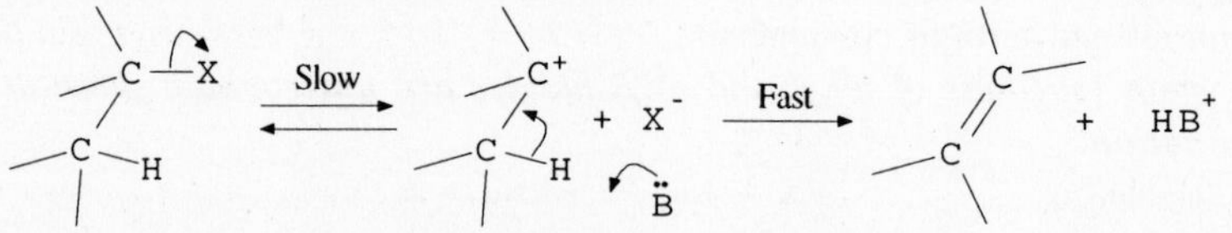

(21) E1 mechanisms are favoured by compounds that can form stable carbocations (e.g., tertiary halides) and polar solvents.

(22) E2 mechanisms have the general form:

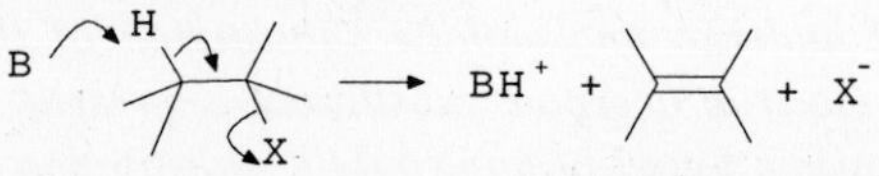

(23) E2 mechanisms are favoured by structures that cannot form stable carbocation and non-polar solvents.

(24) Reactions proceeding by an E2 mechanism are faster if the hydrogen and the halogen are in an anticonformation or configuration (trans to each other).

(25) Organohalides are toxic to insects, plants, fish and mammals. Their slow rate of metabolism in man means that organohalides accumulate in lipid material.

(26) Organohalides are identified by converting the covalently bound halogen into the corresponding halide ion. This ion is often detected or assayed by procedures that involve a reaction with silver nitrate.

14.13 QUESTIONS

(1) Classify the halogen functional groups in the structural formulae shown in Figs 14.3 and 14.4 as either primary, secondary and tertiary alkyl or allyl or alkenyl or aryl halides.

(2) Predict, by means of equations, the most likely product(s) that would be formed in each of the following reactions.

(a) Benzyl bromide with sodium ethoxide in ethanol at room temperature.

(b) 2-Phenylethylbromide with a hot aqueous ethanolic solution of sodium hydroxide.

(c) 1-Bromopentane with sodium hydroxide at about 70°C.

(d) 2-Chloropentane with hot ethanolic sodium ethoxide.

(3) Predict the probable relative reactivities of the halogen atoms in each of the compounds:

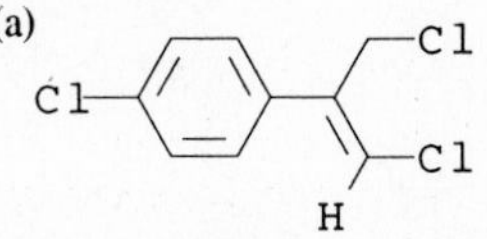

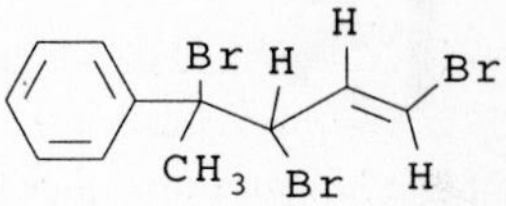

(4) Describe, giving reasons and using equations, the most feasible mechanisms for each of the following reactions:

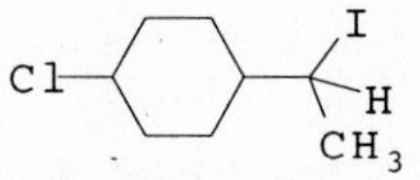

(a) Bromobutane with sodium ethoxide to yield ethoxybutane.

(b) 3-Bromopentane with hot aqueous ethanolic sodium hydroxide to produce mainly pent-2-ene.

(c) 4-Chloropyridine with sodium hydroxide to form 4-hydroxypyridine.

(d) Anti-1,2-diphenylbromoethane with ethanolic potassium hydroxide to yield only E-1,2-diphenylethene.

(5) $CH_2(NH_2)CONHCH(CH_3)COOH$ $CH_3CH(NH_2)CONHCH_2COOH$
                      (A)   (B)

Compounds A and B are structural isomers. Outline, by means of chemical equations a possible way of distinguishing between these compounds. Give brief details of any techniques that are required by the procedure.

(6) Suggest a reason for 2,4,6-trinitrochlorobenzene being more reactive towards nucleophiles than chlorobenzene.

15

Alcoholic and phenolic hydroxyl groups

15.1 INTRODUCTION

The hydroxyl functional group has the structural formula, O–H. This group can occur as either a discrete structure or part of a larger functional group as in carboxylic acids. This chapter discusses the properties of the hydroxyl group when it occurs as a discrete structure in alcohols and phenols. Its properties when it occurs as part of the structure of another functional group will be modified by the other structures in that group and will be described in the chapter for that functional group.

Hydroxyl groups bonded to an aliphatic carbon are called alcohols whilst those bonded to an aromatic carbon, such as the carbon of a benzene ring, are referred to as phenols (Fig. 15.1). Alcohols and phenols have many similar properties but they also have a significant number of different properties.

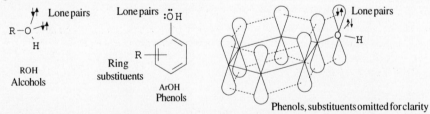

Fig. 15.1. The molecular structures of the hydroxy groups of alcohols and phenols. The oxygen of the hydroxy group of alcohols is tetrahedrally bonded. This can be explained by sp^3 hybridisation of the oxygen atom, the lone pairs occupying two of the sp^3 hybrid orbitals. The structure of the oxygen of phenolic hydroxy groups can be explained by it being sp^2 hybridised. One of the lone pairs occupies a sp^2 hybrid orbital, the other the p_z orbital. As a result, the latter lone pair is an integral part of the conjugated structure of the molecule. This will affect the chemical behaviour of both the ring and the hydroxyl group.

Alcohols are subdivided into primary, secondary and tertiary alcohols depending on the number of hydrogen atoms bonded to the carbon atom to which the hydroxyl group is attached (Fig. 15.2). This is because these classes of alcohol exhibit a significant number of different as well as similar chemical properties.

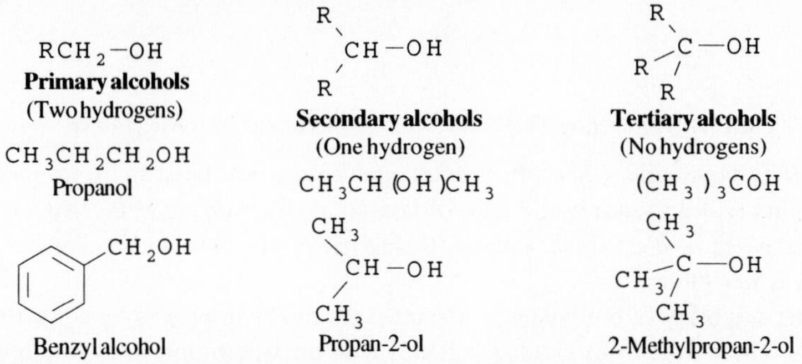

RCH_2—OH
Primary alcohols
(Two hydrogens)

$CH_3CH_2CH_2OH$
Propanol

(benzyl) CH_2OH
Benzyl alcohol

$\underset{R}{\overset{R}{\diagdown}}$CH—OH
Secondary alcohols
(One hydrogen)

$CH_3CH(OH)CH_3$

$\underset{CH_3}{\overset{CH_3}{\diagdown}}$CH—OH
Propan-2-ol

$\underset{R}{\overset{R}{\diagdown}}\underset{R}{\overset{}{\diagup}}$C—OH
Tertiary alcohols
(No hydrogens)

$(CH_3)_3COH$

$\underset{CH_3}{\overset{CH_3}{\diagdown}}\underset{CH_3}{\overset{}{\diagup}}$C—OH
2-Methylpropan-2-ol

Fig. 15.2. Primary, secondary and tertiary alcohols. Note the alternative ways of drawing the structures of propan-2-ol and 2-methylpropan-2-ol.

Alcoholic and phenolic hydroxyl groups are important features of the structures of many different classes of naturally occurring compounds and artificial drugs, for example, steroids (26.9), terpenes (26.8), carbohydrates (Chapter 27), paracetamol and procainamide (25.6). All simple alcohols and phenols should be regarded as toxic. All phenols are corrosive and act as cytotoxic agents.

15.2   NOMENCLATURE

The prefix **hydroxy-** and the suffix **-ol** are used to name both alcohols and phenols. Trivial names are also in common use (Fig. 15.3). The use of the parent name **phenol** indicates that the structure of the compound is based on the structure of phenol and the hydroxyl group of phenol is attached to carbon 1 of the benzene ring.

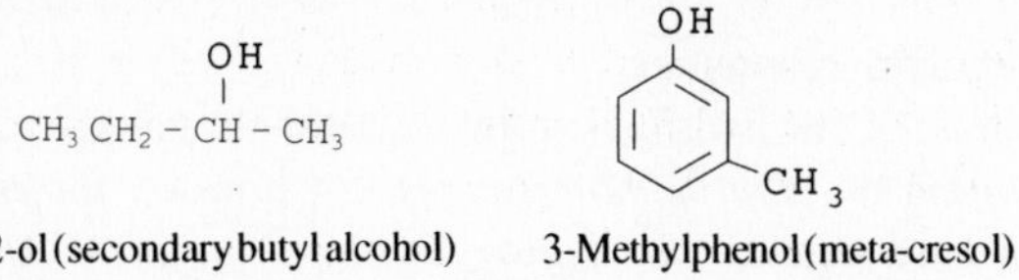

$$CH_3\,CH_2 - \underset{|}{\overset{OH}{C}}H - CH_3$$

Butan-2-ol (secondary butyl alcohol)

(phenol ring with OH and CH_3)
3-Methylphenol (meta-cresol)

Fig. 15.3. Examples of the nomenclature of alcohols and phenols. Trivial names are given in brackets. The abbreviations sec- or s- is often used for secondary whilst *m*- is used for meta.

15.3   PHYSICAL   PROPERTIES

15.3.1 Solubility

Hydroxyl groups can hydrogen bond with water. In theory each hydroxyl group could form three hydrogen bonds, two involving the lone pairs and one the hydrogen atom of the group (Fig. 15.4). However, it is very unlikely that a hydroxyl group would form all three hydrogen bonds at room temperature.

The presence of one or more hydroxyl groups in a molecule will result in that molecule having some water solubility. However, it is not possible to predict the exact degree of this

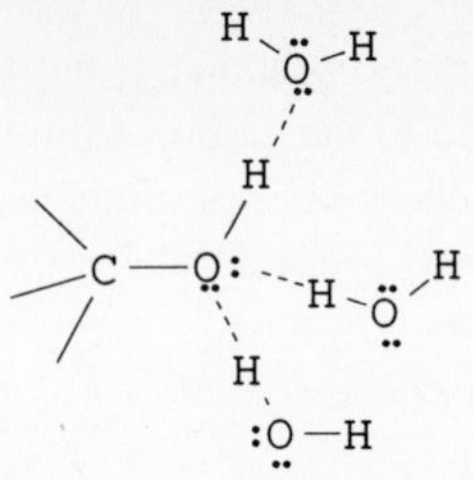

Fig. 15.4. Hydrogen bonding between water molecules and a hydroxyl group.

water solubility especially when other functional groups are present. As a general rule water solubility will decrease as the ratio of non-polar structures to polar structures in the molecule increases. For example, ethanol (C_2H_5OH) is miscible with water but decanol ($C_{10}H_{22}OH$) is insoluble.

The water solubility of both phenols and alcohols can be improved by converting them to their salts. These salts form ion-dipole bonds with the water molecules which increases their water solubility. However, alcohols unlike phenols, are not acidic enough to react with alkalis.

$$R-OH \xrightarrow{Na} R-O^-Na^+ \qquad\qquad Ar-OH \xrightarrow{Either\ Na\ or\ NaOH} Ar-O^-Na^+$$

Alcohols   Phenols

15.3.2 Infra-red
Alcoholic and phenolic hydroxyl groups exhibit strong absorption bands in the 3600-3590 (O–H stretching), 1410-1260 (C–O–H bending) and 1150-1040 (C–OH stretching) cm^{-1} regions. Hydrogen bonding increases the intensity and width of the 3600-3590 cm^{-1} band. Very strong hydrogen bonding can reduce the lower limit of this absorption band to 2500 cm^{-1}. In these cases it is easy to confuse alcoholic and phenolic O–H absorptions with those of other functional groups (e.g. COOH) whose structures contain hydroxyl groups. A strong infra-red absorption in the 3600-3200 cm^{-1} region is taken to indicate the *possible* presence of either alcoholic and/or phenolic hydroxyl groups.

15.3.3 Proton magnetic resonance
The hydrogens (protons) of the hydroxyl groups of alcohols and phenols normally produce one singlet signal within the *approximate ranges*, δ 1.0 to 4.5 for alcohols and δ 4.5 to 11.0 for phenols. In pure samples the hydroxyl groups' proton will show coupling with neighbouring protons. These signals may be broad or sharp depending on the molecular environment of the hydroxyl group. For example, the proton of the hydroxyl group in pure ethanol produces a signal at δ 4.0 to 4.3 due to such coupling (Fig. 15.5).

The hydroxyl groups' proton is sufficiently acidic to exchange with the deuterium of deuterium oxide. Consequently, any reduction in size or the disappearance of a signal after a D_2O shake (10.6.4) is an indication of the possible presence of such a proton. For example, the disappearance of the signal at δ 4.1 in the spectrum of ethanol shows that this signal is due to the hydroxyl groups' proton (Fig. 15.5).

Protons in many other functional groups will also exchange with deuterium so the disappearance of a signal after a D_2O shake is not necessarily indicative of an alcoholic or phenolic hydroxyl group. However, it can be used as additional evidence for the possible

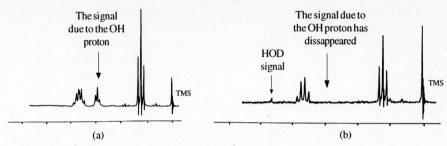

Fig.15.5. (a) ^{1}H n.m.r. spectrum of ethanol. (b) The ^{1}H n.m.r. spectrum of ethanol after a D$_2$O shake.

existence of such a structure in a compound. Detection of the hydroxyl groups of alcohols and phenols in analysis is usually made from the compounds' p.m.r. and IR spectra. There are no reliable specific spot tests for alcoholic hydroxyl groups but phenolic groups may be detected by the iron III chloride test (15.11.1). However, it should be noted that this test does not give a positive result for all phenols.

15.4 GENERAL CHEMICAL PROPERTIES

Alcohols and phenols have many similar chemical properties as can be seen in the summary of their general chemical properties given in Table 15.1.

Table 15.1. A summary of the general properties of alcohols and phenols.

Functional group	Acidity	Nucleophile	Oxidised	Carbocation source
Phenols	Weak to strong	Weak	Yes, easy	No, not possible
Primary alcohols	Very weak	Weak	Yes, easy	Yes, possible
Secondary alcohols	Very weak	Weak	Yes, easy	Yes, possible
Tertiary alcohols	Very weak	Weak	Only with difficulty	Yes, possible

The major differences in the general chemical properties of alcohols and phenols can be explained by the differing natures of the molecular structures of their carbon-hydrogen skeletons. These explanations are discussed later in this chapter.

15.5 ACIDIC NATURE

Alcohols and phenols act as acids because they ionise in water and polar solvents.

$$R{-}OH \rightleftharpoons R{-}O^- + H^+ \qquad Ar{-}OH \rightleftharpoons Ar{-}O^- + H^+$$

Alcohols   Alcoholate ion   Phenols   Phenolate ion

Alcohols are extremely weak acids whilst the acidity of phenols varies from weak to strong. The acid strength of alcohols and phenols depends on the strength of the O–H bond, the stability of the conjugate base, the solvent and the temperature.

15.5.1 The strength of the O–H bond

The O–H bond is polarised in both cases but in phenols this polarisation is greater and so the bond is weaker and ionisation easier. The increased polarisation in phenols occurs because the oxygen of the hydroxyl group of phenols is more electron deficient due to the delocalisation of one of the lone pairs of the oxygen atom (Figs 15.1 and 15.6). This deficiency enhances the inductive effect of the O–H bond and weakens it sufficiently for a significant degree of ionisation of the O–H to occur in aqueous solution and other polar solvents. The lone pairs of electrons of the oxygen atoms of the hydroxyl groups of alcohols are not delocalised and so are not as electron deficient as those of phenols. Consequently, the polarisation of the O–H bonds of alcohols is less than the O–H bonds of phenolic hydroxyl groups. As a result, the O–H bonds of alcoholic hydroxyl groups are stronger and do not ionise significantly in water and other polar solvents.

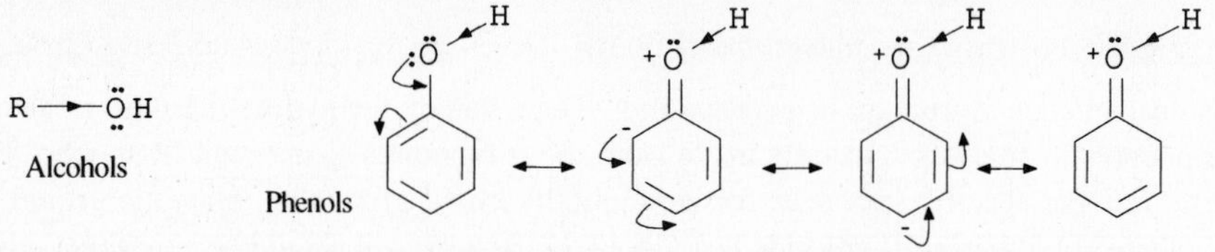

Fig. 15.6. The effect of structures on the O-H bond strength of alcohols and phenols. The electron deficient oxygen increases the polarisation of the O-H bond in phenols.

15.5.2 The stability of the conjugate base

The phenolate anion is more stable than the alcoholate anion. The stability of the phenolate ion is due to the ability of the ion to delocalise its negative charge over the benzene ring via its conjugated structure. This delocalisation means that the negative charge is spread over a wide area making the ion less attractive to protons (Fig. 15.7). The structure of the alcoholate ion does not allow delocalisation of the negative charge to occur (Fig. 15.7). This means the negative charge is concentrated on the oxygen and so the ion readily reacts with a hydrogen ion to reform the alcohol. Subsequently the position of equilibrium for the ionisation moves to the right resulting in alcohols being much less acidic than phenols.

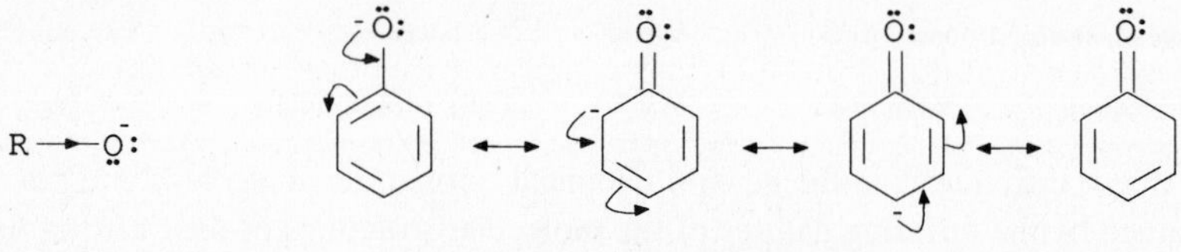

Fig. 15.7. The negative charge of phenolate ions is stabilised by delocalisation but this is not possible in alcoholate ions.

The hydroxyl groups of the vinyl alcohols (enols) are acidic because they have a conjugated structure similar to that of the hydroxyl group of phenols (Fig. 15.8).

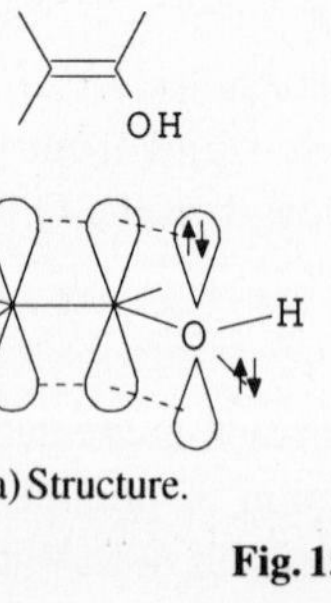

(a) Structure.

$CH_2{=}CH{-}\ddot{O}H \longleftrightarrow \bar{C}H_2{-}CH{=}\overset{+}{O}H$

The O-H bond is weakened by the delocalisation of the lone pair.

$CH_2{=}CH{-}\ddot{O}{:}^- \longleftrightarrow \bar{C}H_2{-}CH{=}O{:}$

The conjugate base is stabilised by delocalisation of the negative charge.

$CH_2{=}CH{-}OH \rightleftharpoons CH_2{=}CH{-}O^- + H^+$

(b) Acidic nature. The position of equilibrium favours the formation of a significant concentration of hydrogen ions.

Fig. 15.8. (a) The structure and (b) acidic nature of vinyl alcohols.

15.6 RING SUBSTITUENTS AND THE ACIDITY OF PHENOLS

Substituents on a benzene ring will effect the pi electron density of that ring. Electron donors will increase the electron density whilst electron acceptors will decrease it. These changes in electron density will affect the ease of ionisation of phenolic hydroxyl groups and the stability of their corresponding phenolate ions. Phenols with electron donor substituents are usually weaker acids whilst those with electron acceptor substituents are generally stronger acids than phenol (Table 15.2). However, there are many exceptions to these general statements.

Table 15.2. Examples of the acid strengths of phenols with electron donor and acceptor substituents. * Chlorophenol is one of many exceptions to the general rules discussed in this section. The pK_as of phenol, ethanoic acid and ethanol are included for comparison purposes.

Phenols with electron donor substituents	pK_a	Reference compounds (pK_a)	Phenols with electron acceptor substituents	pK_a
4-Chlorophenol*	9.2	Ethanoic acid (4.8)	2-Nitrophenol	7.2
2-Methylphenol	10.3	Phenol (9.9)	3-Nitrophenol	8.3
3-Methylphenol	10.1	Ethanol (~16)	4-Nitrophenol	7.2
4-Methylphenol	10.1		2,4,6-Trinitrophenol	0.38

Electron donor groups strengthen the O–H bond and reduce the stability of the phenolate by reducing the delocalisation of the negative charge. Conversely, electron acceptors weaken the O–H bond and increase the stability of the phenolate ion by increasing the delocalisation of the negative charge (Fig. 15.9).

Electron donor groups

Z opposes the O-H bond weakening effect of the mesomeric effect of the hydroxyl group by making the ring electron rich

Resultant electron distribution of Z

Z opposes the delocalisa-tion of the negative charge of the hydroxyl group by making the ring electron rich

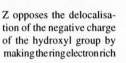

Resultant electron distribution of Z

Electron withdrawing groups

Z enhances the O-H bond weakening effect of the mesomeric effect of the hydroxyl group by making the ring electron deficient

Resultant electron distribution of Z

Z enhances the delocalisation of the negative charge by mak-ing the ring electron deficient.

Resultant electron distribution of Z

Fig. 15.9. The effect of substituents on the acid strengths of phenols. Z is either an electron acceptor or an electron donor group as appropriate.

15.7 ACID REACTIONS OF THE HYDROXYL GROUP

15.7.1 Alcohols

Alcohols will react as acids only with very reactive metals and very strong bases.

$$ROH \xrightarrow{\;Na\;} RO^- \; Na^+ + H_2$$

$$ROH \xrightarrow[\text{Dimsyl sodium}]{CH_3SO\overset{-}{C}H_2 \; Na^+} RO^- \; Na^+ + CH_3SOCH_3$$

$$\text{Dimethyl sulphoxide}$$

The reaction with dimsyl sodium is the basis of a number of volumetric analysis methods for assaying alcohols.

15.7.2 Phenols

In theory, a phenol should exhibit all the reactions of a typical acid, that is, change the colour of indicators, react with electropositive metals to produce hydrogen, form salts with alkalis and bases and liberate carbon dioxide from carbonates and hydrogen carbonates. In practice the reactions of weakly acidic phenols are less comprehensive than those of strongly acidic phenols, for example, some phenols do not react with sodium bicarbonate

Many phenols are sufficiently acidic to be assayed by titration with sodium hydroxide, for example, the antiseptic hexachlorophane which is used in various pharmaceutical and household products. Ethanol is used as the solvent in this assay because hexachlorophane is almost insoluble in water. The indicator for the titration is phenolphthalein.

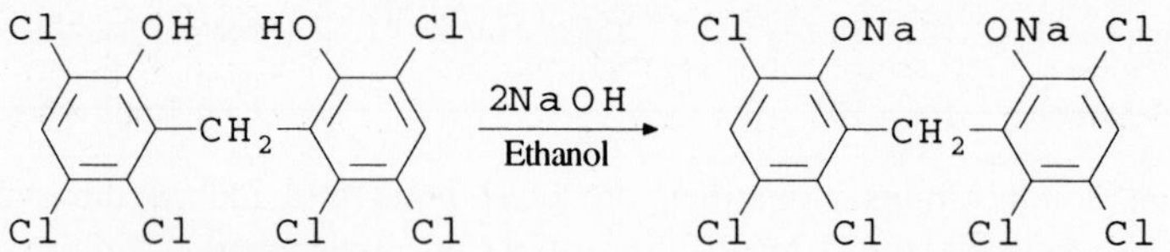

Hexachlorophane

2,4,6-Trinitrophenol (picric acid) forms yellow crystalline salts with many bases. These salts often have sharp melting points and so may be used as derivatives in identification tests and qualitative analysis. For example, picrate formation is the basis of identification test C in the 1993 British Pharmacopoeia monograph of the drug chloroquine.

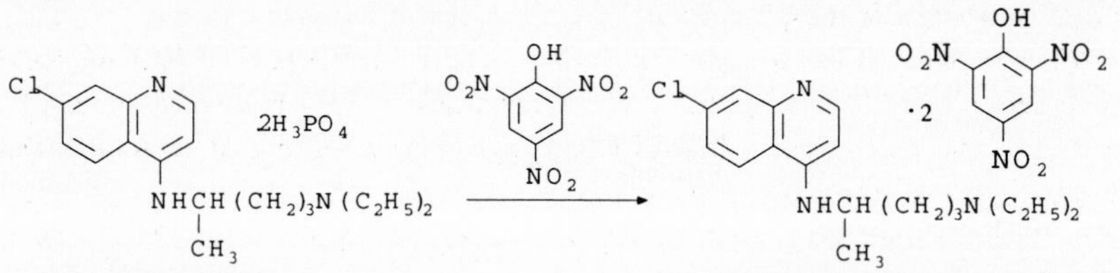

Chloroquine phosphate   Chloroquine picrate, mp 207°C

2,4,6-Trinitrophenol is also used in the Jaffe reaction for the quantitative determination of creatinine. Creatinine, alkali and 2,4,6-trinitrophenol react to form a yellow-brown water soluble complex which can be estimated by colorimetric analysis. The chemistry of the reaction is not understood but the reaction may be used to estimate the efficiency of glomerular filtration in the kidney.

$$\underset{\text{Creatinine}}{} NH_2-\overset{\overset{\displaystyle NH_2^+}{\|}}{C}-\underset{\underset{\displaystyle CH_3}{|}}{N}-CH_2COO^-$$

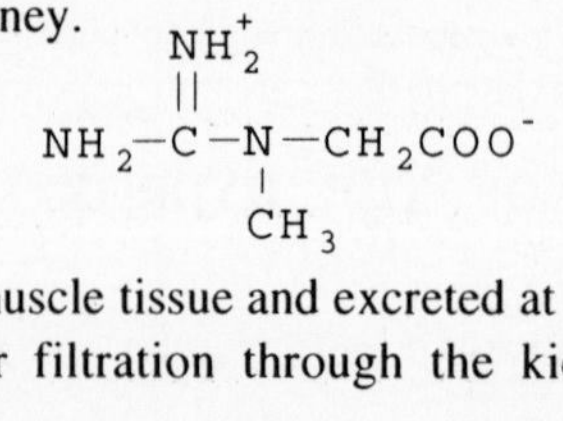

Creatinine is produced by muscle tissue and excreted at a virtually constant daily level almost entirely by glomerular filtration through the kidneys. A comparison of the

creatinine level in the serum and urine of a patient enables one to determine the efficiency
of this function of the patient's kidney.

15.8 REACTION AS A NUCLEOPHILE

Alcohols and phenols can react as nucleophiles in nucleophilic substitution, displacement
and addition reactions. The lone pairs of their hydroxyl groups act as the nucleophilic
centres in these reactions. However, the high electronegativity of oxygen means that the
oxygen is reluctant to allow the lone pairs to form bonds with other atoms. This means that
the hydroxyl groups of alcohols and phenols are usually weak nucleophiles. However,
nucleophilic activity can be improved by the use of suitable catalysts or conversion to the
appropriate alcoholate or phenolate ions. The negative charges of these species make
them stronger nucleophiles than their parent compounds.

$$R-OH \xrightarrow{Na} R-O^- Na^+ \qquad Ar-OH \xrightarrow{\text{Either Na or NaOH}} Ar-O^- Na^+$$

15.8.1 Nucleophilic substitution

All types of alcohol and phenol react in a similar fashion. They do not usually react directly
but have to be converted to the corresponding alcoholate or phenolate ion. For example,
dichlorophenol is converted to its phenolate before being reacted with chloroacetic acid
in the manufacture of the selective herbicide, 2,4-dichlorophenoxyacetic acid.

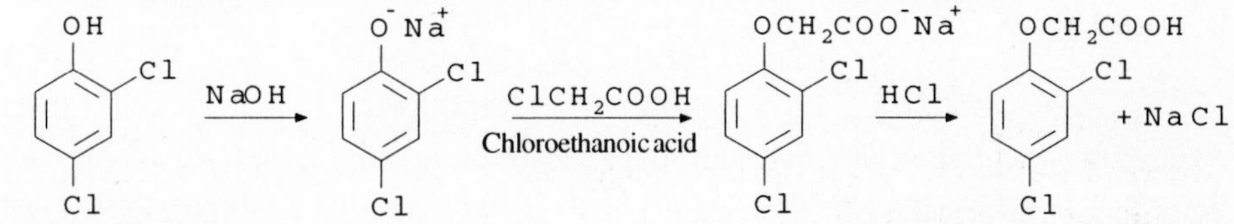

2,4-Dichlorophenol   2,4-Dichlorophenoxyethanoic acid

15.8.2 Nucleophilic displacement

Alcohols and phenols, with some exceptions, act as nucleophiles (Nu) in nucleophilic
displacement reactions (Table 15.3). These reactions usually involve compounds having
the general structural formula RCOZ where Z can be a variety of structures (8.4). The
reactions often take the general form:

$$RCOZ + Nu \longrightarrow RCONu + HZ$$

and frequently require an acid or base catalyst to assist the attack of the weak nucleophile.
It should be noted that phenols do not react directly with carboxylic acids to form esters.

Table 15.3. General examples of some of the nucleophilic displacement reactions of alcohols and phenols

Reactant	General equation	Comments
Acid anhydrides	$RCOOCOR + R'OH \longrightarrow RCOOR' + RCOOH$	Both alcohols and phenols react
Carboxylic acid halides	$RCOX + R'OH \longrightarrow RCOOR' + HX$	
Sulphonyl halides	$RSO_2X + R'OH \longrightarrow RSO_2OR' + HX$	
Carboxylic acids	$RCOOH + R'OH \longrightarrow RCOOR' + H_2O$	Only alcohols react

The reactions of phenols with benzoyl chloride and alcohols with 3,5-dinitrobenzoyl chloride are used to form derivatives for use in analysis and identification tests.

2-Naphthol (mp 123°C)   Benzoyl chloride   2-Naphthyl benzoate (mp 106°C)

Isopropyl alcohol (bp 82°C) NO_2 NO_2
                          3,5-Dinitrobenzoyl chloride   Isopropyl 3,5-dinitrobenzoate (mp 122°C)

The reaction of alcohols with ethanoic anhydride is used as the basis of assay procedures for some alcohols. For example, the antiseptic/anaesthetic benzyl alcohol is heated with excess ethanoic anhydride in the presence of pyridine which acts as a catalyst for the reaction.

$$PhCH_2OH + CH_3COOCOCH_3 \longrightarrow PhOCOCH_3 + CH_3COOH + [excess\ CH_3COOCOCH_3]$$

The reaction mixture is titrated with sodium hydroxide using phenolphthalein as indicator. This indicator changes colour when both the ethanoic acid produced in the reaction and the excess ethanoic anhydride have reacted. A blank determination must also be carried out.

$$CH_3COOH + NaOH \longrightarrow CH_3COONa \qquad (15.1)$$

$$CH_3COOCOCH_{3(unreacted)} + 2NaOH \longrightarrow 2CH_3COONa \qquad (15.2)$$

In most back titration procedures involving a blank the titrating agent C reacts only with the reagent B that reacted with the analyte A (11.6.2). Consequently, the difference between the assay and blank titration values is the volume of the titrating agent C that is equivalent to the reagent B that reacted with the analyte A. In this assay, the titrating agent C also reacts with the ethanoic acid produced by the reaction of the benzyl alcohol and the ethanoic anhydride. Suppose that a sample contained x moles of benzyl alcohol. The equation for the process shows that x moles of ethanoic anhydride would be required to react with these x moles of benzyl alcohol. Suppose the quantity of excess ethanoic anhydride used in the assay is y moles making the total amount of ethanoic anhydride used $x+y$ moles. Therefore, in the blank determination $2x+2y$ moles of sodium hydroxide would be required (equation 15.2). In the assay x moles of benzyl alcohol would react to form x moles of ethanoic acid and so only $x+2y$ moles of sodium hydroxide (equations 15.1 and 15.2). The difference between the blank and assay titres is therefore x moles of sodium hydroxide which is equal to the number of moles of benzyl alcohol in the sample. Consequently, in this assay the difference between the blank and assay titration readings is the volume of sodium hydroxide that is equivalent to the quantity of benzyl alcohol in the sample. It should be noted that in other back titration of this type where a product of the reaction reacts with the titrating agent there may not be a 1:1 relationship between the analyte and the titrating agent. Each case must be considered on its own chemical merits.

Pyridine is widely used as a catalyst for acylation reactions. Although it is a base it is too weak to interfere in titrations when phenolphthalein is used as indicator.

The preparation of a number of drugs involves nucleophilic displacement reactions of alcohols and phenols, for example:

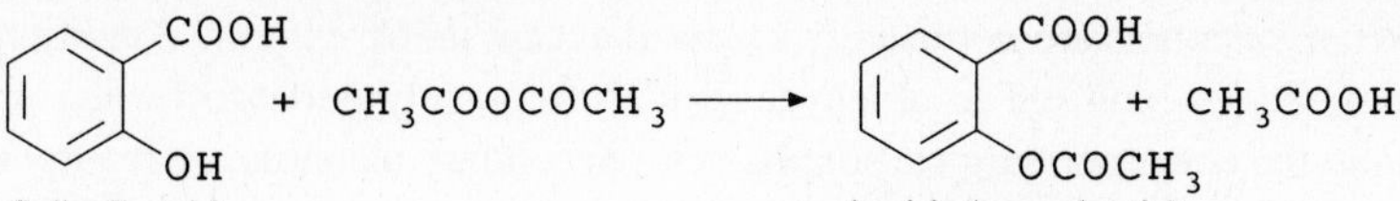

Salicylic acid + $CH_3COOCOCH_3$ ⟶ Aspirin (an analgesic) + CH_3COOH

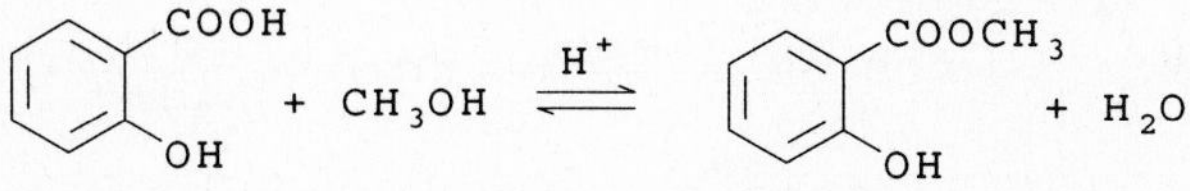

Salicylic acid + CH_3OH ⇌ (H⁺) Methyl salicylate (a counter-irritant) + H_2O

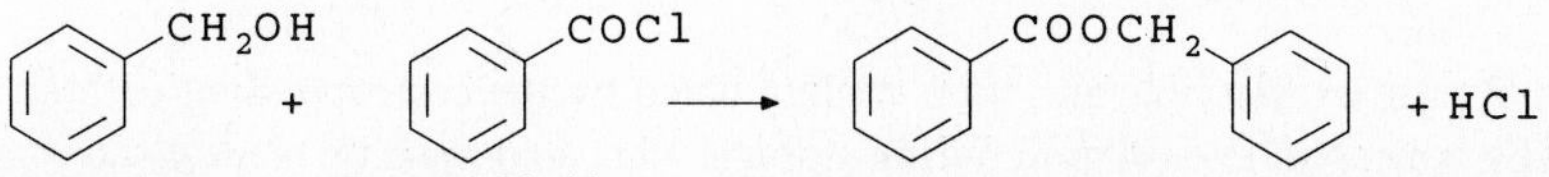

Benzyl alcohol + Benzoyl chloride ⟶ Benzyl benzoate (used in the topical treatment of scabies) + HCl

15.8.3 Nucleophilic addition

Alcohols react with aldehydes and ketones under dry conditions in the presence of an acid catalyst to form hemiacetals. The hemiacetals of ketones are known as hemiketals in older literature. The equilibrium favours the aldehyde or ketone but is forced to completion by excess alcohol. Unsymmetrical hemiacetals possess a chiral centre.

$$R_1CHO + R_2OH \underset{}{\overset{H^+}{\rightleftharpoons}} R_1CH(OH)OR_2$$ 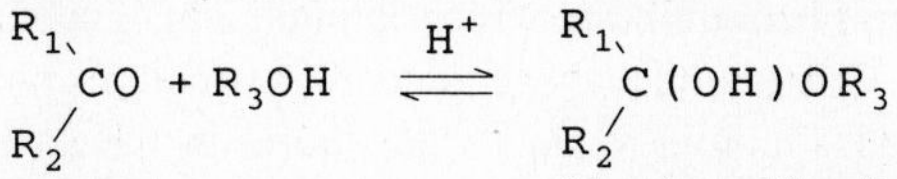
Aldehyde   A hemiacetal

$$\begin{array}{c} R_1 \\ \diagdown \\ CO + R_3OH \\ \diagup \\ R_2 \end{array} \overset{H^+}{\rightleftharpoons} \begin{array}{c} R_1 \\ \diagdown \\ C(OH)OR_3 \\ \diagup \\ R_2 \end{array}$$
Ketone   A hemiacetal (hemiketal)

The cyclic structures of monosaccharides are hemiacetals. The rings are formed by an internal (intramolecular) nucleophilic addition of an alcoholic hydroxyl group to the carbonyl group of the sugar (Chapter 27). This ring closure gives rise to optical isomers which are referred to as the α and β isomers of the monosaccharide (27.2).

Hemiacetals and hemiketals can react further with excess alcohol under dry acid conditions to produce acetals and ketals. These reactions are also equilibria and to obtain reasonable yields of the product it is necessary to move the position of equilibrium to the right by removing the water as soon as it is formed. This is usually achieved in the laboratory by distilling the water out of the reaction mixture. The conversion of hemiacetals into acetals is not a nucleophilic addition but involves the formation of carbocation intermediates (15.13). Oxygen glycosides (27.10) are the acetals of monosaccharides.

$$R_1CH(OH)OR_2 + R_2OH \overset{H^+}{\rightleftharpoons} R_1CH(OR_2) + H_2O$$
Hemiacetal   Acetal

$$\begin{array}{c} R_1 \\ \diagdown \\ C(OH)OR_3 + R_3OH \\ \diagup \\ R_2 \end{array} \overset{H^+}{\rightleftharpoons} \begin{array}{c} R_1 \diagdown \; \diagup OR_3 \\ C \\ R_2 \diagup \; \diagdown OR_3 \end{array} + H_2O$$ 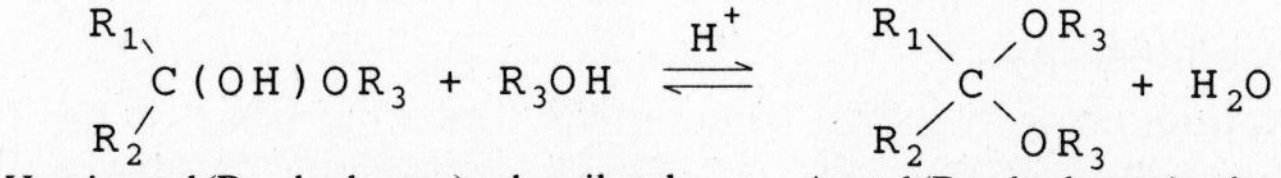
Hemiacetal (R_2 = hydrogen) or hemiketal   Acetal (R_2 = hydrogen) or hemiketal

15.9 OXIDATION

Primary and secondary alcohols are readily oxidised by most oxidising agents including air but tertiary alcohols are not usually oxidised except under extreme conditions. Primary alcohols are easily oxidised to aldehydes which frequently undergo further spontaneous oxidation to the corresponding carboxylic acid. Secondary alcohols are readily oxidised to ketones which do not normally undergo further oxidation.

Primary alcohols $\quad RCH_2OH \xrightarrow{\text{Oxidation}} RCHO \xrightarrow[\text{oxidation}]{\text{Further}} RCOOH$

Secondary alcohols $\quad \begin{array}{c} R \\ \diagdown \\ \quad CHOH \\ \diagup \\ R_1 \end{array} \xrightarrow{\text{Oxidation}} \begin{array}{c} R \\ \diagdown \\ \quad CO \\ \diagup \\ R_1 \end{array}$

Liquid primary alcohols are often contaminated by the corresponding carboxylic acid formed by atmospheric oxidation during storage. This acid must be neutralised if acid free conditions are required. For example, the ethanol used as a solvent in many acid-base assay procedures can contain ethanoic acid formed by atmospheric oxidation on storage. This acid must either be neutralised before the ethanol is used or a blank determination must be carried out.

$$CH_2CH_2OH \xrightarrow[\text{oxidation}]{\text{Atmospheric}} CH_3CHO \xrightarrow[\text{oxidation}]{\text{Atmospheric}} CH_3COOH$$
$$\text{Ethanol} \qquad\qquad \text{Ethanal} \qquad\qquad \text{Ethanoic acid}$$

Phenols have an electron rich ring which makes them very susceptible to attack by oxidising agents. Oxidation of phenols is relatively easy and frequently yields quinones. For example, phenol itself slowly oxidises in air to paraquinone which hydrogen bonds with unreacted phenol to form phenoquinone which is pink. This accounts for the pink tint of older samples of phenol. Compounds whose molecular structures contain quinone moieties are often highly coloured. Quinone residues are found in the structures of many of the pigments responsible for the colours of plants and insects (19.14).

Phenol $\qquad\qquad$ Paraquinone $\qquad\qquad$ Phenoquinone

The oxidation of phenols makes them valuable antioxidants. Antioxidants are substances that oxidise preferentially and in doing so protect another substance from oxidation. Oxidation is retarded but seldom prevented by the presence of an antioxidant. Some tocopherols (vitamins E) and phenols (Fig. 15.11) are permitted antioxidants in the food and pharmaceutical industries. The ease of oxidation of L-ascorbic acid (vitamin C) is the reason for the use of it and its salts as the E300, E301 and E302, permitted antioxidants in the food industry.

Oxidation by titration with iodine solution VS forms the basis of the 1993 British Pharmacopoeia assay for ascorbic acid. However, this assay must be carried out in air free

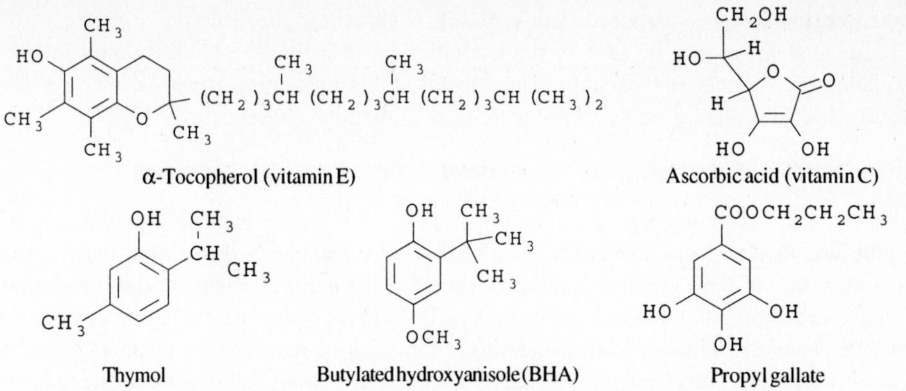

Fig. 15.11. Examples of phenols used as permitted antioxidants in the food and pharmaceutical industries. Thymol, butylated hydroxyanisole (BHA) and propyl gallate are used to stabilise pharmaceuticals.

water since any dissolved oxygen will readily oxidise significant amounts of the ascorbic acid. Starch is used as indicator.

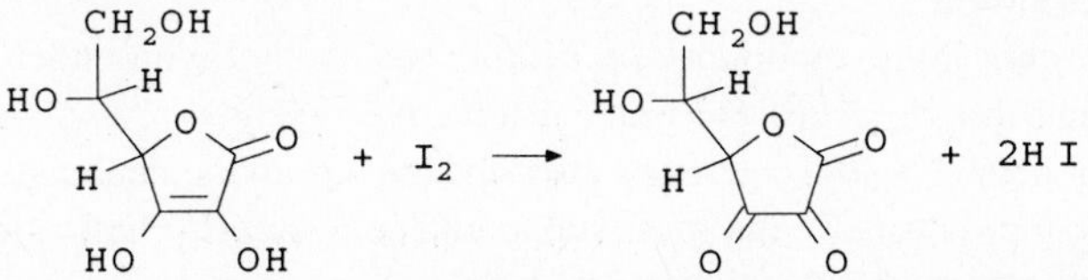

Ascorbic acid is also assayed by titration with Tillman's reagent, the sodium salt of 2,6-dichloro-4-(4-hydroxyphenylamino)cyclohexa-2,5-dien-1-one. The ascorbic acid reduces the solution of the blue/green Tillman's reagent to a colourless hydroxyl compound.

Oxidation of alcohols is a common metabolic process. It generally gives the same products as the corresponding laboratory oxidation. For example, the metabolic oxidation of ethanol in the liver first yields ethanal which is then oxidised further to ethanoic acid.

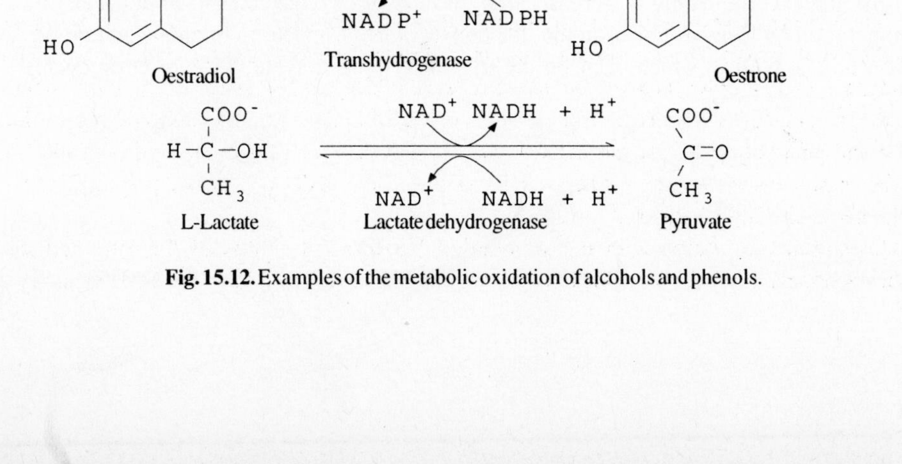

Fig. 15.12. Examples of the metabolic oxidation of alcohols and phenols.

However, metabolic oxidation differs in that it is enzyme-controlled and is often part of a **redox system** as may be seen in the examples given in Fig. 15.12. Easy atmospheric oxidation of alcohols and phenols means that suitable precautions must be taken when storing and formulating many drugs with these functional groups.

15.10 THE REACTIONS OF ALCOHOLS IN ACID CONDITIONS

Alcohols can either undergo elimination or react with some nucleophiles under acid conditions. Reaction mechanisms appear to involve either the hydroxonium ion formed by protonation of the alcohol or the carbocation formed by the subsequent decomposition of the hydroxonium ion in some cases. This carbocation can also rearrange to form a more stable carbocation which can also undergo elimination or react with a nucleophile. The type of reaction and mechanism favoured by a particular alcohol will depend on the reagent used and the prevailing reaction conditions.

$$\underset{\text{Hydroxonium ion}}{\overset{H^+}{\ce{>C-\ddot{O}H ->}} \underset{}{\ce{>C-\overset{+}{O}H_2}}} \longrightarrow \underset{\text{Carbocation}}{\ce{>C^+}} + H_2O$$

15.10.1 Elimination

Elimination occurs in the laboratory on heating the alcohol with either hot concentrated sulphuric or phosphoric acids, the order of reactivity being:

benzyl > allyl > tertiary > secondary > primary alcohols

The major product is usually the most stable alkene (Saytzeff elimination). This is often the alkene with the most substituted C=C bond.

$$\underset{\text{2-Methyl-2-butanol}}{\ce{CH_3-\underset{CH_3}{\overset{OH}{C}}-CH_2CH_3}} \xrightarrow[\text{THF/25°C}]{H^+} \underset{\substack{\text{2-Methyl-2-butene}\\ \text{(major product) trisubstituted C=C bond}}}{\ce{\underset{CH_3}{C}=CH_2CH_3}} + \underset{\substack{\text{2-Methyl-1-butene}\\ \text{disubstituted C=C bond}}}{\ce{\underset{CH_3}{C}-CH_2CH_3}}$$

The mechanism of this type of laboratory reaction is believed to involve the attack of a base, often water, on the appropriate proton of an intermediate carbocation (I).

$$\ce{-\overset{H}{\underset{}{C}}-\overset{\ddot{O}H}{\underset{}{C}}- ->[H^+] -\overset{H}{C}-\overset{\overset{+}{O}H_2}{C}- ->[H_2\ddot{O}] -\overset{}{C}-\overset{+}{C}- -> C=C + H_3O^+}$$

(I)

Biological elimination is stereospecific because of enzyme control. Both cis and trans isomers can be formed. For example, the conversions of malate to fumarate and citrate to cis-aconitate both of which occur in the tricarboxylic acid cycle (TCA or citric cycle).

$$\underset{\text{Malate}}{\ce{\overset{COO^-}{\underset{COO^-}{\overset{|}{\underset{|}{CH_2}} \atop CHOH}}}} \xrightarrow[\text{synthetase}]{\substack{-H_2O\\ \text{Fumarate}}} \underset{\text{Fumarate}}{\ce{^-OOC\overset{H}{\underset{H}{C}}=\overset{COO^-}{C}}} \qquad \underset{\text{Citrate}}{\ce{\overset{COO^-}{\underset{COO^-}{\overset{|}{\underset{|}{CH_2}} \atop HO-C-COO^- \atop CH_2}}}} \xrightarrow[\text{Aconitase/Fe^{2+}}]{-H_2O} \underset{\text{Cis-aconitate}}{\ce{\overset{H}{\underset{CH_2 \atop COO^-}{C}}\overset{COO^-}{\underset{COO^-}{C}}}}$$

15.10.2   Reaction with nucleophiles

Reaction is thought to involve either the carbocation or the hydroxonium ion. In the former case the anion of the acid used to form the carbocation from the alcohol often acts as the nucleophile (Fig. 15.13). Two notable exceptions are phosphoric and hot concentrated sulphuric acids (15.9.1). Reactions involving attack on the hydroxonium ion appear to be less common.

General mechanisms. HA represents the acid.

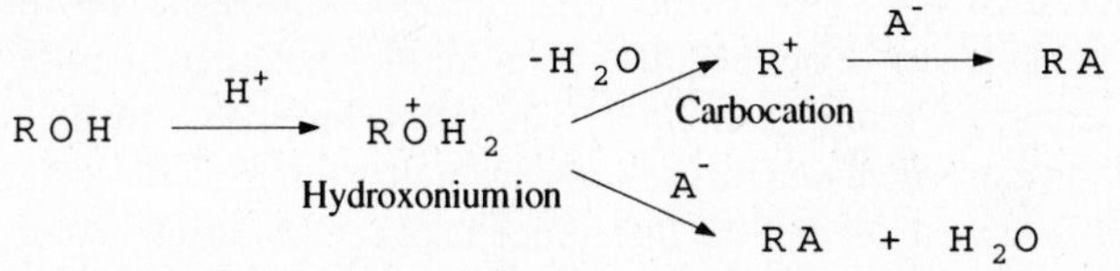

$$ROH \xrightarrow{H^+} R\overset{+}{O}H_2 \begin{array}{c} \xrightarrow{-H_2O} R^+ \xrightarrow{A^-} RA \\ \text{Carbocation} \\ \xrightarrow{A^-} RA + H_2O \end{array}$$

Hydroxonium ion

Cold concentrated hydrochloric acid. The reactivity of alcohols is in the order: benzyl > allyl > t > s > p.

$$ROH \xrightleftharpoons{HCl/ZnCl_2} RCl + H_2O$$

Alkyl halide

Cold concentrated sulphuric acid. Primary alcohols are believed to react by the S_N2 mechanism involving attack of the hydrogen sulphate ion on the hydroxonium ion and not the carbocation.

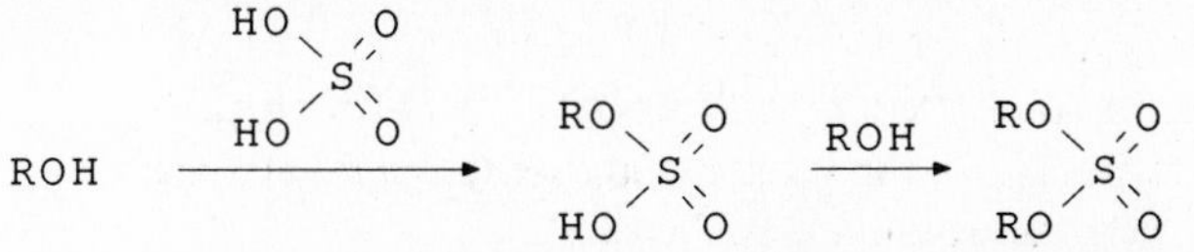

$$ROH \longrightarrow \underset{\text{Alkyl hydrogen sulphate}}{\begin{array}{c} RO \quad O \\ \diagdown S \diagup \\ HO \quad O \end{array}} \xrightarrow{ROH} \underset{\text{Dialkyl sulphate}}{\begin{array}{c} RO \quad O \\ \diagdown S \diagup \\ RO \quad O \end{array}}$$

Fig. 15.13. Examples of the reactions of acids with alcohols.

Glycerol trinitrate which is used to treat angina pectoris, is prepared by the action of a mixture of concentrated nitric and sulphuric acids at $10^{\circ}C$ on glycerol.

$$\begin{array}{c} CH_2OH \\ | \\ CHOH \\ | \\ CH_2OH \end{array} \xrightarrow{HNO_3/H_2SO_4} \begin{array}{c} CH_2ONO_2 \\ | \\ CHONO_2 \\ | \\ CH_2ONO_2 \end{array} + 3H_2O$$

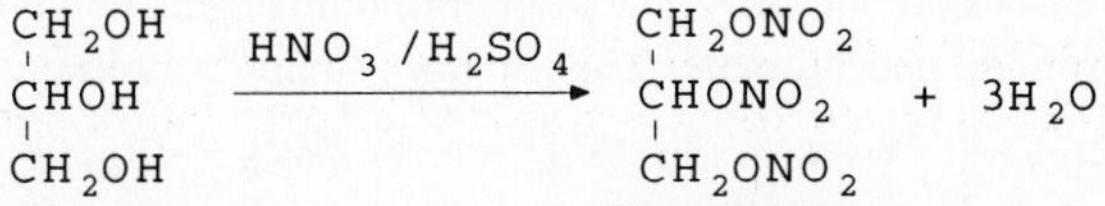

Glycerol   Glycerol trinitrate

Sterols and other alcohols are metabolised to their hydrogen sulphates in the liver. These acidic salts are more water soluble than the parent alcohols and so are more rapidly excreted from the body. Phenols can be metabolised to their hydrogen sulphates or, if sufficiently water soluble, be excreted as such. For example, phenol is excreted as both phenol and phenyl hydrogen sulphate.

The precise route taken by the reaction of an alcohol under acid conditions will depend on the reaction conditions and the chemical species involved. At 140°C concentrated sulphuric acid reacts with excess ethanol to form diethyl ether. The mechanism is believed to involve an S_N2 type of attack of the alcohol on the hydroxonium ion.

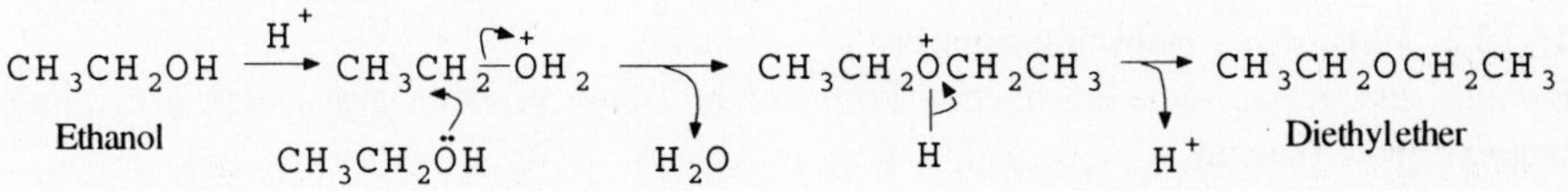

$$CH_3CH_2OH \xrightarrow{H^+} CH_3CH_2\overset{+}{O}H_2 \longrightarrow CH_3CH_2\overset{+}{O}CH_2CH_3 \longrightarrow CH_3CH_2OCH_2CH_3$$

Ethanol $CH_3CH_2\overset{..}{O}H$ H_2O H H^+ Diethyl ether

15.11 ELECTROPHILIC SUBSTITUTION OF PHENOLS

The phenolic hydroxyl group is a strong electron donor (Fig. 15.6). This increases the pi electron density of the ring making aromatic electrophilic substitution comparatively easy. Substitution takes place mainly at the ortho and para positions to the hydroxyl group and is thought to occur by the normal aromatic electrophilic substitution mechanisms (12.6.1c). Nucleophilic substitution of the benzene ring of phenols does not occur.

Bromination is used in the 1993 British Pharmacopoeia as the basis of the assay of some weakly acidic phenols, for example, the anhelmintic, hexylresorcinol. The phenol is reacted with an excess of 'bromine solution'. The excess bromine is treated with potassium iodide to liberate an equivalent quantity of iodine which is titrated with sodium thiosulphate. A blank titration is carried out and the difference between the blank and assay titres is the amount of sodium thiosulphate that is equivalent to the concentration of the phenol (11.6.2).

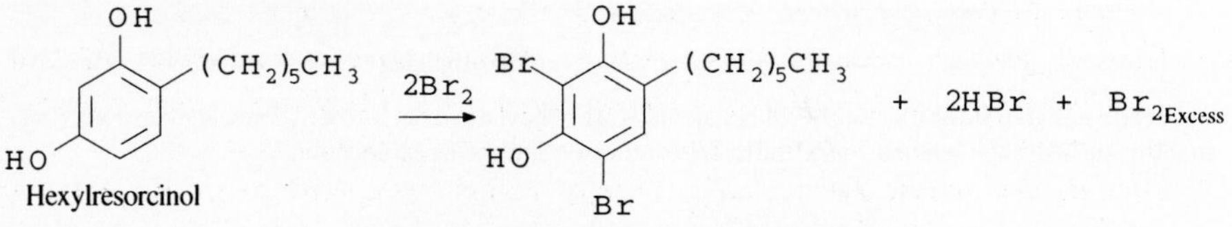

2,4-Dibromo-1,3-dihydroxy-6-hexylbenzene

$$Br_2 + 2KI \longrightarrow 2KBr + I_2$$
Excess

$$I_2 + 2Na_2S_2O_3 \longrightarrow 2NaI + Na_2S_4O_6$$

Therefore, for the bromine that reacts with the hexylresorcinol:

$$1C_{12}H_{18}O_2 \equiv 2Br_2 \equiv 2I_2 \equiv 4Na_2SO_3$$

Bromine solutions are unstable and so the bromine is generated *in situ* from a mixture of potassium bromide and bromate. This mixture liberates bromine when acidified.

$$KBrO_3 + 5KBr + 3H_2SO_4 \longrightarrow 3Br_2 + 3H_2O + 3K_2SO_4$$

15.12 MISCELLANEOUS REACTIONS

15.12.1 Complex formation

Most phenols form coloured complexes with acidic aqueous iron III chloride solution. Colours include green which is often transient, violet and blue. The chemistry of the reaction is not completely understood. However, the reaction is used as the basis of identification and limit tests, a test for phenols in systematic organic analysis and colorimetric assays of phenolic compounds. However, not all phenols will give a coloured complex. 2,4,6-Trichlorophenol, 2-nitrophenol and 2,4,6-tribromophenol amongst others do not form coloured complexes.

15.12.2 Phenols as reducing agents

Phenols can act as reducing agents. For example, they produce gray-black precipitates with Tollen's reagent.

15.13 WHAT YOU NEED TO KNOW

(1) Alcohols and phenols are toxic and in addition phenols are corrosive.

(2) *Alcohols and phenols are compounds whose molecular structures contain the hydroxyl functional group: -O-H. Phenols have the hydroxyl group directly bonded to an aromatic system. The hydroxyl group of alcohols is directly bonded to a saturated carbon atom.*

(3) *Alcohols are classified as primary, secondary and tertiary according to the nature of the carbon skeleton to which the hydroxyl group is bonded.*
         primary, RCH$_2$OH;   secondary, R$_2$CHOH;   tertiary, R$_3$COH.

(4) *The hydroxyl group of alcohols and phenols usually exhibit strong absorption bands in the 3300-3600 cm^{-1} region of the infra-red. Strong hydrogen bonding will lower and broaden the band.*

(5) *The hydroxyl group proton of alcohols and phenols has one ^{1}H n.m.r. signal in the δ1.0 to 11.0 region which is reduced or disappears in a D$_2$O shake.*

(6) *Phenols, primary, secondary and tertiary alcohols have many similar but also a significant number of different chemical properties. These differences are due to the nature of the structure to which the hydroxyl group is attached.*

(7) *Phenols are acidic. Their acidic strength varies from weak to strong. Electron-withdrawing substituents increase the acid strength of phenols whilst electron donors decrease their acid strength relative to phenol.*

(8) *Phenols exhibit many of the characteristic properties of acidic substances. They form salts, acidic phenols liberate carbon dioxide from some carbonates and hydrogen carbonates and can change the colours of acid-base indicators.*

(9) *Strongly acidic phenols may be assayed by titration with sodium hydroxide.*

$$ArOH + NaOH \longrightarrow ArONa + H_2O$$

(10) Picric acid is used to prepare picrates which can be used as derivatives in qualitative organic analysis and identification tests.

(11) Alcohols are almost non-acidic.

(12) *Both alcohols and phenols can act as weak nucleophiles in substitution and displacement reactions. In substitution reactions, alcohols and phenols are too weak to react directly and so have to be converted to their alcoholate anion prior to reaction.*

Substitution $\quad ROH \xrightarrow{Na} RO^- \xrightarrow{R'X} ROR' \quad ArOH \xrightarrow{Na} ArO^- \xrightarrow{R'X} ArOR'$

Nucleophilic displacement $\quad ROH \xrightarrow{R_1COZ} R_1COOR \quad ArOH \xrightarrow{R_1COZ} R_1COOAr$

(13) *Alcohols can also act as nucleophiles in addition reactions. The reaction requires dry conditions and an acid catalyst.*

$$ROH \xrightarrow{\ \overset{|}{C}=O \big/ H^+\ } \overset{|}{\underset{|}{C}}\!\!\begin{array}{c} OH \\ OR \end{array} \xrightarrow{\ ROH \big/ H^+\ } \overset{|}{\underset{|}{C}}\!\!\begin{array}{c} OR \\ OR \end{array}$$

(14) Alcohols may be assayed by a volumetric method based on their nucleophilic displacement reaction with ethanoic anhydride.

(15) Oxidation of primary alcohols usually yield aldehydes which are oxidised further to the corresponding acid. Secondary alcohols are oxidised to ketones but tertiary alcohols are not easily oxidised.

$$RCH_2OH \longrightarrow \underset{\text{Aldehyde}}{RCHO} \qquad R_2CHOH \longrightarrow \underset{\text{Ketone}}{RCOR}$$

(16) Phenols are usually oxidised to quinones:

$$ArOH \longrightarrow \quad \text{or}$$

where R represents the substituents of the appropriate benzene ring.

(17) Alcohols are protonated under acid conditions to form a hydroxonium ion which can decompose to the corresponding carbocation. Both the hydroxonium ion and the carbocation can undergo Saytzeff elimination to the corresponding alkene by E1 and E2 mechanisms respectively. Alternatively they may react with a nucleophile.

Elimination

$$ROH \overset{H^+}{\nearrow} R\overset{+}{O}H_2 \longrightarrow R^+$$
$$\downarrow E2 \qquad \downarrow E1$$
$$\text{Alkene} \qquad \text{Alkene}$$

Reaction with a nucleophile

$$ROH \overset{H^+}{\nearrow} R\overset{+}{O}H_2 \longrightarrow R^+$$
$$\downarrow Nu \qquad \downarrow Nu$$
$$RNu \qquad RNu$$

(18) The carbocations produced from an alcohol under acid conditions can rearrange to a more stable carbocation which can either undergo elimination to form an alkene or react with a nucleophile.

(19) The benzene ring of phenols will undergo electrophilic substitution. Substitution is very much easier than for the corresponding benzene reactions.

(20) Weakly acidic phenols may be assayed by a volumetric back titration method based on the bromination of the benzene ring.

(21) Aqueous iron III chloride is used to test for phenols.

(22) There is no specific chemical test for an alcohol other than derivative formation. IR and ^1H n.m.r. spectroscopy are used to identify alcoholic hydroxyl groups.

15.14 QUESTIONS

(1) Classify the hydroxyl groups in the compounds given as either primary, secondary and tertiary or phenols.

(a)   (b)   (c)   (d)   (e)   (f)

(1) Cont.

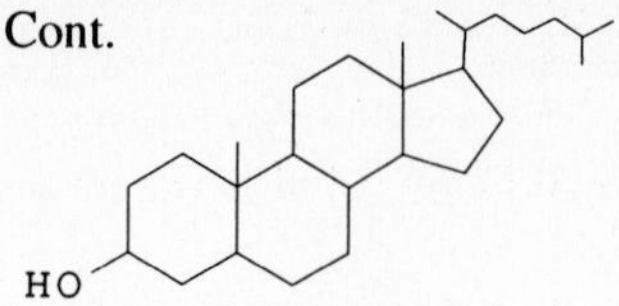

(g) 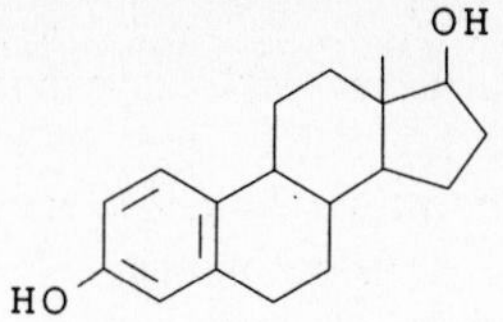   (h) 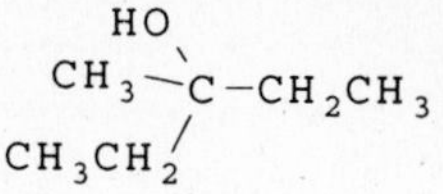   (i)

(2) Which of the following general properties are most likely to be exhibited by (a) primary alcohols, (b) secondary alcohols, (c) tertiary alcohols and (d) phenols? Record the letter for the property against the letter for the compound.

A, Undergo nucleophilic substitution; B, act as a nucleophiles; C, undergo electrophilic substitution; D, act as electrophiles; E, are acids; F, are bases; G, can be oxidised; H, act as oxidising agents; I, can be reduced; J, act as reducing agents; K, can form aliphatic carbonium ions; L, can form aliphatic carboanions.

(3) Predict the relative acid strengths of 4-nitrophenol, 4-methylphenol and 2,4,6-trinitrophenol. Give the reasoning behind your prediction.

(4)

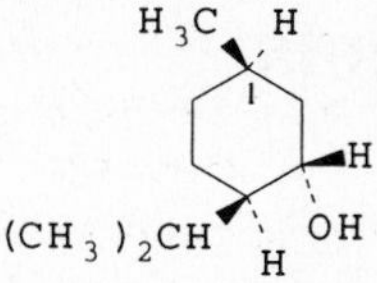

(A)

Compound A is menthol. (a) Specify the type of hydroxyl group found in A. (b) What is the significance of ⸗ and ➤ in the structure drawn for A? (c) Draw the two major conformations that the ring could theoretically assume. (d) Determine whether carbon 1 has an R or S configuration. (e) Predict by means of chemical equations the possible courses of the reaction of A with each of the following reagents: (i) pyruvic acid ($CH_3COCOOH$); (ii) an oxidising agent and (iii) hot concentrated phosphoric acid. (f) Suggest a feasible electronic mechanism for reaction e (iii).

(5)

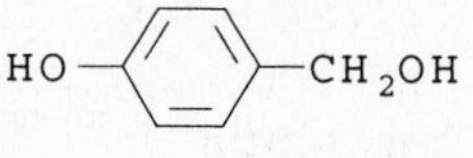

(B)

Compound B is thought to have some pharmacological activity. (a) Classify the hydroxyl groups in B and give one general pharmacological action that could be expected from this compound. (b) Comment on the relative solubilities of B in water, dilute aqueous acid and alkali solutions. (c) Outline chemical reactions that could be made the basis of two identification tests for B. Give equations where possible. (d) Suggest, by means of equations, reactions that might be made the basis of an assay procedure for B. (e) Show, by means of equations, how B might be expected to react with each of the following: (i) sodium hydroxide, (ii) ethanoyl chloride and (iii) ethanoic acid.

16

Thiols and thiophenols

16.1 INTRODUCTION

Thiols and thiophenols are functional groups that have the general structural formula SH. This functional group is attached to a saturated carbon in thiols (RSH) and a ring carbon of an aromatic system in thiophenols (ArSH). They are the sulphur analogues of alcohols and phenols. Thiol and thiophenol functional groups play an important role in the activities of a number of biologically important compounds and drugs (Fig. 16.1).

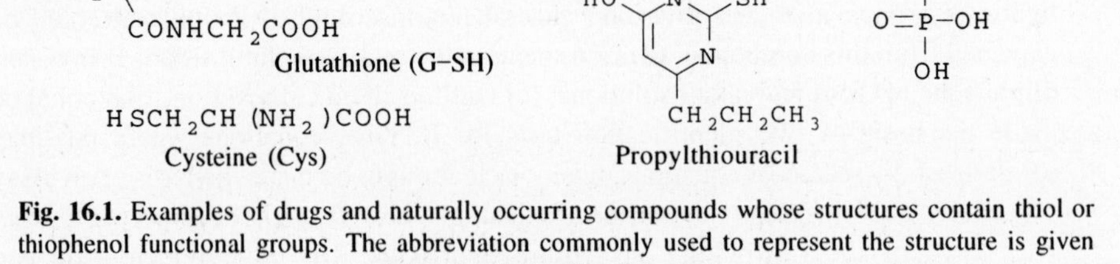

$$HS(CH_2)_2NHCO(CH_2)_2NHCOCH(OH)CCH_2OPOPOCH_2$$

Co-enzyme A (HS–CoA)

Glutathione (G–SH)

Cysteine (Cys)

Propylthiouracil

Fig. 16.1. Examples of drugs and naturally occurring compounds whose structures contain thiol or thiophenol functional groups. The abbreviation commonly used to represent the structure is given in the brackets after the name.

All thiols and thiophenols are toxic and in addition thiophenols are corrosive. As a general rule, when using thiols and thiophenols, gloves should be used and all compounds should be handled in a fume cupboard.

16.2   NOMENCLATURE

16.2.1 Thiols

Compounds whose structures contain thiol functional groups are also known as thioalcohols and mercaptans. **Mercaptan** or **thiol** are used as suffixes whilst **sulphydryl** and **mercapto** are used as prefixes in names to indicate the presence of a thiol group in a molecule.

$$CH_3-SH \hspace{5cm} HS-CH_2COOH$$

Methanethiol or methyl mercaptan   Mercaptoethanoic acid or sulphydrylethanoic acid

The trivial name *thioglycollic acid* is also used for mercaptoethanoic acid.

16.2.2 Thiophenols

The presence of a thiophenol group is indicated by the use of thiophenol as the parent name of the compound, for example:

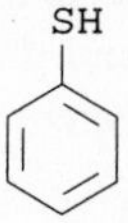

Thiophenol 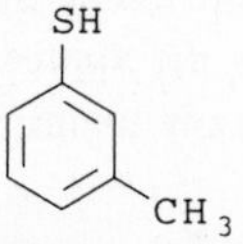   3-Methylthiophenol

The trivial name *m-thiocresol* is also used for 3-methylthiophenol.

16.3   PHYSICAL PROPERTIES

Simple thiols and thiophenols are vile smelling colourless or pale yellow gases, liquids and solids. The solids do not smell as strongly as the gases and liquids. The liquid sprayed by agitated skunks contains butanethiol ($CH_3CH_2CH_2CH_2SH$).

Thiols and thiophenols are less soluble in water than the corresponding alcohols and phenols because sulphur does not hydrogen-bond as well as oxygen. The larger atomic radius of sulphur, 0.104 nm (oxygen 0.066 nm), means that the lone pairs of sulphur are more diffused and not so effective in acting as the negative end of a hydrogen bond.

16.4   GENERAL CHEMICAL PROPERTIES

Thiols and thiophenols behave in a similar fashion to alcohols and phenols. Their general reactivity varies in degree rather than kind. In general they are stronger acids, better nucleophiles and more easily oxidised than the corresponding alcohols and phenols.

16.5   ACIDIC NATURE

Thiols and thiophenols are stronger acids than the corresponding alcohols and phenols because the S–H bond (20.93 kJ mol^{-1}) is weaker than the O–H bond (26.36 kJ mol^{-1}). This is attributed to a poorer orbital overlap in the case of the S–H bond because of the increased size of the orbitals of sulphur atom (Fig. 16.2).

The acidic strengths of thiols and thiophenols are recorded as pK_a values (6.13). These values vary with the solvent and temperature. Consequently, pK_a values are quoted together with the temperature and solvent. If no solvent or temperature is listed, it is usually safe to assume that the solvent is water and the temperature is 25°C.

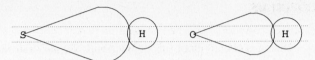

Fig. 16.2. An approximately scale representation of the orbital overlap in the O-H and S-H bonds.

The relative acidic strengths (pK_a values) of thiols and thiophenols may be predicted in the same way as those of phenols (15.5 and 15.6). The predictions are limited to compounds of the same size, shape and type. Electron acceptor groups tend to increase acidity whilst electron donors reduce acid strength. The accuracy of the predictions is limited and there are many exceptions due to factors the theory does not consider.

Thiols and thiophenols behave as typical acids. They form salts with reactive metals, hydroxides, oxides, hydrogen carbonates, carbonates and the salts of heavy metals to form the mercaptides (Fig. 16.3). The alkali metal mercaptides are ionic compounds that are decomposed by water to the appropriate thiol or thiophenol and the corresponding alkali metal hydroxide. The heavy metal mercaptides tend to be covalent compounds.

$$2RSH + 2Na \longrightarrow 2RS^- Na^+ + H_2$$

$$2RSH + 2NaOH \longrightarrow 2RS^- Na^+ + H_2O$$

$$2RSH + (CH_3COO)_2Pb \longrightarrow (RS)_2Pb + 2CH_3COOH$$

$$2RSH + HgO \longrightarrow (RS)_2Hg + 2CH_3COOH$$

Fig. 16.3. General reactions of thiols and thiophenols. RSH may be either a thiol or a thiophenol.

The heavy metal poison antidote dimercaprol (I) is reported to act by forming cyclic dimercaptides which mask the toxic action of the metal and act as a carrier for its excretion. 2,3-Dimercaptosuccinic acid (II) is believed to act in a similar manner. However, unlike dimercaprol it penetrates the blood-brain barrier and removes methylmercury from the brain. Furthermore, 2,3-dimercaptosuccinic acid may be administered orally whereas dimercaprol is administered intramuscularly.

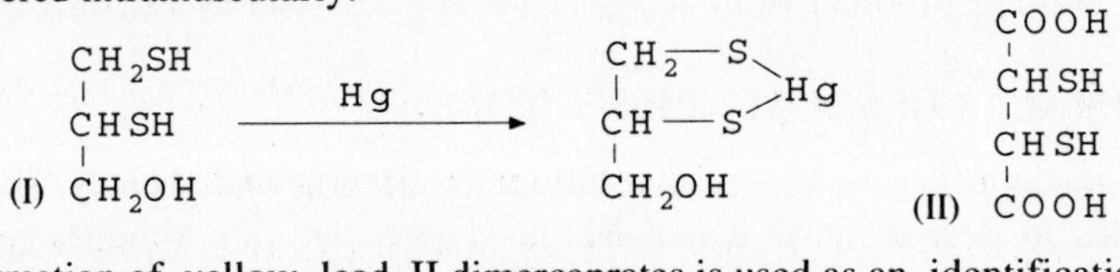

$$
\begin{array}{ccc}
\text{CH}_2\text{SH} & & \text{CH}_2\!\!-\!\!\text{S} \\
| & \xrightarrow{\text{Hg}} & | \quad\quad \text{>Hg} \\
\text{CHSH} & & \text{CH}\!\!-\!\!\text{S} \\
| & & | \\
\text{(I) CH}_2\text{OH} & & \text{CH}_2\text{OH}
\end{array}
\qquad
\begin{array}{c}
\text{COOH} \\
| \\
\text{CHSH} \\
| \\
\text{CHSH} \\
| \\
\text{(II) COOH}
\end{array}
$$

The formation of yellow lead II dimercaprates is used as an identification test for the dimercaprol in the European Pharmacopoeia.

The antithyroid agent, propylthiouracil (III) which suppresses thyroid hormone production can be assayed by titration with mercury II ethanoate after treatment with sodium hydroxide. The indicator for the titration is a freshly prepared solution of diphenylcarbazone which forms a rose-violet mercury-diphenylcarbazone complex at the end point.

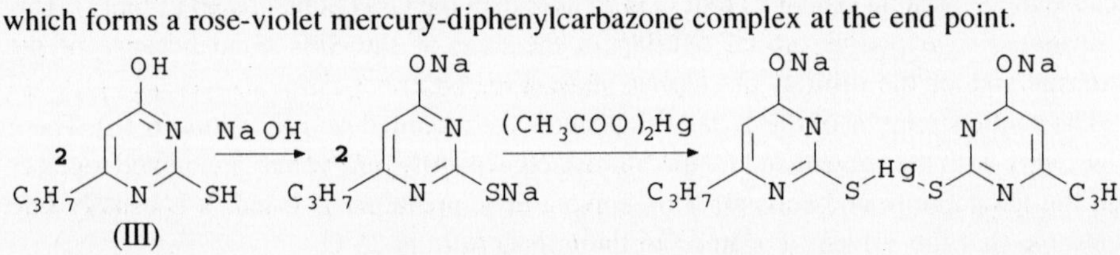

16.6 NUCLEOPHILES

Thiols and thiophenols react in a similar fashion to alcohols and phenols. Thiols and thiophenols are normally better nucleophiles than the corresponding hydroxyl compounds because sulphur has a lower electronegativity than oxygen. This means that the lone pairs of sulphur are not held as strongly as those of oxygen and so the sulphur is more willing to release its lone pairs to form bonds with other atoms.

Thiols and thiophenols act as nucleophiles in substitution, displacement and addition reactions. The mechanisms of these reactions are believed to be similar to those of the corresponding reactions of alcohols and phenols.

16.6.1 Nucleophilic substitution

Thiols and thiophenols, in the form of their thiolate anions, react with alkyl halides and other suitable compounds. Substitution is believed to occur by either S_N1, S_N2 or S_N2Ar mechanisms as appropriate, the rate of reaction being considerably faster than that of the corresponding reactions of alcohols and phenols.

Ethanethiol Ethyl methyl sulphide

Thiophenol Phenyl methyl sulphide

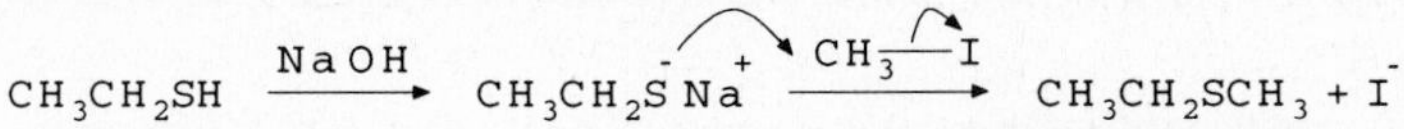

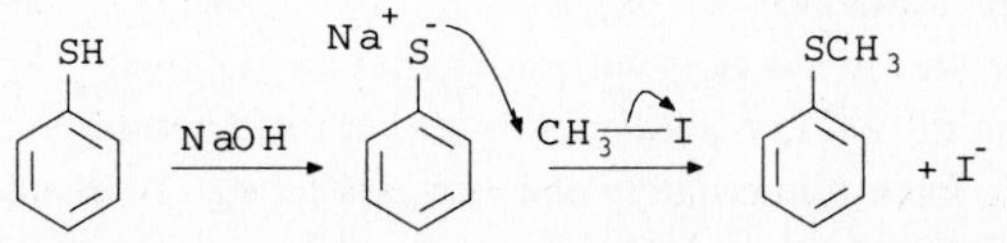

2,4-Dinitrochlorobenzene is used to form derivatives of thiols and thiophenols in identification tests and organic analysis.

Thiophenol 2,4-Dinitrochlorophenyl phenyl sulphide, mp 121°

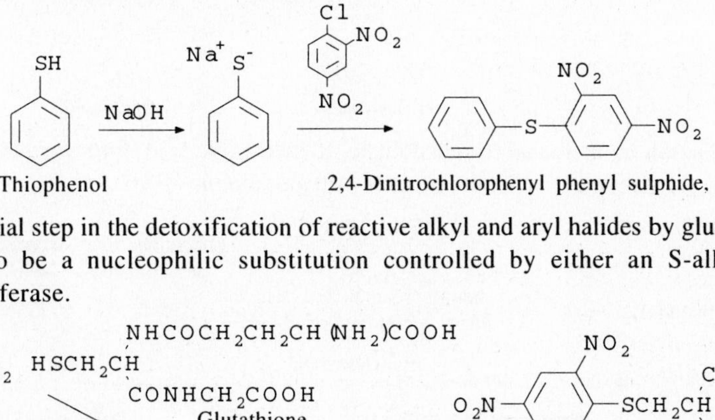

The initial step in the detoxification of reactive alkyl and aryl halides by glutathione is believed to be a nucleophilic substitution controlled by either an S-alkyl or an S-aryltransferase.

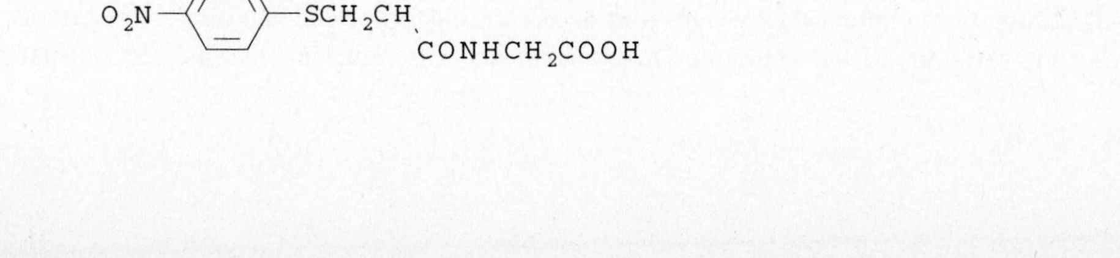

16.6.2 Nucleophilic displacement

The displacement reactions of thiols and thiophenols are similar to those of alcohols and phenols (15.7.2). A notable exception is the reaction of thiols with carboxylic acids. The equilibrium position of this reaction lies well to the left and so it is not usually possible to prepare alkyl thioesters using this reaction. However, S-thioesters may be prepared by reaction of the thiol or thiophenol with the corresponding acid halide or anhydride.

$$CH_3CH_3SH + CH_3COCl \longrightarrow CH_3COSCH_2CH_3 + HCl$$
$$\text{Ethyl ethanethiolate}$$

Ethanoic anhydride, benzoyl chloride and 3,5-dinitrobenzoyl chloride have been used to prepare derivatives of thiols and thiophenols in organic analysis and identification tests as they frequently form crystalline thioesters with sharp melting points.

In biological systems the formation of acetylcoenzyme A, which is the key substance in fatty acid and other biogenetic pathways, involves a nucleophilic displacement reaction between ethanoic acid and coenzyme A. The reaction is catalysed by acetyl coenzyme A synthase which requires magnesium ions and ATP as co-enzymes. The overall equation for the process is:

$$CH_3COOH + ATP + CoASH \xrightarrow{Mg^{2+}} CH_3COSCoA + AMP + H_2O$$

16.6.2 Nucleophilic addition

Thiols react in a similar way to alcohols with aldehydes and ketones (15.7.3). They undergo acid catalysed nucleophilic addition in dry conditions to form hemimercaptals which usually react with the thiol to form immediately the corresponding mercaptals.

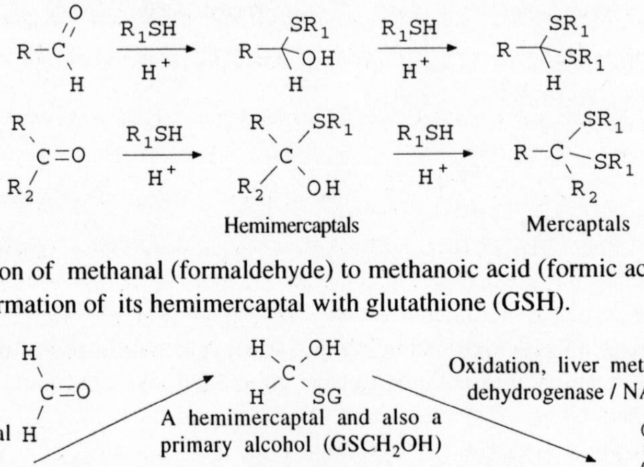

Hemimercaptals   Mercaptals

The oxidation of methanal (formaldehyde) to methanoic acid (formic acid) in the liver involves the formation of its hemimercaptal with glutathione (GSH).

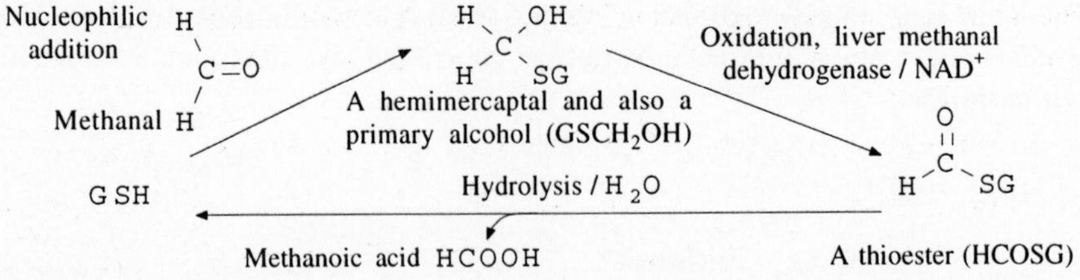

Hemimercaptals and mercaptals are stable to aqueous solutions of alkali but usually hydrolyse in dilute aqueous acid solutions. The rate of hydrolysis will vary from very slow to very rapid depending on the structure and type of mercaptal. Consequently, compounds whose structures contain thiol groups may need careful storage and handling. For example, aqueous solutions of the antibiotics *Lincomycin* and *Clindamycin* must not be sterilised by autoclaving because heating increases their rate of hydrolysis (6.6) to inactive compounds to an unacceptable level. It should be noted that the products of this hydrolysis can exhibit mutarotation (27.3.2). This is represented in structure of the products by the use of (H,OH).

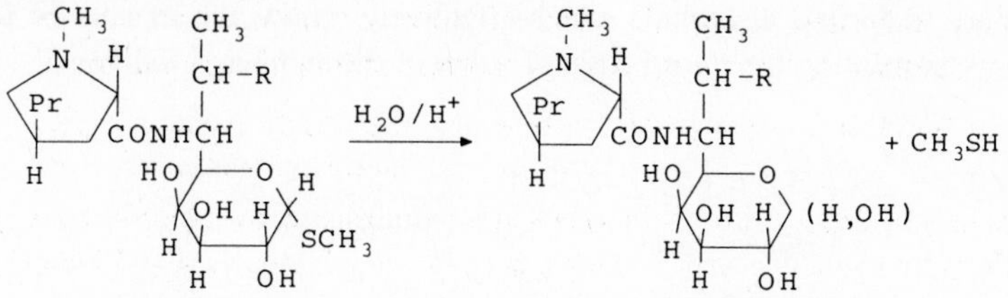

R=Cl, Clindamycin; R=OH, Lincomycin; Pr=$CH_3CH_2CH_2$-.

16.7   OXIDATION

Thiols and thiophenols are easily oxidised to the corresponding disulphide by many oxidising agents including air.

$$2R\text{-}SH \xrightarrow{\text{Oxidising agents}} R\text{-}S\text{-}S\text{-}R \quad \text{Dialkyl disulphide}$$

The ease of oxidation is in the general order:

thiophenols > primary alkylthiols > secondary alkylthiols > tertiary alkylthiols

Easy atmospheric oxidation of thiophenols and primary alkylthiols means that compounds whose structures contain these functional groups may need to be protected from oxidation during storage and in formulation. Atmospheric oxidation is catalysed by Fe^{3+} and Cu^{2+} amongst other transition metals. Thiophenols and thiols are particularly susceptible to oxidation in alkaline solution.

Stronger oxidising agents, such as hot permanganate or hot concentrated nitric acid, yield the appropriate sulphonic acid.

$$R\text{-}SH \xrightarrow[\text{oxidising conditions}]{\text{Stronger}} R\text{-}SO_3H \quad \text{Aryl or alkyl sulphonic acid}$$

Oxidation by iodine is used in the 1993 British Pharmacopoeia to assay the heavy metal poison antidote, Dimercaprol. The Dimercaprol is dissolved in 0.1 M hydrochloric acid and oxidised with a known volume of 0.05 M iodine solution. The excess iodine is determined by back titration with sodium thiosulphate. A blank titration is necessary in this assay to allow for any loss of iodine by evaporation during the titration.

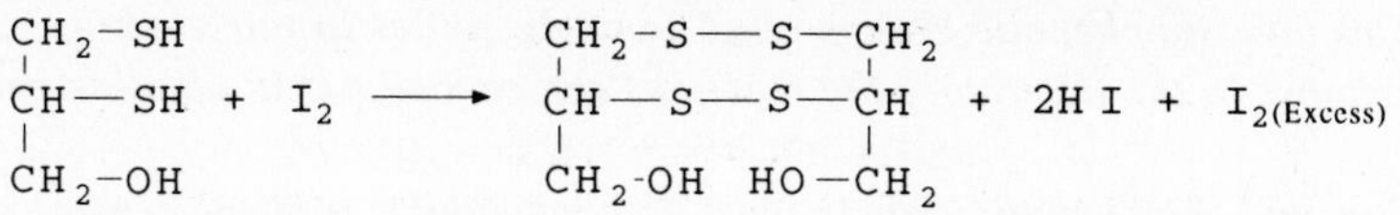

$$2Na_2S_2O_3 + I_{2(Excess)} \longrightarrow 2Na_2S_4O_6 + 2NaI$$

The mechanism of iodine oxidation is thought to involve a sulphenyl iodide intermediate.

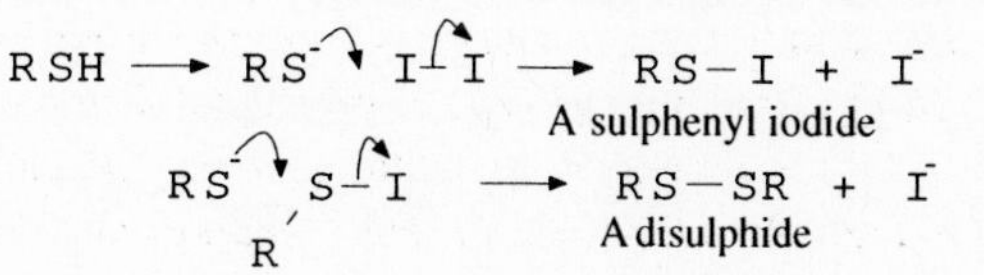

A sulphenyl iodide intermediate is thought by some workers to be involved in the conversion of tyrosine to 3,5-diiodotyrosine in the operation of the thyroid. It has been suggested that drugs such as *Methimazole* (not used in the UK), *Carbimazole* and *Propylthiouracil* that are used to control thyroid action react with the sulphenyl intermediate to form a disulphide which effectively retards the transfer of iodine to the tyrosine by preventing the reformation of some of the sulphenyl iodide.

$HO-$⟨⟩$-CH_2CH(NH_2)COOH$

3,5-Diiodotyrosine

Thyroglobulin — SH I_2

Thyroid peroxidase

— SI I^-

$HO-$⟨⟩$-CH_2CH(NH_2)COOH$

Tyrosine

Drug action retards the iodination by preventing the regeneration of the thyroglobulin

Methimazole

Key: ∿∿∿∿∿
represents a thyroglobulin residue

16.8 WHAT YOU NEED TO KNOW

(1) Thiol and thiophenol functional groups have the structure SH. In thiols (RSH) this functional group if bound to a saturated carbon atom whilst in thiophenols (ArSH) it is attached directly to an aromatic system.

(2) Thiols and thiophenolic groups do not readily hydrogen bond.

(3) Thiols and thiophenolic groups do not improve the water solubility of compounds.

(4) Thiols and thiophenolic groups are usually more reactive than alcohols and phenols.

(5) Thiols and thiophenolic groups are more acidic than the corresponding alcohols and phenols because the S-H bond is weaker than the O-H bond.

(6) The acid strengths of thiols and thiophenols are recorded as pK_a values.

(7) Thiols and thiophenols react to form salts with reactive metals, oxides, alkaline hydroxides and the salts of heavy metals. The formation of coloured heavy metal salts can be used as the basis of identification tests of thiols and thiophenols. The formation of heavy metal salts is also used as the basis of limit tests and assays.

(8) Thiols and thiophenolic groups react as nucleophiles in nucleophilic substitution reactions. In the laboratory the reactions are carried out in alkaline conditions.

$$RSH + R'X \longrightarrow R'SR + HX$$

(9) Thiols and thiophenolic groups react as nucleophiles in displacement reactions.

$$RSH + R'COZ \longrightarrow R'COSR + HZ$$

(10) Thiols and thiophenolic groups react as nucleophiles in addition reactions.

$$\text{C=O} + RSH \longrightarrow \text{C}_{OH}^{SR} \xrightarrow{RSH} \text{C}_{SR}^{SR}$$

(11) Thiols and thiophenolic groups are readily oxidised to disulphides.

$$2\text{R-SH} \longrightarrow \text{R-S-S-R}$$

(12) Thiols and thiophenols may need to be protected from oxidation during storage and on formulation.

(13) The thiol and thiophenolic groups are the reactive groups in many biologically important molecules.

16.9 QUESTIONS

(1) Draw structural formulae for each of the following: (a) ethane-1,2-dithiol, (b) mercaptoethanoic acid, (c) n-butyl mercaptan, (d) calcium ethanethiolate, (e) 2-naphthalenethiol, (f) 2-mercaptobenzoic acid and (g) 2-methylthiophenol.

(2) Predict, by means of chemical equations, the products of the following reactions:
(a) Thiophenol with 3,5-dinitrobenzoyl chloride in alkaline conditions.
(b) Methanethiol with 2,4-dinitrochlorobenzene in alkaline conditions.
(c) Propanethiol with iodine.
(d) Cyclohexanone with ethanethiol in acid conditions.
(e) Propanone and ethane-1,2-dithiol.

(3) Suggest feasible electronic mechanisms for the reactions (a), (b) and (d) in question 2.

(4) Suggest, by means of equations and notes, two identification tests and an assay procedure for ethanethiol. Ethanethiol can be prepared by the reaction of ethyl bromide with sodium hydrogen sulphide. Devise a limit test for ethyl bromide in ethanethiol. Give all relevant practical details and the chemistry underlying the procedure. (You will need to consider the chemistry of alkyl halides (Chapter 14) and thiols in order to answer this question).

17

Ethers and thioethers

17.1 INTRODUCTION

Ether is the name given to the functional group with the structure: C$-$O$-$C, while thioether or **sulphide** is the name given to its sulphur analogue: C$-$S$-$C. Both functional group names are used as class names for simple compounds that contain these structures. Simple ethers and thioethers (sulphides) have the general structures R$-$O$-$R' and R$-$S$-$R' respectively where R may or may not be the same as R' and both these structures may be an aliphatic, alicyclic, aromatic or heterocyclic residue. Ether and thioethers functional groups are frequently found in the structures of naturally occurring molecules and drugs (Fig. 17.1).

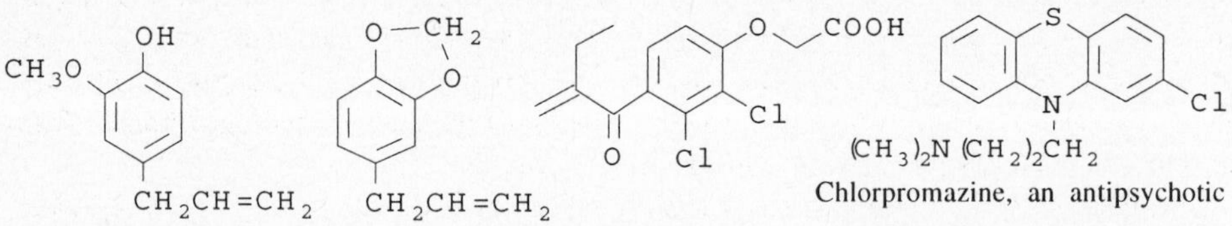

Eugenol a constituent of clove oil   Safrole, a constituent of Sassafras oil   Ethacrynic acid, a diuretic   Chlorpromazine, an antipsychotic   Methionine, an amino acid

Fig. 17.1. Examples of drugs and naturally occurring compounds whose structures contain ether and thioether functional groups.

Differences between the physical and chemical properties of ethers and thioethers with the same carbon skeletons are believed to arise because sulphur has a lower electronegativity and is significantly larger than oxygen. As a result, the bond between an atom A and a sulphur atom is less polar than O$-$A bonds and so would be expected to be less reactive. However, when A is a small atom such as carbon or hydrogen, the size discrepancy results in a poor orbital overlap (Fig. 16.2) with a subsequent increase in the reactivity of the bond.

Consequently, the reactivity of either an S–A or an O–A bond, depends to some extent on the balance struck between the opposing factors of relative size and electronegativity.

17.2   NOMENCLATURE

The prefixes **oxy** and **thio** and the suffixes **ether**, **oxide** and **sulphide** are used to indicate the presence of ether and thioether functional groups in the molecular structure of a compound. When suffixes are used, spaces are incorporated into the name to show that a structure is not a substituent of another structure. For example, the spaces in the name ethyl propyl sulphide (Fig. 17.2), show that the ethyl group is not a substituent of the propyl structure.   Cyclic ethers and thioethers where the functional group is part of the ring system have their own nomenclature systems (13.2).

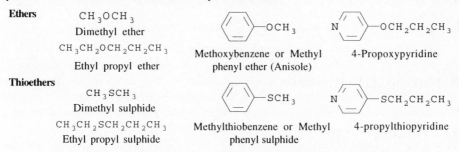

Ethers

CH_3OCH_3
Dimethyl ether

$CH_3CH_2OCH_2CH_2CH_3$
Ethyl propyl ether

Methoxybenzene or Methyl
phenyl ether (Anisole)

4-Propoxypyridine

Thioethers

CH_3SCH_3
Dimethyl sulphide

$CH_3CH_2SCH_2CH_2CH_3$
Ethyl propyl sulphide

Methylthiobenzene or Methyl
phenyl sulphide

4-propylthiopyridine

Fig. 17.2. Examples of the nomenclature of simple ethers and thioethers.

17.3   PHYSICAL   PROPERTIES

Simple ethers are sweet-smelling volatile liquids and colourless solids while simple thioethers are colourless to pale yellow liquids and solids with unpleasant smells. Simple ethers are slightly soluble in water but thioethers are insoluble. This difference in water solubility is attributed to the fact that the sulphur atoms of thioethers are not able to hydrogen bond. It is believed that the larger size of the atomic orbitals of the sulphur atoms results in the lone pairs of the sulphur atoms being too diffuse to act as the negative ends of hydrogen bonds. Conversely, the lone pairs occupying the smaller atomic orbitals of the oxygen atoms of ethers are able to act as the negative ends of hydrogen bonds. This accounts for the partial solubility of many ethers in water.

Simple ethers are good solvents for many types of organic compound. They are used to extract lipids and other naturally occurring compounds from plant and animal material. Methyl tertiary-butyl ether may be used to dissolve gallstones which can occur in either the gall bladder or duct. A main constituent of these 'stones' is cholesterol. This is readily soluble in methyl tertiary-butyl ether which remains liquid at body temperature and is introduced into the gall bladder through a catheter. Treatment only takes a few hours and there appears to be no serious side-effects. Previously, the standard treatment was surgical removal of the gall bladder.

Diethyl and divinyl ethers may be used as general anaesthetics. Their mode of action is not understood but appears to depend on obtaining a sufficient concentration of the ether in the lipid material of the central nervous system (compare with the mode of action of cyclopropane; 12.2). They are not commonly used as anaesthetics because better alternatives are available.

17.4   CHEMICAL PROPERTIES

Ethers and thioethers can be oxidised, act as bases and undergo C−O and C−S bond cleavage. Thioethers can also act as nucleophiles. Ethers are normally less reactive than thioethers and are normally regarded as being chemically inert towards most reagents. However, three- and four-membered cyclic ethers are an exception. These cyclic ethers are very reactive because of weak covalent bonding between the atoms forming their ring structures.

17.4.1 Oxidation

Ethers are usually resistant to oxidation but in air some ethers are slowly converted to peroxides.

$$\text{R-O-R'} \xrightarrow{\text{Air}} \text{R-O-O-R'} \quad \text{Peroxide}$$

The peroxides of some low molecular mass ethers are highly explosive and, in view of the volatility and inflammability of low molecular mass ethers, the presence of small concentrations of peroxides can be extremely dangerous. It is essential that older stocks of diethyl and other liquid ethers are tested for peroxides before use, especially if they are to be distilled. A simple test for peroxides in many ethers is to mix a sample of the ether with a mixture of potassium iodide and starch and allow the mixture to stand in the dark for a specified period of time. The ether is safe to use if no colour is produced after the time has elapsed. A significant concentration of peroxides would oxidise the iodide to sufficient iodine to give a colour with the starch. Starch/iodide paper can also be used for this purpose.

Thioethers are easily oxidised by a variety of oxidising agents, including air, to either the sulphoxide or the sulphone depending on the nature of the oxidising agent and reaction conditions. Milder conditions and oxidising agents usually produce the sulphoxide. For example, diethyl sulphide is oxidised by acidic hydrogen peroxide to the sulphoxide at room temperature but to the sulphone at $100^\circ C$.

$$\underset{\text{Diethylsulphoxide}}{\overset{\displaystyle O}{\overset{\displaystyle \|}{C_2H_5-S-C_2H_5}}} \xleftarrow[25^0]{H_2O_2/H^+} \underset{\text{Diethyl sulphide}}{C_2H_5-S-C_2H_5} \xrightarrow[100^0]{H_2O_2/H^+} \underset{\text{Diethyl sulphone}}{\overset{\displaystyle O}{\overset{\displaystyle \|}{\underset{\displaystyle \|}{\underset{\displaystyle O}{C_2H_5-S-C_2H_5}}}}}$$

Drugs whose structures contain thioether groups are often oxidised by the oxygen in air. This reaction is catalysed by light and can pose problems for the storage of these drugs. For example, the antipsychotic chlorpromazine is slowly oxidised in air to its inactive sulphoxide. The rate of loss of activity is increased if the drug is stored as a liquid preparation in half full or badly stoppered containers.

Chlorpromazine

Chlorpromazine-S-oxide

Thioethers are oxidised in the liver to the corresponding sulphoxides and sulphones during drug metabolism. For example, oxidation of the antipsychotic thioridazine in the liver produces the monosulphoxides which are subsequently oxidised to the disulphone.

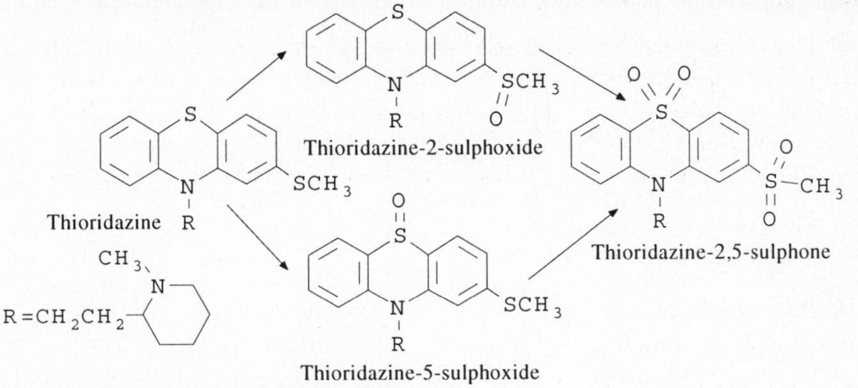

Thioridazine-2-sulphoxide

Thioridazine   R

Thioridazine-2,5-sulphone

Thioridazine-5-sulphoxide

$R = CH_2CH_2-$

17.4.2 Basic nature and bond cleavage

Both ethers and thioethers can act as weak bases. Ethers are soluble in some concentrated mineral acids forming an oxonium ion.

$$R-O-R' \quad \underset{}{\overset{H^+}{\rightleftharpoons}} \quad R-\overset{+}{\underset{|}{O}}-R' \quad \text{Oxonium ion}$$

Oxonium ion formation may occur in the acid conditions of the stomach and contribute to the acid stability of compounds such as phenoxymethylpenicillin.

Hydrogen bromide and hydrogen iodide cleave a C–O bond of the oxonium ion to yield a mixture of alcohols (or phenols) and the appropriate halide. The reaction essentially follows an S_N2 mechanism, protonation being followed by the attack of the halide.

$$R-O-R' \quad \underset{}{\overset{H^+}{\rightleftharpoons}} \quad R-\overset{+}{\underset{|}{O}}-R' \quad \overset{X^-}{\longrightarrow} \quad ROH + R'X \qquad X = Cl \text{ or } Br$$

The exact nature of the mixture of the products produced in these reactions appears to be governed by the stereochemistry of the substrate, the nucleophile attacking the less sterically hindered residue. For example, butyl isopropyl ether reacts with hydrogen iodide to form only propan-2-ol and iodobutane.

$$\underset{CH_3}{\overset{CH_3}{\diagdown}}CH-O-CH_2CH_2CH_2CH_3 \quad \overset{HI}{\longrightarrow} \quad \underset{CH_3}{\overset{CH_3}{\diagdown}}CH-OH + CH_3CH_2CH_2CH_2 I$$

Thioethers are not as basic as ethers because the S–H bond is much weaker than the corresponding O–H bond. Cleavage of the C–S bonds is difficult but can be achieved with cyanogen bromide (BrCN).

$$RSR' + Br\,CN \quad \longrightarrow \quad RSCN + R'Br$$

Dealkylation of ethers and thioethers has been found to be a route for the metabolism in mammals of compounds whose structures contain these functional groups. For example, phenacetin is dealkylated to paracetamol and *Methitural* to the corresponding thiol, with the release of methanal in both cases. However, it should be noted that not all ethers and thioethers are metabolised in this way and that dealkylation is not the only metabolic route followed by such compounds. Phenacetin is also metabolised to *p*-phenetidine which is believed to be the precursor of substances that are responsible for methaemoglobinaemia, hence the withdrawal of phenacetin from the approved drug list.

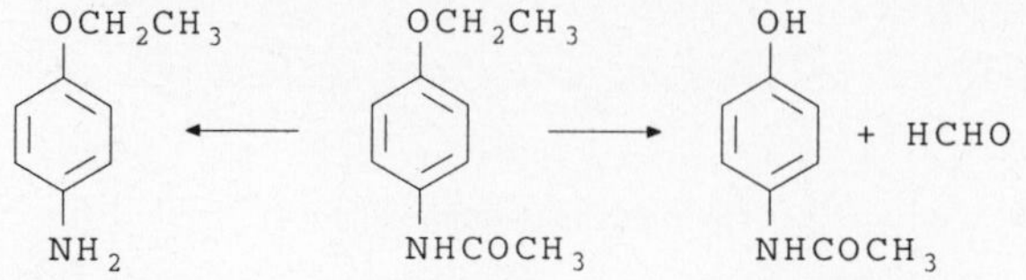

p-Phenetidine   Phenacetin   Paracetamol

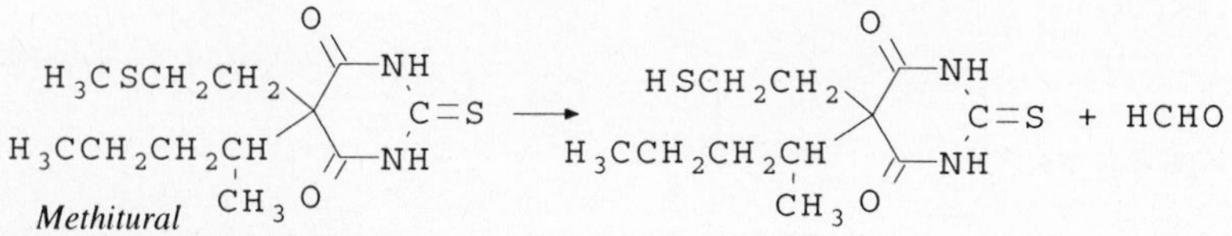

Methitural

17.4.3 Nucleophilic nature

Unlike ethers, thioethers act as nucleophiles, forming sulphonium salts with a wide variety of reagents, such as mercury salts, halogen and alkyl halides. Thioethers are believed to act as nucleophiles because the lone pairs of electrons of the sulphur are further from the nucleus than those of the oxygen. As a result, they are not so strongly held and can take part in chemical reactions more easily.

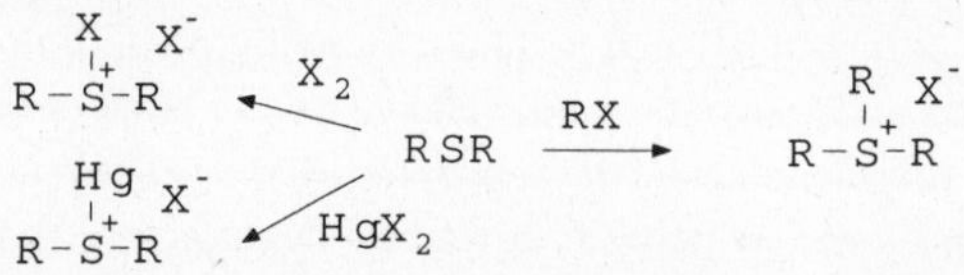

Unsymmetrical sulphonium salts are frequently unstable, readily decomposing to a mixture of all the possible symmetrical compounds. However, unsymmetrical sulphonium salts are potentially optically active because of the tetrahedral shape of their structures, the lone pair occupying one corner of the tetrahedron. For example, ethyl hydroxycarbonylmethyl methyl sulphonium chloride which is prepared from ethyl methyl sulphide has been resolved into its enantiomers.

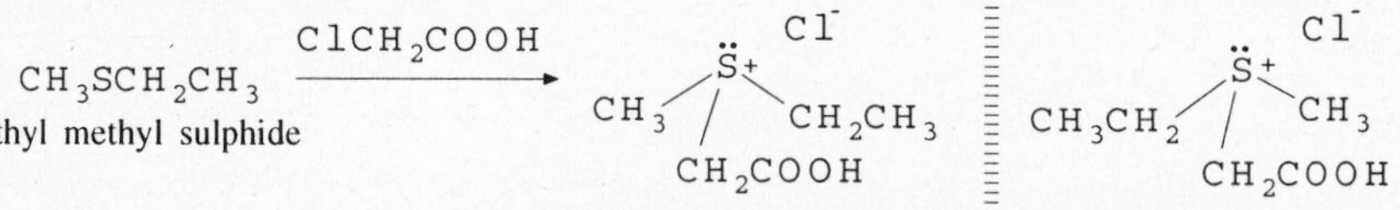

Ethyl methyl sulphide

Ethyl hydroxycarbonylmethyl methyl sulphonium chloride

Sulphonium salts are important intermediates in a number of biological processes. For example, the biogenesis of adrenaline involves the formation of the sulphonium salt S-adenosylmethionine which transfers a methyl group to noradrenaline (norepinephrin) by what is essentially an S_N2 mechanism to form adrenalin (Fig.17.3). In biological systems the various phosphate residue are the equivalent of the halide leaving group.

Sulphonium salts are acetylcholine antagonists inhibiting nervous impulse transmissions.

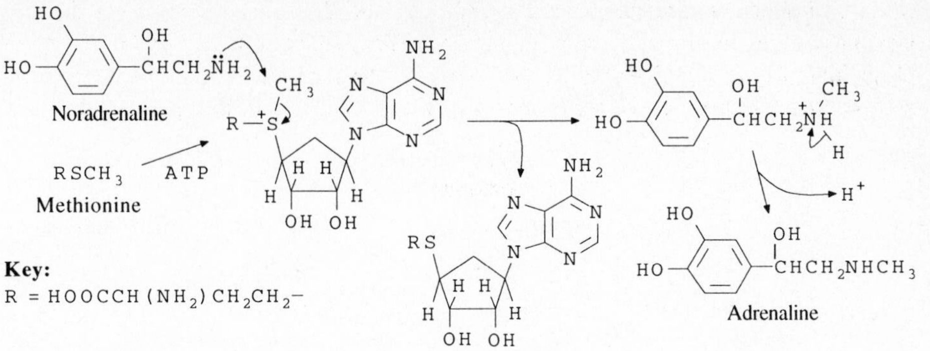

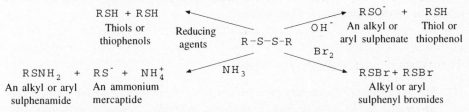

HO
    OH   NH₂
HO—⟨ ⟩—CHCH₂N̈H₂   N   N
    Noradrenaline   CH₃
                    R—⁺S   N
    RSCH₃   ATP   N
    Methionine   H H
                      H   H
                        OH OH

    Key:
    R = HOOCCH(NH₂)CH₂CH₂—

                                NH₂
                            N
        RS   N   N
                      H H
                      H   H
                        OH OH

HO
                        OH   CH₃
HO—⟨ ⟩—CHCH₂N⁺H
                                H
                            →H⁺
HO
                        OH
HO—⟨ ⟩—CHCH₂NHCH₃
                    Adrenaline

Fig. 17.3. Sulphonium salt involvement in the biosynthesis of adrenaline.

17.5   DISULPHIDES

Disulphide functional groups have the structure −S−S−. These functional groups are formed by the oxidation of thiols (16.7) and occur in the structures of many biologically important molecules, such as insulin (Fig.1.1), cystine (I) and lipoic acid (II).

$$CH_2\text{—}S\text{—}S\text{—}CH_2$$
$$CHNH_2 \qquad CHNH_2$$
(I) 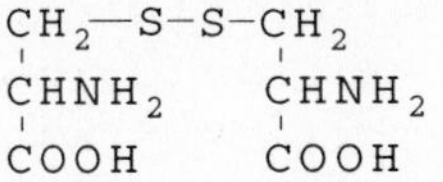 $\qquad$ COOH $\qquad$ COOH $\qquad$ (II) 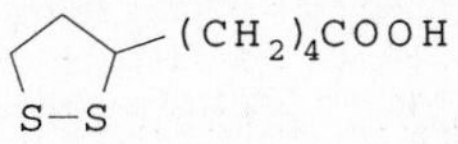 $\qquad$ S—S—$(CH_2)_4COOH$

The prefix **dithio** and the suffix **disulphide** are used in systematic nomenclature to indicate the presence of a disulphide functional group in the structure of a molecule. These prefix and suffix are used in the same way as those for ethers and thioethers. For example, $C_2H_5\text{—}S\text{—}S\text{—}CH_3$ is named as either methyldithioethane or ethyl methyl disulphide.

The main reaction of the disulphide functional group is cleavage of the −S−S− bond which can be achieved by a variety of reagents (Fig. 17.4).

RSH + RSH $\qquad\qquad\qquad\qquad$ RSO⁻ + RSH
Thiols or $\qquad$ Reducing $\qquad$ OH⁻ $\qquad$ An alkyl or $\qquad$ Thiol or
thiophenols $\qquad$ agents $\qquad$ R−S−S−R $\qquad$ aryl sulphenate $\qquad$ thiophenol
$\qquad\qquad\qquad\qquad\qquad\qquad$ Br₂

RSNH₂ + RS⁻ + NH₄⁺ $\qquad$ NH₃ $\qquad\qquad$ RSBr + RSBr
An alkyl or aryl $\quad$ An ammonium $\qquad\qquad\qquad\qquad$ Alkyl or aryl
sulphenamide $\qquad$ mercaptide $\qquad\qquad\qquad\qquad$ sulphenyl bromides

Fig. 17.4. Methods of cleaving the -S-S- bonds of disulphides. R may be aromatic or aliphatic.

Reductive disulphide cleavage is known to occur in a number of biological processes. For example, the reductive cleavage of the disulphide bond of lipoic acid is a key step in the decarboxylation of pyruvate (Fig. 17.5) and other alpha-ketoacids. The decarboxylation of pyruvate results in the formation of acetylcoenzyme A in mammals and bacteria. It is a multistage process which is controlled by a variety of enzyme systems.

The reductive cleavage of the disulphide bridges of keratin, the main protein constituent of hair, is the basis of the permanent waving. The disulphides of the cystine residues in the protein's original structure are cleaved by the use of a solution of a weak reducing agent, such as sodium thioglycollate. This allows the molecules in the hair to be physically

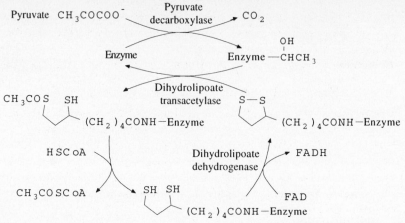

Fig. 17.5. An outline of the enzyme processes involved in decarboxylation of pyruvic acid.

manipulated into a new configuration. It is then 'set' in this new shape by using a solution of a weak oxidising agent to form new disulphide bonds in new positions in the molecule. This is possible because there are large numbers of cystine residues in keratin, the main constituent of hair.

Reductive cleavage is also the main metabolic pathway for drugs such as *Antabuse*, a drug used in the treatment of alcoholics.

$$\textit{Antabuse}\quad (C_2H_5)_2N-\overset{\overset{S}{\|}}{C}-S-S-\overset{\overset{S}{\|}}{C}-N-(C_2H_5)_2 \longrightarrow 2C_2H_5)_2N-\overset{\overset{S}{\|}}{C}-SH$$

17.6 WHAT YOU NEED TO KNOW

(1) Ethers and thioethers have the general structures: R_1-O-R_2 and R_1-S-R_2 respectively. where R_1 and R_2 can be either aliphatic, alicyclic, aromatic or heterocyclic structures which may or may not be the same.

(2) The prefix oxy and the suffix ether are used for ether functional groups in systematic nomenclature.

(3) The prefix thio and the suffix sulphide are used in systematic nomenclature for thioether functional groups

(4) The lone pairs of the oxygen of ethers can form hydrogen bonds but those of the sulphur of thioethers do not because they are too diffused in the larger orbitals of sulphur.

(5) Simple ethers are often partially soluble in water but simple thiols are usually insoluble.

(6) Ethers are good solvents for lipid and non-polar compounds.

(7)   Ethers are resistant to oxidation but some are slowly oxidised by the air to peroxides on storage. These peroxides are explosive and so samples of volatile liquid ethers should be tested for their presence before use.

*(8)   **Thioethers are readily oxidised by oxidising agents to sulphoxides and sulphones.***

$$RSOR \xleftarrow{\text{Mild oxidation}} R{-}S{-}R \xrightarrow{\text{Strong oxidation}} RSO_2R$$

*(9)   **Oxidation of thioethers to sulphoxides and sulphones occurs during the metabolism of some drugs. It can also reduce the activity of drugs during storage.***

(10) Both ethers and thioethers are very weak bases. They are soluble in concentrated strong acids but are insoluble in dilute and weak acids.

*(11) **Ethers are cleaved by hydrogen bromide and hydrogen iodide. Methyl and ethyl ethers usually form the methyl and ethyl halide whilst the other radical is converted to an alcohol or phenol.***

$$R{-}O{-}R' \xrightarrow{\text{HX}} RX + R'OH$$

*(12) **Thioethers can act as nucleophiles. They react to form sulphonium salts.***

$$R{-}S{-}R \xrightarrow{\text{AB}} R{-}\overset{+}{\underset{}{S}}{-}R \quad \overset{A\ B^-}{}$$

(13) Unsymmetrical sulphoxides and sulphonium salts are optically active.

(14) Sulphonium salts are intermediates in biological processes.

*(15) **The disulphide functional group has the structure −S−S−.***

*(16) **Disulphides are easily reduced to thiols in the laboratory and in biological systems.***

$$R{-}S{-}S{-}R \xrightarrow{\text{Reductive cleavage}} R{-}SH + R{-}SH$$

17.7 QUESTIONS

(1)   Draw the structures of the compounds given in Fig. 17.1 and ring the ether and thioether functional groups on each of the structures you have drawn.

(2)   Draw structural formula for each of the following compounds: (a) 2-methoxyphenol, (b) 2-propyl methyl ether, (c) 3-ethoxypyridine, (d) butyl ethyl sulphide, (e) 2-methyl-thiobutane, (f) ethyl propyl disulphide, and (g) 2-(2-methylpropyl) ethyl sulphide.

(3)   Suggest feasible explanations for each of the following observations:
(a) Diethyl sulphide reacts with methyl iodide but diethyl ether does not react.
(b) Diethylether is more soluble in water whilst diethyl sulphide is insoluble.

(4)   Predict the possible course of the following reactions. Illustrate the answer by means of chemical equations. (a) Ethyl propyl sulphide with cyanogen bromide, (b) diethyl sulphide with perbenzoic acid at 25 °C, (c) 4-methylthiomethylcyclohexane with methyl iodide, and (d) diethyldisulphide with sodium thioglycollate.

(5)   Show, by means of equations, how the following compounds may be metabolised by reactions involving their sulphur functional groups.

(a)

$$(CH_3)_2 N \text{—}\underset{N}{\overset{S}{\bigcirc\bigcirc}}\text{—} \overset{Cl^-}{N}(CH_3)_2$$

(b)

$$\underset{}{\overset{S-S}{\bigcirc}}\text{—}CH_2COOH$$

18

Sulphoxides and sulphones

18.1 INTRODUCTION

Sulphoxides and sulphones are functional groups with the formulae $-SO-$ and $-SO_2-$ respectively. Practical evidence suggests that the S—O bonds in these functional groups are basically dative bonds with some double bond character. This double bond character is believed to be due to a pi bond formed by the overlap of a d orbital of sulphur with a p orbital of oxygen (Fig. 2. 12). Consequently, the molecular structures of sulphoxides and sulphones are usually represented in the literature by one of the resonance forms shown in Fig. 18.1. Physical data suggest that structure I is the best representation of the actual structure of the sulphoxide group.

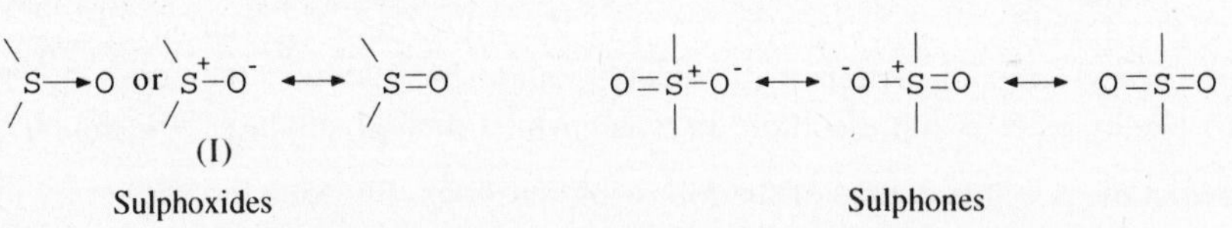

Fig. 18.1. Representations of the structures of sulphoxides and sulphones.

Sulphoxide and sulphone functional groups occur in a number of drugs (Fig. 18.2). A structure/action study on compounds with the structures similar to that of sulphonal showed that the presence of an ethyl group was essential for hypnotic activity.

18.2 NOMENCLATURE

The prefixes **sulphinyl** and **sulphonyl** and suffixes **sulphoxide** and **sulphone** are used by the IUPAC system to indicate the presence of the sulphoxide and sulphone functional

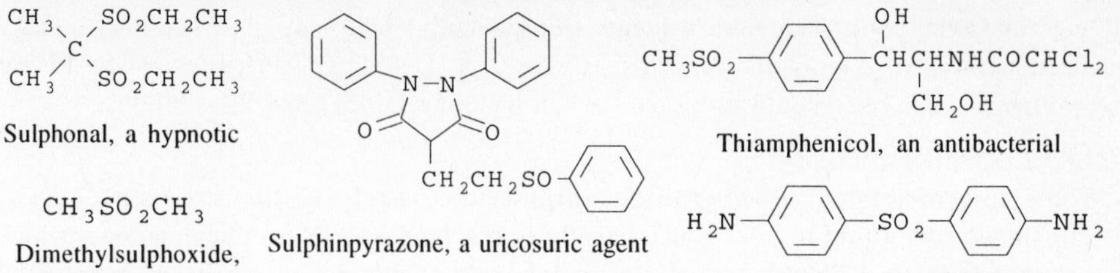

$$CH_3 \underset{CH_3}{\overset{}{C}} \underset{SO_2CH_2CH_3}{\overset{SO_2CH_2CH_3}{}}$$

Sulphonal, a hypnotic

$$CH_3SO_2CH_3$$

Dimethylsulphoxide,
a pharmaceutical aid

Sulphinpyrazone, a uricosuric agent

CH_2CH_2SO—

Thiamphenicol, an antibacterial

$$CH_3SO_2-\text{⬡}-\underset{CH_2OH}{\overset{OH}{\underset{|}{\overset{|}{CH}}}}CHNHCOCHCl_2$$

$$H_2N-\text{⬡}-SO_2-\text{⬡}-NH_2$$

Dapsone, an antileprotic

Fig 18.2. Examples of drugs whose structures contain sulphoxide and sulphone functional groups.

groups respectively. The suffixes are used as separate entities whilst the prefixes are combined with the appropriate substituent name (Fig.18.3).

$$CH_3SOCH_3$$
Dimethyl sulphoxide
Methylsulphinylmethane

$$CH_3SO_2CH_3$$
Dimethyl sulphone
Methylsulphonylmethane

$$CH_3SOC_2H_5$$
Ethyl methyl sulphoxide
Methylsulphinylethane

$$CH_3SO_2C_2H_5$$
Ethyl methyl sulphone
Methylsulphonylethane

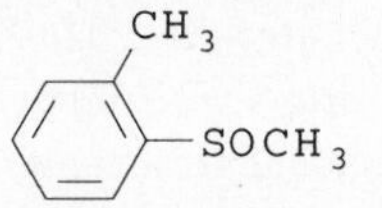

$$\text{⬡}\overset{CH_3}{}-SOCH_3$$

Methyl 2-methylphenyl sulphoxide
2-Methylsulphinylmethylbenzene

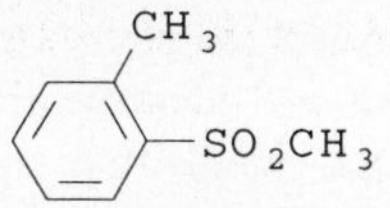

$$\text{⬡}\overset{CH_3}{}-SO_2CH_3$$

Methyl 2-methylphenyl sulphone
2-Methylsulphonylmethylbenzene

Fig 18.3. Examples of the nomenclature of sulphoxides and sulphones.

18.3 PHYSICAL PROPERTIES

Simple sulphoxides and sulphones are almost odourless colourless liquids and solids.

18.3.1 Solubility

Low molecular mass compounds are soluble in water because of hydrogen bonding between the oxygen atoms and the water molecules. Sulphur does not usually form hydrogen bonds. Aliphatic compounds are usually more water soluble than aromatic compounds. Low molecular mass compounds are hygroscopic and should be stored under anhydrous conditions.

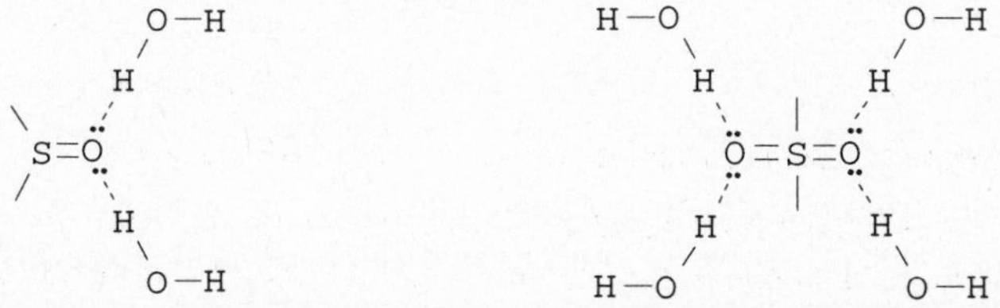

Fig 18.5. The possibilities for hydrogen bonding between water and the sulphoxide and sulphone functional groups. It is unlikely that all the hydrogen bonds shown in the diagram occur at room temperature and above.

Dimethyl sulphoxide (DMSO) is an important solvent. It owes this importance to the fact that it will dissolve both inorganic and organic compounds and is completely miscible with water and many common organic solvents. Consequently, it is a very good medium for many reactions involving inorganic reagents. However, as DMSO can instantaneously diffuse through the skin one should use gloves when handling DMSO and its solutions.

18.3.2   Optical activity

Some unsymmetrically substituted sulphoxides, such as the uricosuric agent sulphinpyrazone (used to treat gout), have been resolved into their optical isomers which suggests that the sulphur atom of the sulphoxide functional group has a tetrahedral structure.

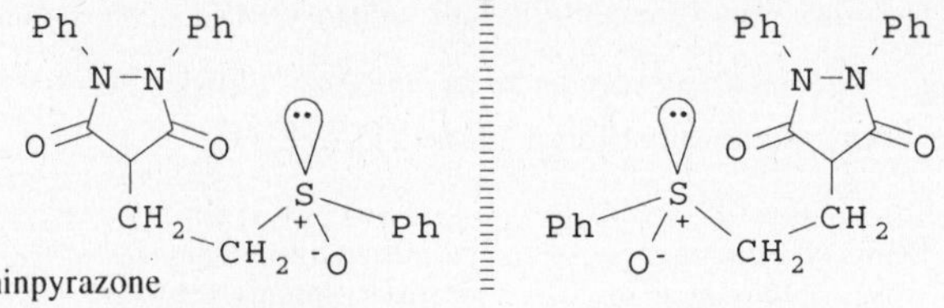

Sulphinpyrazone

18.4   CHEMICAL   PROPERTIES

Sulphoxides can act as weak bases, nucleophiles, electron acceptors and undergo oxidation and reduction. However, sulphones are relatively unreactive. This means that chemically based assay methods and identification tests for drugs containing sulphone group(s) are normally based on the chemistry of the other functional groups present in their molecules. For example, the 1993 British Pharmacopoeia assay for the antileprotic dapsone is based on the reaction of the aromatic amine groups in the molecule with sodium nitrite (24.8.2).

18.4.1   Basic nature

Sulphoxides act as weak bases forming salts by protonation of the oxygen.

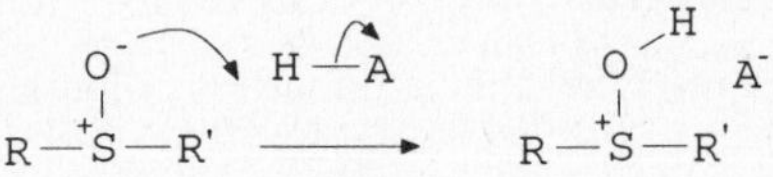

   Salt formation means that sulphoxides are more soluble in aqueous acid than water. Consequently, the excretion of sulphoxides will be helped if the tubular lumen of the kidney exhibits its usual acid pH but will be hindered if the urine should become alkaline.

   The sulphoxide group is also able to form salts and complexes with the transition metal salts. This probably accounts for the solubility of many transition metal salts in DMSO.

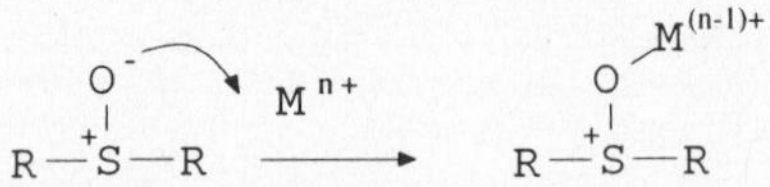

   Sulphones are weaker bases than sulphoxides. They do not protonate so readily as sulphoxides but usually dissolve in concentrated sulphuric acid. Experimental evidence suggests that a protonated species is formed in this acid solution.

18.4.2   Nucleophilic nature

A lone pair of one of the oxygen atoms acts as a nucleophilic centre reacting with labile alkyl halides to form adducts which usually decompose to form a mixture consisting of a hydrogen halide, a sulphide and an aldehyde.

Dimethyl sulphoxide   Phenylglyoxal

Sulphones are less reactive and do not react in this manner.

18.4.3 Action as an electron acceptor

Both the sulphoxide and sulphone functional groups acts as electron acceptors. In aromatic structures they are meta directing to electrophilic substitution whilst in aliphatic structures they weaken the α C–H bonds making these bonds sufficiently acidic to form carboanions in basic conditions. For example, dimethyl sulphoxide reacts with sodium hydride to form sodium methylsulphinylmethide (dimsyl sodium).

sodium methylsulphinylmethide

The sulphinyl carboanions produced by this type of reaction can act as both nucleophiles and bases. For example, dimsyl sodium is a strong base that can be used to titrate weak acids such as alcohols, amines, alkynes and ketones (via their enolic forms). Titration is carried out in a suitable solvent using triphenylmethane as indicator. At the end point of the reaction the triphenylmethane immediately reacts with the excess dimsyl sodium to form the red triphenylmethyl carboanion.

$$CH_3SO\overset{-}{C}H_2\ N\overset{+}{a} + C_2H_5OH \longrightarrow CH_3SOCH_3 + C_2H_5\overset{-}{O}\ N\overset{+}{a}$$

$$CH_3SO\overset{-}{C}H_2\ N\overset{+}{a} + Ph_3CH \longrightarrow CH_3SOCH_3 + Ph_3\overset{-}{C}\ N\overset{+}{a}$$
$$\phantom{CH_3SO\overset{-}{C}H_2\ N\overset{+}{a} + }\text{Triphenylmethane} \phantom{\longrightarrow CH_3SOCH_3 + Ph_3\overset{-}{C}\ }\text{Red}$$

Dimsyl sodium is a strong enough base to form carboanions from compounds with weakly acidic C–H bonds, for example, ketones and alkynes. These carboanions are used as intermediates in synthesis. For example, in the preparation of acetylenic carboxylic acids.

$$CH_3SO\overset{-}{C}H_2\ N\overset{+}{a} + RC\equiv CH \longrightarrow CH_3SOCH_3 + RC\equiv C^-\ N\overset{+}{a}$$

$$RC\equiv C^-\ N\overset{+}{a} + CO_2 \longrightarrow RC\equiv C-COOH$$

Dimsyl sodium also reacts as a nucleophile. For example, it acts as the nucleophile in the nucleophilic displacement reactions of esters to form acyl sulphoxides.

$$2CH_3SO\overset{-}{C}H_2\ N\overset{+}{a} + RCOOR' \longrightarrow RCO\overset{-}{C}HSOCH_3\ N\overset{+}{a} + R'\overset{-}{O}\ N\overset{+}{a} + CH_3SOCH_3$$

$$RCO\overset{-}{C}HSOCH_3\ N\overset{+}{a} \xrightarrow{\ H^+\ } RCOCH_2SOCH_3$$
$$\phantom{RCO\overset{-}{C}HSOCH_3\ N\overset{+}{a} \xrightarrow{\ H^+\ }\ }\text{Acyl sulphoxides}$$

18.4.4 Oxidation and reduction

Sulphoxides are usually oxidised in the laboratory to sulphones and reduced to sulphides.

$$RSR \xleftarrow{\text{Reduction}} RSOR \xrightarrow{\text{Oxidation}} RSO_2R$$

Sulphide   Sulphoxide   Sulphone

Oxidation and reduction of sulphide also follows these routes in some biological systems. For example, dimethyl sulphoxide is oxidised in rats, rabbits and guinea pigs to dimethyl sulphone and it is reduced in cats to dimethyl sulphide.

Sulphones are resistant to both oxidation and reduction.

18.5 WHAT YOU NEED TO KNOW

(1)   The sulphoxide and sulphone functional groups are represented by the structural formulae:

Sulphoxides   Sulphone

(2)   The prefix sulphinyl and the suffix sulphoxide are used to indicate the presence of a sulphoxide group in a structure.

(3)   The prefix sulphonyl and the suffix sulphone are used to indicate the presence of a sulphone group in a molecule.

(4)   Both the sulphoxide and sulphone functional groups can hydrogen bond with water.

(5)   Simple sulphoxides are usually soluble in water. Their solubility increases in acidic aqueous solutions but decreases in basic aqueous solutions.

(6)   Simple sulphones are water soluble.

(7)   Dimethyl sulphoxide (DMSO) is a good solvent for both inorganic and organic compounds.

(8)   Unsymmetrically substituted sulphoxides are optically active.

(9)   Sulphoxides are more reactive than sulphones.

(10) Sulphoxides can act as bases (via the oxygen), nucleophiles (via the oxygen) and electron acceptors. They may be oxidised and reduced.

(11) The carboanions produced by treating a sulphoxide with a base can act as both nucleophiles and bases.

(12) Sulphones can act as electron acceptors but are resistant to oxidation and reduction.

18.6 QUESTIONS

(1)   Draw structural formulae for each of the following compounds: (a) ethyl methyl sulphoxide, (b) ethyl 2-aminophenyl sulphone, (c) 2-ethylsulphonylpropane and (d) 3-methylsulphinylpyridine.

(2)

(A)

Compound A is the insecticide 5-[2-(octylsulphinyl)propyl]-1,3benzodioxole. (a) Predict the relative solubilities of A in (i) water, (ii) hydrochloric acid and (iii) aqueous sodium hydroxide solution and dimethylsulphoxide. (b) Deduce, giving reasons for your deduction where necessary, the products of the reactions of A with each of the following: (i) methyl iodide, (ii) hydrogen in the presence of a platinum catalyst, and (iii) perethanoic acid.

19

Aldehydes and ketones

19.1 INTRODUCTION

Aldehydes and ketones are functional groups with the formulae −CHO and −CO− respectively. Both these functional groups contain the carbonyl group (−CO−) which is planar and has the structure shown in Fig. 19.1. Simple compounds containing aldehyde and ketone functional groups are usually referred to as aldehydes and ketones.

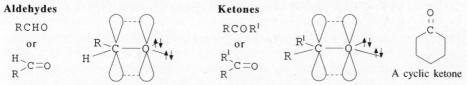

Aldehydes

RCHO

or

$$\begin{array}{c} H \\ R \end{array}\!\!>\!\!C\!=\!O$$

Ketones

RCOR^{I}

or

$$\begin{array}{c} R^{\mathrm{I}} \\ R \end{array}\!\!>\!\!C\!=\!O$$

A cyclic ketone

Fig. 19.1. The structural formulae and orbital structures of the aldehyde and ketone carbonyl groups. The carbon and oxygen atoms of the carbonyl group are sp^2 hybridised. The ketone functional group can form part of a ring structure when R and R^{I} form part of the same structure.

The carbonyl group structure forms part of the structures of many other functional groups, such as the carboxylic acid (−COOH) and ester (−COOR) functional groups. These functional groups have their own distinctive chemical properties but they also exhibit some properties that are similar to those of aldehydes and ketones. Consequently, care should be taken not to confuse the properties of the carbonyl groups of aldehydes and ketones with those of the carbonyl group when it forms part of the structure of other functional groups.

Aldehydes and ketones functional groups are found in the molecular structures of many naturally occurring compounds (Fig. 19.2). They are important functional groups in the metabolism of all living creatures and plants. Small amounts of acetone and acetoacetic acid together with 3-hydroxybutanoic acid (β-hydroxybutyric acid) which is not a ketone, are normally produced and metabolised by the body. However, sometimes when there is insufficient glucose in a cell, as for example in insulin-dependent diabetes, the production of ketones is increased (**ketogenesis**). Hence the use of the term **diabetes ketosis** as found in **hyperglycaemic coma**. Furthermore, the increased production of 3-hydroxybutanoic acid and acetoacetic acid is responsible for the severe acidosis which is sometimes seen in diabeteic ketoacidosis which may occur in untreated diabetes.

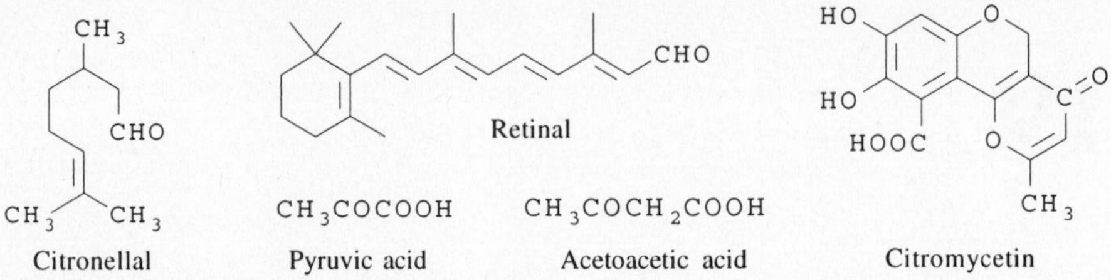

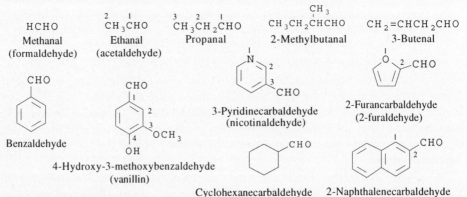

$CH_3COCOOH$ CH_3COCH_2COOH

Citronellal   Pyruvic acid   Acetoacetic acid   Citromycetin

Fig. 19.2. Examples of naturally occurring compounds containing aldehyde and ketone functional groups. Citronellal is the main constituent of citronella oil, retinal is vitamin A, citromycetin is an antibiotic, pyruvic acid and acetoacetic acids are components of the human metabolic systems.

19.2   NOMENCLATURE

19.2.1   Aldehydes

The names of simple aliphatic aldehydes are usually obtained by replacing the ending **-e** of the name of the hydrocarbon with the same number of carbon atoms by the suffix **-al** (Fig. 19.3). The carbon of the aldehyde group is always number 1 of the numbering system when the suffix -al is used. In the case of branched chains, the longest chain ending with the aldehyde group is normally taken to be the parent structure for the compound.

The suffix **aldehyde**, together with an appropriate locant, is normally used to indicate the presence of an aldehyde group which is directly bonded to an aromatic system (Fig. 19.3). If no locant is given, the aldehyde group is attached at position 1 of the aromatic system. In some simple aldehydes and more complex aliphatic and aromatic compounds the IUPAC suffix **carbaldehyde** and the CA suffix **carboxaldehyde** may be used.

HCHO CH_3CHO CH_3CH_2CHO CH_3CH_2CHCHO $CH_2=CHCH_2CHO$

Methanal   Ethanal   Propanal   2-Methylbutanal   3-Butenal
(formaldehyde)   (acetaldehyde)

3-Pyridinecarbaldehyde   2-Furancarbaldehyde
(nicotinaldehyde)   (2-furaldehyde)

Benzaldehyde

4-Hydroxy-3-methoxybenzaldehyde
(vanillin)

Cyclohexanecarbaldehyde   2-Naphthalenecarbaldehyde

Fig. 19.3. Examples of the systematic names of aldehydes. Trivial names are given in brackets.

19.2.2 Ketones

The presence of a ketone functional group is shown by the suffixes **one** and a locant indicating the position of this functional group in the molecule (Fig. 19.4). Alternatively, the prefix **oxo** together with the appropriate locant may be used. It should be remembered that the locant is often dropped if there is no possibility of ambiguity. Simple ketones are sometimes named by using the radical names of the carbonyl groups' substituents with the ending ketone. Each section of the name is separated by a gap to show that it is a separate entity.

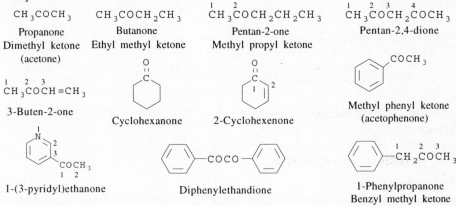

Fig.19.4. Examples of the systematic nomenclature of ketones. Trivial names are given in brackets.

19.3 TAUTOMERISM

Aldehydes and ketones with an α C–H bond exhibit a phenomenon known as **tautomerism when in solution or the liquid state.** In these states the compounds exist as **dynamic equilibrium** mixtures of two types of isomer known as the **keto** and **enol** forms of the molecule. These individual isomers are known as **tautomers**. Their existence is due to a rearrangement reaction involving the appropriate hydrogen atom.

Tautomers differ from the isomers that have been discussed in previous chapters in that tautomers do not usually have an independent existence. They are often found in the form of an equilibrium mixture, the position of equilibrium depending on the relative stabilities of the tautomers, the solvent system and the temperature (Chapter 6). Most simple aldehydes and ketones exist almost exclusively in the keto form. For example, propanone contains about 0.00025% of its enolic isomer. However, a larger quantity of the enol may be present at equilibrium when, for example, the enol structure is stabilised by conjugation and/or other factors such as hydrogen bonding (Fig. 19.5).

The formation of keto-enol equilibrium mixtures is catalysed by both acids and bases. Their presence in the form of a dynamic equilibrium means that aldehydes and ketones will

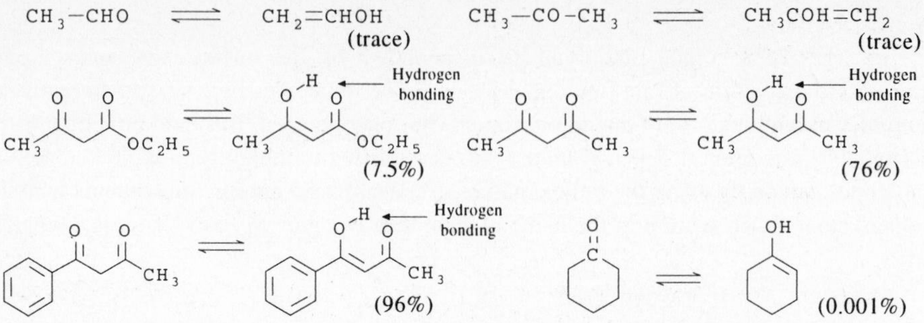

Fig. 19.5. The keto-enol tautomerism of some aldehydes and ketones. The percentages given are those for the enol present at equilibrium in ethanolic solution except for cyclohexanone where the value is for the pure compound. The percentage of enol present at equilibrium usually increases with increasing conjugation in the structure of the enol and intramolecular hydrogen bonding between the hydroxyl and the carbonyl groups (not shown for clarity).

exhibit the chemical and physical properties of the enol group as well as those of their carbonyl groups.

19.4   PHYSICAL PROPERTIES

19.4.1   Water solubility

The carbonyl group has a permanent dipole (Chapter 3) and so is able to undergo permanent dipole-dipole interactions with suitable molecules. The lone pairs of the oxygen atom are also available to form hydrogen bonds. These features explain the solubility of aldehydes and ketones in a wide variety of polar solvents such as water, ethanol and diethyl ether. However, it should be noted that water solubility rapidly decreases as the ratio of the size of the non-polar carbon-hydrogen skeleton to the polar carbonyl group(s) increases. In addition, many of the simpler aldehydes form hydrates in aqueous solution (19.8).

Hydrogen bonding involving the ketonic carbonyl groups of certain steroids is believed to be responsible for the solubility of these potentially insoluble compounds in the aqueous media of biological systems.

19.4.2 Spectra

The carbonyl groups of both aldehydes and ketones absorb in both the infra-red (Fig. 19.6) and ultra-violet regions of the electromagnetic spectrum. The presence of a significant concentration of the enol tautomer results in a broad band in the 3400 to 2850 cm^{-1} region of the infra-red. This absorption band can be used to confirm the presence of an enolic tautomer as well as the basis of an assay procedure for that tautomer.

Infra-red absorptions are used to identify of the presence of aldehyde and ketone functional groups in a molecule. However, it should be noted that other carbonyl-containing functional groups can also absorb in the same regions as aldehydes and ketones and so infra-red spectroscopy cannot be used in isolation to identify the presence of an aldehydic or ketonic carbonyl group in a molecule.

The position and absorbance of aldehydes and ketones in the ultra-violet region will depend on the nature of the carbon hydrogen skeleton of the molecule and the solvent. Compounds in which the carbonyl group is in conjugation will absorb at a longer

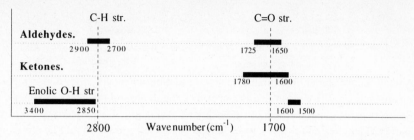

Fig 19.6. The general regions in which aldehydes and ketones absorb in the infra-red region that are important for diagnostic purposes.

wavelength than those in which the carbonyl group is not conjugated. If the conjugation is great enough the compound will absorb in the visible region and will be coloured.

The addition of acid or base will change the wavelength of absorption in the ultra-violet because it changes the position of equilibrium of the keto-enol system. Keto-enol systems normally undergo reversible bathochromic-hyperchromic shifts (10.2) when the pH of the solution is changed. This observation is used to detect the presence of keto-enol tautomeric systems. However, it should be noted that many aldehydes and ketones do not exhibit strong absorptions in aqueous media. This is because these aldehydes and ketones exist in both aqueous acidic and basic solutions as equilibrium mixtures of the hydrate and the aldehyde or ketone (19.3).

19.5   CHEMICAL PROPERTIES

The chemical properties of aldehydes and ketones may be classified for convenience into six main categories: oxidation, reduction, nucleophilic addition, condensation, aromatic substitution and miscellaneous reactions. This last category includes reactions that are not covered by the other classes.

19.6   OXIDATION

Aldehydes are readily oxidised in the laboratory by both mild and strong oxidising agents to the corresponding carboxylic acid.

$$RCHO \xrightarrow[\text{e.g. } KMnO_4/H^+]{\text{Most oxidising agents}} RCOOH$$

Ketones, are not as susceptible to oxidation as aldehydes. They usually react under more vigorous conditions and with stronger oxidising agents to form mixtures of carboxylic acids. However, in some instances a single product is produced.

Cyclopentanone $\xrightarrow{50\% \ HNO_3 \ / \ V_2O_5}$ Glutaric acid (85%)

The oxidations with Tollen's reagent and Fehling's solution are used to distinguish between simple aldehydes and ketones in organic analysis since simple ketones do not normally react with these reagents. Aldehydes yield a silver mirror with Tollen's reagent and a red-brown precipitate with Fehling's solution but there are exceptions to this rule.

$$RCHO + \underset{\text{Tollen's reagent}}{2Ag(NH_3)_2OH} \longrightarrow RCOONH_4 + 2Ag + 3NH_3 + H_2O$$

$$\underset{\text{Silver mirror}}{}$$

$$RCHO + \underset{\text{Fehling's solution}}{\underbrace{2Cu^{2+} + NaOH + H_2O}} \longrightarrow RCOONa + Cu_2O + 4H^+$$

$$\underset{\text{Red-brown solid}}{}$$

Aldehydes are susceptible to atmospheric oxidation when stored at room temperature. For example, the white deposit that is sometimes found around the necks of benzaldehyde bottles is benzoic acid formed by atmospheric oxidation. Consequently, preparations containing aldehydes and drugs with aldehyde functional groups should be stored at low temperatures and in air tight containers.

$$PhCHO \xrightarrow[\text{Slow at room temperature}]{\text{Air}} PhCOOH$$

Metabolic oxidation of aldehydes produces the corresponding carboxylic acid. It occurs in the liver and is mainly catalysed by liver aldehyde dehydrogenase. The co-enzyme for the process is NAD^+ which is converted to NADH. The mechanism of the process is not clear but probably includes the attack of a hydroxide ion in the oxidation.

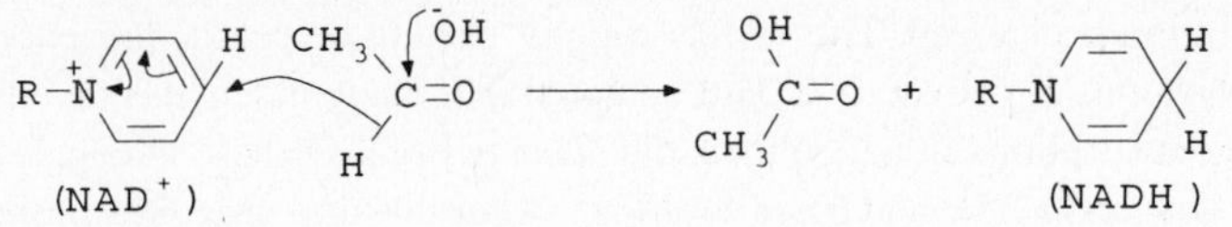

19.7 REDUCTION

Both aldehydes and ketones are readily reduced in the laboratory by a wide variety of reagents. Reduction usually produces either an alcohol or a methylene group.

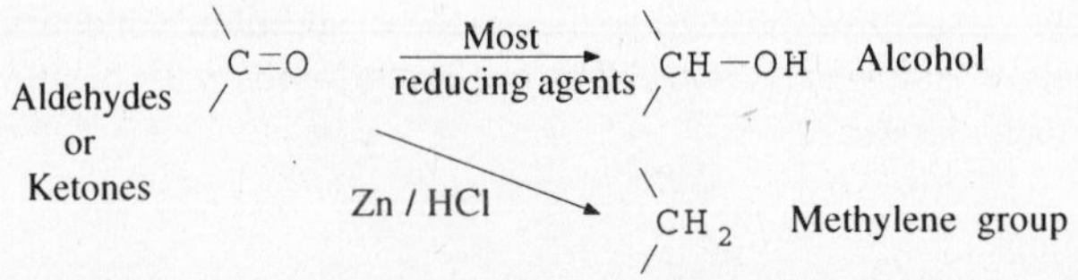

Metabolic reductions usually yield the corresponding alcohols. The reactions are usually stereospecific. The co-enzymes NADPH and NADH frequently act as sources of hydride ions in these reactions. The reductions are pH-dependent, being completed by hydrogen ions supplied from the enzyme for the process. They are in effect nucleophilic additions. The process may be summarised as follows:

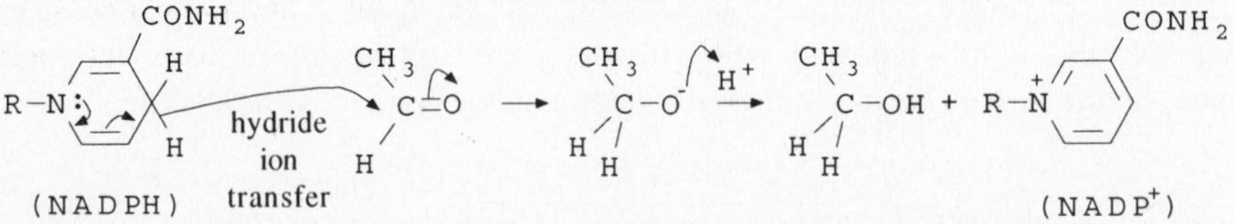

Reductions of this type play an important part in a number of metabolic processes. For example, the oestrone-oestradiol redox system acts as a co-enzyme for the transfer of hydrogen from NADPH to NAD^+ during the action of placental transhydrogenase.

Ketone-secondary alcohol redox systems are often an important feature of the action of steroids.

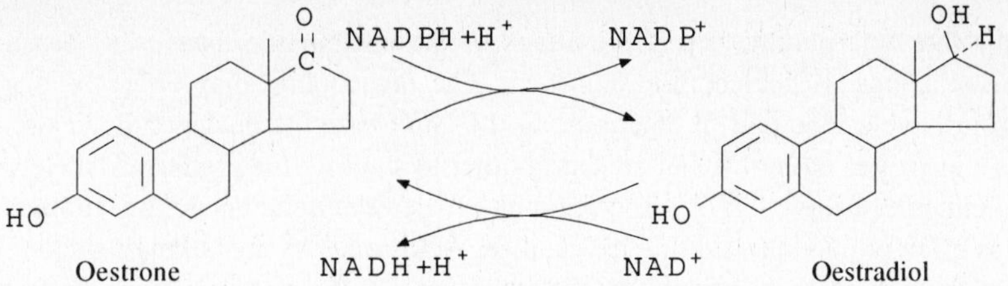

| Oestrone | NADH+H$^+$ NAD$^+$ | Oestradiol |

The carbonyl groups of carbohydrates and other compounds are reduced by NADH. For example, pyruvic acid is reduced to S-(+)-lactic acid in skeletal and heart muscle whilst D-ribose-5-phosphate is converted to D-ribitol-5-phosphate. These conversions are normally catalysed by lactate dehydrogenase.

CH$_3$COCOOH

Pyruvic acid

NADH+H$^+$ NAD$^+$

Lactate
dehydrogenase

COOH
|
HO-C-H
|
CH$_3$

S-(+)-Lactic acid

CHO
|
H-C-OH
|
H-C-OH
|
H-C-OH
|
CH$_2$O PO$_3^{2-}$

D-Ribose-5-phosphate

NADH+H$^+$ NAD$^+$

Lactate
dehydrogenase

CH$_2$OH
|
H-C-OH
|
H-C-OH
|
H-C-OH
|
CH$_2$O PO$_3^{2-}$

D-Ribitol-5-phosphate

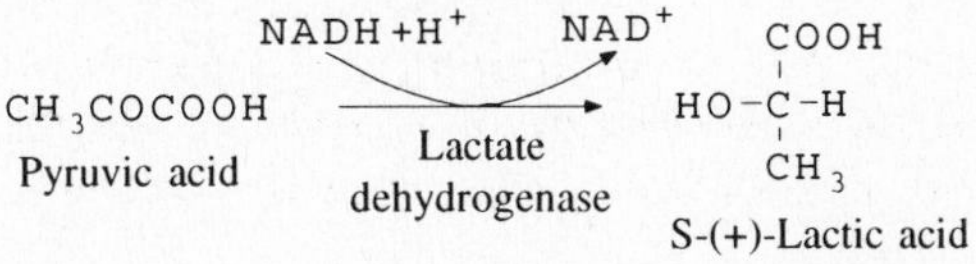

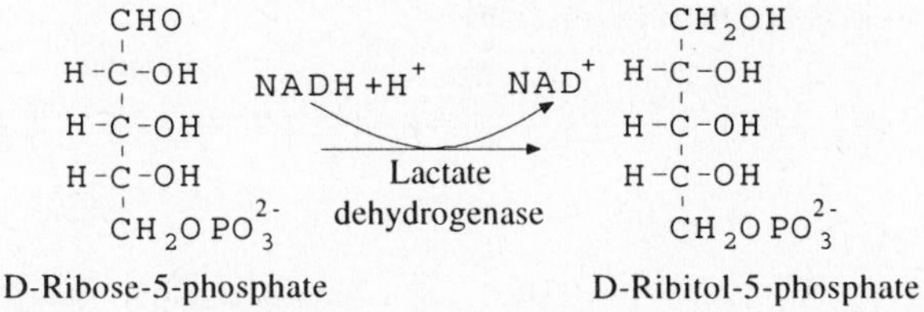

19.8   NUCLEOPHILIC ADDITION

Both aldehydes and ketones undergo nucleophilic addition reactions with a wide variety of reagents. Aldehydes often react more easily than ketones. Aryl aldehydes and ketones usually react more slowly than the corresponding alkyl compounds.

General reaction

$$\text{\textbackslash}C=O + HZ \longrightarrow \text{\textbackslash}C{<}^Z_{OH}$$

Examples

$$\text{\textbackslash}C=O + HCN \longrightarrow \text{\textbackslash}C{<}^{CN}_{OH} \qquad \text{A cyanohydrin}$$

$$\text{\textbackslash}C=O + ROH \underset{}{\overset{H^+}{\rightleftharpoons}} \text{\textbackslash}C{<}^{OR}_{OH} \qquad \text{A hemiacetal or hemiketal}$$

$$\text{\textbackslash}C=O + H_2O \underset{}{\overset{H^+}{\rightleftharpoons}} \text{\textbackslash}C{<}^{OH}_{OH} \qquad \text{A hydrate}$$

Addition occurs because the polar nature of the carbonyl group allows nucleophiles to attack the electron-deficient carbon and electrophiles to attack the electron-rich oxygen atom.

$$\text{\textbackslash}C{=}O \longleftrightarrow \text{\textbackslash}\overset{+}{C}{-}\overset{-}{O}$$

Attack may be **initiated** by either a nucleophile or an electrophile (Fig. 19.7). In both cases the rate-determining step is the attack of the nucleophile, hence the classification of these reactions as nucleophilic additions. The nucleophile may either be negatively charged or have a lone pair of electrons. In the latter case the nucleophilic centre should possess a hydrogen atom that can be lost in order to stabilise the product. The electrophile may be either a positive ion or a species with an electron deficient centre. In practice the reactions often require an acid or base catalyst. Acid catalysts are believed to increase the polarisation of the carbonyl group by either protonating the carbonyl group or, to a lesser extent, hydrogen bonding to the oxygen of the carbonyl group. Base catalysts are usually required to form the appropriate nucleophile for the reaction.

Base-catalysed nucleophilic addition

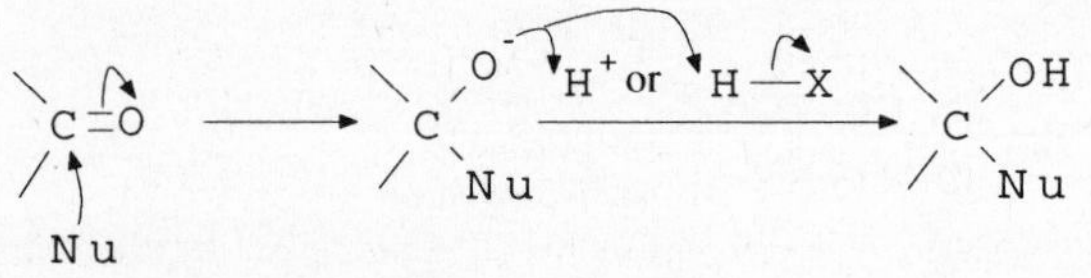

Acid-catalysed nucleophilic addition mechanisms

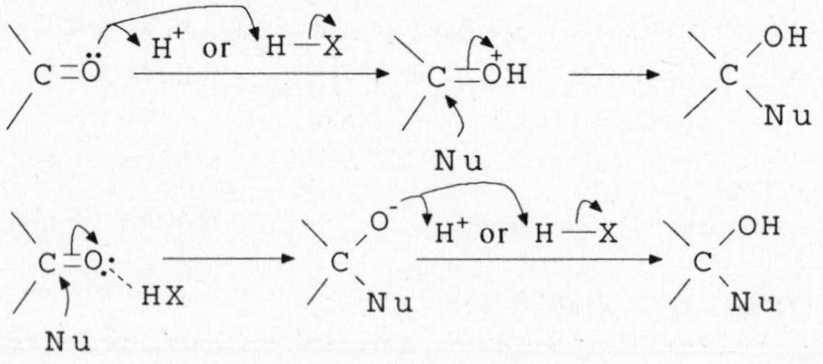

Fig. 19.7. General mechanisms for the nucleophilic addition reactions of aldehydes and ketones. HX may be an acid or a solvent molecule.

Two main factors appear to be responsible for ketones being less reactive than aldehydes. The carbon atoms of the carbonyl groups of ketones are usually less attractive to nucleophiles because the polarisation of the group is normally reduced by the inductive effects of the two R groups. Furthermore, the presence of two R groups on the carbonyl group of ketones leads to a greater degree of steric hindrance between the approaching nucleophile and the carbon as well as a more crowded transition state (Figs 19.8 a and b). This means that a greater quantity of energy will be required to bring the nucleophile close enough to the carbon to react and pass through the transition state (1.4) to produce the product.

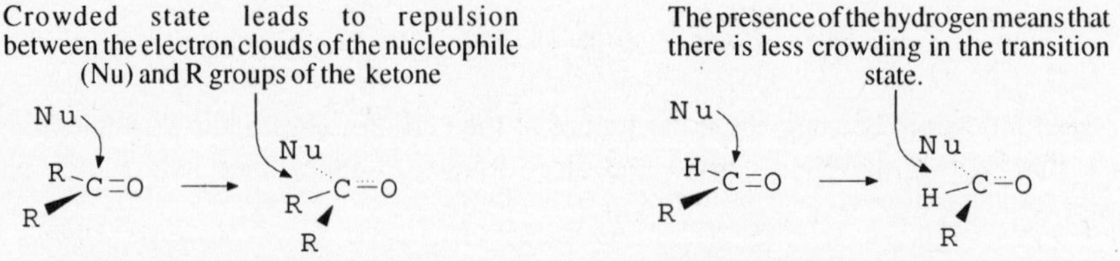

Crowded state leads to repulsion between the electron clouds of the nucleophile (Nu) and R groups of the ketone

The presence of the hydrogen means that there is less crowding in the transition state.

(a) Ketone transition state

(b) Aldehyde transition state

Fig. 19.8. Steric hindrance in the nucleophilic addition reactions of aldehydes and ketones.

19.9 EXAMPLES OF NUCLEOPHILIC ADDITIONS

19.9.1   The nucleophilic addition of hydrogen cyanide

Aldehydes and sterically unhindered ketones react to form hydroxy cyanides commonly known as cyanohydrins. Hydrogen cyanide is not a nucleophile but is the source of the nucleophilic cyanide ions that are responsible for the reaction. However, pure hydrogen cyanide is a poor source of cyanide ions since it is a weak acid and so is only slightly ionised. As a result, reaction is slow unless a base catalyst is used to liberate a significant quantity of cyanide ions and so promote the reaction. The reaction's mechanism follows the normal route for base-catalysed nucleophilic additions.

$$B = \text{a base} \qquad HCN + B \longrightarrow \overset{+}{B}H + \overset{-}{C}N$$

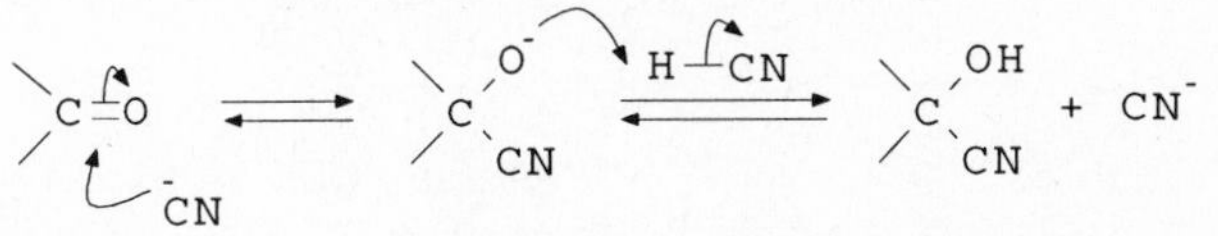

Cyanohydrins occur in nature in a number of plants and insects. They are often bound to a sugar residue by a special type of ether linkage known as a glycosidic link (27.10). For example, mandelonitrile occurs bound to glucose or other simple sugars in almond, apricot and peach pits and the leaves of some African plants.

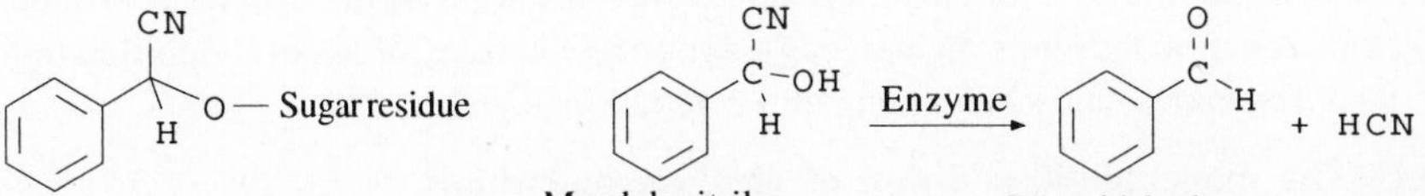

Mandelonitrile   Benzaldehyde

These cyanohydrins form part of the plant or insects defence mechanism. Mandelonitrile, for example, is used by the millipede, *Apheloria corrugatas*, to protect itself against predators. When attacked, the insect uses an enzyme to catalyse the breakdown of the mandelonitrile to benzaldehyde and hydrogen cyanide which it discharges at the predator.

19.9.2 The nucleophilic addition of terminal alkynes

Terminal alkynes will react with aldehydes and ketones in the presence of a strong base. The base reacts with the acidic terminal C–H of the alkyne to form a carboanion (carbanion) which acts as the nucleophile in the addition reaction. The subsequent addition is believed to follow the normal mechanistic route for base-catalysed nucleophilic addition but it is necessary to add aqueous acid to the reaction mixture to liberate the final product.

Ethyl methyl ketone   Methylpentynol

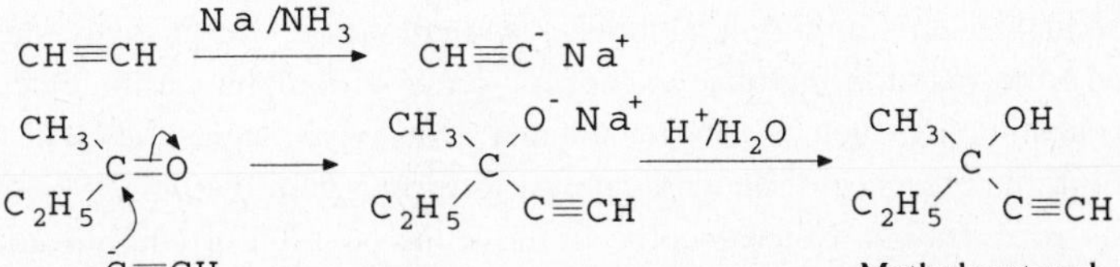

Nucleophilic additions of this type form the basis of the synthesis of methylpentynol and other sedatives whose molecular structures contain an alpha hydroxyalkyne structure.

19.9.3 The nucleophilic addition of water

Both aldehydes and ketones react with water to form hydrates. For example, the hypnotic chloral reacts with water to form a stable crystalline hydrate.

$$\text{Chloral} \quad Cl_3CHO \xrightarrow{\;H_2O\;} Cl_3CH(OH)_2 \quad \text{Chloral hydrate}$$

Very few stable hydrates are known but most aldehydes and ketones exist in aqueous solutions as equilibrium mixtures of the corresponding hydrate and the carbonyl compound. Aldehydes usually form hydrates more readily than ketones but in both cases the position of equilibrium will depend on the nature of the structures bonded to the carbonyl group, the solvent and the temperature. For example, methanal is almost completely hydrated, ethanal is 58% hydrated while propanone (acetone) is only hydrated if acid or base is present.

$$\text{Methanal} \quad \overset{H}{\underset{H}{\diagdown}}C{=}O + H_2O \rightleftharpoons \overset{H}{\underset{H}{\diagdown}}C\overset{OH}{\underset{OH}{\diagup}} \qquad \text{Ethanal} \quad \overset{CH_3}{\underset{H}{\diagdown}}C{=}O + H_2O \rightleftharpoons \overset{CH_3}{\underset{H}{\diagdown}}C\overset{OH}{\underset{OH}{\diagup}}$$

The addition of water to aldehydes and ketones is catalysed by both acids and bases and may be rationalised in terms of the appropriate mechanism (Fig. 19.7).

The hydrates formed by saturated aldehydes and ketones do not absorb in the ultra-violet. Consequently, it is often impossible to use the ultra-violet spectrum of aldehyde and ketone functional groups in aqueous solution as the basis of assays, identification and limit tests for compounds containing these functional groups.

19.9.4 The nucleophilic addition of alcohols and thiols

Alcohols add to aldehydes and ketones under dry conditions in the presence of either an acid or enzyme as catalyst to form moderately stable hemiacetals. In both cases the reactions are reversible, the position of equilibrium usually favouring the carbonyl compound and so the reaction conditions must be adjusted to give a reasonable yield of the hemiacetal. Ketones do not react as readily as aldehydes and often require specific procedures. In both cases the reactions are believed to proceed by mechanisms similar to those shown in Fig. 19.7.

Hemiacetal, $R_1 = H$, $R =$ an alkyl or aryl group

$$\text{For aldehydes, } R_1 = H. \quad \overset{R}{\underset{R_1}{\diagdown}}C{=}O + R'OH \underset{}{\overset{H^+}{\rightleftharpoons}} \overset{R}{\underset{R_1}{\diagdown}}C\overset{OR'}{\underset{OH}{\diagup}}$$

Hemiketal, R_1 and $R =$ an alkyl or aryl group

Alcohol

Aldedydes and ketones whose structures contain a hydroxyl group in a suitable position relative to the carbonyl group of the aldehyde or ketone can form stable cyclic hemiacetals. For example, the ring structures of monosaccharides are cyclic hemiacetals.

The hemiacetals of simple aldehydes and ketones are not usually stable enough to be isolated from reaction mixtures except for some with alpha carbon electron withdrawing group substituents, such as tribromoethanal. However, hemiacetals will react with excess alcohol in the presence of an acid catalyst to form a gem-diether known as an acetal. The reaction is reversible. Consequently, to move the position of equilibrium to the right, the reaction is normally carried out in the laboratory using an excess of the alcohol and the water is removed as it is formed either by azeotropic distillation or the use of excess of an anhydrous acid catalyst which removes the water as the H_3^+O ion.

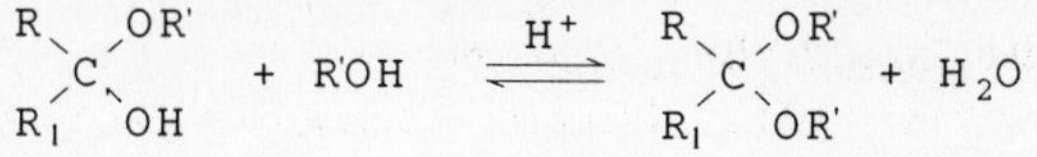

$$\underset{R_1}{\overset{R}{>}}C\underset{OH}{\overset{OR'}{<}} \; + \; R'OH \; \rightleftharpoons \; \overset{H^+}{} \; \underset{R_1}{\overset{R}{>}}C\underset{OR'}{\overset{OR'}{<}} \; + \; H_2O$$

For aldehydes, R_1 = H.

Acetal, R_1 = H, R = an alkyl or aryl group
Ketal, R_1 and R = an alkyl or aryl group

Thiols react in a similar manner to alcohols with aldehydes and ketones to form hemi-thioacetals and thioacetals which are more stable than the corresponding aldehydes and ketones. The reactions occur more readily than those of alcohols because sulphur is a better nucleophilic centre than oxygen.

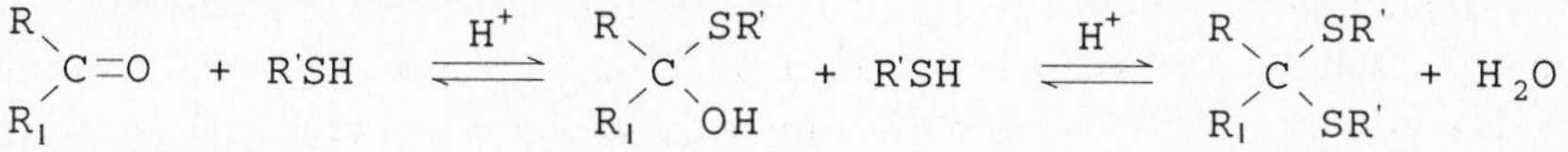

$$\underset{R_1}{\overset{R}{>}}C{=}O \; + \; R'SH \; \overset{H^+}{\rightleftharpoons} \; \underset{R_1}{\overset{R}{>}}C\underset{OH}{\overset{SR'}{<}} \; + \; R'SH \; \overset{H^+}{\rightleftharpoons} \; \underset{R_1}{\overset{R}{>}}C\underset{SR'}{\overset{SR'}{<}} \; + \; H_2O$$

For aldehydes, R_1 = H   Hemithioacetal   Thioacetal

Acetals and ketals are usually stable under alkaline conditions but easily hydrolysed in aqueous acid conditions to the original carbonyl compound and alcohol. Enzyme-controlled hydrolysis is an important feature of the metabolism of mono and polysaccharides (27.8 and 27.10). The structures of these carbohydrates contain hemiacetal and acetal groups. The ease of aqueous hydrolysis also makes acetals very useful as aldehyde- and ketone-protecting groups in the synthesis of drugs and other organic compounds.

Acetal formation has been used to decrease the water solubility and increase the lipid solubility of some drugs. For example, the corticosteroid, triamcinolone, has been converted to its acetonide to increase its lipid solubility.

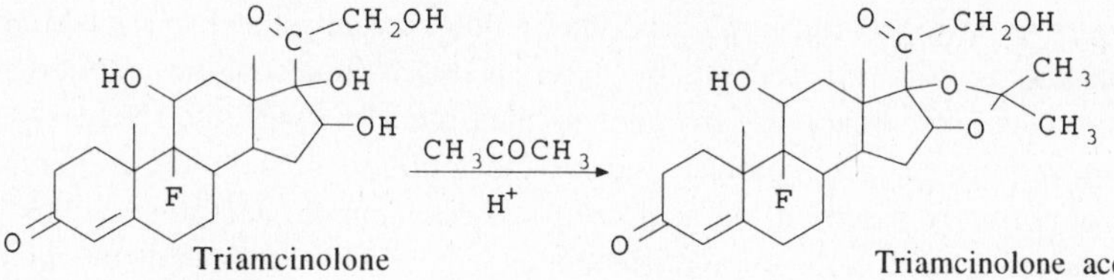

Triamcinolone $\xrightarrow[H^+]{CH_3COCH_3}$ Triamcinolone acetonide

19.9.5   Polymerisation

Aldehydes often polymerise under acid conditions to cyclic trimers. For example, ethanal rapidly polymerises to the mild hypnotic paraldehyde in the presence of acid whilst methanal yields trioxan when a 60% solution is distilled in the presence of sulphuric acid.

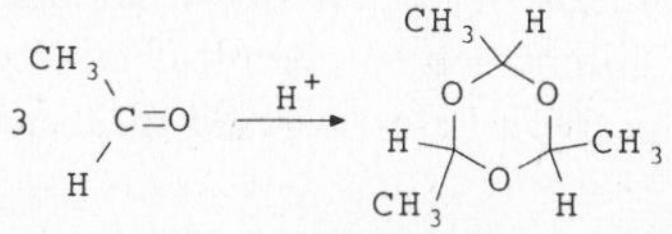

Ethanal   Paraldehyde (bp 124°C) 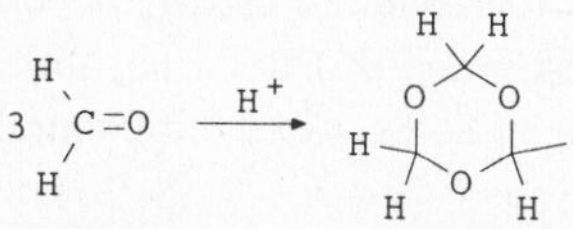   Methanal   Trioxan (mp 62°C)

Aldehydes will also polymerise to other products under different conditions. For example, ethanal will form the cyclic tetramer metaldehyde (slug and snail killer), if the polymerisation is carried out at 20°C in the presence of hydrogen bromide and calcium nitrate.

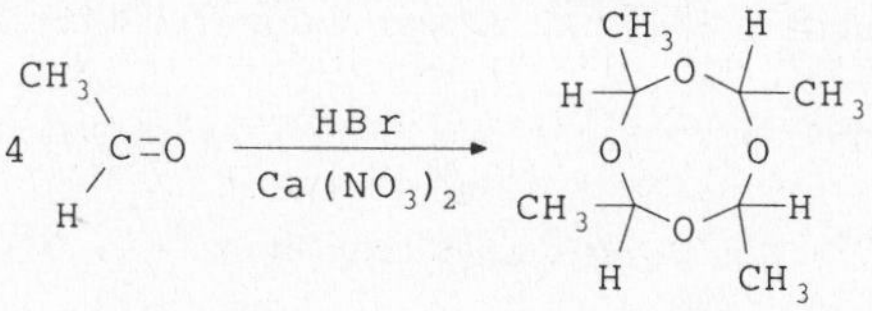

Metaldehyde

Ketones do not normally polymerise.

19.10   CONDENSATION

Condensation reactions are nucleophilic additions that are followed by the elimination of
water from the addition product (Fig. 19.9). Consequently, condensation reactions are
also known as **addition/elimination** reactions.

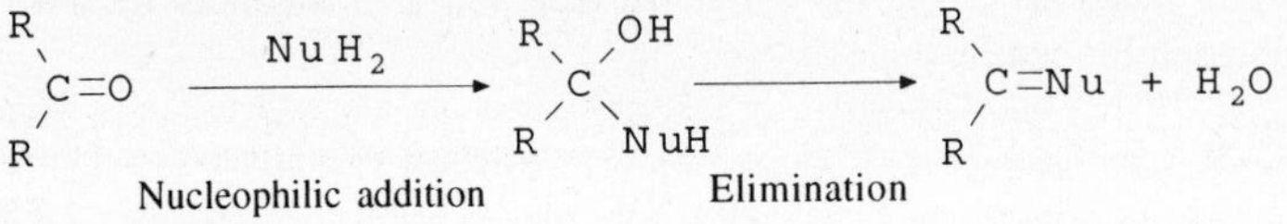

$$\underset{R}{\overset{R}{>}}C=O \quad \xrightarrow{\text{Nu}\,H_2} \quad \underset{R}{\overset{R}{>}}\underset{\text{NuH}}{\overset{OH}{C}} \quad \longrightarrow \quad \underset{R}{\overset{R}{>}}C=Nu \;+\; H_2O$$

Nucleophilic addition   Elimination

Overall general equation:

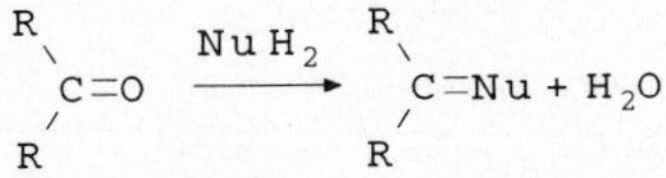

$$\underset{R}{\overset{R}{>}}C=O \quad \xrightarrow{\text{Nu}\,H_2} \quad \underset{R}{\overset{R}{>}}C=Nu + H_2O$$

Fig. 19.9. General equations for the condensation reactions of aldehydes and ketones. The nucleophile
(NuH_2) in these equations possesses two hydrogens directly attached to the nucleophilic centre, hence
the inclusion of H_2 in the general structural formula used for the nucleophile.

The condensation reactions of aldehydes and ketones usually involve nitrogen or
carboanion nucleophiles. The mechanisms of these reactions may be divided into two
stages. The first stage is normally believed to follow a route similar to that of either acid-
or base-catalysed nucleophilic addition reactions of aldehydes and ketones (19.7). For
example, strong nitrogen nucleophiles do not require a catalyst and so the addition step
of the reaction is usually based on the 'base-catalysed' nucleophilic addition reaction
mechanism. However, weak nitrogen nucleophiles, such as hydrazine (NH_2NH_2) and its
derivatives require an acid-catalyst and so the addition step in these condensation
reactions is normally based on the acid-catalysed nucleophilic addition reaction mechanism. On the other hand, carboanion nucleophiles require a base to liberate the nucleophile
and so the mechanism of the addition step of these condensation reactions is usually
similar to the base-catalysed nucleophilic addition mechanism. The mechanism of the
second stage of most condensation reactions is thought to be similar to the acid-catalysed
elimination mechanism of alcohols (15.9.1). However, it should be noted that carboanion
nucleophiles (carbanions) usually need the addition of an acid (the technique is referred
to as an **acid work-up**) to bring about the elimination to complete the condensation.
Furthermore, in some cases elimination occurs by a base-catalysed mechanism.

19.11   EXAMPLES OF CONDENSATION REACTIONS

19.11.1 Hydrazine and its derivatives

Hydrazine and its derivatives react in the presence of an acid catalyst with both aldehydes
and ketones. Aldehydes usually give the E (syn) isomer (5.8) whilst unsymmetrical
ketones usually yield a mixture of E and Z (syn and anti) isomers.

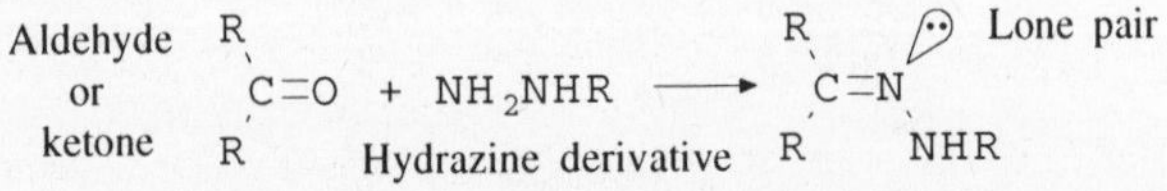

Aldehyde $\underset{R}{\overset{R}{>}}C=O$ + NH_2NHR $\longrightarrow$ $\underset{R}{\overset{R}{>}}C=\underset{NHR}{\overset{\curvearrowleft}{N}}$ Lone pair

or
ketone   Hydrazine derivative

The reactions of many hydrazine derivatives with aldehydes and ketones are used as the basis of identification tests and in qualitative organic analysis. For example, semicarbazide, thiosemicarbazide, phenylhydrazine and 4-nitrophenylhydrazine are used to form derivatives for use in identification tests and qualitative organic analysis while a solution of 2,4-dinitrophenylhydrazine in concentrated sulphuric acid (Brady's reagent) can also be used as a spot test reagent for aldehydes and ketones in qualitative organic analysis.

$$NH_2CONHNH_2 \longrightarrow \underset{\diagup}{\overset{\diagdown}{C}}{=}NHNHCONH_2$$

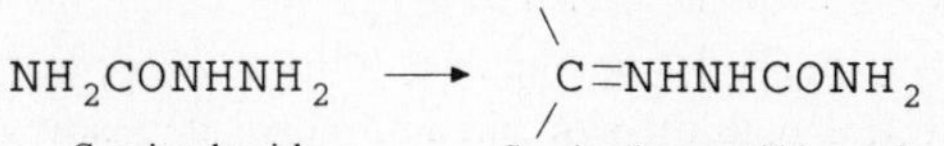

Semicarbazide   Semicarbazone (thiosemicarbazide, $NH_2CSNHNH_2$, gives the corresponding thiosemicarbazone).

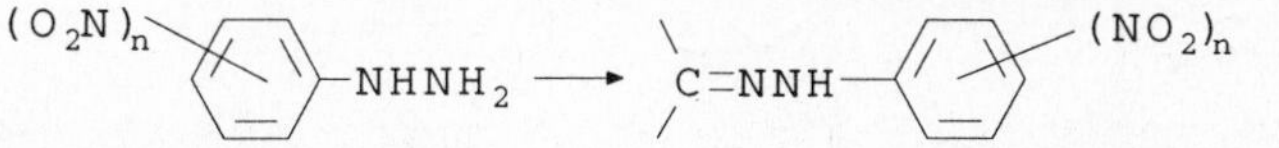

$$(O_2N)_n\text{—}\langle\!\langle\text{—}\rangle\!\rangle\text{—}NHNH_2 \longrightarrow \underset{\diagup}{\overset{\diagdown}{C}}{=}NNH\text{—}\langle\!\langle\text{—}\rangle\!\rangle\text{—}(NO_2)_n$$

Phenyl and nitrophenylhydrazines   Phenyl and nitrophenylhydrazones

The mechanisms of these reactions are believed to have the general form shown in Fig. 19.10. Protonation of the oxygen atom increases the polarisation of the carbonyl group activating it towards nucleophilic attack by the weak hydrazine nucleophiles.

Nucleophilic addition step

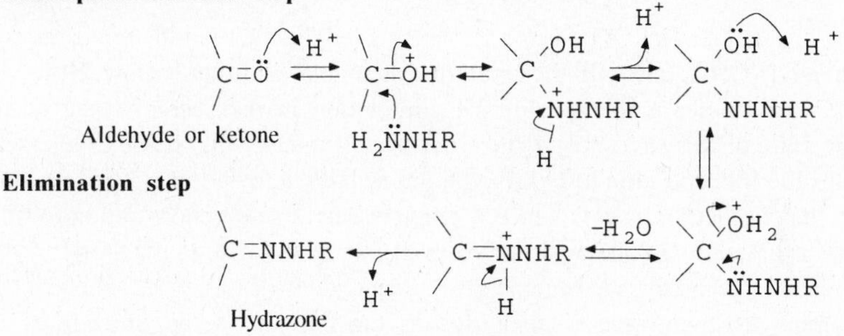

Aldehyde or ketone

Elimination step

Hydrazone

Fig. 19.10. The general mechanism of the reaction of hydrazine and hydrazine derivatives with aldehydes and ketones.

The thiosemicarbazone, thiacetazone is used to treat tuberculosis. It has proved to be particularly useful in cases that exhibit resistance to other antitubercular drugs.

Thiacetazone $CH_3CONH\text{—}\langle\!\langle\text{—}\rangle\!\rangle\text{—}CH{=}NNHCSNH_2$

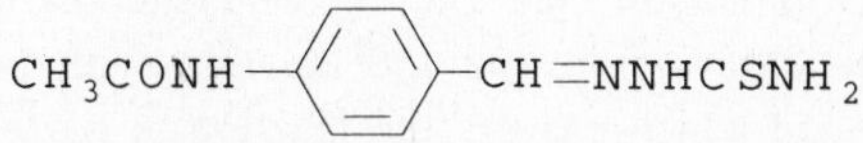

Aldehydes and ketones can be isolated from biological fluids as their quaternary ammonium acyl hydrazones. These quaternary ammonium acyl hydrazones are very water

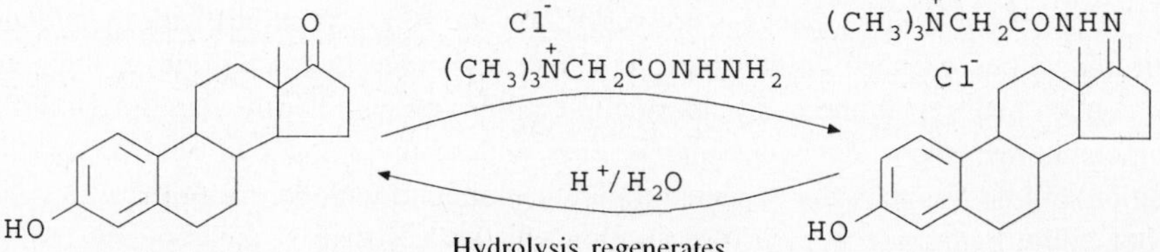

Oestrone   Hydrolysis regenerates the aldehyde or ketone   Oestrone-Girard T complex

soluble. This allows organic impurities to be extracted from aqueous solutions of the quaternary ammonium acetohydrazones by organic solvents. The original aldehyde or ketone is regenerated by acid hydrolysis. For example, the ammonium acyl hydrazide, Girard reagent-T, is used to isolate steroidal hormones from urine by forming the corresponding steroid ammonium acyl hydrazone.

19.11. 2   Hydroxylamine

Hydroxylamine condenses with aldehydes and ketones to form the corresponding oximes. For example, benzaldehyde forms both the syn and antioximes depending on the reaction conditions employed.

$$
\begin{array}{ccccc}
\overset{H}{\underset{Ph}{\diagdown}}C=O & \xrightarrow{NH_2OH} & \overset{H}{\underset{Ph}{\diagdown}}C=N^{OH} & \xrightarrow{HCl} & \overset{H}{\underset{Ph}{\diagdown}}C=N^{OH}
\end{array}
$$

Benzaldehyde   E-Benzaldoxime mp 35^{0} Z-Benzaldoxime mp 130^{0}

The reaction is acid-catalysed, the rate reaching a maximum under weakly acidic conditions. In both low pH (highly acidic) and high pH (highly basic) conditions the reaction is slow. This can be readily explained in terms of the law of mass action. At low pH values the equilibrium:

$$H^+ \curvearrowright \overset{..}{N}H_2OH \;\rightleftharpoons\; \overset{+}{N}H_3OH$$

lies to the right. The cation, $\overset{+}{N}H_3OH$ cannot act as a nucleophile and so the concentration of the nucleophile NH_2OH is low with the result that the rate of addition is slow. However, the high acidity does increase the rate of the elimination in the second stage of the condensation. In spite of this fast second stage the overall rate is still slow. Conversely, at high pH values the concentration of hydroxylamine will be high and the rate of addition high. However, the elimination is slow since it requires a high acid concentration to form the H_2O leaving group and so once again the overall rate of the reaction is slow.

Oximes are used as derivatives in organic analysis and identification tests. The reaction with hydroxylamine hydrochloride is also used as the basis of assay procedures for aldehydes and ketones. For example, cinnamon oil contains a number of different aldehydes, the main one being cinnamaldehyde. The aldehyde content of cinnamon oil is determined in the 1993 British Pharmacopoeia by reacting all the aldehydes in the oil with hydroxylamine hydrochloride and titrating the hydrochloric acid liberated with 0.5 M ethanolic potassium hydroxide solution. The reaction of the aldehyde with hydroxylamine hydrochloride is reversible but the titration with potassium hydroxide removes the hydrochloric acid which enables the reaction to go to completion.

$$RCHO \;+\; NH_2OH\cdot HCl \;\rightleftharpoons\; RCH=NOH \;+\; HCl + H_2O$$

$$HCl + KOH \;\longrightarrow\; KCl + H_2O$$

The end point in these assays is not very sharp and so it is often difficult to get a good indicator colour change. Furthermore, the hydroxylamine hydrochloride solution used will contain hydroxylamine hydrochloride in equilibrium with hydroxylamine. Both these compounds react with aldehydes and ketones and so the amount of acid produced by reaction will be less than the expected theoretical amount with the result that the volume of potassium hydroxide used in the titration will be less than is theoretically required. Consequently, the volume of potassium hydroxide is either multiplied by an empirical correction factor or the results are calculated using the equivalence given in the literature. In the case of cinnamon oil the empirical factor is 1.008.

19.11.3   Ammonia

Aldehydes, with the exception of formaldehyde, usually condense with ammonia to form unstable **imines** although some aromatic imines are stable. These unstable imines often polymerise to cyclic compounds of varying sizes.

$$RCHO \xrightarrow{NH_3} \underset{\text{Imine}}{RCH=NH} \xrightarrow{\text{Polymerisation}} \text{[cyclic compound]}$$

Methanal reacts with ammonia to form a tricyclic compound, urotropine (hexamine or hexamethylenetetramine) which is used as a urinary antiseptic (*de Witt's* liver pills). The urotropine passes into the urine where it decomposes under the acid conditions to methanal which is the active agent. It should be noted that antibiotic treatments are far more effective than the hexamine in dealing with cystitis.

$$HCHO \xrightarrow{NH_3} \text{[Urotropine]}$$

Ketones react with ammonia in a variety of ways. For example, propanone reacts with ammonia to form a number of products. The simplest of these of these products is 2-methyl-2-aminopent-4-one (diacetonamine). This compound is formed by an aldol condensation (19.11.6) which is followed by the addition of ammonia to the C=C bond of the unsaturated ketone formed by the condensation.

$$2CH_3COCH_3 \xrightarrow{NH_3} \underset{\text{Mesityl oxide}}{(CH_3)_2C=CHCOCH_3} \xrightarrow{NH_3} \underset{\text{2-Methyl-2-aminopent-4-one}}{(CH_3)_2\overset{NH_2}{C}CH_2COCH_3}$$

The synthesis of alanine in biological systems is believed to involve the reductive amination of alpha-keto acids. The ammonia condenses with the keto acid and the resultant imine is thought to be reduced by an enzyme to the amino acid.

$$CH_3COCOOH \xrightarrow{NH_3} \underset{\text{Imine}}{CH_3\overset{NH}{C}COOH} \xrightarrow{\text{Reduction}} \underset{\text{Alanine}}{CH_3\overset{NH_2}{C}HCOOH}$$

19.11.4   Amines

Primary amines react with aldehydes and ketones to form **imines**. Imine formation is believed to follow the expected mechanistic route of acid-catalysed nucleophilic addition followed by elimination.

$$\underset{\substack{\text{Aldehyde} \\ \text{or} \\ \text{ketone}}}{\diagdown C=O} + RNH_2 \longrightarrow \underset{\text{Imine}}{\diagdown C=NHR} + H_2O$$

All imines are basic, the lone pair of the nitrogen acting as the basic centre. Imines are unstable unless the carbon atom of the imine group has at least one aromatic substituent (ArCH=NR). These aromatic imines are commonly called **Schiff's bases.** They are often highly coloured solids that are used as derivatives in analysis and identification tests.

Aliphatic imines are believed to be formed as intermediates in the biosynthesis of porphobilinogen (Fig. 19.11) the precursor of porphyrin molecules (Fig. 13.9). This synthetic pathway is also thought to involve the condensation of a carbanion with a ketone (19.11.5) and a Schiff's base type of condensation where an amine attacks the carbon of the C=NH bond in a similar way to the attack of amines on the carbon of the carbonyl groups of aldehydes and ketones.

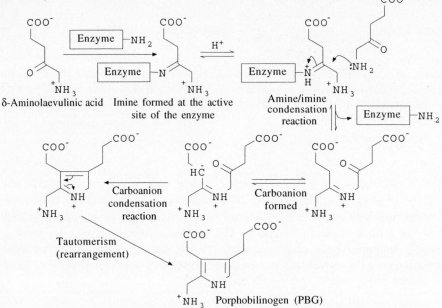

Fig 19.11. An outline of a proposed route for the biogenesis of porphobilinogen (PBG).

Imine formation is also believed to be involved in the transamination action of pyridoxyl phosphate, the antiseptic action of methanal, sight and purine biosynthesis (Fig. 19.11). An understanding of the biogenesis of purines is particularly useful since the purine residues in the nucleic acids are a prime target for anticancer drugs.

Secondary amines react in acid conditions with aldehydes and ketones to produce what are usually unstable intermediate addition products which decompose under the reaction conditions to **enamines** provided the compound has a C–H bond at a position α to the carbonyl group.

Aldehyde or ketone   Unstable intermediate addition product   Enamine

The first stage of the reaction is believed to follow an acid-catalysed nucleophilic addition mechanism (Fig. 19.6). However, the expected second-stage elimination to an imine cannot occur as the single hydrogen of the secondary amine was lost during the nucleophilic addition. As a result, elimination involves a hydrogen atom bonded to a carbon in an α position to the carbonyl group. This elimination probably proceeds by the same type of mechanism as the acid-catalysed eliminations of alcohols (15.10.1)

If no α-hydrogen atom is present the addition product is converted into the **aminal** (compare with acetals 19.9.4).

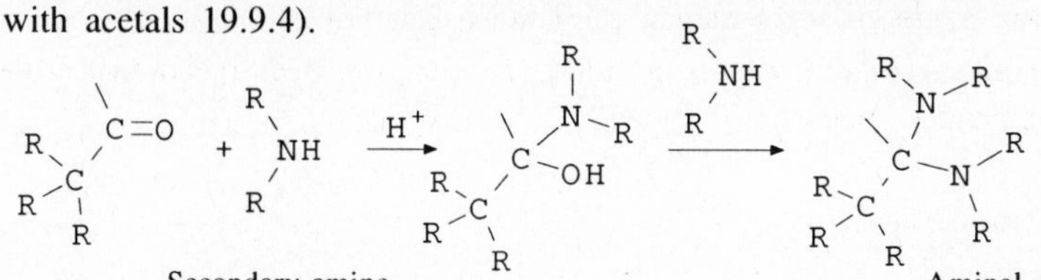

Secondary amine   Aminal

Unlike imines, enamines do not normally function as bases since the lone pair of the nitrogen is delocalised (lone pair in conjugation structure). This activates the beta carbon to electrophilic attack and so enamines are very useful as intermediates in synthesis.

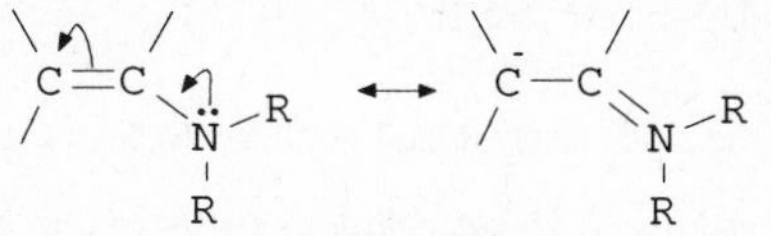

In the synthesis of the narcotic analgesic tilidine, crotonaldehyde is reacted with dimethylamine to form the intermediate 1-dimethylaminobutadiene which is used as the dieneophile in a Diels-Alder reaction to produce a mixture of the racemates of cis and trans tilidine.

$$CH_3CH=CHCHO \xrightarrow{HN(CH_3)_2} CH_2=CHCH=CHN(CH_3)_2$$
Crotonaldehyde   1-Dimethylaminobutadiene

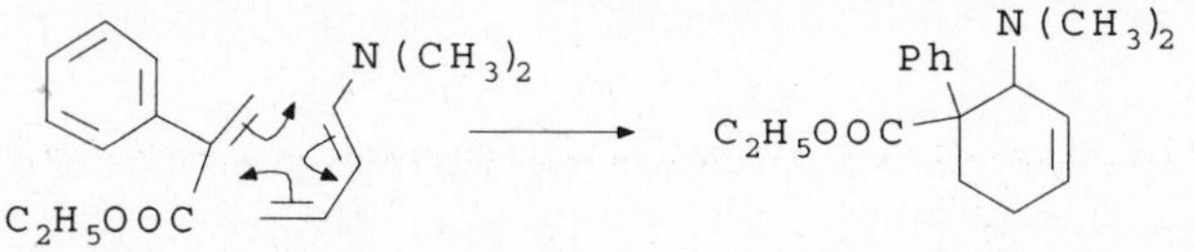

Ethyl 2-phenylacrylate   1-Dimethylaminobutadiene   (±) Cis and trans tilidine

Tertiary amines do not usually react with aldehydes and ketones.

19.11.5   Carboanion condensations

Carboanions are formed by treating compounds whose structures contain a C–H bond in an α position to an electron-withdrawing group(s) with a suitable base (Fig. 19.12). The electron acceptor group(s) withdraw electrons from the carbon which in turn increases the polarity and hence reactivity of the C–H bond. This is an example of the transmission of the inductive effect of a structure having a significant effect on the course of chemical reactions.

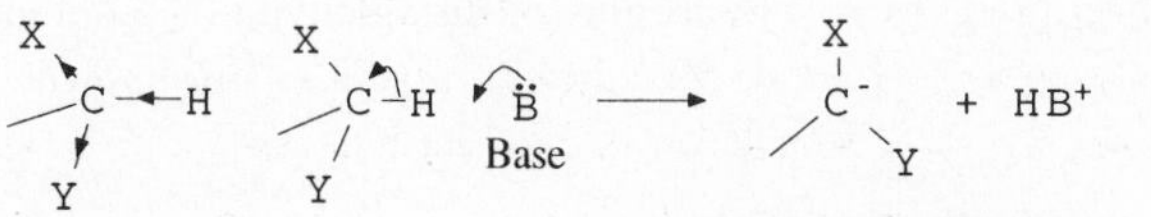

A planar carboanion

Fig. 19.12. The formation of carboanions. X and Y are electron-acceptor groups such as -COOR, -CN, -NO₂, -SO₃H, -SO₂⁻, and -SO⁻, etc. where X may or may not be the same as Y. In some cases Y may be hydrogen. The planar carboanion is stabilised by delocalisation of the negative charge.

Aldehydes and ketones with no α C–H bonds can condense with a wide variety of carboanions (Fig. 19.13). In many cases the elimination of water that completes the condensation only occurs after treatment of the addition product with acid. Many of these condensation reactions were named after their discoverers.

Knoevenagel reaction

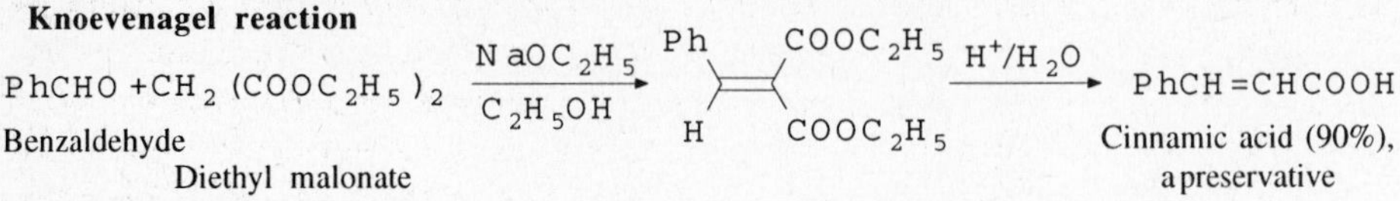

$$PhCHO + CH_2(COOC_2H_5)_2 \xrightarrow[C_2H_5OH]{NaOC_2H_5} \begin{array}{c} Ph \\ \\ H \end{array}\!\!=\!\!\begin{array}{c} COOC_2H_5 \\ \\ COOC_2H_5 \end{array} \xrightarrow{H^+/H_2O} PhCH=CHCOOH$$

Benzaldehyde
  Diethyl malonate
  Cinnamic acid (90%), a preservative

Perkin reaction

$$PhCHO + CH_3COOCOCH_3 \xrightarrow{CH_3COONa} PhCH=CHCOOH + CH_3COOH$$

Benzaldehyde   Ethanoic anhydride   Cinnamic acid (60%)

Claisen condensation

$$PhCHO + CH_3COOC_2H_5 \xrightarrow{Na, 0\text{-}5^0C} PhCH=CHCOOC_2H_5$$

Benzaldehyde Ethyl ethanoate   Ethyl cinnamate (70%)

Fig. 19.13. Examples of the condensation reactions of aldehydes and ketones involving carboanions.

The mechanisms of many of these condensation reactions are believed to follow the general route for aldehyde and ketone condensations (Fig. 19.14). The nucleophilic addition follows the base-catalysed mechanism, its product being completed by the abstraction of a proton from the solvent which is usually water or an alcohol. Subsequent elimination is dependent on the nature of the addition product and may be acid- or base-catalysed.

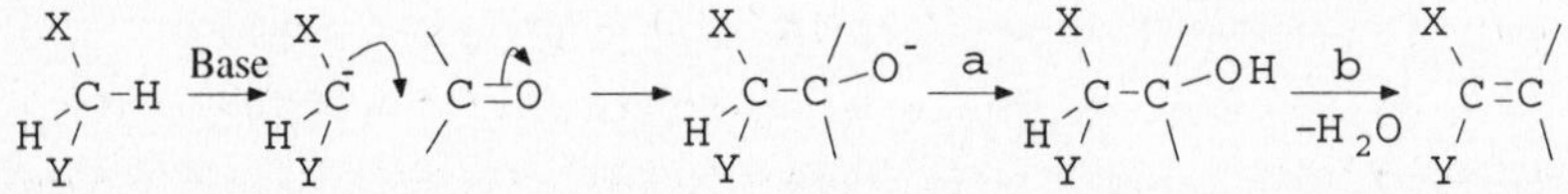

$$\begin{array}{ccccccc}
X & & X & & X & & X & & X \\
\backslash & & \backslash & & \backslash & & \backslash & & \backslash & \diagup \\
C-H & \xrightarrow{Base} & C \quad C=O & \longrightarrow & C-C-O^- & \xrightarrow{a} & C-C-OH & \xrightarrow[-H_2O]{b} & C=C \\
H\diagup & & H\diagup & & H\diagup & & H\diagup & & \diagup \quad \backslash \\
Y & & Y & & Y & & Y & & Y
\end{array}$$

Fig. 19.14. A general mechanism for the condensation reactions of carboanions with aldehydes and ketones. The proton for stage (a) is supplied by either the solvent or water. Stage (b) is brought about by the solvent, an acid or a base.

The carboanion condensations of aldehydes and ketones are used extensively in synthesis of organic compounds. For example, the CNS stimulant amphetamine is prepared by the condensation of nitroethane with benzaldehyde followed by reduction of the condensation product

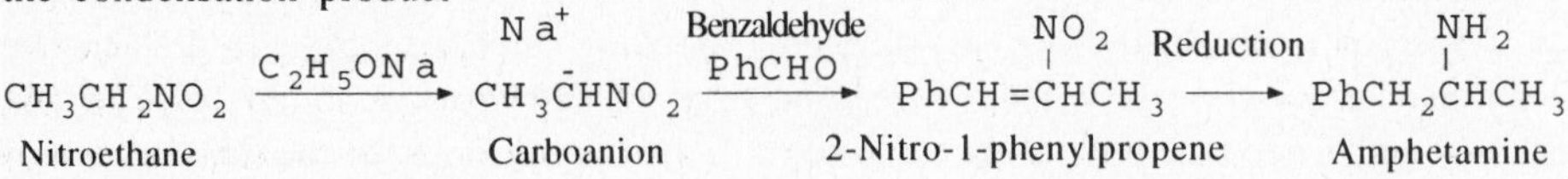

$$CH_3CH_2NO_2 \xrightarrow{C_2H_5ONa} CH_3\overset{-}{C}HNO_2 \xrightarrow[PhCHO]{Benzaldehyde} PhCH=CHCH_3 \xrightarrow{Reduction} PhCH_2CHCH_3$$

Nitroethane   Carboanion   2-Nitro-1-phenylpropene   Amphetamine

19.11.6 Aldol reaction

The Adol reaction involves two molecules of an aldehyde or a ketone whose molecular structures contain an a C–H bond. Mixed aldol reactions are known.

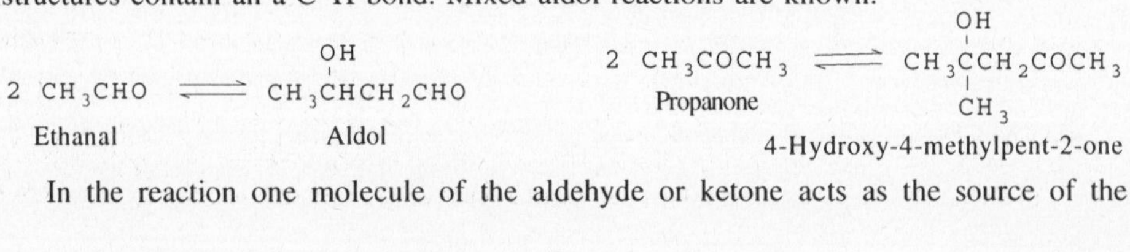

$$2\ CH_3CHO \rightleftharpoons CH_3\overset{OH}{\underset{|}{C}}HCH_2CHO$$

Ethanal   Aldol

$$2\ CH_3COCH_3 \rightleftharpoons CH_3\overset{OH}{\underset{|}{\underset{|}{C}}}CH_2COCH_3$$
  Propanone CH_3

4-Hydroxy-4-methylpent-2-one

In the reaction one molecule of the aldehyde or ketone acts as the source of the carboanion which reacts with the second aldehyde or ketone molecule. The carboanion is formed by the attack of a base on an α H. This attack is possible because the carbonyl group of the aldehyde or ketone acts as the X or Y electron-withdrawing group increasing the polarisation of the α C–H bond (Fig. 19.12). The X and Y groups also stabilise the resultant carboanion by electron delocalisation. The carboanion formed by the action of the base reacts with the carbon of the carbonyl group of the second aldehyde or ketone molecule to form the nucleophilic addition product. The reactions are reversible, the position of equilibrium depending on the nature of the carbonyl compound. For example, with ethanal the equilibrium lies to the right but in the case of propanone it lies to the left but will move almost completely to the right under suitable conditions and in the presence of a catalyst. The catalyst reduces the time to attain equilibrium but not its position.

OH⁻ H—CH$_2$CHO
Ethanal
                    → H$_2$O   O⁻ H—OH   OH
     O |
     ‖
CH$_3$ C—H   CH$_2$CHO ⇌ CH$_3$CHCH$_2$CHO ⇌ CH$_3$CHCH$_2$CHO + OH⁻
Ethanal   Aldol
              Position of equilibrium ————————→

OH⁻ H—CH$_2$COCH$_3$
Propanone
                    → H$_2$O   O⁻ H—OH   OH
     O |
     ‖
CH$_3$ C—CH$_3$ CH$_2$COCH$_3$ ⇌ CH$_3$CCH$_2$COCH$_3$ ⇌ CH$_3$CCH$_2$COCH$_3$ + OH⁻
Propanone | |
                                   CH$_3$ CH$_3$
              ←———————— **Position of equilibrium**   4-Hydroxy-4-methylpent-2-one

In both cases, treatment of the reaction mixtures with acid almost always results in the elimination of water from the hydroxyaldehyde (aldol) or hydroxyketone to yield the more stable unsaturated conjugated aldehyde or ketone.

$$CH_3CHOHCHCHO \xrightarrow{H^+} CH_3CH=CHCHO + H_2O$$
But-2-enal (crotonaldehyde)

$$CH_3CH(OH)CHCOCH_3 \xrightarrow{H^+} CH_3CH=CHCOCH_3 + H_2O$$
Pent-3-ene-2-one

Base-catalysed eliminations are not common. However, in the presence of a strong base, aldol itself loses water to form but-2-enal (crotonaldehyde). It was this loss of water that was responsible for the classification of the aldol reaction as a condensation.

19.12 AROMATIC SUBSTITUTION

The carbonyl groups of aromatic aldehydes and ketones are moderately strong electron withdrawing groups. Consequently, these functional groups are meta directing and deactivating in aromatic electrophilic substitution. For example, reaction of benzaldehyde with a cold mixture of fuming nitric and sulphuric acids yields mainly 3-nitrobenzaldehyde. The resistance of the aldehyde group to oxidation in this reaction is unexpected especially as dilute nitric acid readily oxidises benzaldehyde to benzoic acid.

Benzaldehyde $\xrightarrow{\quad HNO_3\ /\ H_2SO_4\quad}$ 3-Nitrobenzaldehyde

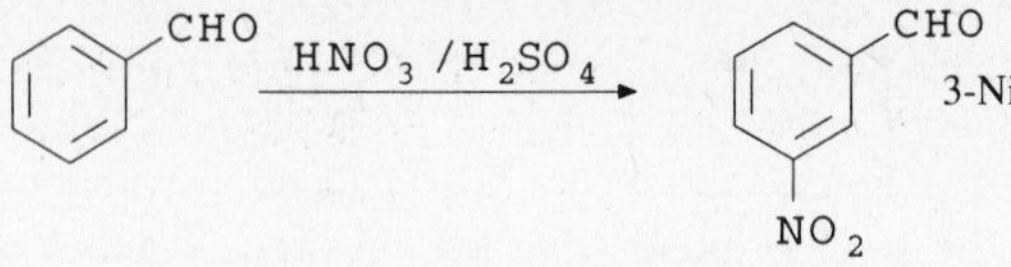

The aldehyde and keto groups are not strong enough electron-withdrawing groups to allow nucleophilic substitution of the ring hydrogen atoms of simple aromatic aldehydes and ketones. However, they are strong enough electron acceptors to allow the nucleophilic substitution of the aromatic halo groups in ortho and para positions relative to the aldehyde or ketone group. For example, the aromatic chloro groups of both 2-chlorobenzaldehyde and 4-chlorobenzaldehyde may be replaced by a sulphonic acid group by treatment of the compounds with aqueous sodium hydrogen sulphite followed by acidification.

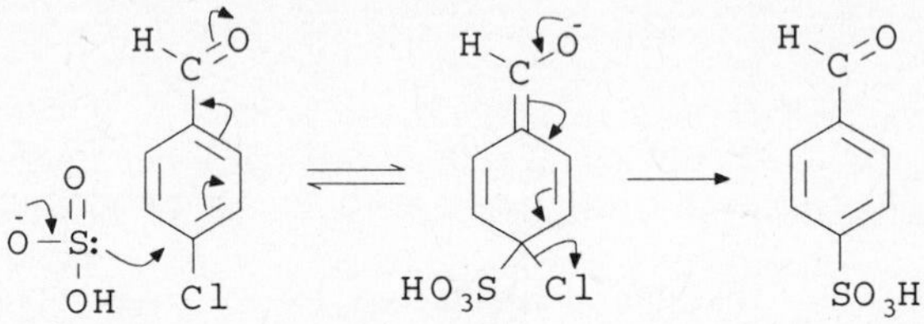

19.13   MISCELLANEOUS   REACTIONS

Aldehydes and ketones with α C–H bonds will react with halogens to form halo substituted aldehydes and ketones. Mono, di and trihalo derivatives may be obtained with acid catalysts but reactions using basic catalysts cannot usually be stopped until all the α C–H bonds have been substituted. Chlorination usually results in substitution occuring only on one side of the carbonyl group of ketones.

$$CH_3CHO \longrightarrow CH_2ClCHO \longrightarrow CHCl_2CHO \longrightarrow CCl_3CHO$$

Ethanal   Chloroethanal   Dichloroethanal   Trichloroethanal

Reaction proceeds by the electrophilic addition of the halogen to the C=C of the enol (acidic catalyst) or the enolate (basic catalyst) of the aldehyde or ketone. The mechanism is believed to be similar to that for electrophilic addition to the C=C of alkenes.

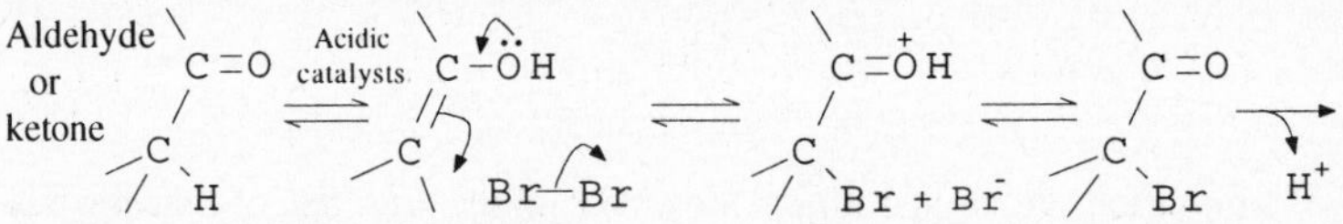

Aldehyde or ketone

This sequence is repeated until all the relevant hydrogens are replaced

The introduction of one electron accepting halogen atom into the molecule increases the polarisation of adjacent C–H bonds which results in selective enol formation and further rapid halogenation.

Increased polarisation due to the electron-withdrawing effect of the bromine atom

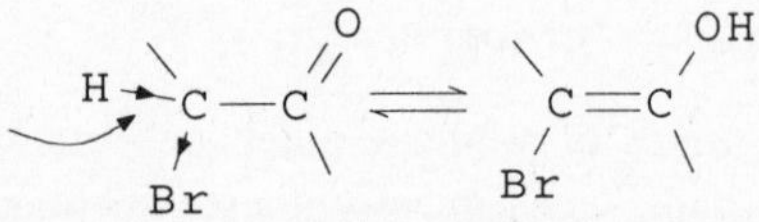

The reaction is the basis of the **haloform reaction** which is used as a test for the presence of a terminal methyl ketone structure ($CH_3CO–$) in a molecule. The compound is dissolved in a suitable solvent such as dioxan and treated with aqueous sodium hydroxide followed by iodine dissolved in aqueous potassium iodide solution. The formation of a yellow precipitate of triiodomethane (iodoform, an early synthetic antibacterial dusting powder) indicates the presence of a terminal methyl ketone or a structure that can be oxidised by the halogen to a terminal methyl ketone.

$$RCH(OH)CH_3 \xrightarrow{I_2} RCOCH_3 \xrightarrow{I_2/OH^-} RCOCI_3 \xrightarrow{OH^-} CHI_3 + RCOO^-$$
A methylcarbinol   Iodoform

Iodoform is produced because the strong electron-withdrawing effect of the CI_3 group makes the carbon of the carbonyl group very electron-deficient which allows nucleophilic attack of the hydroxide ion to occur.

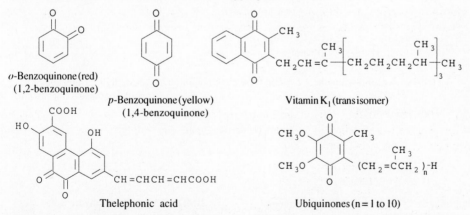

Chlorine and bromine in aqueous alkali will also react in the same way. However, the products of these reactions are liquids, namely trichloromethane (chloroform) and tribromomethane (bromoform) respectively.

19.14 QUINONES

Quinones are compounds that contain the *ortho*-benzoquinone and *para*-benzoquinone ring structures shown in Fig. 19. 15. No *meta*-benzoquinone ring structure is possible. The presence of a quinone structure in a molecule is indicated in the systematic names of compounds by the suffix **quinone** and the appropriate locants.

o-Benzoquinone (red)
(1,2-benzoquinone)

p-Benzoquinone (yellow)
(1,4-benzoquinone)

Vitamin K_1 (trans isomer)

Thelephonic acid

Ubiquinones (n = 1 to 10)

Fig. 19.15. *Para* and *ortho*-benzoquinones. Vitamin K_1 (trans isomer), a yellow viscous oil found in the liver. Vitamin K_1 regulates the formation of thrombin and is involved in blood-clotting. Thelephonic acid (reddish-black) occurs in some *Thelephora* fungi. Ubiquinones (n = 1 to 10) are compounds involved in the control of electron transfer processes in energy production in cells.

Quinones occur widely in nature and are often brightly coloured (Fig. 19.15). They often have characteristic ultra-violet and visible spectra which can be used as the basis of assay and identification tests.

Quinones are reactive structures. They undergo reduction, addition and free radical substitution but as these systems are not aromatic, substitution is uncommon.

19.14.1 Reduction

Quinones are readily reduced by many different reducing agents to the corresponding phenols. The reaction is easily reversed by many oxidising agents.

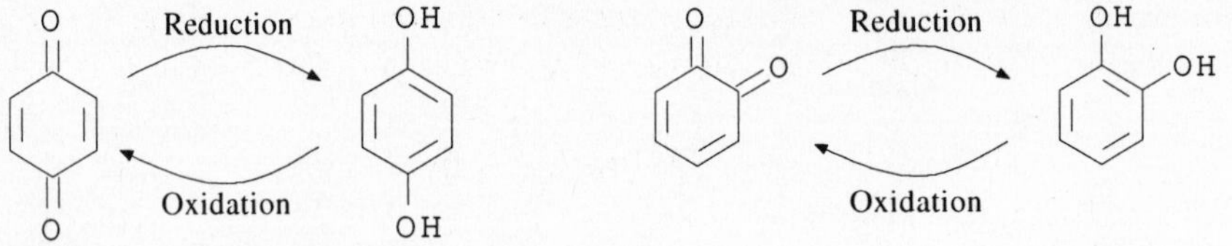

p-Benzoquinone   Hydroquinone (quinol)   o-Benzoquinone   Catechol

Quinone/hydroquinone redox systems are found in a number of biological processes. For example, a quinone/hydroquinone redox system involving ubiquinones occurs as part of an electron transfer processes controlling energy release in cells.

19.14.2 Addition

The addition reactions of quinones can be broadly classified into two types: C=C bond additions similar to those that occur in the alkenes and 1,4-additions involving one of the carbonyl groups. It should be noted that the numbers in the latter case refer to the relative positions of the added groups and are not locants.

Alkene-type addition. Both simple and Diels-Alder additions are known.

Simple alkene-type addition:

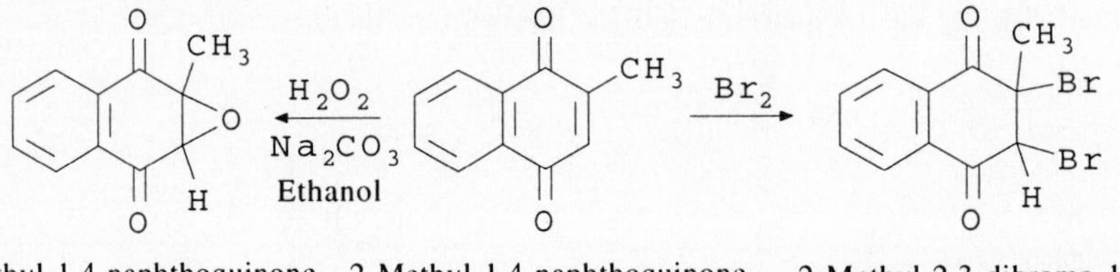

2-Methyl-1,4-naphthoquinone   2-Methyl-1,4-naphthoquinone   2-Methyl-2,3-dibromo-1,4-
oxide   naphthoquinone (mp 107°C)

Diels-Alder reaction:

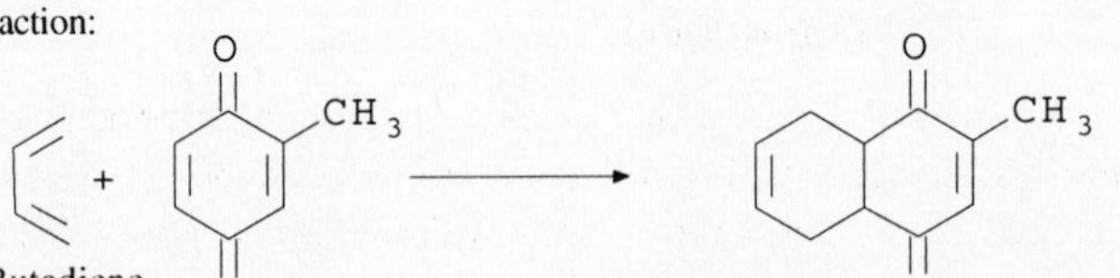

1,3-Butadiene

2-Methyl-1,4-benzoquinone   2-Methyl-5,6,9,10-tetrahydro-1,4-
                                                                          naphthoquinone (mp 81°C)

1,4-Addition. This type of addition involves the carbonyl group. It may include conversion of the product to the appropriate hydroquinone or quinone.

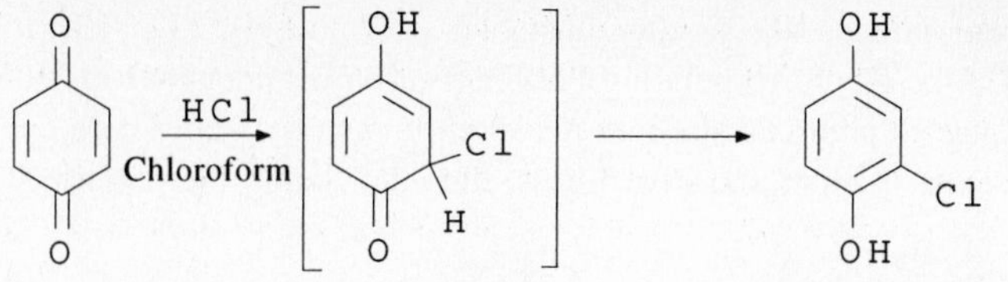

p-Benzoquinone   Unstable 1,4-addition product   2-Chloroquinol 74% (mp 106°C)

Thiols react via a 1,4-addition and a redox reaction with all the unsubstituted (free) positions of a quinoid ring to form thio-substituted quinones. Peptides and proteins with thiol groups have also been known to react with quinones in this manner. This reaction introduces a structure capable of undergoing redox reactions into a biological molecule.

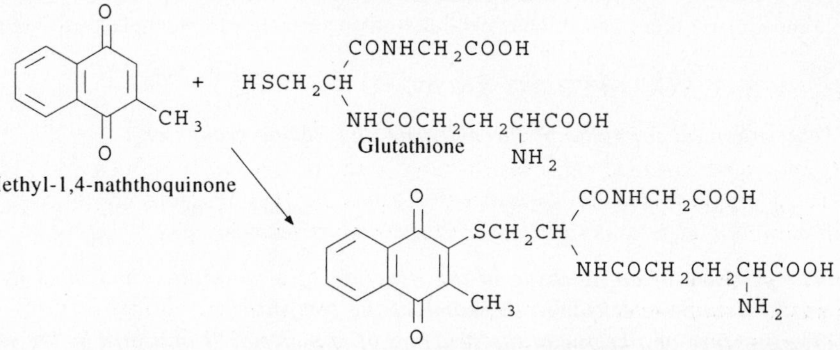

2-Methyl-1,4-naththoquinone

S-(2-Methyl-1,4-naththoquinonyl-3-)-glutathione

19.14.3   Substitution

Quinones rarely undergo traditional electrophilic aromatic substitution but they are prone to free radical substitutions. For example, benzenediazonium chloride is believed to react with 1,4-benzoquinone by a free radical mechanism.

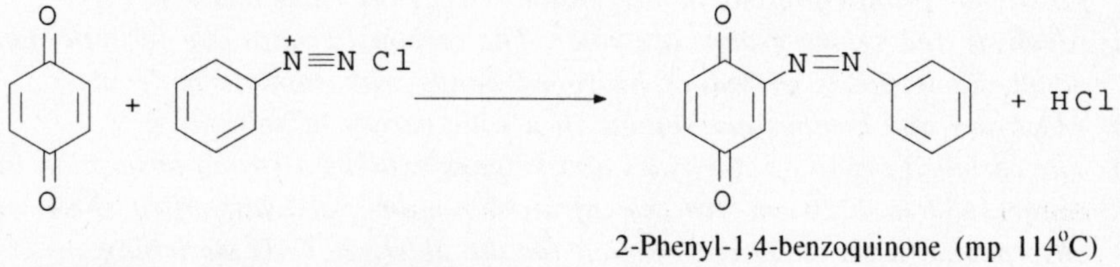

2-Phenyl-1,4-benzoquinone (mp 114°C)

19.15 THIALS AND THIONES

Thials (thioaldehydes) and thiones (thioketones) are the sulphur analogues of aldehydes and ketones (Fig. 19.16). The suffix **thial** is used to indicate the presence of a thioaldehyde group in a molecule whilst either the suffix **thione** or the prefix **thioxo** are used for thioketones.

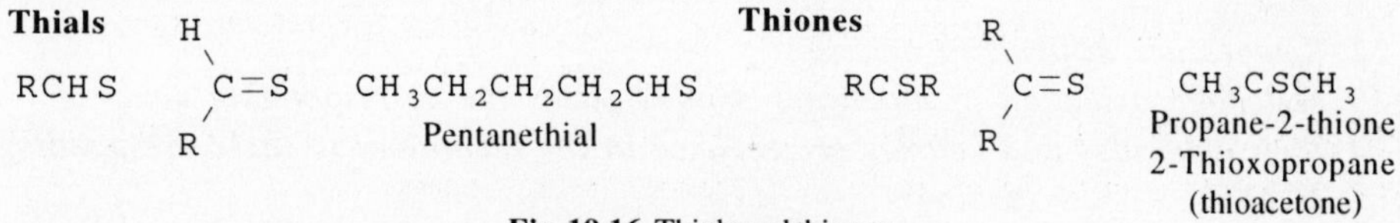

Fig. 19.16. Thials and thiones.

Thials and thiones are usually unstable, polymerising to cyclic trimers known as trithianes. Unlike the corresponding polymerisation of aldehydes (19.9.5), no catalyst is needed because sulphur is a stronger nucleophile than oxygen. These cyclic trithianes are believed to exist in chair and boat forms, the chair being the preferred conformation.

$3CH_3CHS \longrightarrow$
Ethanethial
(thioacetaldehyde)

$3CH_3CSCH_3 \longrightarrow$
2-Propanethione
(thioacetone)

2,4,6-Trimethylcyclohexa-1,3,5-thiane   2,4,6-Hexamethylcyclohexa-1,3,5-thiane

Low molecular mass thials and thiones are red oils with extremely unpleasant smells. Stable monomeric thials and thiones exhibit similar properties to aldehydes and ketones.

19.16   WHAT YOU NEED TO KNOW

(1) The structural formulae of the aldehyde and ketone groups are:

Aldehydes $RCHO$ or $\overset{R}{\underset{H}{\diagdown}}C=O$ *Ketones* $RCOR$ or $\overset{R}{\underset{R}{\diagdown}}C=O$

(2) The presence of an aldehyde in the structure of a molecule is indicated by the suffixes -al or -aldehyde in the name of the compound.

(3) The presence of a ketone in the structure of a molecule is indicated by the suffix -one or the prefix oxo- in the name of the compound.

(4) Both aldehydes and ketones exist in solution and the liquid state as equilibrium mixtures of two or more isomers known as tautomers. In simple compounds the position of equilibrium usually favours the keto form.

(5) The properties of aldehydes and ketones are a mixture of the properties of all the functional groups present in the tautomeric equilibrium mixture.

(6) Aldehyde and ketone groups are polar. The carbonyl groups can form permanent dipole-dipole bonds including hydrogen bonds with suitable molecules.

(7) Aldehydes and ketones are soluble in a wide variety of solvents.

(8) The carbonyl group of aldehydes and ketones exhibits a strong absorption in the range 1650 to 1720 cm^{-1} in the infra-red region. Aldehydes often show strong absorptions in the 2800 cm^{-1} region for the aldehyde C–H stretching.

(9) Aldehydes and ketones show a weak absorption in the ultra-violet whilst their hydrates do not absorb in the ultra-violet.

(10) The general types of chemical reaction undergone by aldehydes and ketones are: oxidation, reduction, nucleophilic addition and condensation.

(11) Aldehydes are readily oxidised in the laboratory to the corresponding carboxylic acid: $RCHO \longrightarrow RCOOH$

(12) Ketones are difficult to oxidise.

(13) Metabolic oxidation of aldehydes usually yields the corresponding acid.

(14) Both aldehydes and ketones are reduced in the laboratory to the corresponding alcohol.

RCHO ——→ RCH$_2$OH   RCOR ——→ RCH(OH)R
Aldehyde   Primary alcohol   Ketone   Secondary alcohol

(15) Metabolic reduction of aldehydes and ketones usually yields the corresponding alcohol.

(16) Both aldehydes and ketones undergo nucleophilic addition.

$$\backslash C=O \xrightarrow{\text{Nu}-\text{E}} \backslash C \overset{OE}{\underset{Nu}{}}$$

(17) The mechanisms of nucleophilic additions are usually based on the general routes shown in Fig. 19.7.

(18) Both aldehydes and ketones undergo condensation reactions usually with nitrogen nucleophiles and carboanions.

$$\backslash C=O + H_2N- \longrightarrow \backslash C=N- + H_2O \qquad \backslash C=O + HC\overset{Y}{\underset{Z}{}} \longrightarrow \backslash C=C\overset{Y}{\underset{Z}{}} + H_2O$$

Nitrogen nucleophiles   Carboanion

(19) The condensation reactions of aldehydes and ketones consist of a nucleophilic addition followed by an elimination. The elimination stage is not always spontaneous but may only proceed after the addition of an acid catalyst.

(20) The condensation reaction of 2,4-dinitrophenylhydrazine with aldehydes and ketones is used as a spot test in organic analysis for these functional groups.

Aldehyde
or ketone $\backslash C=O$ + H$_2$NNH—⟨NO$_2$⟩—NO$_2$ ——→ $\backslash C=$NNH—⟨NO$_2$⟩—NO$_2$ + H$_2$O

(21) Aldehydic and ketonic groups are meta directing to electrophiles in aromatic electrophilic substitution.

(22) Aldehydes and ketones with the structure CH$_3$CO- and compounds that can be oxidised by iodine to this structure will react with iodine in aqueous alkali to form yellow crystals of iodoform (CHI$_3$). The reaction is used as a test for the presence of these structures in a compound.

Compound $\xrightarrow[\text{by iodine}]{\text{Oxidation}}$ RCOCH$_3$ $\xrightarrow{\text{I}_2/\text{OH}^-}$ RCOOH + CHI$_3$

(23) Quinones are compounds whose molecular structures contain the structural units:

1,2-Benzoquinone   1,4-Benzoquinone

(24) Quinones have a characteristic ultra-violet and visible spectrum that can be made the basis of analytical procedures.

(25) Quinone rings are not aromatic, as their structures do not obey the 4n+2 rule.

(26) Quinones exhibit the properties of both ketones and conjugated alkenes. They may be reduced, undergo 1,2- and 1,4-addition and free radical substitution.

(27) Thials and thiones are the sulphur analogues of aldehydes and ketones. They are usually unstable, polymerising to cyclic trimers.

19.17 QUESTIONS

(1) Draw the structural formulae of each of the following compounds: (a) 2-ethyl-3-methylhexanal, (b) 3-methylcyclohexanone, (c) 2-methylpyridine-4-carbaldehyde, (d) 4-hydroxy-3-methoxybenzaldehyde, (e) 2-propenal, (f) penta-2,4-dione, (g) diphenylethanedione, (h) 3-oxopentane, (i) 3-phenyl-1,2-benzoquinone, (j) 2-methyl-1,4-napthaquinone, (k) 1,8-dihydroxy-9,10-anthraquinone, (l) 4-methylpentane-2-thione and (m) 2-thioxo-3-methylbutane.

(2) Compound A is 3,5-dimethylbenzaldehyde. (a) Draw the structural formula of A and describe the three-dimensional shape of a molecule of A. (b) List the principal general chemical and physical properties you would expect A to exhibit. (c) Suggest chemical reactions that could be made the basis of three identification tests for A. (d) Outline, by means of equations, a method of assaying A.

(3) Name the general types of reaction that you would expect 2-hexanone to undergo with each of the reagents (a) to (h). Show, by means of equations, the expected products of each of these reactions. (a) Lithium aluminium hydride, (b) a dilute aqueous acid solution, (c) a mixture of methanol and dry hydrogen chloride, (d) zinc and hydrochloric acid, (e) hydrazine in the presence of an acid catalyst, (f) ethyl ethanoate after treatment of the ester with sodium ethanoate, (g) hydroxylamine and (h) ethylamine.

(4) Draw the structure of the hemiacetals formed when each of the pairs of reagents react: (a) ethanol with propanal, (b) the hydroxy group of 5-hydroxypentanal with the carbonyl group of this molecule and (c) propanol with propanone.

(5) The following conversions are stages in various metabolic processes. Name the general type and suggest a feasible electronic mechanism for each of these reactions. (a) The oxidation of methanal in the liver.

$$HCHO \quad \xrightarrow{\text{Glutathione (GSH)}} \quad GSCH_2OH$$

(b) The biosynthesis of the alkaloid trachelanthamidine.

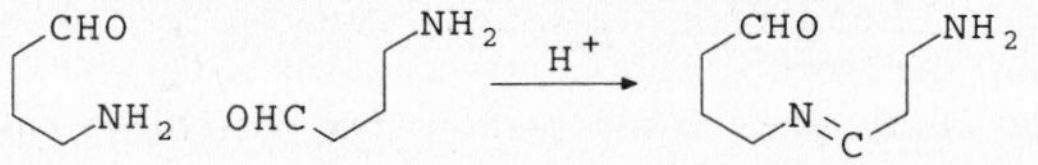

(6)

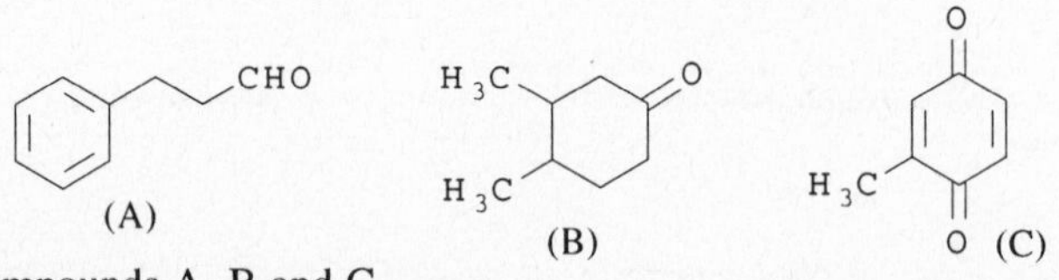

(A)   (B)   (C)

(a) Name compounds A, B and C.

(b) Suggest chemical reactions that could be used to distinguish between compounds A, B and C.

(c) Comment on the stereochemistry of the products of the reaction of compounds A and B with (i) cyclohexanol and (ii) phenylhydrazine.

20

Organic acids

20.1   INTRODUCTION

Acids are defined as compounds that can act as proton donors (6.13). This broad definition covers a wide variety of organic compounds. Many of these compounds are not classified as acids but can, under some conditions, behave as acids. For instance, the aldehydes and ketones discussed in Chapter 19 can donate protons to a base to form a carboanion (19.11.6). Many other types of compound also behave in this manner (e.g.18.4.3, 21.6.2 and 22.6.2).

Table 20.1. Examples of some of the classes of organic compound that are normally referred to as acids. Sulphinic and sulphenic acids have not been isolated but their derivatives are well characterised.

Acid	Structure	Acid	Structure
Carboxylic acids	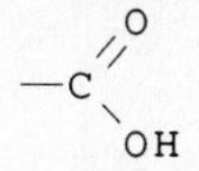 —C⟨$_{OH}^{O}$	Thio acids	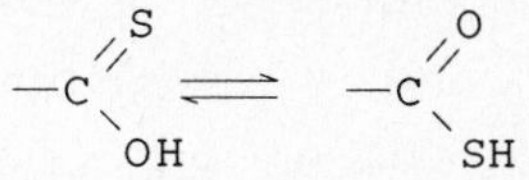 —C⟨$_{OH}^{S}$ ⇌ —C⟨$_{SH}^{O}$
Dithio acids	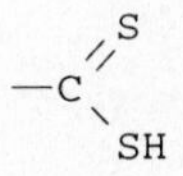 —C⟨$_{SH}^{S}$	Sulphonic acids	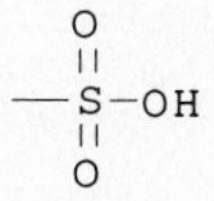 —S—OH (with O above and O below)
Sulphinic acids	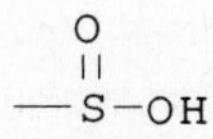 —S—OH (with O above)	Phosphoric acids	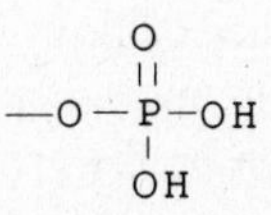 —O—P—OH (with O above and OH below)
Sulphenic acids	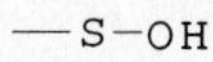 —S—OH		

The main organic compounds that are traditionally classified as acids are the carboxylic acids and some sulphur- and phosphorus-based functional groups whose structures contain -O-H and -S-H bonds (Table 20.1). Carboxylic acids occur widely in nature

(Fig. 20.1) and play important parts in many metabolic processes whilst long chain monocarboxylic acids such as stearic, oleic and palmitic acids are important constituents of natural fats. This fact led to the use of the term **fatty acid** for long chain carboxylic acids. Sulphur- and phosphorus-based acids are not as common but their derivatives are of considerable biological importance.

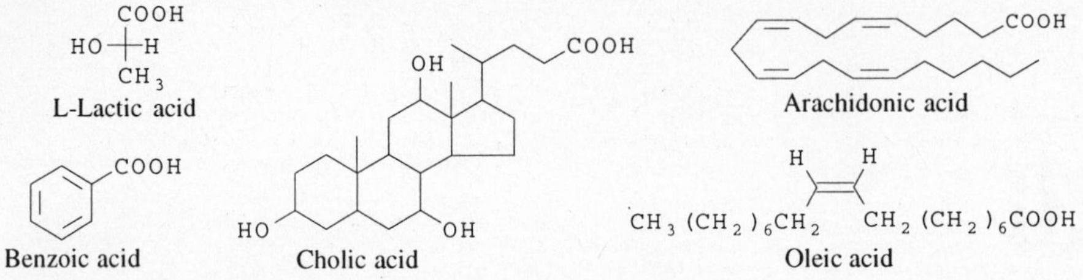

Fig. 20.1. Examples of naturally occurring carboxylic acids. L-Lactic acid occurs in the citric cycle of mammals. Cholic acid is a component of human bile. Benzoic acid occurs in berries and gum benzoin. Arachidonic acid is a precursor of prostaglandins, thromboxanes and leukotrienes. Oleic acid is a component of animal fats.

20.2 NOMENCLATURE

20.2.1 Carboxylic acids

The prefix **hydroxylcarbonyl** and the suffix **oic** acid are used in systematic nomenclature to indicate the presence of a carboxylic acid functional group in a molecule (Fig. 20.2). In the latter instance the carbon of the carboxylic acid group is designated as number 1. Alternatively, the suffix **carboxylic acid** may be used. In this case the locant used to indicate the position of the carboxyl group depends on the nature of the parent structure. It should be noted that the IUPAC rules allow a large number of trivial names to be used for simple carboxylic acids.

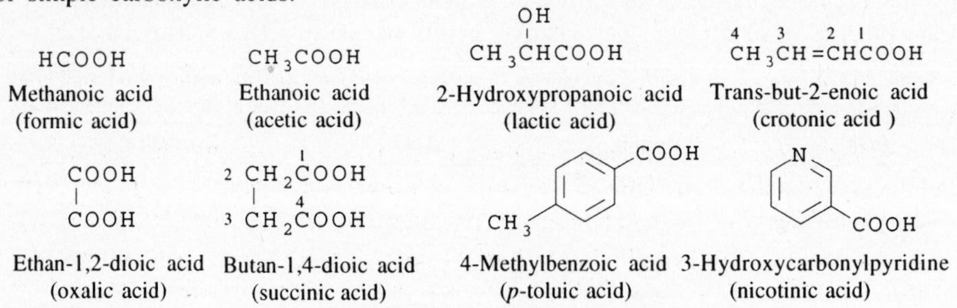

Fig. 20.2. Examples of the systematic and trivial nomenclature of carboxylic acids. Trivial names in common use are given in the brackets.

Substituted carboxylic acids are often classified in terms of the relative position of the substituent to the carboxylic acid group. The position of the substituent is indicated in the classification by the appropriate Greek letter. For example, keto acids are often referred to as α–, β–, γ–, etc. keto acids.

α	β α	γ β α
RCOCOOH	$RCOCH_2COOH$	$RCOCH_2CH_2COOH$
α-Keto acid	β-Keto acid	γ-Keto acid

20.2.2 Sulphur acids

The prefixes and suffixes used for the stable sulphur acids are given in Table 20.2. It should be noted that as sulphenic and sulphinic acids are unstable they are not included in this table. In general, the rules governing the use of locants for sulphur acids are the same as those used for the prefixes and suffixes of the corresponding carboxylic acid structure.

Table 20.2. The prefixes and suffixes used with sulphur acids.

Type of acid	Examples	
	Prefix	Suffix
Thio acid	thio....S or O acid	thionic or thiolic acid
	$CH_3-C{\overset{O}{\underset{SH}{}}} \rightleftharpoons CH_3-C{\overset{S}{\underset{OH}{}}}$	$CH_3-C{\overset{O}{\underset{SH}{}}} \rightleftharpoons CH_3-C{\overset{S}{\underset{OH}{}}}$
	Thioethanoic S-acid Thioethanoic O-acid	Ethanthiolic acid Ethanthionic acid
Dithio acids	dithio	carbothiolic, thionthiolic
	(phenyl)$-C{\overset{S}{\underset{SH}{}}}$	(phenyl)$-C{\overset{S}{\underset{SH}{}}}$
	Dithiobenzoic acid	Benzenecarbothiolic acid
Sulphonic acids	none	sulphonic acid
		(phenyl)$-\overset{O}{\underset{O}{S}}-OH$
		Benzenesulphonic acid
Hydrogen sulphates	none	hydrogen sulphate
		$CH_3-O-\overset{O}{\underset{O}{S}}-OH$
		Methyl hydrogen sulphate

20.2.3   Phosphorus acids

A confusing variety of names has been used in the past for the phosphorus oxy-acids, and so when consulting older literature it is wise to check the names against the structural formula of the acid. The modern systematic names of monophosphorus oxy-acids whose structures contain one phosphorus atom are based on the stem **phosph-** which is combined with a suffix indicating the valency of the phosphorus and basicity of the acid (Table 20.3). For example, the suffix **-onous** indicates that the acid has a trivalent phosphorus that is dibasic. It should be noted that the names of all phosphorus oxy-acids containing a three valent phosphorus atom end in either **-us** or **-ous** whilst the names of all those possessing a pentavalent phosphorus end in **-ic**. The names of polyphosphorus oxy-acids whose structures contain more than one phosphorus atom have a prefix indicating the number of phosphorus atoms in the structure.

Organic phosphorus acids are named by prefixing the name of the appropriate phosphorus oxy-acid with the radical name of the residue bonded to the phosphorus atom (Fig. 20.3). It is important to realise that the appropriate oxy-acid name is selected on the basis of the basicity of the organophosphorus acid. Trivial names are also in common use.

Table 20.3. The nomenclature of the principal phosphorus oxy-acids. The names in brackets are names that may be found in older literature.

Basicity	Tervalent (3)	Pentavalent (5)
Tribasic	Suffix **-orus** or **-orous**	Suffix **-oric**
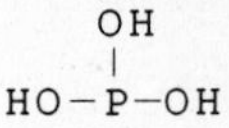	OH | HO−P−OH Phosphorus acid (orthophosphorous)	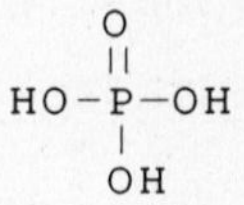 O || HO−P−OH Phosphoric acid (orthophosphoric) | OH
Dibasic	Suffix **-onous** or **-onus**	Suffix **-onic**
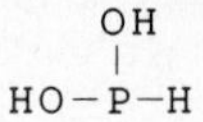	OH | HO−P−H Phosphonus acid (hypophosphorus)	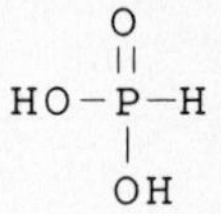 O || HO−P−H Phosphonic acid (in some texts phosphorous acid which is incorrect) | OH
Monobasic	Suffix **-inous** or **-inus**	Suffix **-inic**
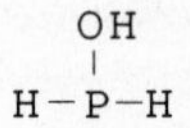	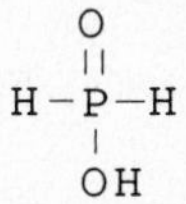 OH | H−P−H Phosphinous acid	O || H−P−H Phosphinic acid (in some texts hypophosphorous acid) | OH

Five valent polyphosphorus-oxy acids

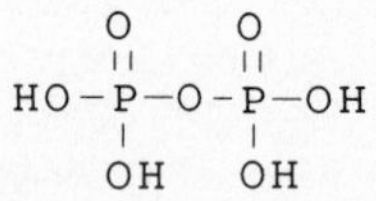

O   O
\|\| \|\|
HO−P−O−P−OH
\| \|
OH OH

Dipolyphosphoric acid
(pyrophosphoric acid)

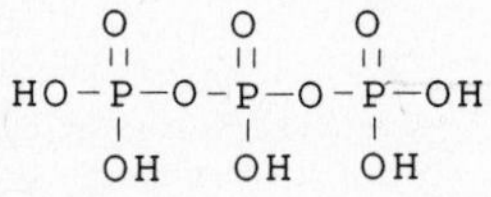

O   O   O
\|\| \|\| \|\|
HO−P−O−P−O−P−OH
\| \| \|
OH OH OH

Tripolyphosphoric acid

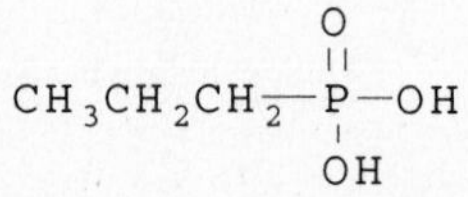

O
\|\|
$CH_3CH_2CH_2$−P−OH
\|
OH

Propylphosphonic acid (a dibasic
phosphorus oxy-acid group)

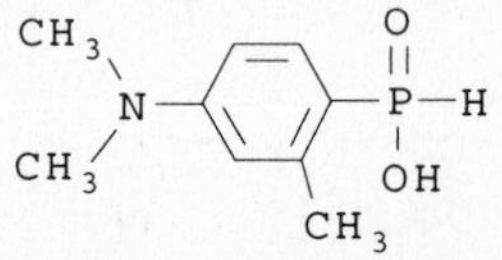

4-Dimethylamino-2-methylphenylphosphinic acid
(a monobasic phosphorus oxy-acid group)

Fig. 20.3. Examples of the names of organic phosphorus acids.

20.3 PHYSICAL PROPERTIES

The low molecular mass organic acids tend to be colourless liquids, whilst higher molecular mass compounds are white solids. Most liquid organic acids have characteristic odours, the sulphur acids being particularly unpleasant. Liquid carboxylic acids and solid carboxylic acids in solution may exist as dimers due to intermolecular hydrogen bonding (Fig. 20.4). It is thought that some phosphorus oxy-acids may also exist as dimers. However, as sulphur is reluctant to form hydrogen bonds, dimerisation is unlikely with the sulphur oxy-acids. The reluctance of sulphur to form hydrogen bonds may be simply explained in terms of the large 'size' of the orbitals containing the lone pairs which results in the lone pair being too diffused to act as the negative end of a hydrogen bond.

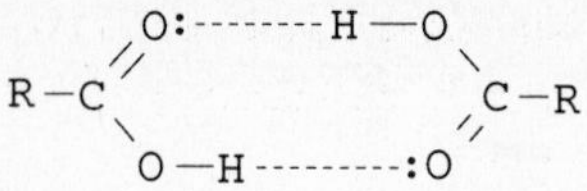

$$R-C \overset{\text{O:}\cdots\cdots\text{H}-\text{O}}{\underset{\text{O}-\text{H}\cdots\cdots\text{:O}}{}} C-R$$

Fig. 20.4. Dimerisation of carboxylic acids.

The nature of higher molecular mass carboxylic acids ranges from crystalline to wax-like solids. For example, benzoic acid is a white crystalline solid whilst stearic acid is a white waxy solid. This, and the fact that stearic acid occurs as a constituent of human fat which renders it relatively harmless, is why it is often used as a lubricant in tableting.

Liquid and solid thio acids exhibit tautomerism, the latter when placed in a suitable solvent.

$$-\text{C SOH} \rightleftharpoons -\text{CO SH} \qquad -\overset{\text{S}}{\underset{\text{OH}}{C}} \rightleftharpoons -\overset{\text{O}}{\underset{\text{SH}}{C}}$$

Thionic form   Thiolic form   Thionic form   Thiolic form

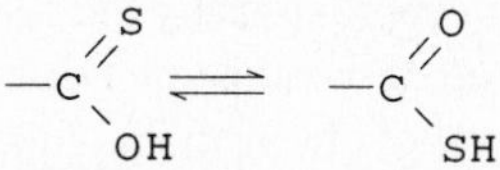

Simple sulphonic acids are hygroscopic liquids and solids. Consequently, drugs and other compounds which contain sulphonic acid functional groups must be stored under anhydrous conditions. Sulphonic acid groups are often introduced into a molecule either to increase its water solubility or, by making the molecule more acidic, to increase its rate of excretion via the kidneys.

The monophosphorus oxy-acids may or may not exhibit tautomerism. For example, experimental work has shown that phosphinic acid does not have a tautomeric form in solution. However, practical evidence suggests that there is a possibility of phosphorous acid existing as a tautomeric equilibrium mixture.

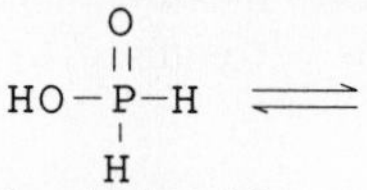

$$\overset{\text{O}}{\underset{\text{H}}{\text{HO}-\text{P}-\text{H}}} \rightleftharpoons \text{Tautomer not detected} \qquad \overset{\text{OH}}{\underset{\text{OH}}{\text{HO}-\text{P}}} \rightleftharpoons \overset{\text{O}}{\underset{\text{OH}}{\text{HO}-\text{P}-\text{H}}}$$

Phosphinic acid   Phosphorous acid ←——— Position of equilibrium, 99%

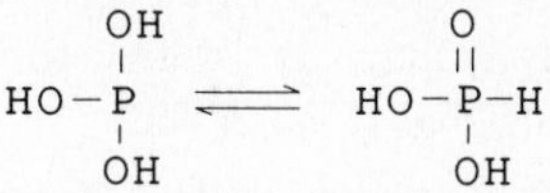

20.3.1 Solubility

Organic acids are able to form hydrogen bonds and, after ionisation, ion-dipole bonds with water molecules. Consequently, simple low molecular mass acids are usually quite water soluble, but as the molecular mass increases the solubility drops unless other functional groups that improve water solubility are present in the molecule. Simple high molecular mass acids are usually insoluble in water. In general, sulphonic and phosphorus oxy-acids are more soluble than carboxylic acids. Phosphorous oxy-acids residues are often introduced into the structure of molecule in order to improve its water solubility. Thio and dithio acids decompose irreversibly in water to hydrogen sulphide and the corresponding carboxylic acid.

$$\text{RCOSH} + \text{H}_2\text{O} \longrightarrow \text{RCOOH} + \text{H}_2\text{S}$$

Aliphatic organic acids are usually more water soluble than aromatic acids. Low molecular mass aliphatic carboxylic acids are soluble in cold water but aromatic carboxylic acids with similar molecular masses are appreciably soluble only in hot water.

20.3.2   Spectroscopic properties

The absorptions of carboxylic, thio and sulphonic acid groups in the infra-red region (Fig. 20.5) can be used with other information to identify these groups in analysis.

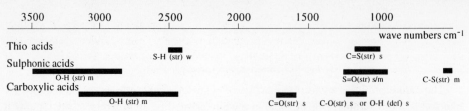

Fig. 20.5. The approximate regions of the infra-red in which carboxylic, thio and sulphonic acids absorb.

The ^{1}H magnetic resonance spectra signals of the acidic protons of organic acids are normally found in approximately the 12 to 13 ppm region of the spectrum. The signals are singlets that are often broad but are usually too far from the other protons in the molecule to couple with and be split by them. Acidic protons readily exchange with D_2O and so their signals disappear after a D_2O shake, which makes it relatively easy to identify their signals.

The ^{13}C magnetic resonance spectra signals for the carbon of the carbonyl group of carboxylic acids are usually found in the 160 to 190 ppm region of the spectrum. Both ^{1}H and ^{13}C spectra are used to identify the possible presence of acid functional groups in a molecule.

20.4 CHEMICAL PROPERTIES

The functional groups traditionally classified as being an acid have a number of general types of property in common. They all act as acids, undergo nucleophilic displacement and act as electron acceptors. However, individual types of acid functional group will also exhibit characteristic properties. For example, thio-acids can react in either of their tautomeric forms but normally react in the thiolic form. In this form thio-acids can be oxidised to the corresponding diacyl sulphide.

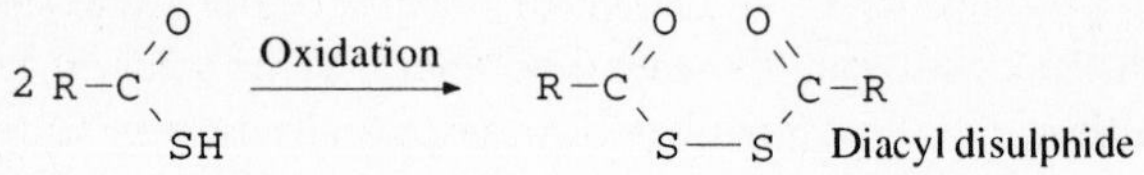

$$2 \ R-C \overset{O}{\underset{SH}{\diagdown}} \quad \xrightarrow{\text{Oxidation}} \quad R-C \overset{O}{\diagdown} \quad \overset{O}{\diagup} C-R \quad \text{Diacyl disulphide}$$

20.5 ACIDITY

The acidic nature of organic acid functional groups can be explained in terms of the electron-withdrawing effect of the adjacent C=O, S=O and P=O structures. These electron-withdrawing effects affect the degree of ionisation of the acid in two ways. First, they weaken the O–H and S–H bonds of the acids (Fig. 20.6) by withdrawing electrons from the oxygen or sulphur atoms of these bonds. This increases the electronegative

$$HA \quad + \quad H_2O \ \rightleftharpoons \ H_3^+O \quad + \quad A^- \qquad\qquad (20.1)$$
$$\text{Acid} \qquad\quad \text{Base} \quad \text{Conjugate acid} \quad \text{Conjugate base}$$

Carboxylic acids   Thio acids   Sulphonic acids   All types of phosphorus oxy-acids

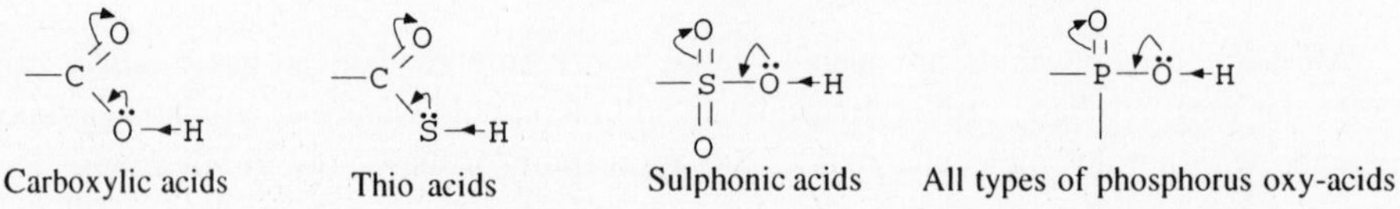

Fig. 20.6. The polarisation of the O–H and S–H bonds of organic acids by adjacent groups. Increased polarity, and hence a weaker bond, is due to the increase in the electronegative character of the oxygen or sulphur atom which is in turn due to the electron-withdrawing effect of the adjacent structure.

character of these atoms which results in an increase in the polarity of the O–H and S–H bonds. This weakens these bonds to the extent that, under the right conditions, they are able to ionise and donate a proton to another chemical species (equation 20.1, Fig. 20.6). It should be remembered that S–H bonds are weaker than O–H bonds (16.5) in this respect.

The second way in which the adjacent C=O, P=O and S=O structures affect the acidity of organic acid functional groups is that they contribute to the stabilisation of the conjugate base by delocalisation of the negative charge of this base. For example, ethanoic acid ionises in water:

$$CH_3COOH + H_2O \longrightarrow H_3^+O + CH_3COO^-$$

Ethanoic acid   Ethanoate ion

The negative charge of the ethanoate ion is spread over the whole of the COO⁻ structure. This makes the ethanoate ion less attractive to protons (Fig. 20.7). As a result, the protons formed by the ionisation are less likely to react with the ethanoate ion and more likely to react with the other chemical species present including the solvent, whereby the acidity of the acid is increased. Stabilisation of the conjugate base appears to be the most important **structural** feature that affects the acidity of an acid. However, the solvent can also have a significant effect on the ionisation of the acid and the stability of the conjugate base.

$$CH_3-C\overset{O}{\underset{O^-}{\diagup}} \rightleftharpoons CH_3-C\overset{O^-}{\underset{O}{\diagup}} \qquad CH_3-C\overset{O}{\underset{O}{\diagup}}^-$$

Fig. 20.7. Stabilisation of the ethanoate ion. (a) Canonical forms. (b) The negative charge is spread over the whole of the COO⁻ structure.

The strength of an acid is usually recorded as a pK_a value (6.13); however, it can also be defined as a pH value. Both pK_a and pH are a measure of the extent of the ionisation of the acid in a particular solvent. In general, sulphur organic acids are stronger, that is, more highly ionised than the corresponding carboxylic and phosphorus organic acids.

Acids with structures that can donate electrons to the acid group are weaker than those of a similar size and type with structures that can withdraw electrons from the acid group provided the solvent effects are not greatly different. For example, 4-methoxybenzoic acid (pK_a 4.46 water) with its electron donor substituent is a weaker acid than 4-nitrobenzoic acid (pK_a 3.41 water) with its electron withdrawing substituent.

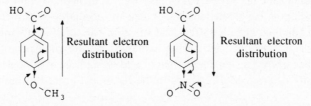

The donation of electrons to the acid group by a structure increases the strength of the O–H or S–H bond by reducing the polarity of the oxygen or the sulphur atom (Fig. 20.8). This increase in strength reduces the ease of ionisation which results in a weaker acid. Moreover, electron donor structures destabilise the conjugate base by reducing the extent of the delocalisation of the negative charge of the oxygen. This increases the attraction of the anion for protons with the result that the equilibrium for the ionisation of the acid (equation 20.1) lies to the left. This also has the effect of making the acid weaker.

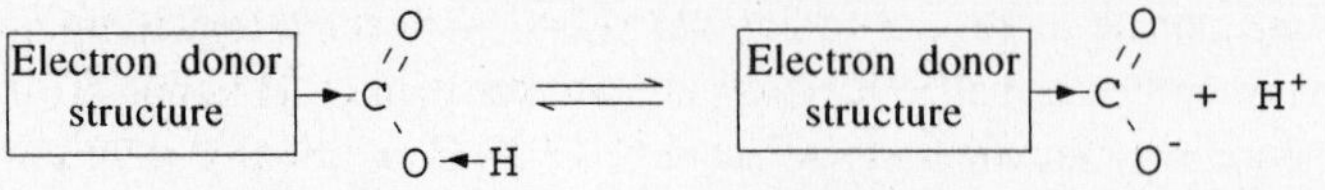

| Electrons donated to the carboxylic acid by the appropriate I and M effects reduces the polarisation of the O-H bond and strengthens the bond | Electrons donated to the carboxyate group by the appropriate I and M effects increases the negative character of the carboxylate ion and reduces its stability |

Fig. 20.8. The effect of an electron donor structure on acid strength of a carboxylic acid. The same arguments are also valid for the other types of acid.

Structures that withdraw electrons from the acid group increase the polarisation of the O–H and S–H bonds of that acid (Fig. 20.9). This decreases the strength of the bond and increases the ease of ionisation with a subsequent increase in acid strength. Furthermore, electron-withdrawing structures stabilise the conjugate base by increasing the extent of the delocalisation of the negative charge on the oxygen. This decreases the attraction of the conjugate base for protons and so the equilibrium for the ionisation of the acid (equation 20.1) lies to the right , that is, the acid is stronger.

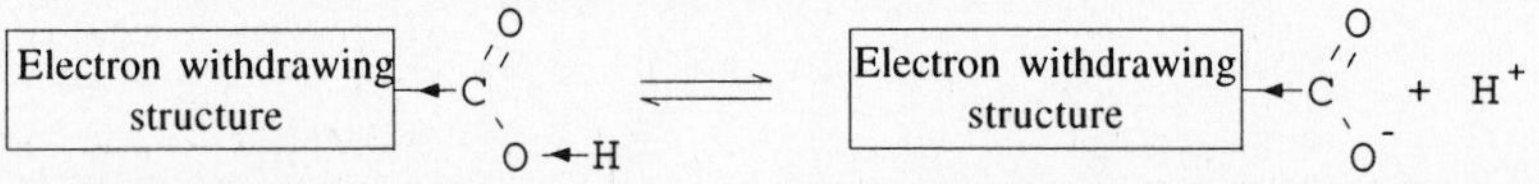

| Electrons withdrawn from the carboxylic acid by the appropriate I and M effects increase the electronegativity of the oxygen and the polarisation of the O-H bond which weakens the O-H bond | Electrons withdrawn from the carboxylate by the appropriate I and M effects reduce the negative charge of the carboxylate ion and so increase the stability of the anion |

Fig. 20.9. The effect of an electron-withdrawing structure on the strength of a carboxylic acid. The same argument is valid for the other types of acid.

At constant temperature the relative strengths of acids with the similar types of structures may be predicted by considering:

(1)   The relative strengths of the S–H or O–H bond of the acid.

(2)   The relative stabilities of their conjugate bases.

These predictions are not very accurate because it is difficult to allow for the effect of the solvent. In addition, the predictions do not take into account other structural features, such as hydrogen bonding, that may affect acid strength. The significantly higher than expected acid strength of salicylic acid can be explained by intramolecular hydrogen bonding stabilising the carboxylate ion. Predictions of acid strength are of limited use when assigning experimentally determined pK_a values to the functional groups of multifunctional group molecules.

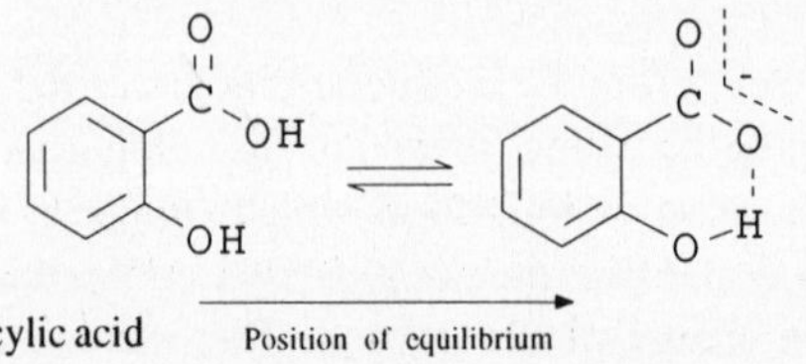

Isomer	pK_a
o	2.98
m	4.08
p	4.58

Salicylic acid   Position of equilibrium

The pK_a value of an acid is an important guide to the ease of movement of acids through biological membranes. Experimental evidence shows that acids pass through biological membranes in their non-ionised form. Factors that affect the position of equilibrium of the ionisation of an acid will affect its ease of absorption. An important factor in biological systems is the pH of the solutions on either side of the membrane. For example, the weak acid aspirin (pK_a 3.6, water), exists almost completely in the non-ionised form in buffer solutions with a pH of about 1 to 3 but is fully ionised in buffer solutions with a pH value of approximately 6 or more (Fig. 20.10). Consequently, aspirin will be unionised in the stomach which is acidic but will be fully ionised in the blood stream which has a pH in the region of 7.4. Full ionisation occurs because the acid probably forms salts with the protein bases in the blood and this drives the equilibrium of the ionisation to the right. The result is that aspirin in its unionised form will be transported through the membranes of the stomach into the blood. Once it enters the blood it does not diffuse back into the stomach as it is now fully ionised.

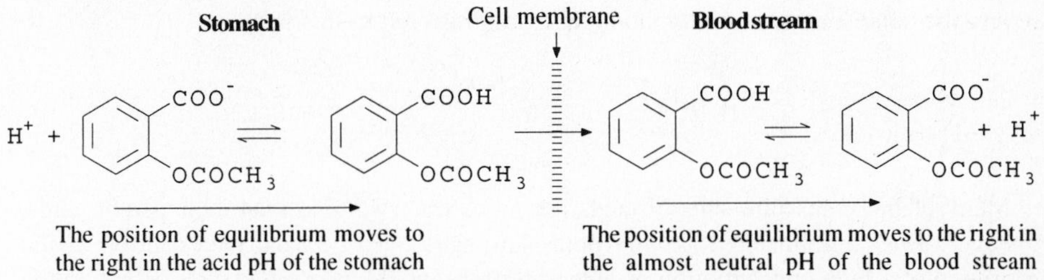

Fig. 20.10. A schematic representation of the transport of aspirin through the membranes of the stomach. The aspirin passes through the cell membrane as the unionised drug.

The strength of an acid will also affect its rate of excretion via the urine which is usually slightly acidic. Consequently, the presence of a strongly acid functional group in a drug will tend to suppress the excretion of that drug since it will tend to remain in its unionised form in the membrane.

20.6 REACTION AS AN ACID

All the types of acid discussed in this chapter form salts, liberate hydrogen from metals, carbon dioxide from carbonates and hydrogen carbonates, and change the colours of indicators.

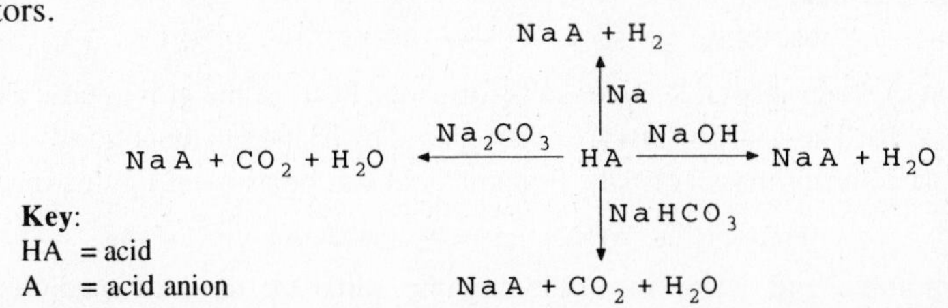

Key:
HA = acid
A = acid anion

The reaction with sodium hydrogen carbonate is used as a test for the presence of an acid group in qualitative organic analysis while titration with alkalis forms the basis of many volumetric analysis methods for assaying compounds that contain carboxylic and sulphonic acidic functional groups (Fig. 20.11).

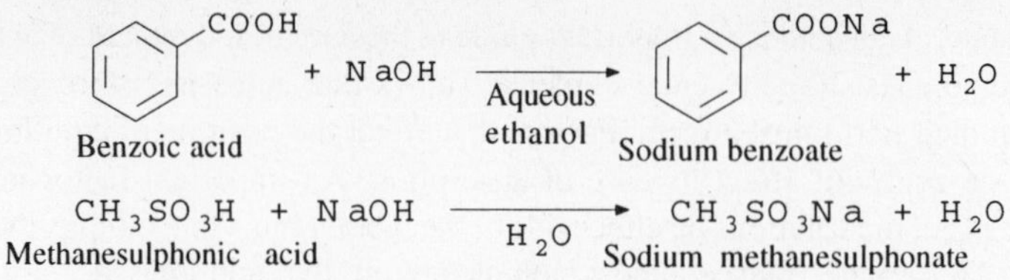

Fig. 20.11. Examples of the assay of acids by titration with sodium hydroxide. Both these examples use phenolphthalein as the indicator.

The sodium salts of many acids are more soluble in water than their parent acids because of the ion-dipole bonds formed between the ions and the water molecules (9.4). These salts normally have the same pharmaceutical action as their parent acids and so drugs with acid functional groups are sometimes administered in this form in order to take advantage of this increased water solubility and, as a consequence, better absorption. The sodium salts of sulphonic acids are less hygroscopic than their parent acids and so drugs such as the antileprotic solapsone are usually stored and administered in this form.

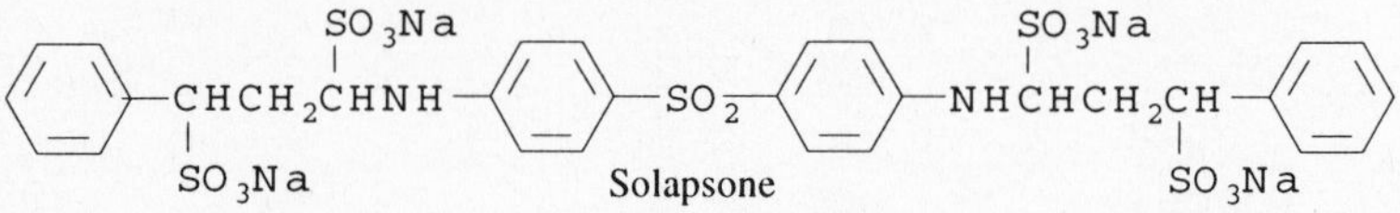

Many of the potassium salts of acids are more water soluble than their parent acids. However, drugs administered as their potassium salts must be used with caution as the absorption of a high concentration of potassium ions would have an adverse affect on the heart (30.2.2).

The groups 1 and 2 salts of 'long chain' aliphatic carboxylic acids are commonly known as soaps. Soaps whose chains contain 6 to 22 carbon atoms are used as surfactants (9.6.5), detergents and lubricants in pharmaceutical preparations. For example, calcium and magnesium stearates are waxy solids which are added to tablet mixtures in order to facilitate the ejection of the tablets from tableting machines.

The heavy metal salts of organic acids, with the exception of some sulphonates are usually insoluble in water. Their formation has been used as the basis of identification tests. For example, the 1993 British Pharmacopoeia identification test for the preservative benzoic acid depends on its reaction with iron III chloride solution to form a buff precipitate of insoluble iron III benzoate:

$$3PhCOOH \ + \ FeCl_3 \longrightarrow (PhCOO)_3Fe \ + \ 3HCl$$

The formation of water insoluble silver salts forms the basis of the gravimetric assay of some carboxylic acids. These silver salts are decomposed by heating to metallic silver which is then weighed to constant mass (11.6.4). Benzoic acid can be assayed by this method.

$$PhCOOH \ + \ AgNO_3 \longrightarrow PhCOOAg \xrightarrow{heat} Ag$$

Both water soluble and water insoluble organic salts are used in medicine. For example, the salts of tartaric, citric and other very water soluble organic acids make it possible for drugs that are sparingly soluble or insoluble in water but whose structures contain basic groups, to be administered by injection in aqueous solution (Fig. 20.12). It should be remembered that these salts have the same pharmacological action as the base.

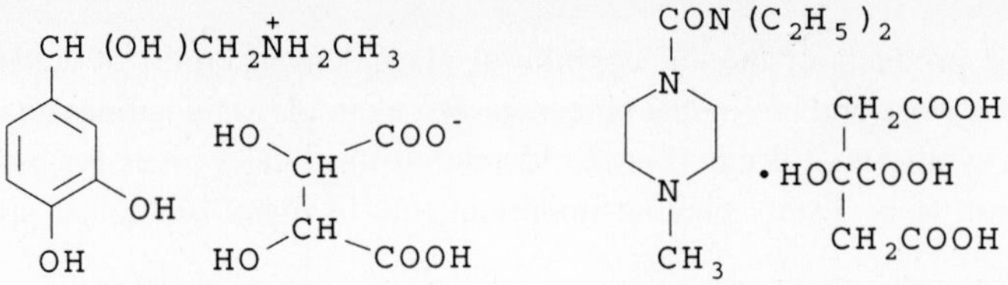

CH(OH)CH$_2$NH$_2$CH$_3$

Adrenaline tartrate

CON(C$_2$H$_5$)$_2$

Diethylcarbamazine citrate

Fig. 20.12. Examples of drugs that may be administered by injection in aqueous solution. It should be noted that a dot is used in the structural formula when it is not obvious which of the functional groups forms the salt. Both the adrenaline tartrate and diethylcarbamazine citrate are very water soluble, unlike their parent bases.

Water insoluble organic salts of embonic and other acids (Fig. 20.13) have been used to treat hookworm and pinworm infestations of the gastrointestinal tract. The poor water solubility of these salts results in poor absorption in the gut which means that the greater part of the drug passes through into the gastrointestinal tract where it acts on the infestations (Fig. 20 13).

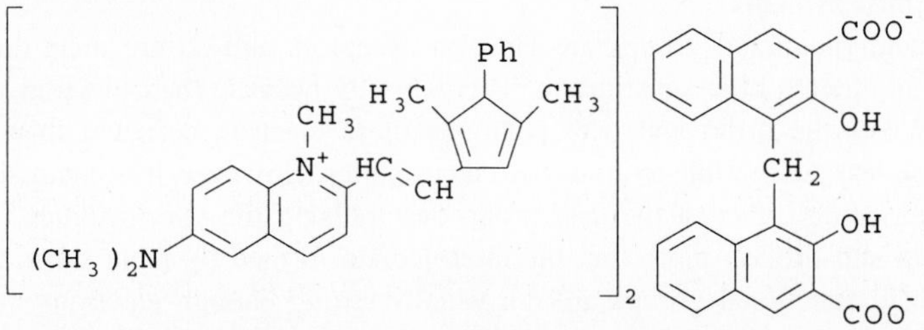

Fig. 20.13. Viprynium embonate, used to treat worm infestations (nematodes) of the gastrointestinal tract.

Salt formation is also important in the binding of acids to carrier and receptor proteins.

20.7 NUCLEOPHILIC DISPLACEMENT REACTIONS

The hydroxyl and thiol groups of some organic acids may be replaced by a variety of nucleophiles (Fig. 20.14). These displacements proceed by a variety of mechanisms.

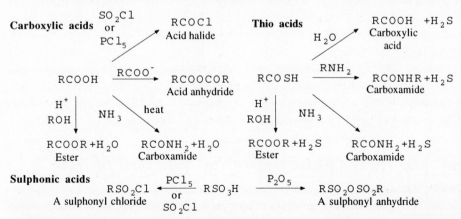

Fig. 20.14. General examples of the nucleophilic displacement reactions of carboxylic and thio acids.

Many of the products of the nucleophilic displacement reactions of acids are important in laboratory work and biological processes. For example, acid halides and anhydrides are used in the synthesis of drugs (Fig. 20.15) and to form derivatives for use in analysis and identification tests. Esters play an important role in many biological processes.

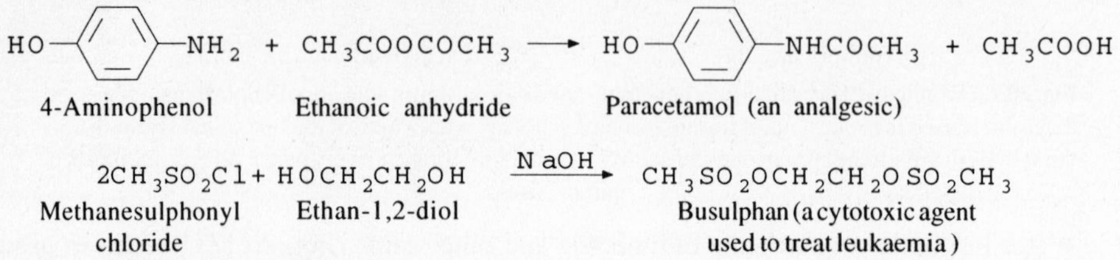

$$HO-\langle\!\!\!=\!\!\!\rangle-NH_2 \ + \ CH_3COOCOCH_3 \longrightarrow HO-\langle\!\!\!=\!\!\!\rangle-NHCOCH_3 \ + \ CH_3COOH$$

4-Aminophenol   Ethanoic anhydride   Paracetamol (an analgesic)

$$2CH_3SO_2Cl + HOCH_2CH_2OH \xrightarrow{NaOH} CH_3SO_2OCH_2CH_2OSO_2CH_3$$

Methanesulphonyl   Ethan-1,2-diol   Busulphan (a cytotoxic agent
chloride   used to treat leukaemia)

Fig. 20.15. Examples of the use of acid anhydrides and halides in the synthesis of drugs.

20.8 AS ELECTRON ACCEPTORS

20.8.1 Aromatic systems

All organic acid functional groups are electron acceptors and so are meta directing to electrophiles in electrophilic substitution. This is partly because their electron-withdrawing effects makes the ortho and para positions more electron deficient than the meta position and so less susceptible to attack by electrophiles. However, it is due mainly to the electron-withdrawing effect of the acid group destabilising the intermediates formed by ortho and para substitution more that the intermediate formed by meta substitution.

Organic acid functional groups are not usually strong enough electron-withdrawing groups to promote the nucleophilic substitution of aromatic systems.

20.8.2 Aliphatic systems

Acid functional groups increase the polarisation of any α C−H bonds which weakens these bonds and allows these hydrogen atoms to be replaced by some electrophiles. For example, chlorine and bromine will successively substitute the α hydrogens of most carboxylic acids. The mechanism of these replacements is believed to be an addition-elimination process involving either an enol or an enol formed by the transfer of α hydrogen atom.

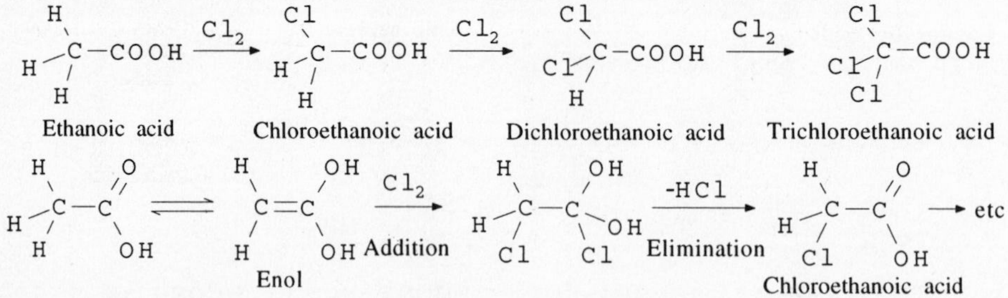

Ethanoic acid   Chloroethanoic acid   Dichloroethanoic acid   Trichloroethanoic acid

Enol   Chloroethanoic acid

20.9 REACTIONS CHARACTERISTIC OF THE TYPE OF ACID

20.9.1 Oxidation

Oxidation of acid functional groups is not common. However, many acids can be oxidised under laboratory conditions (Fig. 20. 16). Carboxylic acids are oxidised by hydrogen peroxide to the corresponding peracid whilst thio acids are converted to the diacyldisulphide derivative. It should be noted that thio acids will slowly oxidise in air and so must be stored correctly.

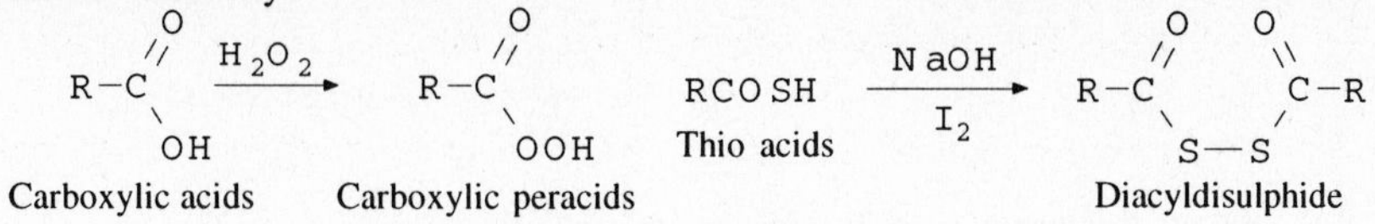

$$R-C\overset{O}{\underset{OH}{<}} \xrightarrow{H_2O_2} R-C\overset{O}{\underset{OOH}{<}} \qquad RCOSH \xrightarrow[I_2]{NaOH} R-C\overset{O}{\underset{S-S}{<}}\overset{O}{\underset{}{>}}C-R$$

Carboxylic acids   Carboxylic peracids   Thio acids   Diacyldisulphide

Fig. 20.16. Examples of the oxidation of organic acids.

20.9.2 Reducing agents

Phosphorus oxy-acids whose structures contain P–H bonds can act as reducing agents. For example, phosphonic acid (phosphorous) and phosphinic (hypophosphorous) acid reduce Tollen's reagent (ammonical silver nitrate) to form a silver mirror. These reactions are the basis of the limit test for these acids in the 1993 British Pharmacopoeia monograph for phosphoric acid. This test states that the appearance of a mixture of aqueous solutions of phosphoric acid and silver nitrate should not change after heating at $100^{\circ}C$ for five minutes.

20.9.3 Reduction

All types of organic acid functional groups are resistant to reduction by standard laboratory reagents. However, carboxylic acids can be reduced, the product depending on the reducing agent.

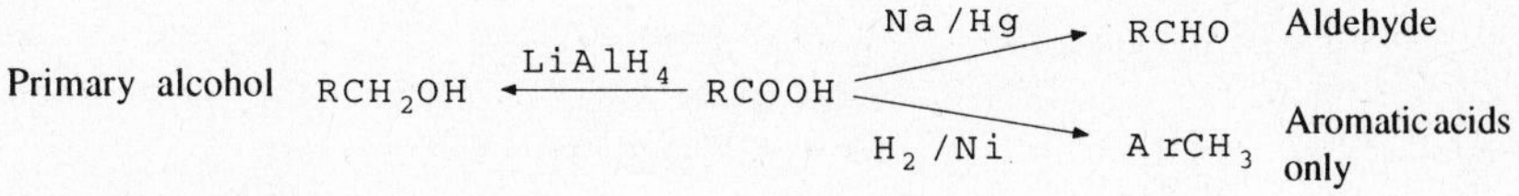

Primary alcohol $\quad RCH_2OH \xleftarrow{LiAlH_4} RCOOH$
$\quad\quad\xrightarrow{Na/Hg} RCHO$ Aldehyde
$\quad\quad\xrightarrow{H_2/Ni} ArCH_3$ Aromatic acids only

20.9.4 Decarboxylation

Carboxylic acids lose carbon dioxide either when heated to high temperatures or when heated with a base (B) to yield the corresponding alkane structure.

$$RCOOH \longrightarrow RH + CO_2$$

The mechanisms of these reactions are believed to involve the slow formation of a carboanion which is followed by the rapid reaction of this carboanion with a proton either supplied by the other compounds in the reaction mixture or by the solvent. This mechanism is supported by the rate expression and trapping experiments.

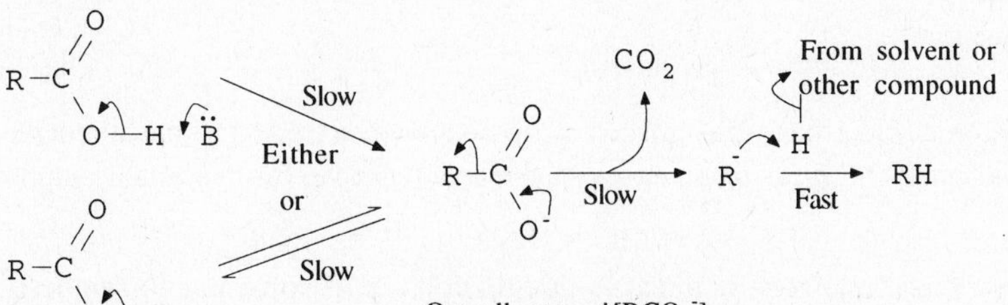

Overall rate = $k[RCO_2^-]$

Metabolic decarboxylations are believed to occur by a similar mechanism. However, the course of these biological decarboxylations may be modified by the presence of other functional groups in the molecule. Consider, for example, the conversion of mevalonic acid pyrophosphate to isopentenyl pyrophosphate, the biological precursor of many terpenes. The involvement of the phosphate ester intermediate results in the formation of the C=C bond instead of the expected C–H bond.

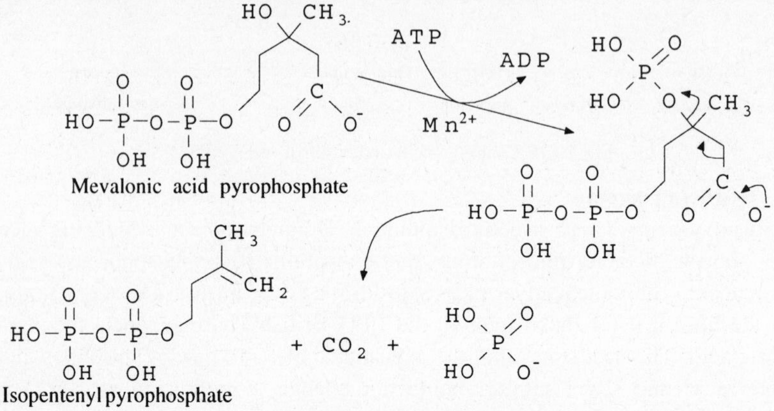

Mevalonic acid pyrophosphate

Isopentenyl pyrophosphate

20.9.5 Replacement reactions of aromatic sulphonic acids

The sulphonic acid group of aromatic sulphonic acids may be replaced by a wide variety of nucleophilic reagents. The reactions usually require the fusion of the sulphonic acid with the relevant reagent except in the case of the reaction with water which is carried out at 150°C.

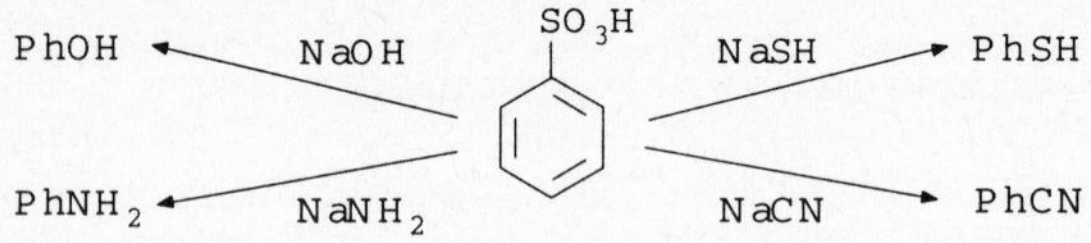

Electrophiles will also replace the sulphonic acid group of aromatic sulphonic acids provided the ring has an electron donor group substituent such as an amino or hydroxyl group.

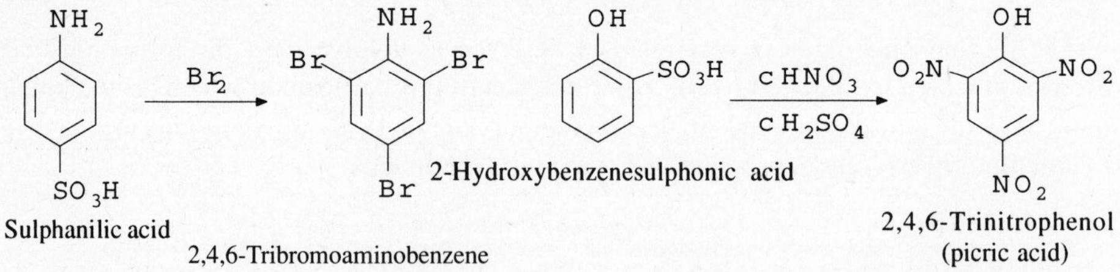

Sulphanilic acid

2,4,6-Tribromoaminobenzene

2-Hydroxybenzenesulphonic acid

2,4,6-Trinitrophenol (picric acid)

Sulphonic acid groups are replaced by hydrogen when treated with water in the presence of an acid catalyst. This is the reverse of the reaction used to form aromatic sulphonic acids.

$$PhSO_3H + H_2O \xrightarrow{H^+} PhH$$

Many of the reactions outlined in this section are of some use in synthesis.

20.10   PROSTAGLANDINS

The prostaglandins are a biologically important group of naturally occurring compounds whose molecular structures are based on that of prostanoic acid.

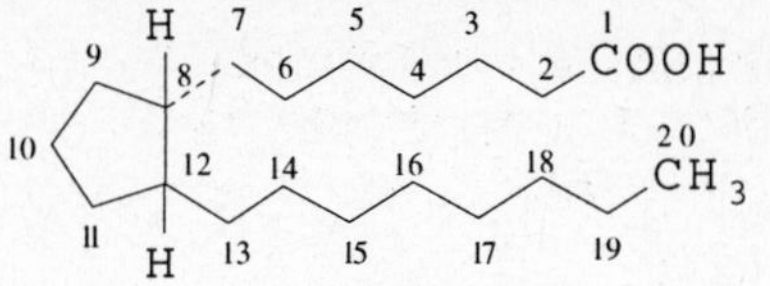

Prostanoic acid

Prostaglandins were originally found in the prostate gland but are now known to occur throughout the body. They may be named using the IUPAC system but are more commonly referred to by a series of abbreviations (Fig. 20.17). These abbreviations are based on the classification of the prostaglandins into families based on the nature of their substituents. The abbreviations also take into account the stereochemistry of each of the compounds.

Table 20.4. The trivial abbreviation nomenclature system of the prostaglandins. Table (a) shows the substituents found in a particular family whilst table (b) indicates the position and configuration of the C=C bonds found in the different members of each family. Note, -- represents a bond directed behind the plane of the paper or away from
the observer whilst - represents a bond directed in front of the plane of the ring or towards the observer.

(a)

Locant / Family	8	9	10	11	12	15
PGA		C=O	C=C			—OH
PGB	C=C	C=O			C=C	—OH
PGC		C=O		C=C	C=C	—OH
PGD		---OH		C=O		—OH
PGE		C=O		---OH		—OH
PGF$_\alpha$		---OH		---OH		—OH
PGF$_\beta$		—OH		---OH		—OH
PGG						—O-OH
PGH		---O-O---		---O-O---		
PGR		---O-O---		---O-O---		—OH

(b)

Subscript	C=C position and (configuration)
1	$C_{13}=C_{14}$ (trans)
2	$C_5=C_6$ (cis) $C_{13}=C_{14}$ (trans)
3	$C_5=C_6$ (cis) $C_{13}=C_{14}$ (trans) $C_{17}=C_{18}$ (trans)

The system uses the capital letters PG to indicate a prostaglandin and the capitals, A, B, C, etc. to identify the nature of the principal substituents (Table 20.4). Compounds that have the same principal substituents are called a family and are described by the same three letter code. The members of a family differ only in the number, position and configuration of

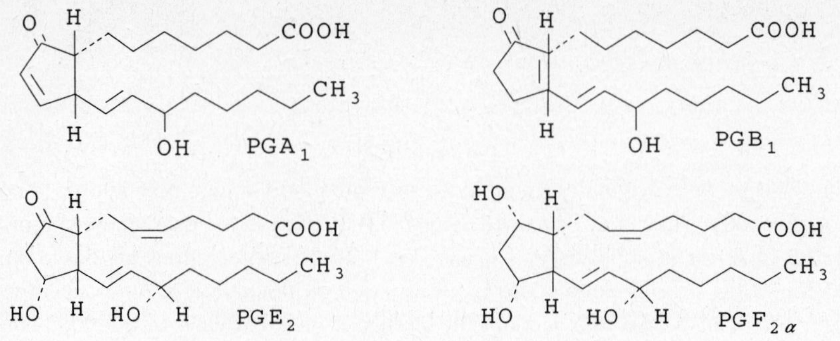

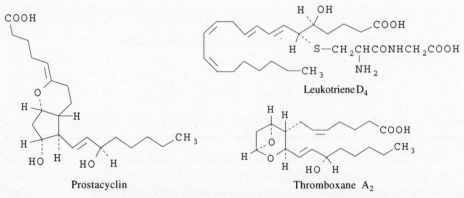

Fig. 20.17. Examples of the nomenclature of prostaglandins.

the alkene C=C bonds in the side chains. Numerical subscripts following the final letter indicate the number and configuration of the alkene double bonds that occur in the chains attached to the five-membered ring. For example, if the compound is classified as PGA_1 the substituent chains contain an alkene C=C bond with a trans configuration between carbons 13 and 14. The subscripts α and β are used to distinguish between the 9α and 9β isomers of the PGF family.

Prostaglandins appear to function as **local hormones**, that is, their action occurs in or near the cells in which they were produced. The limited field of action of prostaglandins appears to be due to their being synthesised on demand and rapidly metabolised after use. This is in contrast to **hormones** which often act in regions far from the sites of their production.

Prostaglandins appear to modify the chemical messages that hormones transmit to cells. Consequently, they have a wide range of effects: for example, prostaglandins enhance inflammation in tissue and cause headaches. Production of prostaglandins is inhibited by aspirin and other anti-inflammatories. These drugs are thought to act by blocking the production of prostaglandins. Prostaglandins also affect blood pressure and kidney function, induce labour at the end of pregnancy and stop the flow of gastric juices. This allows the body to heal gastric ulcers. Compounds related to the prostaglandins, such as the thromboxanes, leukotrienes and prostacyclin are also biologically active (Fig. 20.18). For example, the release of leukotrienes in the body appears to trigger asthma attacks.

Fig. 20.18. Examples of compounds related to the prostaglandins.

20.11 WHAT YOU NEED TO KNOW

(1) *Acids are functional groups that are proton donors.*

$$HA \rightleftharpoons H^+ + A^-$$

(2) *The structures and names of the most common types of organic acid are given in Figs 20.1, 20.2 and 20.3.*

(3) *The main naturally occurring organic acids are the carboxylic acids.*

(4) *The prefix hydroxycarbonyl and the suffix -oic indicate the presence of a carboxylic acid in a molecule.*

(5) A variety of names has been used for the phosphorus oxy-acids. It is important to check the name against the structures.

(6) Substituted acids are often classified into subgroups designated by the appropriate Greek letter.

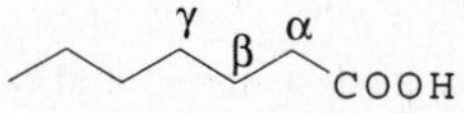

(7) *Thioacids exist as tautomers and can react in both forms.*

$$RCOSH \rightleftharpoons RCSOH$$

(8) Carboxylic, sulphonic and phosphorus oxy-acids are able to hydrogen bond with suitable molecules.

(9) Aromatic sulphonic acids are hygroscopic.

(10) *Aliphatic acids, with the exception of aromatic sulphonic acids, are generally more water soluble than aromatic acids.*

(11) *Low molecular mass aliphatic acids are water soluble because of hydrogen and ion-dipole bonding with water molecules. However, higher molecular mass aliphatic acids are usually insoluble in water unless they possess sufficient water solubilising functional groups.*

(12) All organic acid functional groups exhibit characteristic absorption bands in the infra-red region which can be used to detect the possible presence of these functional groups in a molecule.

(13) The ^1H nuclear magnetic resonance spectra of the acidic protons of all acids are normally singlets which will either reduce in size or disappear after a D_2O shake.

(14) *Strong acids are good proton donors, the position of equilibrium lying to the right.*

$$HA \rightleftharpoons H^+ + A^-$$

(15) *Factors that move this equilibrium to the left will increase the strength of the acid. The commonest of these factors are the strength of the O–H or S–H bond, the stability of the conjugate base and the nature of the solvent.*

(16) *In acidic solvents, ionisation will be depressed, whilst in basic solvents it will be enhanced.*

(17) *Electron-donating groups conjugated or adjacent to the acid group decrease acid strength, whilst electron-withdrawing groups in similar structural situations increase acid strength*

(18) *The strength of an acid is recorded as a pK_a value, where:*

$$pK_a = -\log K_a = -\log \frac{[H^+][A^-]}{[HA]}$$

(19) The smaller the pK$_a$ value the stronger the acid.

(20) The pK$_a$ value of an acid is a guide to its ease of transport across biological membranes. Acids are usually transported in their unionised form and so will usually move from an acid medium to a less acid or basic medium depending on the method of transport through the membrane.

(21) The types of acid discussed in this chapter form salts, liberate hydrogen from metals, carbon dioxide from carbonates and hydrogen carbonates, and change the colours of indicators.

(22) The sodium and ammonium salts of all acids are more water soluble than their parents. Many drugs are administered in this form to take advantage of this improved solubility.

(23) Potassium salts should be used cautiously because of their adverse effect on the heart.

(24) Water soluble organic acids, such as citric and tartaric acids, are used to form water soluble salts of organic bases used as drugs in order to allow them to be administered by injection.

(25) Water insoluble salts of drugs are used to treat infestations of the gastrointestinal tract because their poor solubility prevents absorption which means they remain in the tract.

(26) The hydroxyl and thiol groups of organic acids may be replaced by a variety of nucleophiles. The general equation for these nucleophilic displacement reactions of carboxylic acids is:

$$RCOOH \quad + \quad Nu \quad \longrightarrow \quad RCONu$$

(27) All types of organic acid functional groups act as electron-withdrawing groups. In the electrophilic substitution of aromatic systems they are meta directing. They are not usually strong enough to promote nucleophilic substitution of aromatic systems.

(28) Organic acids undergo limited oxidation.

(29) Organic acids are resistant to reduction. Carboxylic acids may be reduced, the product depending on the reducing agent and the reaction conditions.

(30) Carboxylic acids decarboxylate, the product in biological systems depending on the nature of the other functional groups present.

(31) Prostaglandins are a group of biologically important molecules whose structures are based on that of prostanoic acid (page 355).

(32) Prostaglandins are subdivided into families. Members of a family differ in the number and positions of C=C bonds in their structures.

(33) Prostaglandins are referred to using a three letter system where PG stands for prostaglandin, and the last letter indicates the position, stereochemistry and nature of the substituents in the molecule.

(34) Numerical subscripts indicate the position and stereochemistry of the unsaturated C=C bonds in the different members of a family. Greek letter subscripts are used in the standard manner to indicate differences in the stereochemistry of carbon 9.

(35) Prostaglandins appear to act as local hormones. They are produced on demand and are rapidly metabolised after use.

(36) Prostaglandins have a wide range of biological effects.

20.12   QUESTIONS

(1) Draw structural formulae for each of the following compounds:
(a) 4-aminocyclohexanecarboxylic acid, (b) pyridine-2,3-dicarboxylic acid,
(c) ethanedioic acid, (d) propynoic acid, (e) trans-2-butenoic acid, (f) Z-2-methyl-
2-butenoic acid, (g) 2-hydroxydithiobenzoic acid, (h) propanethiolic acid,
(i) benzylphosphonic acid, (j) p-toluenesulphonic acid, (k) 4-aminophenylphosphinic
acid, and (l) PGD_3

(2) (a) Write the names and draw the structures of four functional groups that act as
traditional acids. (b) List the chemical and physical factors that influence the strengths
of acids. (c) How is the strength of an acid recorded in the literature? (d) Predict the
relative acidic strengths of: (i) 3-nitrobenzoic acid, (ii) 3-hydroxybenzoic acid,
(iii) 3,5-dinitrobenzoic acid, and (iv) 3,5-dinitrobenzenesulphonic acid.

(3) 3-Phenylpropenoic acid (cinnamic acid) is used as a preservative. Draw the structural
formula of this compound. Discuss the stereochemistry of the structure you have
drawn. Show by means of equations how this molecule would be expected to react
with each of the following compounds. State the general type of each of these
reactions. (a) An acidic solution of potassium manganate VII, (b) a mixture of dry
hydrogen chloride and ethanol, (c) calcium carbonate, (d) thionyl chloride, and
(e) cold fuming nitric acid.

(4) Describe the water solubilities of (a) ethanethiolic acid, (b) dithioethanoic acid,
and (c) ethanesulphonic acid. Comment on the stabilities of aqueous solutions of these
acids. List the general chemical properties you would expect each of these acids to
exhibit.

(5) List the general physical and chemical properties that propylphosphinic acid would
be expected to exhibit. Show by means of equations how this acid might be expected
to react with; (a) sodium carbonate, (b) ammonical silver nitrate, and (c) ethanol.
Mention any essential reaction conditions in your answer.

21

Esters

21.1 INTRODUCTION

Esters are derivatives of organic and inorganic acids where the acidic proton(s) of the acid have been replaced by an alkyl or aryl residue, R' (Fig. 21.1). These residues may or may not possess other functional groups.

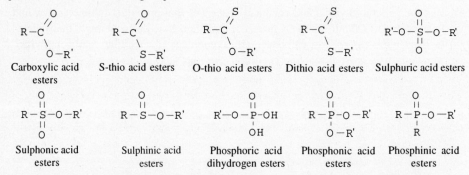

Fig. 21.1. The general structural formulae of a selection of the esters of inorganic and organic acids.

The structural formulae of *some types* of ester functional groups may be written in different ways. For example, in the case of carboxylic acid esters (Fig. 21.2) the carbonyl group of the ester is bonded to the acid residue (R). This carbonyl group is *always* written as CO and is either written adjacent to or drawn bonded to the acid residue R. The structural formulae of the other types of acid ester are related to their parent acids in a similar fashion and so it is necessary to be familiar with the internal structure of an ester group in order to make sense of its structural formula.

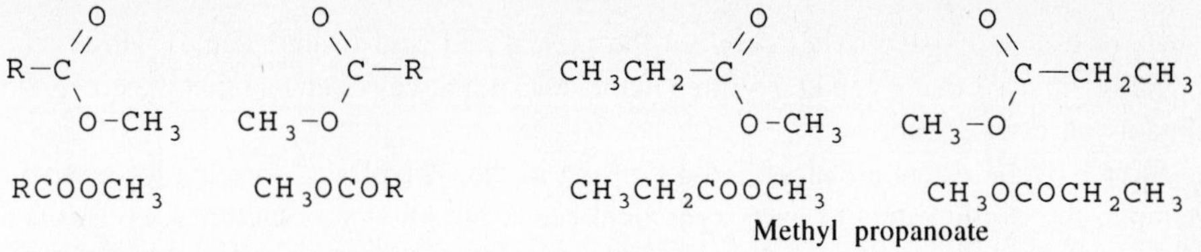

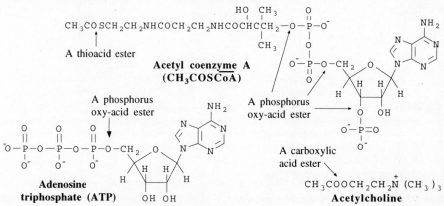

Fig. 21.2. Examples of the interpretation of the formulae used for carboxylic acid esters.

Many types of ester occur naturally, the most common being the esters of the carboxylic, thio and phosphorus oxy-acids. The esters of these acids often have important roles in biological processes (Fig. 21.3). It should be noted that very few examples of phosphorus acid esters with a C–P bond are known.

Fig. 21.3. Examples of biologically important esters.

21.2 NOMENCLATURE

Simple esters are normally named by placing the radical name of the R' group (Fig. 21.1) before the name of the acid residue which has its suffix **-oic** changed to **-ate** (Fig. 21.4). This suffix is used for all types of ester including those of inorganic acids. Incompletely esterified acids are named using the appropriate number of hydrogens as an insert between the radical and acid residue names for the ester. The prefix **alkoxycarbonyl** and the suffix **carboxylate** are also used to name the esters of carboxylic acids. When writing the

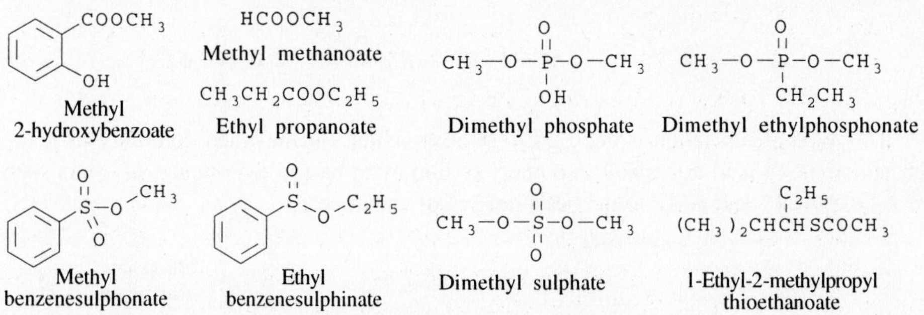

Fig. 21.4. Examples of the nomenclature of esters.

names of esters, a space is left between the radical and acid residue names. However, it should be realised that a gap in a written name does not always indicate that the compound contains an ester group.

Simple cyclic esters are classified according to the type of acid forming the ester. For example, the cyclic esters of carboxylic acid esters are known as **lactones** whilst cyclic sulphonates are classified as **sultones**. The names of lactones contain either the suffix **-lactone** or **-olide** and are based on the corresponding straight chain carboxylic acid, the size of the ring being indicated by the use of locants or a Greek letter (Fig. 21.5). The presence of a sultone group is indicated by the use of the suffix **-sultone** together with the appropriate locants or Greek letter. The names of other cyclic esters are usually based on that of the corresponding hydrocarbon.

Lactones   **Sultones**

Cyclic carboxylic   5-Pentanolide   Cyclic sulphonic   1,4-Butyrosultone
acid esters   (δ-valerolactone)   acid esters   (δ-Butyrosultone)

Fig. 21.5. The general structure and examples of the nomenclature of sultones. It should be noted that the stem names and locants of sultones do not indicate the same sized ring as in the similar name for a lactone because, in the case of lactones, the carbon of the carbonyl group is included in both the stem name designation and is the first carbon in the number system for the ring (Fig. 21.4).

21.3   PHYSICAL PROPERTIES

Low molecular mass simple esters are usually colourless liquids, whilst higher molecular mass compounds are normally white solids. Liquid esters have characteristic odours. For example, carboxylic acid esters often have a sweetish smell whilst thioesters have very penetrating, unpleasant odours. This is partly due to instability and their subsequent decomposition to thiols.

21.3.1   Solubility

The acidic esters of **inorganic acids**, such as sulphuric and the phosphorus oxy-acids are often water soluble. This is believed to be due to hydrogen bonding between the oxygen atoms of the ester and the water molecules and ion-dipole bonding due to the ionisation of any unesterified acidic O–H groups.

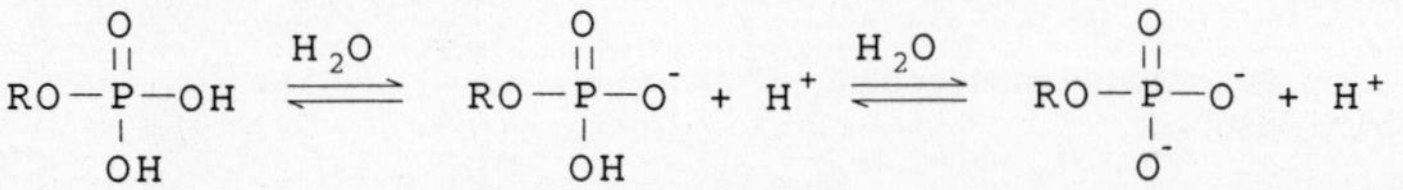

This type of ionisation is believed to be responsible for the water solubility of many phosphorus oxy-acid esters and also plays an important part in the binding of esters such as ATP, NADP$^+$ and AMP to the basic centres of enzymes.

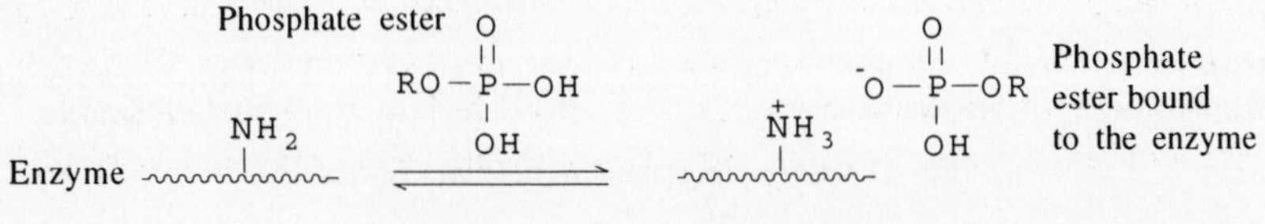

The incorporation of a phosphorus oxy-acid residue into the structure of a drug is used to improve the water solubility of that drug.

Most organic acid ester groups are not as polar as those of the corresponding acid. Consequently, the simple esters of **organic acids** are not usually very soluble in water although some hydrogen bonding is possible between the non-conjugated lone pairs of any oxygens and the water molecules. Methyl and ethyl methanoates are notable exceptions being completely miscible with water. Simple esters are usually soluble in lipids and other non-polar solvents.

A number of liquid esters are useful solvents for many non-polar compounds. For example, ethyl ethanoate and dimethyl sulphate are particularly useful as solvents in the manufacture of drugs, ethyl ethanoate being used as a solvent in the manufacture of the antidepressant *Amoxapine*. However, it is somewhat difficult to remove all the ethyl ethanoate from the product since traces of it are often trapped in the crystals of the *Amoxapine*.

Oils and fats which consist mainly of mixtures of the esters of fatty acids and glycerol (Chapter 26) are used as solvents in the formulation of drugs because since they are harmless and are readily metabolised by enzymes in the body. Oils are used as solvents for the injection of steroids whilst both oils and fats are the basis of ointments and creams.

21.3.2 Spectroscopy
The absorptions of the esters of carboxylic and thioacids in the infra-red region (Fig. 21.6) are characteristic enough to be used to detect the possible presence of these esters in a molecule. However, the main absorptions of other sulphur and phosphorus esters occur largely in the fingerprint region, which means that it is not usually feasible to use them to identify the presence of these esters in a molecule.

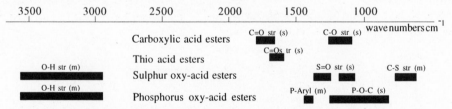

Fig. 21.6. The main regions in which esters absorb in the infra-red.

Nuclear magnetic resonance spectroscopy is of limited use in identifying esters. [13]C nuclear magnetic resonance spectroscopy will readily identify the presence of a carbonyl group in a molecule but will not easily identify the type of carbonyl group. This is because the carbon atoms of the carbonyl groups of many different functional groups exhibit resonance within the range δ 165 to 210 . [1]H nuclear magnetic resonance spectroscopy is of some value in detecting the presence of esters that contain O–H, S–H and P–H bonds. The signals due to the protons of the hydroxyl and thiol groups will either disappear or be reduced in size after a D_2O shake because these protons will exchange with deuterium. This is a useful aid to the location of these signals in the spectrum.

21.4 CHEMICAL PROPERTIES

The reactions of the different types of ester are very diverse but there are some common denominators. Most types of ester react with nucleophiles in what are essentially nucleophilic displacement and /or substitution reactions. Carboxylic and sulphur acid esters act as electron-withdrawing groups. However, phosphorus oxy-acid esters do not act in this manner. Esters also exhibit many properties that are characteristic of the type of ester. For example, the acid residues of phosphorus oxy-acid esters act as good leaving groups in biological elimination reactions, whilst carboxylic acid esters can be reduced (21.8).

Cyclic esters usually react in a similar manner to the corresponding acyclic compound. However, four-membered cyclic esters are more reactive than the same type of acyclic ester but larger rings exhibit about the same degree of reactivity as the corresponding acyclic structures.

21.5 REACTIONS WITH NUCLEOPHILES

21.5.1 Carboxylic acid esters

The alcoholic and phenolic residues of acylic esters may be replaced by suitable nucleophiles.

$$RCOOR^1 \xrightarrow[\text{Hydrolysis}]{H_2O /H^+} RCOOH + R^1OH$$

$$RCOOR^1 \xrightarrow[\text{Hydrolysis}]{NaOH} RCOONa + R^1OH$$

$$RCOOR^1 \xrightarrow[\text{Ammonolysis}]{NH_3} RCONH_2 + R^1OH$$

$$RCOOR^2 \xrightarrow[\text{Ammonolysis}]{R^1NH_2} RCONHR^1 + R^2OH$$

$$RCOOR^2 \xrightarrow[\text{Alcoholysis}]{R^1OH} RCOOR^1 + R^2OH$$

$$RCOOR^1 \xrightarrow[\text{Transesterification}]{R^2COOR^3} RCOOR^3 + R^2COOR^1$$

$$RCOOR \xrightarrow[\substack{\text{Hydroxylamine} \\ \text{hydrochloride}}]{NH_2OH .HCl} \underset{\substack{\text{A hydroxamic acid}}}{R-\overset{\overset{\displaystyle O}{\|}}{C}-NHOH} \overset{\text{Tautomerism}}{\rightleftharpoons} R-\overset{\overset{\displaystyle OH}{|}}{C}=NOH$$

Lactones react with nucleophiles to yield the substituted acid. However, lactones are usually stable to acid hydrolysis.

$$HOCH_2CH_2COOCH_3 \xleftarrow{\quad} \quad \xrightarrow{\quad} H_2NCH_2CH_2COOH$$
Methyl 3-hydroxypropanoate $\quad CH_3ONa \quad$ $\begin{array}{c} CO \\ | \\ O \end{array}$ $\quad NH_3 \quad$ 3-Aminopropanoic acid

$$HSCH_2CH_2COOH \xleftarrow{\quad NaSH \quad} \qquad \xrightarrow{\quad NaOH \quad} HOCH_2CH_2COONa$$
3-Mercaptopropanoic acid $\qquad\qquad\qquad$ Sodium 3-hydroxypropanoate

Practical evidence suggests that the mechanisms of the ester nucleophilic displacement reactions are based on two general routes. Route (b) is a modification of route (a) which is believed to occur when hydrolysis is carried out under acid conditions (Fig. 21.7).

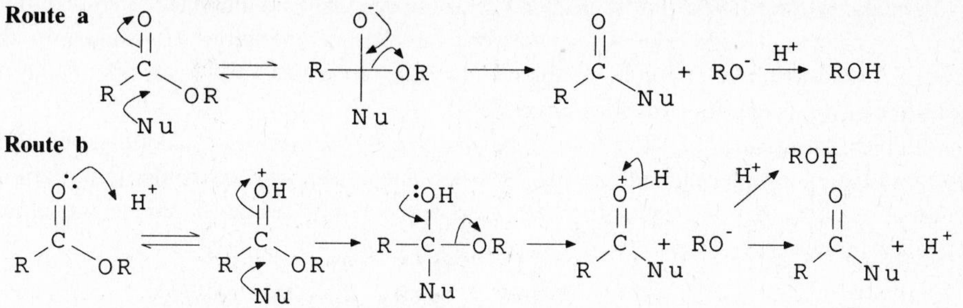

Route a

Route b

Fig. 21.7. General mechanisms for the nucleophilic reactions of carboxylic acid esters. Many other types of compound with the general formula RCOZ are believed to react with nucleophiles by these mechanisms.

The reactions of nucleophiles with carboxylates are important in pharmacy and biological processes. For example, the reaction with hydroxylamine hydrochloride is the basis of a qualitative test for carboxylic acid esters. In this test, the hydroxamic acid formed by the reaction of the ester with the hydroxylamine hydrochloride is reacted with iron III chloride to form a purple complex. The colour of this complex should not be confused with the cherry red colour produced by some nitro compounds.

Alcoholysis and transesterification are used to modify the structures of fats and oils (26.3) in order to change their physical properties. Alcoholysis involving benzocaine is the preferred route for the manufacture of the local anaesthetic procaine.

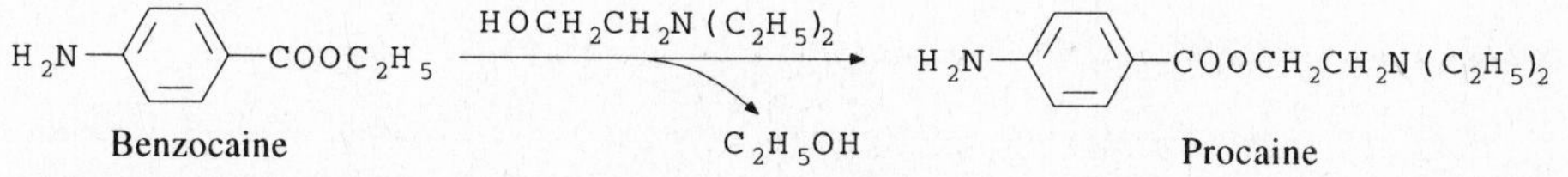

$$H_2N-\underset{\text{Benzocaine}}{\underset{|}{\bigcirc}}-COOC_2H_5 \xrightarrow[\text{C_2H_5OH}]{HOCH_2CH_2N(C_2H_5)_2} H_2N-\underset{\text{Procaine}}{\underset{|}{\bigcirc}}-COOCH_2CH_2N(C_2H_5)_2$$

Hydrolysis has a variety of uses and is one of the most important of all the nucleophilic substitution reactions of the esters of carboxylic acids. It is utilised in drug administration and forms the basis of many assays and identification tests for esters. Hydrolysis by atmospheric moisture is often the cause of the slow deactivation of 'ester' drugs on storage whilst enzyme-catalysed hydrolysis forms part of the main route for their metabolism. For example, procaine is hydrolysed in blood plasma by pseudoesterases to 4-aminobenzoic acid and 2-(N,N-diethylamino)ethanol.

$$H_2N-\underset{\text{Procaine}}{\underset{|}{\bigcirc}}-COOCH_2CH_2N(C_2H_5)_2 \xrightarrow[\text{Pseudoesterases}]{H_2O} \underset{\text{2-(N,N-diethylamino)ethanol}}{HOCH_2CH_2N(C_2H_5)_2} + \underset{\text{4-Aminobenzoic acid}}{H_2N-\underset{|}{\bigcirc}-COOH}$$

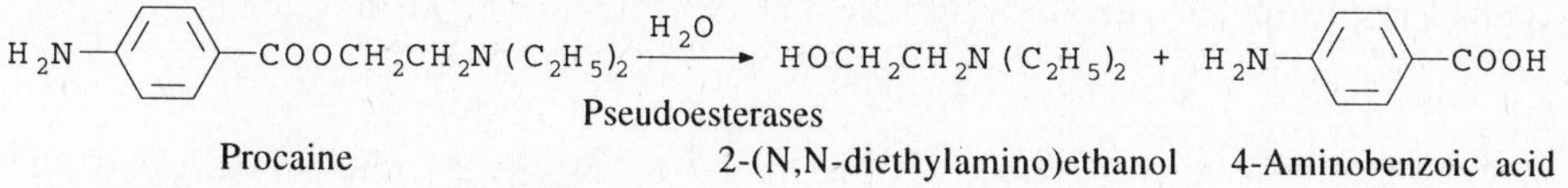

Hydrolysis sometimes makes it necessary to protect 'ester' drugs with surfactants in order that they reach their site of action.

Drug administration. Ester hydrolysis provides a mechanism for maintaining the level of the concentration of some drugs in the blood because both acid- and base-catalysed hydrolysis is slow, the reactions taking place in hours rather than minutes. The drug is administered to the patient in the form of a suitable ester. The good lipid solubility of esters results in a build-up of the ester in the patient's body fat. This acts as a depot and the drug is slowly released from the ester by hydrolysis.

The partly esterified cellulose derivative, cellulose ethanoate phthalate (*Cellacephate*), is stable under acid conditions but is hydrolysed in alkaline conditions. Consequently, this ester is used as an enteric coating for tablets either to release drugs specifically in the intestine or to protect the stomach from the drug.

Unpleasant-tasting drugs that have to be taken orally are sometimes administered in the form of their esters because the esters have a bland taste. For example, the antibiotic chloramphenicol has a persistent bitter taste and so it is administered as the almost tasteless palmitate.

$$O_2N-\langle \rangle-\overset{HO}{\underset{H}{\overset{|}{C}}}H\overset{H}{\underset{NHCOCHCl_2}{\overset{|}{C}}}CH_2OCO(CH_2)_{14}CH_3$$

Chloramphenicol palmitate

Assays. Hydrolysis is the basis of a number of assay methods for esters. The ester is completely hydrolysed by heating under reflux with an excess of either aqueous sodium hydroxide or ethanolic potassium hydroxide. After hydrolysis the amount of excess alkali is determined by back titration with hydrochloric or other suitable acid (11.6.2). A blank titration is usually carried out to allow for the extraction of alkali from the glass of the flask used for the hydrolysis. Benzyl benzoate, used for the topical treatment of scabies, may be assayed by this method. Ethanolic potassium hydroxide is used for the hydrolysis because benzyl benzoate is insoluble in water but completely soluble in ethanol.

$$PhCOOCH_2Ph + KOH \xrightarrow[\text{heat}]{\text{Ethanol}} PhCOOK + PhCH_2OH \quad [+ KOH_{(Excess)}]$$

$$KOH_{(Excess)} + HCl \xrightarrow[\text{phenolphthalein}]{\text{Indicator}} KCl + H_2O$$

Many other drugs containing ester groups may be assayed by this method, including the analgesic, aspirin, the insect repellent dimethyl phthalate, and the counter-irritant methyl salicylate. Ester hydrolysis is also the basis of the determination of the saponification value of fats and oils (26.10).

Identification tests. The ester is hydrolysed using either an acid- or base-catalysed method and the products are identified by appropriate methods. For example, the identification test B for the analgesic aspirin in the 1993 British Pharmacopoeia is based on hydrolysing the drug by heating with aqueous sodium hydroxide. The reaction mixture is acidified to precipitate salicylic acid which is identified by its melting point.

Aspirin $\xrightarrow[\text{Heat}]{\text{NaOH}\atop H_2O}$ + CH_3COONa $\xrightarrow{H^+/H_2O}$ Salicylic acid, mp 156°C + CH_3COOH

21.5.2 Sulphur acid esters

Acyclic and cyclic esters of sulphonic and thioacids tend to react in the same general manner as carboxylic acid esters but are usually more reactive than the corresponding type of carboxylic acid ester and an appreciable degree of hydrolysis of these esters can rapidly occur in moist conditions at room temperature. Consequently, drugs such as the antileprotic Ditophal, which contains two thioester groups, and the cytotoxic agent Busulphan, which is a disulphonate ester, must be stored under dry conditions.

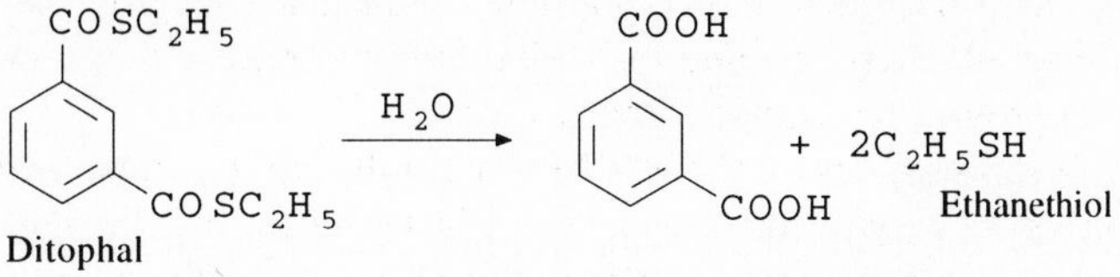

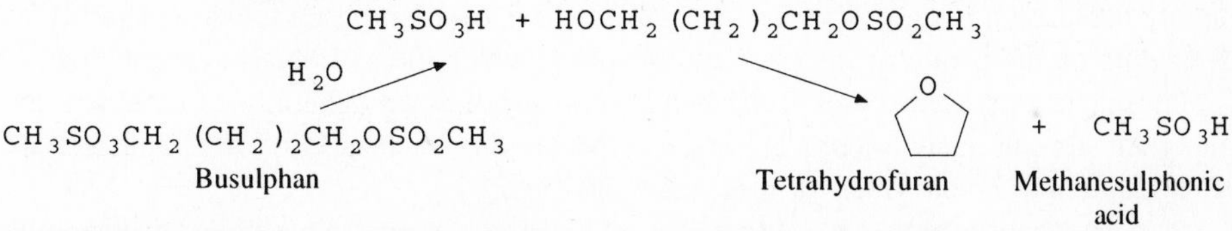

It is interesting to note that the active agent of Ditophal is the ethanethiol produced by the hydrolysis of the esters when the drug is applied to the sores caused by the leprosy. Since this compound has an antisocial odour the drug is formulated with a strong perfume.

The mechanisms of the nucleophilic reactions of the sulphur acid esters are different for the different types of these esters. The reactions of the thioesters are nucleophilic displacements, nucleophilic attack occurring at the unsaturated C atom of the ester group, but the reactions of sulphonates are substitutions, nucleophilic attack taking place on the carbon adjacent to the oxygen of the sulphonate group. This difference is due to the sulphonate being a strong electron-withdrawing group and a good leaving group. The sulphonate group's very strong electron-withdrawing effect induces a positive charge on the carbon atom causing it to act as an electrophilic centre. The sulphonate anion is a good leaving group because it is stabilised by charge delocalisation in the sulphonate structure.

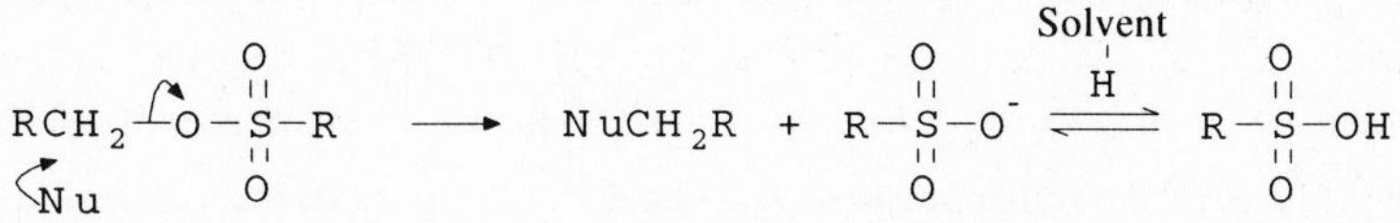

Nucleophilic displacement reactions involving thioesters, especially those of co-enzyme A, occur in many metabolic pathways. For example, the formation of acetylcholine (Fig. 21.8 a) and the conjugation of glycine with aromatic acids in excretion (Fig. 21.8 b).

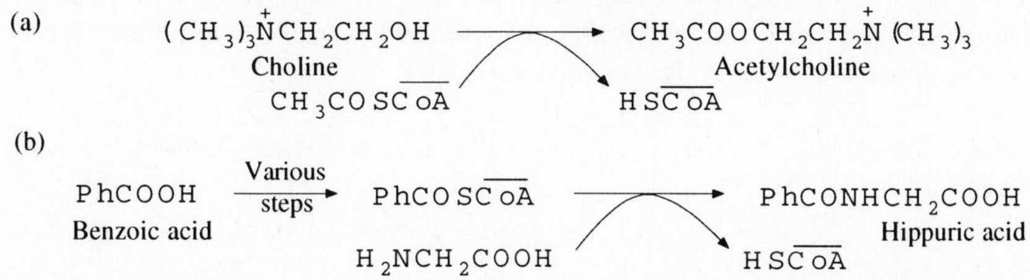

Fig. 21.8. (a) The formation of acetylcholine. This reaction is in essence an alcoholysis, the choline replacing the co-enzyme A residue of the ethanoate ester. (b) The conjugation of carboxylic acids with glycine in excretion processes. The nucleophilic displacement illustrated is an ammonolysis.

21.5.3 Phosphorus acid esters

The most important of the nucleophilic displacement reactions of the acyclic and cyclic esters of the oxy-acids of phosphorus is hydrolysis. All types of phosphorus oxy-acid ester will hydrolyse to an alcohol and the corresponding acid.

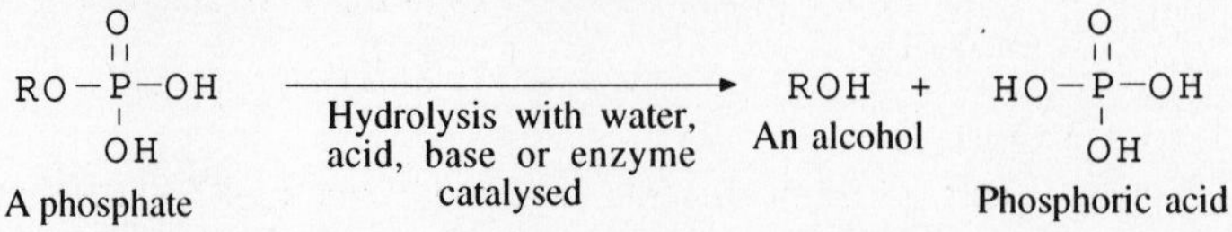

$$
\underset{\text{A phosphate}}{RO-\overset{\overset{\displaystyle O}{\|}}{\underset{\underset{\displaystyle OH}{|}}{P}}-OH}
\xrightarrow[\substack{\text{acid, base or enzyme} \\ \text{catalysed}}]{\text{Hydrolysis with water,}}
\underset{\text{An alcohol}}{ROH} \; + \;
\underset{\text{Phosphoric acid}}{HO-\overset{\overset{\displaystyle O}{\|}}{\underset{\underset{\displaystyle OH}{|}}{P}}-OH}
$$

Cyclic esters, especially those with five-membered ring structures, hydrolyse more rapidly than acyclic esters. Various mechanisms have been proposed for these hydrolyses depending on the nature of the ester and the pH at which the hydrolysis is carried out.

Phosphate ester hydrolyses, catalysed by enzymes known generally as esterases, are important steps in many biological processes. Metal ions, such as Mn^{2+}, Mg^{2+} and Zn^{2+} are often required as co-enzymes for these reactions (21.7.3).

Phosphorus oxy-acid ester residues also act as leaving groups in biological nucleophilic substitutions that are the equivalent of the laboratory nucleophilic substitutions of alkyl halides (14.5). The biosynthesis of the terpene geraniol involves the nucleophilic substitution of dimethylallyl pyrophosphate (DMAPP) by isopentenyl pyrophosphate (IPP). The reaction is thought to be controlled by the participation of a nucleophilic centre in the enzyme for the reaction. This involvement is believed to be responsible for the formation of the C_2 double bond of the geraniol pyrophosphate.

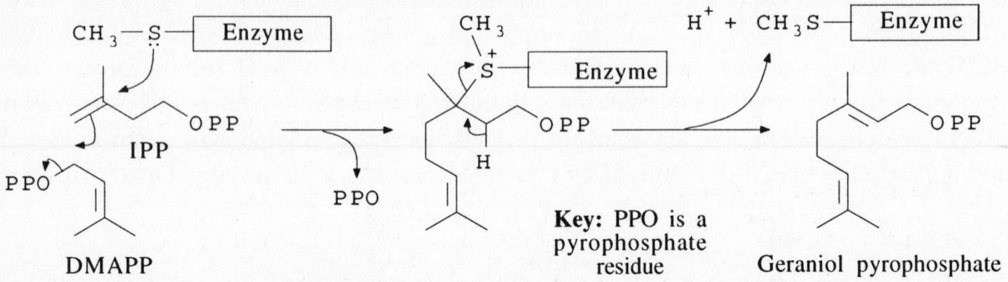

Key: PPO is a pyrophosphate residue

The hydrolysis of the phosphorus oxy-acid ester ATP to ADP is an essential source of energy for the human body (21.6.4). An adult human requires about 40 kg of ATP a day just to keep the body ticking over. When we exercise the consumption of ATP rises dramatically and can be as high as 0.5 kg per minute. Fortunately, the body continuously manufactures ATP to replace that used to provide energy for human activities. Compounds such as creatine phosphate, acetyl phosphate, 1,3-diphosphoglycerate and phosphoenolpyruvate are used as sources of phosphate units in the conversion of ADP back to ATP.

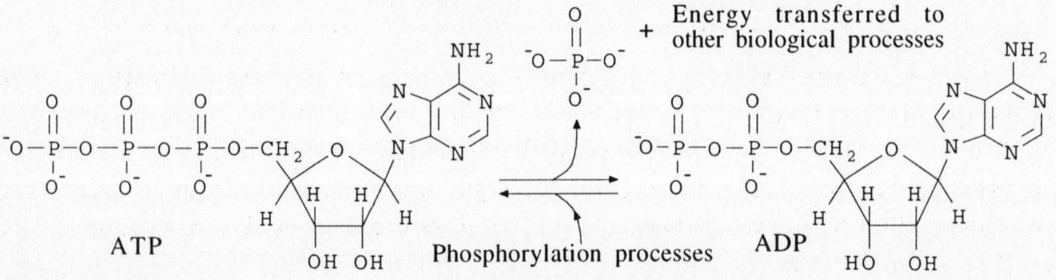

21.6 AS ELECTRON ACCEPTORS

Esters can act as electron-withdrawing groups in both aromatic and aliphatic structures. The strength of the withdrawing effect and the types of reaction the compound will undergo depend on the nature of the ester. Phosphorus oxy-acid esters tend to be the weakest electron acceptors.

21.6.1 Aromatic systems

Carboxylic and sulphur acid esters are meta directing in electrophilic substitution. However, these ester groups are not usually strong enough electron-withdrawing groups to allow nucleophilic substitution of the ring to occur.

21.6.2 Aliphatic systems

The withdrawal of electrons from the α carbon by the ester group increases the electronegative character of this carbon atom. As a result, the α carbon atom exerts a strong attraction for the electrons of its remaining three bonds and so increases the polarisation of these bonds. This increase in polarisation can weaken α C–H bonds sufficiently for them to react with bases to form a carbanion in suitable conditions. These carbanions can act as nucleophiles in substitution, displacement and condensation reactions. The mechanisms of these reactions are usually based on the general mechanisms for these types of reaction which are described in sections 14.5, 15.7.2 and 19.10 respectively.

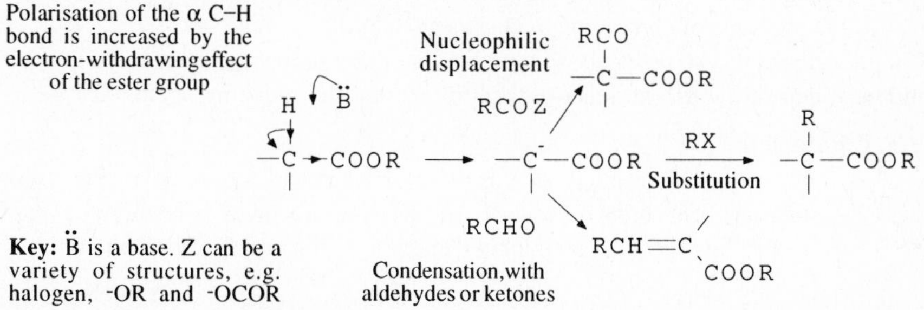

Polarisation of the α C–H bond is increased by the electron-withdrawing effect of the ester group

Nucleophilic displacement

RCOZ

RX — Substitution

RCHO — Condensation, with aldehydes or ketones

Key: B̈ is a base. Z can be a variety of structures, e.g. halogen, -OR and -OCOR

The carbanions in many of these reactions are produced from **active methylene groups** where a second electron-withdrawing group adds to the polarisation of the α C–H bond (Fig. 21.9).

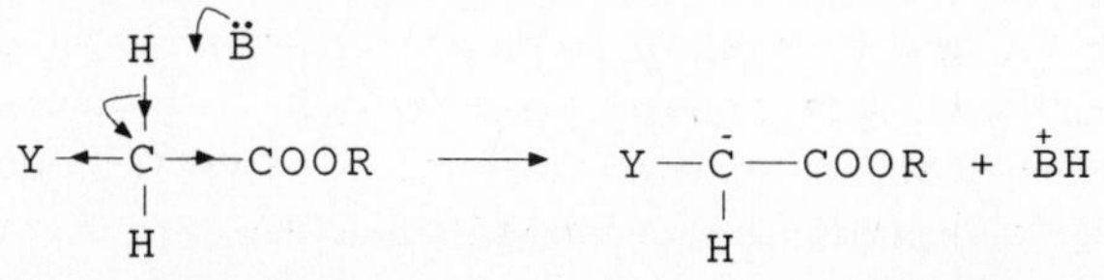

Fig. 21.9. The formation of a carboanion from an active methylene group (YCH_2COOR). Examples of the second electron-withdrawing group Y are carboxylic acid ester, thioester, nitro or cyanide groups.

Reactions in which carboxylic acid esters act as sources of carboanions are important in the synthesis of drugs whilst a number of biological reactions are believed to involve carboanion formation in their mechanisms. For example, the biosynthesis of S-malic acid in plants and micro-organisms is thought to involve a carboanion formed from acetylcoenzyme A by the action of a basic group of the enzyme for the process, malate synthase (Fig. 21.10).

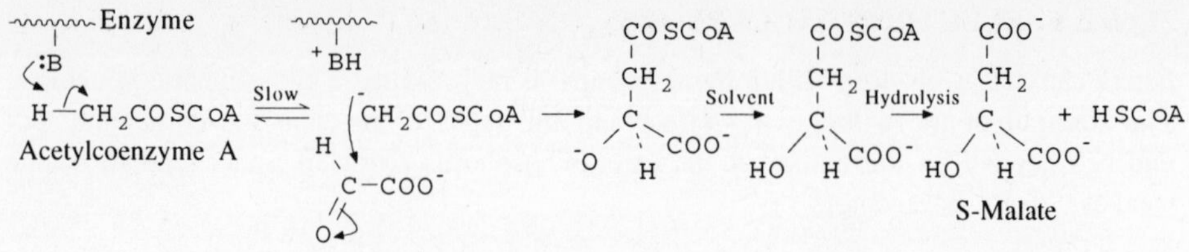

Fig. 21.10. A proposed mechanism for the biosynthesis of S-malic acid in plants and micro-organisms.

21.7 REACTIONS CHARACTERISTIC OF THE TYPE OF ESTER

21.7.1 Oxidation and reduction

Esters are usually resistant to oxidation and reduction. However, it is relatively easy to reduce carboxylic acid esters in the laboratory, the products depending on the reagent and the reaction conditions.

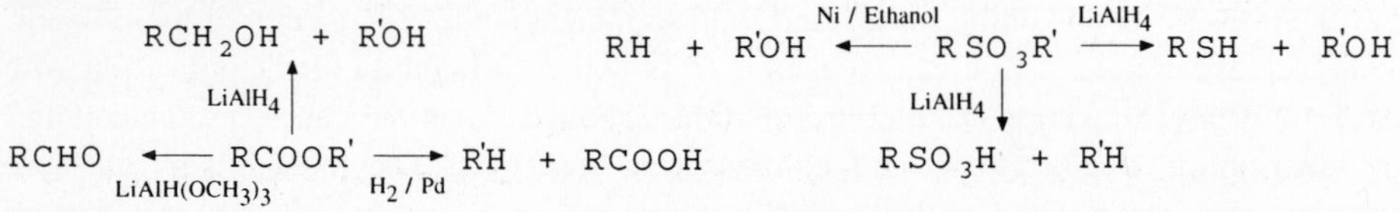

Carboxylic acid esters are not reduced by hydrogen in the presence of metallic catalysts under **mild** conditions. This allows the selective reduction of more sensitive functional groups in the presence of carboxylic acid esters (26.3.6).

Sulphonates may also be cleaved in the laboratory by reducing agents to yield a variety of products depending on the reaction conditions and the reducing agent.

21.7.2 Elimination

The phosphorus oxy-acid residues of this type of ester can act as a leaving group in elimination reactions. The biosynthesis of the terpene precursor isopentenyl pyrophosphate is believed to involve such a coupled decarboxylation and E_2 elimination.

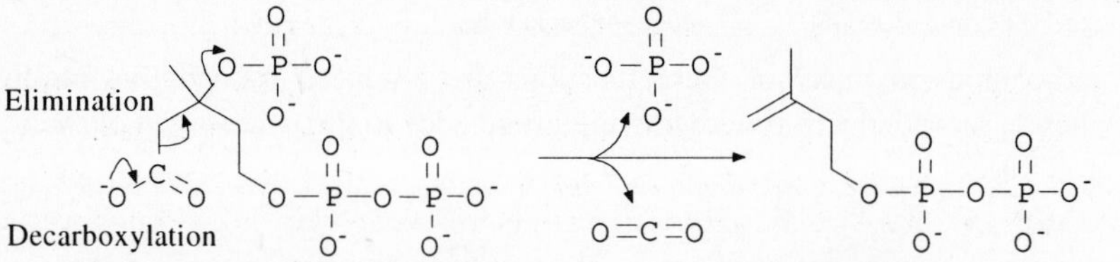

Mevalonate phosphate derivative   Isopentenyl pyrophosphate

21.7.3 Coordination of phosphorus oxy-acid esters with metals

Divalent metal ions are known to act as catalysts in both laboratory and biological reactions involving phosphorus oxy-acid esters. For example the rate of hydrolysis of salicyl phosphate is increased by a factor of seven when the reaction is carried out in the presence of Cu^{2+} ions.

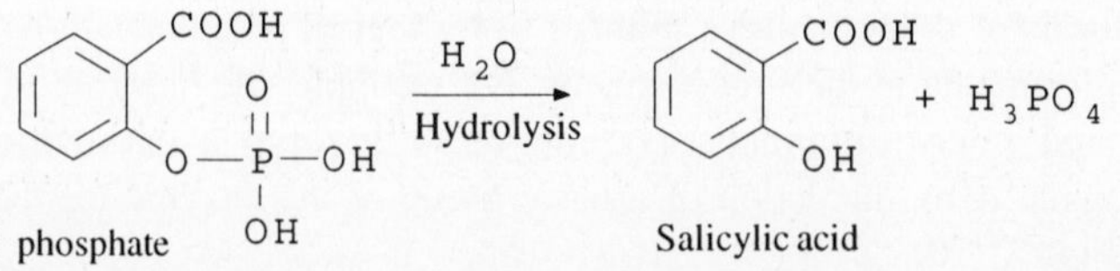

Salicyl phosphate   Salicylic acid

The action of these divalent metal ions is believed to be due to their ability to coordinate with some of the oxygen atoms of the ester.

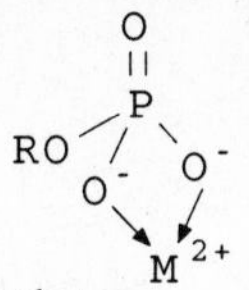

Phosphate coordination 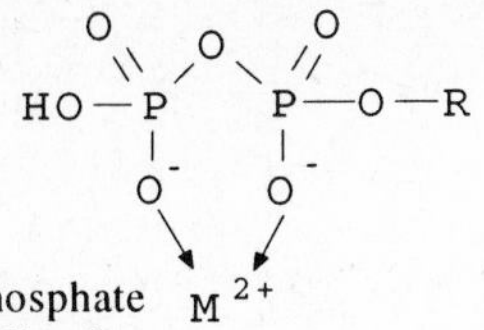   Diphosphate coordination 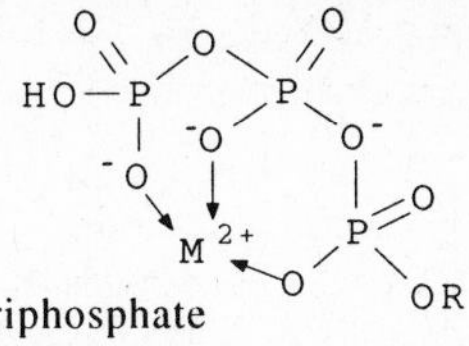   Triphosphate coordination

This coordination is thought to mask the negative charge of the oxygen atoms which in turn increases the chance of nucleophilic attack on the phosphorus atom. The electron-withdrawing effect of the positive metal ion is also believed to enhance the polarisation of the uncoordinated P=O and P−O− bonds and so increases their reactivity. This explains why the presence of a metal ion is essential in many biological processes involving the reaction of a phosphorus oxy-acid ester if the reaction is to proceed.

21.7.4 Hydrolysis of the phosphoric anhydride group

The phosphoric anhydride group (−PO$_2$−O−PO$_2$−) found in polyphosphoric oxy-acid esters is an energy-rich structure. This high energy content is required to counterbalance the strong repulsive forces between the electron clouds of the oxygen atoms in the structure (Fig. 21.11). In other words, the strong bond strength of the −PO$_2$−O−PO$_2$− system is necessary to stabilise this structure. Consequently, the breaking of the −PO$_2$−O−PO$_2$− bonds releases large amounts of energy.

Key: ◄────► Indicates repulsion between the electron clouds of the adjacent oxygen atoms

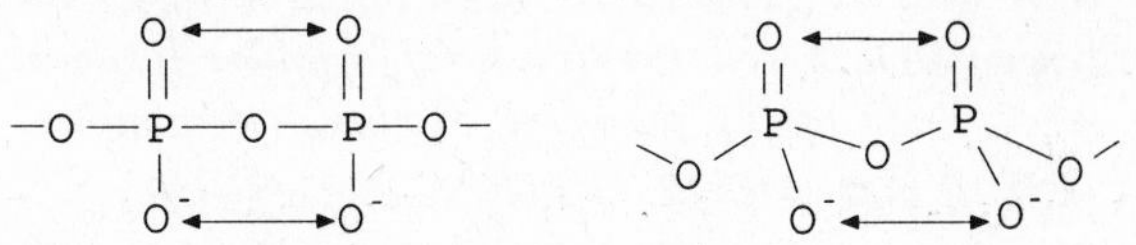

Fig. 21.11. Repulsive forces in the phosphoric anhydride structure. It should be remembered that polyphosphoric esters are fully ionised at physiological pHs.

The body uses this structure as an energy store because the structure does not hydrolyse appreciably at body temperatures except in the presence of suitable enzymes. This feature allows nature to control the release of the energy stored in the phosphoric anhydride structure. Two of the principal sources of phosphoric anhydride structures in living matter are adenosine triphosphate (ATP) and adenosine diphosphate (ADP) (21.5.3). It is emphasised that these molecules will only release their energy in the presence of suitable enzymes. For example, ADP exists in the aqueous media of cells because there are *no* enzymes in the cells that catalyse the nucleophilic attack of the water on ADP.

Adenosine triphosphate is probably the most widely distributed triphosphate in the human body. Not only is it involved in the release of energy but it is also appears in most biosynthetic and metabolic processes. Its wide range of reactivity is due to the fact that there appear to be four positions in the molecule where hydrolysis can occur (Fig. 21.12). It is not possible to predict in advance where hydrolysis will occur but it is believed that the coordination with divalent metal ions plays a key role in the operation of the enzymes controlling the reactions (21.7.3).

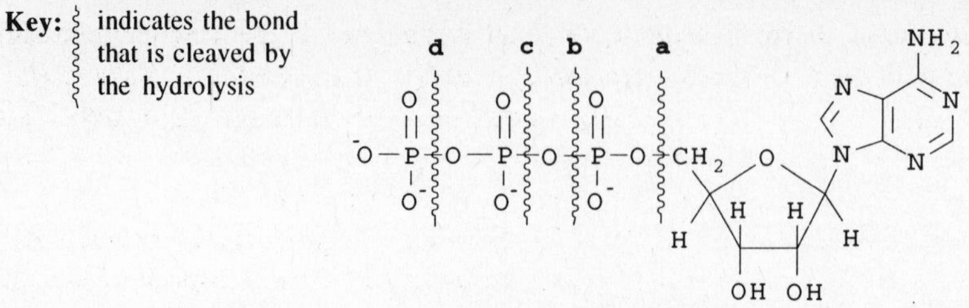

Key: ⌇ indicates the bond
     ⌇ that is cleaved by
     ⌇ the hydrolysis

Fig. 21.12. The modes of hydrolysis of ATP. Fission at position (d) leads to the formation of ADP. Position (c) is involved in diphosphate transfers. Reaction at position (c) occurs in acetyl adenylate production whilst fission at position (a) occurs in the synthesis of S-adenosylmethionine.

The $-PO_2-O-PO_2-$ structure is normally more easily hydrolysed in the laboratory than the $RO-PO_2-$ bond of monophosphoric oxy-acid esters. Consequently, it is possible to estimate the concentration of polyphosphoric oxy-acid residues in tissue by treating the sample with 1 M hydrochloric acid at 100° for 6-7 minutes. These conditions convert most polyphosphoric oxy-acid residues to phosphate but do not normally cause the hydrolysis of monophosphorus oxy-acid residues. The phosphate produced by the hydrolysis can be determined by gravimetric analysis or colorimetrically by converting to a blue phosphomolybdate complex.

21.8 WHAT YOU NEED TO KNOW

(1) Esters are the organic derivatives of organic and inorganic acids in which either the O–H or S–H groups have been replaced by an alcoholic or phenolic residue.

(2) It is possible to write the structural formulae of some types of ester in several different ways. Each of the following groups of structures represent the same ester:
–COOR and –OCOR;   RCOS– and –SCOR;   –SO₃R, RSO₂O– and –OSO₂R

(3) The presence of an ester functional group in a molecule is commonly indicated by the use of a radical name of the alcohol or phenol from which the ester is derived followed by a space and the name of the parent acid with the suffix -ate. The prefix alkoxycarbonyl and the suffix carboxylate are used for the esters of carboxylic acids.

(4) Lactones are cyclic carboxylic acid esters.

(5) The esters of inorganic acids, such as sulphuric and the phosphorus oxy-acids are often water soluble. The introduction of this type of ester group is used to increase the water solubility of some drugs.

(6) The esters of carboxylic acids are not normally soluble in water. They are lipid soluble.

(7) The presence of an absorption in the 1740 to 1710 cm^{-1} region of the infra-red indicates the possible presence of a carbonyl group of either a carboxylic or thioacid esters in the structure of a compound.

(8) ^{13}C nuclear magnetic resonance is of limited use in identifying the presence of the carbonyl group of a carboxylic or thio ester in a molecule.

(9) The reactions of an ester depends largely on the type of ester but all types of ester will react with nucleophiles and act as electron-withdrawing groups.

(10) Sulphonates, sulphates and phosphorus oxy-acid esters will react with

nucleophiles in both nucleophilic displacement and substitution reactions. Carboxylic and thioacid esters only undergo nucleophilic displacements.

(11) *The nucleophilic displacement reactions of carboxylic acid esters may be summarised by the general equation:*

$$RCOOR^1 + HNu \longrightarrow RCOONu + R^1OH$$

The esters of thioacids, dithioacids, phosphorus oxy-acids and sulphur oxy-acids can also act with nucleophiles in this manner to produce the appropriate acid or its derivative plus an alcohol, phenol or thiol.

(12) *The nucleophilic substitution reactions of the esters of the phosphorus oxy-acids may be summarised by the general equation, where OP is a phosphate residue:*

$$ROP + Nu \longrightarrow RNu + \text{Phosphorus oxy-acid residue}$$

(13) Cyclic esters usually react in a similar fashion to the corresponding acylic ester. Four-membered rings are more reactive than larger ring systems and the corresponding acylic compound.

(14) Hydrolysis is frequently used as the basis of assay and identification tests for esters.

(15) *All types of aromatic ester group are meta directing in the electrophilic substitution of aromatic systems.*

(16) *Compounds with alpha C–H bonds can give rise to carboanions in basic conditions. These carboanions are good nucleophiles.*

(17) Esters are usually resistant to oxidation and reduction. However, carboxylic acid and sulphonic acid esters can be reduced in the laboratory.

(18) *The phosphorus oxy-acid residues of phosphorus oxy-acid esters can act as a leaving group in biological elimination reactions.*

(19) *Phosphorus oxy-acid esters are able to coordinate with metal ions. This is believed to be the basis of the catalytic action of metal ions in the biological reactions of these esters.*

(20) *The phosphoric anhydride structure ($-PO_2-O-PO_2-$) is used by living matter as an energy store because it will only hydrolyse appreciably in the presence of an enzyme.*

21.9 QUESTIONS

(1) Draw the structural formulae of all the **ester** isomers of each of the following compounds. The structures you draw should possess only ester functional groups.
(a) Ethyl propanoate, (b) ethyl ethanesulphonate, (c) propyl phosphate, (d) 3-methyl-5-pentanolide, (e) propyl ethylphosphonate, (f) ethyl hydrogen sulphate,

(g)   methoxycarbonylcyclohexane and (h) 2-methyl-1,4-butyrosultone.

(2) $PhCH_2CH_2COOCH_3$ (A)

Compound A was isolated from biological material. (a) List the general chemical and physical properties you would expect this compound to exhibit. (b) Comment on the general solubility of A in water and lipids. (c) Predict, by means of chemical equations, how A could react with (i) ethanolic sodium hydroxide, (ii) ethyl ethanoate, (iii) sodium ethoxide followed by methyl bromide, (iv) sodium ethoxide followed by propanone and an acid work up. (v) ethanol, Give details of any essential reaction conditions.

(3)   Describe, with the aid of equations, the general similarities and differences in the reactivities of sulphur and phosphorus oxy-acid esters.

(4)   Suggest a feasible electronic mechanism for each of the reactions:

(a)

$$(CH_3)_3\overset{+}{N}CH_2CH_2OH \longrightarrow CH_3COOCH_2CH_2\overset{+}{N}(CH_3)_3$$

$$CH_3CO\,S\overline{C\,oA} \qquad\qquad H\,S\overline{C\,oA}$$

(b)

$$\underset{H}{\overset{O}{\underset{\|}{C}}}\underset{S}{\diagdown}{}^{G} \xrightarrow{\;H^+/H_2O\;} HCOOH + HSG$$

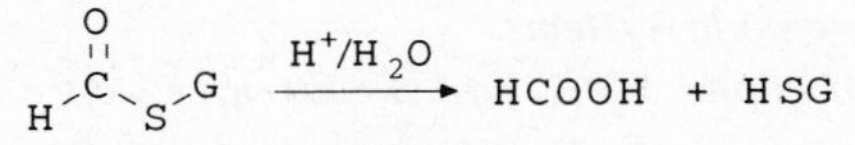

G = Glutathione

(5)

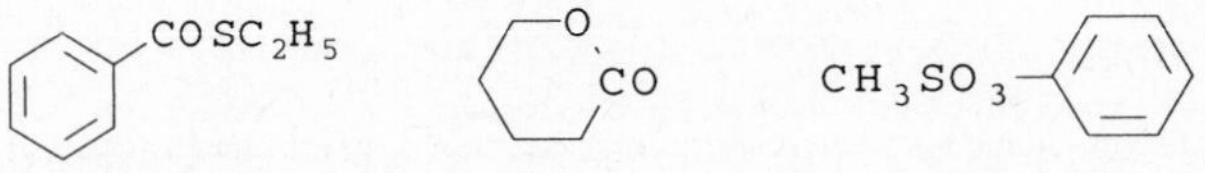

(A)   (B)   (C)

(a)   State the general type of ester illustrated by the examples A, B and C. Comment on their stability in (i) H^+/H_2O and (ii) OH^-/H_2O.

(b)   Describe by means of equations, reactions which could be made the basis of one identification test for compounds A and B.

(c)   Devise a chemical method of assaying C.

(d)   Outline *two* reasons for drugs being administered in the form of their esters.

(6)

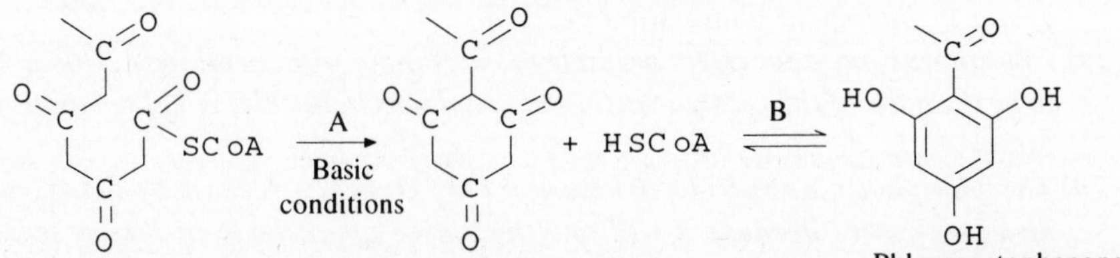

Phloroacetophenone

The step A shown above has been suggested as part of a route for the biogenesis of phloroacetophenone. Suggest a feasible electronic mechanism for this step. What is the general name given to the phenomenon illustrated by step B?

22

Nitrile and nitro functional groups

22.1 INTRODUCTION

The structural formulae and orbital structures of the nitro and nitrile (cyanide) functional groups are shown in Fig. 22.1. The nitro group is a resonance hybrid whilst the orbital structure of the nitrile group is similar to that of the alkyne group but with the nitrogen atom's lone pair occupying one of the sp hybrid atomic orbitals of the nitrogen atom.

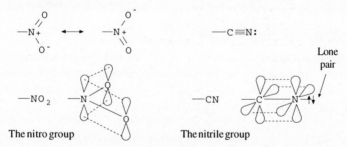

Fig. 22.1. The orbital structures and structural formulae of the nitro and nitrile functional groups.

The nitro and nitrile functional groups are found in some naturally occurring compounds but they are not widely distributed in nature. For example, the nitro group occurs in the broad spectrum antibiotic chloramphenicol which was isolated from *Streptomyces venezuelae* whilst the nitrile group is found in amygdalin which occurs in the seeds of the bitter almond (*Prunus amygdalus*).

The main uses of nitriles and nitro compounds are as intermediates in the synthesis and manufacture of organic compounds including drugs.

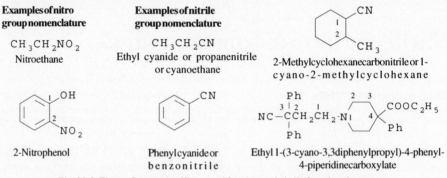

Chloramphenicol   Amygdalin

22.2   NOMENCLATURE

The presence of a nitro group in a molecule is indicated by the use of the prefix **nitro** in the name of the compound. There is no suffix for the nitro group. The prefix used to indicate the presence of a nitrile group in a molecule is **cyano** whilst the suffixes used are **nitrile, carbonitrile** and **cyanide** (Fig. 22.2). It should be noted that when either cyano or cyanide are used, the carbon of the −CN group is not considered to be part of the carbon skeleton on which the stem name of the molecule is based. However, the carbon of the −CN group is included in the stem name and numbering of an aliphatic molecule when the suffix nitrile is used. This is not the case when the nitrile is a substituent of a ring system.

Examples of nitro group nomenclature

$CH_3CH_2NO_2$
Nitroethane

2-Nitrophenol

Examples of nitrile group nomenclature

CH_3CH_2CN
Ethyl cyanide or propanenitrile
or cyanoethane

Phenyl cyanide or
benzonitrile

2-Methylcyclohexanecarbonitrile or 1-
cyano-2-methylcyclohexane

Ethyl 1-(3-cyano-3,3diphenylpropyl)-4-phenyl-
4-piperidinecarboxylate

Fig. 22.2. The prefixes and suffixes used for nitro and nitrile functional groups.

22.3   PHYSICAL PROPERTIES

22.3.1   Physical nature

Simple nitriles and aliphatic nitro compounds are colourless liquids and solids. Aromatic nitro compounds are often yellow or red liquids and solids. Most nitriles and nitro compounds are toxic and should be handled with care.

Aliphatic nitro alkanes with an α C−H bond exhibit tautomerism in solution and the liquid state. The position of equilibrium usually favours the **nitro** form of the molecule. However, the **aci-nitro** form is acidic enough to form salts with bases (22.6) and some compounds such as 2-nitropropane are acidic enough to ionise in water. Aromatic nitro compounds do not exhibit tautomerism because they do not possess an α C−H bond.

Primary nitro compounds   Secondary nitro compounds

nitro form   aci-nitro or nitronic acid form

The nitrile and nitro functional groups have strong dipole moments (about 3 to 3.5 D). Consequently, strong attractive forces exist between the molecules in the liquid state. These strong attractive forces account for the relatively high boiling points of the simple aliphatic and aromatic nitriles and nitro compounds (Table 22.1). These attractive forces would also play a part in the binding of molecules containing these functional groups to active and receptor sites.

Table 22.1. The boiling points of some simple nitriles and nitro compounds contrasted with those of the corresponding hydrocarbons.

Nitriles	bp $^{\circ}$C	Nitro compounds	bp $^{\circ}$C	Hydrocarbon	bp $^{\circ}$C
Methyl cyanide	81.6	Nitromethane	101.2	Methane	-162
Ethyl cyanide	97.2	Nitroethane	114-115	Ethane	-87
Propyl cyanide	117.5	Nitropropane	120.3	Propane	-42
Phenyl cyanide	190.7	Nitrobenzene	210-211	Benzene	80

The strong attractive forces caused by nitrile and nitro groups partly explains the formation of charge transfer complexes. For example, 1,3,5-trinitrobenzene is thought to form a stable charge transfer complex (3. 7) with 1,3,5-trimethylbenzene (mesitylene) in which the two benzene rings lie parallel to each other (Fig. 22.3). The strong electron-withdrawing effect of the nitro groups result in the benzene ring being electron deficient and able to accept electrons from the electron rich ring of the 1,3,5-trimethylbenzene.

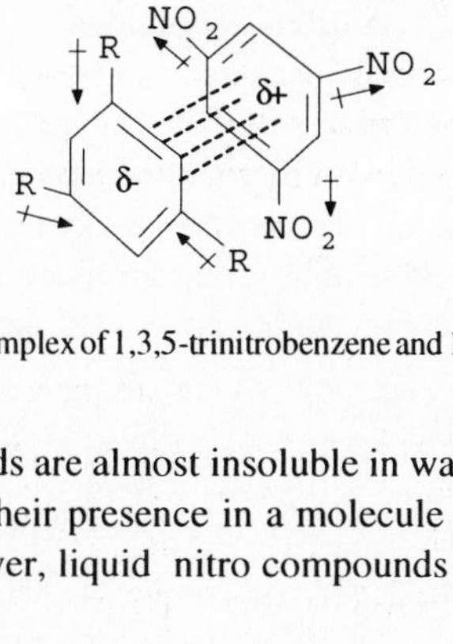

Key: R=CH$_3$

Fig. 22.3. The charge transfer complex of 1,3,5-trinitrobenzene and 1,3,5-trimethylbenzene.

22.3.2 Solubility

Simple nitriles and nitro compounds are almost insoluble in water. Although the nitrile and nitro functional groups are polar their presence in a molecule does not improve the water solubility of that molecule. However, liquid nitro compounds are good solvents for many organic compounds.

22.4 CHEMICAL PROPERTIES

The nitro and nitrile functional groups are stable to a wide variety of reagents. Nevertheless, both groups are readily reduced and act as electron-withdrawing groups. Both groups will also exhibit characteristic reactions. For example, nitriles are hydrolysed to amides and under vigorous conditions, acids. Nitriles also react with alcohols in acid conditions to form alkyl imidates. Nitroalkanes are acidic. For example, nitromethane and dinitromethane have pK_a values of 10.2 and 3.6 respectively.

22.5   REDUCTION

Nitro compounds are easily reduced under both laboratory and biological conditions. The most common product is the corresponding amine but oximes and other compounds can also be produced (Fig. 22.4).

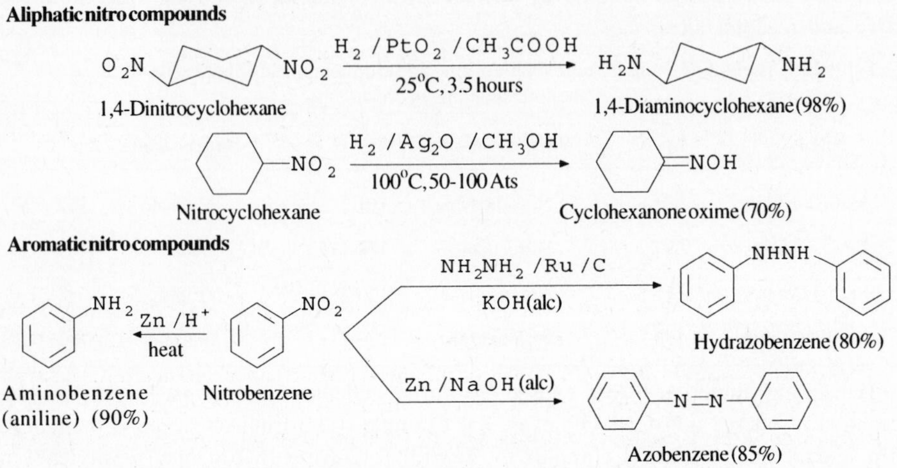

Aliphatic nitro compounds

1,4-Dinitrocyclohexane

$H_2 / PtO_2 / CH_3COOH$
25°C, 3.5 hours

1,4-Diaminocyclohexane (98%)

Nitrocyclohexane

$H_2 / Ag_2O / CH_3OH$
100°C, 50-100 Ats

Cyclohexanone oxime (70%)

Aromatic nitro compounds

Aminobenzene
(aniline) (90%)

Zn / H^+
heat

Nitrobenzene

$NH_2NH_2 / Ru / C$
KOH(alc)

Hydrazobenzene (80%)

$Zn / NaOH$ (alc)

Azobenzene (85%)

Fig. 22.4. Examples of the reduction of nitro compounds. Note azobenzene should not be confused with azabenzenes (pyridines) where the nitrogen is located in the ring.

Biological reduction of nitro compounds usually yields the amine. The reaction is catalysed by enzymes known as **nitro reductases.**

The reduction of nitroarenes is used as a source of aromatic amines in drug synthesis and manufacture. This reaction is used in the preparation of the local anaesthetics benzocaine and procaine. In this synthesis the acetyl group introduced at stage 2 protects the amino group from oxidation. The acetyl group is known as a **protecting group.** Any structure may be used as a protecting group provided it is stable under the conditions of the reactions being employed but can be removed at the appropriate point in the synthesis.

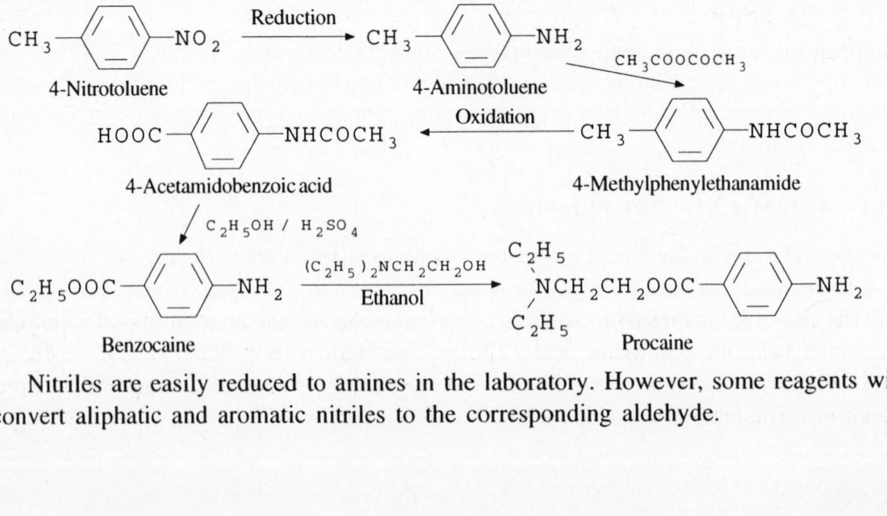

4-Nitrotoluene

Reduction

4-Aminotoluene

$CH_3COOCOCH_3$

4-Acetamidobenzoic acid

Oxidation

4-Methylphenylethanamide

C_2H_5OH / H_2SO_4

Benzocaine

$(C_2H_5)_2NCH_2CH_2OH$
Ethanol

Procaine

Nitriles are easily reduced to amines in the laboratory. However, some reagents will convert aliphatic and aromatic nitriles to the corresponding aldehyde.

$$PhCN \xrightarrow{\text{LiAlH}_4} PhCH_2NH_2$$

Phenyl   Benzylamine
cyanide   (72%)

$$PhCN \xrightarrow[\text{(C}_2\text{H}_5)_2\text{O} \sim -5^0\text{C}]{\text{LiAlH (OC}_2\text{H}_5)_3} PhCHO$$

Phenyl   Benzaldehyde
cyanide   (76%)

$$CH_3CH_2CH_2CN \xrightarrow[\text{C}_2\text{H}_5\text{OH /HCl}]{\text{H}_2 \text{ /PtO}_2} CH_3CH_2CH_2CH_2\overset{+}{N}H_3 \; \overset{-}{Cl}$$

Butanenitrile   Butylammonium chloride(95%)

22.6 ACTION AS ELECTRON-WITHDRAWING GROUPS

22.6.1 Aromatic systems

Nitro and nitrile groups are meta directing to electrophiles in the electrophilic substitution of aromatic systems.

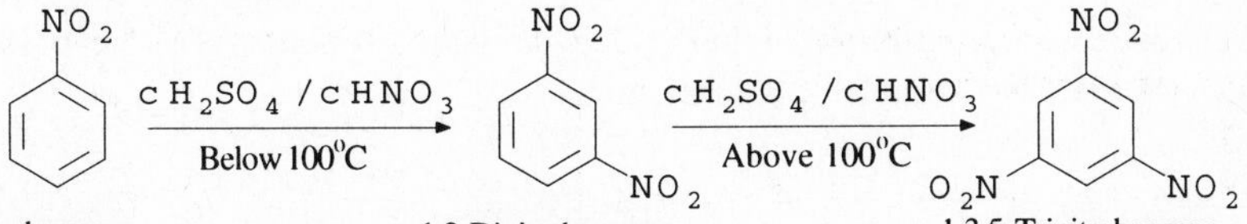

Nitrobenzene   1,3-Dinitrobenzene   1,3,5-Trinitrobenzene

The electron-withdrawing effect of the nitro group is strong enough to allow strong nucleophiles to attack the rings of simple nitroarenes. If more than one nitro group is present, nucleophilic attack may result in the substitution of a nitro group in either an ortho or para position relative to the other nitro group. The strong electron-withdrawing effect of nitro groups also increases the reactivity of other substituents on the aromatic system. This also explains for the high acidity of picric acid (15.6) and the increased reactivity of nitroaryl halides (14.7).

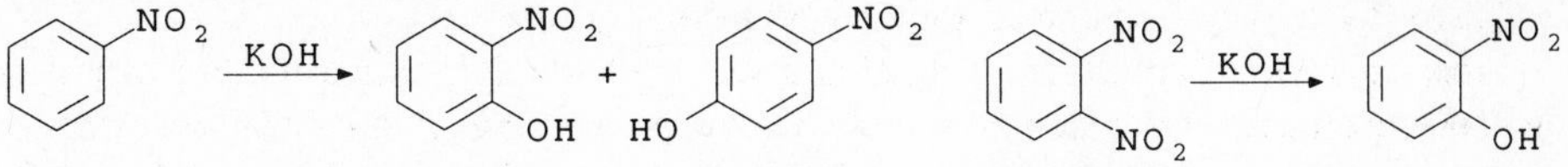

Nitrobenzene   2-Nitrophenol   4-Nitrophenol   1,2-Dinitrobenzene   2-Nitrophenol

22.6.2 Aliphatic systems

The electron-withdrawing effects of both nitriles and nitro groups weaken α C–H bonds to the extent that primary and secondary structures are able to form planar carboanions in basic conditions. These carboanions act as strong nucleophiles because they are stabilised by resonance (electron delocalisation).

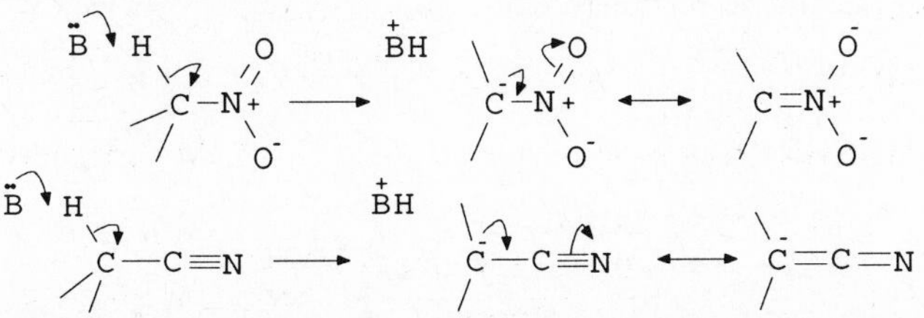

The carboanions formed from primary and secondary nitriles and nitro compounds act as nucleophiles in substitution, addition and condensation reactions (Fig. 22.5). Most of these reactions are believed to proceed by mechanisms based on the general schemes given in the appropriate sections of the text.

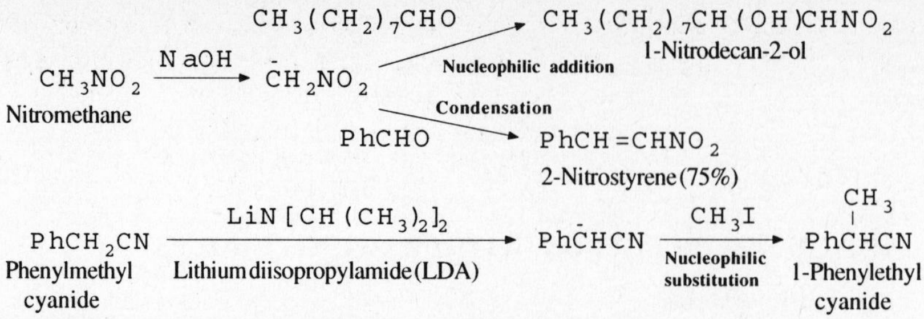

$$CH_3(CH_2)_7CHO \longrightarrow CH_3(CH_2)_7CH(OH)CHNO_2$$

1-Nitrodecan-2-ol

$$CH_3NO_2 \xrightarrow{NaOH} \bar{C}H_2NO_2 \quad \text{Nucleophilic addition}$$

Nitromethane

Condensation

$$PhCHO \longrightarrow PhCH=CHNO_2$$

2-Nitrostyrene (75%)

$$PhCH_2CN \xrightarrow{LiN[CH(CH_3)_2]_2} Ph\bar{C}HCN \xrightarrow[\text{substitution}]{CH_3I} \overset{CH_3}{\underset{|}{PhCHCN}}$$

Phenylmethyl   Lithium diisopropylamide (LDA)   Nucleophilic   1-Phenylethyl

cyanide   substitution   cyanide

Fig. 22.5. Examples of the nucleophilic addition, substitution and condensation carboanion reactions of nitriles and nitro compounds.

The condensation reaction of benzaldehyde with nitroethane is the basis of the synthesis of the CNS stimulant, amphetamine.

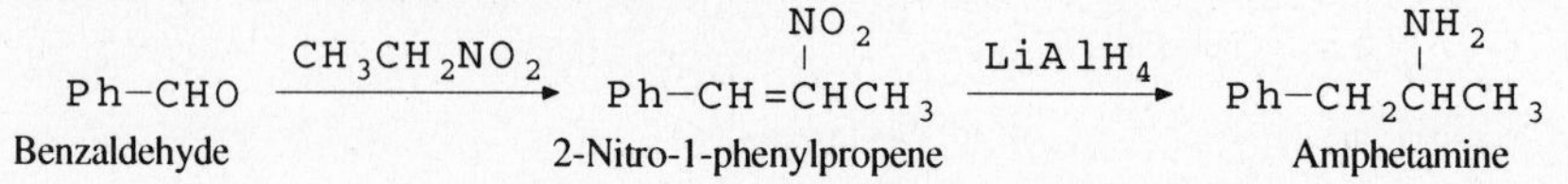

$$Ph-CHO \xrightarrow{CH_3CH_2NO_2} \overset{NO_2}{\underset{|}{Ph-CH=CHCH_3}} \xrightarrow{LiAlH_4} \overset{NH_2}{\underset{|}{Ph-CH_2CHCH_3}}$$

Benzaldehyde   2-Nitro-1-phenylpropene   Amphetamine

It is interesting to note that lithium aluminium hydride does not usually reduce alkene C=C bonds because the pi electrons of the bond repel the hydride ions from this reagent. However, in this case the polarisation of the C=C bond by the electron-withdrawing effect of the nitro group is sufficient to allow the nucleophilic attack on the C=C bond by a hydride ion from the lithium aluminium hydride.

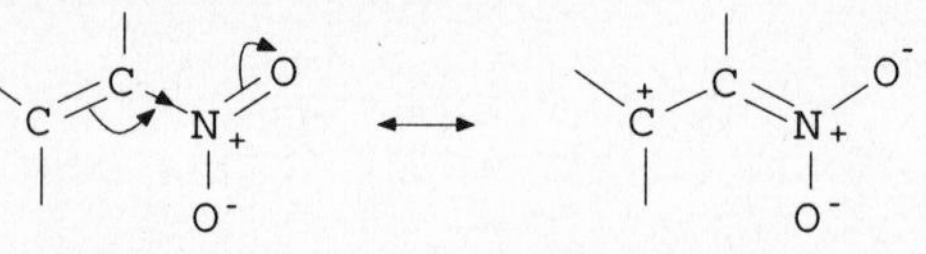

22.7 REACTIONS OF THE TYPE OF FUNCTIONAL GROUP

22.7.1 Hydrolysis of nitriles

Nitriles can be hydrolysed by aqueous acid to the corresponding amide. The reaction is usually slow and is normally carried out by heating the nitrile with either sulphuric or hydrochloric acids for several hours in order to obtain an appreciable yield of the amide. The mechanism of the reaction is essentially an acid catalysed nucleophilic addition followed by a tautomeric rearrangement.

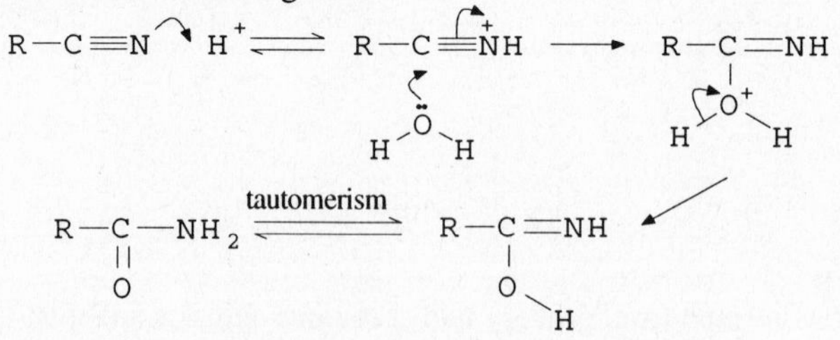

$$R-C\equiv N \quad H^+ \rightleftharpoons R-C\equiv \overset{+}{N}H \longrightarrow R-C=NH$$

$$R-\underset{\underset{O}{\|}}{C}-NH_2 \underset{\xrightarrow{\text{tautomerism}}}{\rightleftharpoons} R-C=NH$$

Further hydrolysis converts the amide into the carboxylic acid.

$$RCONH_2 \xrightarrow{H^+/H_2O} RCOOH + NH_4^+$$

Nitriles are hydrolysed under basic conditions. The reaction does not stop, except under controlled conditions, at the amide stage. It usually yields the corresponding carboxylic acid.

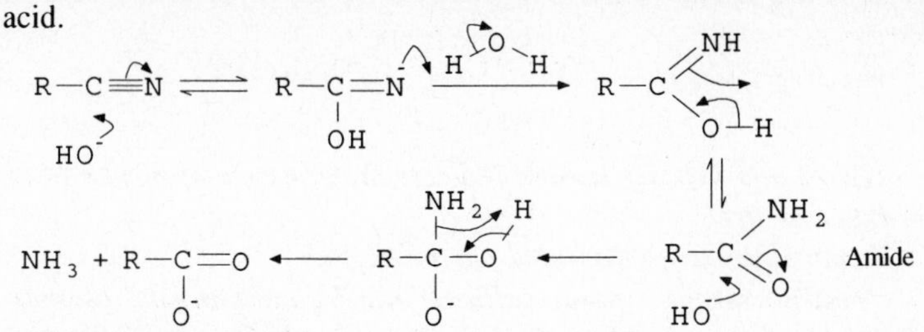

22.7.2   Reaction with alcohols

Nitriles react with alcohols in the presence of either hydrochloric or sulphuric acids to form alkyl imidate salts which can be converted into the alkyl imidate by treatment with sodium carbonate. The mechanism of the reaction is similar to that for the acid hydrolysis of nitriles given in 22.7.1.

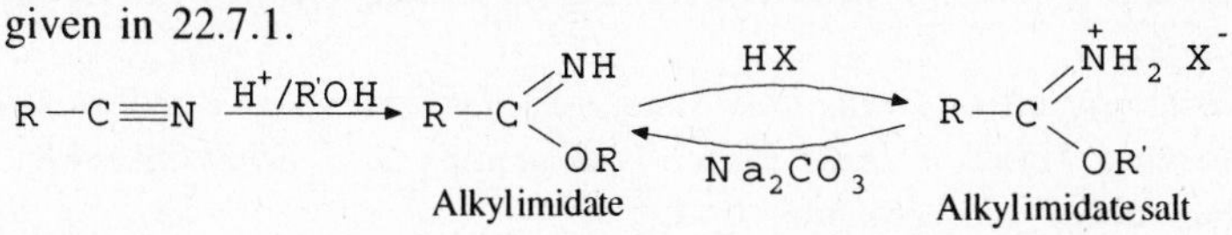

Alkyl imidates are strong bases and are useful intermediates in the synthesis of organic compounds.

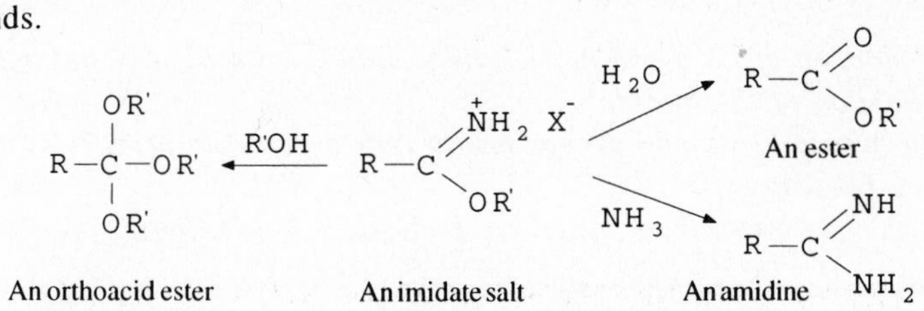

22.7.3   Reaction of nitro compounds with nitrous acid

Nitrous acid reacts with primary nitroalkanes to form nitrolic acids whilst secondary nitroalkanes react to form pseudonitroles. However, tertiary nitroalkanes do not react. In the primary and secondary nitroalkane reactions the nitrous acid acts as a base because the nitroalkanes are stronger acids. These reactions may be used to distinguish between primary, secondary and tertiary nitro compounds since nitrolic acids form red potassium salts in aqueous potassium hydroxide solution and pseudonitroles form a blue solution in chloroform.

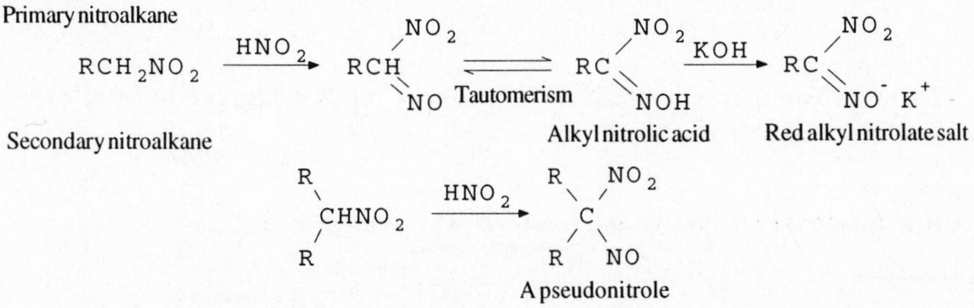

22.8 WHAT YOU NEED TO KNOW

(1) The nitro and nitrile functional groups have the structural formulae:

$$\text{Nitro group} \quad -NO_2 \quad -\overset{\overset{\displaystyle O}{\diagup\!\!\diagup}}{\underset{\underset{\displaystyle O^-}{\diagdown}}{N^+}} \qquad\qquad \text{Nitrile} \quad -CN \quad -C\equiv N$$

(2) The prefixes and suffixes used to indicate the presence of nitro and nitrile groups in a molecule are:

> *nitro group: prefix nitro but no suffix;*
> *nitrile: prefix cyano; suffixes: nitrile, carbonitrile, cyanide.*

(3) The carbon of the nitrile group is included in the numbering system of acyclic compounds when the suffix nitrile is used but is not included when the other suffixes are used.

(4) Compounds containing nitro groups are often yellow or red.

(5) Primary and secondary nitroalkanes exhibit tautomerism in the liquid state and in solution.

(6) The nitrile and nitro groups have strong dipoles which can result in strong forces of attraction between these functional groups and other suitable structures including active and receptor sites.

(7) Nitroarenes are believed to form charge transfer compounds with electron rich species.

(8) The introduction of nitro and nitrile groups into a molecule will not improve the water solubility of the molecule.

(9) Nitro and nitrile functional groups are usually reduced to primary amines by a variety of reagents.

$$RNO_2 \longrightarrow RNH_2 \qquad RCN \longrightarrow RCH_2NH_2$$

(10) Nitro groups are sometimes reduced to oximes.

(11) Nitrile and nitro groups are strong electron acceptors.

(12) Nitrile and nitro groups are meta directing to electrophiles in the electrophilic substitution of aromatic systems.

(13) Nitro and nitrile functional groups are strong enough electron-withdrawing groups to allow nucleophilic attack on an aromatic system. Substitution usually occurs in the ortho and/or para positions relative to the nitro or nitrile group.

(14) C–H bonds in an alpha position to a nitrile or nitro group are usually acidic enough to react with a base and form a carboanion.

$$\overset{H}{\underset{\diagup}{\diagdown}}C-\overset{\overset{\displaystyle O}{\diagup\!\!\diagup}}{\underset{\underset{\displaystyle O^-}{\diagdown}}{N^+}} \xrightarrow{\text{base}} \overset{-}{\underset{\diagup}{\diagdown}}C-\overset{\overset{\displaystyle O}{\diagup\!\!\diagup}}{\underset{\underset{\displaystyle O^-}{\diagdown}}{N^+}} \qquad \overset{H}{\underset{\diagup}{\diagdown}}C-CN \xrightarrow{\text{base}} \overset{-}{\underset{\diagup}{\diagdown}}C-CN$$

These carboanions may act as nucleophiles in nucleophilic substitution, addition and condensation reactions.

Key: Z= a nitro or a nitrile group

$$R-\overset{\overset{\displaystyle |}{C-Z}}{\underset{\underset{\displaystyle H}{|}}{C}}-OH \xleftarrow[\text{Nucleophilic addition}]{RCHO} \overset{-}{C}-Z \xrightarrow[\text{carboanion}]{\overset{\text{Nucleophilic substitution}}{RX}} \overset{R}{\underset{\underset{\displaystyle |}{|}}{-C-Z}}$$

$$\xrightarrow[\text{Condensation}]{ArCHO} ArCH=\overset{|}{C}-Z$$

(15) Primary and secondary nitroalkanes are often referred to in the literature as being acidic because of their weak α C–H bonds.

(16) Nitriles are hydrolysed by aqueous acid to the corresponding amide. Continued hydrolysis yields the corresponding carboxylic acid.

$$RCN \xrightarrow{\quad H^+/H_2O \quad} RCONH_2 \xrightarrow{\quad H^+/H_2O \quad} RCOOH$$

(17) Nitriles are hydrolysed by alkali to the salt of corresponding carboxylic acid.

$$RCN \xrightarrow{\quad H_2O/NaOH \quad} RCOONa$$

(18) Nitrous acid is used to distinguish between primary, secondary and tertiary nitroalkanes.

22.9 QUESTIONS

(1) Draw structural formulae for each of the following compounds: (a) 2-cyanopropanoic acid, (b) propanedinitrile, (c) 2-nitrobutane, (d) 3-nitropyridine, and (e) 3-nitrobenzonitrile.

(2) Determine which of the compounds in question 1 are optically active. Draw the R and S isomers of these compounds.

(3) List the *general* chemical and physical properties you would associate with the nitro group. Predict the products from the reaction of nitrobutane with each of the following reagents: (a) LDA followed by ethyl iodide, (b) sodium ethoxide followed by benzaldehyde, (c) zinc and hydrochloric acid, and (d) nitrous acid followed by potassium hydroxide.

(4) List the *general* chemical and physical properties you would associate with the nitrile group. Predict the products from the reaction of benzonitrile with each of the following reagents. (a) dilute aqueous sodium hydroxide, (b) lithium aluminium hydride, and (c) a mixture of concentrated nitric and sulphuric acids at 25°C.

(5) (a) Suggest by means of chemical equations one chemical identification test for each of the following compounds:

A   B

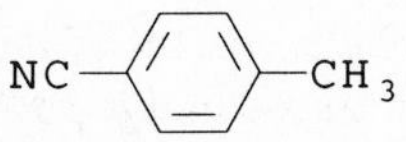

O_2N—⟨benzene ring⟩—CH_3 NC—⟨benzene ring⟩—CH_3

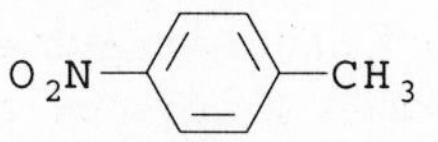

(b) Outline a physical method of distinguishing between compounds A and B.

(c) Describe, with the aid of chemical equations, a chemical method of distinguishing between compounds A and B.

23

Carbonic acid and related compounds

23.1 INTRODUCTION

Carbonic acid (H_2CO_3) has never been isolated but very small amounts of the acid are believed to exist in equilibrium with dissolved carbon dioxide and water in aqueous solution.

$$H_2CO_3 \rightleftharpoons CO_2 + H_2O$$
Carbonic acid

The organic derivatives of carbonic acid are well known (Fig. 23.1).

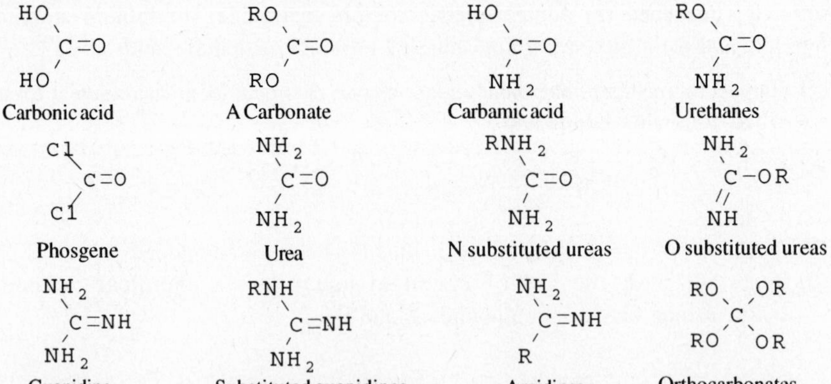

HO
  \
   C=O
  /
HO

Carbonic acid

RO
  \
   C=O
  /
RO

A Carbonate

HO
  \
   C=O
  /
NH_2

Carbamic acid

RO
  \
   C=O
  /
NH_2

Urethanes

Cl
  \
   C=O
  /
Cl

Phosgene

NH_2
  \
   C=O
  /
NH_2

Urea

RNH_2
  \
   C=O
  /
NH_2

N substituted ureas

NH_2
  \
   C−OR
  //
NH

O substituted ureas

NH_2
  \
   C=NH
  /
NH_2

Guanidine

RNH
  \
   C=NH
  /
NH_2

Substituted guanidines

NH_2
  \
   C=NH
  /
R

Amidines

RO   OR
  \ /
   C
  / \
RO   OR

Orthocarbonates

Fig. 23.1. The structure of carbonic acid and related compounds. Note that in this figure R can be either an aliphatic, aromatic or heterocyclic structure.

Compounds containing the functional groups listed in Fig. 23.1 are found in many biological processes and drugs. For example, citrulline, a substituted urea and arginine, a substituted guanidine occur in the **urea cycle** which is responsible for the synthesis of urea in terrestrial vertebrates (Fig. 23.2). The urea cycle was the first cyclic metabolic pathway to be discovered.

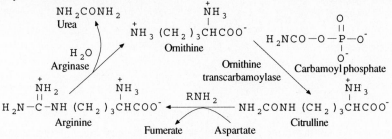

Fig. 23.2. A simplified outline of the urea cycle.

23.2   NOMENCLATURE

Table 23.1. The prefixes and suffixes commonly used in the systematic nomenclature of carbonic acid derivatives and related compounds.

Functional group	Prefix	Suffix	Functional group	Prefix	Suffix
Carbonates		carbonate	Orthocarbonates		orthocarbonate
Urethanes	carbamoyl	carbamate	Guanidines	guanidino	guanidine
Amidines	amidino	amidine carboxamidine	Ureas	ureido	urea carbamide

The presence of a carbonic acid related functional group in a molecule is usually indicated in systematic names by the prefixes and suffixes given in Table 23.1. Some examples of the use of these prefixes and suffixes are given in Fig. 23.3.

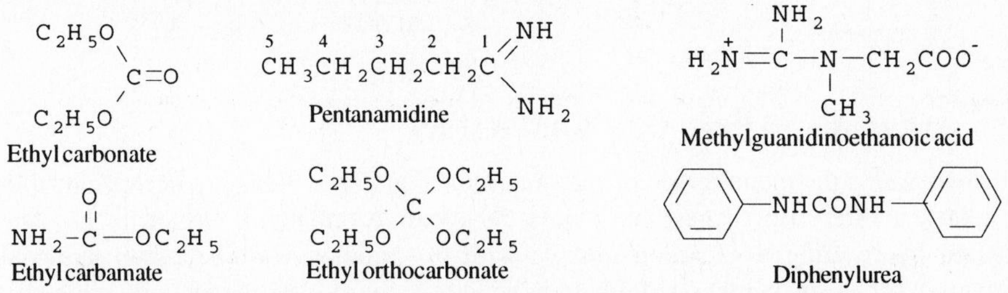

$$C_2H_5O{\diagdown}\atop C_2H_5O{\diagup}\!\!C{=}O$$

Ethyl carbonate

$$\overset{5}{C}H_3\overset{4}{C}H_2\overset{3}{C}H_2\overset{2}{C}H_2\overset{1}{C}{\diagup}^{NH}_{\diagdown NH_2}$$

Pentanamidine

Methylguanidinoethanoic acid

$$\overset{O}{\overset{\|}{NH_2{-}C}}{-}OC_2H_5$$

Ethyl carbamate

$$C_2H_5O{\diagdown}\atop C_2H_5O{\diagup}\!\!C{\diagdown}^{OC_2H_5}_{OC_2H_5}$$

Ethyl orthocarbonate

Diphenylurea

Fig. 23.3. Examples of the use of the prefixes and suffixes commonly used for carbonic acid derivatives and related compounds.

23.3   GENERAL   CHEMICAL   PROPERTIES

The chemical properties of carbonic acid derivatives are similar to those of the corresponding carboxylic acid derivatives because they have common structural features. For example, phosgene reacts with water, ammonia, alcohols, primary and secondary amines in a similar fashion to acid chlorides because the structure of phosgene contains a $-COCl$ group. Like the reactions of the acid chlorides, all of these reactions involve the liberation of hydrogen chloride.

$$
\begin{array}{c}
NH_2CONH_2 \\
\text{Urea}
\end{array}
\xleftarrow{NH_3}
\qquad
\begin{array}{c}
RNH_2 \\
\end{array}
$$

RNHCONHR
A substituted urea

$$
O=C \overset{Cl}{\underset{Cl}{}} \quad \text{Phosgene}
$$

R_2NH

R_2NCONR_2
A substituted urea

$$
\begin{array}{c}
\text{Cl} \quad H_2O \\
\end{array}
\quad
O=C \overset{OH}{\underset{OH}{}} \rightleftharpoons CO_2 + H_2O
$$
Carbonic acid

$$
\text{ROH} \quad O=C \overset{OR}{\underset{Cl}{}} \xrightarrow{ROH} O=C \overset{OR}{\underset{OR}{}}
$$
A chloro carbonate   A carbonate

Compounds containing amino structures whose lone pairs are not conjugated are usually medium to strong bases and will form salts. However, compounds with amino structures whose lone pairs are conjugated will either be weak bases or not basic and so will be lesslikely or unable to form salts. Those compounds whose structures contain an amide like structure will behave like amides and undergo hydrolysis and other nucleophilic displacement reactions.

23.4 ALKYL CARBONATES AND ORTHOCARBONATES

Alkyl carbonates and orthocarbonates have the general structural formulae shown in Fig. 23.1. They are the esters of carbonic and orthocarbonic acids respectively. Orthocarbonic acid may be considered to be the fully hydrated form of carbonic acid, that is, water has been added across the carbonyl group of the carbonic acid. Like carbonic acid, orthocarbonic acid has never been isolated but its derivatives are well characterised.

Simple alkyl carbonates are readily soluble in water. They are used in organic synthesis to introduce $-COOR$ into molecules with ketone and nitrile functional groups.

$$
\begin{array}{c}
CN \\
| \\
RCHCOOR \\
\alpha\text{-cyanoester}
\end{array}
\xleftarrow[\text{(ii)} \quad H^+]{\text{(i)} \quad NaOR}
\begin{array}{c}
RCH_2CN \\
\end{array}
(RO)_2CO
\xrightarrow[\text{(ii)} \quad H^+]{\text{(i)} \quad NaOR}
\begin{array}{c}
RCOCH_3 \\
\end{array}
\begin{array}{c}
RCOCH_2COOR \\
\beta\text{-ketoester}
\end{array}
$$

23.5 CARBAMIC ACID AND URETHANES

Carbamic acid is the mono-amide of carbonic acid (Fig. 23.1). It has not been isolated but its salts and esters (urethanes) have been prepared. Ammonium carbamate has been prepared by reacting dry carbon dioxide with dry ammonia whilst ethyl carbamate (urethane) has been synthesised by treating ethyl chloromethanoate with ammonia. N-substituted urethanes have also been synthesised in the laboratory. Urethane has been shown to cause cancer in mice, rats and hamsters.

$$
CO_2 + 2NH_3 \longrightarrow
O=C \overset{O^- \; NH_4^+}{\underset{NH_2}{}}
\qquad
O=C \overset{OC_2H_5}{\underset{Cl}{}}
\xrightarrow{NH_3}
O=C \overset{OC_2H_5}{\underset{NH_2}{}}
$$

Ammonium
carbamate

Ethyl
chloromethanoate

Ethyl carbamate
(urethane)

Urethanes and the salts of carbamic acids are usually soluble in water. However, aqueous solutions of urethanes are often unstable, hydrolysis being catalysed by acids, bases and enzymes.

$$NH_2COOC_2H_5 \begin{array}{l} \xrightarrow{HCl/H_2O} NH_4Cl + C_2H_5OH + CO_2 \\ \xrightarrow{NaOH/H_2O} Na_2CO_3 + C_2H_5OH + NH_3 \end{array}$$

It should be noted that some urethane derivatives, such as carbachol and neostigmine, are sufficiently stable to hydrolysis to be sterilised in aqueous solution by autoclaving.

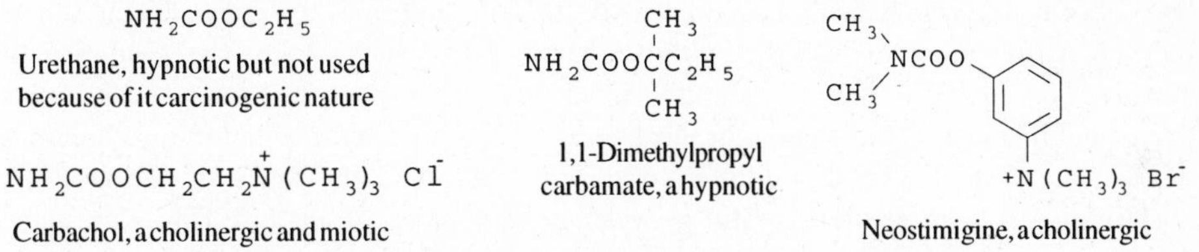

$NH_2COOC_2H_5$

Urethane, hypnotic but not used because of it carcinogenic nature

$NH_2COOCH_2CH_2\overset{+}{N}(CH_3)_3\ \overset{-}{Cl}$

Carbachol, a cholinergic and miotic

$NH_2COOCC_2H_5$ (with CH_3 groups)

1,1-Dimethylpropyl carbamate, a hypnotic

$(CH_3)_2NCOO$— (aromatic ring with $^+N(CH_3)_3\ Br^-$)

Neostimigine, a cholinergic

Fig. 23.4. Examples of drugs with urethane residues in their structures.

The structures of a number of drugs contain urethane residues (Fig. 23.4). These residues are usually hydrolysed in aqueous body fluids to either inactive or active metabolites. For example, the antithyroid drug carbimazole whose structure contains an N-substituted urethane residue, is hydrolysed by plasma esterases to ethanol and the corresponding N-substituted carbamic acid which subsequently decarboxylates to methimazole which is also an antithyroid agent.

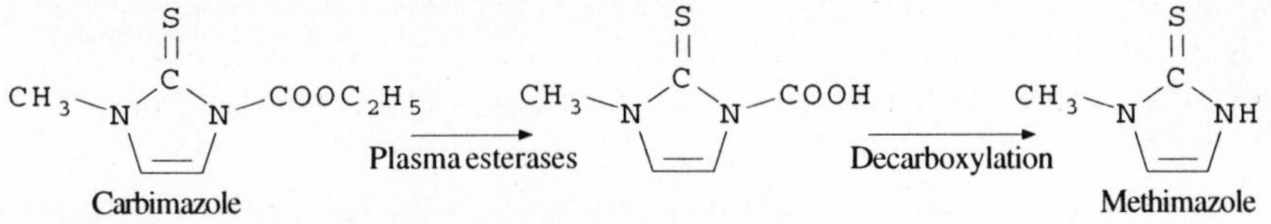

Carbimazole $\xrightarrow{\text{Plasma esterases}}$ (N-substituted carbamic acid) $\xrightarrow{\text{Decarboxylation}}$ Methimazole

Urethanes are also metabolised by N-hydroxylation.

$$NH_2COOC_2H_5 \xrightarrow{\text{Hydroxylation}} HONHCOOC_2H_5$$

23.6 DITHIOCARBAMIC ACIDS

Dithiocarbamic acid (Fig. 23.5) is the unstable sulphur analogue of carbamic acid, however; its salts, derivatives and salt derivatives are well characterised. The carbamic acid derivatives are easily oxidised to the corresponding disulphide often known as thiuram

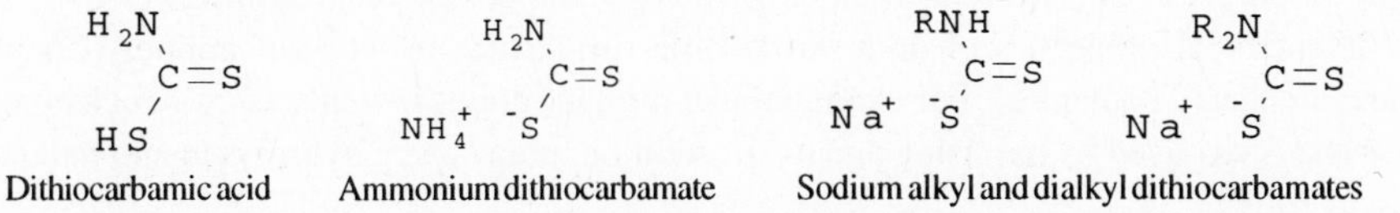

Dithiocarbamic acid

Ammonium dithiocarbamate

Sodium alkyl and dialkyl dithiocarbamates

Fig. 23.5. The structural formula of dithiocarbamic acid and the general structural formulae of dithiocarbamates. It should be noted that substituted dithiocarbamates may contain positive ions other than the sodium ions shown in the above examples. These ions may be metallic or organic in nature.

disulphides. For example, oxidation of N,N-diethylthiocarbamic acid yields tetraethylthiuram disulphide (*Disulfiram, Antabuse*) which is used to treat alcoholism.

$$\underset{\substack{C_2H_5 \\ \diagup}}{\overset{\substack{C_2H_5 \\ \diagdown}}{N}}-\underset{\overset{\|}{S}}{C}-SH \quad \xrightarrow{\text{Oxidation}} \quad \underset{\substack{C_2H_5 \\ \diagup}}{\overset{\substack{C_2H_5 \\ \diagdown}}{N}}-\underset{\overset{\|}{S}}{C}-S-S-\underset{\overset{\|}{S}}{C}-\underset{\substack{\diagdown \\ C_2H_5}}{\overset{\diagup C_2H_5}{N}}$$

N,N-Diethylthiocarbamic acid 　　　　　 *Disulfiram (Antabuse)*

Disulfiram does not cure alcoholism but is used to discourage alcoholics from drinking alcohol. Patients suffer very unpleasant reactions, such as nausea, headaches, palpitations and fainting, if they drink whilst taking the drug. Ethanol is metabolised in the body to ethanal which itself is oxidised to ethanoic acid and excreted. Disulfiram is reduced in the body to N,N-diethylthiocarbamic acid which blocks the oxidation of ethanal. Consequently, the concentration of ethanal in the body increases and it is this compound that causes the nausea, headaches, etc. The only way to avoid these symptoms when being treated with this drug is to abstain from alcohol. Since its introduction, Disulfiram has been very successful in weaning alcoholics off alcohol. However, many doctors recommend that patients being treated with Disulfiram also attend suitable counselling sessions.

Metallic salts of dithiocarbonates are used as insecticides and fungicides. For example, zinc bis(dimethyldithiocarbamoyl) disulphide (*Zerlate*) and disodium ethylenebisdithio- carbamate (*Nabam*) are used as a fungicides.

$$\underset{\substack{CH_3 \\ \diagup}}{\overset{\substack{CH_3 \\ \diagdown}}{N}}-\underset{\substack{\diagdown \\ S}}{\overset{\diagup S}{C}}\quad Zn \quad \underset{\substack{\diagup \\ S}}{\overset{\diagdown S}{C}}-\underset{\substack{\diagdown \\ CH_3}}{\overset{\diagup CH_3}{N}}$$

Zinc bis(dimethyldithiocarbamoyl) disulphide (*Zerlate*)

$$Na^+ \; {}^-S-\underset{\overset{\|}{S}}{C}-NHCH_2CH_2NH-\underset{\overset{\|}{S}}{C}-S^- \; Na^+$$

Disodium ethylenebisdithiocarbamate (*Nabam*)

23.7   UREA

Urea is the diamide of carbonic acid. Its properties indicate that the C–N bonds are both the same length and possess a significant degree of double bond character. Furthermore, both nitrogens have identical properties. Consequently the structure of urea is described as being a resonance hybrid of the three canonical forms:

$$H_2N\underset{\overset{\|}{O}}{\diagdown}\underset{C}{}\diagup NH_2 \quad \longleftrightarrow \quad H_2\overset{+}{N}\underset{\overset{|}{O^-}}{\diagdown}\underset{C}{}\diagup NH_2 \quad \longleftrightarrow \quad H_2N\underset{\overset{|}{O^-}}{\diagdown}\underset{C}{}\diagup \overset{+}{N}H_2$$

Urea is the end product of nitrogen metabolism in man. It is synthesised in the liver and about 30g is excreted daily in the urine by the average adult human being. A low urine and a high blood urea concentration is often a strong indicator of renal failure.

The chemical properties of urea show some similarities to those of amines (Chapter 24) and amides (Chapter 25). For example, like primary amines, urea acts as a nucleophile, a base and is oxidised by oxidising agents. In addition, urea can be hydrolysed and reduced.

23.7.1   Reactions similar to those of primary amines
As a base. Urea acts as a '**monoacidic**' base forming salts with acids. [1]H n.m.r. indicates that it is the oxygen that accepts the proton from the acid.

$$NH_2CONH_2 \xrightarrow{HNO_3} NH_2CONH_2 \cdot HNO_3$$
Urea nitrate

As a nucleophile. Urea acts as a nucleophile in nucleophilic displacement reactions with acid chlorides, anhydrides and carboxylic acid esters. The products of these reactions are **ureides**.

$$RCOCl + NH_2CONH_2 \longrightarrow RCONHCONH_2 + HCl$$
A ureide

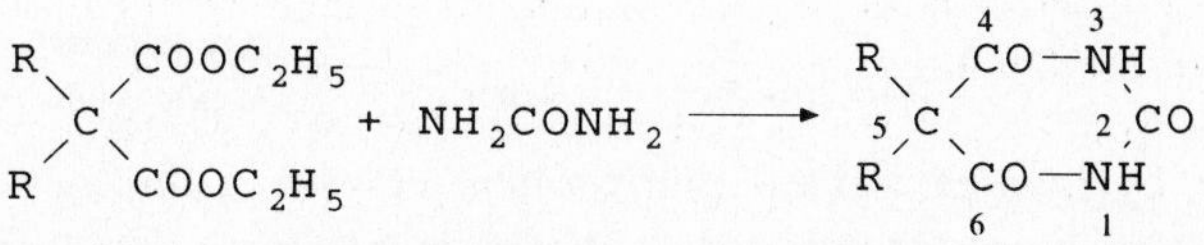

Many ureides are pharmacologically active. For example, carbromal and bromovaletone are hypnotics.

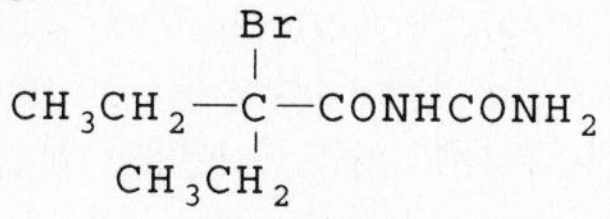

Carbromal 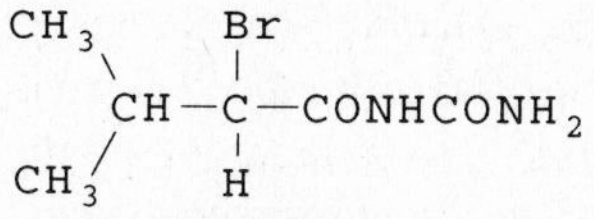　　　　　　　　　　Bromovaletone

An important group of cyclic ureides that are pharmacologically active are the barbiturates. These compounds have a variety of therapeutic uses, for example, hypnotics, sedatives, anticonvulsants and antiepileptics. It is interesting to note that pharmacologically useful barbiturates have structures that can be classified into two main types (Table 23.2). Another important group of cyclic ureides are the purines and pyrimidines (Chapter 13).

Table 23.2. Examples of the structure-action of barbiturates. It should be noted that both the suffixes -one and -al can be used in the names of barbitones.

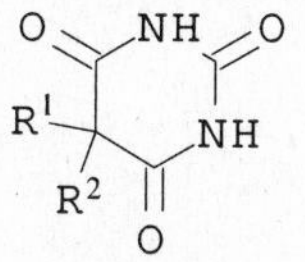

Type 1 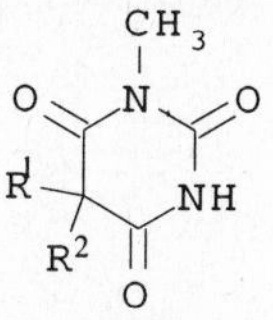　　　　　　　　　　　Type 2

Compound and (type)	R^1	R^2	Use	Trade name
Pentobarbitone (1)	$-C_2H_5$	$-CH(CH_2)_2CH_3$ with CH_3	Hypnotic	*Nembutal*
Phenobarbitone (1)	$-C_2H_5$	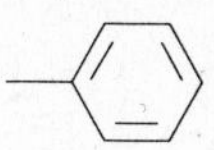 (phenyl)	Hypnotic, anticonvulsant	*Luminal*
Hexobarbitone (2)	$-CH_3$	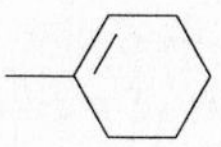 (cyclohexenyl)	Hypnotic	*Evidorm*
Methohexital (2)	$-CH_2CH=CH_2$	$-CHC\equiv CCH_2CH_3$ with CH_3	Narcotic	*Brevital*

Barbiturates are not very soluble in water. However, in the form of their hydroxyl tautomers, they act as weak acids, forming salts with cold aqueous solutions of the group 1 and 2 metals hydroxides and in some cases their carbonates.

| Phenobarbitone | Three tautomers of the type | Phenobarbitone sodium, a mixture of the salts of the various tautomers |

$C_{12}H_{11}N_2NaO_3$

The sodium salts are hygroscopic and are readily soluble in water. Consequently, these salts are used to prepare solutions for injection. It should be noted that dissolved carbon dioxide is acidic enough to convert these salts to the parent barbitone which, because of its poor water solubility precipitates from solution. Furthermore, the sodium salts of barbiturates hydrolyse in aqueous solution especially when heated. Consequently, good storage conditions are required if solutions are to be kept for any length of time and they should not be sterilised by autoclaving.

Oxidation. Urea is oxidised by both nitrous acid and alkaline sodium hypobromite solution. In both cases the yield of nitrogen is not quantitative and so cannot be made the basis of assay procedures.

$$CO_2 + H_2O + N_2 \xleftarrow{HNO_2} \underset{NH_2}{\overset{NH_2}{C=O}} \xrightarrow[NaOH]{NaOBr} Na_2CO_3 + NaBr + H_2O + N_2$$

23.7.2   Reactions similar to those of amides
Hydrolysis. Urea, like most amides, is hydrolysed by water, the reaction being catalysed by both acids, bases and enzymes.

$$NH_4^+ + CO_2 \xleftarrow{H_2O /H^+} \underset{NH_2}{\overset{NH_2}{C=O}} \begin{array}{c} \xrightarrow{H_2O /OH^-} NH_3 + CO_3^{2-} \\ \xrightarrow[Urease]{H_2O} CO_2 + 2NH_3 \end{array}$$

The enzyme catalysed hydrolysis of urea is the basis of a method for its determination in blood samples. An electrode coated with urease immobilised in a special polymer is placed in the sample. Urea and water diffuse into the polymer and hydrolysis of the urea occurs within the structure of the polymer which causes changes in the nature of the electrode. These changes can be correlated with the concentration of urea in the sample.

Biuret reaction. Urea decomposes on heating to biuret.

$$2NH_2CONH_2 \xrightarrow{Heat} NH_2CONH_2CONH_2 + NH_3$$
$$\text{Urea} \qquad\qquad \text{Biuret}$$

Biuret reacts with aqueous sodium hydroxide and copper II sulphate to form a purple complex. A similar reaction is the basis of a colorimetric method used to determine the total concentration of peptides and proteins in serum.

23.7.3   Properties characteristic of urea
Urea forms **channel (canal) inclusion complexes** when it is crystallised in the presence

of certain *straight chain* molecules. Inclusion complexes are formed when a molecule, known as the **guest** molecule, is trapped inside the crystal lattice of a second compound known as the **host**. The guest molecule is not chemically bonded to the host but it is likely that weak forces of attraction (Chapter 3) will exist between the guest and the host. In urea inclusion complexes, the guest molecules are trapped in channels that occur in the loosely packed urea crystal. These channels have a diameter of about 0.5 nm and can contain several trapped molecules depending on their size.

Many different types of straight chain molecule are able to form channel complexes with urea, for example, alkanes, alcohols, carboxylic acids and esters. However, complexes will not form unless the trapped molecule possesses *more than* a specific minimum number of carbon atoms. This minimum depends on the type of trapped molecule, for example, it is five for alkanes and six for alcohols. Branched chain and cyclic molecules do not normally form channel complexes because their shape makes them too large to fit in the channels.

Channel complexes have been used to resolve racemic modifications and separate geometric isomers. For example, 2-chlorooctane has been resolved by fractional crystallisation with urea. Other compounds that form inclusion complexes have also been used to resolve racemic modifications.

23.8 BIOTIN

Biotin

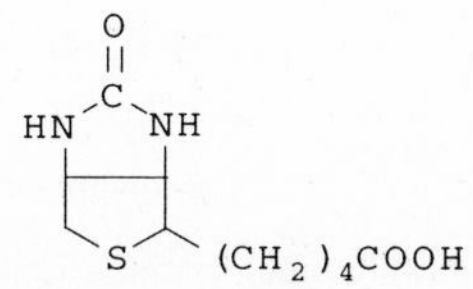

Biotin is a bicyclic urea derivative that is essential for the growth of mammals. It is obtained from the diet and synthesised by the intestinal flora. Biotin is usually required by biological reactions in which carbon dioxide is the reagent (carboxylation). For example, it acts as co-enzyme in the conversion of pyruvate to oxaloacetate in glucogenesis.

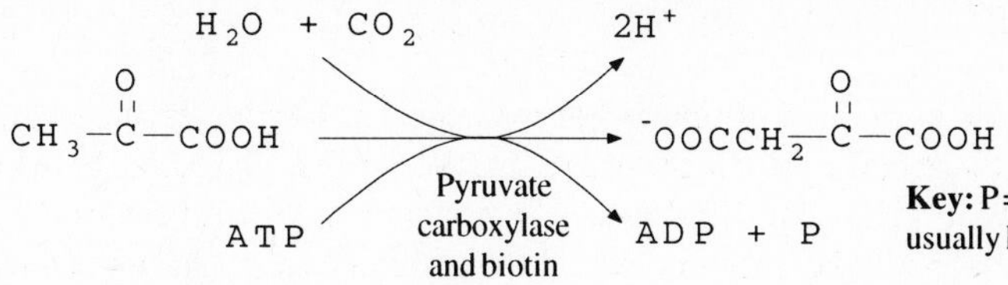

23.9 THIOUREA

Thiourea is the sulphur analogue of urea. Like urea, its properties indicate that its structure is a resonance hybrid. Thiourea acts as a monobasic acid and exhibits some of the general properties of an amine and a thioamide. For example, thiourea acts as a nucleophile, can be hydrolysed and oxidised. However, the sulphur atom modifies the course of some of these reactions because sulphur is in general a better nucleophile than either oxygen or nitrogen (Fig. 23.6). Thiourea also forms channel inclusion compounds with many different types of molecule. It differs from urea in that the channels are large enough to accommodate some branched chain and cyclic molecules.

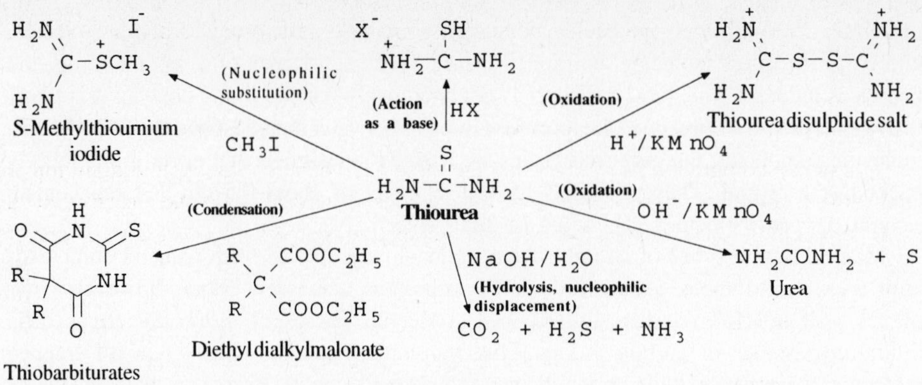

Fig. 23.6. Examples of the reactions of thiourea, the general type of the reaction being given in brackets

Thiourea is not involved in human metabolism. However, the structures of a number of drugs contain thiourea residues (Fig. 23.7).

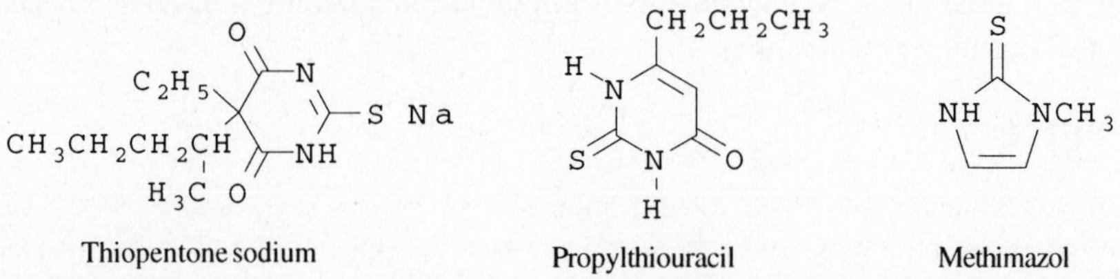

Thiopentone sodium   Propylthiouracil   Methimazol

Fig. 23.7. Examples of drugs containing thiourea residues. Thiopentone sodium is a general anaesthetic whilst propyluracil and methimazol are antithyroid drugs that inhibit iodine uptake.

The metabolism of these drugs often involves the loss of the sulphur by a mechanism that is not understood. For example, thiopentone sodium is largely metabolised to pentobarbitone.

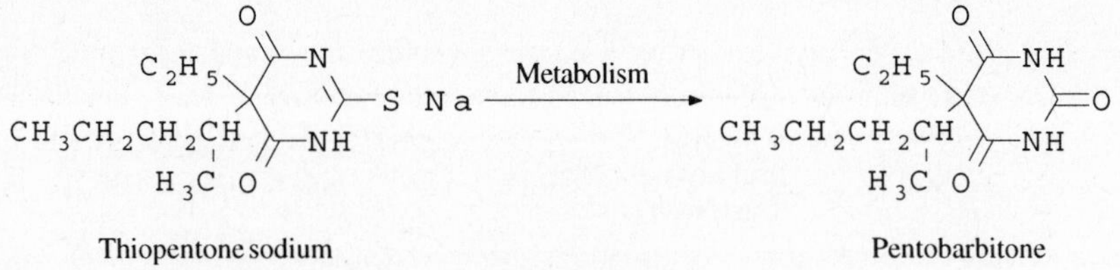

Thiopentone sodium   Pentobarbitone

23.10 GUANIDINE AND ITS DERIVATIVES

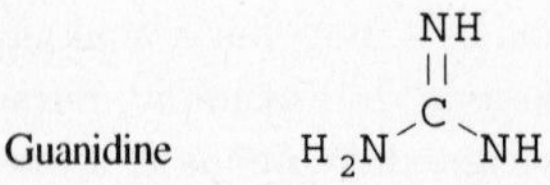

Guanidine is a deliquescent crystalline solid, melting about $50\,^{\circ}C$. It is soluble in water and ethanol. Guanidine will act as a strong mono-acidic base ($pK_a \sim 13.6$). For example, guanidine forms forms stable salts with both inorganic and organic acids.

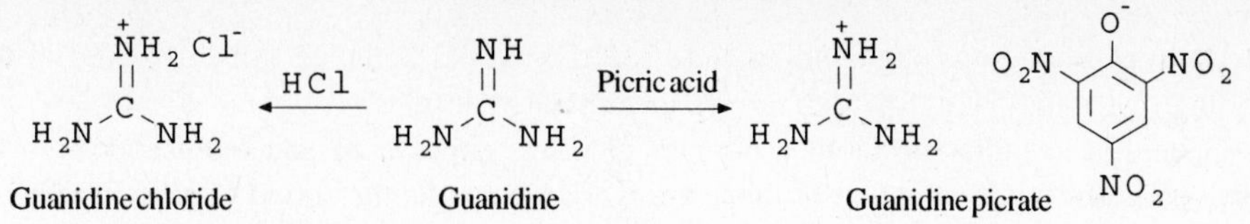

Guanidine chloride   Guanidine   Guanidine picrate

X-ray crystallography has shown that the three C−N bonds of the guanidinium ion are the same length. This observation and other evidence indicates that the structure of the guanidinium ion is a resonance hybrid with the positive charge concentrated mainly on the carbon atom.

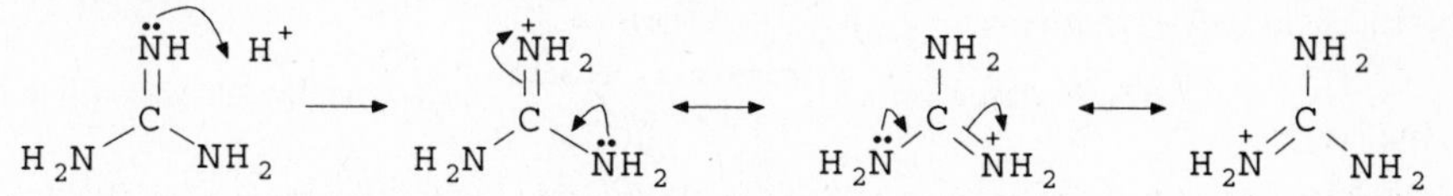

Guanidine is hydrolysed by barium hydroxide to urea but there is no evidence that enzymic hydrolysis occurs in mammalian systems.

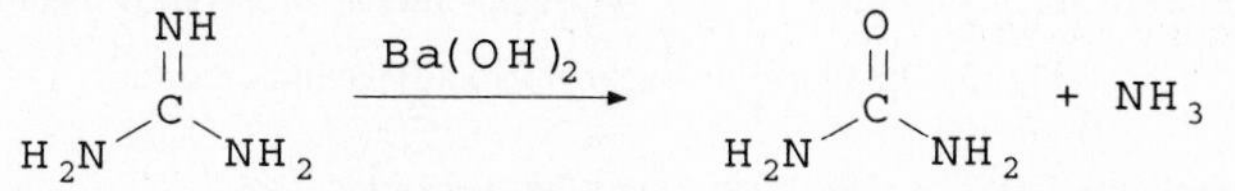

Guanidine will also form coloured complexes with many transition metal salts, such as copper II sulphate and cobalt II chloride.

Substituted guanidines have similar chemical properties to guanidine although these may be modified by the presence of other functional groups in the molecule. For example, creatine (pK_a 11.0) is not such a strong base as guanidine (pK_a ~13.6) because of zwitterion formation between the guanidino group and the carboxylic acid.

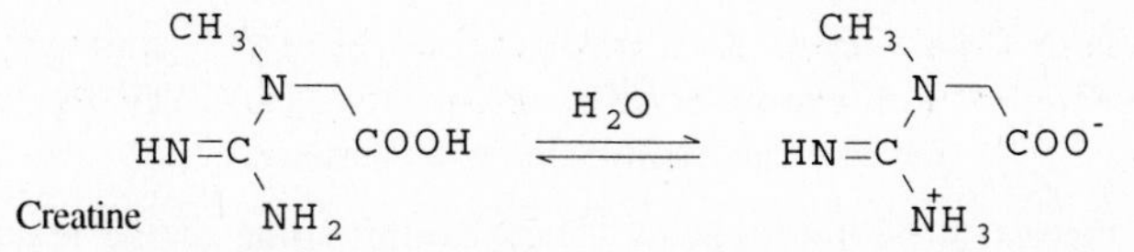

Creatine

Guanidine and its derivatives occur almost exclusively in the salt form in biological fluids. They are important constituents of many metabolic processes. For example, arginine provides the guanidino residue in the biosynthesis of creatine phosphate which is involved with energy storage in muscles.

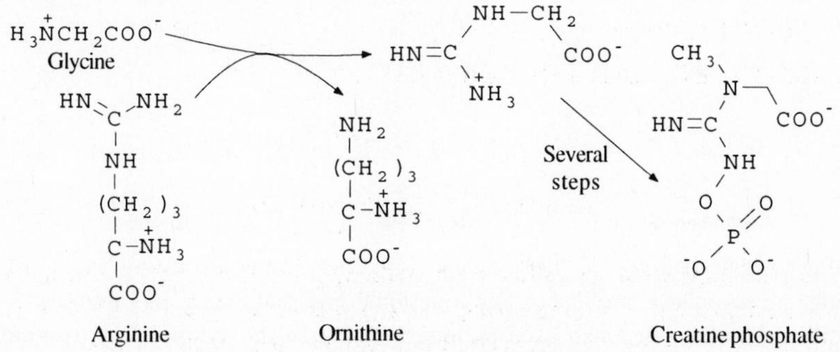

Arginine   Ornithine   Creatine phosphate

Hydrogen bonding by guanidine and its derivatives is believed to be involved in the binding of guanidine and its derivatives to protein molecules.

Guanidine residues occur in a number of drugs (Fig. 23.8) and often appear to be involved in the mode of action of those drugs. For example, the action of guanethidine as an adrenergic neurone blocking agent is believed to be due to its guanidine residue.

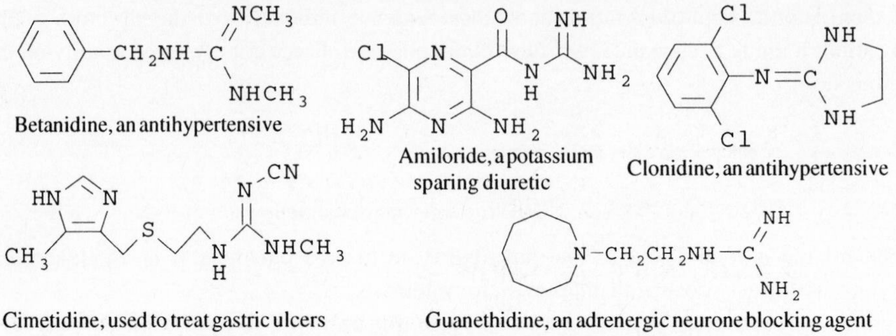

Betanidine, an antihypertensive

Amiloride, a potassium sparing diuretic

Clonidine, an antihypertensive

Cimetidine, used to treat gastric ulcers

Guanethidine, an adrenergic neurone blocking agent

Fig 23.8. Examples of drugs containing guanidine residues.

23.11 IMIDIC ACID AND ITS DERIVATIVES

Imidic acids have the general structural formula shown in Fig. 23.9. They are the tautomeric forms of the corresponding amide (25.3.1) and have never been isolated in the form shown in Fig. 23.9. However, their imidate, amidine and other derivatives are well characterised.

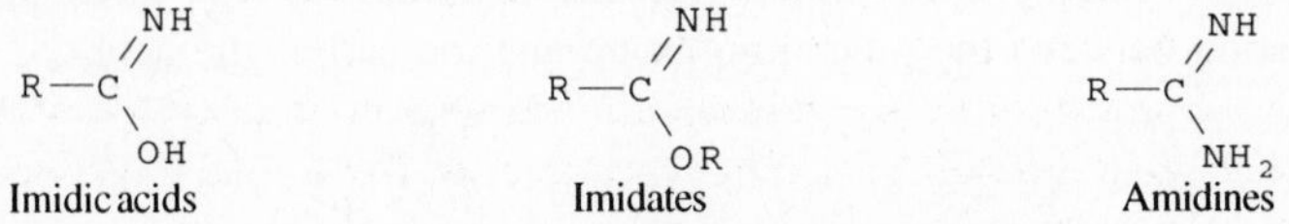

Imidic acids

Imidates

Amidines

Fig. 23.9. Imidic acids, imidates and amidines.

Imidates are the esters of imidic acids. They exhibit some of the reactions of both esters and bases. For example, imidates form salts with acids, undergo ammonolysis with ammonia and hydrolysis with water in the presence of an acid.

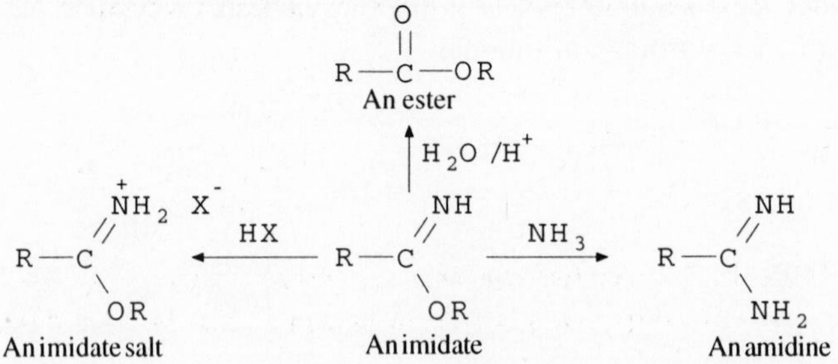

An ester

An imidate salt

An imidate

An amidine

Amidines are derivatives of imidic acids (Fig. 23.9). They are strong bases (pK_a ~12) whose salts are resonance stabilised. The functional group is found in a number of drugs (Fig. 23.10.). Amidines are also used as intermediates in the synthesis of pyrimidines.

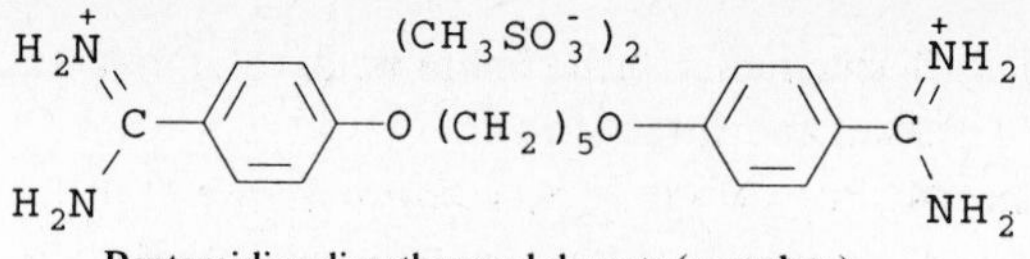

Pentamidine dimethanesulphonate (mesylate),
used to treat trypanosomiasis and AIDS

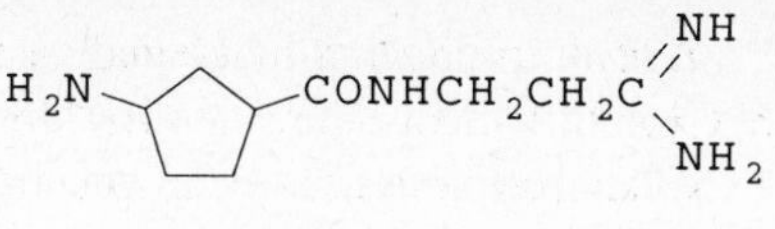

Amidinomycin, an antiviral

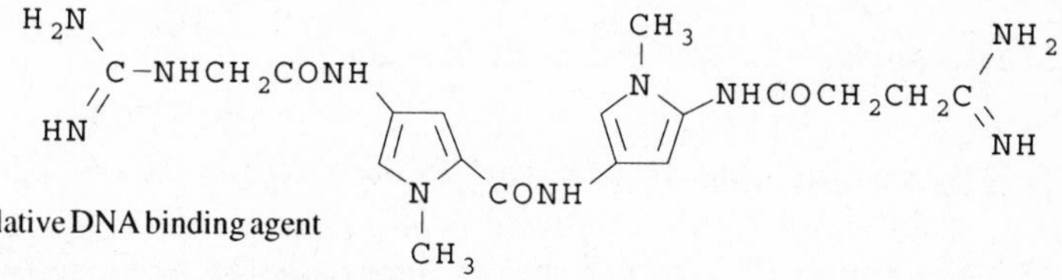

Netropsin, a non-intercalative DNA binding agent

Fig. 23.10. Examples of drugs containing amidine structures.

23.12   WHAT YOU NEED TO KNOW

(1) The structures of the derivatives of carbonic acid and the prefixes and suffixes used to indicate the presence of these structures (Fig. 23.1 and Table 23.1).

(2) The simple derivatives of carbonic acid and their related compounds are usually soluble in water.

(3) Urethanes are hydrolysed in aqueous solution, the hydrolysis being catalysed by acids, bases and enzymes.

(4) Dithiocarbamates are oxidised to the corresponding disulphides (cf. thiols).

(5) Urea is the main metabolic end product of nitrogen in humans and other mammals.

(6) The reactions of urea are often similar to those of amines and amides.

(7) Urea acts as a base forming salts with acids.

(8) Urea acts as a nucleophile in nucleophilic displacement reactions with esters, acid chlorides and anhydrides. The products of these reactions are known generally as a ureides.

$$RCOZ + H_2NCONH_2 \longrightarrow RCONHCONH_2$$
$$\text{A ureide}$$

Key: Z = OR (esters), Cl (acid halides) and OCOR (anhydrides)

(9) Purines and pyrimidines are cyclic ureides.

(10) Barbiturates may be classified as either cyclic ureides or pyrimidines. Barbiturates used as drugs usually have the general structures:

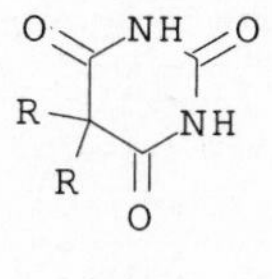

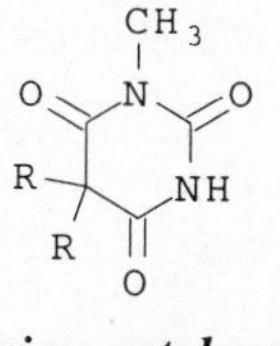

(11) Urea is hydrolysed by water, the reaction being catalysed by acids, bases and enzymes, the products being either ammonia and carbon dioxide or their derivatives.

(12) Enzyme-catalysed hydrolysis using special electrodes as detectors to follow the hydrolysis is the basis of the estimation of urea in some biological fluids.

(13) Urea forms channel inclusion compounds.

(14) Thiourea is the sulphur analogue of urea. Its reactions are similar to those of urea but modified to some extent by the presence of the sulphur atom.

(15) Thiobarbitones have the general formula:

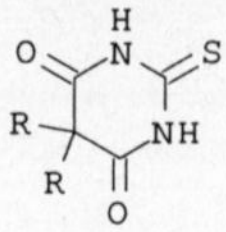

(16) *Guanidine has the structural formula :*

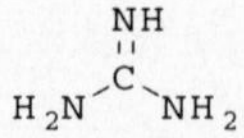

(17) *Guanidine and its derivatives are strong bases because of resonance stabilisation of the conjugate acid.*

(18) *Guanidine and its derivatives occur almost exclusively in the salt form in biological fluids.*

(19) *Guanidine and its derivatives are believed to bind to proteins by means of hydrogen bonds.*

(20) Imidates and amidines have the general structural formulae:

Imidates 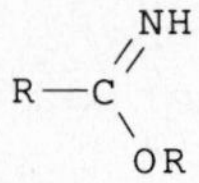   Amidines

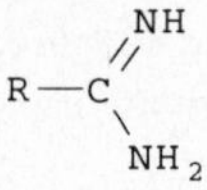

(21) Imidates and amidines are bases.

23.13 QUESTIONS

(1) Classify the type(s) of functional group found in each of the following compounds:

(a) $(CH_3O)_2CO$ (b) $CH_3\overset{NH}{\overset{\|}{C}}NH_2$ (c) $CH_3CONHCONH_2$ (d) $(C_2H_5)_2NCSSH$

(2) List the general chemical properties you would expect each of the following compounds to exhibit: (a) butanamidine, (b) diethylurea and (c) methylguanidine.

(3) Predict the most likely product(s) of the reaction of each of the following:
(a) phosgene with excess ethylamine, (b) ethyl carbonate with propanone mixed with sodium ethoxide. Suggest a structure for the compound that would be formed if the reaction mixture was treated with very dilute aqueous acid at room temperature, (c) N-ethylthiocarbamate with iodine, (d) thiourea with diethyl malonate, (e) methylurea with an aqueous solution of dilute hydrochloric acid, (f) guanidine with ethanoicacid, (h) ethyl ethanimidate with ammonia.

(4)

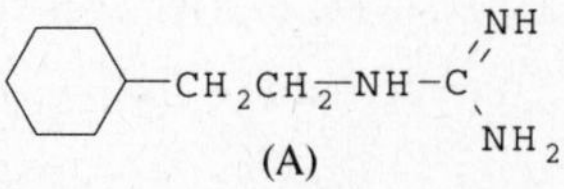

(A) 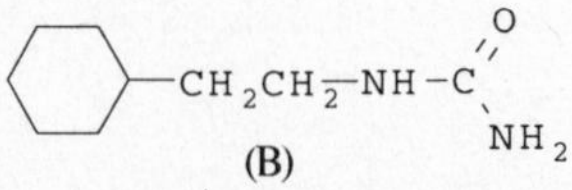   (B)

Compound A is a potential drug. Outline chemical reactions that may be suitable for: (a) one identification test for A, (b) one assay for A, and (c) a limit test for B in A. Your answers may include reference to physical methods that are used in conjunction with the chemical reactions.

(5) Predict, giving reasons, the relative basic strengths of methylurea (A) and methylguanidine (B).

24

Amines

24.1 INTRODUCTION

Amine functional groups have the general structures shown in Table 24.1. They are classified as either primary, secondary or tertiary amines depending on the number of hydrogen atoms bonded to the nitrogen. This classification is necessary because these structures exhibit different as well as similar properties. For the same reason, it is also necessary to distinguish between aromatic and aliphatic amines. **Aromatic amines have the amino group directly bonded to the aromatic ring system.** Mixed aromatic and aliphatic amines are also known.

Table 24.1. The classification and general formulae of amine functional groups.

Classification (symbols)	Structure	Examples		
Primary amines (p or 1' or 1^0) **(Two hydrogens)**	$-NH_2$	$CH_3CH_2NH_2$ Ethylamine	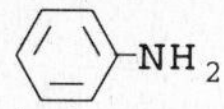 ⬡$-NH_2$ Aminobenzene	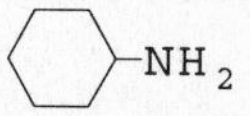 ⬡$-NH_2$ Cyclohexylamine
Secondary amines (s or 2' or 2^0) **(One hydrogen)**	$-NH-$	CH_3NHCH_3 Dimethylamine	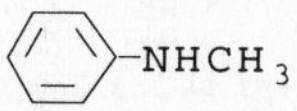 ⬡$-NHCH_3$ N-Methylaminobenzene	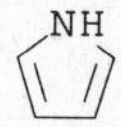 Pyrrole
Tertiary amines (t or 3' or 3^0) **(No hydrogens)**	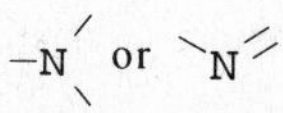 $-N\diagup$ or $\diagdown N\diagup$	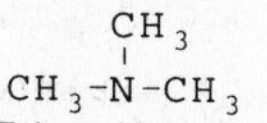 $CH_3-\underset{\underset{CH_3}{\mid}}{N}-CH_3$ Trimethylamine N,N-Dimethylaminobenzene	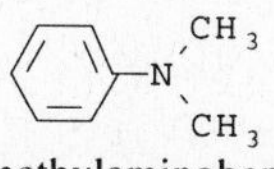 ⬡$-N\underset{CH_3}{\overset{CH_3}{\diagup}}$	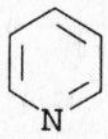 Pyridine

The ring nitrogens of heterocyclic compounds are either secondary or tertiary amines. Saturated ring nitrogen atoms are secondary amines whilst unsaturated ring nitrogen atoms are tertiary amines (Table 24.1).

24.2 NOMENCLATURE

The prefix **amino** and the suffix **amine** are used to denote the presence of all types of amino group in the name of a compound. Primary amines are named in the IUPAC system by either prefixing the stem name of a compound with amino plus the appropriate locant or using the suffix amine with the radical name of the compound (Fig. 24.1). However, some trivial names, such as aniline for aminobenzene and *o*-, *m*- or *p*-toluidine for the various isomers of methylaminobenzene are still in common use.

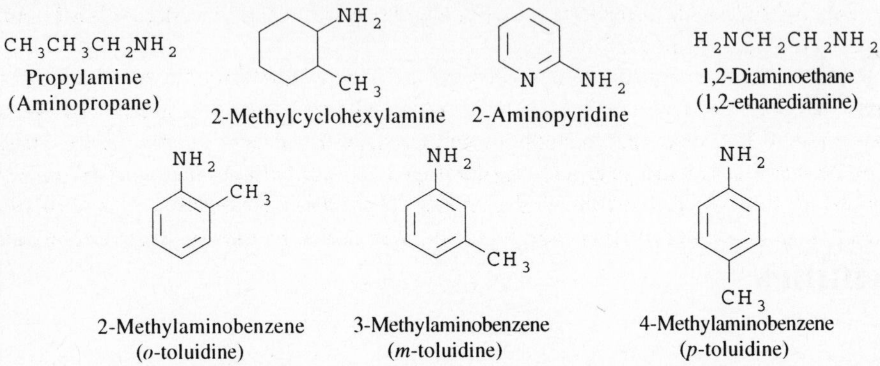

$CH_3CH_2CH_2NH_2$

Propylamine
(Aminopropane)

2-Methylcyclohexylamine   2-Aminopyridine

$H_2NCH_2CH_2NH_2$

1,2-Diaminoethane
(1,2-ethanediamine)

2-Methylaminobenzene   3-Methylaminobenzene   4-Methylaminobenzene
(*o*-toluidine)   (*m*-toluidine)   (*p*-toluidine)

Fig. 24.2. Examples of the systematic and trivial nomenclature of primary amines.

Symmetrical secondary and tertiary amines are usually named by prefixing the radical name of the substituent with **di** or **tri** respectively This is followed by the suffix **amine** (Figs 24.2 and 24.3). In unsymmetrical secondary and tertiary amines the largest structure is usually taken to be the parent structure. The other structures are treated as substituents and often have their names prefixed by **N** to denote that they are bonded to the nitrogen of an amino group. In all cases the terms amino or amine are used as appropriate. It should be noted that amino is prefixed with the relevant alkyl group name when using substituted amines as substituents of other structures.

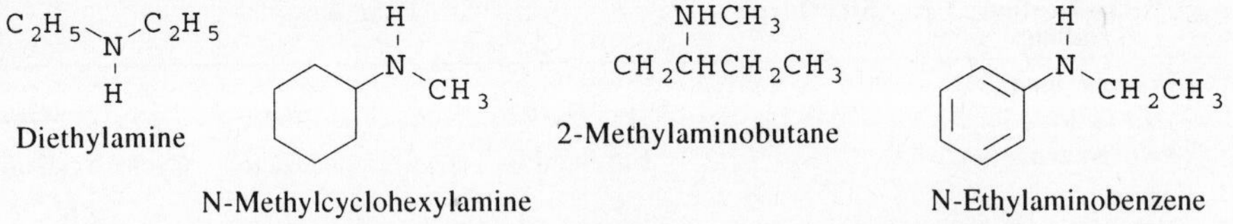

Diethylamine

N-Methylcyclohexylamine

2-Methylaminobutane

N-Ethylaminobenzene

Fig. 24.2. Examples of the nomenclature used for secondary amines.

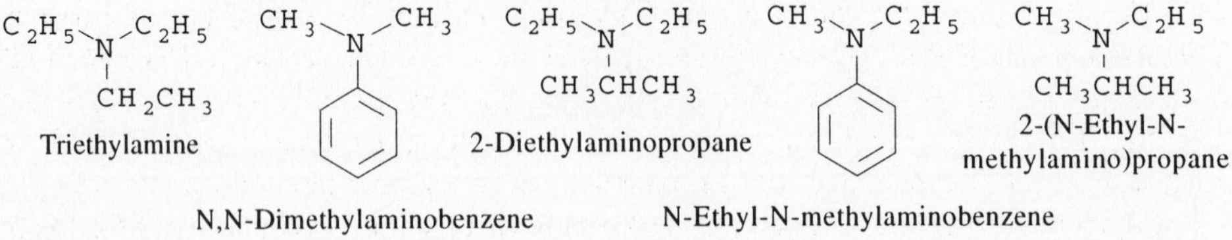

Triethylamine

N,N-Dimethylaminobenzene

2-Diethylaminopropane

N-Ethyl-N-methylaminobenzene

2-(N-Ethyl-N-methylamino)propane

Fig. 24.3. Examples of the systematic nomenclature used for tertiary amines. Note the use of two Ns in cases where ambiguities could arise due to there being alternative positions for a substituent.

The nomenclature system of heterocyclic ring nitrogen compounds is discussed in Chapter 13.

24.3   STRUCTURE

The nitrogen atom of aliphatic amines has a tetrahedral configuration. Alkyl groups or a mixture of alkyl groups and hydrogen atoms occupy three of the corners and the lone pair, the fourth. In practice, the bond angles are close to the theoretically predicted value of 109° for an sp^3 hybridised nitrogen atom. For example, experimental work has shown that the bond angles in trimethylamine are 108.7°.

The molecules of aliphatic amines with three different groups bonded to the nitrogen atom are chiral because of their tetrahedral configuration. However, these chiral amines have not been resolved into their enantiomers at room temperature because the practical evidence suggests that the structure is rapidly oscillating (inverting) between the two enantiomeric forms (Fig. 24.4). It has been shown that at room temperature the barrier to this oscillation is only about 24 kJ mol^{-1} which is of the same order as that required to cause free rotation in C–C sigma bonds.

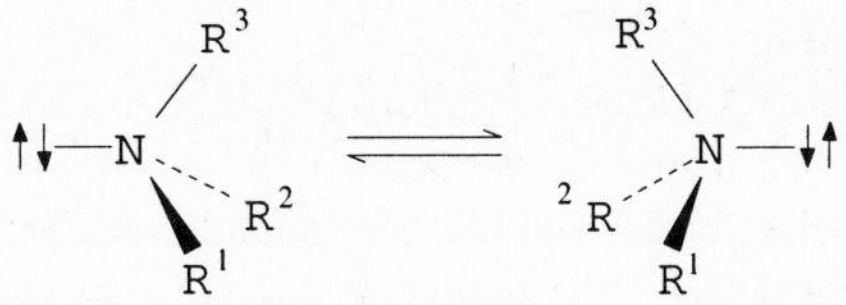

Fig. 24.4. The interconversion of tertiary amines.

Alicyclic tertiary ring nitrogens have a tetrahedral configuration which means that unsymmetrically substituted amines will be chiral. However, at room temperature few of these compounds have been resolved into their enantiomers. Once again this is explained as being due to rapid oscillation between the two enantiomeric forms. A compound that has been resolved is Troger's base which has a twisted chiral molecule (Fig. 24.5). It is believed that the bridgehead positions of the nitrogens prevents them inverting thus leading to separate existence of the two enantiomers.

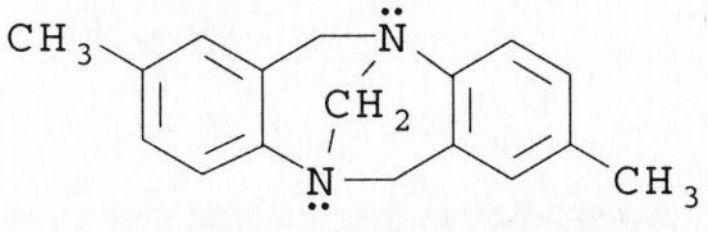

Fig. 24.5. Troger's base. This compound was resolved into its enantiomers by repeated column chromatography using specially prepared (+)-lactose hydrate.

The nitrogen atom of aromatic amines tends to have only a slight pyramidal configuration. For example, the C–N–C bond angles in triphenylamine have been calculated as being 114° which gives the molecule a low pyramidal configuration (Fig. 24.6). This result is supported by the fact that triphenylamine has an experimentally determined dipole of 0.47 D. However, it should be noted that for simplicity in theoretical discussions, the nitrogen atoms of aromatic amines is usually regarded as having a flat trigonal planar configuration which is explained by sp^2 hybridisation even though the observed bond angles suggest that the structure should be explained by a hybridisation between the extremes of sp^2 and sp^3.

Ph–N̈–Ph
|
Ph

Ph–N̈–Ph   Ph 114°
Ph

Ph = ⟨benzene ring⟩–

Fig. 24.6. Triphenylamine.

24.4 PHYSICAL PROPERTIES

Low molecular mass amines are colourless gases, liquids and solids. They have high vapour pressures and have strong distinctive ammoniacal or fishy smells. The odour of decomposing tissue is due to the presence of amines such as putrescene (1,4-diaminobutane). Amines with higher molecular masses are solids with lower vapour pressures and so do not have such strong odours.

24.4.1 Solubility

Simple low molecular mass amines are often very water soluble, but as molecular mass increases this solubility decreases. Simple aliphatic amines are more water soluble than simple aromatic amines. Solubility of both simple aliphatic and aromatic amines in organic solvents and lipids is usually good.

24.4.2 Hydrogen bonding

All amines are capable of hydrogen bonding to other molecules including other amine molecules (Fig. 24.7). For example, they can hydrogen bond to water but this does not usually have a significant effect on the compound's water solubility. Liquid amines are often highly associated.

24.4.3 Infra-red spectroscopy

Primary and secondary amines have a characteristically sharp N–H bond stretching absorptions in approximately the 3300 to 3500 cm^{-1} region of the electromagnetic spectrum. Tertiary amines cannot absorb in this region because they do not have N–H bonds. The

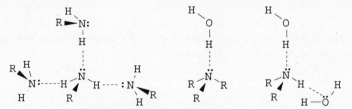

Fig. 24.7. Examples of the hydrogen bonding between amine molecules and also between amine molecules and water molecules.

O–H bonds of alcohols and phenols also exhibit absorb in this region. However, their absorptions bands are usually stronger than those of amines. They are also usually broader because of extensive hydrogen bonding. It is possible to distinguish between N–H and these O–H stretching absorptions by adding a few drops of a mineral acid to the compound. If a primary or secondary amine is present the absorption bands will broaden and move to approximately the 2500 to 3000 cm^{-1} region because amines react to form the corresponding salts with subsequent changes in their spectra (Fig. 24.8). The absorption bands of alcohols and phenols do not change because these compounds have no corresponding reaction with mineral acids.

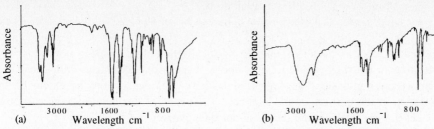

Fig. 24.8. The infra-red spectra of (a) aminobenzene and (b) benzene ammonium bromide.

24.4.4 ^{1}H n.m.r. spectroscopy

The N–H protons of primary and secondary aliphatic amines have chemical shifts of approximately δ 0.5 to 4.5 whilst the chemical shifts of the corresponding protons of aromatic compounds usually lie in the δ 3 to 6 region. Addition of a few drops of mineral acid moves the absorption of these protons towards the 7 to 10 ppm region. This movement is due to the formation of amine salts whose protons absorb downfield to those of amines. The protons of both primary and secondary amines are exchangeable with deuterium so their signals either disappear or are considerably reduced when the sample is shaken with D_2O.

24.5 BASIC NATURE

The lone pairs of electrons of amines allow them to act as bases and form salts with a variety of acidic substances. These salts can be converted back to the amine by treatment with a stronger base such as sodium hydroxide. Picrates are used as derivatives in organic analysis and identification tests.

$$\underset{\text{Amphetamine}}{\overset{\overset{\displaystyle NH_2}{|}}{PhCH_2CHCH_3}} \underset{\text{NaOH}}{\overset{\text{HCl}}{\rightleftharpoons}} \underset{\text{Amphetamine hydrochloride}}{\overset{\overset{\displaystyle \overset{+}{N}H_3\ Cl^-}{|}}{PhCH_2CHCH_3}}$$

$$2\ \underset{\text{Amphetamine}}{\overset{\overset{\displaystyle NH_2}{|}}{PhCH_2CHCH_3}} \underset{\text{NaOH}}{\rightleftharpoons} \left[\underset{\text{Amphetamine tartrate}}{\overset{\overset{\displaystyle \overset{+}{N}H_3}{|}}{PhCH_2CHCH_3}} \quad \begin{matrix} HO-CH-COO^- \\ | \\ HO-CH-COO^- \end{matrix}\right]_2$$

$$\underset{\text{Amphetamine}}{\overset{\overset{\displaystyle NH_2}{|}}{PhCH_2CHCH_3}} \underset{\text{NaOH}}{\rightleftharpoons} \underset{\text{Amphetamine picrate}}{\overset{\overset{\displaystyle \overset{+}{N}H_3}{|}}{PhCH_2CHCH_3}}$$

Molecules with both acid and amine functional groups can form internal amine salts in aqueous solution. These salts are often referred to as either **zwitterions** or **dipolar ions**.

$$\underset{\text{Alanine}}{\overset{\overset{\displaystyle NH_2}{|}}{CH_3-CH-COOH}} \overset{H_2O}{\rightleftharpoons} \overset{\overset{\displaystyle \overset{+}{N}H_3}{|}}{CH_3-CH-COO^-}$$

4-Aminobenzenesulphonic acid (sulphanilic acid)

Some amines, such as pyrrole, do not behave as bases which means that their lone pairs are not available for this type of reaction. In the case of pyrrole, this lack of availability is explained in theory by the almost complete delocalisation of the nitrogen atom's lone pair of electrons in the ring. However, this does not mean that all amines with conjugated structures are non-basic. The degree of basicity will depend on the extent of the conjugation of the nitrogen atoms lone pair. The extent of this delocalisation cannot be predicted because mechanistic theory is not quantitative.

The lone pair of electrons are part of the conjugated structure of the molecule and are not readily available for reaction

Pyrrole

The more readily an amine reacts with an acid the stronger the amines basic strength. For a strong base the position of equilibrium in the reaction should lie well to the right. In order for this to occur the lone pair of a base must be readily available and its conjugate acid stable.

Base   Conjugate acid

Basic strengths of amines are recorded as pK_a values for the dissociation of the conjugate acid (6.13.1 and 6.13.2) so that the strengths of bases can be directly compared with those of acids (Chapter 18).

$$\text{Conjugate acid} \rightleftharpoons \text{Base} + H^+$$

If a base is strong, the position of this equilibrium will lie to the left which means that the concentration of the base will be low and the concentration of its conjugate acid will be high. Consequently, strong bases have high pK_a values (equations 6.13 and 6.15). Conversely, for weak bases the concentration of the base will be high and that of the conjugate acid low with the result that weak bases have low pK_a values. The pK_a values of some simple amines are given in Table 6.4.

Aromatic amines are usually weaker bases than aliphatic amines. It is believed that this is because both the positive charge of the ammonium group of their conjugated acids is not

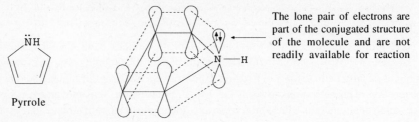

(a)

(b)

Fig. 24.9. The delocalisation of electrons in (a) aminobenzene and (b) pyrrole.

usually stabilised by delocalisation and the availability of the lone pair is restricted because it is partly delocalised in the conjugated aromatic system (Fig. 24.9). In the case of aminobenzene, which is a weak base, the lone pair is not fully delocalised, but in the case of pyrrole, where the amine is not basic, the lone pair is fully delocalised.

The reduced stability of the conjugated acids of aromatic amines is believed to be due to the inability of the structure of the conjugate acid to stabilise the positive charge by delocalisation. Once the conjugate acid is formed the nitrogen's configuration changes to a tetrahedron corresponding to sp^3 hybridisation. This means that in theory there is no orbital route for the delocalisation of the positive charge (Fig. 24.10).

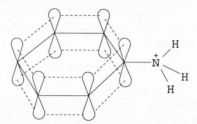

Fig. 24.10. The structure of the phenylammonium ion. There is no p orbital route for the delocalisation of the positive charge. The only stabilising factor is the C-N bonds inductive effect.

The basic strengths of all amines depends on the temperature, solvent, availability of the lone pair and the stability of the conjugate acid. The effect of temperature and solvents cannot be predicted. However, at constant temperature, it is possible to predict and explain the relative basic strengths of amines having similar structures with some degree of success provided no solvation occurs. Compounds whose structures act as electron donors to the nitrogen will be stronger bases than those whose structures act as electron acceptors from the nitrogen. This is because electron donors increase the availability of the lone pair and the stability of the corresponding conjugate acid (by reducing the size of the positive charge, Figs 8.8 and 8.9) whilst electron acceptors reduce the availability of the lone pair and the stability of the conjugate acid (by increasing the size of the positive charge). Consider, for example, the relative basic strengths of aminobenzene, 4-methyl-aminobenzene and 4-nitroaminobenzene (Fig. 24.11). The methyl group of 4-methylami-nobenzene increases the electron density of the benzene ring. This increase opposes the delocalisation of the lone pair into the ring making it more available than in the case of aminobenzene. Furthermore, this increase in negative character of the ring makes the inductive effect of the ammonium group more effective in reducing and so stabilising the positive charge than in the case of aminobenzene. The net result is that 4-methylaminoben-zene is a stronger base than aminobenzene. The nitro group of 4-nitroaminobenzene has the opposite effect. It removes electrons from the benzene ring making it electron deficient by comparison with the ring of aminobenzene. This improves the delocalisation of the lone pair into the benzene ring since the repulsion between the pi electrons of the ring and the lone pair is reduced. It also reduces the stability of the ammonium group by decreasing the effectiveness of its inductive effect in reducing the positive charge.

It is not possible to predict the relative strengths of bases when solvation of the conjugate acid occurs. However, it is possible to explain irregularities in the expected

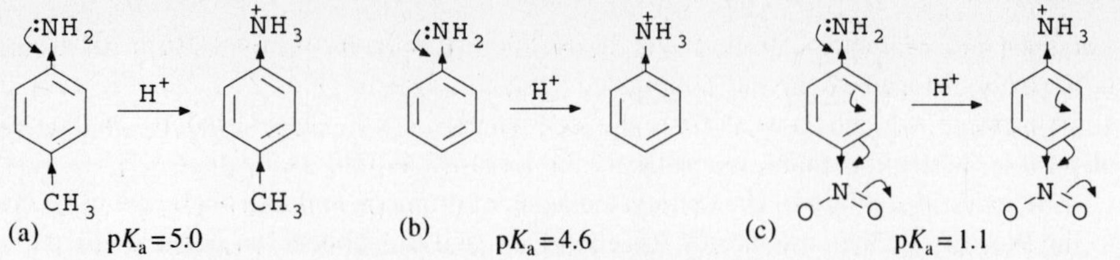

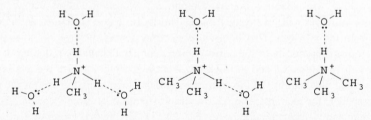

Fig. 24.11. The effect of the substituent on the relative basic strengths of (a) 4-methylaminobenzene and (c) 4-nitroaminobenzene. Aminobenzene (b) is used as a standard for comparison purposes.

order of basic strength in terms of solvation. Consider, for example, the relative basic strengths of methylamine, dimethylamine and trimethylamine in water. The methyl group is an electron donor. Therefore, in the absence of solvation, one would expect the order of basic strength of these amines to be:

$$\text{trimethylamine} > \text{dimethylamine} > \text{methylamine}$$

However, in practice it is found that the order is:

$$\text{dimethylamine} > \text{methylamine} > \text{trimethylamine} \qquad (24.1)$$
$$(pK_a\ 10.7) \qquad (pK_a\ 10.6) \qquad (pK_a\ 9.8)$$

This experimental observation is explained by differences in the solvation of the conjugate acids of these bases. In water, solvation mainly takes the form of hydrogen bonding between the water molecules and the conjugate acid. As the number of hydrogen atoms attached to the nitrogen increases so does the degree of hydrogen bonding and solvation (Fig. 24.12). This increase in solvation increases the stability of the conjugated acid with a subsequent increase in basic strength.

Fig 24.12. The theoretically possible hydrogen bonding of the conjugate acids of the methylamines to water. It is unlikely that hydrogen bonding takes the precise form shown and that all the possible hydrogen bond options are taken up at room temperature and above by any one conjugate acid.

The increased hydrogen bonding and consequently the increased stability of the conjugated acids of methylamine and dimethylamine explains why they are stronger bases in water than simple tertiary amines. However, the degree of hydrogen bonding must be balanced against the other parameters governing basic strength. Although methylamine has the highest degree of hydrogen bonding, dimethylamine is the stronger base because the greater availability of its lone pair and the increased structural stability of its conjugate acid outweigh the stronger solvation effect of the conjugate acid of methylamine. The increased availability of methylamine's lone pair and the greater structural stability of its conjugate acid are explained in theory by the presence of two inductive effects in the structure of dimethylamine as against one inductive effect in the structure of the primary amine.

Arguments of the type described in the preceding paragraphs are moderately successful in explaining the relative basic strengths of many amines.

The basic strengths of amines in water at 25° can be used to calculate the extent to which they ionise under biological conditions. Their degree of ionisation influences their transport across membranes, distribution, pharmacological effect and excretion in living materials. For example, the effectiveness of a local anaesthetic will depend partly on its ability to cross lipid membranes to get to its site of action. Experimental evidence shows that the uncharged form of the compound crosses the lipid membrane more easily than the charged form. However, once the drug has penetrated the membrane, evidence suggests that it is the charged form that is the most effective in producing the anaesthetic effect. For example, in aqueous solution, the local anaesthetic procaine has a pK_a of 8.7. Using this value in Equation 24.2 it is possible to calculate the percentage of the free base present in solutions with different pH values. At pH 7 (approximately blood pH) calculation shows that the percentage of free base present is about 2% whereas at pH 9 the percentage of free base is about 66%.

$$\mathrm{pH} = \mathrm{p}K_a - \log \frac{[\text{conjugate acid}]}{[\text{base}]} \qquad (24.2)$$

Many amines can be assayed by titration with an acid, such as perchloric or hydrochloric, under the appropriate reaction conditions. For example, adrenaline may be assayed by a potentiometric titration with a solution perchloric acid in ethanoic acid (11.6 .4), triethanolamine by direct titration (11.6.2) with hydrochloric acid and ephedrine by reaction with an excess of hydrochloric acid followed by back titration (11.6.3) of the excess acid with sodium hydroxide (Fig. 24.13).

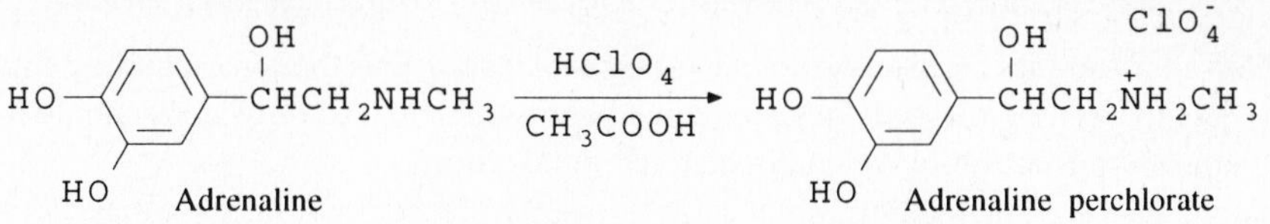

$$(HOCH_2CH_2)_3N + HCl \xrightarrow{\text{Methyl red}} (HOCH_2CH_2)_3\overset{+}{N}H\ \overset{-}{Cl}$$

Triethanolamine → Triethanolamine hydrochloride

$$PhCH(OH)CH(CH_3)NHCH_3 \xrightarrow[\text{Ethanol}]{HCl_{\text{Excess}}} PhCH(OH)CH(CH_3)\overset{+}{N}H_2CH_3\overset{-}{Cl} + [HCl_{\text{Excess}}]$$

Ephedrine → Ephedrine hydrochloride

$$HCl_{\text{Excess}} + NaOH \xrightarrow{\text{Methyl red}} NaCl + H_2O$$

Fig. 24.13. The chemistry underlying the assay of some amines.

24.6 AMINE SALTS

Aliphatic amine salts have the general structure:

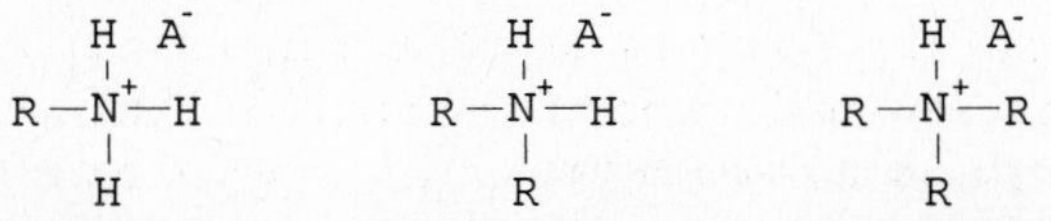

A primary amine ammonium salt — A secondary amine ammonium salt — A tertiary amine ammonium salt where the substituent R groups can be hydrogens and/or different or similar organic residues. They are named either as derivatives of ammonia or as a combination of the names of the amine and the acid (Fig. 24.14). The latter method is often used when there are several amino groups in a molecule and it is not certain which one is involved in the salt formation. In these cases separate structures of the amine and acid separated by a dot are used to represent the structure of the salt as can be seen in the structure of naphthalene-1,6-diamine monohydrochloride (Fig. 24.14). Because two nomenclature systems are in current use care must be taken when relating the structures of the amine salts of dibasic acids such as amine sulphates and hydrogen sulphates to their names.

Aromatic and heterocyclic amine salts are named as derivatives of the amine by changing the ending of the name of the amine to **-inium** (Fig. 24.14).

$$CH_3\overset{+}{N}H_3 \ \overset{-}{Cl}$$

Methylammonium chloride
(methylamine hydrochloride)

$$(CH_3)_2\overset{+}{N}H_2 \ \overset{-}{Cl}$$

Dimethylammonium chloride
(dimethylamine hydrochloride)

$$(CH_3)_3\overset{+}{N}H \ \overset{-}{Cl}$$

Trimethylammonium chloride
(trimethylamine hydrochloride)

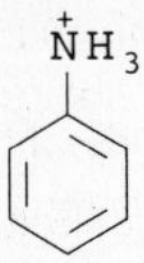

Anilinium chloride 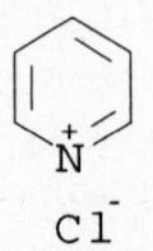   Pyridinium chloride

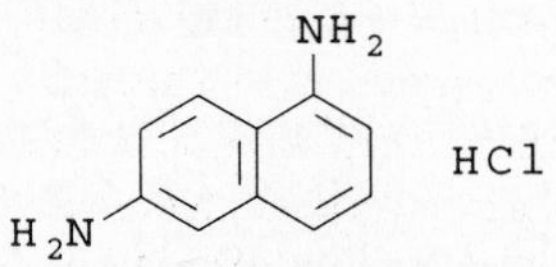

Naphthalene-1,6-diamine
monohydrochloride

$$Ph\overset{+}{N}H_3 \ H\overset{-}{S}O_4$$

Anilinium hydrogen sulphate
(anilinium sulphate)

$$(Ph\overset{+}{N}H_3)_2 \ SO_4^{2-}$$

Dianilinium sulphate
(anilinium sulphate)

Fig. 24.14. Examples of the systematic nomenclature used for amines. Alternative names, not in common use, are given in brackets. Note the confusing use of anilinium sulphate.

The salts of amines are usually more resistant to oxidation than the corresponding amines. Consequently, as they normally have the same pharmacological activity as the amine itself, many amine drugs are stored and administered in this form.

Amine salts are usually more water soluble than the parent amine because they can form ion-dipole bonds with water molecules. The change in solubility is more marked with aliphatic than aromatic amines. However, it should be noted that different salts of the same amine will have different water solubilities.

Many amine drugs are administered as their salts in order to take advantage of the increase in water solubility. Salts used for this purpose are mainly chlorides, sulphates and hydrogen sulphates. Citrates and tartrates are frequently used for very insoluble amines because hydrogen bonding between the carboxylic acid and alcohol groups of these acid residues and water molecules is believed to be responsible for the improved water solubility of these salts. The ionisation of the carboxylic acid groups also give rise to ion-dipole bonds which

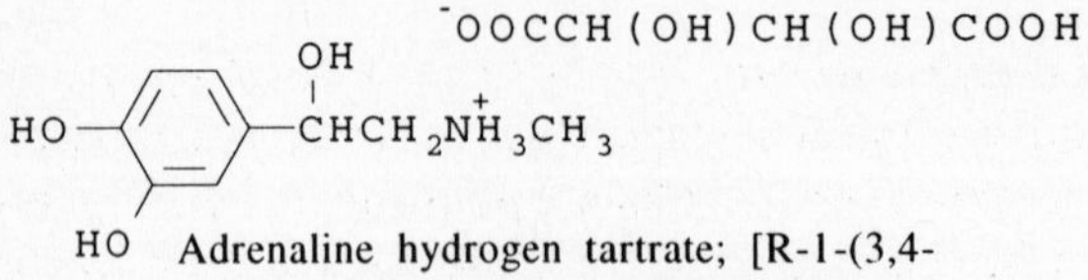

Adrenaline hydrogen tartrate; [R-1-(3,4-dihydroxyphenyl)-2-methylaminoethanol hydrogen tartarate]

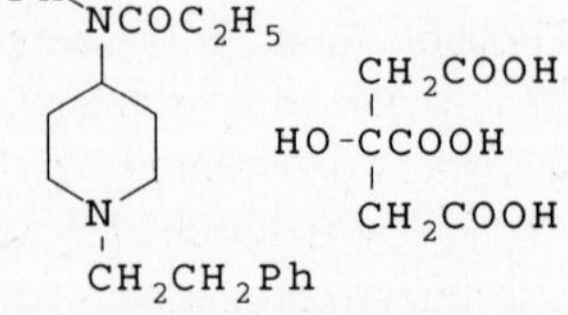

Fentanyl citrate; [N-(1-phenethyl-4-piperidyl)-propananilide dihydrogen citrate]

also improves water solubility. However, organic acids such as maleic acid whose structures do not contain hydroxyl groups are also used to improve water solubility of amines.

Amine picrates and other salts that are easy to purify by recrystallisation and have sharp melting points are used as derivatives in the identification of amines. For example, the 1993 British Pharmacopoeia uses picrate formation as one of its identification tests for the local anaesthetic lignocaine hydrochloride and other drugs with amine functional groups.

Lignocaine hydrochloride   Lignocaine picrate

The water solubility of amine salts can be used to isolate amines from mixtures and natural sources. The sample containing the amine is dissolved in a suitable organic solvent, such as diethyl ether and the solution shaken in a separating funnel with an aqueous solution of a suitable acid. Alternatively, the sample is extracted directly with an aqueous solution of an acid. In the former case the acid reacts with the amine to form the more water soluble amine salt which partitions between the aqueous and organic phases, the aqueous phase usually containing a significantly higher concentration of the amine salt than the organic phase. The aqueous solution is separated from the organic phase and neutralised with a base to convert the salt back to the free base which is then extracted into a suitable organic solvent. The organic solution is dried and the solvent removed from the crude amine which is either identified or purified further depending on the circumstances. This type of procedure is used to isolate many alkaloids (24.12) from natural sources. Ethanolic aqueous acid solutions are often used to improve the efficiency of alkaloid extraction.

Precipitation of an insoluble sulphosalicylate is the basis of a semi-quantitative screening test for protein in urine. The density of the precipitate is compared with standards to determine whether the protein content of the urine requires further investigation.

Peptide or protein   Sulphosalicylic acid   Water insoluble sulphosalicylate

24.7 REACTIONS AS A NUCLEOPHILE

The lone pair of electrons of the nitrogen enables amines to act as nucleophiles in substitution, displacement and addition-elimination (condensation) reactions. Primary, secondary and tertiary amines often react in a similar fashion. It is usually the N–H bonds of primary and secondary amines that are responsible for any differences in the reactions of primary secondary and tertiary amines. In general, aliphatic amines are usually better nucleophiles than aromatic amines because the lone pairs of aromatic amines are partly delocalised in the aromatic system and therefore not so readily available for reaction.

All types of amine undergo nucleophilic substitution. These reactions often involve S_N1, S_N2 and $S_N2(aromatic)$ mechanisms. However, only primary and secondary amines react in nucleophilic displacement and addition elimination (condensation) reactions. This is because tertiary amines do not have a hydrogen atom attached to the nitrogen which can be lost to stabilise the product.

24.7.1　Nucleophilic substitution of aliphatic compounds

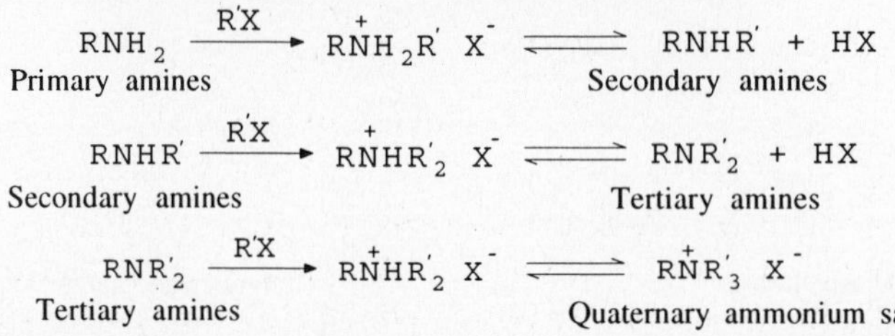

$$RNH_2 \xrightarrow{R'X} R\overset{+}{N}H_2R'\ X^- \rightleftharpoons RNHR' + HX$$

Primary amines　　　　　　　　　　　　　　Secondary amines

$$RNHR' \xrightarrow{R'X} R\overset{+}{N}HR'_2\ X^- \rightleftharpoons RNR'_2 + HX$$

Secondary amines　　　　　　　　　　　　　Tertiary amines

$$RNR'_2 \xrightarrow{R'X} R\overset{+}{N}HR'_2\ X^- \rightleftharpoons R\overset{+}{N}R'_3\ X^-$$

Tertiary amines　　　　　　　　　　　Quaternary ammonium salts

Fig. 24.15. N-alkylation of primary, secondary and tertiary amines. The products of the primary and secondary amine reactions will react further with R'X to give a mixture of all the possible products.

Amines are able to substitute a number of functional groups including aliphatic halogens and sulphates (Fig. 24.15). The reaction is generally referred to as **N-alkylation** and usually involves S_N1 and S_N2 types of mechanism. N-Alkylation of primary and secondary amines in the laboratory usually yields a mixture of the appropriate amines, their salts and the quaternary ammonium salt. In these cases, the reaction is of very little practical use. However, tertiary amines give only the quaternary ammonium salt and so N-alkylation is used to prepare and manufacture these compounds.

N-Alkylation of amines is an important reaction in biological systems. For example, the biosynthesis of choline involves an N-methylation, the methyl being transferred from S-adenosylmethionine while the action of a number of cytotoxic agents used in the treatment of cancer involves the nucleophilic substitution of a quaternary salt by an amine (Fig. 14.12)

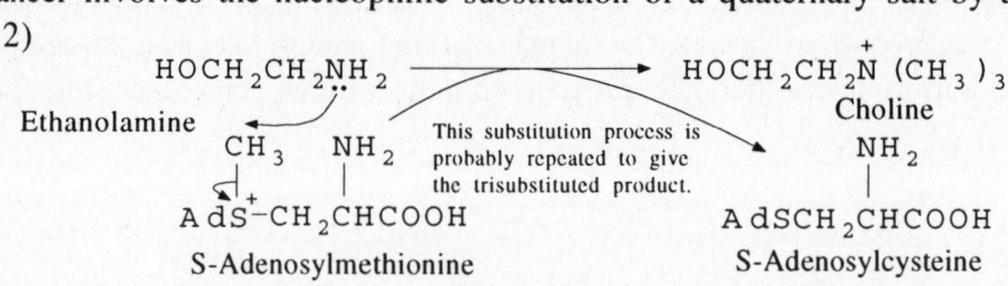

$$HOCH_2CH_2NH_2 \longrightarrow HOCH_2CH_2\overset{+}{N}(CH_3)_3\ X^-$$

Ethanolamine　　　　　　　　　　　　　　　　　　Choline

This substitution process is probably repeated to give the trisubstituted product.

$$\underset{\text{S-Adenosylmethionine}}{A\,d\overset{+}{S}{-}CH_2CHCOOH}$$　　$$\underset{\text{S-Adenosylcysteine}}{A\,dSCH_2CHCOOH}$$

The reverse of N-alkylation, **N-dealkylation**, is also an important reaction in biological processes. For example, N-dealkylation of an amine is a common route for N-methyl and N-ethyl substituted amines in xenobiotic metabolism. For instance, the antidepressant imipramine is dealkylated to the more powerful antidepressant desipramine by an oxidative process.

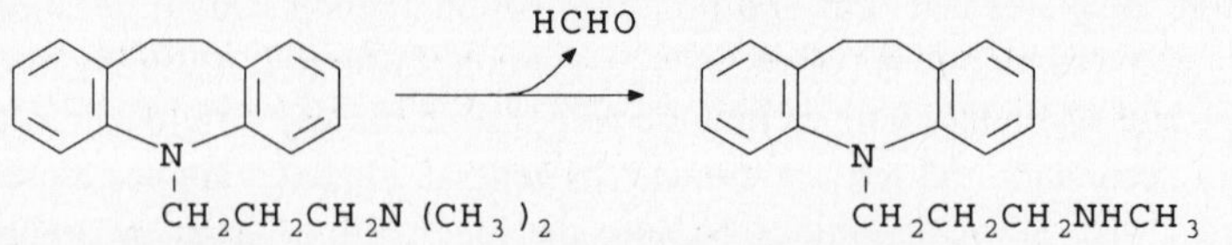

　　　　　　　　　　　　　　　　HCHO

$$\underset{\text{Imipramine}}{CH_2CH_2CH_2N(CH_3)_2}$$　　　　　　$$\underset{\text{Desipramine}}{CH_2CH_2CH_2NHCH_3}$$

24.7.2 Nucleophilic substitution of aromatic compounds

This only occurs with activated aryl halides, such as 2,4-dinitrochlorobenzene (14.7).

24.7.3 Nucleophilic displacement

Primary and secondary amines react with acyl halides and acid anhydrides to form amides. This reaction is generally referred to as **N-acylation**. Tertiary amines do not normally react.

$$R'COZ \; + \; RNH_2 \; \longrightarrow \; R'CONHR \; + \; HZ$$
$$\text{an N-alkylamide}$$

$$R'COZ \; + \; RNHR \; \longrightarrow \; R'CONHR \; + \; HZ$$
$$\text{an N,N-dialkylamide}$$

Many drugs are prepared by the reaction of an amine with either an acid halide or anhydride. For example, the analgesic, paracetamol, is manufactured by the reaction of 4-aminophenol with ethanoic anhydride. Reaction takes place selectively at the amino group when the reactants are used in an approximately 1:1 molar ratio even though both the hydroxyl and the amino groups can react as nucleophiles with ethanoic anhydride. However, nitrogen nucleophiles are usually more reactive than oxygen nucleophiles. This can be explained simply in terms of the electronegativities of nitrogen and oxygen. Nitrogen has the lower electronegativity and so its lone pair of electrons is not held so tightly by the atom as those of oxygen. Consequently, the lone pairs of nitrogen nucleophiles are more readily available to form bonds than those of oxygen nucleophiles.

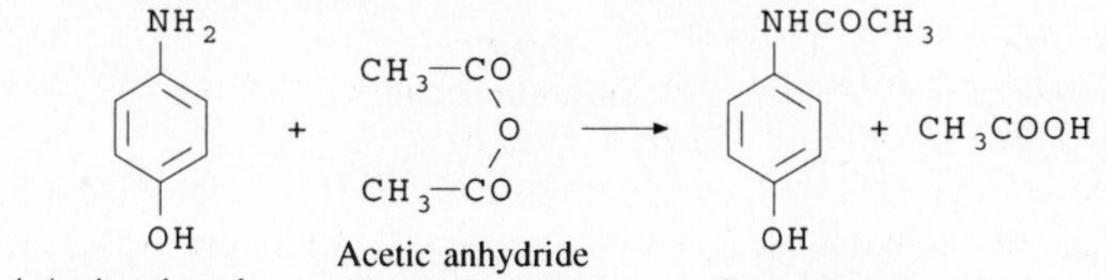

4-Aminophenol   Acetic anhydride   Paracetamol
      (CH₃COOCOCH₃)

N-Acetylation is also an important step in the metabolism of aromatic amines in mammals (Fig. 24.16). The incorporation of an N-acetyl group into the molecule improves its water solubility which leads to easier excretion. Drugs, such as paracetamol, that contain N-acetyl groups are excreted in the urine.

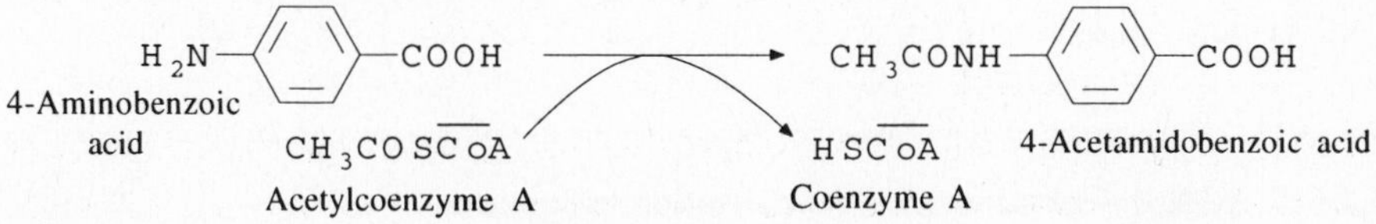

Fig. 24.16. The acetylation of 4-aminobenzoic acid (a constituent of some suncreams). The enzyme arylamine acetyltransferase which catalyses this reaction is found mainly in the liver and kidneys of most mammals.

Sulphonyl chlorides react with primary and secondary amines to form sulphonamides but react with tertiary amines to form unstable salts which decompose in the presence of water to the amine and the corresponding sulphonic acid.

$$R'SO_2Cl \; + \; RNH_2 \; \longrightarrow \; R'SO_2NHR \; + \; HCl$$
$$\text{Primary amine} \qquad \text{A sulphonamide}$$

$$R'SO_2Cl + RNHR'' \longrightarrow R'SO_2NRR'' + HCl$$

Secondary amine   A sulphonamide

$$R'SO_2Cl + R_3N \longrightarrow R'SO_2\overset{+}{N}R_2 \ \ Cl^- \xrightarrow{H_2O} R'SO_3H + R_3N$$

Tertiary amine   A sulphonic acid

The reaction of primary and secondary amines with sulphonyl chloride is an essential stage in the synthesis of sulphonamide drugs. These multistage syntheses often involve the use of **protecting groups** which prevent unwanted reactions occurring. The introduction of a protecting group into a molecule changes a functional group to one that cannot take part in a specific reaction. However, this change must be reversible at a later stage in the synthesis. Consider, for example, the preparation of sulphapyridine which is used to treat dermatitis herpetformis (Fig. 24.17). If the acetyl group had not been introduced in stage one of this synthesis the amino group would have reacted with the sulphonyl chloride group introduced into the ring in stage two. The use of an acetyl group as a protecting group in stage one prevents this happening because the N-acetyl derivative is unable to react with the sulphonyl chloride. In the last stage of the synthesis the acetyl protecting group is removed by acid catalysed aqueous hydrolysis of the amide. It should be noted that sulphonamides, unlike amides, do not hydrolyse under dilute aqueous acid conditions.

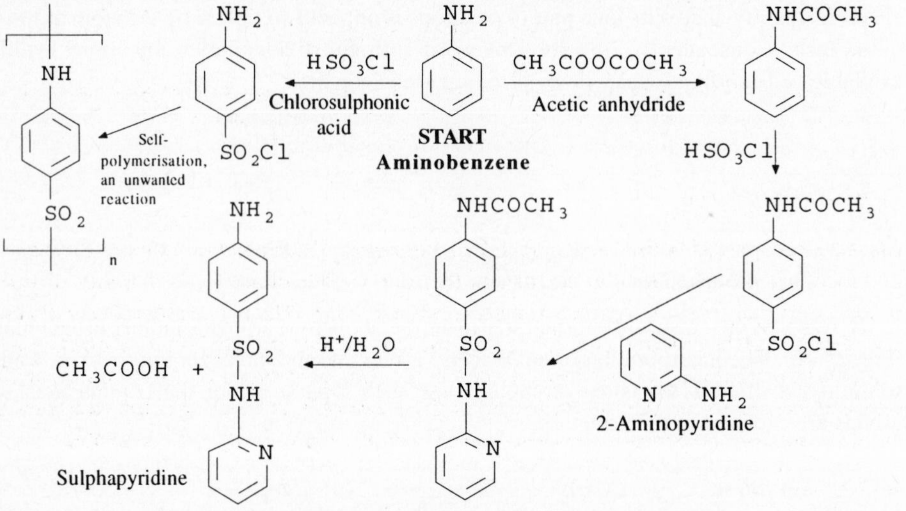

Fig 24.17. The synthesis of sulphapyridine, illustrating the use of a protecting group in a synthesis.

24.7.3   Addition elimination (condensation) reactions

Amine condensation reactions (19.11.4) take the general forms:

Primary amines $RNH_2 + O{=}C\big\langle \xrightarrow{H^+} RNH{=}C\big\langle + H_2O$

Aldehyde or ketone   An imine

Secondary amines $\overset{R}{\underset{R}{>}}NH + O{=}C\big\langle \xrightarrow{H^+} \overset{R}{\underset{R}{>}}N{-}C\big\langle + H_2O$

An enamine

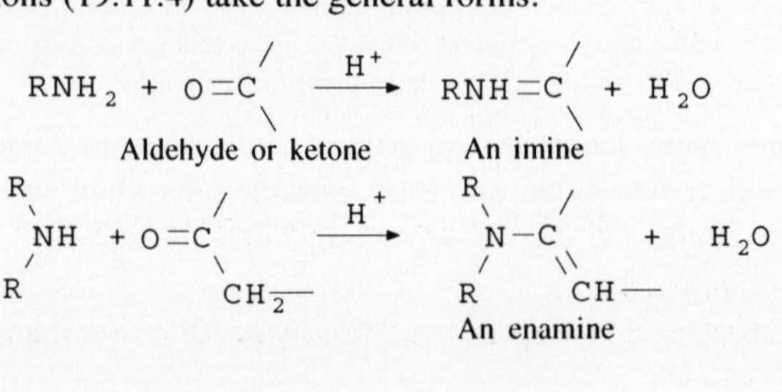

These reactions have been discussed in some detail in 19.11.4 but it should be remembered that aliphatic imines are unstable and usually polymerise to cyclic compounds.

A consequence of these reactions is that the activity of drugs which have primary or secondary amino groups can be considerably diminished if they are used in preparations containing compounds possessing aldehyde or ketone groups. For example, intravenous infusions of dextrose (aqueous glucose solutions) and laevulose (aqueous fructose solutions) are sterilised by autoclaving. During this process some of the sugar is converted to 2-hydroxymethylfurfural, some of which oxidises to methanoic and laevulinic acids (Fig. 24.18). The resulting solutions (pH 3.0 to 5.5) are acidic enough to catalyse the reaction of amine drugs, such as ampicillin, with the 2-hydroxymethylfurfural. This reaction considerably reduces their activity. The presence of phosphates, tartrates, citrates and ethanoates should also be avoided as the Schiff's base reaction is catalysed by these anions.

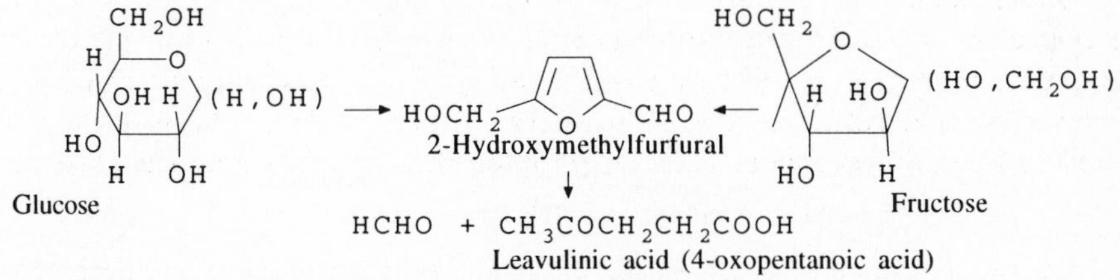

Fig. 24.18. The effect of heating aqueous solutions of glucose and fructose. It should be noted that the brackets around sections of the formulae of glucose and fructose signifies that both the α and β isomers (27.2.1) of these sugars react in the same way.

24.8 OXIDATION

Most amines are readily oxidised under both laboratory and biological conditions to a variety of functional groups. This variety means that one cannot classify the majority of these oxidations simply in terms of a few general reactions. The product of an individual oxidation will depend on the reaction conditions, the oxidising agent and the amine. However, it is possible to predict the course of the oxidation of a particular type of amine with a specific oxidising agent. Consequently, this section discusses only selected specific examples of amine oxidation. In general, aromatic amines oxidise more easily than aliphatic.

24.8.1 Air

Most amines oxidise in air to a variety of products (Fig. 24.19). The oxidation is catalysed by light and so amines should be stored in airtight, lightproof containers.

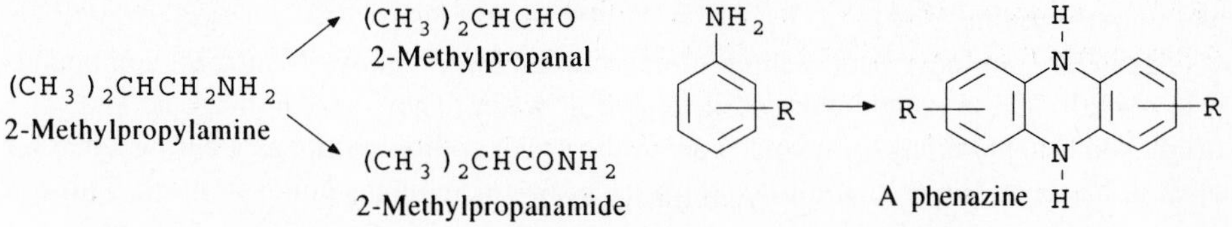

Fig. 24.19. Examples of the atmospheric oxidation of amines.

24.8.2 Nitrous acid

Nitrous acid reacts under acidic conditions with all types of amine to form a wide variety of products. The nitrous acid must be generated immediately prior to use in these reactions by treating sodium nitrite with a dilute acid. Dilute hydrochloric acid is commonly used but other mineral and organic acids are used in some circumstances.

All types of primary amines react with nitrous acid at room temperature to form the diazonium salt. Aliphatic diazonium salts are unstable and spontaneously decompose via a carbocation intermediate to give nitrogen and mixtures of other compounds. Propylamine, for example, reacts to yield nitrogen and a mixture consisting of 7% propanol, 32% propan-2-ol and 28% of propene. Methylamine, on the other hand, yields mainly methyl nitrite and a very small amount of methanol.

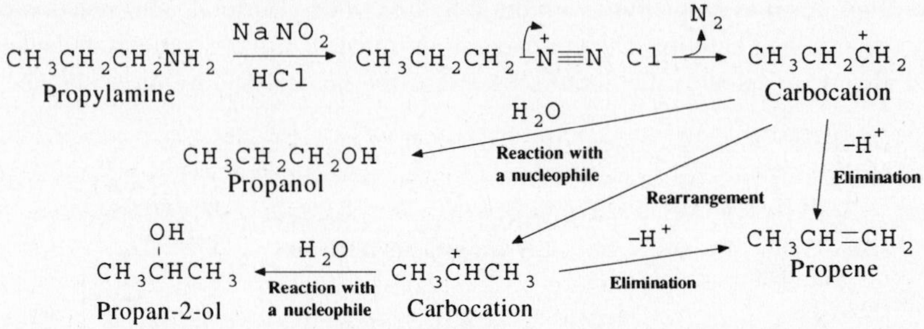

At room temperature most aromatic diazonium salts are unstable and decompose to nitrogen and the corresponding phenol. However, aromatic diazonium salts are usually stable at 0°C and below but are not normally isolateable from solution. In spite of this they are important intermediates in the chemical, pharmaceutical and food industries (24.9).

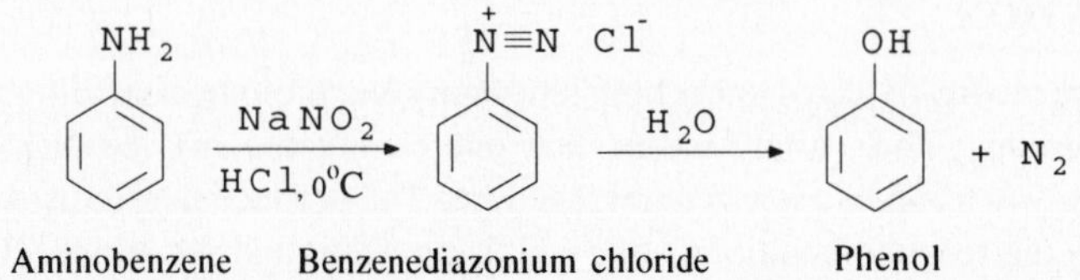

Aminobenzene   Benzenediazonium chloride   Phenol

The evolution of nitrogen when a compound is treated with nitrous acid is used to indicate the presence of a primary amine. However, the test is not always reliable as other functional groups, such as simple amides ($RCONH_2$), can produce nitrogen with nitrous acid.

Some primary aromatic amines, such as the local anaesthetic benzocaine, are assayed by titration with sodium nitrite using the dead stop indicator method. Other compounds that can be assayed by this method are sodium aminosalicylate, dapsone, procaine hydrochloride and sulphapyridine.

Secondary amines react with nitrous acid to form yellow N-nitroso compounds (Fig. 24.20). These compounds are insoluble in water. They are carcinogenic and, as a result, some concern has been voiced about the use of sodium nitrite as a curing agent for meat. It has been shown that nitrites form traces of nitrosamines in the stomach. This can cause cancers in animals. However, no such cancers have yet been detected in man.

Tertiary amines form colourless salts with nitrous acid. Consequently, the reaction of nitrous acid with amines can be used to distinguish between primary, secondary and tertiary amines (Fig. 24.20).

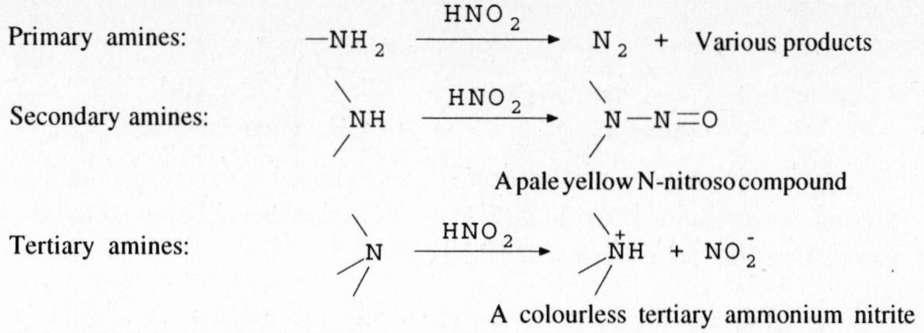

Primary amines: $-NH_2$ $\xrightarrow{HNO_2}$ N_2 + Various products

Secondary amines: $\diagdown NH$ $\xrightarrow{HNO_2}$ $\diagdown N-N=O$

A pale yellow N-nitroso compound

Tertiary amines: $\diagdown N$ $\xrightarrow{HNO_2}$ $\diagdown \overset{+}{N}H$ + NO_2^-

A colourless tertiary ammonium nitrite

Fig. 24.20. The reaction of primary, secondary and tertiary amines with nitrous acid.

24.8.3   Concentrated sulphuric acid

Hot fuming sulphuric acid completely oxidises all classes of amine to water, carbon dioxide and ammonium sulphate. The yield of ammonium sulphate is quantitative and is the basis of Kjeldahl's method of assaying protein in serum and other amino compounds (Fig. 24.21). The amine is heated with fuming sulphuric acid. Yellow mercury II oxide is used as a catalyst and anhydrous sodium sulphate is added to elevate the boiling point of the sulphuric acid. The reaction mixture is treated with a concentrated aqueous solution of sodium hydroxide which neutralises the residual sulphuric acid and then reacts with the ammonium sulphate to liberate ammonia. The ammonia is either distilled into a known volume of hydrochloric acid which is back titrated with sodium hydroxide or it is distilled into boric acid containing Tashiro's indicator which is titrated with hydrochloric acid until it is grey or just red.

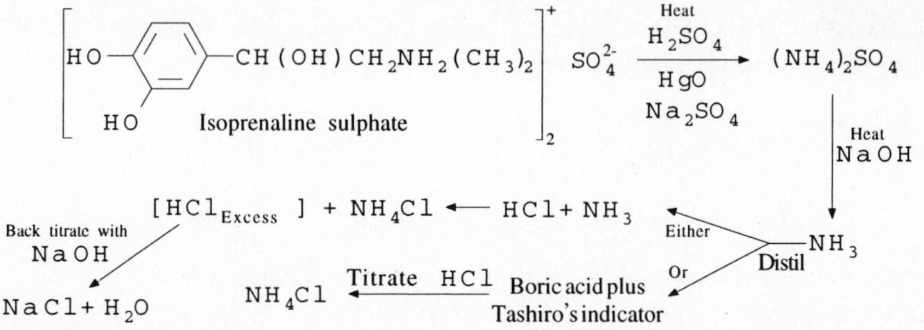

$$\left[HO-\langle\rangle-CH(OH)CH_2NH_2(CH_3)_2 \ SO_4^{2-} \right]_2 \xrightarrow[\underset{Na_2SO_4}{HgO}]{\overset{Heat}{H_2SO_4}} (NH_4)_2SO_4$$

HO   Isoprenaline sulphate

$\xrightarrow[NaOH]{Heat}$

Back titrate with
NaOH   [HCl$_{Excess}$] + NH$_4$Cl $\longleftarrow$ HCl + NH$_3$ Either $\diagup NH_3$
$\diagdown$ Distil
NaCl + H$_2$O   NH$_4$Cl $\xrightarrow{\text{Titrate HCl}}$ Boric acid plus   Or
Tashiro's indicator

Fig 24.21. An outline of the assay of the sympathomimetic isoprenaline sulphate by Kjeldahl's method.

24.8.4 Biological oxidation

Primary aliphatic amines with no alpha carbon substituents are often oxidised in mammals to the corresponding aldehyde by the enzyme, monoamine oxidase (MAO), which occurs in the kidneys, liver, heart, lungs and central nervous system. Monoamine oxidase is also believed to oxidise some secondary aliphatic amines to the appropriate aldehyde.

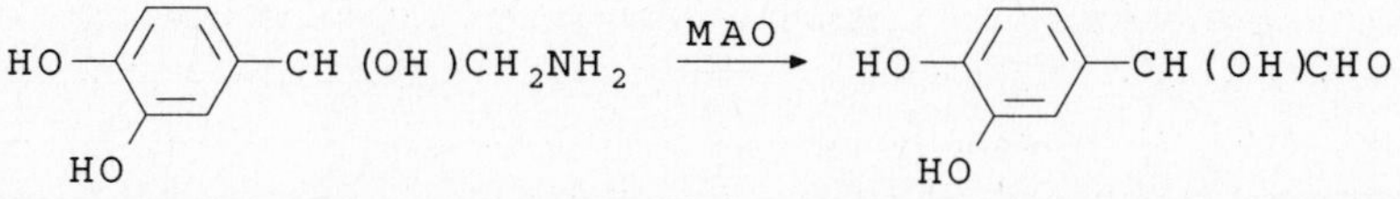

$$HO-\langle\rangle-CH(OH)CH_2NH_2 \xrightarrow{MAO} HO-\langle\rangle-CH(OH)CHO$$

HO   HO

Primary amines are also metabolised to aldehydes by a number of other enzymes. For example, histamine is oxidised by histaminase to 4-iminazolylethanal.

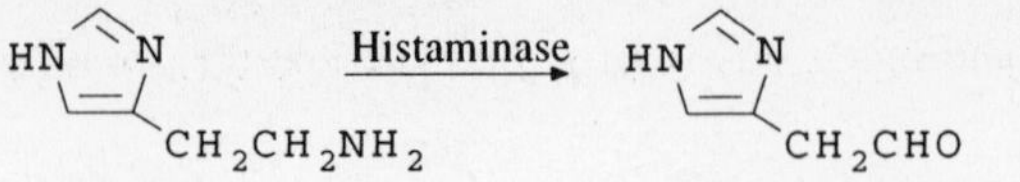

Histamine   4-Iminazolylethanal

Both primary and secondary amines are sometimes oxidised to the corresponding nitroso and hydroxylamino compounds (Fig. 24.22). These compounds are believed to be carcinogenic and also responsible for other toxic effects.

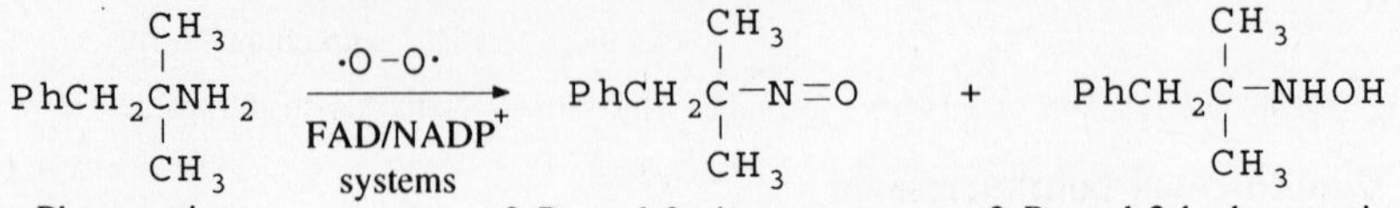

Phentermine   systems   2-Benzyl-2-nitrosopropane   2-Benzyl-2-hydroxyaminopropane

Fig. 24.22. The biological oxidation of the anorexigenic agent, phentermine.

Tertiary amines are sometimes oxidised in humans to the N-oxide. For example, the tranquilliser chlorpromazine is oxidised to chlorpromazine-N-oxide amongst other metabolic products.

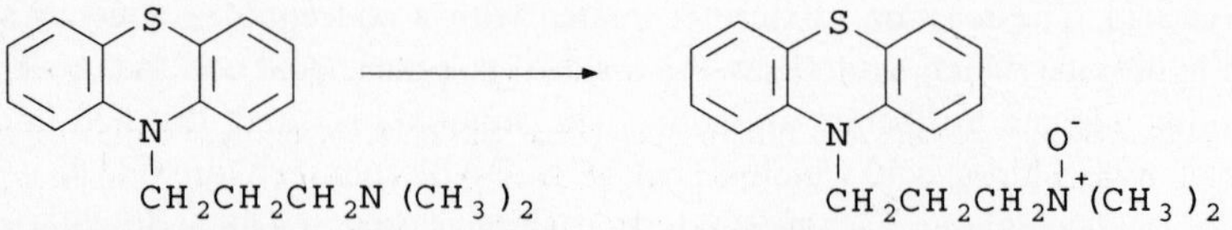

Chlorpromazine   Chlorpromazine-N-oxide

Microsomal oxidation of tertiary amines can involve hydroxylation of a carbon atom adjacent to the amine. For example the long-acting tranquilliser diazepam is metabolised to the short-acting tranquilliser temazepam.

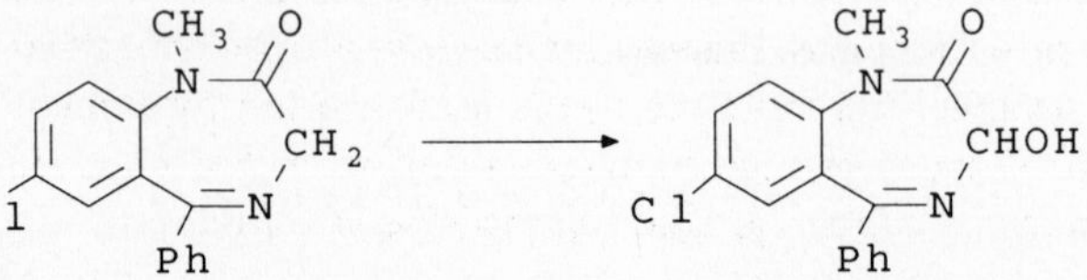

Diazepam   Temazepam

24.9   AROMATIC DIAZONIUM SALTS

Aromatic diazonium salts have the general formula. $Ar-\overset{+}{N}\equiv N$ A^-. They are stable in cold aqueous solutions but it is not usually possible to isolate the actual compounds. Their stability is believed to be due to the stabilisation of the positive charge by delocalisation (Fig. 24.23). Aromatic diazonium salts are important intermediates in the chemical, food and pharmaceutical industries. Their reaction can be divided into two

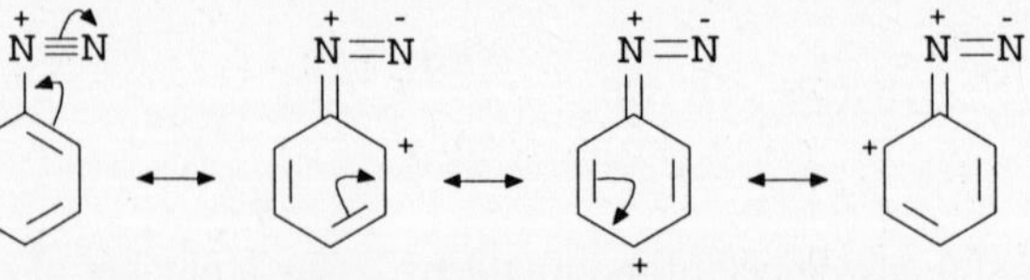

Fig. 24.23. The delocalisation of the positive charge in aromatic diazonium salts as exemplified by the delocalisation in benzenediazonium ions.

types. Those which involve the loss of nitrogen and those which do not involve a loss of nitrogen.

24.9.1 Reactions involving a loss of nitrogen

The diazonium group is replaced by a wide variety of groups (Fig. 24.24). These reactions are of use in the synthesis of many compounds.

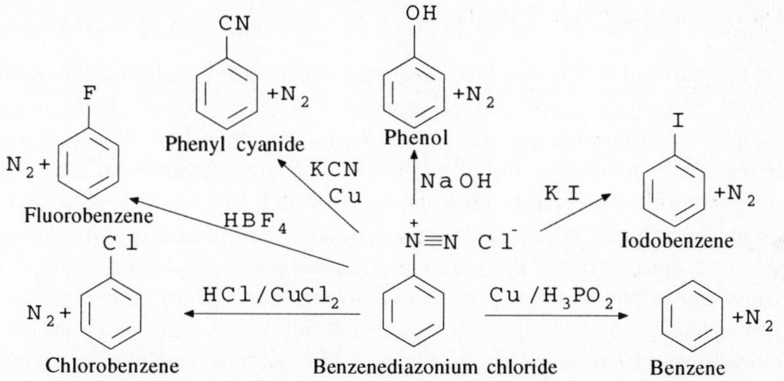

Fig. 24.24. Examples of some of the reactions of benzenediazonium chloride in which nitrogen is lost. It should be noted that the reactions illustrated will also be shown by other aromatic diazonium salts and not just benzenediazonium chloride.

24.9.2 Reactions that do not involve the loss of nitrogen

The diazonium cation can act as an electrophile in aromatic substitutions. However, it attacks only aromatic compounds, such as aromatic amines and phenols with electron rich rings, to form compounds whose structures contain the azo (−N=N−) functional group. The conversion of an aromatic primary amine to its diazonium salt and its subsequent reaction with 2-naphthol to form a red dye is used as a test for primary aromatic amino functional groups in many organic analysis procedures and some identification tests for drugs whose structures contain primary aromatic amino groups.

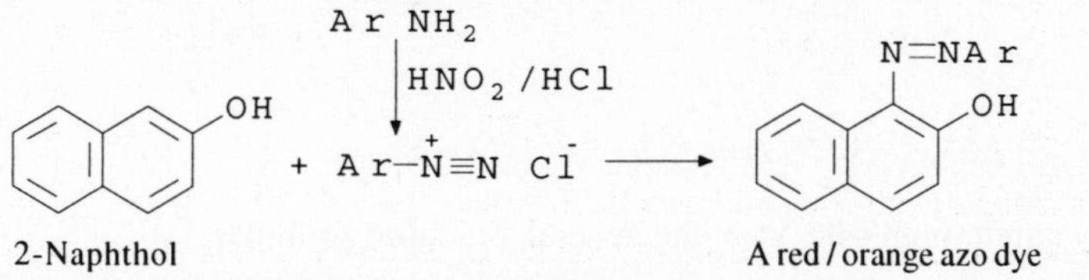

Many azo compounds are important dyes and some are used in the food and pharmaceutical industry as colourings, for example, *Tartrazine, Azovan Blue* and *Amaranth.*

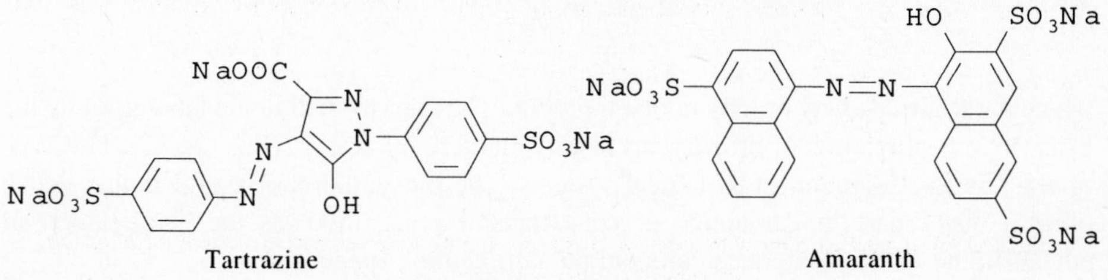

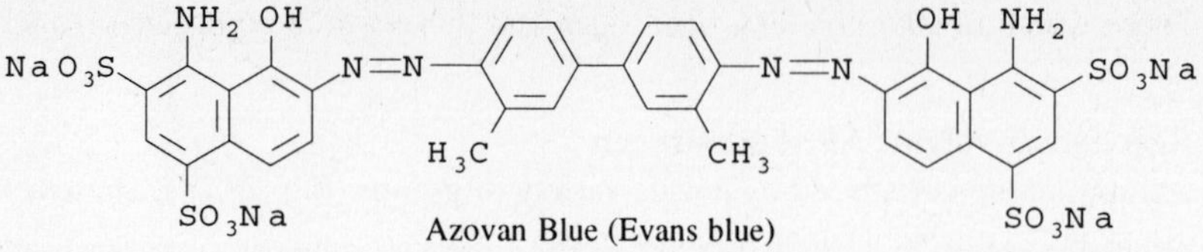

Azovan Blue (Evans blue)

24.10 ELECTROPHILIC SUBSTITUTION OF AROMATIC AMINES

The general course of the electrophilic substitution of aromatic amines will depend on the nature of the reagents. Normally amino groups act as electron donors increasing the ease of electrophilic substitution and are ortho and para directing. For example, bromination of aminobenzene in aqueous solution yields 2,4,6-tribromoaminobenzene. However, acidic reagents form the corresponding amine salt which acts as an electron-withdrawing group, reducing the ease of electrophilic substitution and is mainly meta directing. For example, the main product of the nitration of aminobenzene using a mixture of concentrated nitric and sulphuric acids is the meta-nitro derivative. Many other reactions carried out under acidic conditions will also result in mainly meta rather than ortho and para substitution although this is not always the case. For example, sulphonation of aminobenzene at $180^{\circ}C$ results in the formation of only the para-sulphonic acid derivative.

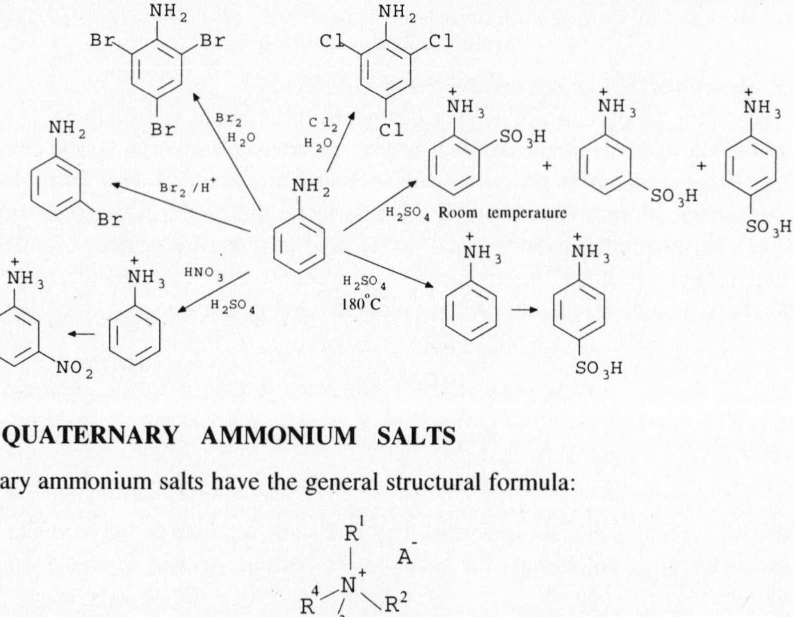

24.11 QUATERNARY AMMONIUM SALTS

Quaternary ammonium salts have the general structural formula:

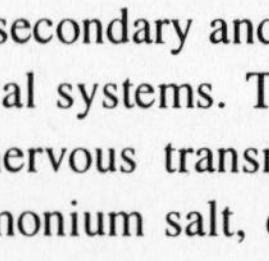

where the R groups may or may not be the same. They are formed in the laboratory by the exhaustive N-alkylation of primary, secondary and tertiary amines (Fig. 24.25). This type of reaction is also found in biological systems. The biosynthesis of acetylcholine which plays a major part in autonomic nervous transmissions, involves the N-alkylation of ethanolamine to the quaternary ammonium salt, choline (page 408).

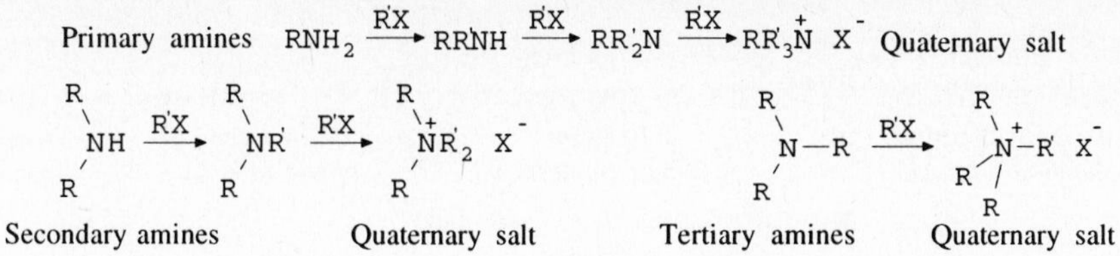

Primary amines RNH_2 $\xrightarrow{R'X}$ $RR'NH$ $\xrightarrow{R'X}$ RR'_2N $\xrightarrow{R'X}$ $RR'_3\overset{+}{N}$ X^- Quaternary salt

Secondary amines   Quaternary salt   Tertiary amines   Quaternary salt

Fig. 24.25. The formation of quaternary salts by the exhaustive alkylation of amines.

Many quaternary ammonium salts are hygroscopic and most are water soluble. Low molecular mass compounds exhibit some lipid solubility but larger molecules usually have either a low solubility or are insoluble in lipids. Some quaternary salts such as viprynium embonate (Fig. 24.26), which is used to treat some worm infections of the lower bowel, owe their mode of action to their low solubilities in both water and lipids. This low solubility means that the drug is not absorbed when taken orally but passes through the stomach and the intestines into the lower bowel.

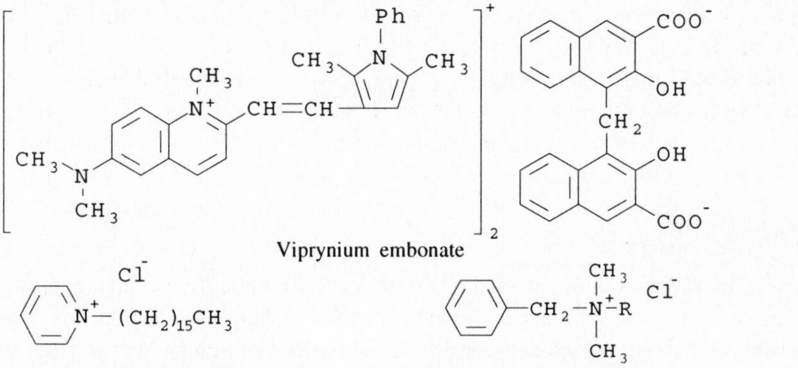

Viprynium embonate

Cetylpyridinium chloride   Benzalkonium chloride

Fig. 24.26. The structures of viprynium embonate, cetylpyridinium chloride and benzalkonium chloride. Benzalkonium chloride is a mixture of compounds with the general formula shown in which the R group ranges from C_8H_{17} to $C_{18}H_{37}$.

Long chain quaternary ammonium salts such as cetylpyridinium chloride and benzalkonium chloride (Fig. 24.26) are used as antiseptic detergents because of their surfactant (9.6.5) and antibacterial actions.

Quaternary ammmonium salts with a β C–H bond undergo elimination on reaction with most bases to form an alkene. The reaction is believed to proceed by an E2 mechanism.

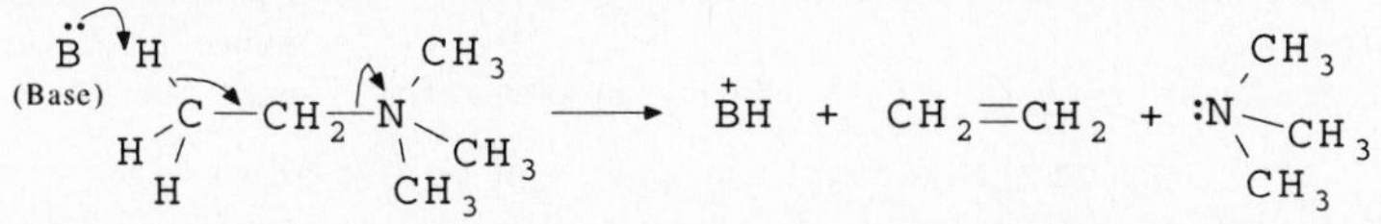

In cases where there is a possibility of more than one product, the least substituted alkene usually predominates (Fig. 24.27). The reason for this selectivity is not fully understood but is believed to be mainly due to steric factors. This type of elimination is referred to as the Hofmann elimination in order to distinguish it from the Saytzeff type of elimination which normally produces the most substituted alkene (14.8).

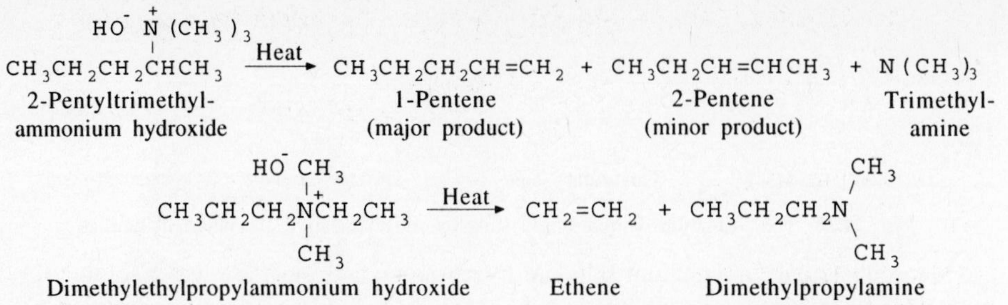

$$\underset{\substack{\text{2-Pentyltrimethyl-}\\\text{ammonium hydroxide}}}{CH_3CH_2CH_2\overset{\overset{\overset{+}{N}(CH_3)_3}{|}}{C}HCH_3} \xrightarrow{\text{Heat}} \underset{\substack{\text{1-Pentene}\\\text{(major product)}}}{CH_3CH_2CH_2CH=CH_2} + \underset{\substack{\text{2-Pentene}\\\text{(minor product)}}}{CH_3CH_2CH=CHCH_3} + \underset{\substack{\text{Trimethyl-}\\\text{amine}}}{N(CH_3)_3}$$

$$\underset{\text{Dimethylethylpropylammonium hydroxide}}{CH_3CH_2CH_2\overset{\overset{CH_3}{\overset{+}{|}}}{\underset{\underset{CH_3}{|}}{N}}CH_2CH_3} \xrightarrow{\text{Heat}} \underset{\text{Ethene}}{CH_2=CH_2} + \underset{\text{Dimethylpropylamine}}{CH_3CH_2CH_2\overset{CH_3}{\underset{CH_3}{N}}}$$

Fig. 24.27. Examples of the elimination reactions of quaternary ammonium salts.

Few examples of Hofmann elimination of quaternary ammonium salts have been found in biological processes. However, some workers think that the vinylic ether group found in plasmalogens (26.4.2) is due to a Hofmann type elimination of trimethylamine from one of the choline residues found in a suitable phospholipid.

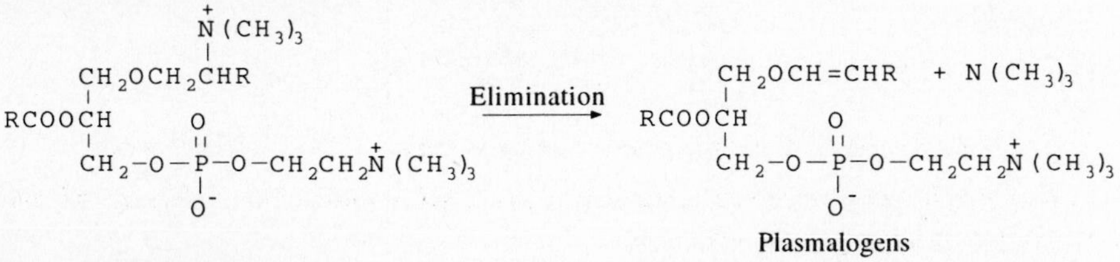

$$\begin{array}{c}\overset{+}{N}(CH_3)_3\\|\\CH_2OCH_2CHR\\|\\RCOOCH \quad\quad O\\|\quad\quad\quad||\\CH_2-O-P-O-CH_2CH_2\overset{+}{N}(CH_3)_3\\|\\O^-\end{array} \xrightarrow{\text{Elimination}} \begin{array}{c}CH_2OCH=CHR \quad + \; N(CH_3)_3\\|\\RCOOCH \quad\quad O\\|\quad\quad\quad||\\CH_2-O-P-O-CH_2CH_2\overset{+}{N}(CH_3)_3\\|\\O^-\end{array}$$

Plasmalogens

24.12 ALKALOIDS

The group of compounds referred to as alkaloids are naturally occurring amines that are extracted mainly from plants, shrubs and trees. Alkaloids occur in the free state, as appropriate salts and as quaternary ammonium salts. A large number have been isolated and many of them have been found to have pharmacological activity (Fig. 24.28). Unfortunately a number of the more useful medicinal compounds and their synthetic derivatives have addictive properties which considerably reduces their usefulness.

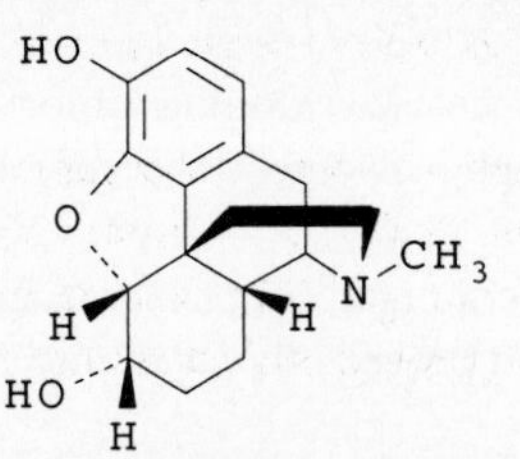

Morphine, an addictive analgesic

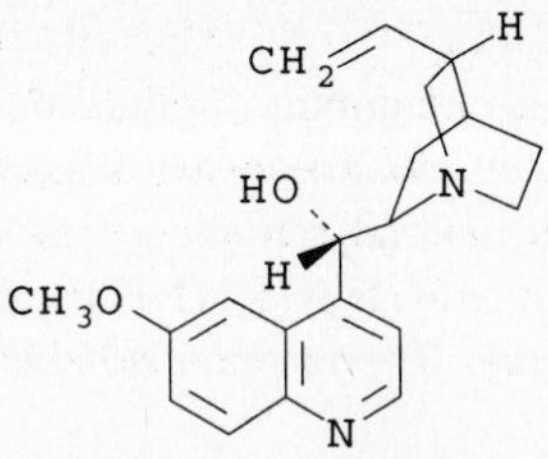

Quinine, an antimalarial

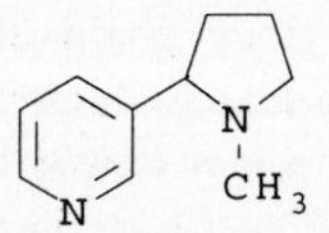

Nicotine, a toxic alkaloid found in tobacco which is used as an insecticide

Fig. 24.28. Examples of pharmacologically active alkaloids.

The pharmacological activity of the alkaloids has generated a great deal of interest over the years. It led to the discovery of many compounds, such as the local anaesthetic benzocaine and the narcotic analgesic pethidine, amongst others used in medicine today (Fig. 24.29). Unfortunately it also led to the discovery of the narcotic analgesic heroin and other compounds that are subject to abuse.

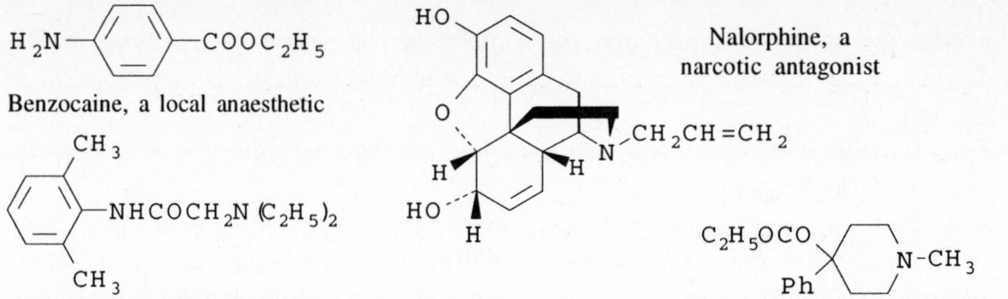

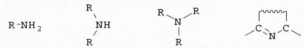

Benzocaine, a local anaesthetic

Lignocaine, a local anaesthetic
and antiarrhythmic

Nalorphine, a
narcotic antagonist

Pethidine, a narcotic analgesic

Fig. 24.29. Some examples of pharmaceutically active compounds developed from alkaloids.

24.13   WHAT YOU NEED TO KNOW

(1)   Amines are classified as either primary, secondary or tertiary depending on the number of hydrogen atoms bonded to the nitrogen atom.

$R-NH_2$ Primary amines   Secondary amines   Tertiary amines   Cyclic tertiary amines

(2)   The prefix amino and the suffix amine are used to indicate the presence of amine functional groups in a molecule.

(3)   N is used in the names of amines to indicate that the following structure is a substituent bonded to the nitrogen of the amino group.

(4)   The nitrogen of amino functional groups does not usually act as a chiral centre.

(5)   In theory the nitrogen atom of an aliphatic amine is sp^3 hybridised whilst that of an aromatic amine is usually taken to be sp^2 hybridised.

(6)   Amines with low molecular masses are water soluble but those with higher molecular masses are not very water soluble.

(7)   The N–H bonds of primary and secondary amines absorb about 3300-3500 cm^{-1} in the infra-red region. Addition of a few drops of acid moves this absorption to approximately 2500-3000 cm^{-1}.

(8)   The protons of primary and secondary aliphatic amines have chemical shifts at approximately 0.5 to 4 while the corresponding aromatic compounds have chemical shifts about $\delta 3$ to 6 in ^1H n.m.r. spectroscopy. The signals of these protons either disappear or are reduced in intensity when the sample is shaken with D_2O.

(9)   Amines act as bases, nucleophiles and are oxidised.

(10) Amines act as bases (proton acceptors), forming salts with acids.

(11) The basic strength of an amine is recorded as a pK_a value. Its value depends on the temperature, solvent, the availability of the lone pair and the stability of the conjugate acid.

(12) At constant temperature, electron donor structures increase the availability of the lone pair and the stability of the conjugate acid and so increase basic strength.

(13) At constant temperature, electron-withdrawing structures decrease the

availability of the lone pair and the stability of the conjugate acid and so they decrease basic strength provided there is no solvation of the conjugated acid by the solvent.

(14) The pK_a value of an amine is a measure of the extent to which it ionises in aqueous biological fluids.

(15) Amine salts are usually more water soluble than the corresponding base and so drugs are often administered in this form as the salts have the same action as the base.

(16) The water solubility of amine salts is the basis of an extraction method for isolating amines from mixtures and naturally occurring materials. Naturally occurring amines are also precipitated from aqueous solutions as their water insoluble amine salts.

(17) Picrates are used as derivatives in qualitative analysis of amines.

(18) All amines act as nucleophiles in substitution reactions. Primary and secondary amines also act as nucleophiles in displacement and addition-elimination (condensation) reactions. The general equations for these reactions are:

Substitution:

$$\backslash N: + RX \longrightarrow \overset{+}{\backslash NR} \; X^- \; \underset{}{\overset{Base}{\rightleftharpoons}} \; \backslash N:$$

Displacement:
Primary and secondary amines

$$RCOZ + H\backslash N \longrightarrow RCON\backslash + HZ$$
An amide

Condensation:

$$RNH_2 + O{=}C \overset{H^+}{\longrightarrow} RN{=}C + H_2O$$
Primary amines   Imine

$$\backslash NH + O{=}C \underset{CH-}{\overset{H^+}{\longrightarrow}} \backslash N{-}C + H_2O$$
Secondary amines   Aldehydes and ketones   Enamine

(19) The formulation of amine drugs with aldehydes and ketones could reduce their activity especially if acids, phosphates, citrates and ethanoates are present.

(20) Amines, especially aromatic amines, are oxidised by air and other oxidising agents to yield a wide range of products.

(21) Nitrous acid reacts at room temperature with all types of primary amine to give nitrogen and other products, secondary amines to yield yellow N-nitroso compounds and tertiary amines to form nitrites. This reaction is used as a test for primary amino groups and to distinguish between the different classes of amine.

(22) Nitrous acid reacts under acidic conditions with aromatic primary amines at 0^oC to form diazonium salts which are used to manufacture azo dyes.

(23) The formation of an aromatic diazonium salt and its subsequent reaction with 2-naphthol to form a red dye is used as a test for an aromatic amine group.

(24) Some aromatic primary amines can be assayed by titration with sodium nitrite using the dead stop indicator method.

(25) All types of amine may be assayed by Kjeldahl's method of nitrogen determination.

(26) Electrophilic substitution of aromatic amines under non-acidic conditions usually

occurs in the ortho and para positions but under acidic conditions it can occur in the meta positions.

(27) Quaternary ammonium salts have the general formula: $R_4N^+ \ A^-$, where the R groups may or may not be the same.

(28) Quaternary ammonium salts are usually water soluble.

(29) Quaternary ammonium salts undergo Hofmann elimination to form mainly the least substituted and simplest alkenes in the presence of a base. Drugs containing this functional group should not be stored in glass (alkaline surface) containers.

(30) Alkaloids are naturally occurring nitrogenous compounds that are isolated from plants, shrubs and trees. They are often pharmacologically active.

24.14 QUESTIONS

(1) Draw the structural formula of each of the following compounds. (a) 1,6-Diaminohexane, (b) 3-Aminopyridine, (c) Diethylmethylamine, (d) 2-(2-Methylaminoethyl)pyridine, (e) N,N-Dimethylcyclohexylamine, (f) Diphenylamine, (g) Ethylpropylamine, (h) 2-Dimethylaminopropylbenzene, and (i) 2-Methylaminocyclohexylamine.

(2) Classify the amino groups of each of the compounds given in question 1 as either a primary, secondary, tertiary or quaternary amine.

(3) Construct systematic names for each of the following amines:

(a) CH_3 / $C_2H_5-C-NH_2$ / CH_3 (b) $C_2H_5-N-CH_3$ (cyclohexyl)   (c) pyridine-NH_2, NH_2 (d) CH_3-N-CH_3 / $CH_3CH_2CHCH_3$

(4) Predict the relative basic strengths of the members of each of the following groups of amines in a solvent which does not solvate with their conjugated acids:
(a) N-methylaminobenzene (A), aminobenzene (B) and N,N-dimethylaminobenzene (C).
(b) Aminobenzene (A), 4-nitroaminobenzene (B) and 4-aminophenol (C).
(c) Pyrrole (A), pyrrolidine (B) and N-methylpyrrolidine (C).

(5) An accurately weighed mass of amphetamine sulphate was heated with sodium hydroxide solution. The reaction mixture was distilled into 50 cm^3 of a 0.5 M solution of hydrochloric acid until the solution had doubled in volume. This solution was titrated with a 0.5 M solution of sodium hydroxide using methyl red as indicator. Outline, by means of equations, the chemistry of this assay. If 6.8754 g of a sample of amphetamine sulphate and a 0.51 M hydrochloric acid solution were used, the titration required 15.7 cm^{-3} of 0.49 M sodium hydroxide for neutralisation. Calculate, (a) the number moles of $C_9H_{13}N$ and (b) %w/w of $C_9H_{13}N$ the sample.

(6) Classify each of the amino functional groups in nicotine (Fig. 24.28) as either primary, secondary or tertiary amines. List the general chemical properties that each of these functional groups would be expected to exhibit. Indicate which properties are associated with which functional group.

(7) Write equations for the most likely reaction of piperidine with each of the following compounds: (a) hydrochloric acid, (b) methyl iodide.

25

Acid amides

25.1 INTRODUCTION

Acid amides are derivatives of organic acids in which one of the hydroxyl or thiol group(s) have been replaced by either an amino (simple amides) or N substituted amino group (N substituted amides). Both acyclic and cyclic structures are known (Fig. 25.1).

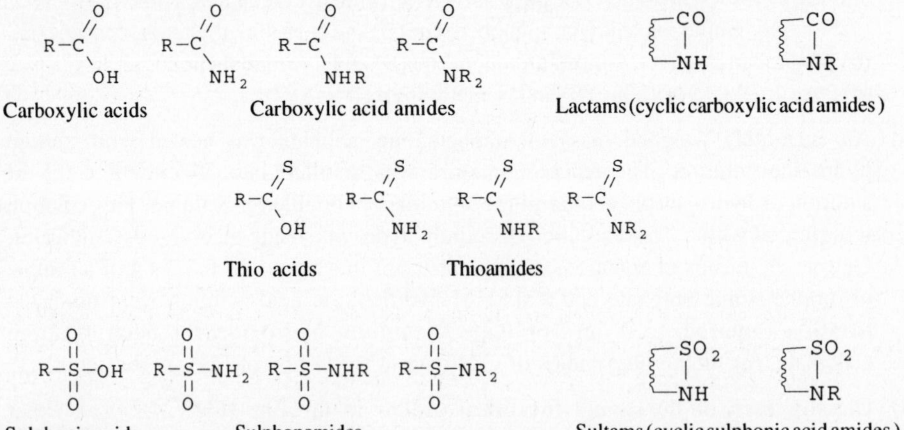

Fig. 25.1. Examples of some amides and their relationship to their corresponding acids.

Acid amide functional groups are found in many drugs (Fig. 25.2) and naturally occurring compounds. For example, amide functional groups are the **peptide links** that are the characteristic feature of peptide and protein structures (28.6).

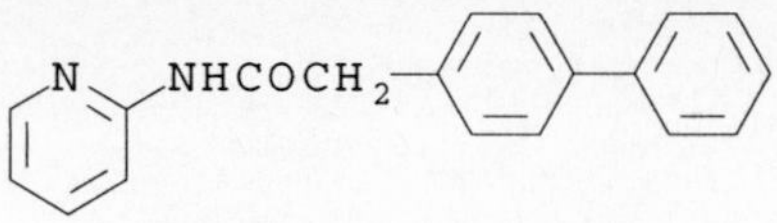

Difenpiramide (anti-inflammatory)

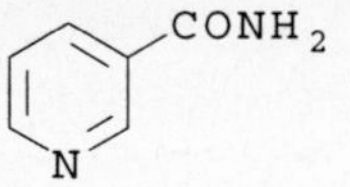

Nicotinamide
(vitamin B supplement)

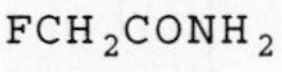

Fluoroacetamide
(insecticide)

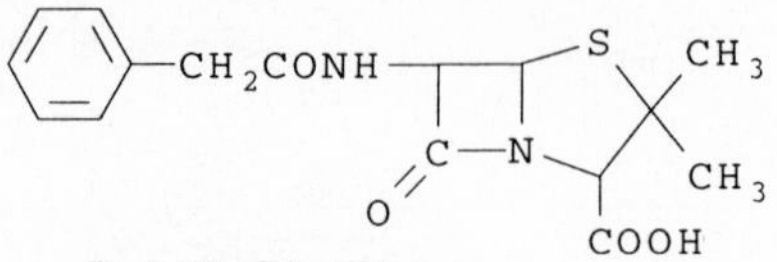

Penicillin G (antibiotic)

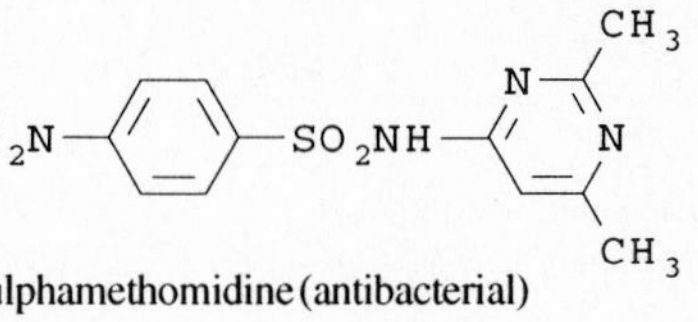

Sulphamethomidine (antibacterial)

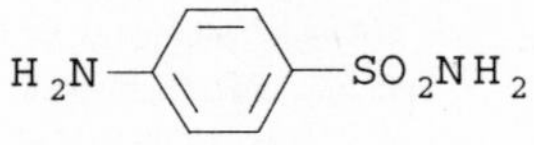

Sulphanilamide (antibacterial)

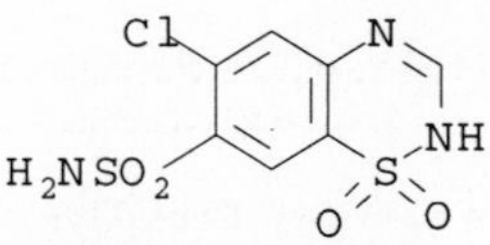

Chlorothiazide (diuretic)

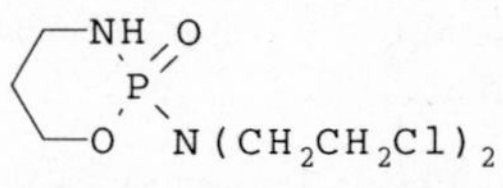

Cylophosphamide (cytsostatic agent)

Fig. 25.2. Examples of drugs whose structures contain acid amide functional groups.

25.2 NOMENCLATURE

The prefixes and suffixes that are used to indicate the presence of the different types of common acid amide functional group in a molecule are given in Table 25.1. Examples of the use of these prefixes and suffixes are given in Fig. 25.3. The trivial names given in brackets are commonly used for many of these molecules.

Table 25.1. The prefixes and suffixes used to indicate the presence of common acid amide structures in molecules. It should be noted that lactam and sultam names are usually based on that of the appropriate nitrogen and sulphur heterocyclic and so do not include the suffix lactam or sultam.

Type of amide	Prefix	Suffix	Type of amide	Prefix	Suffix
Carboxylic acid amides	amido carbamoyl	amide carboxamide carboxamido	Sulphonic acid amides	sulphamoyl	sulphonamide sulphonamido
Cyclic carboxylic acid amides	--	lactam	Cyclic sulphonic acid amides	--	sultam
Thioamides	thio...amide Thiocarbamoyl	--			

The ring sizes of all types of cyclic acid amide are indicated in their names by a prefix. These prefixes may be either the locants of the carbon atoms separated by the nitrogen atom or a Greek letter acting as a locant. These prefixes are dropped if there is no ambiguity. In the case of sultams, carbon number 1 is bonded to the sulphur of the sulphonamide group.

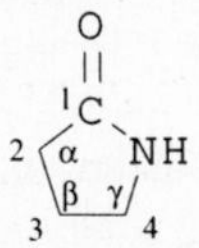

1,4-Butrolactam or γ-Butrolactam (pyrrolidone)

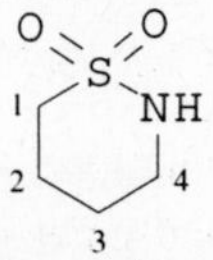

Butanesultam

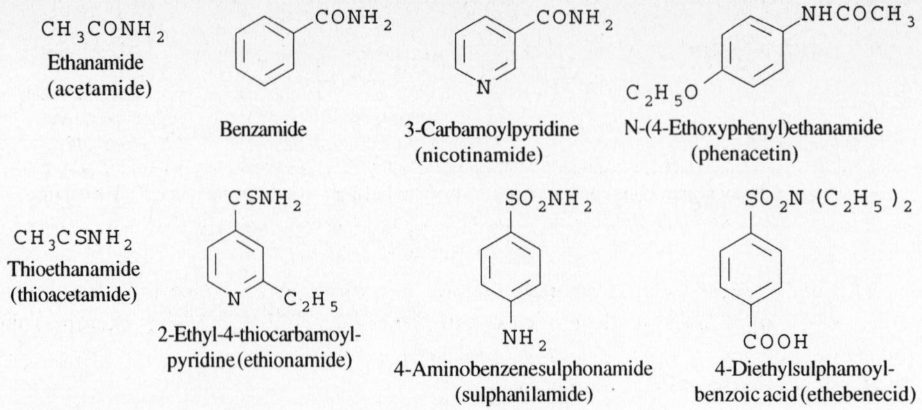
CH3CONH2
Ethanamide
(acetamide)
CONH2
Benzamide
CONH2
N
3-Carbamoylpyridine
(nicotinamide)
NHCOCH3
C2H5O
N-(4-Ethoxyphenyl)ethanamide
(phenacetin)
CH3CSNH2
Thioethanamide
(thioacetamide)
CSNH2
N
C2H5
2-Ethyl-4-thiocarbamoyl-
pyridine (ethionamide)
SO2NH2
NH2
4-Aminobenzenesulphonamide
(sulphanilamide)
SO2N (C2H5)2
COOH
4-Diethylsulphamoyl-
benzoic acid (ethebenecid)

Fig. 25.3. Examples of the use of the systematic and (trivial) nomenclature of acid amides.

25.3 PHYSICAL PROPERTIES

25.3.1 Structure

The properties of the carboxamide functional group indicates that the lone pair of electrons of the nitrogen atom are conjugated with the pi bond of the carbonyl group (Fig. 25.4). Thioamides have a similar structure while in the sulphonamide group it appears that conjugation exists between the lone pair of electrons of the nitrogen atom and one of the S=O bonds.

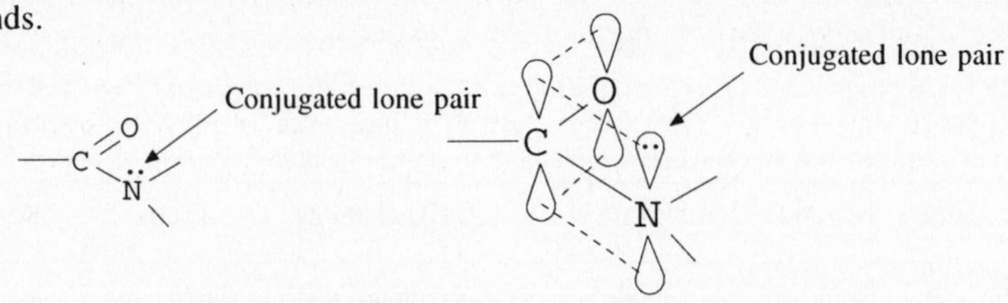
Conjugated lone pair
Conjugated lone pair
C
O
N
C
O
N

Fig. 25.4. The structure of the amide functional group.

Acid amides with at least one N–H bond exhibit tautomerism in the liquid and solution states, the position of equilibrium for both simple acyclic and cyclic acid amides lying almost exclusively to the left. However, many purines and pyrimidines that could exist as phenolic tautomers prefer to exist in the amide form.

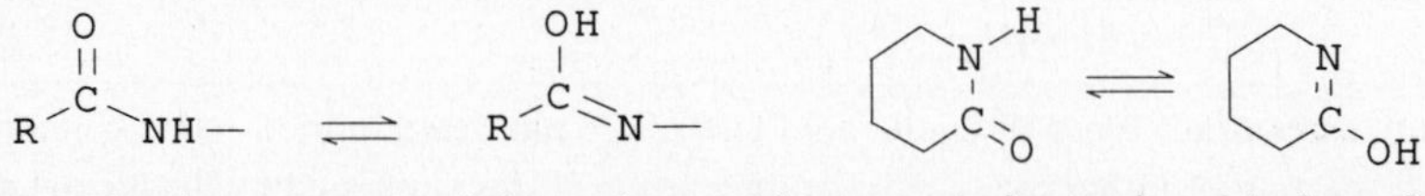
O
R
C
NH—
OH
R
C
N—
H
N
C
O
N
C
OH

The molecular structure of the carboxamide group is **isosteric** with that of the ester carboxylic acid ester group (Fig. 25.5). Structures or atoms that are isosteric have similar electronic configurations in the outer electronic shells of the principal atoms of the structure. This definition does not take into account the relative sizes and shapes of groups and so isosteres are also classified in terms of their valency. For example, the class I isosteric groups, hydroxyl, thiol, primary amino and methyl have a valency of one whilst the class II isoteric atoms and groups, oxygen, sulphur, selenium secondary amine and methylene have a valency of two.

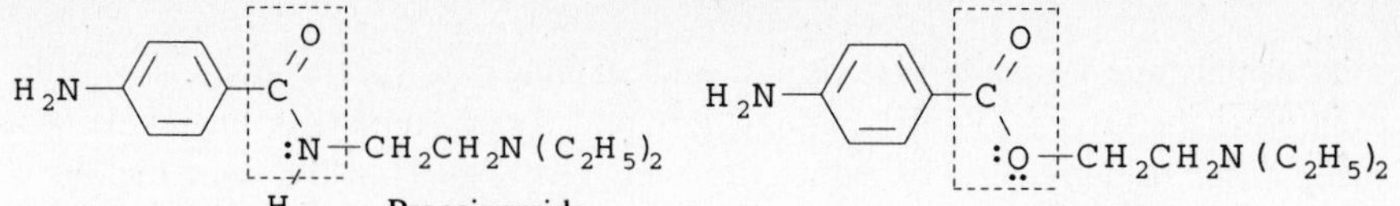

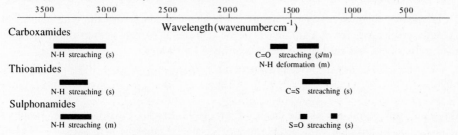

H_2N—⟨benzene⟩—C(=O)—$:N$(H)—$CH_2CH_2N(C_2H_5)_2$
H
Procainamide

H_2N—⟨benzene⟩—C(=O)—$:O$—$CH_2CH_2N(C_2H_5)_2$
Procaine

Fig. 25.5. The isosteric nature of the carboxamide and carboxylic acid ester groups as exemplified by the structures of the local anaesthetic procaine and the myocardial depressant procainamide. The structures in the dotted boxes are isosteric, the pair of electrons in the N-H bond corresponding to the second lone pair of the oxygen.

The replacement of an atom or structure in a biologically active molecule by an isosteric group is used to design new antimetabolites and drugs. For example, the replacement of the hydroxyl group of hypoxanthine by its isostere, the thiol group, lead to antitumour agent 6-mercaptopurine.

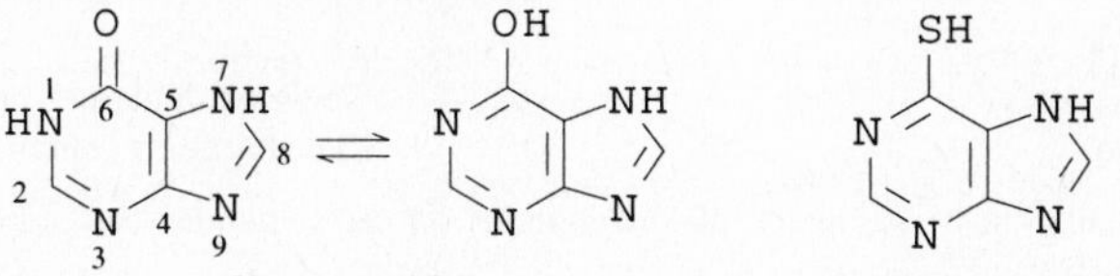

Hypoxanthine ⇌ (OH structure) ⇌ 6-Mercaptopurine

25.3.2 Solubility

Simple aliphatic carboxamides with low molecular masses are readily soluble in water but this solubility decreases as the relative molecular mass of the compound increases. Aromatic carboxamides are less soluble in water than aliphatic compounds of similar size and relative molecular mass. Thioamides and sulphonamides are not usually very soluble in water.

Liquid low relative molecular mass carboxamides, such as dimethylformamide, are good solvents for a variety of reagents and extensive use is made of this property in the synthesis of organic compounds.

25.3.3 Spectra

The regions in which acid amides absorb in the infra-red region are given in Fig. 25.6. A strong broad absorption band in the 3500 to 3300 cm^{-1} region may be taken to indicate the possible presence of a carboxamide or thioamide group in a molecule. It must be remembered that other functional groups absorb in this region and so any deduction of this nature must be confirmed by other evidence,

3500	3000	2500	2000	1500	1000	500

Wavelength (wavenumber cm^{-1})

Carboxamides
N-H streaching (s)   C=O streaching (s/m)
                       N-H deformation (m)

Thioamides
N-H streaching (s)   C=S streaching (s)

Sulphonamides
N-H streaching (m)   S=O streaching (s)

Fig. 25.6. The approximate absorption regions of acid amides in the infra-red.

Acid amides with N–H bonds usually exhibit broad proton nuclear magnetic resonance signals. Acid amide protons are exchangeable with deuterium. This is used to identify signals due to amide protons. ^{13}C nuclear magnetic spectroscopy is of little use in identifying the presence of the carbonyl group of a carboxamide or thioamide because so many different derivatives of carboxylic acids give signals in the same region.

25.4   GENERAL CHEMICAL PROPERTIES

Acid amides can act as very weak bases and they can also act as acids. Acid amides are hydrolysed but sulphonamides are far less reactive than carboxamides and thioamides. This reduced reactivity is probably due to steric hindrance and the repulsion of the nucleophile by the electron clouds of the two S=O groups (Fig. 25.7).

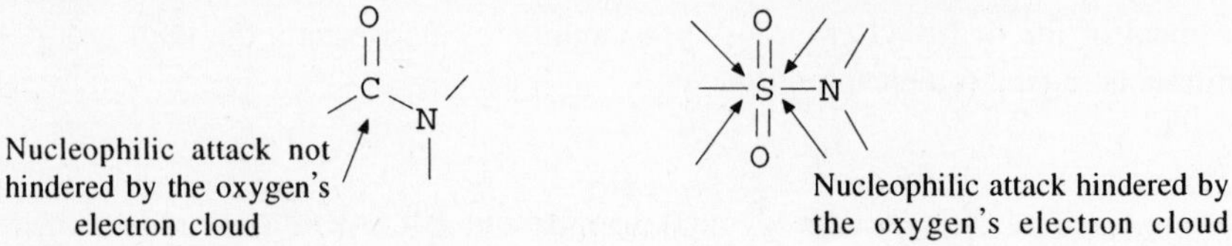

Nucleophilic attack not hindered by the oxygen's electron cloud

Nucleophilic attack hindered by the oxygen's electron cloud

Fig. 25.7. A representation of the attack of nucleophiles on carboxamides and sulphonamides.

Acid amides may be oxidised and reduced. All types of acid amide also exhibit reactions that are characteristic of their type. For example, unsubstituted carboxamides usually react with alkaline solutions of bromine to form the corresponding amine and /or nitrile.

RCH_2NH_2 R has five or less carbon atoms

RCH_2CONH_2 ⟵ Br_2 /KOH

RCN R has five or more carbon atoms

25.5   AMPHOTERIC NATURE

Acid amides are weak bases. The lone pair of the nitrogen is not readily available to accept protons because of the strong electron-withdrawing effect of the adjacent C=O, C=S or S=O groups.

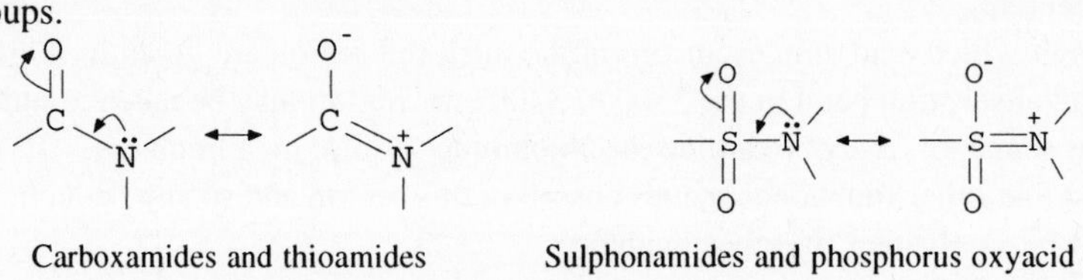

Carboxamides and thioamides

Sulphonamides and phosphorus oxyacid amides

In spite of their weakly basic nature, carboxamides can form salts with acids. For example, ethanamide forms a nitrate with nitric acid and an oxalate with oxalic acid. These salts are unstable and are usually easily hydrolysed by water.

CH_3CONH_2 + HNO_3 ⟶ $CH_3CONH_2 .HNO_3$

Ethanamide H_2O Ethanamide nitrate

However, salt formation is the basis of the 1993 British Pharmacopoeia assay of nicotinamide by titration with perchloric acid in glacial ethanoic acid (11.6.4).

Acid amides whose structures contain at least one N−H bond can act as acids. This acidic behaviour occurs because the C=O, C=S and SO_2 electron-withdrawing groups weaken the N−H bond and, more importantly, the resultant anion is resonance stabilised.

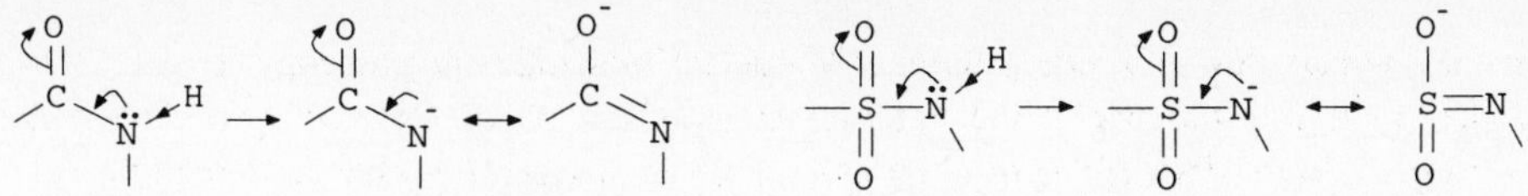

Carboxamides can have pK_a values of about -1 to -2. They form ionic alkali metal salts. These salts are unstable being easily hydrolysed in aqueous solution. Stable mercury salts have been produced in which it is thought that the mercury is covalently bonded to the nitrogen of the carboxamide.

Sulphonamides and sultams readily form stable well characterised sodium salts with sodium hydroxide but not with sodium carbonate as it is not a sufficiently strong base.

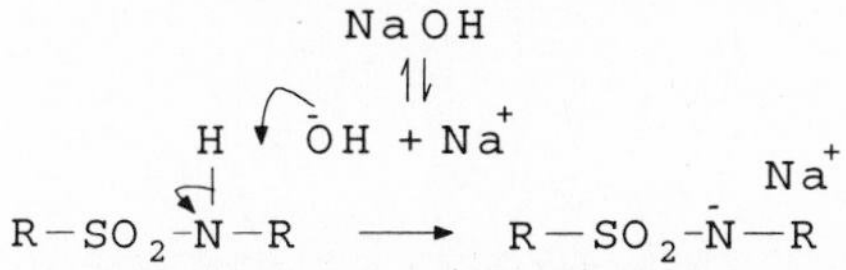

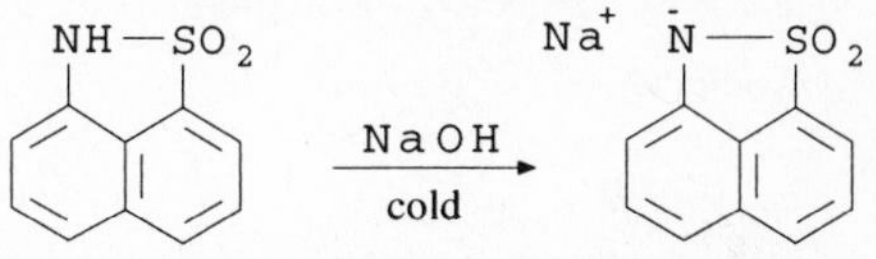

As expected, the sodium salts of sulphonamide and sultams are more soluble in water than their parent compounds. Consequently, a number of drugs containing sulphonamide and sultam groups are administered in the form of aqueous solutions of their sodium salts, for instance, the antibacterial sulphadimidine is administered by injection as an aqueous solution of sulphadimidine sodium. However, it should be noted that these aqueous solutions are often susceptible to light-catalysed atmospheric oxidation Consequently, they must be prepared with water that contains an antioxidant but no dissolved air. It is interesting to note that most workers think that it is the sulphonamide ion that is responsible for the antibacterial action, but the drug penetrates the bacterial cell wall in its unionised form. It should be noted that the solubility of thiazide diuretics, such as chlorothiazide (Fig. 25.2), in alkaline solutions is probably due to the formation of the salts of both the sulphonamide and sultam groups.

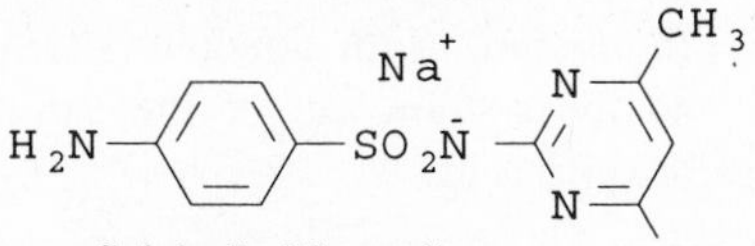

Sulphadimidine sodium

25.6 HYDROLYSIS

Acid amides are hydrolysed by water, the reaction being catalysed by acids, enzymes and bases. The ease of hydrolysis is generally in the order:

thioamide > carboxamide > sulphonamide

Sulphonamides are relatively stable to hydrolysis reacting only under vigorous conditions, such as heating with 10 M hydrochloric acid in the laboratory. They are stable to alkali-catalysed hydrolysis. However, sultams are hydrolysed by hot alkali.

The products of the hydrolysis of carboxamides will depend on the nature of the acid amide and the reaction conditions. All types of carboxamide yield either the corresponding carboxylic acid or its salt. N-Substituted carboxamides also produce the amine or its salt but unsubstituted carboxamides yield ammonia or ammonium salts. Acid conditions produce the amine salts whilst base conditions yield the salt of the carboxylic acid.

$$RCOO^- + NH_3 \xleftarrow{\ ^-OH/H_2O\ } RCONH_2 \xrightarrow{\ H^+/H_2O\ } RCOOH + NH_4^+$$

Simple carboxamides

$$RCOO^- + RNH_2 \xleftarrow{\ ^-OH/H_2O\ } RCONHR \xrightarrow{\ H^+/H_2O\ } RCOOH + RNH_3^+$$

N Substituted carboxamides

The mechanisms of these reactions follow the general route outlined in Chapter 21 (Fig. 21.7). However, in the acid catalysed hydrolysis the nitrogen atom gains a proton from the oxonium ion (**a** in Fig. 25.8) before being displaced whilst a similar transfer of a proton from the hydroxyl group of the intermediate (**b**) to the NH$_2$ leaving group occurs in the base-catalysed mechanism (Fig. 25.8).

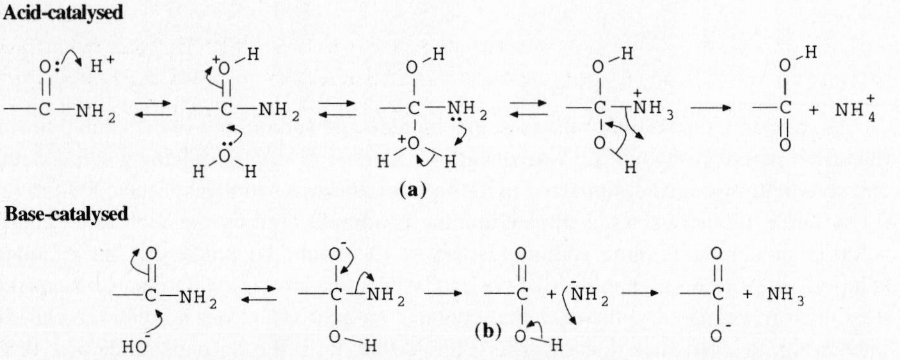

Acid-catalysed

(a)

Base-catalysed

(b)

Fig. 25.8. The mechanisms of the acid- and base-catalysed hydrolyses of carboxamides by water.

Thioamides react in a similar fashion but usually produce the corresponding carboxylic acid or its salt and not the thioacid as might be expected. In some cases a mixture of the corresponding carboxamide and carboxylic acid or its salt are formed. For example, hydrolysis of the antitubercular drug *Ethionamide* with aqueous sodium hydroxide yields the corresponding carboxamide and the sodium salt of the carboxylic acid whereas acid hydrolysis yields only the carboxylic acid.

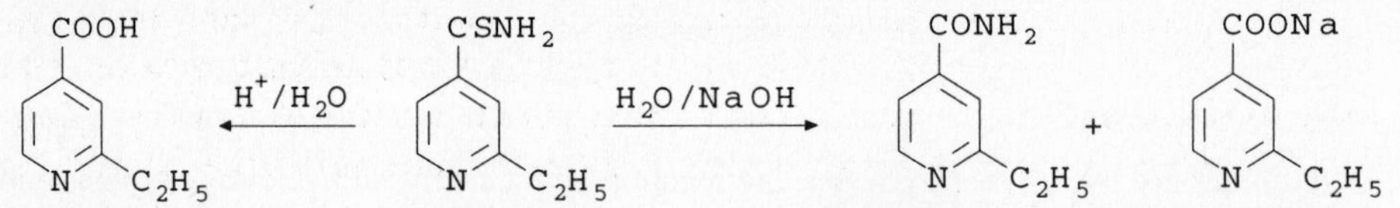

2-Ethylpyridine-4-carboxylic acid   *Ethionamide*   2-Ethylpyridine-4-carboxamide   Sodium 2-ethylpyridine-4-carboxylate

Primary carboxamides (RCONH$_2$) can be assayed by hydrolysis. The carboxamide is hydrolysed with alkali and the ammonia liberated during hydrolysis distilled into a known volume of acid. The excess acid is back titrated with sodium hydroxide using methyl red as indicator.

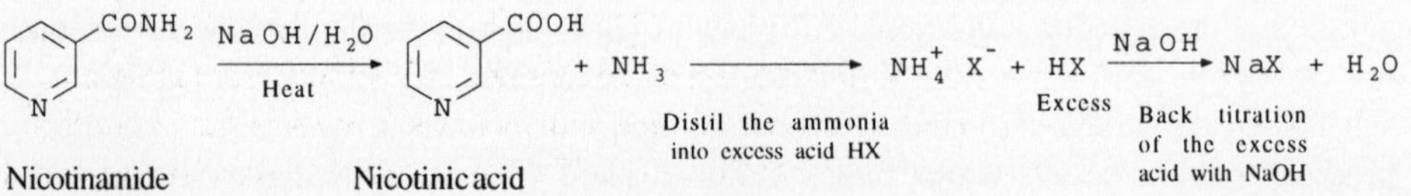

Nicotinamide   Nicotinic acid

$$+ NH_3 \longrightarrow NH_4^+ \; X^- + HX \xrightarrow{\ NaOH\ } NaX + H_2O$$

Distil the ammonia into excess acid HX   Excess   Back titration of the excess acid with NaOH

Hydrolysis is the basis of identification tests for all types of acid amides. All tests are based on identifying, by chemical or physical methods, one or more of the products of the hydrolysis. Consequently, the precise nature of the procedure used in the identification test will depend on the molecular structure of the amide. Consider, for example, identification test C in the 1993 British Pharmacopoeia for the respiratory stimulant nikethamide. The compound is heated with aqueous sodium hydroxide and the formation of diethylamine, which is identified by its smell and effect on red litmus paper, is taken as a positive result.

Nikethamide $\xrightarrow{\text{NaOH/H}_2\text{O}}$ Sodium nicotinate   + $(\text{C}_2\text{H}_5)_2\text{NH}$ Diethylamine

Sulphonamides are difficult to hydrolyse but can be decomposed by heating with a mixture of phosphoric and sulphuric acids. The amine and sulphonic acid are separated and identified by making suitable derivatives such as the benzamide for the amine and the S-benzylthiuronium salt of the acid.

$$\text{RNH}_2 \xrightarrow[\text{NaOH}]{\text{PhCOCl}} \text{RNHCOPh} + \text{HCl}$$
A benzamide

$$\text{RSO}_3\text{H} \xrightarrow[\text{NaOH}]{\text{PhCH}_2\text{SC}\overset{\overset{+}{\text{NH}_2}\,\text{Cl}^-}{\underset{\text{NH}_2}{\diagup}}} \text{PhCH}_2\text{SC}\overset{\overset{+}{\text{NH}_2}}{\underset{\text{NH}_2}{\diagup}} \text{RSO}_3^- + \text{HCl}$$
S-Benzylthiuronium chloride   S-Benzylthiuronium salt

Traditional qualitative analysis uses hydrolysis with hot sodium hydroxide to identify the presence of primary carboxamide ($-\text{CONH}_2$) and thioamide functional groups ($-\text{CSNH}_2$) in a compound. The liberation of ammonia on hydrolysis is taken as a positive result. Carboxamide hydrolysis also forms the basis of the structure determination of peptides and proteins (28.11).

Hydrolysis is one of the main metabolic reactions of compounds containing carboxamide functional groups. Drugs may be activated or deactivated by metabolic hydrolysis. For instance, metabolic hydrolysis of paracetamol by serum amidases produces 4-aminophenol which is responsible for its analgesic affect. However, hydrolysis of procainamide by serum amidases produces 4-aminobenzoic acid (PABA) which has no local anaesthetic action. In most cases, metabolic hydrolysis of amides is usually too slow for amides to be of much use as prodrugs, one exception being paracetamol.

Paracetamol $\xrightarrow{\text{Serum amidases}}$ 4-Aminophenol

Procainamide $\xrightarrow{\text{Serum amidases}}$ 4-Aminobenzoic acid

Drugs whose structures contain carboxamide and thioamide functional groups must be stored under dry conditions to reduce the risk of hydrolysis. For example, penicillins are rapidly decomposed by moist air to inactive compounds by hydrolysis of the very reactive β-lactam ring. This instability means that penicillins should be stored only as the solid drug and aqueous penicillin preparations must be used within the time stated on the label.

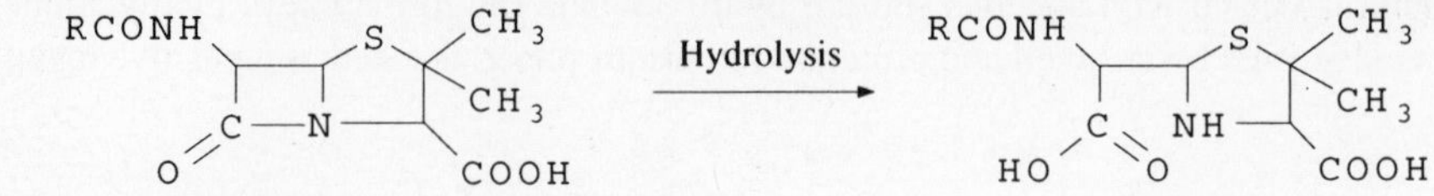

A penicillin (R controls the nature of the drug)   A penicilloic acid

25.7   OXIDATION

Some acid amide functional groups with the notable general exception of the sulphonamide group can be oxidised in the laboratory. The ease of oxidation and the nature of the products depends on the structure of the acid amide, the reagent used and conditions under which the reaction is conducted (Fig. 25.9).

$$CH_3CONH_2 \xrightarrow{HNO_2} CH_3COOH + N_2 + H_2O$$

Ethanamide   Ethanoic acid

$$(CH_3)_3CCONH_2 \xrightarrow[DMF / (CH_3CH_2)_3N]{Pb(OCOCH_3)_2} (CH_3)_3CNCO$$

2,2-Dimethylpropanamide   1,1-Dimethylethyl isocyanate
                                           (t-butyl isocyanate)

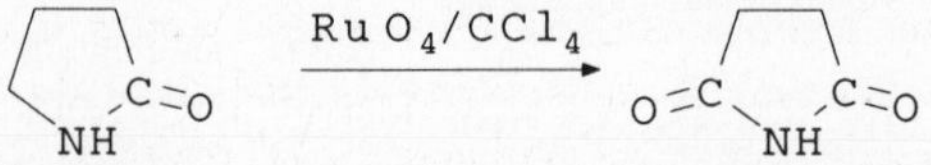

Pyrrolidone $\xrightarrow{RuO_4/CCl_4}$ Succinimide

Fig. 25.9. Examples of the laboratory oxidation of acid amides.

Metabolic oxidation of carboxamide groups does not appear to occur to any significant extent. However, *Ethanioamide* is oxidised in humans to the S-oxide which appears to be active against tuberculosis. Sulphonamide groups do not appear to be oxidised in biological systems.

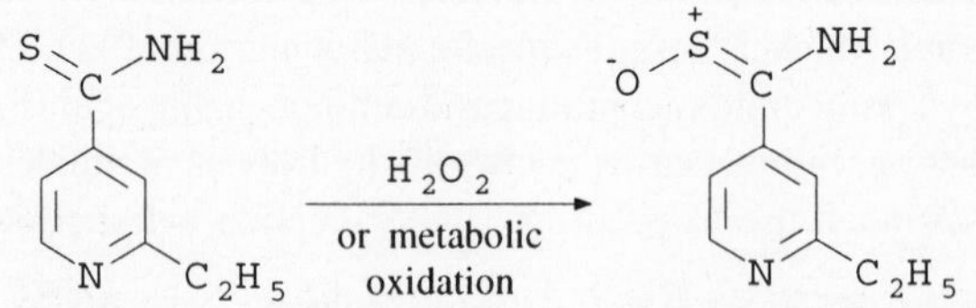

Ethionamide $\xrightarrow[\text{oxidation}]{\substack{H_2O_2 \\ \text{or metabolic}}}$ Ethionamide-S-oxide

25.8   REDUCTION

Acid amide functional groups may be reduced in the laboratory. The ease of reduction and the nature of the products depends on the type of amide, the reagent and the reaction conditions (Fig. 25.10).

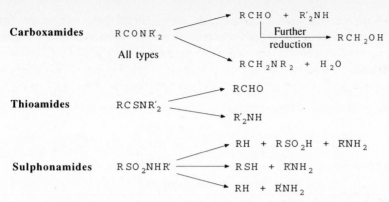

Fig. 25.10. General routes for the reduction of acid amides. The R groups may be the same or different, aliphatic, aromatic or heterocyclic structures.

The reduction of the carboxamide functional group with lithium aluminium hydride is the basis of a method for the assay of compounds whose structures contain this functional group. The method is of limited use because it relies on the formation of a reasonably volatile amine. This amine is steam distilled into a known volume of acid and the excess acid is determined by back titration with alkali.

$$RCONH_2 \xrightarrow{LiAlH_4} RCH_2NH_2 \xrightarrow[\substack{\text{Distil into excess} \\ \text{acid (HX)}}]{} RCH_2\overset{+}{N}H_3 \ X^- + HX \xrightarrow[\substack{\text{Titrate the excess} \\ \text{acid with NaOH}}]{\text{Excess}} NaX + H_2O$$

A further limitation of this method is the presence of either nitro and/or nitrile groups in the compound. Lithium aluminium hydride will reduce these functional groups to amines. The formation of these 'extra' amine groups can lead to erroneous titration values.

25.9 REACTIONS CHARACTERISTIC OF SIMPLE AMIDES (RCONH₂)

25.9.1 Imide formation

Simple unsubstituted carboxamides can react with acid chlorides and anhydrides to form imides. For example, dipropanimide is prepared by treating propanamide with propanoic anhydride in the presence of sulphuric acid.

$$CH_3CH_2CONH_2 \xrightarrow[\substack{\text{Propanoic anhydride} \\ H_2SO_4}]{CH_3CH_2COOCOCH_2CH_3} CH_3CH_2CONHCOCH_2CH_3$$

Propanamide   Dipropanimide (74%)

Cyclic imides are formed by strongly heating either the appropriate dicarboxylic acid ammonium salt or the simple diamide (Fig. 25.11).

25.9.2 Hofmann degradation

Primary carboxamides react with bromine in aqueous alkali solution to form the

$$RCONH_2 \xrightarrow[OH^-]{Br_2} RNCO \xrightarrow{H_2O} RNHCOOH \longrightarrow RNH_2 + CO_2$$

A carboxamide   An isocyanate   A carbamic acid   An amine corresponding amine. The bromine converts the carboxamide to the isocyanate. This is rapidly hydrolysed to the carbamic acid which immediately decarboxylates to the amine.

25.10   IMIDES

Imido functional groups have the structural formula $-CO-NH-CO-$. This group is usually formed by the general methods illustrated in Fig. 25.11 and the reaction given in section 25.9.1.

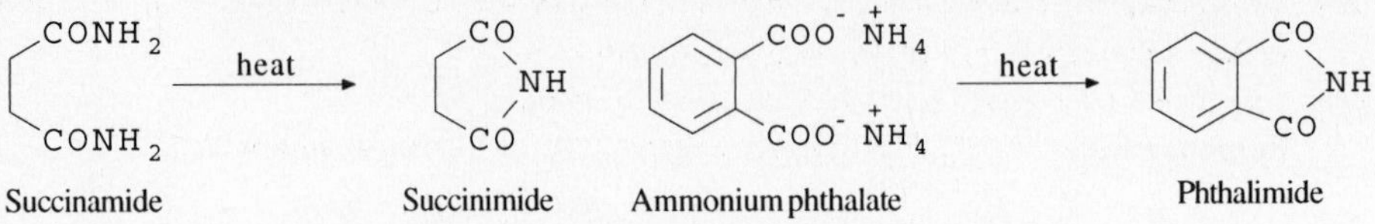

Fig. 25.11. General methods of preparing imides.

A number of cyclic imides act as hypnotics and anticonvulsants (Fig. 25.12).

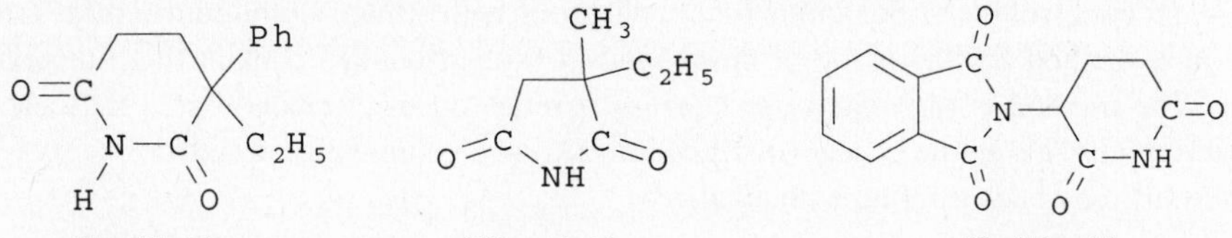

Fig. 25.12. Examples of imides used as hypnotic and anticonvulsant drugs. Thalidomide (Fig. 5.25) has tetratogenic properties (5.11.2) and must not be used in normal circumstances.

Imides act as weak acids forming resonance stabilised salts with alkali.

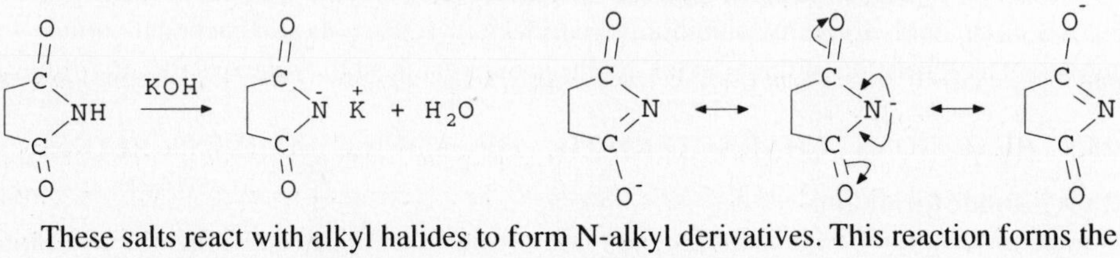

These salts react with alkyl halides to form N-alkyl derivatives. This reaction forms the basis of the Gabriel synthesis of amines and amino acids.

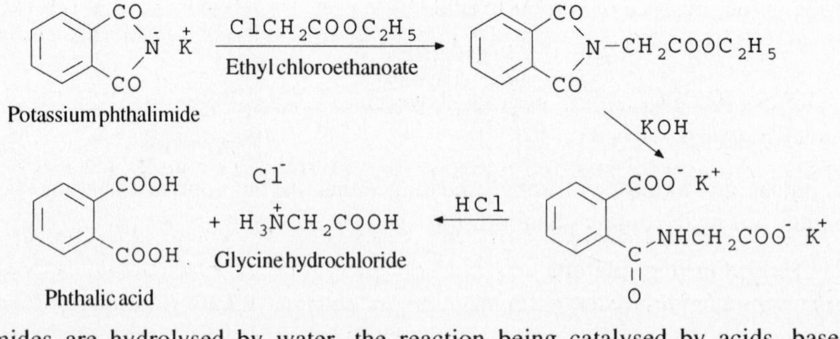

Imides are hydrolysed by water, the reaction being catalysed by acids, bases and enzymes. The hydrolysis with either acid or base can be used to identify the imide by identifying the carboxylic acid(s) formed in the reaction.

$$
\begin{array}{ccccc}
\begin{array}{c}\diagup COO^- \\ + NH_3 \\ \diagdown COO^- \end{array} & \xleftarrow{\ ^-OH\ /H_2O\ } & \begin{array}{c}\diagup CO \\ NH \\ \diagdown CO \end{array} & \xrightarrow[NH_4^+]{H^+/H_2O} & \begin{array}{c}\diagup COOH \\ \\ \diagdown CONH_2 \end{array} & \xrightarrow[NH_4^+]{H^+/H_2O} & \begin{array}{c}\diagup COOH \\ \\ \diagdown COOH \end{array}
\end{array}
$$

Succinate   Succinimide   Succinamide   Succinic acid

25.11   WHAT YOU NEED TO KNOW

(1) Acid amides are derivatives of organic acids in which either a hydroxyl or a thiol group has been replaced by an amino group.

(2) The types and general structures of acid amides are:

Carboxamides: Thioamides: Sulphonamides:

$$-\overset{\overset{\displaystyle O}{\|}}{C}-N\diagup_{\diagdown} \qquad -\overset{\overset{\displaystyle S}{\|}}{C}-N\diagup_{\diagdown} \qquad -\overset{\overset{\displaystyle O}{\|}}{\underset{\underset{\displaystyle O}{\|}}{S}}-N\diagup_{\diagdown}$$

(3) The prefixes and suffixes used to indicate the presence of an acid amide functional group are listed in Table 25.1

(4) Carboamides and thioamides exhibit tautomerism but the position of equilibrium lies almost exclusively to the left.

$$-\overset{\overset{\displaystyle O}{\|}}{C}-\overset{\overset{\displaystyle H}{|}}{N}- \ \rightleftharpoons\ -\overset{\overset{\displaystyle OH}{|}}{C}=N- \qquad\qquad -\overset{\overset{\displaystyle S}{\|}}{C}-\overset{\overset{\displaystyle H}{|}}{N}- \ \rightleftharpoons\ -\overset{\overset{\displaystyle SH}{|}}{C}=N-$$

(5) The lone pair of the nitrogen atom of all types of amide group is conjugated with either a C=O, S=O or P=O groups in the structure.

(6) The carboxamide and carboxylic acid ester groups are isosteric.

(7) The carboxamide functional group will enhance the water solubility of a compound because of hydrogen bonding between the lone pairs of the oxygen atom and the proton of the N–H bonds of the amide with water molecules.

(8) Thioamide and sulphonamide functional groups do not enhance the water solubility of a compound.

(9) The N–H bonds of appropriate carboxamides and thioamides absorb around 3500 to 3300 cm^{-1} in the infra-red region of the electromagnetic spectrum. A strong absorption band in this region could be indicative of the presence of these functional groups.

(10) The protons of the N–H bonds of acid amides can exchange with deuterium. In the ^{1}H n.m.r. spectroscopy this results in either the loss of the signal for this proton(s) or a reduction in its size.

(11) Acid amides are usually very weak bases.

(12) Acid amides are weak acids.

(13) The sodium salts of sulphonamides are more water soluble than the parent compound.

(14) Thioamides and carboxamides are hydrolysed by water (catalysed by acids, bases and enzymes) to the acid and amine or their appropriate salt depending on the nature of the catalyst.

(15) Hydrolysis is the basis of assay procedures for carboxamides of the type RCONH$_2$.

(16) Hydrolysis of an acid amide functional group, followed by identification of the hydrolysis products is the basis of many identification tests for compounds whose

structures contain these functional groups.

(17) Hydrolysis is one of the main metabolic routes for compounds whose molecular structures contain carboxamide groups.

(18) Metabolic hydrolysis of carboxamides is not usually suitable for amides to be of use as prodrugs.

(19) Acid amides should be stored under dry conditions.

(20) Acid amide functional groups may be oxidised to a variety of products.

(21) Acid amide functional groups may be reduced to a variety of products.

(22) Reduction of carboxamides to volatile amines can be used as the basis of an assay procedure for suitable compounds.

(23) The imide functional group has the structure: –CO–NH–CO–.

(24) Imide functional groups are weak acids, forming salts with alkalis.

$$RCONHCOR \xrightarrow{\text{KOH}} RCON^{-}COR \; K^{+}$$

(25) Imide functional groups are easily hydrolysed, the product(s) of the reaction depending on the extent of the hydrolysis. The most likely products are either the carboxylic acids or a mixture of carboxylic acids, amides and ammonia or the corresponding amine if a N-substituted imine is hydrolysed.

25.12 QUESTIONS

(1) Draw structural formulae for each of the following compounds: (a) butanamide, (b) 2-ethanamidobutane, (c) 6-hexanelactam, (d) 3-ethylcarbamoylmethylbenzene, (e) 4-aminobenzenesulphonamide, (f) 2-(4-aminophenylsulphonamido)pyridine, (g) 1,8-naphthosultam, and (h) 2-ethyl-3-methylsuccinimide.

(2) N-(2-Methylphenyl)propanamide (compound A) is thought to have some biological activity. (a) Explain (i) why A is not basic but (ii) could act as a weak acid. (b) Comment on the expected water solubility of A. (c) List the general chemical properties that A would be expected to exhibit. (d) Predict the products of the reaction of A with (i) aqueous hydrochloric acid and (ii) lithium aluminium hydride.

(3) (a) Suggest, by means of equations, chemical reactions which could be made the basis of two identification tests for benzamide (compound B).
(b) Outline chemical reactions that may be made the basis of an assay method for B.
(c) Devise a limit test for benzoic acid in B.
(d) What precautions are necessary when storing chemicals containing carboxamide groups? Explain why these precautions are required.

(4) Explain each of the following observations.
(a) Phthalimide is a weak acid.
(b) Sulphonamides are often administered as their sodium salts.
(c) Aqueous solutions of penicillins are unstable.

(5) Suggest feasible reaction mechanisms for each of the following reactions:
(a) The reaction of 4-ethanamidophenol with aqueous sodium hydroxide.
(b) The reaction of propanamide with propanoic anhydride.
(c) The reaction of potassium phthalimide with 2-bromo-2-methylpropane.

26

Lipids

26.1 INTRODUCTION

Lipid is the traditional collective name given to a wide range of naturally occurring compounds that are isolated from plant and animal material by extraction with non-polar solvents such as chloroform, diethyl ether, propanone and petroleum ethers. The term *lipid* encompasses a wide variety of **different classes of compounds** which are related by the method of their isolation and not their structures. However, only certain of the types of compound extracted from plants and animals are traditionally classified as lipids.

Lipids may be subdivided into two general types: **the saponifiable lipids** that may be hydrolysed by sodium hydroxide and the **non-saponifiable lipids** that are not hydrolysed by sodium hydroxide (Fig. 26.1). The former group may be further subdivided into lipids whose structures do or do not contain a glycerol residue whilst the latter are classified according to their structural type. Lipids may also be classified as **simple, complex** and **derived** lipids. Simple lipids are the fatty acid esters of glycerol and long chain high molecular mass alcohols. Complex lipids are those which contain a fatty acid ester and another residue, for example lipoproteins and glycolipids. Derived lipids are substances that are obtained by the hydrolysis of lipids, for example, fatty acids, high molecular mass alcohols and steroids. However, neither of these classifications is entirely satisfactory.

Lipids are the basic building block of cell membranes (26.11) and are involved as hormones in the control of many biological processes.

26.2 FATTY ACIDS

Fatty acids are the most abundant of all the naturally occurring lipids. However, they are usually found in the form of the acid residue of the esters of glycerol, high molecular mass

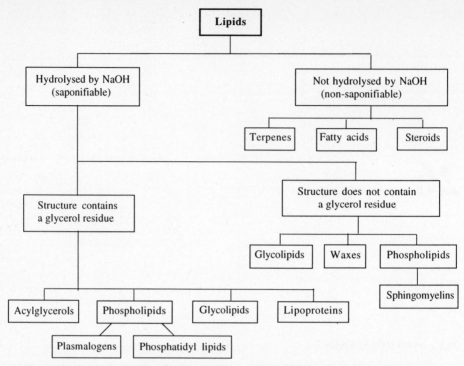

Fig. 26.1. A general classification of the main types of compound that are traditionally regarded as lipids. It should be noted that phospholipids and glycolipids are very diverse groups of compound which may or may not contain a glycerol residue.

alcohols, steroids and sphingosine. Hence their classification as derived lipids.

The fatty acids that are commonly found in lipids usually have straight chain structures that contain an even number of carbon atoms. Most of these even numbered fatty acids have between 14 and 22 carbon atoms, the most common being those with 16 or 18 (Fig. 26.2). Odd numbered fatty acids do occur but in very small quantities. They are usually found in the milk of ruminants.

The fatty acids found in lipids may be saturated (no C=C bonds) or unsaturated (contain C=C bonds) in nature. **Living cells usually contain more unsaturated than saturated fatty acids, animal cells have a relatively higher proportion of saturated fatty acids while plant cells have a higher proportion of unsaturated fatty acids.** The latter is believed to be a factor in the aetiology of coronary heart disease. Unsaturated fatty acids are classified as monoenoic, dienoic, trienoic, tetraenoic, etc. according to the number of C=C bonds in their structures. They can exist in both cis and trans forms but the cis isomers are the more common. A few of the fatty acids found in lipids will contain other functional groups. For example, ricinoleic acid has a hydroxyl group bonded to carbon 12. This acid is optically active, both enantiomers being known.

An important group of fatty acids that are sometimes classified as lipids are the prostaglandins. These compounds and the chemical and physical properties of fatty acids are discussed in Chapter 20.

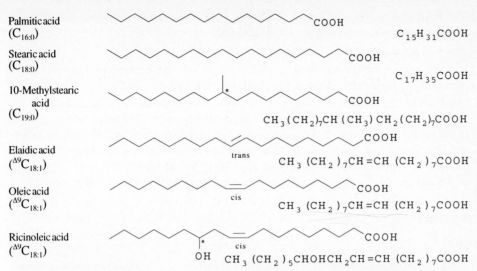

Palmitic acid
($C_{16:0}$)

$C_{15}H_{31}COOH$

Stearic acid
($C_{18:0}$)

$C_{17}H_{35}COOH$

10-Methylstearic
acid
($C_{19:0}$)

$CH_3(CH_2)_7CH(CH_3)CH_2(CH_2)_7COOH$

Elaidic acid
($^{\Delta 9}C_{18:1}$)

trans

$CH_3(CH_2)_7CH=CH(CH_2)_7COOH$

Oleic acid
($^{\Delta 9}C_{18:1}$)

cis

$CH_3(CH_2)_7CH=CH(CH_2)_7COOH$

Ricinoleic acid
($^{\Delta 9}C_{18:1}$)

OH   cis

$CH_3(CH_2)_5CHOHCH_2CH=CH(CH_2)_7COOH$

Fig. 26.2. Examples of the fatty acids commonly found in lipids. Full and abbreviated structures are shown. Chiral centres are marked by *. It should be noted that 10-methylstearic acid is unusual in that it contains an odd number of carbon atoms. The numbers in the brackets show the ratio of the number of carbon atoms to C=C bonds while the number(s) after the Δ indicates the position of the C=C bond.

26.3   ACYLGLYCEROLS

Acylglycerols, or glycerides as they were originally known, are the mono, di and tri esters of the trihydric alcohol glycerol and fatty acids (Fig. 26.3). The most commonly occurring acylglycerols are the triacylglycerols. These compounds are the main constituents of animal and plant fats and oils. Fats are solid whilst oils are liquids at room temperature and pressure. Triacylglycerols form the main energy store in all animals. However, large quantities of the mono and diacylglycerols are produced during the course of the digestion of triacylglycerols by the action of lipase in the intestine.

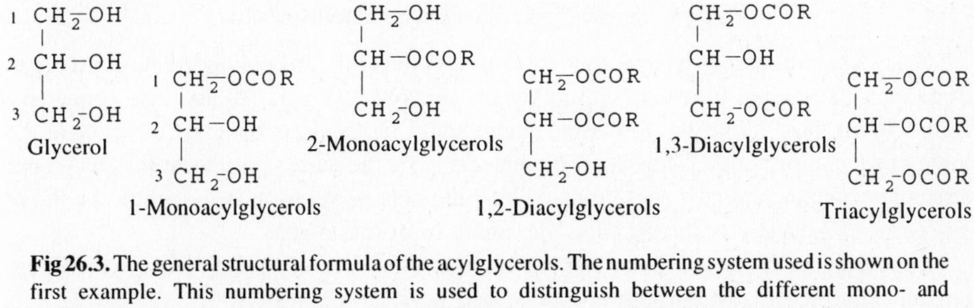

1 CH_2-OH

2 $CH-OH$

3 CH_2-OH
   Glycerol

1 CH_2-OCOR

2 $CH-OH$

3 CH_2-OH
   1-Monoacylglycerols

CH_2-OH

$CH-OCOR$

CH_2-OH
2-Monoacylglycerols

CH_2-OCOR

$CH-OCOR$

CH_2-OH
1,2-Diacylglycerols

CH_2-OCOR

$CH-OH$

CH_2-OCOR
1,3-Diacylglycerols

CH_2-OCOR

$CH-OCOR$

CH_2-OCOR
Triacylglycerols

Fig 26.3. The general structural formula of the acylglycerols. The numbering system used is shown on the first example. This numbering system is used to distinguish between the different mono- and diacylglycerol isomers. The acyl groups (RCO) may or may not have the same structures.

Acylglycerols are the major components of naturally occurring fats and oils. Fats and oils are mixtures of acylglycerols in which are dissolved other compounds including other types of lipid. These mixtures are often complex and it is difficult to identify their component acylglycerols and subsequently the fat or oil. However, a particular fat will tend to contain a small number of 'dominant' acyl residues whose relative concentration and composition can be used to identify the fat.

26.3.1   Nomenclature

A wide variety of nomenclature systems are in use for naming acylglycerols. The commonly used systems are either based on that shown in Figs 26.3 and 26.4. In the latter case if all the acyl radicals are identical the name of the acid radical is prefixed by **tri-** and has the suffix **-in**. If the acyl radicals are different, the name of the acid radical consists of the names of the acids in the order they occur in the structure. In this case the ending **-ic** is replaced by **-o** except for the name of the last acid which is given the suffix **-ate**. In the case of mono and diacylglycerols the appropriate locant is used to indicate the position of the acyl group. The appropriate prefix is used when acyl groups are identical.

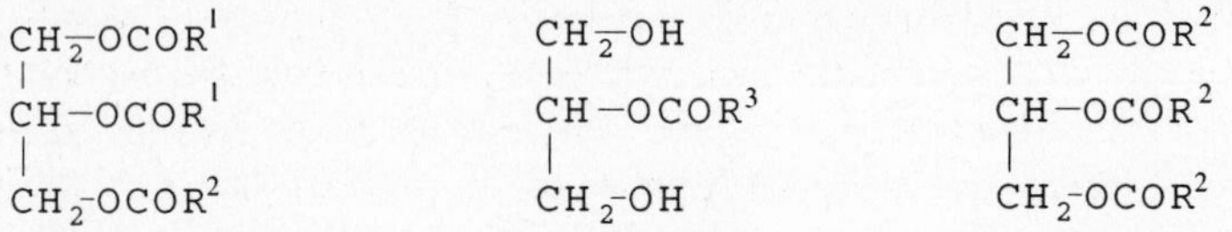

$$
\begin{array}{lll}
CH_2\text{-}OCOR^1 & CH_2\text{-}OH & CH_2\text{-}OCOR^2 \\
| & | & | \\
CH\text{-}OCOR^1 & CH\text{-}OCOR^3 & CH\text{-}OCOR^2 \\
| & | & | \\
CH_2\text{-}OCOR^2 & CH_2\text{-}OH & CH_2\text{-}OCOR^2 \\
\text{Glyceryl dipalmitostearate} & \text{Glyceryl-2-oleate} & \text{Glyceryl tristearin}
\end{array}
$$

Key: R^1 = Palmitic acid, R^2 = Stearic acid, R^3 = Oleic acid.

Fig. 26.4. Examples of the trivial and systematic nomenclature of acylglycerols.

26.3.2   Structure and physical properties

Acylglycerols may be liquids or solids at room temperature. However, the structures of the acyl groups of liquid glycerides (oils) exhibit a higher degree of unsaturation than those of solid glycerides (fats). Free rotation about the C–C bonds of the glycerol residue can occur in the liquid state but in the solid state the triacylglycerols normally assume the so-called tuning fork conformation (Fig. 26.5).

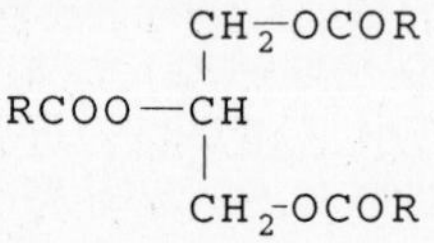

$$
\begin{array}{l}
CH_2\text{-}OCOR \\
| \\
RCOO\text{-}CH \\
| \\
CH_2\text{-}OCOR
\end{array}
$$

Fig. 26.5. The tuning fork conformation of triacylglycerols.

X-ray crystallography shows that solid triacylglycerols are polymorphic, existing in three major allotropic forms which are known as alpha (α), beta (β) and beta prime (β') forms. In all these allotropic forms the acylglycerol molecule is believed to occur in the tuning fork conformation. The beta form appears to be the most stable form for fats whose structures contain saturated acyl groups whilst the beta prime form is thought to be the *in situ* form of naturally occurring fats. The alpha form is unstable.

Triacylglycerols are non-polar and insoluble in water but mono and diacylglycerols are amphipathic, that is exhibit both non-polar and polar properties.

26.3.3   Chemical properties

The general chemical properties exhibited by an acylglycerol will depend on the nature of the functional groups in the molecule. In the first instance one would expect reactions that are characteristic of those functional groups. For example, all acylglycerols will exhibit the chemical properties of esters such as hydrolysis and transesterification (21.5) whilst unsaturated glycerides will also undergo reactions such as oxidation, reduction and addition that are characteristic of alkenes (Chapter 12). Acylglycerols whose structures contain other functional groups would be expected to exhibit the chemical properties that are associated with those functional groups. However, it must be remembered that the presence of other functional groups and the configuration of the carbon-hydrogen skeleton of a molecule may either prevent or modify these reactions.

26.3.4 Hydrolysis

All acylglycerols can be hydrolysed by water, the reaction can be catalysed by acids, bases and enzymes. The products of the hydrolysis will depend on the reagents and conditions used. For example, hydrolysis with aqueous sodium hydroxide will yield glycerol and the sodium salts of the appropriate fatty acids. This reaction is known as **saponification** and is the basis of the determination of the **saponification value** which may be used in the characterisation of lipids (26.10). Saponification can also be used to classify a lipid as being glycerol based, since the presence of glycerol in the reaction mixture can be detected using potassium hydrogen sulphate or chromatographic methods. This reagent converts the glycerol to the highly toxic compound acrolein, which is identified by its characteristic smell. Saponification also forms the basis of the production of soaps from fats.

$$
\begin{array}{llll}
CH_2\text{-}OCOR^1 & & CH_2OH & CHO \\
| & ^1RCOONa & | & | \\
CH\text{-}OCOR^2 \xrightarrow{NaOH/H_2O} & ^2RCOONa + CHOH \xrightarrow{KHSO_4} & CH \\
| & & | & || \\
CH_2\text{-}OCOR^3 & ^3RCOONa & CH_2OH & CH_2 \\
\text{A triacylglycerol} & \text{Fatty acids} & \text{Glycerol} & \text{Acrolein}
\end{array}
$$

Hydrolysis, in the presence of the enzyme pancreatic lipase, results initially in the preferential removal of the outer acyl groups to yield the 2-monoacylglycerol. The reaction will slowly continue to completion if it is not stopped at this point. This reaction has been used to identify triacylglycerol species. The triacylglycerol is preferentially hydrolysed to the monoacylglycerol using pancreatic lipase and the fatty acids released from position 1 and 3 are identified in the form of their methyl esters by GL chlomatography. The monoacylglycerols are isolated, hydrolysed and the fatty acids from position 2 also identified by GL chromatography of their methyl esters. Although the initial hydrolysis with pancreatic lipase is not taken to completion, the deductions from this analytical procedure are reasonably accurate.

$$
\begin{array}{llll}
CH_2\text{-}OCOR^1 & ^1RCOOH + CH_2\text{-}OH & CH_2OH \\
| & & | & | \\
CH\text{-}OCOR^2 \longrightarrow & CH\text{-}OCOR^2 \longrightarrow & CHOH + R^2COOH \\
| & & | & | \\
CH_2\text{-}OCOR^3 & ^3RCOOH + CH_2\text{-}OH & CH_2OH \\
\text{Triacylglycerol} & \text{Fatty acids} \quad \text{Monoacylglycerol} & \text{Glycerol} \quad \text{Fatty acid}
\end{array}
$$

26.3.5 Oxidation

Oxidising agents readily oxidise the C=C bonds of unsaturated acylglycerols to a variety of products. For example, ozone will form ozonides which can be decomposed by oxidative or reductive fission (12.3.3). The identification of the compounds produced by these reactions can be used to identify the positions of the C=C double bonds in the acylglycerol molecule.

Oxidation of the C=C bonds of unsaturated acylglycerols via peroxidation with atmospheric oxygen yields carboxylic acids, aldehydes and ketones. Aldehyde formation may give rise to either a rancid or unpleasant flavour. The reaction has a free radical mechanism and so an uncontrolled reaction would do considerable damage to biological tissues if it were not for the presence of naturally occurring antioxidants such as vitamin E. Oxidation of the lipids in food is undesirable and so it is controlled by the addition of antioxidants such as BHT (3,5-di-t-butyl-4-hydroxytoluene).

Metabolic oxidation of fatty acids can occur at the α, β and ω carbon atoms of the fatty acid. β-Oxidation is the most common type of biological oxidation. In mitochondria, the β-oxidation cycle results in the formation of a β-keto ester whose acid residue contains two carbons less than the original fatty acid and acetyl coenzyme A. This cycle can be repeated until the carbon chain of the fatty acid has been completely oxidised to acetyl coenzyme A. Fatty acids with odd numbers of carbon atoms yield both acetyl coenzyme A and propionoyl coenzyme A as the products of β-oxidation in mitochondria. β-Oxidation is responsible for supplying about 50% of the energy requirements of many cells.

$$
\text{Fat} \begin{cases} \overset{\beta}{\text{RCH}_2\text{CH}_2\text{CH}_2\text{CH}_2\text{COOH}} \xrightarrow{\beta\text{-Oxidation cycle}} \text{RCOCH}_2\text{COSCoA} + \text{CH}_3\text{COSCoA} \\ \underset{\alpha}{\text{RCH}_2\text{CH}(\text{CH}_3)\text{CH}_2\text{COOH}} \xrightarrow{\alpha\text{-Oxidation cycle}} \text{RCH}_2(\text{CH}_3)\text{COOH} \end{cases}
$$

α-Oxidation is important when the structure of the fatty acid prevents β-oxidation. It results in the loss of the α carbon atom from the fatty acid residue and usually the formation of a product that can undergo β-oxidation. α-Oxidation is an important step in the breakdown of branched chain fatty acids in the kidney and liver. Its failure can lead to Refsum's disease.

26.3.6   Reduction

The C=C bonds of unsaturated acylglycerols are easily reduced by hydrogen in the presence of a metal catalyst such as palladium or nickel at room temperature (12.4.3). The reaction is known as **fat hardening** because the loss of C=C bonds increases the density of the fat. Fat hardening is used to convert liquid vegetable and animal oils into solid fats in the production of margarine and cooking fats. The reduction is allowed to proceed until the product reaches the desired physical state.

26.3.7   Transesterification

Transesterification (21.5) of acylglycerols occurs both under laboratory and biological conditions. In the laboratory the reaction yields a mixture containing all the possible products. The reaction is more specific in biological systems.

26.3.8   Addition

Electrophiles will add to the C=C bonds of unsaturated triacylglycerols (12.4.2). Electrophilic addition using iodine monochloride is the basis of the method for determining the **iodine value** used in the characterisation of lipids (26.10).

26.4   PHOSPHOLIPIDS

Phospholipids may be subdivided into: (1) derivatives of phosphatidic acid and its isomers, (2) other phospholipids including the plasmalogens, and (3) the sphingomyelins.

The phosphatidyl compounds, ether phospholipids and the plasmalogens are derivatives of glycerol but the sphingomyelins are derivatives of the amino alcohol, sphingosine.

The phospholipids are important constituents of cell membranes. Plasmalogens occur widely in the membranes of nerve and muscle cells. Sphingomyelins are widely distributed in plants and animals but are particularly abundant in the brain and nerve cells.

26.4.1 Phosphatidyl compounds

The structures of these lipids are based on that of α phosphatidic acid (Fig. 26.6). Phosphatidyl compounds are subclassified according to the nature of the R^3 substituent of the phosphate group (Table 26.1).

Table 26.1. Examples of the subclassification of phosphatidyl lipids.

Sub class	R^1 Substituent	General structure and comments
Phosphatidyl glycerols	$HOCH_2CH(OH)CH_2OH$ Glycerol	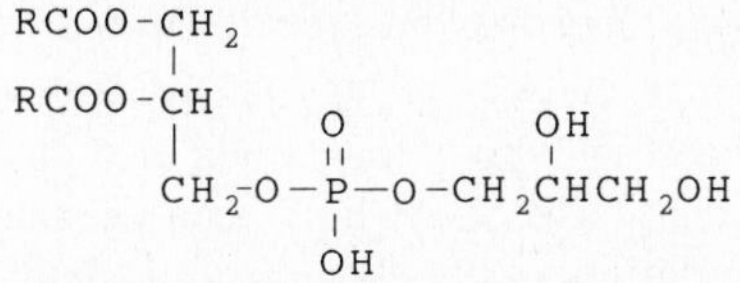 $RCOO-CH_2$ $RCOO-CH$ $CH_2-O-P(=O)(OH)-O-CH_2CHCH_2OH$ (with OH) Phospatidyl glycerol, main plant phospatidyl lipid
Phosphatidyl choline (lecithins)	$HOCH_2CH_2\overset{+}{N}(CH_3)_3$ Choline	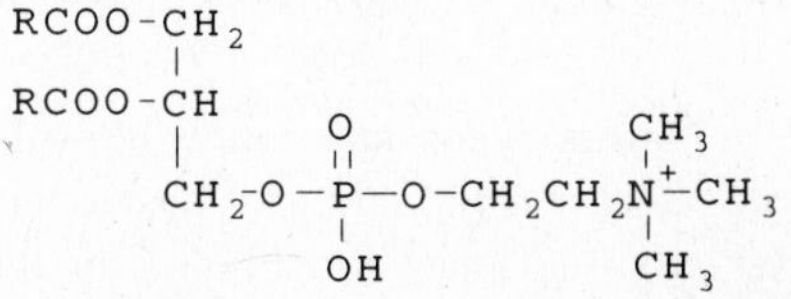 $RCOO-CH_2$ $RCOO-CH$ $CH_2-O-P(=O)(OH)-O-CH_2CH_2\overset{+}{N}(CH_3)_3$ α-Lecithins, the main animal phospolipid
Phosphatidyl ethanolamine (cephalins)	$HOCH_2CH_2NH_2$ Ethanolamine	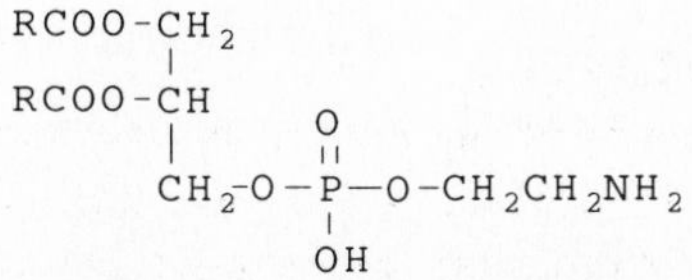 $RCOO-CH_2$ $RCOO-CH$ $CH_2-O-P(=O)(OH)-O-CH_2CH_2NH_2$ α-Cephalins, a major animal phospholipid
Phosphatidyl serine	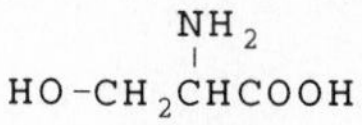 $HO-CH_2CHCOOH$ (with NH_2) Serine	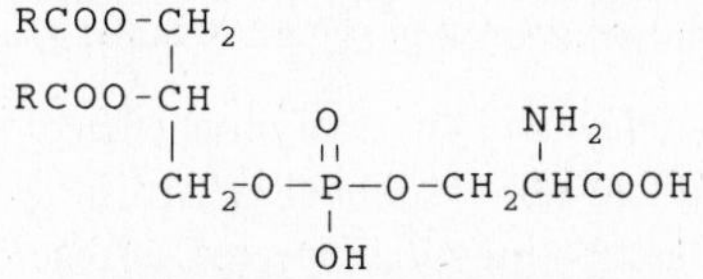 $RCOO-CH_2$ $RCOO-CH$ $CH_2-O-P(=O)(OH)-O-CH_2CHCOOH$ (with NH_2) Small amounts occur in brain and other tissue
Phosphatidyl inositol	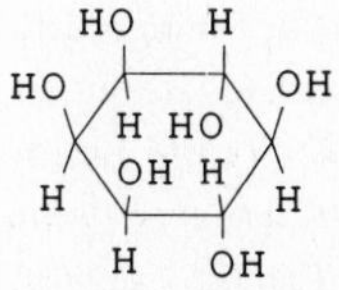 myo-Inositol	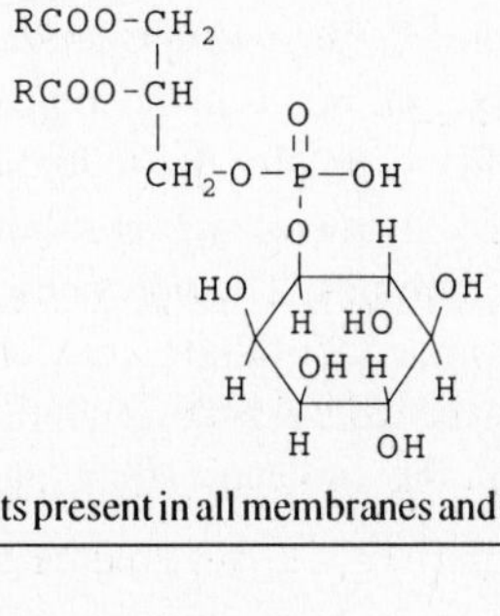 $RCOO-CH_2$ $RCOO-CH$ $CH_2-O-P(=O)-OH$ Small amounts present in all membranes and the brain

Carbon 2 of the glycerol residue of naturally occurring phosphatidyl compounds is a chiral centre and usually has an R configuration.

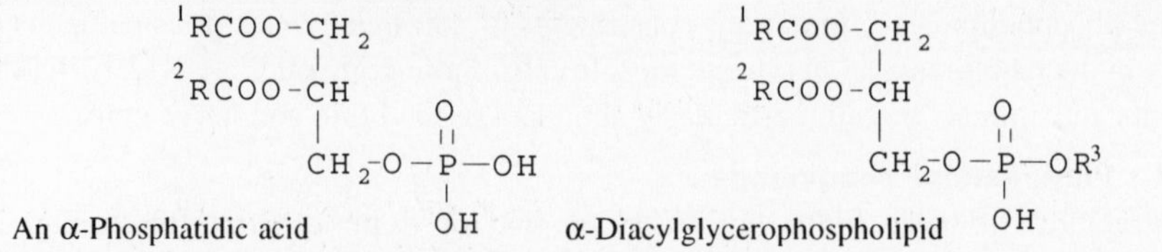

$$
^1RCOO-CH_2 \\
^2RCOO-CH \\
CH_2-O-\overset{\overset{O}{||}}{P}-OH \\
OH
$$

An α-Phosphatidic acid

$$
^1RCOO-CH_2 \\
^2RCOO-CH \\
CH_2-O-\overset{\overset{O}{||}}{P}-OR^3 \\
OH
$$

α-Diacylglycerophospholipid

Fig. 26.6. The structures of α-phosphatidic acids and the phosphatidyl lipids. ^{1}R is usually a saturated while ^{2}R is usually an unsaturated fatty acid residue. R^3 is an fatty acid residue which is not the same as the other R groups.

The chemical and physical properties of a phosphatidyl compound depend on the nature of the functional groups in the molecule. In the first instance each of these functional groups may be expected to react in its own characteristic manner. For example, the presence of ester groups means that all phosphatidyl lipids may be hydrolysed to glycerol and the appropriate acids or their salts depending on the nature of the catalyst, unsaturated phosphatidyl lipids can be oxidised, reduced and will undergo electrophilic addition and phospatidylserines will yield a purple complex with ninhydrin (28.5.4). However, the presence of other functional groups and the nature of the carbon-hydrogen skeleton of the molecule may either prevent or modify these reactions.

26.4.2   Plasmalogens and ether phospholipids

Plasmalogens occur in brain and muscle tissue. They have molecular structures in which the ester group attached to carbon 1 of the glycerol residue of α-phosphatidic acid is replaced by an unsaturated ether (alkenyl) residue (Fig. 26.7).

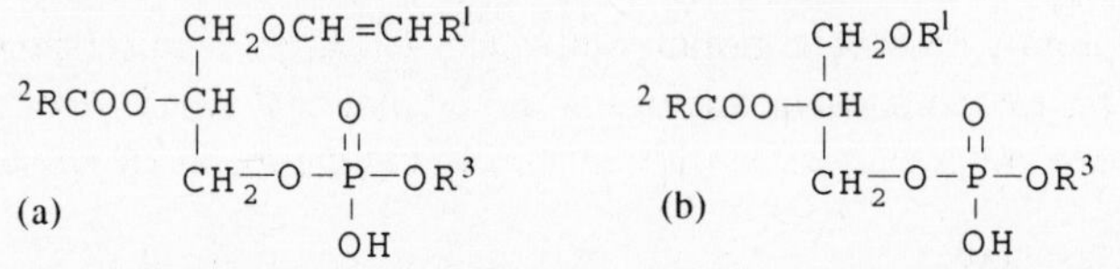

$$
CH_2OCH=CHR^1 \\
^2RCOO-CH \\
CH_2-O-\overset{\overset{O}{||}}{P}-OR^3 \\
OH
$$

(a)

$$
CH_2OR^1 \\
^2RCOO-CH \\
CH_2-O-\overset{\overset{O}{||}}{P}-OR^3 \\
OH
$$

(b)

Fig. 26.7. The general structures of (a) plasmalogens and (b) ether phospholipids. The R^1 groups are the same as those found in the phosphatidyl compounds. ^{2}R and R^3 may or may not be the same.

The chemical and physical properties of plasmalogens and ether phospholipids depend on the nature of the functional groups in the molecule. In the first instance each of these functional groups may be expected to react in the appropriate manner for this functional group. For example, the ester groups may be hydrolysed but in this case hydrolysis will not yield glycerol because of the unreactive ether group attached to carbon 1 of the glycerol residue. The C=C bonds in the molecule may be oxidised, reduced and will undergo addition reactions. For example, the oxidation of the C=C double bond of the ether group of the plasmalogens by mercury II chloride is the basis of a test to distinguish plasmalogens from other phospholipids. The plasmalogen is oxidised to the corresponding aldehyde which is detected using either Schiff's reagent or *o*-dianisidine. However, it should be remembered that the presence of other functional groups and the nature of the carbon-hydrogen skeleton of the molecule may either prevent or modify the expected reactions of a functional group.

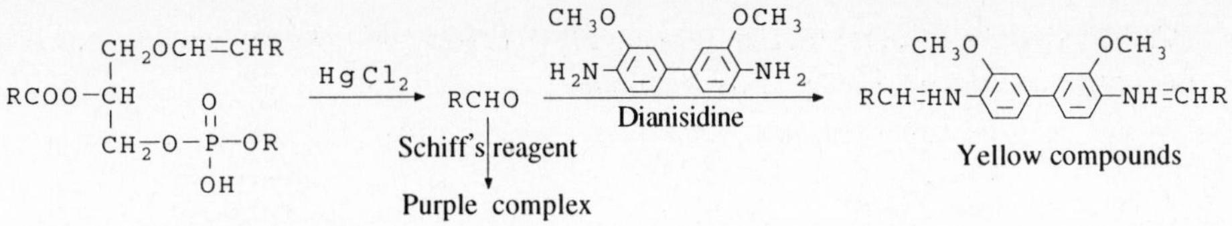

CH$_2$OCH=CHR
|
RCOO−CH   O $\xrightarrow{\text{HgCl}_2}$ RCHO
| || Schiff's reagent
CH$_2$−O−P−OR
|
OH

$$\xrightarrow[\text{Dianisidine}]{\text{H}_2\text{N}-\text{(CH}_3\text{O, OCH}_3\text{)}-\text{NH}_2}$$

RCH=HN−⟨⟩−⟨⟩−NH=CHR

Yellow compounds

Purple complex

26.4.3 Sphingomyelins

Sphingomyelins are the amide derivatives of the choline substituted phosphate esters of the unsaturated amino alcohol, sphingosine (Fig. 26.8). These lipids are widely distributed in plants and animals but are particularly abundant in the brains of animals.

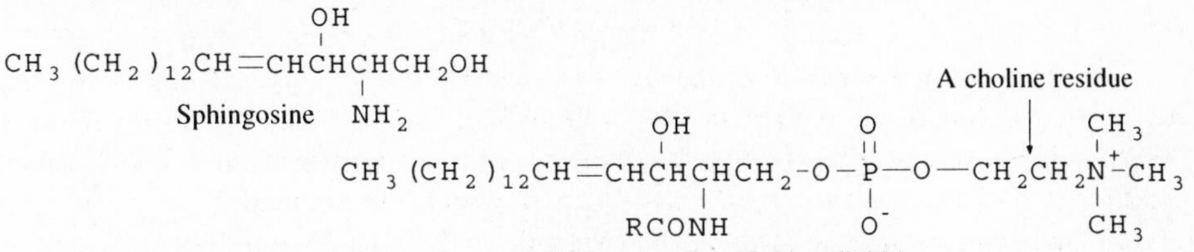

OH
|
CH$_3$(CH$_2$)$_{12}$CH=CHCHCHCH$_2$OH
|
Sphingosine   NH$_2$

A choline residue

CH$_3$(CH$_2$)$_{12}$CH=CHCHCHCH$_2$−O−P−O−CH$_2$CH$_2$N$^+$CH$_3$
(with OH on one carbon, RCONH on another, O$^-$ on phosphate, CH$_3$ groups on N)

Sphingomyelins (Sphingolipids)

Fig. 26.8. The general structural formula of the sphingomyelins. R represents an alkyl residue.

The chemical and physical properties of sphingomyelins depend on the nature of their constituent functional groups. This observation may be used to predict the possible course of the chemical reactions of sphingomyelins. For example, complete hydrolysis would result in the fission of all the amide and ester groups in the molecule with the formation of sphingosine, phosphoric acid, choline and a fatty acid. The exact form of these products will depend on the nature of the catalyst used for the hydrolysis. For example, sodium hydroxide will produce the appropriate sodium salts of the acids formed whilst hydrochloric acid would form the hydrochloride salts of the amino compounds. However, it must be remembered that any predictions based on the characteristics of these functional groups may be modified by either the carbon-hydrogen skeleton or other functional groups in the molecule.

26.5 GLYCOLIPIDS

Most glycolipids are sphingolipids (Fig. 26.8) in which the phosphate residue has been replaced by a sugar moiety with a β-D- configuration (Fig. 26.9). They may also be regarded as the glycosides with a sphingosine derivative acting as the aglycone. Glycolipids may be subclassified as cerebrosides, sulphatides and gangliosides depending on the nature of the sugar residue. Cerebrosides have one glucose or galactose residue (neutral glycolipids), sulphatides, a sugar with a sulphate group and gangliosides (acidic glycolipids) which have a polysaccharide residue that contains a sialic acid residue. The individual members of a class are distinguished by the nature of the fatty acid residue R.

The chemical and physical properties of individual glycolipids will depend on the types of functional group and the carbon-hydrogen skeleton of the molecule. These properties may be predicted by drawing comparisons with the characteristic properties of these functional groups and carbon-hydrogen skeletons. However, it should be remembered that when making predictions of this sort the other functional groups in the molecule can interfere and these predictions may not agree with the practical observations.

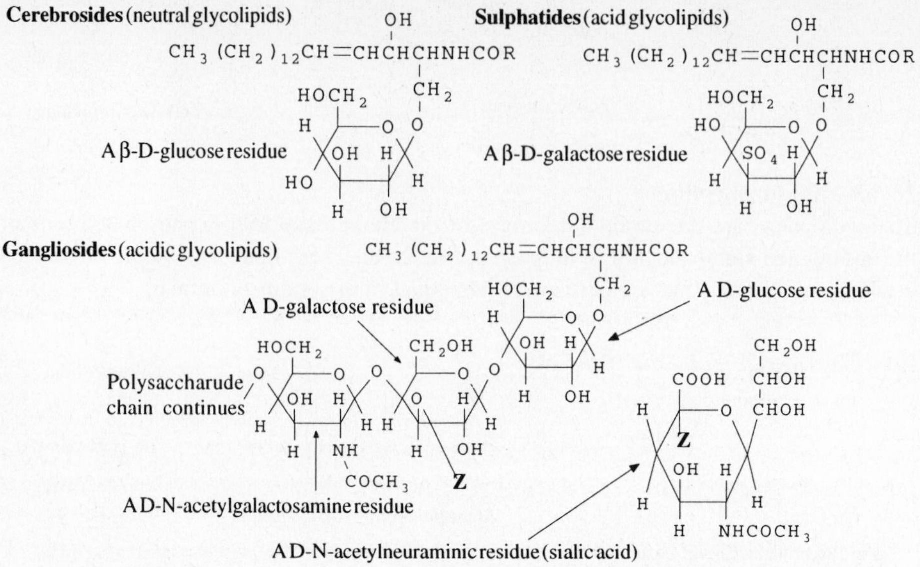

Cerebrosides (neutral glycolipids)   OH   **Sulphatides** (acid glycolipids)   OH

$CH_3 (CH_2)_{12}CH=CHCHCHNHCOR$ $CH_3 (CH_2)_{12}CH=CHCHCHNHCOR$

A β-D-glucose residue

A β-D-galactose residue

Gangliosides (acidic glycolipids) $CH_3 (CH_2)_{12}CH=CHCHCHNHCOR$

A D-galactose residue   A D-glucose residue

Polysaccharide chain continues

A D-N-acetylgalactosamine residue   Z

A D-N-acetylneuraminic residue (sialic acid)

Fig. 26.9. The general structural formulae of the subclasses of glycolipids. In the ganglioside structure the polysaccharide chain branches at the point marked **Z** via an ether link. A typical branch residue is shown at the bottom right of the diagram.

26.6   WAXES

Waxes are solids that are easily moulded when warm. They are insoluble in water. Many biological waxes are the esters of carboxylic acids and high molecular mass alcohols, steroids (26.9) and triterpenes (26.8). The most common alcohol residues found in these compounds are usually derived from cetyl, ceryl and myricyl alcohols and sterols. For example, cholesterol esters are found in blood. Their chemical and physical properties will depend on the functional groups in the molecule and the nature of its carbon-hydrogen skeleton. The nature of these properties may be predicted in the same way as those of all multifunctional group molecules. However, it must be remembered that in practice the results of these predictions may be incorrect because of the unpredictable behaviour of the carbon-hydrogen skeleton and other functional groups in the molecule.

26.7   LIPOPROTEINS

Lipoproteins are believed to have a micellular (9.9) shell-like structure of phospholipids, proteins and cholesterol surrounding a non-polar core of triacylglycerols and cholesterol esters (Fig. 26.10). The shell consists of a monolayer of phospholipid, protein and cholesterol about 0.2 nm thick. The density of the particles increases with increasing protein content and so plasma lipoproteins are referred to as chylomicrons, chylomicron remants, very low density (VLDL), intermediate density (IDL), low density (LDL) and high density (HDL) lipoproteins.

Lipoproteins have a variety of physiological functions. For example, low density lipoproteins transport cholesterol from the liver to other tissues and are implicated as a risk factor in coronary heart disease.

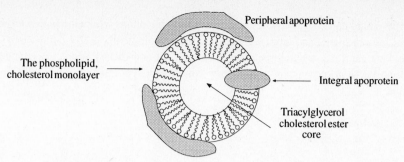

Fig. 26.10. A schematic representation of the micellular structure of lipoproteins.

26.8   TERPENES

Terpenes are a diverse group of lipids that have molecular formulae with five or multiples of five carbon atoms (Fig. 26.11). They are mainly hydrocarbons but can contain oxygen functional groups such as ethers, alcohols, esters and ketones. Terpenes are the main constituents of the so-called essential oils that are extracted from animals and plants. These fragrant oils have been used for thousands of years as medicines, spices and perfumes.

Myrcene   Menthol   Camphor   Eucalyptol   α-Pinene

Fig. 26.11. The structures of some monoterpenes.

26.8.1   Classification of a compound as a terpene

Compounds are classified as terpenes if their structures contain the requisite number of carbon atoms and obey the isoprene rule although there are some exceptions. The rule in effect states that terpene structures can be artificially divided into isoprene units placed *nose* to *tail* (Fig. 26.12).   To decide whether a structure conforms with the rule it is necessary to distort the isoprene structure and ignore all C=C bonds. It is also useful to note that the *nose* of an isoprene unit will probably correspond to a methyl or gem dimethyl group in the structure of the suspected terpene. It is emphasised that isoprene is not involved in the biogenesis of terpenes but the rule is simply the traditional method used to classify compounds as terpenes.

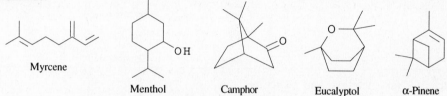

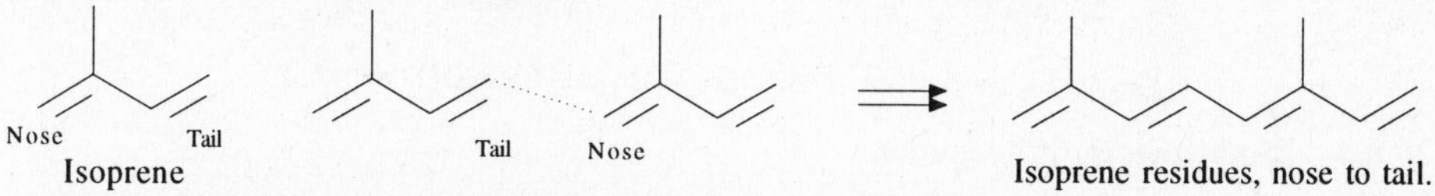

Nose   Tail   Tail   Nose
  Isoprene   Isoprene residues, nose to tail.

Fig. 26.12. Classification of compounds as terpenes using the isoprene rule.

26.8.2   Classification of terpenes

Terpenes are classified according to the number of carbon atoms in their structures (Table 26.2). The illogical nature of this classification is due to the way in which terpenes were

discovered. Mono and sesquiterpenes are mainly found in plants whilst the higher terpenes
are found in both plants and animals. They are often biologically active compounds. For
example, beta-carotene, a tetraterpene is an antioxidant and dietary source of vitamin A_1
while squalene, a triterpene, is a precursor for the biosynthesis of steroids in living
organisms.

β-Carotene

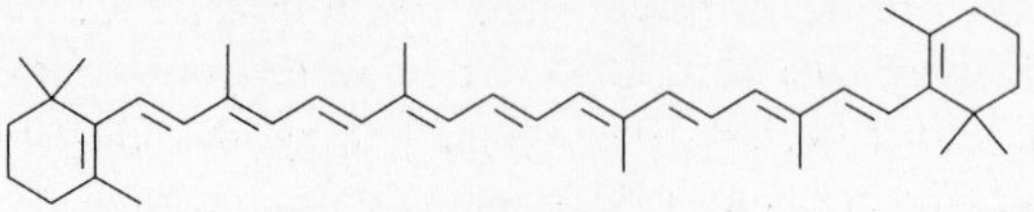

Squalene

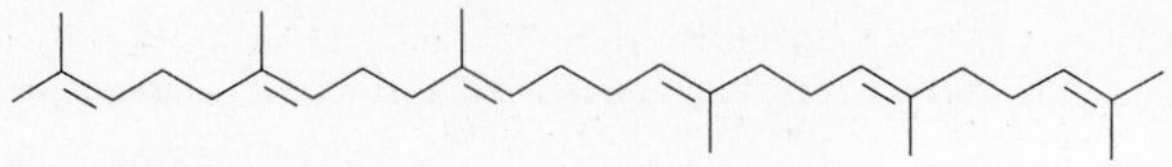

Table 26.2. The classification system for terpenes.

Number of carbon atoms	Number of isoprene units	Classification
10	Two	Monoterpene
15	Three	Sesquiterpene
20	Four	Diterpene
25	Five	Sesterterpene
30	Six	Triterpene
40	Eight	Tetraterpene

26.8.3   Nomenclature and structure

Terpenes are usually known by their trivial names. However, compounds whose structures
are based on mono and bicyclic ring systems may be named using the appropriate systematic
nomenclature.

The ring structures of terpenes will exhibit the appropriate conformations and configurations. For example, the preferred conformation of menthol is a chair form with the ring
substituents occupying equatorial positions. Similarly the preferred conformation of
neomenthol, the diastereoisomer of menthol, is also the chair form with the hydrocarbon
substituents occupying equatorial positions and the hydroxyl group an axial position
(Fig. 26.13).

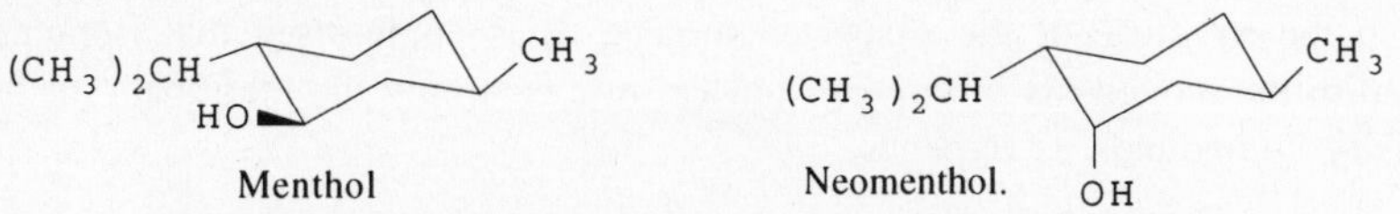

Menthol   Neomenthol.

Fig. 26.13. The preferred conformations of menthol and neomenthol.

26.8.3   Biosynthesis of terpenes

Isopentenyl pyrophosphate and dimethylallyl pyrophosphate are the precursors of terpenes in biological systems. The biosynthesis of many monoterpenes appears to involve the
head to tail addition of dimethylallyl pyrophosphate to isopentenyl pyrophosphate to form
geranyl pyrophosphate. This compound is an important intermediate in the biosynthesis
of many terpenes (Fig. 26.14). The relationship between geraniol pyrophosphate and

Isopentenyl pyrophosphate   Dimethylallyl pyrophosphate

these terpenes can be rationalised using the principles of reaction mechanisms plus appropriate oxidations and reductions. This rationalisation often involves carbocations that either react with a nucleophile, eliminate to an alkene or rearrange to a more stable carbocation which proceeds to undergo either an elimination or react with a nucleophile. Experimental evidence suggests that the rationalisation given in Fig. 26.14 is correct.

26.8.4   Physical and chemical properties

The diverse nature of the terpenes makes it difficult to generalise about their physical and chemical properties. In general their water solubility is poor but their solubility in other lipids

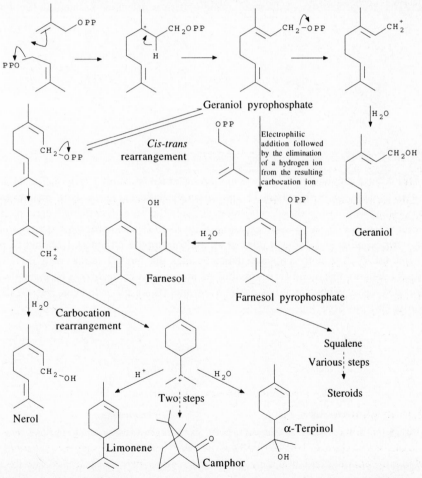

Fig. 26.14. A chemical rationalisation of the biosynthesis of some terpenes. P represents a phosphate residue.

is good. The chemical properties of a terpene will be largely similar to those of alkanes and alkenes (12.2.3 and 12.3.3). However, terpenes whose structures contain other functional groups will also exhibit the chemical properties characteristic of those functional groups. For example, the secondary alcohol group found in some terpenes could act as a nucleophile in nucleophilic substitutions, displacements and additions (15.7). It could also be oxidised and undergo elimination to an alkene (15.8 and 15.9.1).

26.9   STEROIDS

Steroids are a class of compound based on the multi-ring structure shown in Fig. 26.15. The individual rings in a system are referred to by the letters A, B, C and D. They may be saturated, unsaturated, and in the case of the A ring of oestrogens, aromatic in nature. However, the same number system is used for all steroids regardless of the nature of the rings. The stereochemistry of the steroid ring system is discussed in detail in section 4.5.

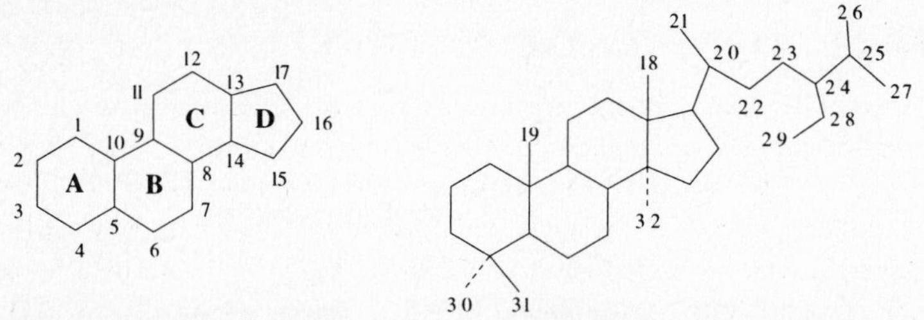

Fig. 26.15. The fundamental steroid ring system and the numbering system used with substituents.

Steroids are usually known by a semi-systematic trivial naming system. Trivial names are used as the basis of the name. The positions are indicated in the usual manner by the appropriate locant. Locants are allocated to both the ring system and certain of the substituents according to the scheme shown in Fig. 26.15. This numbering system is retained even when a compound does not possess all the appropriate carbon(s). It should be noted that the use of *nor* with a locant indicates that the methyl group with the carbon corresponding to that number is missing in the named compound. For example, 5α-androstane has a methyl group at position 18, but 18-nor-5β-androstane has no methyl group in this position (Fig. 26.16).

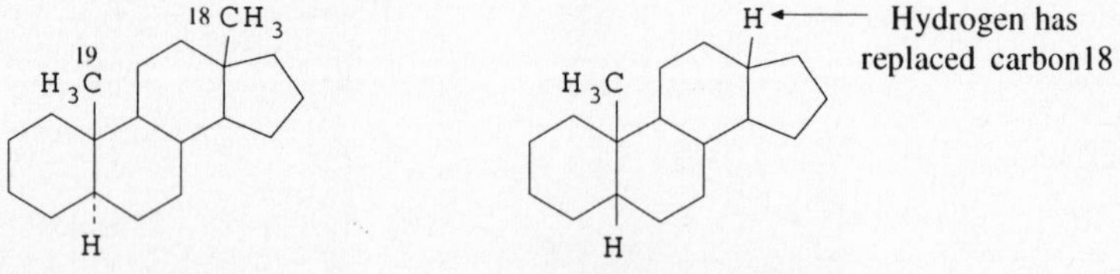

5-α-Androstane   18-Nor-5-β-androstane

Fig. 26.16. The use of *nor* in the names of steroids. The hydrogens at the other ring junctions have been omitted in the interests of clarity.

Many steroids are biologically active (Fig. 26.17). Their chemical and physical properties will depend on the nature of the functional groups in the molecule. For example, steroids are not usually very soluble in water because their structures do not contain sufficient water solubilising groups. Steroids whose structures contain secondary alcohol groups will exhibit the reactions of secondary alcohols (Chapter 15) whilst those whose structures contain ketone groups will exhibit the reactions of ketones. However, these reactions may not give the expected product because many steroid reactions are either accompanied by the rearrangement of the ring system or the migration of alkyl substituents from one position to another.

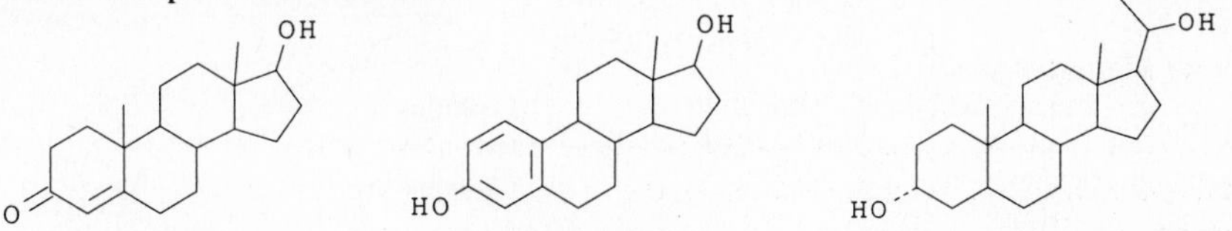

Testosterone, the main male sex hormone

Oestradiol (estradiol), a female sex hormone

Pregnanediol, found in the urine of pregnant women

Fig. 26.17. Examples of biologically important steroids.

26.10   THE CHARACTERISATION OF LIPIDS

The characterisation of lipids is complicated by the fact that most naturally occurring lipids are complex mixtures of compounds. However, plant and animal fats and oils consist mainly of acylglycerols. The composition of these fats and oils varies depending on the source of the material. However, it does appear that the composition of the same type of lipid samples from a particular type of source, say individual human beings, will usually contain the same fatty acids within a definite range of concentrations (Table 26.3). This fact is used as the basis of the specifications used for the naturally occurring fats and oils that are used in the pharmaceutical and food industries.

Table 26.3. The approximate fatty acid composition of some naturally occurring fats and oils. The relative sizes of the acids are represented by the number of carbon atoms in their structures while x indicates that there is less than 1% of the acid present in the sample.

Source	Saturated acid residues					Unsaturated acid residues			
	C12	C14	C16	C18	>C20	C14	C16	C18	>C20
Beef fat	x	3%	29%	20%	x	x	4%	42%	x
Chicken fat		2%	21%	5%			7%	64%	x
Lard	x	2%	26%	13%			5%	52%	
Palm oil		2%	42%	5%			x	51%	
Ground nut oil			16%	8%			1%	75%	
Olive oil			14%	3%			2%	80%	x

Lipids are classified by both chemical and physical methods. Chemical methods are based on a sequential series of chemical tests. Each test leads to further tests until a classification is made. A typical sequence is shown in Fig. 26.18. The chemistry of some of these tests is known but in those cases where the chemistry is not known the reason for their use is that they work and are specific. When testing tissue samples, delipidised samples should be used as blanks (controls).

The main physical methods used to characterise lipids are: thin layer chromatography to separate and identify individual lipids, enzymic hydrolysis followed by gas chromatography to identify the fatty acid residues in lipid material, spectroscopy to identify simple lipids and the nonfatty acid residues in complex lipids. High pressure liquid chromatography (HPLC) is also used to identify fatty acid residues and other lipids.

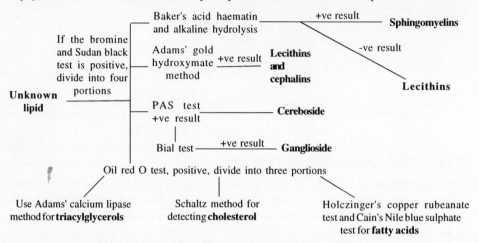

Fig. 26.18. An analytical scheme for the chemical classification of an unknown lipid.

Individual lipids are identified by standard chemical procedures after they have been separated from the naturally occurring lipid material. Separation and purification is usually achieved by chromatography. Identification of the pure lipid is largely based on physical methods such as mass spectroscopy and nuclear magnetic resonance spectroscopy. This is backed up by some limited chemical testing such as hydrolysis and identification of the fatty acids produced by either TLC, HPLC or gas chromatography of the methyl esters of the fatty acids as the parent fatty acids are not sufficiently volatile. These methyl esters are produced by treating the fatty acid mixture with diazomethane:

$$RCOOH \xrightarrow[\text{Diazomethane}]{CH_2N\equiv N} \underset{\text{Methyl ester}}{RCOOCH_3} + N_2$$

The positions of the fatty acid residues in triacylglycerols and some of the other lipids with glycerol-based structures may be determined by selective enzyme hydrolysis. The 1 and 3 residues are liberated first leaving the 2-monoacylglycerol.

Specific lipid mixtures are evaluated and also characterised by the measurement of a number of empirical measurements and chemical tests. Three of the principal ones are **acid value**, **saponification value** and **iodine value** (Table 26.4). It is emphasised that these are not the only measurements of this type that are used to evaluate lipids. Other measurements that are defined in a similar manner are given in the appropriate literature.

The acid value is usually determined by titration with aqueous potassium hydroxide using phenolphthalein as indicator. A mixture of ethanol and diethyl ether is used as a solvent for the lipid.

Saponification values are usually determined by heating the lipid under reflux with a known volume of ethanolic solution of potassium hydroxide which contains excess potassium hydroxide. The excess potassium hydroxide is determined by back titration with hydrochloric acid using phenolphthalein as indicator.

Table 26.4. The definition of some of the physical measurements used to evaluate lipids.

Specification	Definition
Acid value	This is the number of milligrams of potassium hydroxide that will neutralise the free acids in one gram of the lipid
Saponification value	This is the number of milligrams of potassium hydroxide that will neutralise the free acids and hydrolyse all the esters in one gram of the lipid
Iodine value	This is the number of parts by mass of iodine that is absorbed by 100 parts by mass of the lipid

The iodine value of a lipid is a measure of the degree of unsaturation of that lipid. It is normally determined by treating the lipid with a known volume of iodine monochloride solution containing excess iodine monochloride. The excess iodine monochloride is reacted with potassium iodide to form iodine which is titrated with sodium thiosulphate using starch mucilage as indicator.

$$\text{Lipid} \xrightarrow[\text{Iodine monochloride}]{ICl} \text{Product} + ICl_{\text{Excess}}$$

$$ICl_{\text{Excess}} + KI \longrightarrow KCl + I_2$$

$$2\,I_2 + Na_2S_2O_3 \xrightarrow{\text{Titrate}} 2\,NaI + Na_2S_4O_6$$

The reaction is carried out in the dark and a blank determination must be carried out. The former is to reduce the light catalysed decomposition of the iodine monochloride whilst the latter allows for any decomposition and losses due to the monochlorides volatility.

26.11   LIPIDS AND CELL MEMBRANES

Cell membranes are the specialised tissues that separate the contents of cells from their surroundings. The structures of cell membranes are currently believed to consist of a phospholipid bilayer which is made up of mainly phosphatidyl cholines and phosphatidyl ethanolamines in which the molecules are arranged with their polar phosphate ends facing out (Fig. 26.19). Cholesterol is embedded in the surface of the bilayer while proteins are either found on both surfaces of the bilayer (peripheral or extrinsic proteins) or pass right through the bilayer (integral or intrinsic proteins). Carbohydrates always occur on the extracellular surface of the membrane in the form of glycoproteins (proteins which have a polysaccharide chain bonded to them) and glycolipids (26.5).

The molecules of the bilayer are held together by hydrophobic bonding and van der Waals' forces (9.3 and 3.6). Consequently, it will take little energy to force the molecules apart to allow the passage of chemical species through the membrane. Furthermore, the membrane will be self sealing, the molecules moving together to reconstitute the hydrophobic bonds and van der Waals' forces after the passage of the chemical species.

Many peripheral protein molecules are thought to be bound to the bilayer largely by hydrogen and electrostatic bonding. This means that these peripheral proteins are liberated from the membrane by relatively mild changes in their environment such as changes in pH and salt concentration. Their loss leaves the membrane intact. Other peripheral proteins are covalently bound to phosphatidyl inositols in the bilayer. Many peripheral proteins such as cytochrome *c* act as enzymes.

Integral proteins can only be liberated by procedures that disrupt the membrane. Integral proteins contain the receptor sites that trigger the physiological response of the cell. They also act as enzymes, carriers and form ion channels through the membrane.

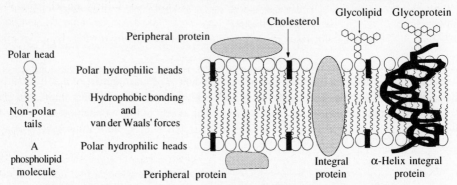

Fig. 26.19. A representation of the fluid mosaic model of a cell membrane. The carbohydrate chains of glycolipids and glycoproteins are found only on the extracellular side of the membrane.

The methods by which chemical species pass through a cell membrane fall into seven basic types: osmosis, filtration, simple diffusion, facilitated diffusion, active transport mechanisms, endocytosis and exocytosis.

26.11.1 Osmosis
Osmosis is the movement of **water** through a membrane because of a difference in solute concentration on the opposite sides of the membrane (9.8).

26.11.2 Filtration
The configurations of the molecules forming a cell membrane result in the structure being transversed by a number of narrow channels. Filtration is the movement of small molecules through these channels under hydrostatic pressure.

26.11.3 Simple diffusion
Diffusion is the movement of a **molecule** or **ion** through a membrane from an area of high concentration to an area of low concentration. Movement will cease when the concentration of the substance is the same on both sides of the membrane. The mechanism of diffusion is of two types: simple lipid diffusion due to lipid solubility and simple aqueous diffusion via protein channels. Many drugs enter a cell by simple lipid diffusion.

The rate of diffusion of a chemical species through a membrane is governed by the size of the molecule, temperature, thickness of the membrane (d), the concentration gradient across the membrane (c_1-c_2), the surface area of the membrane involved in the diffusion (a) and the relative solubilities of the diffusing species in water and the lipid membrane.

The diffusion coefficient (k) which takes into account the chemical and physical state of the diffusing species and the membrane. A number of these factors were summarised by Fick in his first law of diffusion (equation 26.1):

$$\text{Rate of diffusion} \quad = \quad \frac{ka(c_1\text{-}c_2)}{d} \qquad (26.1)$$

Since k, a and d are constant for a given membrane and chemical species the rate of diffusion of that species through the membrane will therefore depend, according to Fick, on the concentration gradient across that membrane, that is, **first order kinetics**.

The rate of diffusion of a species through a membrane also depends on the relative solubilities of that species in water and lipids. These relative solubilities are related to the lipid/water partition coefficient of the diffusing species. In most cases the higher the partition coefficient, the more soluble the species is in the lipid and the faster it will diffuse through the membrane.

It is not possible to measure partition coefficients for biological membranes and so the partition coefficient values determined for model organic solvent/water systems such as the octanol/water system are used instead. The results from these systems are often a useful guide to the behaviour of the species *in situ*.

26.11.4 Facilitated diffusion
Facilitated diffusion involves a carrier protein transporting the substance through the membrane (Fig. 26.20). The process does not require energy supplied by the cell. It is highly specific but is always from a high to a low concentration. For example, glucose is transported through the membranes of most cells in this manner.

26.11.5   Active transport
Active transport occurs when a substance is transported through a membrane usually against the concentration gradient (low to high concentration). The substance combines with a specific carrier protein. The reaction causes the carrier to change its conformation and in doing so it transports the substance from one side of the membrane to the other (Fig. 26.20). The carrier releases the chemical species into the surrounding medium, the complete process requiring the expenditure of considerable quantities of cellular energy.

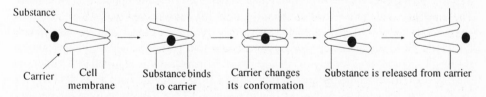

Substance

Carrier   Cell   Substance binds   Carrier changes   Substance is released from carrier
        membrane   to carrier   its conformation

Fig. 26.20. A schematic representation of carrier protein transport.

The rate of active transport is dependent on the concentration of the transported substance. At concentrations lower than that needed to saturate the available carrier it will follow **first order kinetics**. Above this concentration it will exhibit **zero order kinetics**.

The full details of the mechanisms by which active transport occurs are not known. However, active transport is chemically specific and is known to be involved in the transport of drugs, metabolites and minerals across biological membranes in the human body, for example the absorption of melphalan into tumours, amino acids into brain tissue and Na^+ and K^+ ions from the tubular epithelium of the kidney. The main active transport system in all animals is the sodium pump (30.2.3) in the outer cell membrane.

26.11.6 Endocytosis

Endocytosis is the processes by which substances such as large molecules or even whole bacterium, pass through the membrane *into* the cell. The membrane flows around the substance in effect forming a membrane bag (vesicle) around the substance as the membrane closes behind the substance. The vesicle is then released into the cell (Fig. 26.21). The importance of endocytosis in drug absorption is not fully known.

There are two main types of endocytosis: phagocytosis and pinocytosis. Phagocytosis ('cell eating') is concerned with the transport of large particles through the membrane and includes the engulfing of bacteria by white blood cells. Pinocytosis (' cell drinking') involves the transport of large molecules in solution.

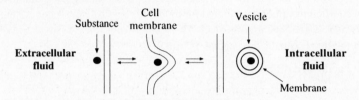

Fig. 26.21. Schematic representations of endocytosis.

26.11.7 Exocytosis

Exocytosis is the processes by which substances are transported out of a cell. It essentially is the reverse of endocytosis. The substance is packaged into a vesicle which fuses with the membrane. The substance is then released on the extracellular side of the membrane.

Exocytosis is used to secrete digestive enzymes into the gastrointestinal tract and release many hormones and neurotransmitters into the extracellular fluid.

26.12 LIPIDS AS A HUMAN ENERGY SOURCE

The triacylglycerols are the main fats found in humans. They are obtained from the diet. The triacylglycerols in the food are hydrolysed in the intestine and the glycerol and fatty acids formed absorbed into the cell where they are reconstituted into triacylglycerols (26.11.1). These reconstituted triacylglycerols are transported mainly as lipoproteins in the blood and deposited largely in adipose tissue. When energy is required by a biological process these triacylglycerols are hydrolysed by a lipase, and the fatty acids liberated are oxidised in various tissues by β-oxidation to produce ATP. ATP acts as the energy source to 'power' the biological processes of the cell. Fats supply about 40% of the energy requirements for humans on a normal diet.

26.13 WHAT YOU NEED TO KNOW

(1) Lipid is the traditional collective name given to a number of naturally occurring classes of compound that are extracted from plant and animal tissues with non-polar solvents such as hexane, chloroform, diethyl ether and propanone.

(2) Lipids may be classified as either saponifiable or non-saponifiable lipids.

(3) Lipids may be classified as either simple, complex or derived lipids.

(4) Simple lipids are the esters of long chain fatty acids with either high molecular alcohols or glycerol.

(5) Complex lipids are lipids that contain a fatty acid ester and some other residue.

(6) Derived lipids are those that are produced by the hydrolysis of other lipids.

(7) Lipids are usually known by trivial names although some systematic names are used.

(8) Fatty acids are classified as derived lipids. The concentration of free fatty acids in a biological system is usually small.

(9) The common fatty acid residues found in lipid molecules have 'straight chain' structures with an even number of carbon atoms. These acid residues may be saturated or unsaturated.

(10) Prostaglandins are not usually classified as lipids.

(11) Acylglycerols are the mono, di and tri fatty acid esters of glycerol (Fig. 26.3). The triesters usually prefer a tuning fork conformation (Fig. 26.5).

(12) Fats are solid lipids at room temperature while oils are liquid lipids at room temperature. Oils have a higher degree of unsaturation than fats.

(13) Triacylglycerols are insoluble in water and are not amphipathic.

(14) Phospholipids have structures which contain a phosphate ester. Phosphatidyl compounds (Fig. 26.6) and plasmalogens (Fig. 26.7) have glycerol residues but the sphingomyelins (Fig. 26.8) are the esters of the alcohol sphingosine.

(15) Phosphatidyl lipids are divided into subgroups based on the nature of the substituent (R^1) of the phosphate group. The most important of these subgroups are the phosphatidyl cholines (R^1 = choline) and the phosphatidyl ethanolamines (R^1 = ethanolamine).

(16) Naturally occurring phosphatidyl lipids are optically active and usually have an **R** configuration.

(17) Glycolipids are sphingolipids which are similar to sphingomyelins except that the choline phosphate residue has been replaced by a sugar residue. They are subdivided according to the nature of this sugar residue.

(18) Waxes are the esters of high molecular mass alcohols, terpenes and steroids.

(19) Lipoproteins are complexes formed by the association of a lipid with a specific apoprotein. The nature of the bonding between the apoprotein and the lipid is not known but is thought to be electrostatic or hydrophobic in nature.

(20) Terpenes are the main constituents of naturally occurring substances known as the essential oils.

(21) Terpenes are compounds whose molecular formula contains either five, or a whole number multiple of five, carbon atoms. Their structures obey the isoprene rule, although there are some exceptions.

(22) Isopentenyl and dimethylallyl pyrophosphates are the precursors of terpenes in biological systems.

(23) Some terpenes are the precursors of steroids in plants and animals.

(24) Steroids are a class of compound based on the ring system which has the non-systematic numbering system shown in Fig. 26.15.

(25) Bonds that project above the plane of the steroid rings are shown as solid lines and

are referred to as β bonds whilst bonds projecting below the plane of the rings are referred to as α bonds and are represented by dotted lines.

(26) The ring systems of naturally occurring steroids usually have one of the configurations shown in Fig. 4.13 (p50).

(27) The chemical properties of all types of lipid are based on those of the functional groups found in the molecule. The chemical properties of a particular lipid may be predicted by assuming that each of these functional groups will react in the first instance in the appropriate manner for that functional group. However, it must be remembered that the other functional groups and the nature of the carbon-hydrogen skeleton of the molecule may prevent or modify the expected reactions of a functional group. Common chemical reactions of lipids are: (a) hydrolysis of all types of ester and amide groups, (b) transesterification of ester groups, (c) oxidation of C=C bonds (peroxidation and rancidity of fats), (d) reduction of C=C and C=C bonds and (e) electrophilic addition to C=C bonds.

(28) Naturally occurring lipids are mixtures of different lipids and other compounds.

(29) Lipids are characterised by specific chemical tests, physical techniques and the measurement of empirical constants.

(30) The main empirical measurements are acid, saponification and iodine values.

(31) Biological membranes consist of a phospholipid bilayer in which are found cholesterol, proteins and other compounds. Proteins embedded through the bilayer are referred to as integral proteins whilst those bound to the surfaces of the bilayer are known as peripheral proteins.

(32) The lipid molecules in a biological membrane are believed to be held together by hydrophobic bonding and van der Waals' forces. Consequently the structure may be forced apart to allow the passage of chemical species through the membrane.

(33) The transport of chemical species through a biological membrane can occur by: osmosis, filtration, simple diffusion, facilitated diffusion active transport, endocytosis and exocytosis.

(34) The metabolism of fats is an important energy source for the human body.

26.14   QUESTIONS

(1) Draw the general structural formulae for each of the following: (a) a triacylglycerol, (b) a plasmalogen, (c) a sulphatide and (d) a sphingolipid.

(2) Outline chemical reactions that may be made the basis of (a) two identification tests and (b) an assay procedure for a triacylglycerol.

(3) (a) Identify and name the functional group in the terpene, menthol (Fig. 26.13). (b) List the general chemical properties you would expect this functional group to exhibit.

(4) Draw the structures of the main phospholipids found in animal cell membranes. What net charge will these phospholipids have at pH 7?

(5) Explain, using suitable examples, the meaning of each of the following: (a) endocytosis, (b) active transport and (c) lipoprotein.

27

Carbohydrates

27.1 INTRODUCTION

Carbohydrates are a group of naturally occurring compounds. Many are essential for life. The group ranges from straight chain polyhydroxy aldehydes and ketones to complex polyhydroxy single (Fig. 27.1) and multi-ring compounds (Fig. 27.22). The class also includes amino compounds such as glucosamine but not compounds such as 2,4-dihydroxycyclohexanone. Simple carbohydrates are referred to as **sugars** or **saccharides**.

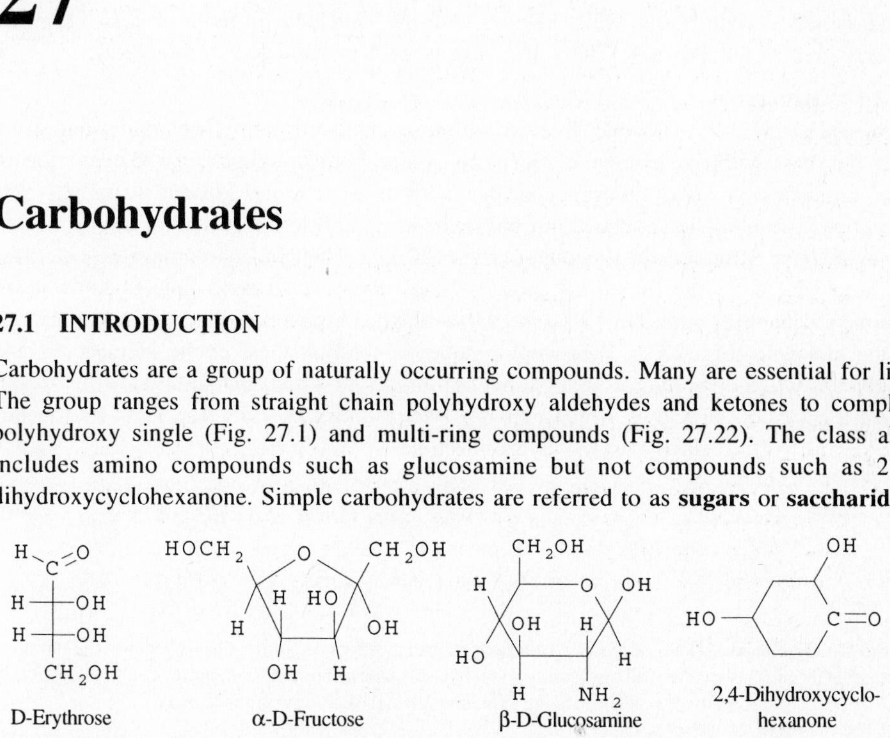

Fig. 27.1. D-Erythrose, α-D-fructose and β-D-glucosamine are classified as carbohydrates but 2,4-dihydroxycyclohexanone is not classified as a carbohydrate.

Carbohydrates are divided into three classes: **monosaccharides**, **oligosaccharides** and **polysaccharides**. Monosaccharides are sugars that cannot be hydrolysed by water to simpler compounds. Oligosaccharides yield a few, usually taken to be 2 to 9, monosaccharide molecules per oligosaccharide molecule on hydrolysis, whilst polysaccharides yield a large number of monosaccharide molecules per polysaccharide molecule. However, the frontier between oligosaccharides and polysaccharides is not clearly defined and carbohydrates whose molecules contain ten or more monosaccharide residues may be referred to as oligosaccharides. Polysaccharides are further subdivided into **homopolysaccharides** whose structures contain one type of monosaccharide residue and **heteropolysaccha-rides** whose structures contain more than one type of monosaccharide residue.

27.2   MONOSACCHARIDES

Simple monosaccharides are usually classified as **aldoses** and **ketoses**. Aldoses are monosaccharides which either have an aldehyde group in their molecular structures or form an aldehyde group on hydrolysis whilst ketoses either have a ketone group in their molecular structures or form a ketone group on hydrolysis. Aldoses and ketoses are further classified as bioses, trioses, tetroses, pentoses, hexoses, etc. according to the number of carbon atoms in their structures. For example, aldoses with the molecular formula $C_3H_6O_3$ would be classified as aldotrioses whilst ketoses with the molecular formula of $C_6H_{12}O_6$ would be classified as ketohexoses. The suffix **-ose** is used to indicate a carbohydrate.

27.2.1   Structure

Monosaccharides with less than five carbon atoms have 'straight chain' molecular structures but those with five or more carbons can exist in both straight chain and cyclic forms. The molecules of solid monosaccharides with five or more carbons usually (some exceptions) have cyclic structures but will exist as equilibrium mixtures of the cyclic and 'straight chain' structures in aqueous solution (27.3.2). The cyclic structures arise because the aldehyde or ketone functional group of the monosaccharide is able to undergo an internal nucleophilic addition (19.8 and 19.9) with one of the alcohol groups in another part of the molecule (Fig. 27.2). However, in aqueous solution these cyclic hemiacetals and hemiketals will slowly hydrolyse to form equilibrium mixtures containing the corresponding straight chain structures which explains why monosaccharides can react in aqueous solution as hydroxyaldehydes and hydroxyketones.

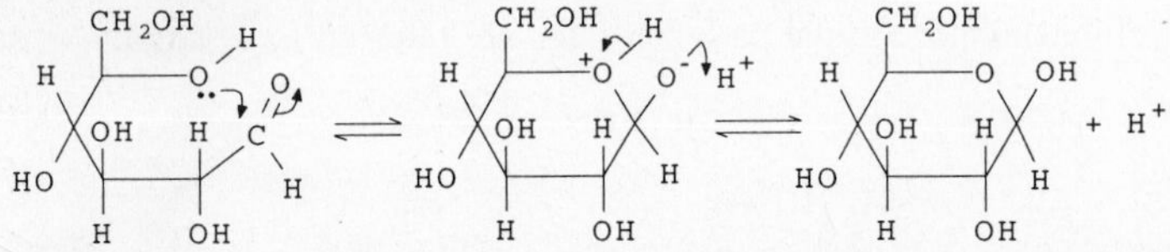

Fig. 27.2. Nucleophilic addition leading to the formation of the ring structure of glucose. Each of the carbon atoms of the ring has a tetrahedral configuration (sp³ hybridisation). Free rotation about the C–C bonds results in the hydroxyl group being close enough to the carbonyl group to react.

The formation of the ring hemiacetal or hemiketal leads to the formation of a new chiral centre in the molecule. The two isomers formed by the ring closure are known as **anomers** and the carbon of the carbonyl group involved in the ring closure, the **anomeric carbon**. Anomers differ only in the relative orientations of the new hydroxyl groups in space (Fig. 27.3). The anomer with the new hydroxyl group on the **same side** of the ring as the terminal CH_2OH is known as the **beta** (β) anomer whilst that with the hydroxyl group situated on the **opposite side** of the ring to the terminal CH_2OH is called the **alpha** (α) anomer.

The rings of monosaccharide structures are not flat but will take up the appropriate conformation for the type of ring system. For example, six-membered ring systems are normally found in a chair conformation (Fig. 27.4) and less commonly the boat forms whilst five-membered rings occur in the envelope conformation.

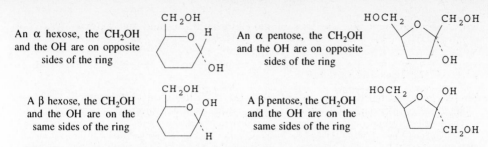

An α hexose, the CH₂OH and the OH are on opposite sides of the ring

A β hexose, the CH₂OH and the OH are on the same sides of the ring

An α pentose, the CH₂OH and the OH are on opposite sides of the ring

A β pentose, the CH₂OH and the OH are on the same sides of the ring

Fig. 27.3. The alpha and beta isomers of aldohexoses and ketohexoses. The rings are drawn using the Haworth convention for monosaccharides where bonds above the plane of the ring are drawn up the page whilst bonds below the plane of the ring are drawn down the page.

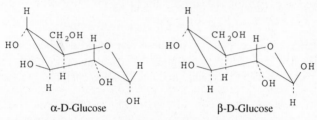

α-D-Glucose   β-D-Glucose

Fig. 27.4. The conformations of α-glucose and β-glucose.

27.2.2   Configuration

Monosaccharides are usually classified as having either a D or L configuration (4.8.1). This classification is based on the configuration of the highest numbered asymmetric carbon in the 'straight chain' form of the molecule (Fig. 27.5). Most naturally occurring monosaccharides have a D configuration.

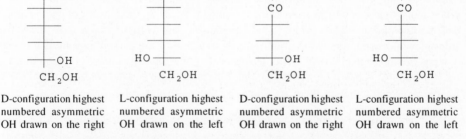

D-configuration highest numbered asymmetric OH drawn on the right

L-configuration highest numbered asymmetric OH drawn on the left

D-configuration highest numbered asymmetric OH drawn on the right

L-configuration highest numbered asymmetric OH drawn on the left

Fig. 27.5. The positions of the hydroxyl groups that decide the D/L configurations of monosaccharides. The carbonyl group must always be drawn at the top of the Fischer projection.

The configurations of the other chiral carbons in the molecule are indicated by the stem name of the monosaccharide (27.2.3). These names indicate the configurations of all the secondary hydroxyl groups in the molecule (Fig. 27.6). They apply irrespective of whether the hydroxyl groups are adjacent or separated by methylene groups. These names can also be used as prefixes to indicate the configurations of multichiral centre structures, in which case they are normally printed in italics.

The R/S system of designating configuration may also be used to indicate the

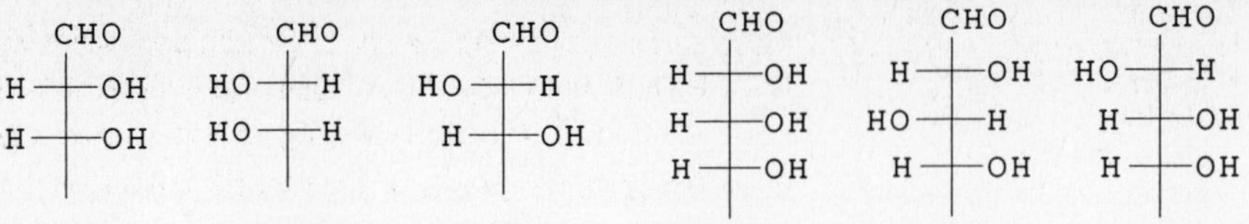

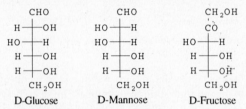

CHO	CHO	CHO	CHO	CHO	CHO
H—OH	HO—H	HO—H	H—OH	H—OH	HO—H
H—OH	HO—H	H—OH	H—OH	HO—H	H—OH
			H—OH	H—OH	H—OH
D-Erythro	L-Erythro	D-Threo	D-Ribo	D-Xylo	D-Arabino

Fig. 27.6. Examples of the configurational prefixes associated with the stem names of monosaccharides. The aldehyde group represents the principal functional group reference point on which the configuration is based. Mainly D configurations are shown since the L configuration with the same name has the mirror image configuration. The same configurational prefixes are used for ketoses except some trivial names such as fructose and psicose are also used as the names of the compounds.

configuration of a monosaccharide but it is somewhat inconvenient to use when naming monosaccharides. For example, D-glucose would be:

$$2R,3S,4R,5R-2,3,4,5,6\text{-pentahydroxyhexanal}$$

Monosaccharides such as glucose, mannose and fructose, whose structures have the same configurations except for the configuration at one chiral centre are **epimers** (5.4.2 and Fig. 27.7) and can yield the same products in some chemical reactions (27.5.2).

CHO	CHO	CH$_2$OH
H—OH	HO—H	CO
HO—H	HO—H	HO—H
H—OH	H—OH	H—OH
H—OH	H—OH	H—OH
CH$_2$OH	CH$_2$OH	CH$_2$OH
D-Glucose	D-Mannose	D-Fructose

Fig. 27.7. Examples of epimeric configurations.

27.2.3 Nomenclature

'Straight chain' monosaccharide molecules may be named using the combination of a prefix indicating the configuration of the molecule (Fig. 27.8), the number of carbon atoms and the appropriate suffix. Aldoses have the suffix **-ose** whilst ketoses have the suffix **-ulose** together with a locant that indicates the position of the carbonyl group if it occurs at a position other than 2 (Fig. 27.8). However, in both cases it is more common to use a trivial name that is based on a combination of the suffix **-ose** and the configurational suffix for the molecule although there are some exceptions to this rule.

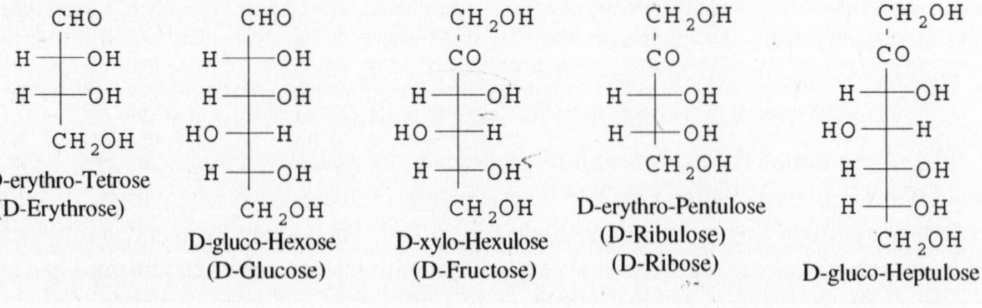

CHO	CHO	CH$_2$OH	CH$_2$OH	CH$_2$OH
H—OH	H—OH	CO	CO	CO
H—OH	H—OH	H—OH	H—OH	H—OH
CH$_2$OH	HO—H	HO—H	H—OH	HO—H
D-erythro-Tetrose	H—OH	H—OH	CH$_2$OH	H—OH
(D-Erythrose)	CH$_2$OH	CH$_2$OH	D-erythro-Pentulose	H—OH
	D-gluco-Hexose	D-xylo-Hexulose	(D-Ribulose)	CH$_2$OH
	(D-Glucose)	(D-Fructose)	(D-Ribose)	D-gluco-Heptulose

Fig. 27.8. Examples of the nomenclature used for straight chain monosaccharide structures. Non-systematic trivial names are given in brackets. It should be noted that dextrose is glucose and sorbose is fructose.

Monosaccharides with ring structures are given specific stem names. **Pyranose** is used for six-membered rings whilst **furanose** is used for five-membered rings. These stem names are combined with the prefixes used to indicate their configurations (Fig. 27.9). Monosaccharides in which a hydroxyl group has been replaced by a hydrogen have the name of the parent sugar prefixed by **deoxy-** and a locant indicating the position of the hydrogen except in aldoses when the hydrogen is in position 2 the locant is usually dropped.

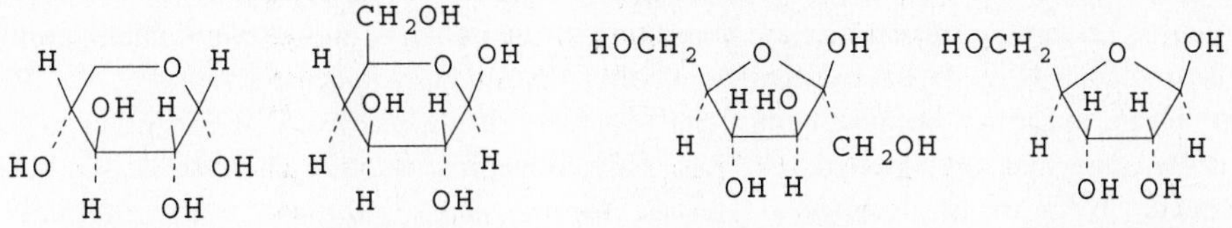

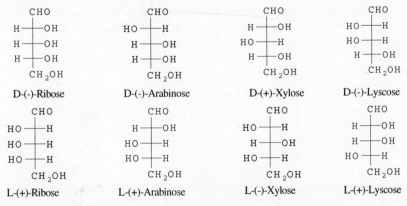

| α-D-Xylopyranose | β-D-Glucopyranose | β-D-Fructofuranose | β-D-Ribofuranose |
| (α-D-xylose) | (β-D-glucose) | (β-D-fructose) | (β-D-ribose) |

Fig. 27.9. Examples of the nomenclature used for monosaccharides with ring structures. The commonly used name is given in brackets.

Substituent monosaccharide rings are named by having the ending **-ose** changed to **-osyl**. For example, pyranose and furanose are changed to **pyranosyl** and **furanosyl** respectively.

27.3   THE PHYSICAL PROPERTIES OF MONOSACCHARIDES

27.3.1   Optical activity

Monosaccharides are optically active because they possess a number of chiral centres. Each of these centres has a different structure and so the theoretical number of isomers of a particular 'straight chain' monosaccharide structure is given by the 2^n rule (5.3.4). For example, four C aldoses have two chiral centres and so have four optically active stereoisomers (Fig. 5.9) whilst five C aldoses have three chiral centres and so in theory have eight optically active straight chain stereoisomers (Fig. 27.10). However, monosaccharides that exist as ring structures can have 2 x 2^n optically active isomers since they can exist as alpha and beta isomers. For example, each of the isomers of ribose has an alpha and a beta form and so in theory ribose has a total of sixteen optically active isomers.

D-(-)-Ribose	D-(-)-Arabinose	D-(+)-Xylose	D-(-)-Lyscose
L-(+)-Ribose	L-(+)-Arabinose	L-(-)-Xylose	L-(+)-Lyscose

Fig. 27.10. The stereoisomers of five carbon atom aldoses.

25.3.2   Solubility

Monosaccharides are very soluble in water. However, the specific rotation (5.2) of an aqueous solution of a monosaccharide with a cyclic structure will change over a period of time from its initial value to a second constant value. For example, the specific rotation of an aqueous solution of α-D-glucose changes from an initial value of +19° when freshly dissolved to +52° whilst the specific rotation of an aqueous solution of β-D-glucose changes from +113° to +52°. This phenomenon is called **mutarotation**. It is due to the formation of an equilibrium mixture consisting of the anomers and 'straight chain' forms of the molecule (Fig. 27.11). At equilibrium this mixture has a specific rotation of 52°. The equilibrium is formed because hemiacetal structures of glucose hydrolyse in water to the *straight chain* hydroxy aldehyde (19.9.4). This allows free rotation about the C_1-C_2 bond to occur. As a result, it is now possible for the ring to reform in the alternative configuration. All types of *monosaccharide* can exhibit mutarotation.

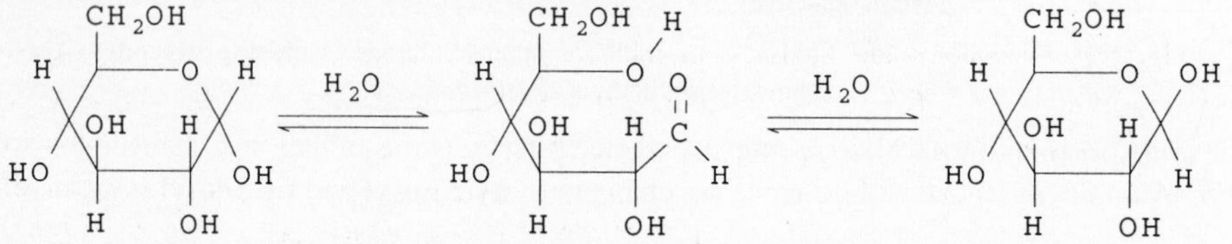

α-D-glucose$[\alpha]_D^{298} = +19°$, water   'Straight chain' form   β-D-glucose$[\alpha]_D^{298} = +113°$, water

Fig. 27.11. The mutarotation of D-glucose

27.4   THE CHEMICAL PROPERTIES OF MONOSACCHARIDES

The chemical properties of monosaccharides are a mixture of those of primary and secondary alcohols (Chapter 15), aldehydes, ketones (Chapter 19), hemiacetals and hemiketals, and characteristic reactions which usually involve more than one of these functional groups. Interpretation of these reactions may be complicated by tautomerism of the straight chain monosaccharide if the reaction is carried out under aqueous alkaline conditions. For example, after a short period of time an aqueous alkaline solution of glucose will contain fructose and mannose amongst other monosaccharides (Fig. 27.12). Tautomerism does not occur in aqueous acid solutions but dilute acid does catalyse the interconversion of the α and β isomers of a monosaccharide. A further point to bear in mind when interpreting monosaccharide reactions is that epimers will often yield the same products.

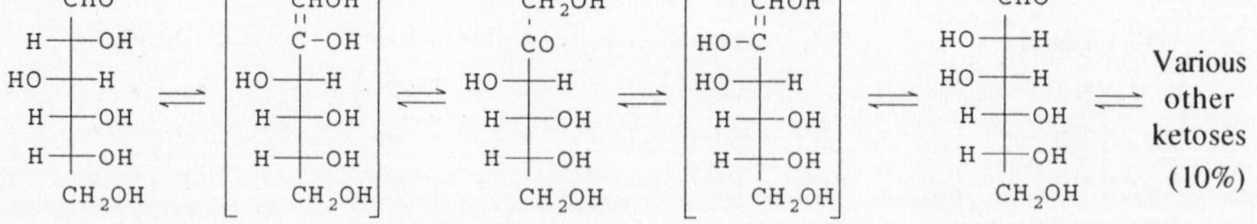

Glucose (63.5%)   Enolic form   Fructose (21%)   Enolic form   Mannose (2.5%)

Fig 27.12. The tautomerism of glucose in a basic aqueous solution. It should be noted that the enolic forms of the molecule can revert to any of the aldose or ketose forms. The figures in brackets are the approximate concentrations of the isomers present at equilibrium.

27.5 REACTIONS CHARACTERISTIC OF MONOSACCHARIDES

27.5.1 Acids and bases

Monosaccharides are stable in dilute aqueous acid at room temperature although some compounds such as glucose and fructose will undergo mutarotation. They are unstable in concentrated acids at all temperatures. For example, at room temperature pentoses form furfural whilst hexoses form 5-hydroxymethylfurfural amongst other compounds.

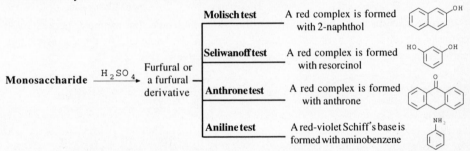

$C_5H_{10}O_5$
Pentoses ⟶ Furfural

$C_6H_{12}O_6$ ⟶ HOCH$_2$—furan—CHO ⟶ $HCHO$ + $CH_3COCH_2CH_2COOH$ + Humins, brown compounds of unknown structure
Hexoses   5-Hydroxymethylfurfural   Laevulinic acid

These reactions are the basis of a number of tests used to detect the presence of carbohydrates (Fig. 27.13). Polysaccharides usually break down under the reaction conditions to yield sufficient monosaccharide for the test.

Monosaccharide $\xrightarrow{H_2SO_4}$ Furfural or a furfural derivative

Molisch test	A red complex is formed with 2-naphthol	
Seliwanoff test	A red complex is formed with resorcinol	
Anthrone test	A red complex is formed with anthrone	
Aniline test	A red-violet Schiff's base is formed with aminobenzene	

Fig. 27.13. The chemical basis of some of the tests used to detect carbohydrates.

In basic aqueous solutions monosaccharides either decompose or undergo tautomerism (Fig. 27.12).

27.5.2 Phenylhydrazine

Aqueous solutions of monosaccharides react with phenylhydrazine hydrochloride to form initially the expected phenylhydrazone. However this compound is unstable and undergoes further reaction with phenylhydrazine to form an osazone.

Glucose $\xrightarrow{PhNHNH_2}$ Glucosazone $\xleftarrow{PhNHNH_2}$ Fructose

Osazones do not have reliable enough melting points for them to be used as derivatives in the normal manner because epimers form the same osazone. However, the time taken for the formation of an osazone under standard conditions is characteristic of the monosaccharide and may help in its identification. For example, the times taken for the formation of glucosazone from glucose and fructose are 5 and 2 minutes respectively. In both cases glucosazone is formed because the monosaccharides are epimers. Sucrose (Fig. 27.21), the common domestic white sugar, also forms glucosazone because it hydrolyses to glucose and fructose under the reaction conditions used to prepare the derivative but the time of formation is 30 minutes.

27.5.3 Oxidation

Oxidation can affect both the aldehyde and alcohol groups of monosaccharides. The products of the reaction will depend on the nature of the oxidising agent. For example, mild oxidising agents usually oxidise either the aldehyde and/or the primary alcohol groups to the corresponding carboxylic acids;

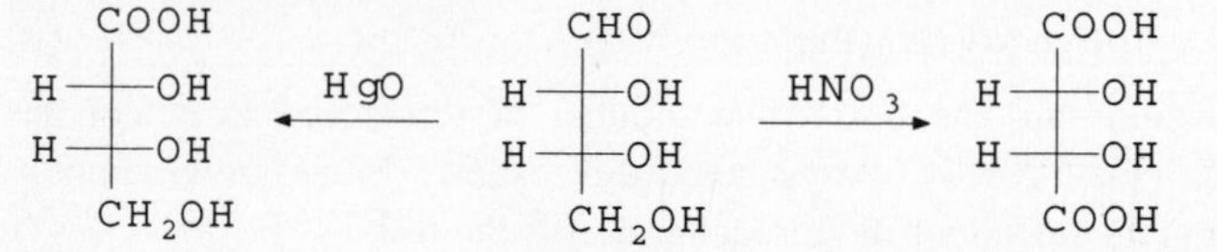 COOH	CHO	COOH
H——OH $\xleftarrow{\text{HgO}}$	H——OH $\xrightarrow{\text{HNO}_3}$	H——OH
H——OH	H——OH	H——OH
CH$_2$OH	CH$_2$OH	COOH
D-Glyceric acid	D-Erythrose	meso-Tartaric acid

Stronger oxidising agents can disrupt the whole molecule. For example, periodic acid oxidises the diols and hydroxy carbonyl groups in the molecule to the corresponding carboxylic acids, aldehydes, methanoic acid and methanal (Fig. 27.14).

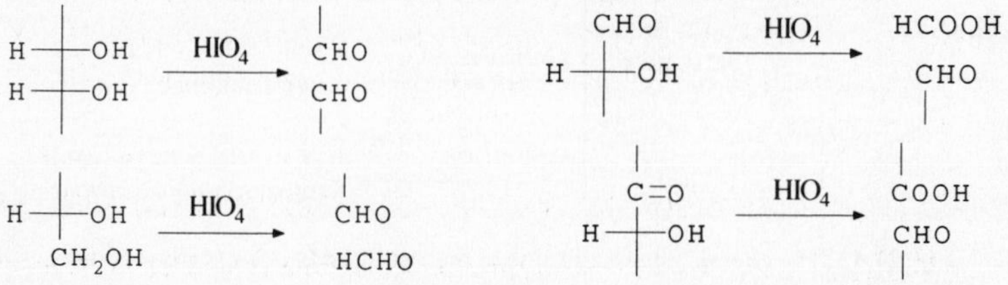

H——OH $\xrightarrow{\text{HIO}_4}$	CHO	CHO $\xrightarrow{\text{HIO}_4}$	HCOOH
H——OH	CHO	H——OH	CHO
H——OH $\xrightarrow{\text{HIO}_4}$	CHO	C=O $\xrightarrow{\text{HIO}_4}$	COOH
CH$_2$OH	HCHO	H——OH	CHO

Fig. 27.14. Models for the oxidation of various diol and hydroxy carbonyl structures by periodic acid. The cleavage of one mole of C-C bond requires one mole of the oxidising agent.

Periodic acid oxidation of the carbohydrates in tissue is the basis of the PAS staining procedure. Any aldehydes, and methanoic acid produced by the oxidation of carbohydrate deposits in the tissue are detected by staining with Schiff's reagent which produces a red-purple colouration in the areas where oxidation has taken place. The colour is not easy to see and counter-staining followed by washing with water is necessary. The type of counter-staining used will depend on the nature of the investigation being carried out.

The oxidation of glucose by glucose oxidase is the basis of the estimation of glucose in biological fluids. The oxidation converts the glucose to hydrogen peroxide and gluconic acid.

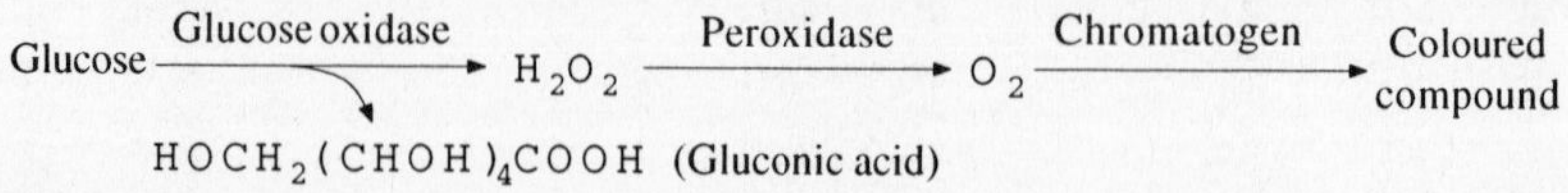

$$\text{Glucose} \xrightarrow{\text{Glucose oxidase}} H_2O_2 \xrightarrow{\text{Peroxidase}} O_2 \xrightarrow{\text{Chromatogen}} \begin{array}{c}\text{Coloured}\\\text{compound}\end{array}$$

$$HOCH_2(CHOH)_4COOH \quad \text{(Gluconic acid)}$$

The hydrogen peroxide is converted to water and oxygen by treatment with a peroxidase. An oxygen acceptor (chromogen) is added which converts the oxygen to a coloured compound which can be estimated colorimetrically (10.2.1).

27.6 REACTIONS SIMILAR TO THOSE OF ALCOHOLS

Monosaccharides exhibit some of the reactions of primary and secondary alcohols. The hydroxyl groups may be oxidised (27.5.3) and they also act as nucleophilic centres in substitution reactions. For example, the hydroxyl groups of glucose react with ethanoic anhydride to form the corresponding pentaethanoate derivative and undergoes nucleophilic substitution with methyl iodide to yield the pentamethyl ether derivative.

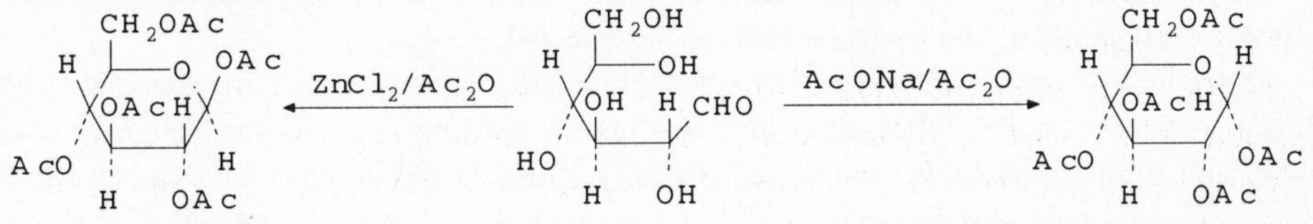

CH$_2$OAc ← ZnCl$_2$/Ac$_2$O ← CH$_2$OH → AcONa/Ac$_2$O → CH$_2$OAc

Penta-O-acetyl-β-D-glucose   D-Glucose   Penta-O-acetyl-α-D-glucose

Key: Ac = acetyl group (CH$_3$CO), Ac$_2$O = ethanoic anhydride, AcONa = sodium ethanoate.

α-D-glucose   CH$_2$OH → Ag$_2$O /MeI → CH$_2$OMe   Penta-O-methyl-α-D-glucose

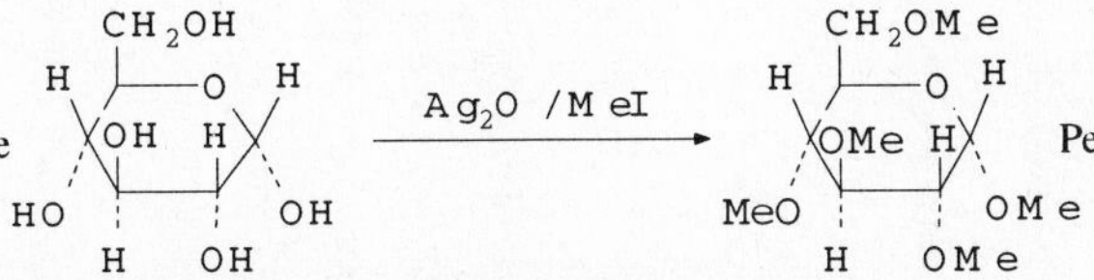

Key: MeI = methyl iodide (CH$_3$I), Me = methyl group (CH$_3$).

The esters of monosaccharides are used as derivatives in organic analysis and identification tests. The formation of ethers is used in the elucidation of the structures of oligosaccharides and polysaccharides. However, it is not possible to form ethers in alkaline solution using the normal synthetic methods unless the monosaccharide is first stabilised by converting it to a glycoside (27.10).

27.7 REACTIONS SIMILAR TO THOSE OF ALDEHYDES AND KETONES

The carbonyl groups of the simple 'straight chain' monosaccharides such as threose and erythrose can exhibit some of the traditional reactions of aldehydes and ketones such as oxidation, reduction, nucleophilic addition and condensation. Those monosaccharides that

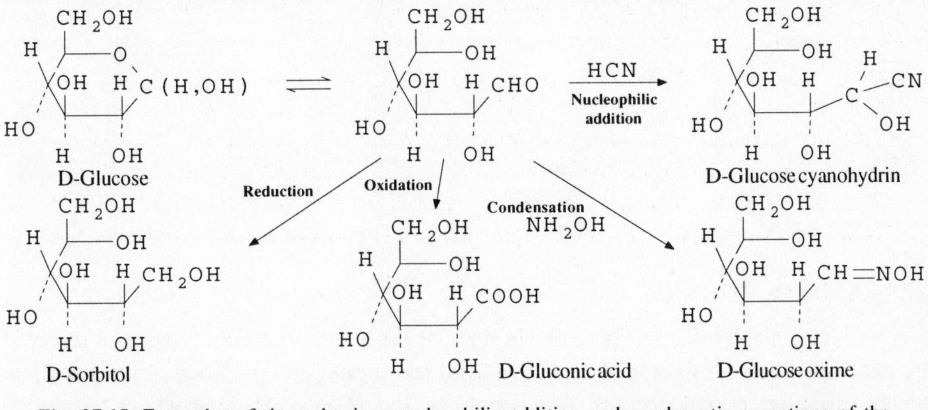

CH$_2$OH C(H,OH) — D-Glucose ⇌ CH$_2$OH CHO → HCN Nucleophilic addition → CH$_2$OH CN OH — D-Glucose cyanohydrin

Reduction → CH$_2$OH CH$_2$OH — D-Sorbitol

Oxidation ↓ CH$_2$OH COOH — D-Gluconic acid

Condensation NH$_2$OH → CH$_2$OH CH=NOH — D-Glucose oxime

Fig. 27.15. Examples of the reduction, nucleophilic addition and condensation reactions of the carbonyl groups of glucose. Other aldoses and ketoses react in a similar fashion.

exist in the form of cyclic structures will also exhibit these properties in aqueous solution because of the presence of the 'straight chain' form of the molecule in solution (Fig. 27.11). The concentration of the 'straight chain' is small but, because one is dealing with an equilibrium system, as the 'straight chain' form reacts, the related ring forms open to produce more of the 'straight chain' molecules. These 'new' molecules react, the equilibrium is disturbed and more of the molecules in the ring form open to form 'straight chain' molecules. This process leads to further reaction of the carbonyl group and the process repeats itself until all the monosaccharide has reacted.

D-sorbitol is used as an artificial sweetener and sugar substitute in many diabetic foods because of its poor absorption. It is also used as the starting point in the manufacture of isosorbide dinitrate which is used to treat angina pectoris. It should be noted that isosorbide dinitrate is in essence the ester of the dihydric alcohol D-isosorbide and nitric acid.

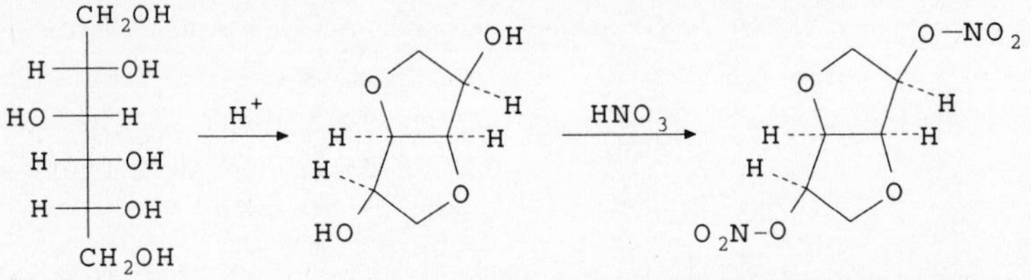

D-sorbitol   D-isosorbide   D-isosorbide dinitrate

The addition of hydrogen cyanide to aldoses is the basis of the Kiliani synthesis of a larger monosaccharide from a smaller compound whilst the reaction of monosaccharides with hydroxylamine is the basis of the Wohl degradation for shortening an aldose chain.

27.8 REACTIONS AS A HEMIACETAL OR HEMIKETAL

Cyclic monosaccharides react as hemiacetals and hemiketals with alcohols under dry conditions in the presence of an acid catalyst to form glycosides. Glycosides are the monosaccharide equivalent of acetals and ketals (19.9.4). Their chemistry is discussed in section 27.10. 'Straight chain' monosaccharides do not possess a hemiacetal or hemiketal functional group and so react under similar conditions to form hemiacetals and hemiketals.

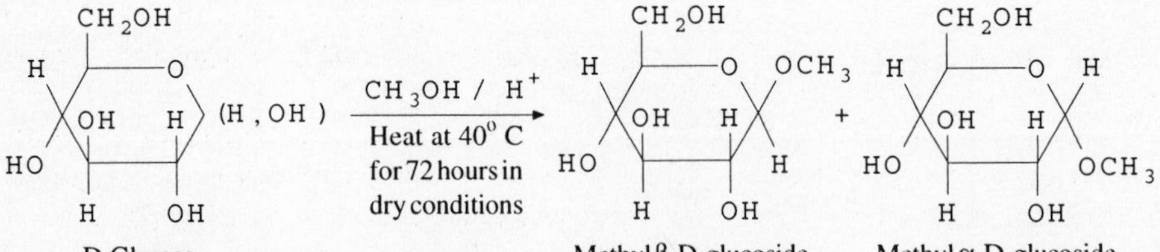

D-Glucose   Methyl β-D-glucoside   Methyl α-D-glucoside

27.9 AMINO SUGARS

Amino sugars are aldoses in which one or more of the hydroxyl groups have been replaced by either primary, alkyl or acyl substituted primary amino groups. They occur widely in nature, the most common being D-glucosamine and D-galactosamine where the amino group is at position 2. Like the monosaccharides, the structures of the amino sugars are cyclic, existing in both α and β forms. However, amino sugars do not exhibit tautomerism if the amino group occupies a position α to the aldehyde group.

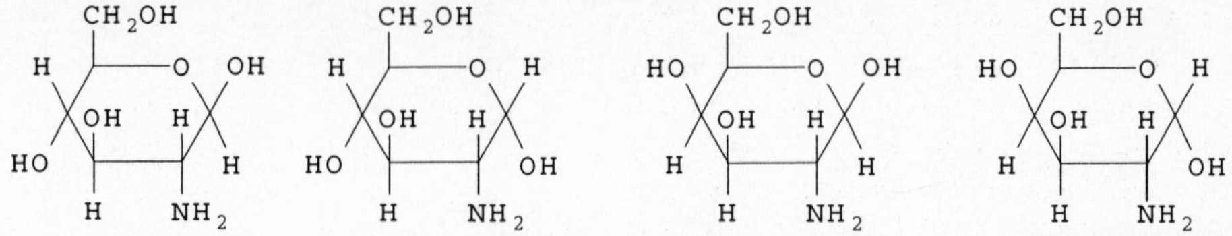

β-D-glucosamine   α-D-glucosamine   β-D-galactosamine   α-D-galactosamine

Amino sugars exhibit many of the chemical properties of monosaccharides and in addition, some of the properties of the appropriate nitrogen functional group. For example, amino sugars with a primary amino group can form salts, undergo condensation and nucleophilic substitution reactions.

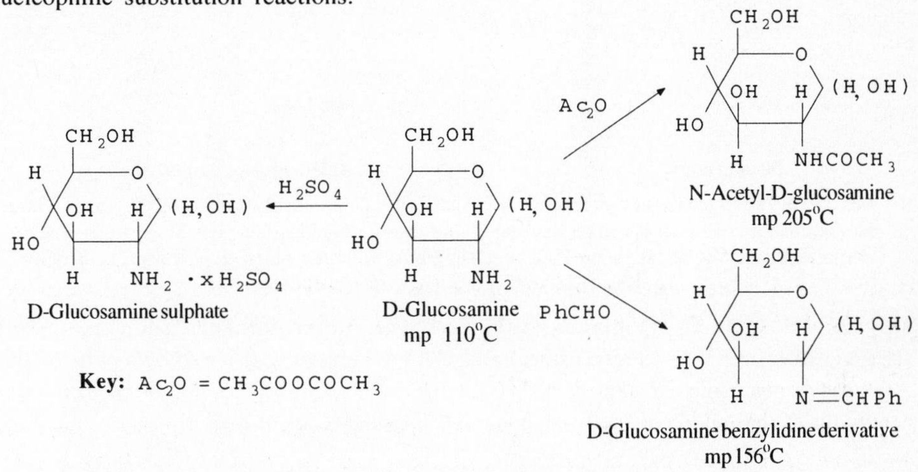

N-Acetyl-D-glucosamine
mp 205°C

D-Glucosamine sulphate

H_2SO_4

D-Glucosamine
mp 110°C

Ac_2O

PhCHO

D-Glucosamine benzylidine derivative
mp 156°C

Key: $Ac_2O = CH_3COOCOCH_3$

The N-acyl derivatives of amino sugars are less susceptible to acid catalysed hydrolysis by water than the corresponding O-acyl derivatives (esters). Furthermore, the glycosides of amino sugars with an amino group at position 2 are also resistant to acid catalysed hydrolysis. It is also difficult to form the glycosides of these amino sugars because the amino group acts as a base and removes the hydrogen ions that would normally catalyse the glycoside formation by reacting with the hydroxyl group.

Amino sugars and their derivatives are found in many important biological structures. D-N-acetylgalactosamine residues occur in the polysaccharides that determine certain blood-group antigens (Fig. 27.16). Muramic acid residues such as N-acetylmuramic acid (Fig. 27.24) are a constituent of the cell walls of bacteria and N-methylglucosamine residues are found in the antibiotic streptomycin.

27.10   GLYCOSIDES

Glycosides consist of a sugar residue (the **glycone**) bonded to a non-sugar residue (the **aglycone**) via the carbon of what was the carbonyl group of the sugar. The most common type of linkage, generally known as the glycosidic link, between the glycone and the

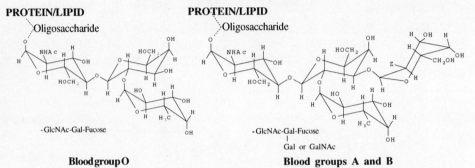

α-D-N-Acetylglucosamine (α-GlcNAc)   α-L-6-Deoxygalactose (Fucose)

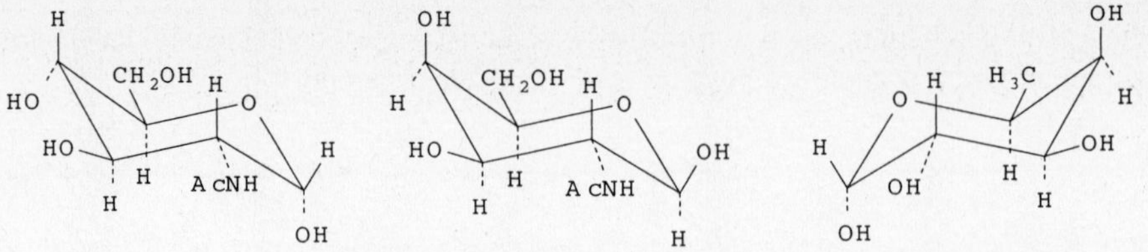

β-D-N-Acetylgalactosamine (β-GalNAc)

PROTEIN/LIPID   **PROTEIN/LIPID**

Oligosaccharide   Oligosaccharide

- GlcNAc-Gal-Fucose   - GlcNAc-Gal-Fucose
                                                       |
                                                  Gal or GalNAc

Blood group O   **Blood groups A and B**

Fig. 27.16. The structures of the carbohydrate residues that determine A, B and O blood groups. The polysaccharide chains act as markers and determine a person's blood group. For blood group A, Z = -NHCOCH₃ and for blood group B, Z = -OH. People with AB blood groups have both types of polysaccharide chain attached to their red blood cells.

aglycone is C–O–C (O–glycosides). However, other linkages such as C–N–C (N–glycosides), C–S–C (S–glycosides) and C–C (C–glycosides) are known (Fig. 27.17). All types of glycoside can exist as both α and β isomers. Glycosides are usually classified according to the nature of their aglycone and pharmacological action.

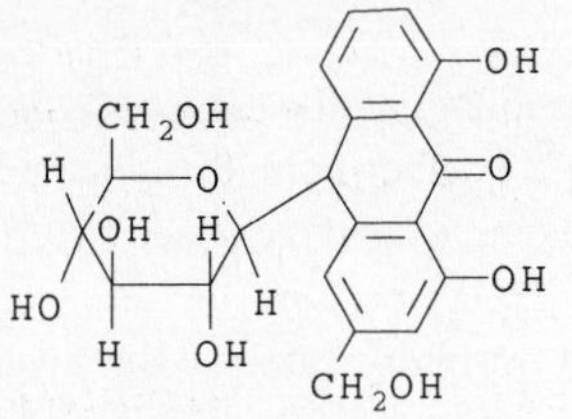

Aloin, ex aloes 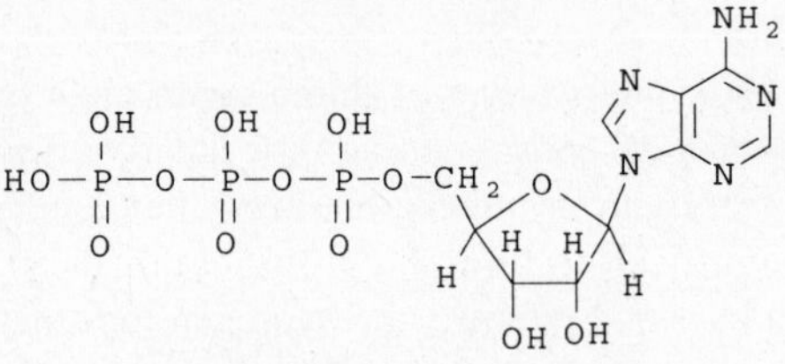   Adenosine triphosphate (ATP), ex mammals

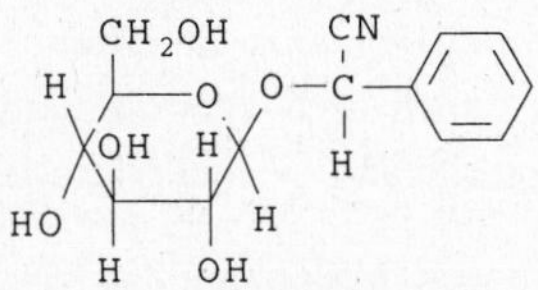

Prunasin, ex wild cherry bark 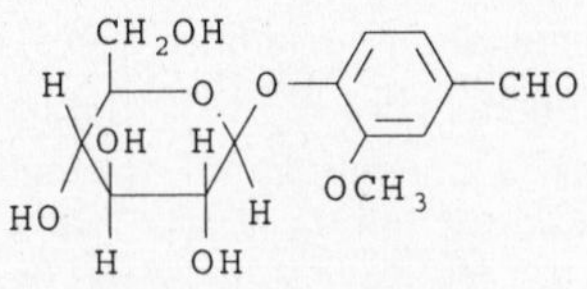   Glucovanillin, ex vanilla

Fig. 27.17. Examples of naturally occurring C-, N- and O-glycosides.

Trivial names are used for most glycosides but O-glycosides may be systematically named using the radical name of the aglycone followed by a name based on that of the sugar moiety but with the ending **-oside** (Fig. 27.18). However, N- and C-glycosides are systematically named using the name of the sugar residue as a substituent of the aglycone.

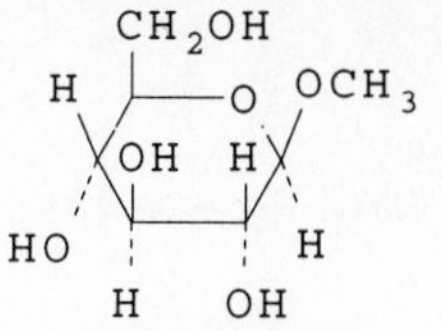

Methyl β-D-glucoside

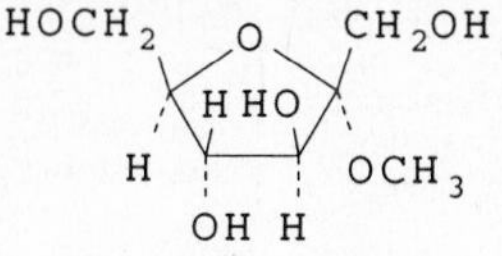

Methyl α-D-fructoside

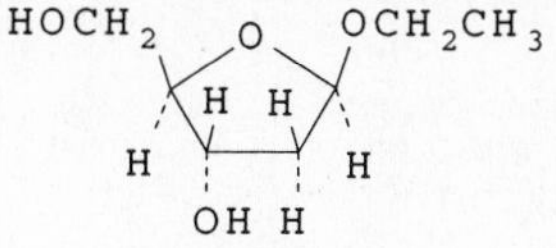

Ethyl β-D-deoxyriboside

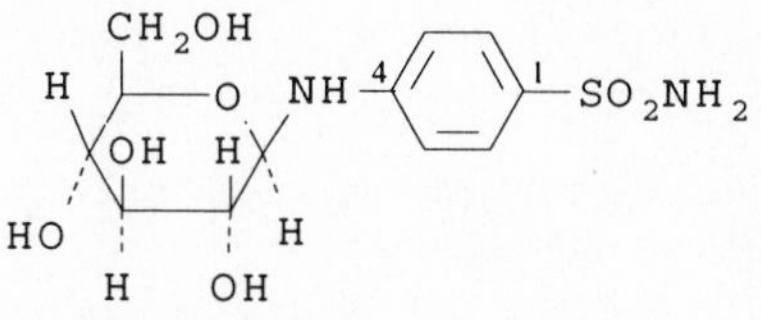

N_4-(β-D-Glucopyranosyl)sulphanilamide

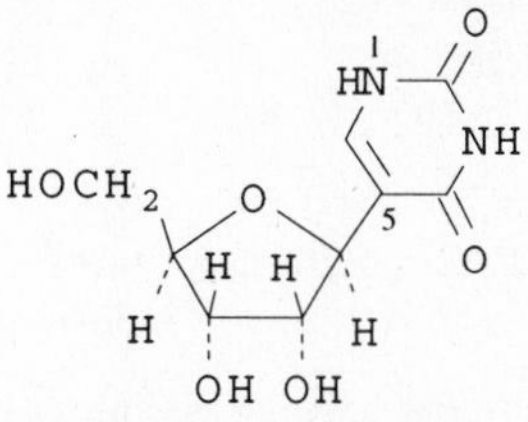

5-(β-D-Ribofuranosyl)uracil

Fig. 27.18. Examples of the nomenclature of glycosides.

Glycosides are usually soluble in water, aqueous media and some polar solvents such as ethanol. They are usually extracted from plant and animal tissue using ethanol or aqueous ethanol.

The chemical properties of glycosides depend on the nature of the functional groups in the molecule and their influence on each other. However, it should be noted that O-, N- and S-glycosides are easily hydrolysed by water in the presence enzymes or by heating with aqueous acid to give the aglycone and either the glycone or a mixture of monosaccharide molecules when the sugar residue is a polysaccharide. C-Glycosides are more difficult to cleave in this manner and need stronger conditions.

There are no specific tests for glycosides. However, they may be identified by hydrolysis and identification of the resultant glycone and aglycone. It is also possible to identify glycosides using ^{1}H and ^{13}C nuclear magnetic resonance spectroscopy.

27.11 THE STRUCTURES OF DISACCHARIDES

Disaccharides are oligosaccharides whose structures consist of two monosaccharide residues linked together by O-glycosidic linkage. In these structures one monosaccharide residue acts as the glycone the other like an aglycone, their structures being drawn using either the Haworth or conformational formulae. For example, in maltose the molecule of α-D-glucose acts as the glycone whilst the other molecule of D-glucose acts as the aglycone, the glycosidic link being formed between C_1 of the α-D-glucose and C_4 of the D-glucose. As it is the α isomer of the glycone that forms the glycoside link, the link is generally referred to as an α-1,4-linkage. The glycosidic links in oligosaccharides and polysaccharides are often classified in terms of the configuration of the monosaccharide residue acting as the glycone for the linkage.

The bonds of a glycosidic link in the structural formula of a disaccharide are drawn so that the functional groups forming the glycosidic link line up with each other.

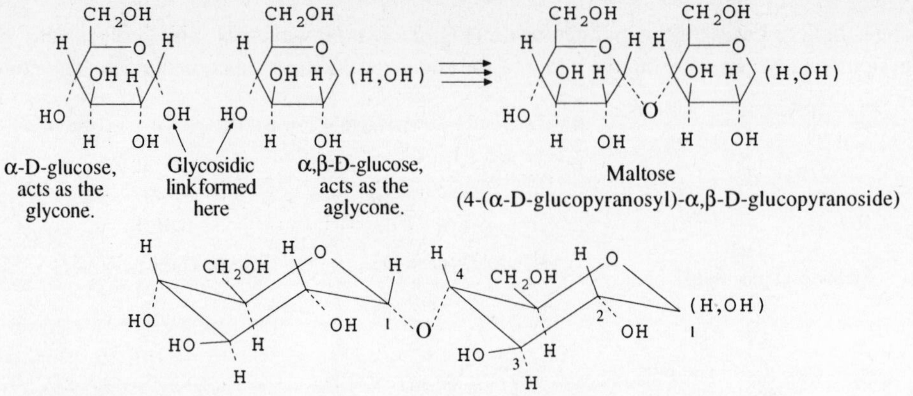

CH₂OH   CH₂OH   CH₂OH   CH₂OH

α-D-glucose, acts as the glycone.   Glycosidic link formed here   α,β-D-glucose, acts as the aglycone.

Maltose
(4-(α-D-glucopyranosyl)-α,β-D-glucopyranoside)

Fig. 27.19. The structure of maltose (4-(α-D-glucopyranosyl)-α,β-D-glucopyranoside) which consists of two glucose residues.

Consequently, the α and β structures of some monosaccharide residues are not drawn in their familiar orientation but are either turned around and/or over in the drawing of the structure of the compound. This fact must be remembered when determining whether it is the α or β isomer of a monosaccharide (27.2.1) that is incorporated into the structure of the compound (Fig. 27.20).

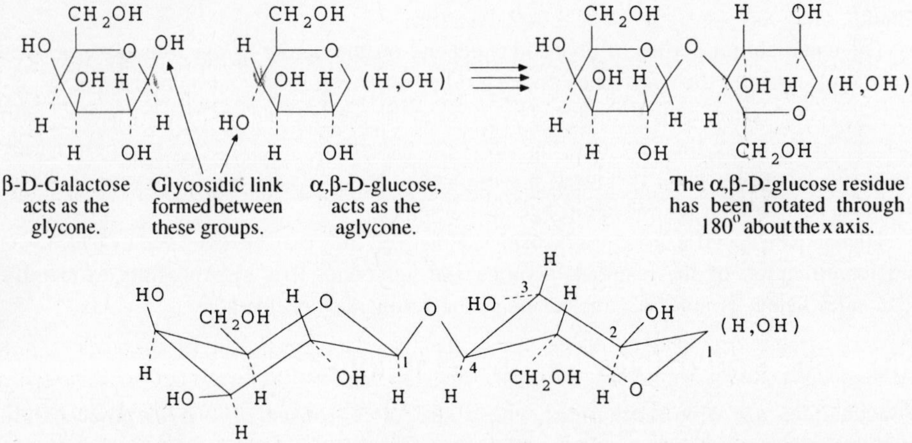

β-D-Galactose acts as the glycone.   Glycosidic link formed between these groups.   α,β-D-glucose, acts as the aglycone.

The α,β-D-glucose residue has been rotated through 180° about the x axis.

Fig. 27.20. The structure of lactose (4-O-β-D-galactopyranosyl-α,β-D-glucopyranose), a disaccharide consisting of one glucose and one galactose residue that occurs in the milk of mammals.

Glycosidic links are also formed between the carbons of what were the carbonyl groups of some monosaccharides. For example, the glycosidic link in sucrose is formed between C_1 of an α-D-glucose residue and C_2 of a β-D-fructose residue (Fig. 27.21). In these types of structure it is not possible to designate one residue as the glycone and so the glycosidic link is classified in terms of both the isomers forming the link. For example, the glycosidic link of sucrose would be referred to as an α,β-1,4-linkage.

The glycosidic link between the monosaccharide residues of disaccharides is hydro-lysed by water. The reaction is catalysed by acids and enzymes. Enzymic hydrolysis is

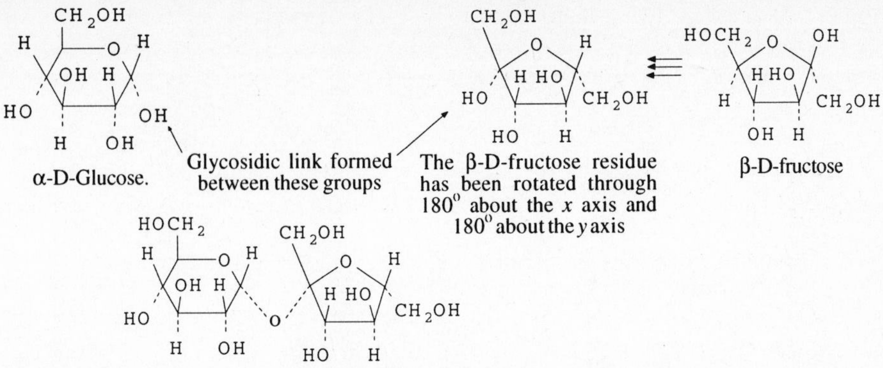

Fig. 27.21. The structure of sucrose (2-(α-D-glucopyranosyl)-β-D-fructofuranoside) whose structure contains one glucose and one fructose residue.

usually specific, a particular enzyme only being capable of catalysing the cleavage of a specific type of glycosidic link. For example, an α-glucosidase would catalyse the hydrolysis of maltose to glucose but would not catalyse the hydrolysis of the disaccharide cellobiose which has a β-1,4-glycosidic link between its two glucose residues (Fig. 27.23). This degree of stereospecificity is very important in the control of the metabolism of disaccharides, other oligosaccharides and polysaccharides.

27.12   OLIGO- AND POLYSACCHARIDES

The structures of oligo- and polysaccharides consist of monosaccharide residues linked by the same types of O-glycosidic linkages that are found in disaccharides. The common linkages are usually 1,4 and 1,6 but others are known. Homopolysaccharides whose structures contain only glucose residues are known collectively as **glycans**.

Naturally occurring oligosaccharides and polysaccharides can be generally divided into those which are involved in biological structures, such as cellulose and the mureins and those which act as metabolites, such as starch and glycogen.

27.12.1 Starch

Plants convert the glucose that is formed by photosynthesis into starch which the plants use as an energy store. These natural starches consists of a mixture of two types of glycan: the amyloses (10 to 20%) and the amylopectins (80 to 90%). Amyloses consist of linear chains of glucose residues joined by α-1,4-glycosidic linkages of the type found in maltose (Figs 27.19 and 27.22a). The chains vary in length, and relative molecular masses of between 150,000 and 6,000,000 have been recorded. Amylopectins have a similar structure but the chains have branches that are joined to the 'main' branch by α-1,6-glycosidic links (Figs 27.22 b and 27.22 c). Branches are usually separated by approximately twenty to thirty glucose units. Amylopectins usually have relative molecular masses of between 1 and 6 million. The amyloses and amylopectins are not very soluble in water. However, starch particles will form colloidal solutions.

The glycosidic links in starch are hydrolysed by water, the reaction being catalysed by acids and enzymes. Partial hydrolysis results in the formation of a mixture of smaller polymer molecules known as **dextrin** which is used as a gum. Complete hydrolysis yields

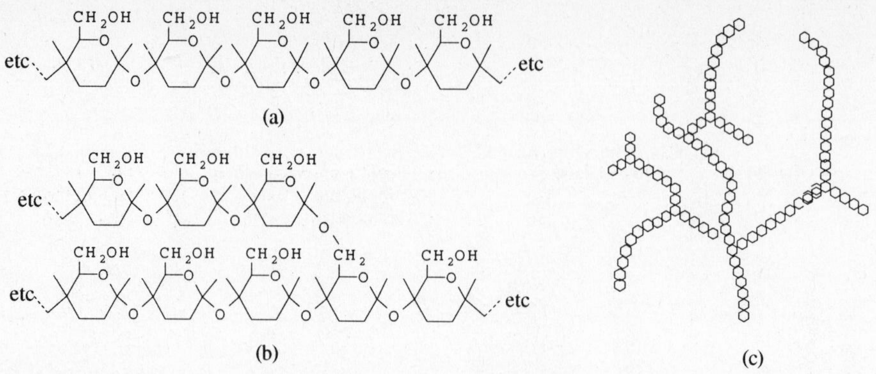

Fig. 27.22. Fragments of the structures of amylose (a) and amylopectins (b) and (c). Each of the hexagons in (c) represents a chain of five α-D-glucose residues.

only glucose. Enzymic hydrolysis that occurs during digestion enables humans to use plant starches as a source of glucose. The glucose that is not immediately used is converted into glycogen which is used as an energy store.

27.12.2　Glycogen

Glycogen is used by mammals as an energy store. It has a similar structure to amylopectin except the branches occur every 8 to 10 glucose residues. Glycogen releases its stored energy by a process known as glycolysis. The glycogen is converted to glucose-6-phosphate which in turn is converted to pyruvate, the sequence of steps being complex but well documented. The complete process produces more ATP than it uses and it is the conversion of this ATP to ADP which provides some of the energy for the organism. The fate of the pyruvate depends on the nature of the organism. For example, in yeast the pyruvate is decarboxylated and reduced to ethanol whilst in higher animals, such as humans, when oxygen is available (aerobic conditions) the pyruvate enters the citric acid cycle which proceeds through the electron transfer chain to produce a high yield of ATP. If no oxygen is available the pyruvate is converted to lactate (anaerobic conditions) which results in a much lower net formation of ATP. It is the local accumulation of lactate that is believed to be responsible for the severe pain of *cramp* following unaccustomed or excessive exercise.

27.12.3　Cellulose

Cellulose is a glycan which is found in the cell walls of higher plants. Its structure consists of chains of β-D-glucose units linked by 1,4-glycosidic linkages (Fig. 27.23). These glucose chains have relative molecular masses that vary from 150,000 to 1,000,000. However, cellulose cannot be used by humans as a source of glucose because the body does not produce a β-1,4-glucosidase which will break the 1,4-glycosidic linkages between the glucose units in cellulose. It passes through the digestive tract and provides the roughage that is essential for proper excretion of the body's waste products.

Cellulose is responsible for the rigidity of the cell walls of higher plants. This rigidity is believed to be due to the large numbers of hydrogen bonds that are formed between adjacent cellulose molecules. This results in the formation of bundles of cellulose molecules that form fibres with considerable strength.

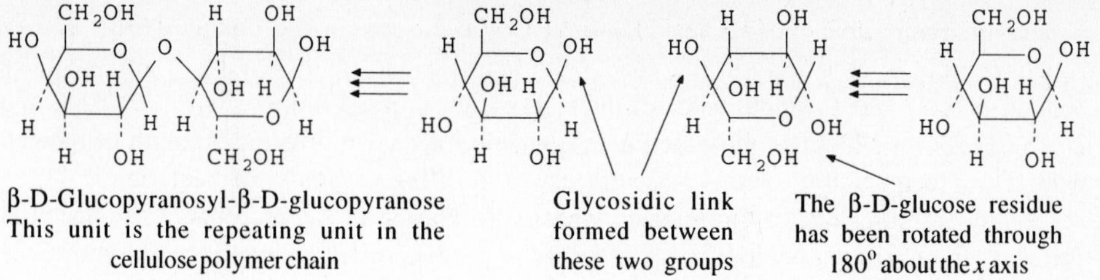

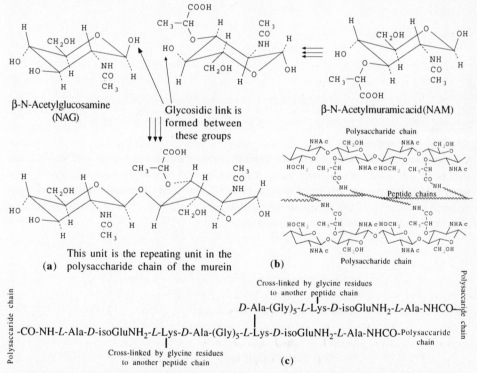

β-D-Glucopyranosyl-β-D-glucopyranose
This unit is the repeating unit in the
cellulose polymer chain

Glycosidic link
formed between
these two groups

The β-D-glucose residue
has been rotated through
180° about the *x* axis

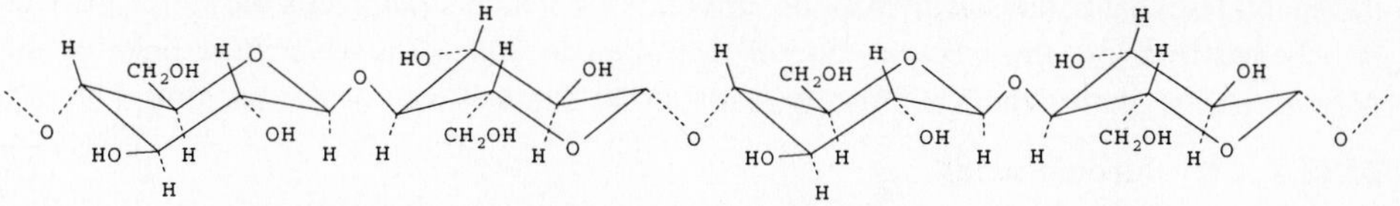

Fig. 27.23. The structure of a fragment of the β-D-glucose chain in cellulose.

27.12.4   Mureins

Mureins are a group of glycoproteins that are found in the bacterial cell wall. Mureins are
based on a polysaccharide that consists of alternating N-acetylglucosamine (NAG) and

β-N-Acetylglucosamine
(NAG)

Glycosidic link is
formed between
these groups

β-N-Acetylmuramic acid (NAM)

This unit is the repeating unit in the
(a)   polysaccharide chain of the murein

(b)

Polysaccharide chain

Peptide chains

Polysaccharide chain

Cross-linked by glycine residues
to another peptide chain

D-Ala-(Gly)$_5$-L-Lys-D-isoGluNH$_2$-L-Ala-NHCO–

-CO-NH-L-Ala-D-isoGluNH$_2$-L-Lys-D-Ala-(Gly)$_5$-L-Lys-D-isoGluNH$_2$-L-Ala-NHCO-Polysaccaride
chain

Cross-linked by glycine residues
to another peptide chain

(c)

Polysaccharide chain (left vertical)

Polysaccaride chain (right vertical)

Fig. 27.24. (a) The repeating unit of the polysaccharide chains of a murein. The sequence right to left shows
how the structural formula of the repeating units are constructed. (b) A schematic representation of a larger
fragment of the structure of a murein. The peptide chains (〜〜〜) cross-link the separate polysaccharide chains
together to form a structure with a lattice like appearance. (c) The structure of the peptide chain linking the
polysaccharide chains in *Staphylococcus aureus*, the groups of letters represent the amino acid residues in
the chain.

N-acetylmuramic acid (NAM) residues linked by β-1,4-glycosidic linkages (Fig. 27.24). A tetrapeptide is attached via the carboxylic acid residue of the NAM. These peptides contain both D and L amino acid residues. They are crossed-linked (28.7.1, 28.13.4 and 28.13.5) to form a latticework which is responsible for the rigidity and strength of the cell wall. The precise nature of this structure varies in different genera of bacteria.

Many antibiotics act by interfering with the formation of the mureins of the bacterial cell wall. For example, penicillins are believed to produce weak cell walls by inhibiting the transpeptidase that catalyses the peptide cross linking which strengthens the cell wall. This is believed to allow the enzyme murein hydrolase to attack the glycosidic links of the polysaccharide chain which ultimately leads to the death (lysis) of the bacteria.

27.12.5   Hyaluronic acids

Hyaluronic acids are a group of polysaccharides that control the viscosity of body jellies and lubricating fluids. Their relative molecular masses vary from 400,000 to 4,000,000. Their structures appear to be a chain that consists of alternating glucuronic acid and glucosamine residues linked by β-1,3-glycosidic links (Fig. 27.25).

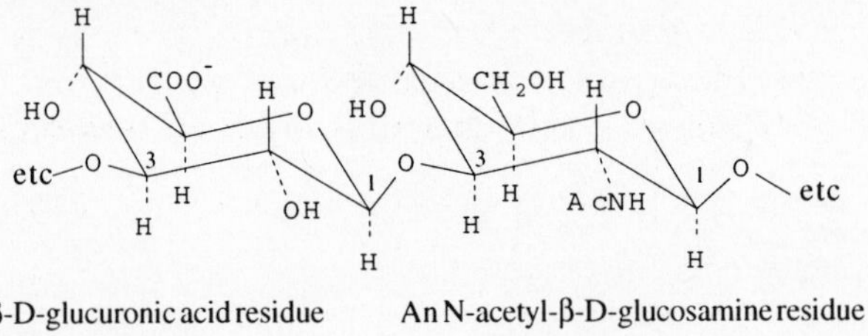

A β-D-glucuronic acid residue   An N-acetyl-β-D-glucosamine residue

Fig. 27.25. The repeating fragment of the polysaccharide chain of a hyaluronic acid.

27.12.6   Chondroitins

Chondroitin are polysaccharides that occur in cartilage, skin and connective tissue. Their structures are similar to the hyaluronic acids except galactosamine replaces glucosamine and at least one of the hydroxyl groups in the disaccharide repeating unit is a charged sulphate ester (Fig.27.26). Heparin which inhibits the clotting of blood, keratan sulphate and dermatan sulphates have a similar structure.

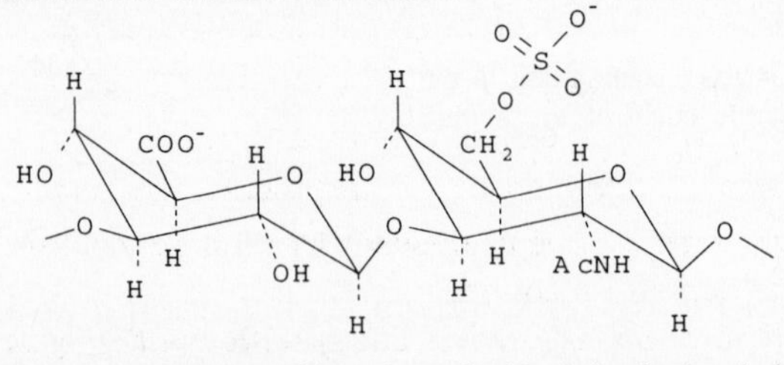

Fig. 27.26. The repeating fragment of the polysaccharide chain of a chondroitin-6-sulphate.

27.13 WHAT YOU NEED TO KNOW

(1) Carbohydrates are classified as: monosaccharides, oligosaccharides and polysaccharides.

(2) Monosaccharides are aliphatic polyhydroxy compounds that either contain or

yield aldehyde or ketone groups on hydrolysis and have structures of the type shown in Fig. 27.1.

(3) Oligosaccharide and polysaccharides are polymers of monosaccharides. An oligosaccharide molecule yields up to approximately nine monosaccharide molecules on hydrolysis while a polysaccharide molecule will yield a large number of monosaccharide molecules on hydrolysis.

(4) *Monosaccharides are classified as either aldoses or ketoses depending on whether their structures are based on aldehyde or ketone functional groups. They are further classified as trioses, tetroses, pentoses, etc. according to the number of carbon atoms in their structures.*

(5) *Trioses have 'straight chain' structures but tetroses, pentoses and larger monosaccharides have ring structures, the ring being closed by a hemiacetal or hemiketal group. Six-membered rings usually exist in the chair form whilst five-membered rings exist in the envelope conformation.*

(6) *Monosaccharides with six-membered ring are generally referred to as pyranoses whilst those with five-membered rings are generally known as furanoses.*

(7) *α Monosaccharide ring structures have the hydroxyl group of the hemiacetal or ketal on the opposite side to the terminal hydroxyl group whilst β monosaccharides have it on the same side of the ring structure as the terminal hydroxyl group.*

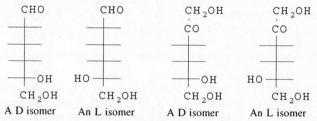

(8) *Monosaccharides are designated as D or L depending on the configuration of the chiral carbon with the highest locant. The configuration of the other chiral centres in the molecule are indicated by the name of the monosaccharide.*

(9) Epimers are multichiral centred compounds which have the same configurations for all of their chiral centres except one.

(10) *Monosaccharides are optically active. The specific rotation of an aqueous solution of a monosaccharide will change with time until it reaches a certain value. This phenomenon is known as mutarotation and is due to the establishment of the equilibrium system:*

$$\alpha\text{-Isomer} \;\rightleftharpoons\; \textbf{Straight chain form} \;\rightleftharpoons\; \beta\text{-Isomer}$$

(11) *Monosaccharides exhibit many of the chemical properties of primary and secondary alcohols, aldehydes, ketones, hemiacetals and hemiketals.*

(12) *Amino sugars are aldoses that have one or more of their hydroxyl groups replaced by amino or substituted amino groups, for example, glucosamine.*

(13) *Amino sugars have similar structures to those of the corresponding monosaccharides.*

(14) *Glycosides consist of a sugar residue (glycone) bonded to a non-sugar residue (aglycone) via a glycosidic link.*

(15) *The common types of linkage between the glycone and the aglycone are C–O–C (O–glycosides), C–N–C (N–glycosides), C–S–C (S–glycosides) and C–C (C–glycosides).*

β-O-glycosides   α-S-glycosides   β-N-glycosides   β-C-glycosides

(16) *Glycosides can exist as α and β isomers.*

(17) Glycosides are soluble in water and ethanol.

(18) *The glycosidic linkages of O-glycosides, N-glycosides and S-glycoside are hydrolysed by enzymes or aqueous acid to form the sugar and the aglycone.*

(19) *Many naturally occurring compounds exist in nature as glycosides.*

(20) *Oligosaccharides and polysaccharides have molecular structures in which monosaccharide units are linked together by O-glycosidic links.*

(21) The common positions of the glycosidic links in oligosaccharides and polysaccharides are 1,4 and 1,6. The links are classified according to the nature of the monosaccharide residue acting as the glycone.

Maltose, an α 1,4 link   An α 1,6 link and β 1,4 links in a fragment of an amylopectin.

(22) *Polysaccharides whose structures contain only glucose residues are known as glycans.*

(23) *The glycosidic links in polysaccharides are hydrolysed by enzymes and aqueous acid. Enzymes are usually specific in their action and a particular enzyme will only hydrolyse a certain type of glycosidic link.*

(24) Starch is a mixture of amyloses and amylopectins. Amyloses consist of chains of glucose residues linked by α-1,4-glycosidic linkages. Amylopectins are glycans whose structure contains branched chains of glucose residues linked by α-1,4-glycosidic links. Branches are attached by α-1,6-glycosidic links and are separated by about every twenty to thirty glucose residues.

(25) Glycogen is used as part of the energy store in animals. It has a similar structure to amylopectins except the branches occur about every eight to ten residues.

(26) Cellulose is a glycan that is responsible for the rigidity of the cell walls of plants. Its molecules consist of chains of glucose residues linked by β-1,4-glycosidic links.

(27) Mureins are glycoproteins that are found in the cell walls of bacteria. They consist of polysaccharides linked by small peptides to form a latticework-like structure.

(28) Polysaccharides play an important part in the structure of the human body, for example, hyaluronic acids govern the viscosity of the body jellies and lubricating fluids and chondroitins form part of the structure of the skin, cartilage and connective tissues.

27.14   QUESTIONS

(1) Classify the following compounds as D or L isomers.

(a)
```
      CHO
  H ——OH
 HO ——H
  H ——OH
    CH₂OH
```
(b)
```
      CHO
 HO ——OH
 HO ——H
  H ——OH
    CH₂OH
```
(c)
```
    CH₂OH
    CO
 HO ——H
 HO ——H
    CH₂OH
```
(d)
```
    CH₂OH
    CO
  H ——OH
  H ——OH
    CH₂OH
```

(2) Classify the following compounds as α or β isomers.

(a), (b), (c), (d) [structural formulae]

(3) State whether the glycosidic linkages in each of the compounds is either α, β or α,β link.

(a), (b), (c) [structural formulae]

(4) Draw the structural formula of α-D-galactosamine. Classify the functional groups in this compound and list under each functional group their general properties. Predict, using chemical equations, the possible products of the reaction of glucosamine with: (a) methyl iodide, (b) ethanoic acid, (c) hydroxylamine, (d) nitrous acid and (e) excess ethanoic anhydride.

(5) Explain why fructose and glucose yield the same product on reaction with phenylhydrazine.

(6) A disaccharide A is hydrolysed in the presence of an α-D-glucosidase to D-glucose and D-galactose. Compound A was completely methylated by reaction with methanol in dry acid conditions followed by further reaction with dimethyl sulphate in the presence of sodium hydroxide. Hydrolysis of the fully methylated compound produced 2,3,4,6-tetra-O-methylglucose and 1,2,3,6-tri-O-methylgalactose. Deduce the structure of A.

28

Amino acids, peptides and proteins

28.1 INTRODUCTION

Amino acids are compounds whose structures contain a carboxylic acid group with an amino group usually but not always α to the carboxylic acid. About twenty-five amino acids constitute the basic building blocks of the peptides and proteins that occur in biological systems. They are found in all cells and are essential for life. For example, in the human body peptides and proteins account for the strength and elasticity of the skin, the strength of bones, act as catalysts for metabolic processes, carry metabolites around the body, are an essential part of the nervous system and, in the form of antibodies, act as the body's defence force. However, the human body cannot synthesise many of the amino acids that it requires and so must obtain them from plant and animal sources in its diet. Dietary deficiencies often lead to severe deficiency diseases and ultimately death. Peptides and proteins obtained from natural sources are used extensively as drugs.

28.2 AMINO ACIDS

Amino acids are generally classified as α, β, γ, etc. amino acids depending on the relative positions of the acid and amino functional groups (Fig. 28.1).

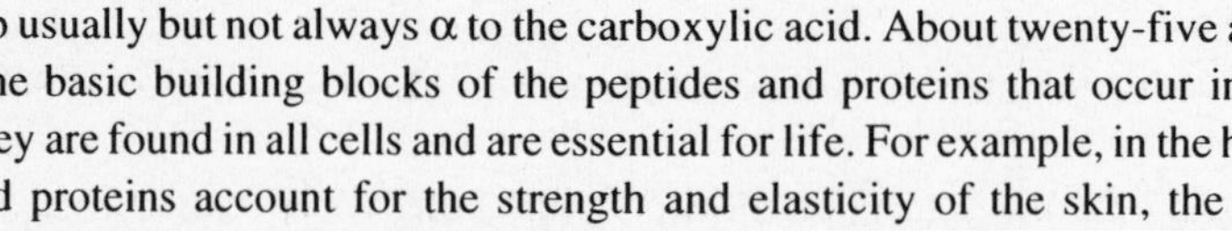

$$\overset{\displaystyle NH_2}{\underset{\alpha}{\underset{|}{RCHCOOH}}}\qquad\overset{\displaystyle NH_2}{\underset{\beta\ \ \alpha}{\underset{|}{RCHCH_2COOH}}}\qquad\overset{\displaystyle NH_2}{\underset{\gamma\ \ \beta\ \ \ \alpha}{\underset{|}{RCHCH_2CH_2COOH}}}$$

α-Amino acid   β-Amino acid   γ-Amino acid

Fig. 28.1. The general structures and classification of amino acids.

The structures of amino acids can contain a variety of functional groups (Table 28.1). Many of these functional groups are either acidic or basic and so amino acids are classified as being neutral, acidic and basic compounds.

Table 28.1. Examples of neutral, acidic and basic amino acids.

Neutral amino acids	Name	Symbol/Letter		Isoelectric point
$CH_2(NH_2)COOH$	Glycine	Gly	G	6.0
$CH_3CH(NH_2)COOH$	Alanine	Ala	A	6.0
$CH_2OHCH(NH_2)COOH$	Serine	Ser	S	5.7
$PhCH_2CH(NH_2)COOH$	Phenylalanine	Phe	F	5.5
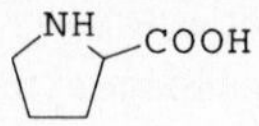	Proline	Pro	P	6.3
$CH_3SCH_2CH_2CH(NH_2)COOH$	Methionine	Met	M	5.7
$H_2NCOCH_2CH_2CH(NH_2)COOH$	Glutamine	Gln	Q	5.7
Acidic amino acids				
$HOOCCH_2CH_2CH(NH_2)COOH$	Glutamic acid	Glu	E	3.2
$HOOCCH_2CH(NH_2)COOH$	Aspartic acid	Asp	D	3.0
Basic amino acids				
	Arginine	Arg	R	10.8
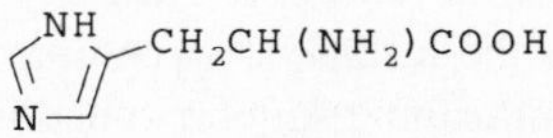	Histidine	His	H	7.6
$H_2NCH_2CH_2CH_2CH_2CH(NH_2)COOH$	Lysine	Lys	K	9.7

All amino acids, with the exception of glycine, have chiral centres and are optically active. The configurations of these centres are classified using both the D/L and R/S nomenclature systems (4.8). Most naturally occurring amino acids have an L or S configuration. One exception is L-cysteine which has an R configuration.

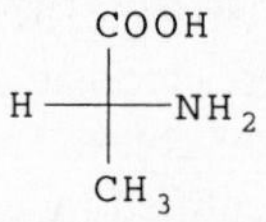

D-Alanine or R-Alanine 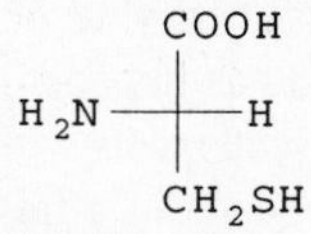   L-Cysteine or R-Cysteine (Cys or C) 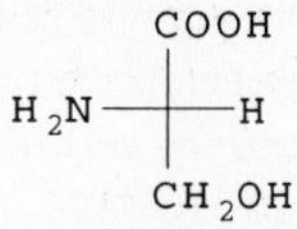   L-Serine or S-Serine

Amino acids exist in the solid state as a dipolar ion known as a **zwitterion** (Fig. 28.2). Zwitterions are in essence internal salts and so exhibit many of the properties of salts. For example, they have high melting points, large dipoles and are soluble in water.

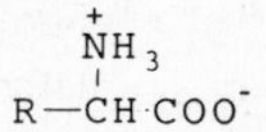

$$\overset{+}{N}H_3$$
$$R-CH\cdot COO^-$$

Fig. 28.2. A zwitterion.

28.3 NOMENCLATURE OF AMINO ACIDS

Amino acids may be named systematically but it is more usual to use the trivial names given in Table 28.1. The symbols/letters listed with the names of the amino acids are often used to represent amino acids residues in the structures of peptides and proteins (Fig. 28.8).

28.4 PHYSICAL PROPERTIES OF AMINO ACIDS

Amino acids are usually high melting point solids which often decompose on heating.

28.4.1 Solubility

Amino acids are usually soluble in water and insoluble in many organic solvents. In aqueous acid solution (low pH) the carboxylate ion acts as a base and accepts a proton to form a cation, whilst in aqueous basic solution (high pH) the ammonium ion acts as an acid and loses a proton to form an anion. These processes form two interdependent dynamic equilibrium systems and so either the addition or the presence of acid or base in an amino acid solution will move the positions of equilibrium in the appropriate direction.

$$\underset{\text{R--CH--COO}^-}{\overset{\overset{\displaystyle NH_2}{|}}{}} \quad \underset{\text{Base}}{\overset{\text{Acid}}{\rightleftharpoons}} \quad \underset{\underset{\textbf{Zwitter ion}}{\text{R--CH--COO}^-}}{\overset{\overset{\displaystyle \overset{+}{N}H_3}{|}}{}} \quad \underset{\text{Base}}{\overset{\text{Acid}}{\rightleftharpoons}} \quad \underset{\text{R--CH--COOH}}{\overset{\overset{\displaystyle \overset{+}{N}H_3}{|}}{}}$$

The pH at which only the zwitterion is present in solution is known as the **isoelectric point** of the amino acid (Table 28.1). The value of the isoelectric point of an amino acid depends on the relative strengths of the acid and basic groups in the molecule (6.13). For example, aspartic acid has an acidic isoelectric point because the acidic strengths of the two carboxylic acid groups outweigh the basic strength of the one amino group

The differences between the isoelectric points of amino acids form the basis of the isolation, separation and identification of amino acids using the technique of **comparative electrophoresis** (9.10.3). The identification is carried out using a support consisting of either chromatography paper soaked in a suitable buffer or an agar gel containing a suitable buffer. Those amino acids that exist as cations at the pH of the buffer will migrate to the cathode while those that exist as anions will migrate to the anode. The different ions usually migrate at different speeds. Amino acids whose isoelectric point corresponds to the pH of the buffer will remain almost stationary at the point of application. Electrophoresis is stopped before the amino acid ions move off the support. The electrochromatogram is stabilised by quickly drying or chemically fixing. The positions of the components of the mixture are found as in paper and thin layer chromatography by treating the paper or gel with ninhydrin which reacts with most amino acids to form purple to violet complexes (28.5.4a). The compounds in a mixture are identified by comparing their positions with those of standard samples which are run at the same time. If the relative positions of a component of the mixture and one of the standards are the same they are assumed to be the same compound.

Electrophoresis is not normally used for the analysis of amino acids; however, it is used for the purification and analysis of peptides, proteins and other macromolecules (28.10). Amino acid analysis methods are normally based on ion exchange chromatography, high pressure liquid chromatography and thin layer chromatography.

28.4.2 Optical activity

All amino acids with the exception of glycine have a chiral centre and usually exist in an optically active form in nature. However, their specific rotations in solution are pH-dependent which makes it difficult to use the specific rotation for identification purposes.

28.5 CHEMICAL PROPERTIES OF AMINO ACIDS

The chemical properties of amino acids may be conveniently classified into those due to the 'carboxylate group', the 'amino group', other functional groups in the molecule (if any), and those characteristic of amino acids.

28.5.1 Properties characteristic of the 'carboxylate group'

Amino acids exhibit the properties of weak carboxylic acids (Chapter 20). For example, they form salts with hydroxides and carbonates, esters with alcohols, decarboxylate on heating above 200° and are reduced to primary alcohols by lithium aluminium hydride (Fig.28.3).

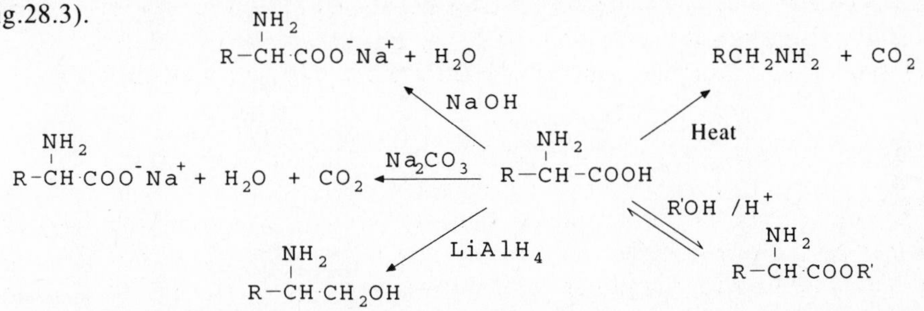

Fig. 28.3. Examples of the reactions of amino acids that are similar to those of carboxylic acids.

Enzyme-catalysed decarboxylation of amino acids is an important step in many biological processes. For example, decarboxylation of levodopa yields dopamine, the precursor of noradrenalin which is responsible for hypertensive effects in humans. The decarboxylation is catalysed by dopa decarboxylase. The anti-hypertensive drug *Methyldopa* acts by successfully competing with the levodopa for this enzyme. However, decarboxylation of the drug gives rise to α-methylnoradrenalin which also acts as a hypertensive agent but not so effectively as the naturally produced noradrenalin. Consequently, the body's hypertensive activity is reduced but not completely suppressed.

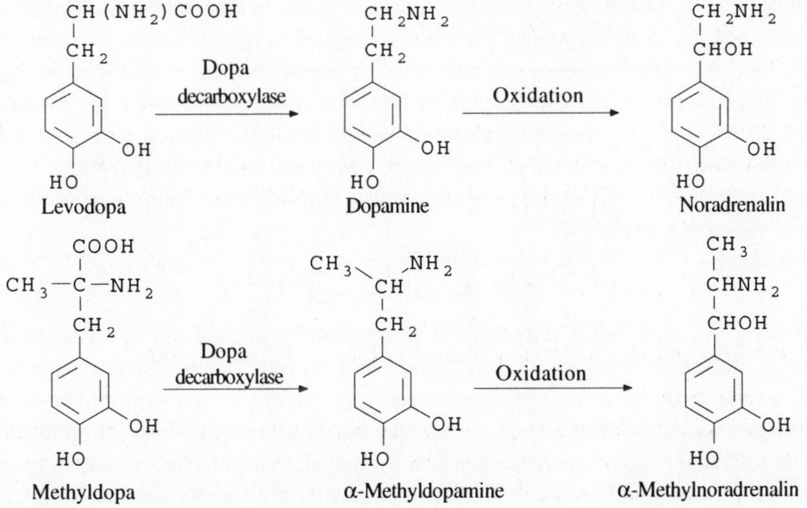

Esterification of amino acids is used extensively to protect the carboxylic acid group in the synthesis of peptides and proteins.

28.5.2   Properties characteristic of the 'amino group'

The 'amino groups' of amino acids exhibit many of the properties of primary amines (Chapter 24). For example, the 'amino group' can form salts with both organic and inorganic acids, react as a nucleophile and be oxidised (Fig. 28.4).

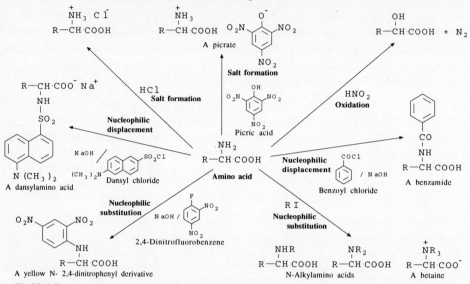

Fig 28.4. Examples of the 'amino' group reactions of amino acids. In the nucleophilic reactions the 'amino' group is acting as the nucleophile in the type of reaction specified.

Many of the reactions of the 'amino group' of amino acids are used for the identification and assay of amino acids. For example, the reactions with picric acid, 2,4-dinitro-fluorobenzene or 2,4-dinitrochlorobenzene, dansyl chloride and benzoyl chlorides are used in identification tests for amino acids and to prepare derivatives of these compounds in organic analysis. These reagents are also used in the determination of the structures of peptides and proteins (28.12). Dansylglycine has been used to investigate the binding of acidic drugs to plasma proteins because it shows an enhanced fluorescence with a blue hypsochromic shift when it binds to substances with a low dielectric constant.

Amino acids may be assayed by titration with perchloric acid using glacial ethanoic acid as solvent (11.6.3).

Overall reaction:

$$\underset{\text{RCHCOOH}}{\overset{NH_2}{|}} \xrightarrow[\text{Ethanoic acid}]{HClO_4} \underset{\text{RCHCOOH}}{\overset{\overset{+}{N}H_3 \quad ClO_4^{-}}{|}}$$

Alternatively, amino acids may be treated with methanal and the resulting solution titrated with sodium hydroxide (the formol titration). In this method the methanal reacts with the 'amino group' but the precise nature of the reaction is not certain. However, it effectively prevents the formation of zwitterions which allows the direct titration of the carboxylic acid with sodium hydroxide which for calculation purposes reacts in a one to one ratio with the original amino acid. Phenolphthalein is normally used as indicator.

N-Acylation of glycine and occasionally glutamic acid to form a hippuric acid is the basis of the excretion of aromatic and heteroaromatic carboxylic acids in mammals. These hippuric acids are excreted in the urine.

$$ArCOOH \xrightarrow{\text{Two steps}} ArCOSCoA \xrightarrow{\hspace{1cm}} ArCONHCH_2COOH$$

$$H_2NCH_2COOH \qquad HSCoA \quad \text{A hippuric acid}$$

$$\text{Glycine} \qquad \text{Co-enzyme A}$$

The excretion of hippuric acid is the basis of the use of sodium benzoate to check the detoxification function of the liver. The liver combines the benzoate with glycine to form hippuric acid which is subsequently excreted in the urine. A reduced amount of hippuric acid is secreted if the liver is damaged. This test has now been superceded.

28.5.3 Properties characteristic of the other functional groups

The other functional groups found in amino acid molecules will usually exhibit many of the chemical properties that are found in the simple compounds containing those functional groups. For example, the thiol group of cysteine is easily oxidised to the disulphide cystine which is readily reduced to cysteine. The cysteine/cystine redox system is of considerable importance in protein chemistry and biochemistry.

$$HSCH_2CHCOOH \underset{\text{Reduction}}{\overset{\text{Oxidation}}{\rightleftarrows}} HOOCCHCH_2\text{-S-S-}CH_2CHCOOH$$

$$\overset{NH_2}{|} \qquad \qquad \overset{NH_2}{|} \qquad \qquad \overset{NH_2}{|}$$

It is interesting to note that the thiol group of cysteine has growth promoting properties. A solution of cysteine hydrochloride in a sodium borate buffer (Squibb) is used to aid the healing of wounds.

Many of the reactions of the other functional groups, if any, in a particular amino acid may be found in the appropriate chapters of this text.

28.5.4 Properties characteristic of amino acids

Amino acids exhibit a number of properties that are due to the joint action of the 'amino' and carboxylate groups.

(a) Ninhydrin reaction. Most amino acids react with ninhydrin to form purple compounds. A few amino acids such as proline form yellow compounds. This reaction is the basis of identification tests and colorimetric assays for amino acids. The ninhydrin reaction is sufficiently characteristic to be used as a reliable spot test for amino acids and to locate amino acids on paper and thin layer chromatograms. The ninhydrin reaction is also the basis of the detection, identification and measurement of the relative concentrations of amino acids in amino acid analysers.

Ninhydrin   Purple complex

(b) Complex salt formation. Amino acids readily form stable metal complexes with a variety of metals. The ratio of metal ion to amino acid residues in these complexes is usually 1:2. Their structures normally involve the bonding of both the acid and 'amino' groups to the metal. However, in the case of histidine, experimental evidence indicates that it is only the basic groups that are bonded to the metal. These histidine complexes are significantly more stable than those formed by other amino acids (32.2). This enhanced stability is thought to be the reason for histidine inhibiting the affect of cobalt on tumours and micro-organisms.

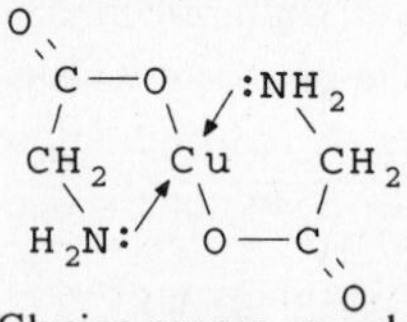

Glycine-copper complex 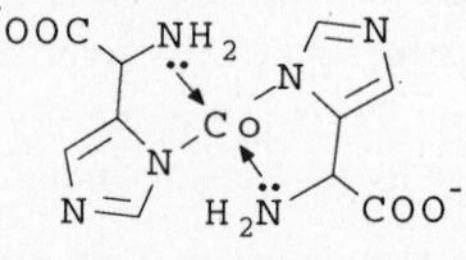   Histidine-cobalt complex

The basis of the treatment of cases of lead and iron poisoning with penicillamine is the formation and subsequent excretion of the appropriate amino acid-metal complexes. The structures of these complexes are still the subject of much discussion.

28.6 PEPTIDES AND PROTEINS

Peptides and proteins are polymers whose structures consist of amino acid molecules linked together by amide functional groups formed between the carboxylic acid of one amino acid molecule and the 'amino' group of a second, usually different, amino acid molecule. For example, the structure of the dipeptide alanylglycine consists of an alanine molecule bonded via an amide link to a glycine molecule (Fig. 28.5). The amide links ($-CONH-$) of peptides are commonly referred to as **peptide links or bonds** and the amino acid molecules in the structures as **residues**. Hydrolysis of all the amide links in a peptide or protein yields a mixture of their constituent amino acids.

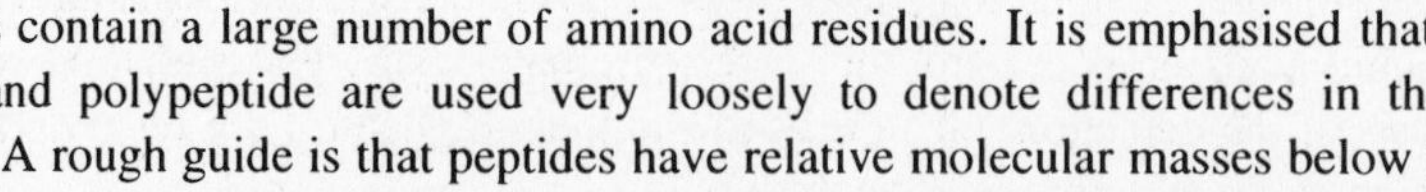

$$NH_2$$
$$CH_3CHCOOH$$
Alanine

$$NH_2 \quad A$$
$$CH_3CH-CO-NH-CH_2COOH$$
The alanine residue   The glycine residue

$$H_2NCH_2COOH$$
Glycine

Fig 28.5. The structure of alanylglycine. The amide functional group (A) is commonly referred to as a peptide link or bond in peptide and protein structures.

The term **peptide** is usually used for compounds whose structures contain a small number of amino acid residues whilst the term **polypeptides** is used for larger peptides whose structures contain a large number of amino acid residues. It is emphasised that the terms peptide and polypeptide are used very loosely to denote differences in the sizes of peptides. A rough guide is that peptides have relative molecular masses below 500 whilst polypeptides have relative molecular masses between 500 and 2000. The term **oligopeptide** is sometimes used for peptides containing 2 to approximately 9 amino acid residues.

Proteins are naturally occurring macromolecules with relative molecular masses greater than 2000. They may be broadly classified into two types: **simple** and **conjugated proteins**. Simple proteins are large polypeptides that yield only amino acids on hydrolysis whilst conjugated proteins are large polypeptides that produce amino acids and other compounds.

on hydrolysis. Both simple and conjugated proteins may be further subdivided into globular and fibrous proteins. The molecules of globular proteins are roughly spherical in shape and tend to be water soluble whilst those of fibrous proteins have either a linear or sheet-like structure and tend to be insoluble in water. These shapes and solubilities reflect their biological activity. Fibrous proteins, for example, tend to be associated with biological structures such as muscles, hair and connective tissue whilst globular proteins are usually involved in metabolism.

28.7 THE STRUCTURES OF PEPTIDES AND PROTEINS

The structures of peptides and proteins are very diverse. However, they exhibit a number of common features which are referred to as the **primary, secondary, tertiary** and **quaternary structures** of peptides and proteins. These terms define aspects of the organisation of peptides and proteins. However, because of the diverse nature of the structures of peptides and proteins, these terms should not be applied rigidly to all structural situations. Alteration of any of these structural features in a peptide or protein will usually destroy the compound's biological activity.

28.7.1 Primary structure

The primary structure of peptides and proteins is the order in which the amino acids appear in the structure. It is the free rotation about the C_1–C_2 and the C_2–N bonds of the amino acid residues in this structure that allows the structure to assume the variety of shapes that characterise peptide and protein structures. The amide bonds linking the amino acid residues have a flat rigid molecular structure which is readily explained by the sp^2 hybridisation of the carbon, oxygen and nitrogen atoms of the amide group (Fig. 28.6).

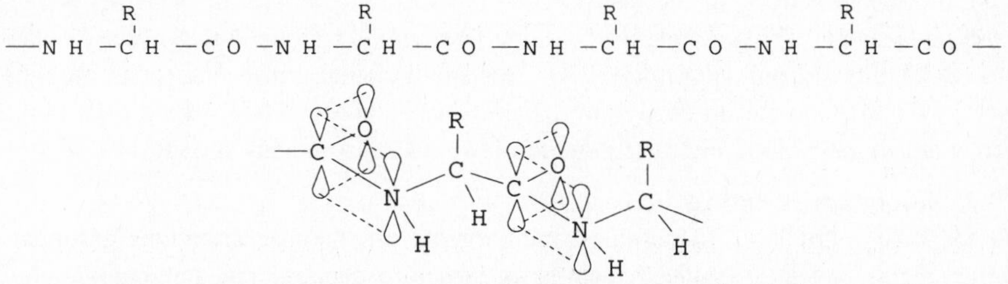

Fig. 28.6. The structure of a fragment of the primary structure of peptides and proteins.

The structures of most peptides and proteins consist of a single chain of amino acids. However, different parts of the same peptide chain or different peptide chains may be linked (**cross-linking**) (Fig. 28.7). For example, the structure of insulin (Fig. 1.1) contains two peptide chains that are held together by two **disulphide links** or **bridges** (−S−S−) due to the presence of cysteine residues in the chains. The structure of insulin also contains a closed loop formed by a disulphide bridge. Disulphide bridges are a very common form of cross-linking in peptides and proteins. Other types of cross-linking are illustrated in Figs 28.21 and 28.23.

It is inconvenient to draw normal structural formula for the primary structures of polypeptides and proteins because of the size of the structures involved. Consequently, either three letter symbols or single letters (Table 28.1) are used to represent the amino

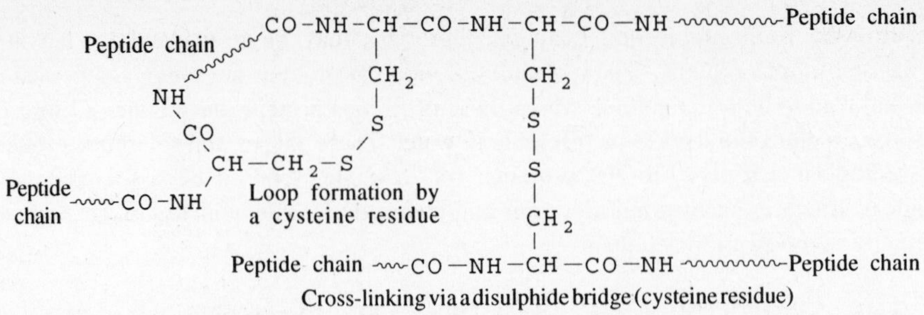

```
                        CO—NH—CH—CO—NH—CH—CO—NH ~~~~~~Peptide chain
  Peptide chain ~~~~           |                |
                              CH₂              CH₂
          NH                   |                |
            \                  S                S
            CO                /                 |
              \     CH—CH₂—S                    S
  Peptide ~~~CO—NH /   Loop formation by        |
  chain            a cysteine residue          CH₂
                                                |
        Peptide chain ~~~CO—NH—CH—CO—NH ~~~~~~Peptide chain
        Cross-linking via a disulphide bridge (cysteine residue)
```

Fig. 28.7. Cross-link and loop formation by disulphide bridges

acid residues in the structures of peptides and proteins (Fig. 28.8). These symbols or letters are usually written with the N-terminal (NH_2 or H) of the peptide chain on the left and its C-terminal (COOH or OH) on the right-hand side of the line of symbols. In more complex compounds, whose primary structures cannot be written in a straight line, a key is used to indicate the N and C terminals in the structure.

(NH_2)Tyr-Gly-Gly-Phe-Met-Thr-Ser-Glu-Lys-Ser-Glu-Thr-Pro-Leu-Val
|
(HOOC)Glu-Gly-Lys-Lys-Tyr-Ala-Asn-Lys-Ile-Ile-Ala-Asn-Lys-Phe-Leu-Thr

(H) YGGFMTSEKSETPLVTLFKNAIIKNATKKGE(OH)

Fig. 28.8. Representations of the structure of β-endorphin, a naturally occurring peptide with opiate activity (anaesthetic and analgesic activities). β-Endorphin is believed to be one of the compounds that are produced in mammals to counteract pain.

Minor changes in the primary structure of a polypeptide will have important consequences on the biological activity of the polypeptide. For example, sickle-cell anaemia is caused by the replacement of a glutamic acid residue by a valine residue in the primary structure of the β-subunit of haemoglobin (28.7.4). This small change causes the round 'doughnut' shape of a red blood cell to change to a sickle shape. These sickle-shaped cells clog the arteries and impair blood circulation which leads to heart attacks and strokes.

28.7.2 Secondary structure
The secondary structures of peptides and proteins are the conformations assumed by peptide chains and the organised sections of structure that occur in polypeptide chains. Secondary structures are mainly formed by non-covalent bonding between sections of the same chain or adjacent peptide chains. The principal form of the non-covalent bonding is hydrogen bonding between the oxygen atom of the carbonyl group of one amide group with a hydrogen attached to the nitrogen of the another amide group further along the chain. Van der Waals' forces and hydrophobic bonding are also responsible for stabilising the secondary structures of peptides and proteins. The main forms of secondary structure are the right-handed α-helix, the β-pleated sheet and the triple helix (Fig. 28.9).

The α-helix and β-pleated sheet secondary structures usually occur in parts of the structures of peptides and protein and not throughout its entire structure. Segments of α-helix secondary structures are found in many peptides and proteins. However, β-pleated sheets are uncommon except in β-keratin, a protein found in human skin and fibroin, a protein that occurs in natural silk. Triple helixes are found in a small group of polypeptides

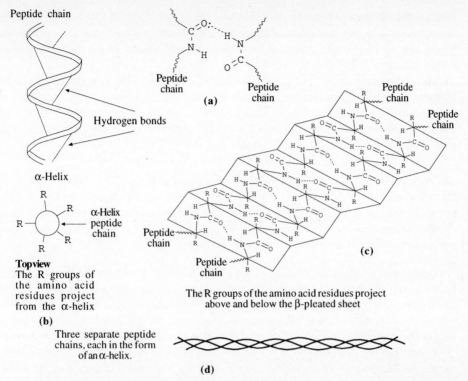

Fig. 28.9. Secondary structures of peptides and proteins. (a) Hydrogen bonding between amide groups.
(b) The α-helix. (c) β-pleated sheet. (d) Triple helix.

polypeptides known as tropocollagens that occur in collagen (28.12), the major fibrous material that occurs in the connective tissue of the human body. Tropocollagen molecules consist of three polypeptide chains, each containing about 1000 amino acid residues, plaited together to form a triple helix. The structure is held together by hydrogen bonds and a few disulphide bridges. Groups of these tropocollagen molecules form very strong microfibres (fibrils) that constitute the basis of the structures of bones and teeth.

28.7.3   Tertiary structure

The tertiary structure of a polypeptide is the overall shape assumed by the folding of the polypeptide chain. It is stabilised by disulphide bridges, hydrogen bonding, van der Waals'

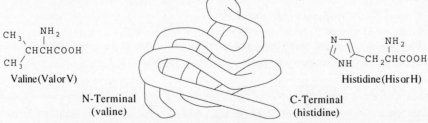

Fig. 28.10. The folded tertiary structure of the β chain of human haemoglobin.

forces and hydrophobic bonding. For example, the folded shape of the β-chain of human haemoglobin is largely held in shape by hydrophobic bonding (Fig. 28.10). It is interesting to note that the hydrophilic groups of the β-chain of haemoglobin are on the surface of the folded structure.

The electrostatic attraction between the free $-COO^-$ and $-\overset{+}{N}H_3$ groups that are found in some amino acid residues at body pH will also stabilise the tertiary structures of peptides and proteins. These electrostatic attractions, which are in effect ionic bonds, are known as **salt bridges**. They are usually formed between acidic and basic amino acid residues such as glutamic acid and lysine.

Conjugated proteins incorporate non-protein residues into their tertiary structures. For example, the β chain of human haemoglobin incorporates haem (Fig. 32.12) into its tertiary structure. These residues are known as **prosthetic groups** and are usually the source of the proteins biological activity. For example, the haem unit in the β chain of human haemoglobin is the part of the molecule that carries the oxygen or carbon dioxide.

The way in which a polypeptide chain is folded plays an important part in determining the physical, chemical and biological properties of the molecule. Disruption of the folded structure usually destroys the polypeptide or protein's biological activity.

28.7.4 Quaternary structure

Quaternary structures occur when polypeptide and protein molecules fit together to form dimers, trimers, etc. that are mainly held together by hydrogen bonds, van der Waals' forces, salt bridges and hydrophobic bonding. Human haemoglobin has a quaternary structure that consists of two types of conjugated polypeptide known as the α and β subunits respectively. Each subunit incorporates a haem molecule and the complete haemoglobin molecule consists of two α and two β subunits held together by hydrogen bonds and salt bridges. In deoxyhaemoglobin (haemoglobin with no oxygen) the four subunits are arranged around a cavity that is large enough to hold the anion 2,3-bisphosphoglycerate (BPG). This BPG ion is squeezed out of the cavity when the first oxygen molecule binds to a haem unit of the haemoglobin.

$$2,3\text{-Bisphosphoglycerate} \qquad {}^-O_3POCH_2\overset{\overset{\displaystyle OPO_3^-}{\displaystyle |}}{C}HCOO^-$$

The separation of the component polypeptide molecules of the quaternary structure of a protein will usually destroy biological activity.

28.8 NOMENCLATURE OF PEPTIDES AND PROTEINS

Trivial names are commonly used for peptides and proteins. However, small peptides may be named systematically. The name of the C-terminal amino acid residue is taken as the parent or stem name of the compound. The other amino acid residues are treated as substituents of this residue. The ending **-ine** of the names of these substituent amino acid residue are changed to **-yl** and they are placed before the stem name in the order that they occur in the primary structure of the peptide (Fig.28.11). This system of nomenclature is only suitable for small peptides.

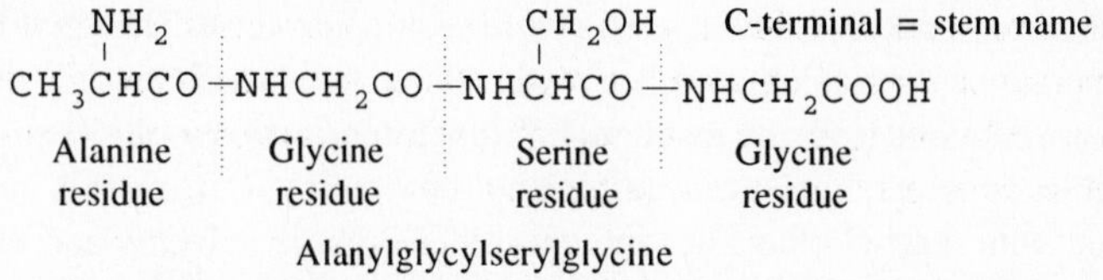

$$\underset{\substack{\text{Alanine}\\ \text{residue}}}{CH_3\overset{\displaystyle NH_2}{\overset{|}{C}HCO}} {-} \underset{\substack{\text{Glycine}\\ \text{residue}}}{NHCH_2CO} {-} \underset{\substack{\text{Serine}\\ \text{residue}}}{NH\overset{\displaystyle CH_2OH}{\overset{|}{C}HCO}} {-} \underset{\substack{\text{Glycine}\\ \text{residue}}}{NHCH_2COOH} \quad \text{C-terminal = stem name}$$

Alanylglycylserylglycine

Fig. 28.11. An example of the nomenclature used for small peptides. The structures CO–NH are the peptide bonds (amide functional groups) in the structure of alanylglycylserylglycine

28.9   DENATURATION AND ISOLATION OF PEPTIDES AND PROTEINS

28.9.1 Denaturation

The secondary, tertiary and quaternary structures of peptides and proteins are altered by changes in their environment. These changes usually destroy biological activity and are normally irreversible. The process is referred to as **denaturation** and may be brought about by heating, changes in pH and high concentrations of reagents such as certain salts, urea and guanidine. Denaturation results in a significant loss of organisation in the compound's structure. It leads to a structure which is usually less soluble in water. This fact has been utilised in a number of tests for the detection of proteins in body fluids. For example, the detection and estimation of protein in urine (Fig. 28.12).

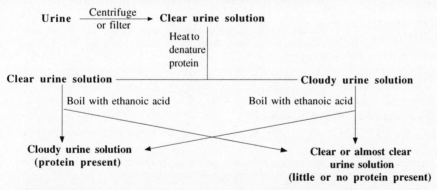

Fig. 28.12. Denaturation and the detection of proteins in urine. Small (10-15 mg dm^{-3}) amounts of protein are normally present in urine, but this quantity is not detected by this method which is sensitive down to about 120-150 mg dm^{-3}. A positive test is a strong indicator of a renal disease although increased protein can sometimes appear in normal young people. The ethanoic acid is used to remove calcium phosphates which would otherwise precipitate and give a false result. This test has been largely replaced by 'dipstick' type methods.

28.9.2   Isolation

The natural forms of polypeptides and proteins are referred to as their **native forms**. The isolation of native forms is complicated by their ease of denaturation. Consequently, isolation must be carried out under mild conditions. However, it is still difficult to know whether the isolated compound is actually in its native or denatured form.

Isolation is carried out in two stages, the first being the separation of the polypeptide and proteins from the cellular material. This involves destroying the cellular material by, for example, either maceration in a blender or homogeniser, alternatively freezing and thawing the material to break down cell walls or ultrasonic vibrations. The conditions under which these operations are conducted are determined by experiment. The polypeptides and proteins are obtained from the resultant mixture by extraction with either water, salt solutions or suitable aqueous buffers. Lipids may be removed from this aqueous extract by extraction with diethyl ether, hexane and other organic solvents and low molecular mass and electrolyte impurities by dialysis (9.12). The isolation of the polypeptide or protein from the aqueous extract may then be carried out by a variety of techniques (Table 28.2). The extracted polypeptide or protein is usually purified by a combination of dialysis, paper chromatography, TLC or HPLC.

Table 28.2. General methods of isolating polypeptides and proteins from aqueous extracts.

Technique	Notes
Solubility control	Addition of either buffer or salt solutions to a solution of the peptide or protein is used to selectively precipitate the required peptide or protein. The solubility of a peptide or protein is lowest at its isoelectric point. Organic solvents can also be used to precipitate peptides and proteins from aqueous solution
Salt formation	Peptides and proteins are precipitated from solution by the formation of insoluble salts with reagents such as sulphosalicylic acid
Partition	Carried out automatically using counter-current and steady state distribution machines. Partition is usually between two buffered solutions
Gel filtration	The peptides and proteins are eluted from the column at different times depending on their size and shape
Chromatography	All forms are used
Electrophoresis	Suitable for small scale separations
Freeze drying	The aqueous solution of the peptide or protein is frozen and the water removed by sublimation under reduced pressure

28.10 GENERAL PHYSICAL PROPERTIES OF PEPTIDES AND PROTEINS

The structures of polypeptides and proteins contain many charged groups such as the carboxylate ($-COO^-$) and ammonium ($-\overset{+}{N}H_3$) groups. These charges do not always cancel out and so the structure often has a net negative or positive charge. These net charges make it possible to separate and identify mixtures of polypeptides and proteins by electrophoresis (Fig. 28.13). The electrophoretic separation of the mixture of proteins in serum is an important diagnostic tool in the identification of disease. The sample under test and a sample of normal serum are applied alongside each other on the support which is usually a strip of cellulose acetate or agarose. The current is passed for a set time. The proteins are detected by staining with suitable reagents and the results from the sample under test compared with those from the sample of normal serum. Differences between the two electrophoretic patterns can be used to indicate either the course of any further investigation or the nature of any disease.

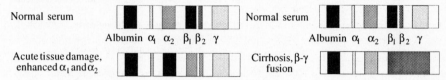

Fig. 28.13. A diagrammatic representation of an electrochromatogram of the proteins in normal and abnormal human serum.

The solubility of polypeptides and proteins in water is dependent on the pH of the solution. It is lowest when the pH of the buffer corresponds to the isoelectric point of the polypeptide or protein. At this pH the polypeptide or protein molecules are electrically neutral and so coagulate into particles that are often too large to be soluble in water. For example, casein, a protein found in milk, is precipitated in the form of white curds when milk turns sour and its pH of drops from its normal value of about 6.5 to 4.7.

Polypeptide and protein structures contain large numbers of chiral centres and so are optically active.

28.11   GENERAL CHEMICAL PROPERTIES OF PEPTIDES AND PROTEINS

The chemical properties of peptides and proteins depend on the nature of the functional groups within the molecule and their interactions if any. For example, all peptides and proteins can be hydrolysed and will also exhibit many of the other reactions of simple amides. Amide hydrolysis forms the basis of the structure determination and identification of peptides and proteins (28.12) while acid and enzymatic hydrolysis in the stomach is the start of the digestion process of peptides and proteins. Peptides and proteins also exhibit many of the properties of the ammonium, carboxylate and any other functional groups, such as, alcohols, disulphides and guanidine groups that occur in the molecule. Peptides and proteins also display characteristic reactions. For example, they usually react with ninhydrin to form a purple-violet complex (compare with amino acids). This reaction may be used to locate peptides and proteins in chromatography. Peptides and proteins form complexes with metal ions. The biuret reaction involving copper ions is the basis of a colorimetric method for the determination of the total protein content of serum. One copper ion is believed to complex with four peptide links but the chemistry of the method is not fully understood.

Protein $\xrightarrow[\text{Incubate at }37^{\circ}\text{C for 10 minutes}]{\text{The working biuret reagent containing copper sulphate}}$ Violet-coloured soluble complex λ_{max} 555 nm

28.12.   STRUCTURE DETERMINATION OF PEPTIDES AND PROTEINS

The general approaches to the determination of the primary structure of a peptide or protein are outlined in Fig. 28.14. The information obtained from each of these routes is put together, very much like the pieces of a jig-saw, to obtain the required structure. Computer programs, such as the *MultAlin* from Cherwell Scientific Publications, are now available to assist in this task.

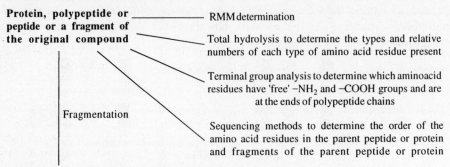

Protein, polypeptide or ——————— RMM determination
peptide or a fragment of
the original compound ——————— Total hydrolysis to determine the types and relative numbers of each type of amino acid residue present

Terminal group analysis to determine which amino acid residues have 'free' $-NH_2$ and $-COOH$ groups and are at the ends of polypeptide chains

Fragmentation

Sequencing methods to determine the order of the amino acid residues in the parent peptide or protein and fragments of the parent peptide or protein

The formation of smaller peptide fragments by chemical and enzymic methods. These fragments are analysed by the same routes that are used for the parent peptide or protein

Fig. 28.14. The general method of approach for determining the structures of peptides and proteins.

The secondary, tertiary and quarternary structures of peptides and proteins are deduced using a variety of physical methods such as X-ray crystallography, ultra-violet spectroscopy and nuclear magnetic resonance. A discussion of the application of these and other techniques is beyond the scope of this text and the reader is advised to consult more specialised texts.

28.12.1   Total hydrolysis
Total hydrolysis of a peptide or protein is normally carried out by heating the compound with 5 M hydrochloric acid at 100°C for about twelve hours. The reaction can result in the partial decomposition of some sensitive amino acids such as tryptophan and cysteine. The hydrolysis mixture is normally analysed using an amino acid analyser (Fig. 28.15). These

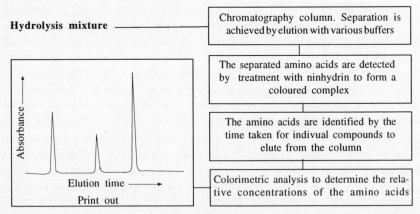

Hydrolysis mixture ——————— Chromatography column. Separation is achieved by elution with various buffers

The separated amino acids are detected by treatment with ninhydrin to form a coloured complex

The amino acids are identified by the time taken for indivual compounds to elute from the column

Colorimetric analysis to determine the relative concentrations of the amino acids

Absorbance

Elution time ——→
Print out

Fig. 28.15. The principle of amino acid analysers. The time taken for an amino acid to elute is characteristic of that acid. The amino acids in a mixture are identified by running standard amino acid mixtures through the analyser and comparing the elution times of the components of these standards with those of the mixture.

machines enable one to identify and determine the molar ratios of the amino acids in the acid hydrolysate. However, the components of simple mixtures may be determined both quantitatively and qualitatively by both paper and thin layer chromatography.

28.12.2   Relative molecular mass determination

The relative molecular masses of peptides and proteins are usually determined by osmometry, ultracentrifuge sedimentation and optical methods. The results are reasonably accurate for the smaller peptides and proteins but are not as accurate for the larger compounds.

28.12.3   Terminal group analysis

Terminal group analysis (end group analysis) is used to identify the amino acid residues in a peptide or protein with free 'carboxylic' and 'amino' groups. These groups are usually at the end of a chain and so their identification can be used to determine the number of polypeptide chains in the protein. For example, N-terminal group analysis of insulin (Fig. 1.1) shows the presence of equal amounts of glycine (Gly) and phenylalanine (Phe) which indicates that insulin has two polypeptide chains.

Terminal group analysis is based on attaching what is in effect a chemical label to the *free* carboxylic and amino groups in the structure. The labelled structure is completely hydrolysed and the liberated labelled amino acids identified by analytical techniques such as thin layer and paper chromatography. These labelled amino acids are the residues with the *free* carboxylic and amino groups in the peptide or protein molecule. For example, 2,4-dinitrofluorobenzene is used to locate *free* amino groups in peptides and proteins by attaching a 2,4-dinitrophenyl residue to the *free* amino group (Fig 28.16).

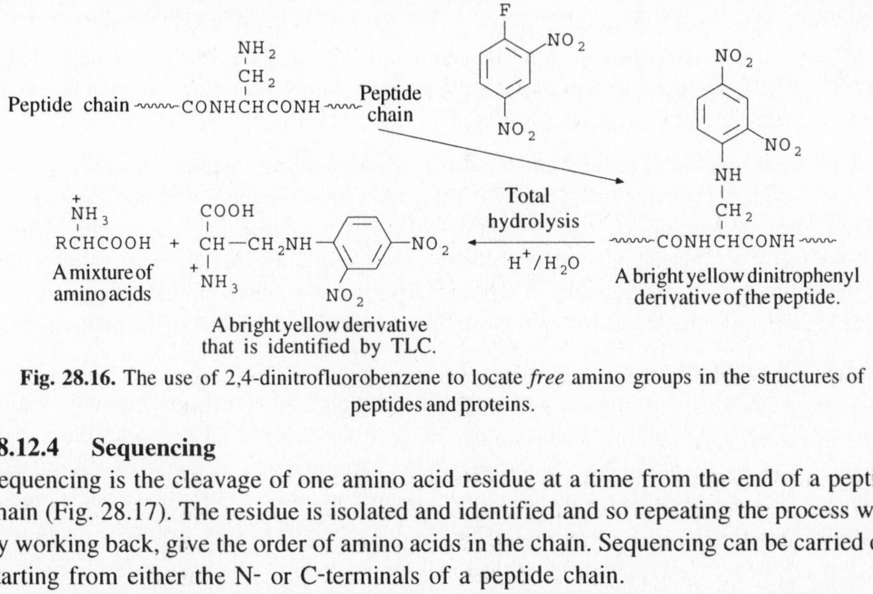

Fig. 28.16. The use of 2,4-dinitrofluorobenzene to locate *free* amino groups in the structures of peptides and proteins.

28.12.4   Sequencing

Sequencing is the cleavage of one amino acid residue at a time from the end of a peptide chain (Fig. 28.17). The residue is isolated and identified and so repeating the process will, by working back, give the order of amino acids in the chain. Sequencing can be carried out starting from either the N- or C-terminals of a peptide chain.

(a) Edman degradation method of sequencing. This method starts at the N-terminal of the peptide chain and so requires a free amino group before it can be used. The peptide is

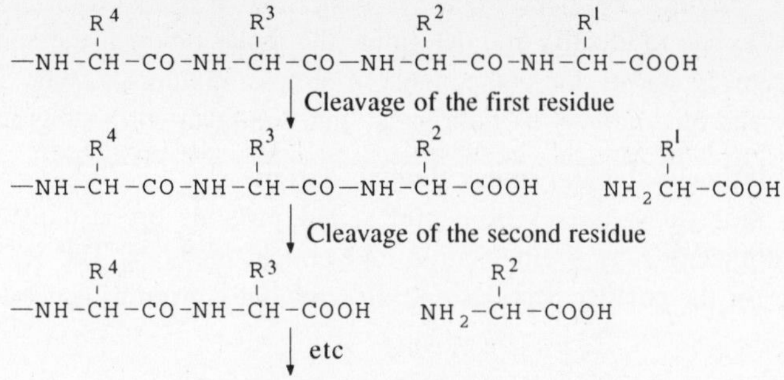

$$R^4 \quad\quad R^3 \quad\quad R^2 \quad\quad R^1$$
$$\text{—NH—CH—CO—NH—CH—CO—NH—CH—CO—NH—CH—COOH}$$

Cleavage of the first residue

$$R^4 \quad\quad R^3 \quad\quad R^2 \quad\quad\quad R^1$$
$$\text{—NH—CH—CO—NH—CH—CO—NH—CH—COOH} \quad \text{NH}_2\text{CH—COOH}$$

Cleavage of the second residue

$$R^4 \quad\quad R^3 \quad\quad\quad R^2$$
$$\text{—NH—CH—CO—NH—CH—COOH} \quad \text{NH}_2\text{—CH—COOH}$$

etc

Fig. 28.17. Sequencing.

reacted with phenylisothiocyanate to form the N-phenylthiourea derivative of the peptide. Controlled hydrolysis of this derivative with dilute hydrochloric acid gives the phenylthiohydantoin derivative of the last amino acid residue and the rest of the peptide chain. The phenylthiohydantoin derivative is separated and identified by thin layer chromatography. Since the hydrolysis has revealed the amino group of the next amino acid residue in the peptide chain the procedure can be repeated to identify the next amino acid residue in the chain and so on until the complete sequence has been determined.

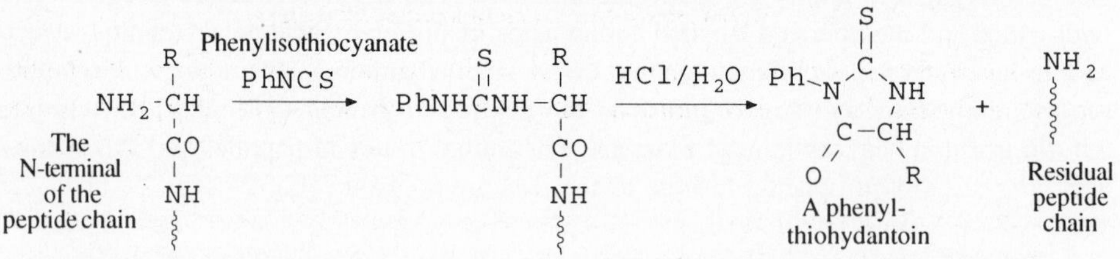

Machines have been developed that can automatically carry out the Edman degradation sequence. These machines are able to carry out up to twenty successive degradation steps before a build-up of by-products interferes with the results.

(b) The use of carboxypeptidase in amino acid sequencing. Incubation of the peptide or protein with carboxypeptidase cleaves the amide bond of the C-terminal amino acid residue. The liberated amino acid is identified by TLC or paper chromatography. Unfortunately, the new C-terminal residue amide bond is likely to be cleaved before all the first residue is liberated. Consequently, it is necessary to continuously monitor the incubation mixture to identify the order in which the amino acids are liberated from the peptide chain.

28.12.5   Fragmentation

A variety of chemical reactions may be used to break a peptide or protein down into smaller fragments. Reactions commonly used are acid hydrolysis using dilute hydrochloric acid for short lengths of time, base hydrolysis with barium hydroxide and oxidation with performic acid (HCOOOH). This last reagent is used to cleave disulphide bridges. Many reagents are specific for particular residues or structures. For example, cyanogen bromide cleaves the peptide links of methionine residues whilst N-bromosuccinamide cleaves peptide links involving tryptophan and tyrosine residues (Fig. 28.18).

Enzymatic hydrolysis is also a valuable tool as enzyme attack is often specific for particular types of peptide link. For example, chymotrypsin rapidly cleaves tyrosine,

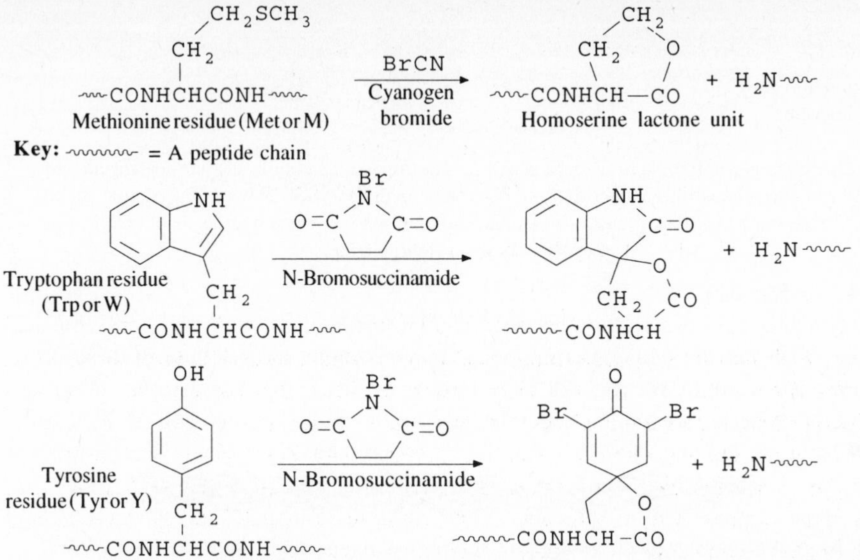

Fig. 28.18. Examples of the cleavage of specific peptide links.

tryptophan, and phenylalanine peptide links as well as slowly cleaving serine, threonine, asparagine, lysine, leucine, histidine and methionine peptide links. However, it should be noted that some enzymes, such as subtilisin and pronase, are not specific in their action.

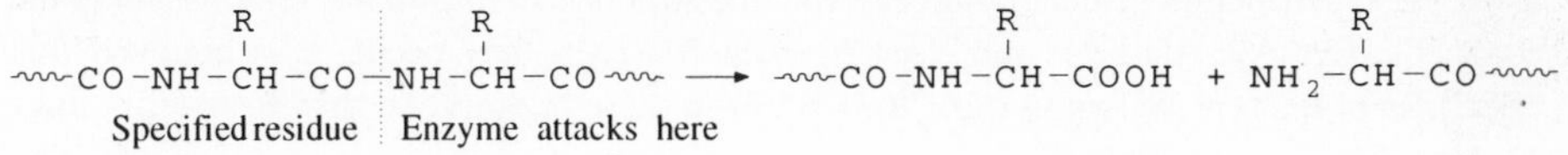

28.13 SOME PEPTIDES AND PROTEINS OF BIOLOGICAL IMPORTANCE

28.13.1 Collagen

Collagen is the most abundant fibrous protein in vertebrates, giving strength to skin, teeth, bones and other tissues. The basic building block of the structure of collagen is the polypeptide tropocollagen. These peptide molecules have a structure that consists of three peptide chains, each chain containing about 1000 amino acid residues (about 30% glycine) in an open left-handed helix. In the tropocollagen molecule three of these left-handed chains are plaited together to form a right-handed triple helix that is held together by hydrogen bonds and a few disulphide bridges. This results in a very strong cable-like molecular structure that is known as a **protofibril**. Groups of tropocollagen protofibrils are held together in a regular array to form microfibres (**fibrils**) by hydrogen bonding and cross links (Fig. 28.19). The resultant structure is very strong, a collagen fibril with a cross-section of 1 mm can support masses of the order of 10 kg. The arrangement of the collagen fibrils in tissue will depend on the nature of the tissue. In skin the collagen microfibres form an irregular latticework.

28.13.2 β-Keratin

Beta-keratin occurs in the skin. Its structure contains a considerable proportion of beta pleated sheets.

Tropocollagen molecule

Fig 28.19. The general structure of a collagen fibril. The tropocollagen molecules are held together by hydrogen bonds and disulphide bridges. In bones and teeth the spaces between the peptide chains contain the inorganic ions, such as calcium and sodium ions. Enzyme hydrolysis of collagen by pepsin and trypsin yield gelatine.

28.13.3 α-Keratin

Alpha-keratin is a fibrous protein that occurs in hair, skin, nails, wool and horn. Practical evidence shows that the peptide chain is coiled into an α helix and that three of these chains are plaited into a protofibril (Fig. 28.11d). Further evidence suggests that parallel groups of these protofibrils are coiled together to form a multistranded helical microfibre (Fig. 28.20).

Fig. 28.20. A representation of the structure of a portion of a microfibre of α-keratin. Each line represents triple helix of three peptide chains.

28.13.4 Elastin

Elastins are the rubber-like proteins that are found in the elastic connective tissue found in the arteries and tendons. The structures of these proteins have not been fully elucidated. However, experimental evidence suggests that the structure of elastin has regions where the sequences -Lys-Ala-Ala-Lys- and -Lys-Ala-Ala-Ala-Lys- may occur. It is believed that cross links of the type shown in Fig. 28.21 occur in these regions and that these cross links

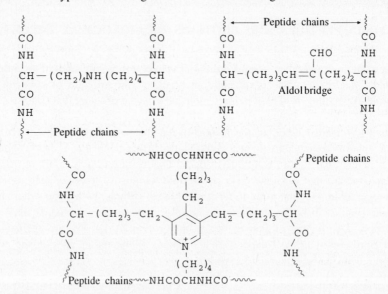

Fig 28.21. Cross links that are believed to occur in the structure of elastin. Each of the cross links is based on a residue that is derived from either a lysine (Lys or K) or a leucine (Leu or L) residue.

may be important in the elasticity of elastin. These cross-linked regions appear to be separated by sections of the peptide chain that are rich in glycine, proline and valine. Some of these regions appear to have a repeating amino acid sequence.

28.13.5   Fibrin

This is a complex material that is formed from fibrinogen during blood clotting. Fribrinogen molecules are very soluble in the plasma. Its structure is not fully elucidated but electron microscope data suggest that fibrinogen molecules consist of three nodules joined by two rods (Fig. 28.22). These molecules contain three pairs of three types of peptide chain known as the α, β and γ peptide chains.

Fig. 28.22. The overall shape of a fibrinogen molecule based on electron microscope data. This structure is about 46 nm in length.

Fibrinogen is involved in the formation of blood clots. Bleeding initiates a series of enzyme reactions known as a *cascade*. Thrombin, an enzyme that is part of this cascade, cleaves four of the arginine-X (X is usually glycine) peptide links in the α and β chains to produce four small peptides. Cleavage of the two α chains produces two A peptides, whilst cleavage of the two β chains result in the formation of two B peptides, each containing twenty residues. The remaining fibrinogen structure is known as the fibrin monomer and contains about 97% of the amino acid residues of the original fibrinogen molecule. Fibrin monomers, unlike fibrinogen molecules, spontaneously polymerise to form a water insoluble fibrous network which forms the basis of the blood clot. This structure is fragile and it is subsequently strengthened by factor XIIIa via the formation of cross links between lysine and glutamine residues (Fig. 28.23).

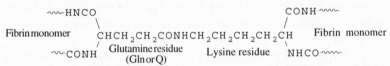

Fig. 28.23. Glutamine-lysine cross links in fibrin networks.

28.13.6   Insulin

Insulin is a globular polypeptide hormone that controls glucose metabolism. The primary structure of insulin consists of two peptide chains, the A chain containing 21 residues and the B chain containing 30 residues (Figs 1.1 and 28.24).

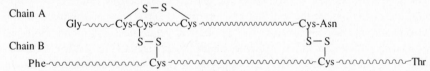

Fig. 28.24. The A and B chains of human insulin. (Asn is an asparagine residue)

Insulin is produced and released by the β-cells of the pancreas in response to increases in blood glucose. It is believed to be synthesised from a larger linear peptide, preproinsulin whose structure contains the A and B chain sequences of insulin separated by a peptide chain (Fig. 28.25). Evidence indicates that enzymatic hydrolysis of preproinsulin followed by the formation of disulphide bridges yields proinsulin. Further hydrolysis of proinsulin results in the formation of insulin plus an inactive peptide known as C-peptide.

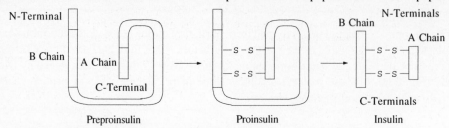

Fig. 28.25. A schematic representation of the formation of insulin from preproinsulin.

Practical evidence suggests that insulin has a compact three-dimensional structure in which the A chain is tucked into the B chain. The structure has a non-polar core consisting of the aliphatic chains of residues from both the A and B chains. It is stabilised by hydrogen bonds, salt bridges and hydrophobic bonding.

28.13.7   Lysozyme

Lyoszyme is a hydrolytic enzyme that cleaves certain of the glycosidic links of the polysaccharides found in the cell walls of some bacteria. This reduces the strength of the cell wall with the result that the high osmotic pressure inside the cell causes it to burst.

The structure of lysozyme consists of a single chain of 129 amino acid residues. This chain contains some α-helix regions and one region where one section of the chain is hydrogen bonded to another section in a structure that is similar to the β-pleated sheet found in fibroin the protein that occurs in natural silk. The complete chain is folded into an approximately ellipsoidal shape whose interior is almost completely composed of the non-polar groups of the amino acid residues. This ellipsoidal structure is stabilised by hydrogen bonding, disulphide bridges and hydrophobic bonding.

The active site of the lysozyme molecule is a cleft in the surface of the enzyme. The active residues involved in this site are residues 59, 62, 63, 101 and 107. Substrates appear to bind to these residues by hydrogen bonding and van der Waals' forces.

28.13.8   Endorphins and enkephalins

Endorphins are polypeptides with analgesic and anaesthetic activity that are found in the brain, pituitary gland and peripheral tissues of all vertebrates. The activity of endorphins and enkephalins resembles that of morphine and so is commonly referred to as **opiate** activity. The main ones are the α, β and γ endorphins, α and β neo-endorphins, dynorphin, met-enkephalin and leu-enkephalin. The amino acid sequences of the endorphins have been found in prohormones. For example, human β-endorphin (Fig. 28.26) is the 61-91 amino acid sequence of the C-terminal fragment of β-lipotropin, a hormone that stimulates the release of fatty acids from adipose tissue. It is interesting to note that other species produce endorphins with primary structures that are very little different from human β-endorphin. This suggests that the complete sequence is necessary for activity.

The enkephalins are pentapeptides (Fig. 28.27) that were isolated from pig brain by Hughes in 1975. They are also found in human brain tissue and exhibit opiate activity that

$$\overset{\mid}{(NH_2)}Tyr\text{-}Gly\text{-}Gly\text{-}Phe\text{-}Met\text{-}Thr\text{-}Ser\text{-}Glu\text{-}Lys\text{-}Ser\text{-}Glu\text{-}Thr\text{-}Pro\text{-}Leu\text{-}Val$$

$$\underset{31}{(HOOC)}Glu\text{-}Gly\text{-}Lys\text{-}Lys\text{-}\underset{27}{Tyr}\text{-}Ala\text{-}Asn\text{-}Lys\text{-}\underset{23}{Ile}\text{-}Ile\text{-}Ala\text{-}Asn\text{-}Lys\text{-}Phe\text{-}Leu\text{-}Thr}$$

Fig. 28.26. The primary structure of human β-endorphin. The β-endorphin of pig is identical except for valine instead of isoleucine at position 23. Rat and bovine β-endorphins have histidine instead of tyrosine at position 27 and glutamine instead of glutamic acid at position 31.

is comparable to that of the endorphins. However, practical evidence suggests that the enkephalins are not synthesised in the brain but in the pituitary gland.

(NH$_2$)Tyr-Gly-Gly-Phe-Met(COOH)   (NH$_2$)Tyr-Gly-Gly-Phe-Leu(COOH)

Met-enkephalin   Leu-enkephalin

Fig. 28.27. The structures of met-enkephalin and leu-enkephalin.

The structure of met-enkephalin corresponds to residues 1 to 5 of the N-terminal end of β-endorphin (Fig. 28.26) and also residues 61 to 65 of the hormone lipotropin. All the amino acid residues in both the met-enkephalin and leu-enkephalins have L configurations. Experimental evidence shows that the opiate activity of the enkephalins depends on their having a phenolic amino acid residue at position 1 and a non-phenolic aromatic residue at position 4. Furthermore, the enkephalins are deactivated by hydrolysis of the Tyr-Gly peptide link and so modifications to this link can prolong the action of the enkephalin.

The physiological roles of the endorphins and enkephalins are not known. However, they are thought to play a part in narcotic addiction. There is some evidence to suggest that people with an endorphin deficiency appear to be more susceptible to addiction than those without this deficiency.

28.13.9   Immunoglobins

Immunoglobins are a group of globular proteins that act as antibodies although not all immunoglobins exhibit this activity. All immunoglobins consist of two short and two long peptide chains known as **light** and **heavy** chains respectively (Fig. 28.28). Each of the heavy

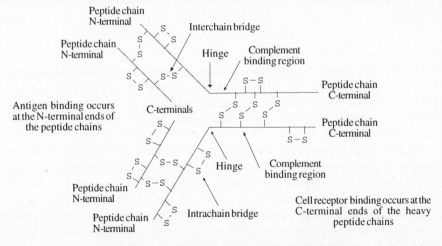

Fig 28.28. A model for the structures of the immunoglobins.

chains has a central flexible region that is commonly referred to as the **hinge**. All the chains are held together by disulphide bridges (**interchain** disulphide bridges) and the molecule also contains disulphide bridges within each chain (**intrachain** disulphide bridges). The amino acid sequence in both light and heavy chains is variable. However, two general types of light chain and five general types of heavy chain are found in immunoglobin molecules. The two general types of light chain are known as the kappa (K) and lamda (L) chains respectively.

The N-terminal segments of the immunoglobin peptide chains are the regions of the molecule which bind to antigens whilst their C-terminal segments interact with the receptors of a variety of cells. The hinge region of the structure is linked to complement binding. The C1 enzyme in the plasma binds to this region of the immunoglobin and in doing so activates the complement system which controls the effects of the antigen-antibody interaction.

28.13.10   Pharmaceuticals
A number of micro-organisms produce peptides that act as antibiotics. For example, vancomycin has been isolated from *Streptomyces orientalis* and Polymixin B_1 from *Bacillus polymyxa*.

28.14   WHAT YOU NEED TO KNOW

(1)   Amino acids are carboxylic acids whose structures contain a nearby amino group. They are classified as α β γ etc. amino acids depending on the relative positions of the carboxyl and amino groups (Fig. 28.1).

(2)   Amino acids exist as zwitterions in solution (Fig. 28.2).

(3)   Amino acids are classified as either acidic, neutral or basic depending on the number and relative strengths of the acidic and basic groups in their structures.

(4)   The configurations of amino acids are usually classified as D or L. The R/S system is not in common use for these compounds.

(5)   The structures of amino acid molecules in aqueous solution depends on the pH of the solution.

$$\underset{\text{Base}}{\overset{\text{Acid}}{\rightleftharpoons}} \quad \underset{\text{Zwitterion}}{\text{R}-\underset{\overset{|}{\text{NH}_3^+}}{\text{CH}}-\text{COO}^-} \quad \underset{\text{Base}}{\overset{\text{Acid}}{\rightleftharpoons}} \quad \text{R}-\underset{\overset{|}{\text{NH}_3^+}}{\text{CH}}-\text{COOH}$$

(6)   The pH of an aqueous solution of an amino acid which contains only the zwitterion form of the acid is known as the isoelectric point of that amino acid.

(7)   Electrophoresis can be used to separate, identify and isolate the components of mixtures of amino acids but the preferred methods are ion exchange chromatography and HPLC.

(8)   Amino acids exhibit many of the chemical properties that their constituent functional groups would exhibit in the corresponding simple compounds as well as properties that are characteristic of amino acids.

(9)   Peptides and proteins are polymers whose structures contain amino acid residues linked together by amide functional groups that are referred to as peptide bonds.

(10) Peptides with RMM values >500 are referred to as polypeptides while proteins have RMM values >2000. These definitions are not rigid.

(11) Proteins are classified as simple and conjugated. Simple proteins yield only amino acids on hydrolysis but conjugated proteins will also produce other compounds.

(12) Simple and conjugated proteins are further classified as globular and fibrous proteins. The molecules of globular proteins have a spherical shape while those of fibrous have a linear or sheet-like appearance.

(13) The primary structure of a peptide or protein is the order in which the amino acid residues occur in the structure. Primary structures may contain branches and cross links between chains known as bridges.

(14) The primary structure of a peptide or protein may be represented by a series of three letter symbols or single letters which are used to represent the amino acid residues in the structure. By convention the N-terminal of the chain is written on the left-hand side of the structure.

(15) The secondary structures of peptides and proteins are the organised regions of structure that appear in some areas of the molecule.

(16) The most common secondary structures are the αhelix, the βpleated sheet and the triple helix.

(17) The tertiary structure of a peptide or protein is the overall shape assumed by the folding of the peptide chain.

(18) Tertiary structures are held in shape by disulphide bridges, salt and other types of bridge and non-covalent bonds such as hydrogen and hydrophobic bonds.

(19) Conjugated proteins incorporate non-protein compounds, known as prosthetic groups, into their tertiary structures.

(20) Quaternary structures are formed when peptide and protein molecules fit together to form dimers, trimers, etc. These quaternary structures are held together by salt bridges and non-covalent bonds such as hydrogen and hydrophobic bonds.

(21) Alterations to either the primary, secondary, tertiary or quaternary structures of a peptide or protein will usually destroy the biological activity of the compound.

(22) Changes in the chemical and physical environment of a peptide or protein can irreversibly change either its secondary, tertiary or quaternary structure. The process of change is known as denaturation and usually results in a more open structure that is less water soluble than the original peptide or protein.

(23) The natural form of a peptide or protein is known as its native form. The isolation method used for a peptide or protein may result in the denaturation of its native form.

(24) The structures of peptides and proteins often have a resultant positive or negative charge. This enables mixtures of peptides and proteins to be separated by electrophoresis.

(25) The solubility of peptides and proteins in aqueous solutions depends on the pH of the solution. The solubility of a peptide or protein is lowest at its isoelectric point.

(26) The chemical and biological properties of peptides and proteins depend largely on the nature of the functional groups on the surface of the molecule.

(27) Acid and enzymatic hydrolysis of the amide bonds of peptides and proteins is the first step in the digestion of these compounds in the stomach.

(28) Total hydrolysis of a peptide or protein is used to identify and determine the molar ratio of the amino acids present in a peptides and proteins.

(29) The amino acid residues with free 'amino' and 'carboxyl' groups in a peptide or protein are found by terminal group analysis.

(30) Sequencing is the stepwise removal, in the order that they occur in the structure, and the identification of the amino acid residues in the peptide chain. This sequential removal may be carried out from either the N-terminal or the C-terminal ends of the peptide chain. N-Terminal sequencing is the more accurate

(31) Fragmentation of a peptide or protein into smaller peptides is used to investigate the structure of the original peptide or protein. Each fragment is analysed using the techniques described in 28.12. The results are put together like the pieces of a jig-saw to obtain the structure of the peptide or protein.

(32) Fibrous proteins have structures in which the peptide chains form either fibrous threads or tough sheets. They are usually insoluble in water and are used as construction materials by living organisms.

(33) Globular proteins are spherical in shape and are usually soluble in water. They act as enzymes, buffers, carriers of other compounds, antibodies, etc.

(34) A number of peptides obtained from natural sources are used as antibiotics and medicines.

28.15   QUESTIONS

(1)   Draw the Fischer projections of S-serine and D-cysteine.

(2)   Explain, by means of chemical equations, the effect of the addition of (a) hydrochloric acid and (b) sodium hydroxide to an aqueous solution of serine in water.

(3)   Classify the functional groups in serine. List the main general chemical reactions that each of these functional groups would be expected to exhibit

(4)   Determine whether the following compounds have either D, L, R and S configurations.

(a)
COOH
H_2N H CH_3

(b)
COOH
H_2N——H
C_2H_5

(c)
COOH
H_2N CH_3 H

(d)
CH_2SH
$HOOC$——H
NH_2

(5)   Show by means of chemical equations how glycine would be expected to react with: (a) sodium carbonate, (b) 2,4-dinitrochlorobenzene, (c) ethanoic acid, (d) benzoyl chloride in the presence of sodium hydroxide, and (e) lithium aluminium hydride.

(6)   Outline the part played by amino acids in the metabolism of aromatic acids.

(7)   Draw the structural formulae of the peptides: (a) (NH_2)Ala-Gly-Lys-Ser(COOH) and (b) ASRKKG. Give the systematic names of both of the structures you have drawn.

(8)   Explain, in general terms only, the meaning of the terms: primary, secondary, tertiary and quaternary structures of peptides and proteins.

(9)   Total hydrolysis of a peptide gave glycine, valine and glutamic acid in the molar ratio 2:1:1. Reaction with 2,4-dinitrofluorobenzene followed by hydrolysis yielded three yellow derivatives that were identified as DNP-Gly, DNP-Gly-Val and DNP-Gly-Val-Glu. Determine the structure of the peptide. (DNP = 2,4-dinitrophenyl)

(10) Draw all the isomers of the tripeptide glycylglycylalanine. Outline a chemical method that would enable one to distinguish between each of the compounds you have drawn.

(11) A synthetic met-enkephalin has a D-alanine residue instead of a glycine residue at position 2. Suggest a reason for this modification having a longer action than natural met-enkephalin.

(12) Explain the meaning of each of the following terms in the context of peptide and protein chemistry: (a) salt bridge, (b) triple helix, (c) globular protein, (d) conjugated protein, (e) disulphide bridge, and (f) sequencing.

(13) The RMM of a peptide (A) was found to be 260. Treatment of A with phenylisothiocyanate followed by dilute hydrochloric acid yielded alanine N-phenylthiohydantoin. Deduce the structure of A if total hydrolysis produced only glycine and alanine in the ratio 3:1.

29

The nucleic acids

29.1   INTRODUCTION

The nucleic acids were isolated in 1869 by Friedrich Miescher from the nuclei of leukocytes (pus cells). It has since been discovered that the nucleic acids are the compounds that carry the cells genetic information. They control the growth, function and reproduction of all types of cell.

The nucleic acids are polymers whose structures are based on a number of monomer units known as **nucleotides** (Fig. 29.1). These nucleotides consist of a purine or pyrimidine base (29.2) bonded to a pentose sugar, this structure being known as a **nucleoside**. These nucleosides are linked by phosphate groups to form a chain which makes up the nucleic acid molecule. The phosphate residues are ionised at physiological pH and so in the body, nucleic acid molecules are giant anions.

There are two general types of nucleic acid, the **deoxyribonucleic acids** (DNA) and the **ribonucleic acids** (RNA). These classes of nucleic acid utilise two different pentoses

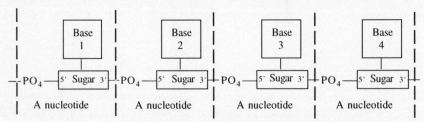

Fig. 29.1. The general arrangement of the bases, pentoses and phosphate units that comprise the structure of a nucleic acid chain. The nature of the genetic information carried by a nucleic acid is known to be related to the order of its bases and the length of its chain.

in their structures. The structures of the deoxyribonucleic acids contain β-D-deoxyribose residues whilst those of the ribonucleic acids contain β-D-ribose residues.

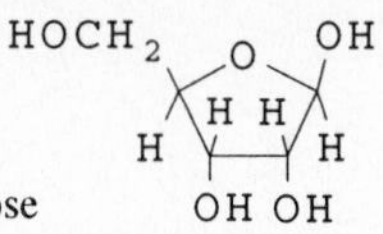

β-D-Ribose 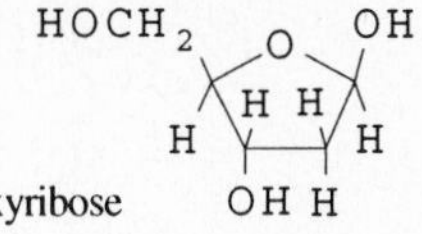   β-D-Deoxyribose

DNA molecules are extremely large having relative molecular masses up to one trillion (10^{-12}) and are usually found in the nucleus of the cell whilst RNA molecules are much smaller having relative molecular masses as low as 35,000. RNA is mainly found in small particles known as ribosomes that are found in the cyctoplasm of a cell. Ribosomes are the site of protein synthesis in the cell.

29.2   DEOXYRIBONUCLEIC ACIDS

The four bases normally found in DNA are the purines adenine (A) and guanine (G) and the pyrimidines cytosine (C) and thymine (T). However, derivatives of these bases, such as N^6-methyladenine and 5-methylcytosine, are found in some DNA molecules. X-ray, n.m.r. and other spectroscopic techniques have shown that the bases with an oxygen function exist in their keto forms in the nucleic acids.

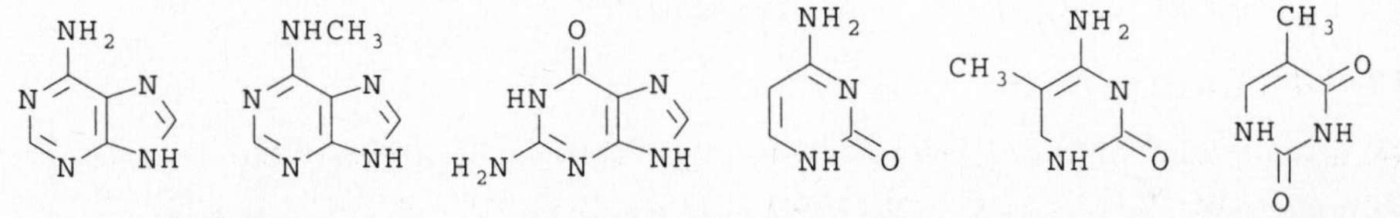

Adenine (A)   N^6-Methyladenine   Guanine (G)   Cytosine (C) 5-Methylcytosine Thymine (T)

The ratio of adenine to thymine and also that of cytosine to guanine in a DNA molecule is always 1:1. This is known as **Chargaff's rule**. Chargaff's rule also applies to DNA that contains derivatives of the four bases provided these base derivatives are counted with their parent bases. Each of the bases in DNA is the aglycone of a N-glycoside formed with the β-D-deoxyribose. These N-glycosides are referred to as nucleosides.

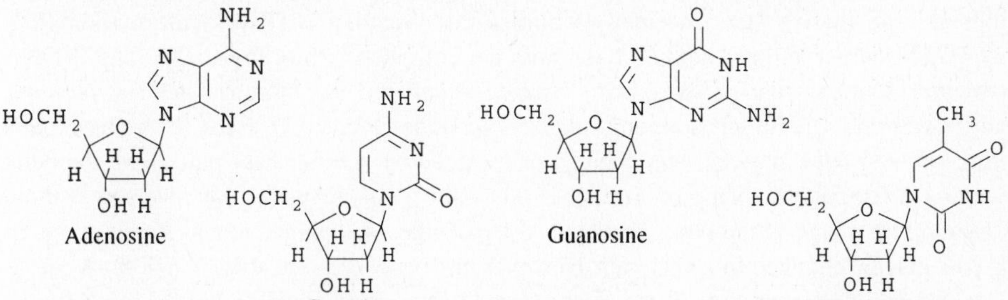

Adenosine   Guanosine

Cytosine   Thymosine

The phosphate residue is bonded to $C_{5'}$ of the nucleoside to form the nucleotide. Each nucleotide in the nucleic acid chain is linked via this phosphate residue to the $C_{3'}$ of the deoxyribose residue of the next nucleotide in the chain (Fig.29.2). It should be noted that by convention primes are used for the locants of the sugar residues and unprimed locants for the bases of nucleic acids.

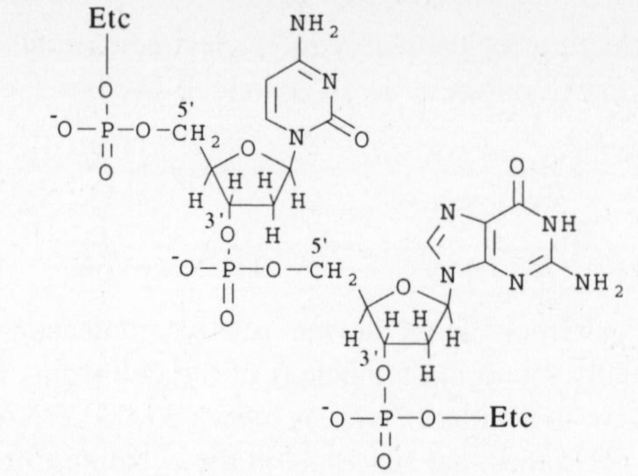

Fig. 29.2. The structure of a fragment of the chain of the nucleic acid found in DNA.

The overall structure of DNA was determined by Crick and Watson in 1953. They suggested that the structure of DNA molecules was based on two nucleic acid strands twisted into a right-handed double helix with the bases lying on the inside of the helix (Fig. 29.3). This double helix is held together by hydrogen bonding that is always between either adenine and thymine (A-T) residues or guanine and cytosine (G-C) residues, which explains Chargaff's observation that the ratio of A:T and G:C residues in any DNA molecule is always 1:1. These pairs of bases are commonly referred to as base pairs (bp) when discussing the structures of nucleic acids.

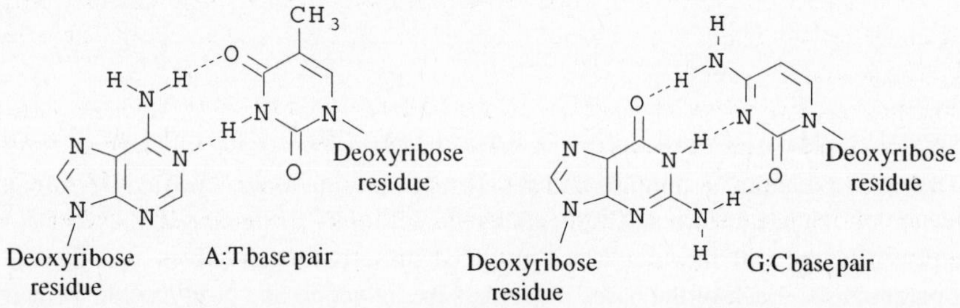

Evidence from X-ray diffraction studies indicates that the double helix of DNA in its natural state (native state) normally assumes the so-called B-DNA conformation (Fig. 29.3). This structure has about 10 base pairs per complete turn of the strand. The B-DNA structure contains two grooves in its surface which are the binding sites for proteins, antibiotics and carcinogens among other compounds (Fig. 29.3). Each of the base pairs in the double helix may be interchanged or replaced by another base pair along the helix without affecting the shape of the helix. The base pairs may also be reversed without changing the shape of the helix, that is, a A-T pair may be changed to a T-A pair and a G-C pair may be changed to a C-G pair. However, the base pairs are always A-T (T-A) or G-C (C-G). In all structures the 5' end of one strand lies opposite the 3' end of the other strand. This has some significance in DNA replication (29.3).

Two other conformational forms of the DNA double helix have been identified . These forms are known as the A-DNA and Z-DNA conformations. However, it is difficult to know whether these forms exist in living biological systems as the DNA may have been altered by the isolation procedure. In fact the presence of A-DNA has not been detected

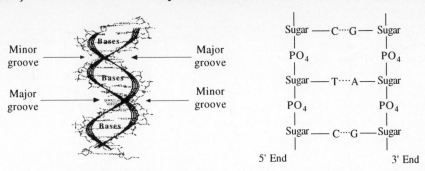

Fig. 29.3. A fragment of the B-DNA double helix. Interchanging (i) the two bases of a base pair and/or (ii) base pair with base pair does not affect the geometry of the double helix.

in biological systems; however, there is some evidence to show the presence of Z-DNA in *Esch. coli*. Furthermore, a change in the humidity of the environment of B-DNA from 92% to 75% will change its conformation to that of A-DNA. This conformational change is reversible. Similarly some DNA molecules will assume a Z-DNA conformation in concentrated salt solutions. Both these changes could occur in living organisms.

The A-DNA conformation takes the form of a right-handed helix with 11 base pairs per turn. This gives it a wider flatter twist with a deep major groove and a shallow minor groove. The base pairs are inclined at an angle of about 20^0 to the axis of the double helix which gives the structure an empty core lined by the bases and their hydrogen bonds. Z-DNA is a left-handed double helix with 12 base pairs per turn. It has a deep minor groove but has no major groove.

Electron microscope evidence shows that the DNA chain may be folded, twisted and coiled into a very compact shape. Many DNA molecules have cyclic tertiary structures that are in the form of a circle or distorted loop. Some of these cyclic DNAs are coiled or twisted into unusual shapes. This twisting and coiling of the molecule is known as either **supercoiling**, **supertwisting** or **superhelicity** as appropriate.

29.3 REPLICATION OF DNA

Replication is believed initially to involve the partial unwinding and separation of the two strands of the double helix (Fig. 29.4). The unwound sections act as a template for the formation of a new (**daughter**) strand, the daughter nucleotides being added in an order that is complementary to the sequence of bases in the parent strand. In other words, a daughter nucleotide will only be added if its base forms a base pair with the appropriate base of the parent strand. For example, for every A or T exposed a T or A respectively will be added to the daughter strands. Similarly, for every C or G exposed a G or C respectively will be added to the daughter strands. Unwinding continues as the daughter strands grow until two separate pairs of DNA molecules are formed, each containing one old strand and one new daughter strand. Experimental evidence shows that both the daughter strands are formed by 5'-nucleotides forming a phosphate ester with the 3' hydroxyl group of the last nucleotide in the daughter strand. In other words, both the daughter strands grow in the 5' to 3' directions, the growth being controlled by enzymes such as DNA polymerase I. However, the two strands are not synthesised in the same way. Okazaki showed that the growth of the daughter strand which started at the 3' end of the DNA parent is a smooth build-up of the strand by the successive addition of nucleotides. He called this daughter the **leading strand**. The other strand is also formed in the 5' to 3' direction but as the DNA continues to unwind it is not possible for the daughter strand to be formed smoothly in the 5' to 3' direction. Consequently it is formed in a series of sections which are known as Okazaki fragments. These sections are joined together by the enzyme DNA ligase to form the second daughter strand which Okazaki called the **lagging strand**. It is interesting to note that DNA ligase also repairs broken DNA strands.

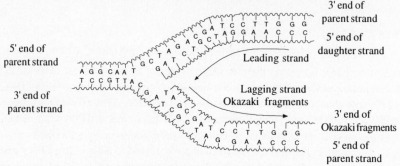

Fig. 29.4. A schematic representation of the replication of DNA. The arrows show the direction of the growth of the leading and lagging strands.

A number of anticancer drugs act by preventing replication of abnormal DNA. These drugs covalently bond to both strands of the abnormal DNA double helix This locks the strands together which prevents their unwinding and replication occurring. For example, the nitrogen mustard melphalan selectively bonds to the purine bases in both strands of a DNA helix (Fig. 14.12). Drugs like melphalan react preferentially with abnormal DNA but they also react with normal DNA and so their use must be carefully controlled.

29.4   RIBONUCLEIC ACIDS

There are three major types of ribonucleic acids: messenger RNA (mRNA), transfer RNA (tRNA) and ribosomal RNA (rRNA). They are named according to their function in protein synthesis. Messenger RNA 'informs' the ribosome of the type of amino acid and the order in which they are required, whilst transfer RNA transports the required amino acid residue to the ribosome for inclusion in the protein. Ribosomal RNA occurs in the ribosome and is involved in the translation of the RNA message.

The structures of the ribonucleic acids have not been elucidated to the same degree as those of the deoxyribonucleic acids. However, it is known that ribonucleic acid structures usually consist of a single nucleic acid strand. (Fig. 29.5). The nucleosides in this strand have similar structures to those found in the DNA strands except a ribose residue replaces the deoxyribose residue found in the DNA strands. The bases found in the RNA strand are the same as those in the DNA strands except the pyrimidine, uracil (U) (Fig. 29.5), replaces thymine. However, the structures do not obey Chargaff's rules, that is, the ratios of U:A and C:G are not usually 1:1. The RNA nucleosides are linked by phosphate units in an identical fashion to those in DNA nucleic acid strands.

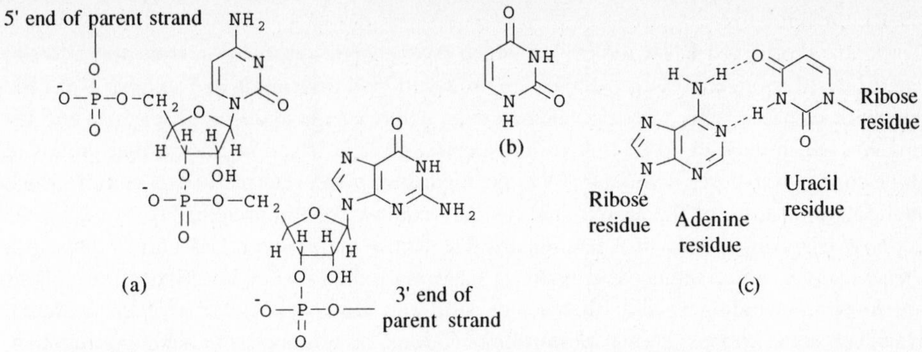

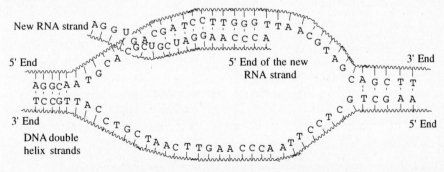

Fig. 29.5. (a) The general structure of a section of the chain of an RNA molecule. (b) Uracil. (c) Hydrogen bonding between uracil and adenine.

Single strands of RNA often have hairpin loops in which base pairing and double helix formations occur. At these and other suitable points in a structure uracil forms a hydrogen bonded base pair with adenine (Fig. 29.5). It should be noted that the RNA molecules of some viruses consist of a double stranded helix similar to that found in DNA.

RNA forms of ribonucleic acid are believed to be formed using specific sections of a DNA molecule as a template. The DNA double helix partially unwinds and the new RNA molecule is formed along this unwound strand in the direction 5' to 3', starting with the 3' end of the first nucleotide bonding to the 5' position of the next nucleotide in the new molecule (Fig. 29.6). The type and order of the bases in the unwound DNA strand will control the type and order of the bases in the new molecule because only a base that can base pair with a particular base in the unwound strand will be incorporated at that point in the new RNA molecule. A will be associated with U, U with A, G with C and C with G. This process which is known as **transcription**, is controlled by enzymes known as RNA polymerases.

The strands of DNA molecules appear to contain stop and start signals in the form of specific base sequences for the transcription of particular RNA molecules. In addition the enzyme, **rho factor**, may be involved in the termination of the synthesis and release of an RNA molecule although, in many cases the RNA chain is released without the intervention of this enzyme. After release the RNA molecules may be cleaved and chemically modified into other forms of RNA.

Fig. 29.6. A schematic representation of a transcription process.

29.4.1 mRNA

These are short-lived RNA molecules which pass genetic information from the DNA to the tRNA. Messenger RNA molecules are produced by transcription and so have a specific sequence of bases. This sequence indicates the nature of the amino acid residues and the order in which they are to be linked in a specific protein. It is now known that groups of three consecutive bases in the mRNA molecule form a code, referred to as a **codon**, which indicates the nature of the amino acid residue required for the protein. For example, the codon CUU corresponds to a leucine residue whilst the codon CUG corresponds to a serine residue. Some amino acid residues have more than one codon. Sixty-four codons are known, including several that act as stop and start signals for protein synthesis. However, some codons, known as **nonsense codons**, do not appear to have any function.

The order in which the codons appear on the messenger RNA corresponds to the sequence of the amino acid residues in the protein. In protein synthesis the mRNA moves to the ribosome. Each codon is 'read' by the complementary tRNA molecule in a process known as **translation** and the appropriate residue is transported to the ribosome to be incorporated at the correct point in the peptide chain. A stop codon on the mRNA halts the protein biosynthesis.

$$\text{DNA} \xrightarrow{\text{Transcription}} \text{RNA} \xrightarrow{\text{Translation}} \text{Protein}$$

If there is an error in a codon, by chance or mutation in the DNA template, a different amino acid may be incorporated into the protein. Such substitutions lead to the formation of a faulty protein. This in turn may result in disease. For example, substitution of valine for glutamic acid leads to the abnormal haemoglobin of sickle-cell anaemia. However, not all mutations are unfavourable, some are advantageous and are the basis of the evolution of the species.

29.4.2 tRNA

Transfer RNA molecules transport the amino acids required for protein synthesis to the ribosome. Each individual type of tRNA molecule is a specific carrier for a particular type of amino acid residue.

Transfer RNA molecules usually contain between 70 and 80 nucleotides. These nucleotides can contain a number of unusual bases, such as 2'-O-methylguanosine (OMG) and inosine (I) as well as uracil, adenine, cytosine and guanine. The single strand of tRNA molecules is often folded into the 'clover leaf' configuration, the 'stems' of which are stabilised by base pairing between either adenine and uracil or cytosine and guanine

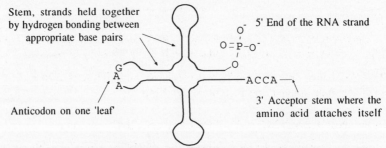

Fig. 29.7. A schematic representation of the general structure of a typical tRNA molecule. This structure is not flat but is folded into the approximate shape of a clover leaf.

(Fig. 29.7). The amino acid being transferred is attached to the 3' end (the acceptor stem) of the tRNA by an ester group.

Each tRNA molecule contains a section that has a three base sequence, known as the **anticodon**, that 'reads' the codons of mRNA. The tRNA is activated when its anticodon sequence is able to base pair with the codon of a mRNA. For example, a proline bearing tRNA with an anticodon of GGA would be induced to add its proline residue to the growing protein in the ribosome by a CCU codon in a mRNA molecule.

29.4.3 rRNA

Ribosomes contain approximately 60% rRNA and 40% polypeptide. The structure of ribosomes is complex and has not yet been elucidated. However, the *Esch. coli* ribosome has been separated into two subunits, one that is shaped something like a mitten and a larger roughly spherical subunit which has three protuberances on one side. The small *Esch. coli* ribosome subunit has been shown to contain a 1542 nucleotide rRNA molecule while the larger subunit has been shown to contain two rRNA molecules, one with 120 and the other with 2094 nucleotides. Both subunits also contain a number of different peptides.

The base sequences of the rRNA molecules found in the ribosomes of *Esch. coli* have been determined. However, less is known about their secondary structures and overall shape. Experimental evidence, coupled with deduction appears to show that the secondary structures of the rRNA molecules contain base paired double helix sections which link single strand areas (Fig. 29.8). Other rRNA molecules isolated from the ribosomes of other species appear to have similar structures.

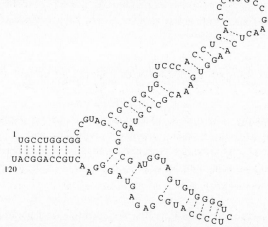

Fig. 29.8. The proposed sequence of the 120 nucleotide rRNA found in E. coli ribosomes. [Reproduced with permission from the *Annual Review of Biochemistry*, **53**, 134 (1984) by Annual Reviews, Inc.]

It is believed that base pairing between mRNA and rRNA is responsible for the selection of the translational site in protein synthesis.

29.5 BASE SEQUENCING OF NUCLEIC ACIDS

The first step in determining the base sequence in a DNA molecule is to divide the DNA chain into smaller fragments by the use of enzymes called *restriction endonucleases*.

These cleave the chain at specific points corresponding to particular base sequences. For example, the restriction enzyme Xho I cleaves the chain between cytosine and thymine on the 5' side of the thymine nucleoside when the 3' to 5' sequence of bases is GAGCTC.

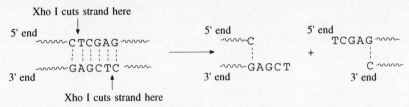

A number of different restriction enzymes (over 200 are known) are used with different samples of the original DNA to produce a large number of different DNA fragments. If a sufficient number of these fragments are obtained they will contain overlapping base sequences from the original DNA molecule (Fig 29.9). Consequently, knowing the base sequences of these fragments, it is possible to fit them together and deduce the base sequence of the original DNA molecule.

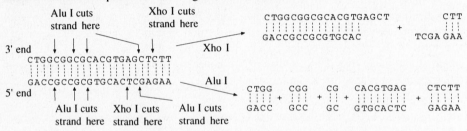

Fig. 29.9. The cleavage of the DNA chain by different restriction enzymes. Each fragment is purified and its base sequence is determined. The structures of the fragments are put together like the pieces of a jig-saw to deduce the base sequence of the original DNA molecule.

The DNA fragments are separated and purified. Each fragment is analysed separately. Its base sequence is usually determined by enzymically attaching a phosphate group containing radioactive P-32 to the 5' ends of the two strands. This acts as a label enabling one to follow the progress of the strand through the analytical procedure. The purified fragment of DNA chain is separated into its constituent strands by heating. These individual strands are isolated.

Each strand is now cleaved at specific points on both sides of a nucleoside using reagents and carefully controlled conditions that are specific for a particular nucleoside (Table 29.1). Cleavage is carried out by dividing the sample of the purified strand into four portions. Each portion is treated with one of the reagents listed in Table 29.1. This treatment gives a very complex mixture of *all* the possible cleavage products. The products of all four reactions are separated on the same plate by gel electrophoresis using a polyacrylamide gel. The positions of the radioactive products are located using a photographic plate. This use of radioactivity to locate the products is known as **autoradiography**. It does not record the positions of the unlabelled fragments. The position of a product in the gel chromatogram depends on the number of bases in the product, the smaller the number the faster, and so the further, the product migrates. This means that the

Table 29.1. The reagents and conditions used for the selective cleavage of single DNA strands.

Cleavage on both sides of a .... nucleoside	Method
Adenine	Methylation with dimethylsulphate (DMS) followed by treatment with dilute acid and heating with aqueous piperidine. Adenine is cleaved faster than guanine. Some guanine cleavage occurs
Guanine	Methylation with DMS followed by heating with aqueous piperidine. Under these conditions guanine is cleaved faster than adenine. Some adenine cleavage occurs
Cytosine	Treatment with hydrazine in 2M sodium chloride followed by heating in aqueous piperidine
Cytosine and thymine	Treatment with hydrazine followed by heating with aqueous piperidine

product that has migrated the furthest, probably has the structure: **phosphate[P-32]-sugar-base** because one of the four cleavage reactions must remove the last nucleotide with the radioactive phosphate from the chain since there are only four types of nucleotide in the chain (Fig. 29.10). Similarly, one of the four cleavage reactions must also result in the formation of a P-32 labelled product with a structure that consists of the last two nucleotides in the strand. Because only four different bases are found in DNA the four specific cleavage reactions will result in products that contain 1,2,3,4, etc. P-32 labelled nucleotides. Since the relative distances these products migrate on electrophoresis is proportional to the number of bases they contain it is possible to read off the sequence of bases from the electrophoresis chromatogram starting with the base that has moved furthest along the chromatogram. In the hypothetical electrochromatogram shown in Fig. 29.10, the product that migrated the furthest is formed by the reagents that cleave the strand at cytosine and cytosine/thymine and so the end 5' of the strand is a cytosine nucleoside. If thymine had been the end nucleoside the chromatogram would have only shown one spot in the C+T column and not spots in both the C and C+T columns. The product which migrated the second furthest along the plate was formed by the reagent that cleaves the strand at guanine containing nucleotides and so the base sequence for the end of the strand is GC. Proceeding in this manner in the direction shown in Fig. 29.10 the complete sequence for the strand must be CGGATCAAGGCTTACGGT.

The identification of guanine is sometimes complicated by the fact that the reagent for adenine causes some cleavage of guanine nucleosides. The amount of this cleavage is usually smaller than that due to the guanine reagent and so the spots on the electrogram in the adenine reagent column are usually much paler than those in the guanine reagent column.

The method used for determining the sequence of bases in RNA molecules is similar to that used to determine the base sequences of DNA.

29.6 DNA IN FORENSIC SCIENCE

DNA 'fingerprinting' is used in police work to identify criminals and in other instances to determine a person's genetic origin. Human and animal genes contain many successively

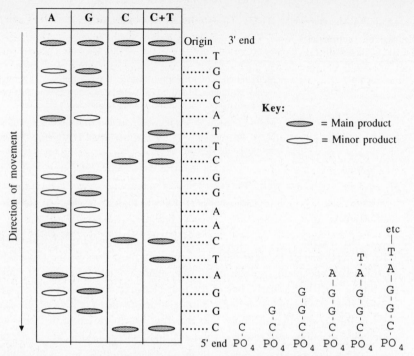

Fig 29.10. The determination of the base sequence of a hypothetical DNA strand by analysis of the fragments formed by selective cleavage of the strand.

repeating nucleotides. The regions in which these repeating sequences occur are known as **minisatellites.** The number of these repeating sequences varies considerably from person to person and so can be used as the basis of a procedure to identify individuals. Suitable DNA-containing material from a suspect is treated with restriction enzymes and the fragments produced separated. These fragments are combined, by base pairing, with a

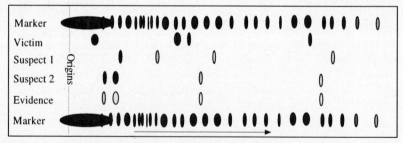

Fig. 29.11. DNA 'fingerprinting'. A simulation of a DNA semen sample taken from a rape victim compared with those taken from two suspects. The pattern taken from suspect 2 matches the sample recovered from the victim but is not identical with the victim's DNA and so this electrochromatogram will be used as evidence to identify suspect 2 as the rapist.

short length of specially prepared radioactive DNA. Gel electrophoresis is used to compare the samples from the suspect with those taken from the victim. The electrochromatograms are visualised by autoradiography as in DNA and RNA sequencing procedures

(29.5). A relative molecular mass marker is used to provide a base line reference. Identification is made by matching the patterns obtained on the electrochromatograms (Fig. 29.11). This type of evidence is very reliable as old samples will either give a true test or none at all. Consequently, there is little chance of a person being wrongly convicted on genetic fingerprinting evidence.

29.7 WHAT YOU NEED TO KNOW

(1) Nucleic acids are the molecules that store and contain the cells genetic material.

(2) There are two types of nucleic acid: deoxyribonucleic acids (DNA) and ribonucleic acids (RNA).

(3) All nucleic acids are polymers whose structures are based on a number of monomer units called nucleotides. These consist of a base residue, a p e n t o s e sugar residue and a phosphate unit bonded in the order shown (Fig. 29.1).

(4) The phosphate units are bonded to the sugar residues at the 3' and 5' positions whilst the base is bonded at position 1'.

(5) The base-sugar residue is known as a nucleoside.

(6) The sugar residue in DNA is deoxyribose while the sugar residue in RNA is ribose. Both sugar residues are bonded to the base by βglycosidic links (page 505).

(7) The principal bases found in DNA are adenine (A), cytosine (C), guanine (G) and thymine (T).

(8) The bases found in RNA are adenine (A), uracil (U), cytosine (C) and guanine (G).

(9) The secondary structure of DNA consists of two nucleic acid strands twisted together to form a double helix held together by hydrogen bonding between suitable bases.

(10) The pairs of bases which can form hydrogen bonds between the strands are:
       DNA, A—T and C—G   RNA, A—U and C—G

(11) The conformation of naturally occurring DNA is thought to be the B-DNA conformation which has a right-handed helix containing 10 bases per turn.

(12) The B-DNA structure has two grooves known as the minor and major grooves. These are the binding sites for many substrates.

(13) Replication is believed to involve the unwinding of the two strands of a DNA molecule and the formation of a new second strand (the daughter strand) alongside the original strand (the parent strand). The order and nature of the bases in the parent strand will dictate the order and nature of the bases in the daughter strand. Daughter strands are formed in the direction 5' to 3' of the parent strand.

(14) Some anticancer drugs act by locking the two strands together, thereby preventing replication of cancer cells.

(15) RNA molecules consist of a single strand of nucleosides linked by phosphate units. The strand often has hairpin loops which result in double stranded sections that are base paired into a helix similar to that found in DNA.

(16) All forms of RNA are formed in the 5' to 3' direction along an unwound DNA strand. This process is known as transcription.

(17) There are three major types of RNA, namely, messenger RNA (mRNA), transfer RNA (tRNA) and ribosomal RNA (rRNA). They are all concerned with

the synthesis of peptides and proteins.

(18) mRNA is a short-lived RNA that passes genetic information to tRNA.

(19) tRNA carries the required amino acid to the ribosome for incorporation into the peptide or protein.

(20) rRNA controls the peptide and protein synthesis.

(21) Certain consecutive sequences of three bases in mRNA form a code known as a codon, This code indicates to the tRNA which amino acid is required in the synthesis of a peptide or protein.

(22) Codons are reported by the three letters representing the three bases forming the codon. By convention the 5' end base of the codon is written on the left.

(23) tRNA molecules are specific for one type of amino acid residue. Each tRNA molecule has a sequence of three bases known as the anticodon that 'reads' the codon and so informs the tRNA molecule when to deliver its amino acid residue to the ribosome. Reading involves the matching, by base pairing, of the codon of a mRNA to the anticodon of a tRNA molecule. The process is known as translation.

(24) tRNA are single stranded nucleic acids. They often have 'clover leaf' conformations in which the stems contain hydrogen bonded base pairs and the leaves the anticodon. The amino acid residue being transferred is attached to the 3' end of the RNA strand by an ester group.

(25) rRNA are single stranded nucleic acids found in the ribosomes of a cell. Their secondary structures are thought to consist of single stranded areas linked by base paired double helixes formed by hairpin loops.

(26) The sequence of bases in a nucleic acid is determined by selective cleavage of the DNA into smaller fragments using enzymes. These fragments are separated and a radio active phosphate unit attached to the 5' ends of each fragment. The fragments are separated into their separate strands. The sample of each strand is divided into four samples. Each sample is treated with a reagent that specifically cleaves the fragment about a particular nucleoside. The reaction mixture is analysed by gel electrophoresis, the smallest fragments moving furthest along the plate. The sequence of bases in the fragment is read from the plate, the smallest fragment being at the 5' end of the structure.

(27) DNA fingerprinting is used to identify people.

29.8   QUESTIONS

(1)   Draw the structures of the sugars and principal bases that would be obtained by the complete hydrolysis of (a) DNA and (b) RNA.

(2)   Explain the meaning of each of the terms (a) nucleotide, and (b) nucleoside.

(3)   What is Chargaff's rule? Outline the principal features of the structure of DNA. What is the structure of the strand of DNA that lies opposite the structure TTAGCTA?

(4)   Explain the replication process of (a) DNA and (b) RNA.

(5)   Explain how the sequence of bases in a fragment of RNA could be determined. Four sections of an RNA fragment were found to have the sequences: ACAGG, GAUUGCCA, UACAGGAU and CCACCGAA. Deduce the complete sequence of the fragment.

30

General inorganic chemistry

30.1 THE PERIODIC TABLE

The periodic table is an attempt to rationalise the chemical and physical properties of the elements and their compounds. The form of the table used in this text is based on the format recommended in 1984 by the International Union of Pure and Applied Chemistry (IUPAC) in which the elements are arranged in order of increasing atomic number (Fig. 30.1).

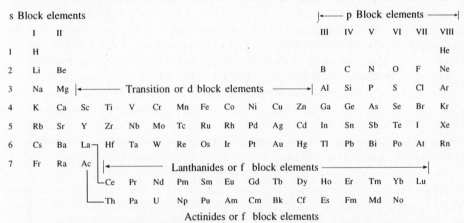

Fig. 30.1. The periodic table. The lanthanoids occupy periods 6a and follow lanthanum whilst the actinoids occupy period 7 and follow actinium.

In this text the main groups (vertical columns) of the periodic table are referred to by the Roman number at the top of the column whilst elements in the central sections of periods 4, 5, 6 and 7 are normally referred to by the general names: **transition metals**, **lanthanides** and **actinides** respectively. However, in the IUPAC table all the groups are numbered 1 to 18 from right to left. The periods (horizontal lines) in the table are identified by a number. Blocks of elements are often classified in terms of their electronic configuration (Fig 30.1). Groups 1 and 2, for example, are known as the s block elements because they have complete or partially complete s orbitals in their outer shells. Similarly, groups III to VIII are known as the **p** block elements since these elements have a partially or fully complete p orbitals in their outer shell.

Elements in the same group of the periodic table will possess many similar chemical properties because they have similar outer shell electronic configurations (Table 30.1). They also exhibit relatively regular changes in their physical properties as one proceeds up or down the group. In some groups there is a distinct change in the nature of the chemical and physical properties of the elements at a specific point in the group. For example, in group IV there is a marked difference between the properties of silicon and germanium. These observations enable one to make general predictions about the nature and reactivity of elements within a group. Elements in the same group will tend to form compounds with similar formulae. This enables one to predict the formulas of either the unfamiliar or unknown compounds in that group. The formula of calcium hydroxide, for example, is $Ca(OH)_2$ and so one would expect the hydroxides of the other elements in this group to have formulas of the form $M(OH)_2$. Similarly, it is possible to predict the reactivity of the elements and their compounds by following the trends within a group. For example, as one goes down group III the elements become more metallic and their chlorides less susceptible to hydrolysis with water. Consequently, one would predict that thallium monochloride is ionic in nature and relatively stable in water.

In contrast, the chemical properties of elements occupying the same period of the periodic table exhibit distinct differences as one progresses from one element to another. These changes are more marked between elements of the main groups in a period since these elements have different outer shell configurations. The physical properties of the elements of a period often show relatively regular changes as one proceeds across a period (Table 30.1). However, some physical properties have a maximum value at some point in the period. Notable exceptions to these general trends are the elements of the transition metals, lanthanoids and actinoids. Adjacent elements in a period in these regions of the table have very similar general chemical and physical properties. In these periods the electronic configurations of the outer shells are usually the same and it is the configuration of an inner shell that changes. These differences in the electronic configurations of inner shells appear to have less of an effect on the properties of the element.

Elements that are necessary for life are found mainly in the first four periods of the periodic table. The most abundant elements in humans are hydrogen, carbon and nitrogen whilst calcium, chlorine, iron, magnesium, phosphorus, potassium, sodium and sulphur are also present in significant amounts. Small amounts of chromium, cobalt, copper, fluorine, iodine, manganese, molybdenum, nickel, selenium and zinc have been found to be essential for healthy human life whilst nickel and vanadium are though to be necessary. The concentrations of the elements required to maintain a healthy human body lies within a specific range. Concentrations lower than the minimum requirement can result in

Table 30.1. Some general trends in the periodic table.

Property	Down a major group	Transition elements across a period	Across a period
Melting point	Decreases	Erratic	Erratic
Boiling point	Decreases	Erratic	Erratic
Atomic radius	Increases	Similar	
Metallic nature	Increases	Mainly metallic in nature	Decreases
Non-metallic nature	Decreases	Mainly metallic in nature	Increases
Electronegativity	Decreases	Remains almost constant	Increases
Ionisation energy	Decreases	Increases slowly	Increases
Electron affinity	Decreases		Increases
Oxidation states	Values usually remain the same	Variable, the number of and values reach a maximum at about the centre of the series	The number of states increases

deficiency diseases whilst concentrations above the maximum of the range can be toxic.

Periods 1, 2, 3 and 4 contain most of the essential **minerals** and **trace elements** required by humans. In biological systems, minerals are defined as elements that the human body requires in concentrations of approximately 100 mg or more per day whilst trace elements are those the body requires at a level of about 20 mg or less per day. The IUPAC defines a trace element as elements that are present at a concentration of 0.01% w/v. The exact amounts of minerals and trace elements required vary from person to person and will depend on body weight and age. Both minerals and trace elements are normally obtained from water and food. Deficiencies arise because of poor diet, disease or a lack of the element in the water and food eaten. This is often due to a lack of the element in the soil on which the food is grown.

In general the elements in periods 5, 6 and 7 tend to be toxic to human beings although some of their compounds have pharmaceutical and other uses. The noble gases of group VIII are not very reactive and apart from helium have no pharmaceutical uses. Helium is used as a diluent for oxygen in deep sea diving.

30.2   GROUP I

The elements of group I are commonly called the alkali metals (Table 30.2). Sodium and potassium in the form of their ions are essential constituents of human metabolism. The balance of intracellular and extracellular sodium and potassium is responsible for maintaining the water levels in human tissue and body fluids. Trace amounts of lithium are found in the body but it does not appear to be an essential element. However, there is evidence to suggest that people whose bodies contain higher than normal levels of lithium are less likely to suffer from heart attacks, mental disorders, gout, rheumatism, gastric and duodenal ulcers. Lithium appears to function as a mood stabiliser and its compounds have been used to treat depression, hyperthyroidism, migraine and epilepsy although prolonged use may have toxic effects. The remaining elements are not essential to health.

Table 30.2. The elements of group I. The most stable isotope of francium has a $t_{1/2}$ of 20 minutes

Elements	Symbol	Electronic configuration
Lithium	Li	$1s^2\ 2s^1$
Sodium	Na	$1s^2\ 2s^2\ 2p^6\ 3s^1$
Potassium	K	$1s^2\ 2s^2\ 2p^6\ 3s^2\ 3p^6\ 3d^{10}\ 4s^1$
Rubidium	Rb	$1s^2\ 2s^2\ 2p^6\ 3s^2\ 3p^6\ 3d^{10}\ 4s^2\ 4p^6\ 4d^{10}\ 4f^{14}\ 5s^1$
Caesium	Cs	$1s^2\ 2s^2\ 2p^6\ 3s^2\ 3p^6\ 3d^{10}\ 4s^2\ 4p^6\ 4d^{10}\ 4f^{14}\ 5s^2\ 5d^{10}\ 5f^{14}\ 6s^1$
Francium	Fr	$1s^2\ 2s^2\ 2p^6\ 3s^2\ 3p^6\ 3d^{10}\ 4s^2\ 4p^6\ 4d^{10}\ 4f^{14}\ 5s^2\ 5d^{10}\ 5f^{14}\ 6s^2\ 6p^6\ 7s^1$

30.2.1 Sodium

The healthy human adult contains on average 4000 mmol (92 g) of sodium in the form of sodium ions. About 38% of this sodium is present in bones and is relatively inactive biologically, about 13% is retained in cells (intracellular fluid) and approximately 50% is found in the fluids surrounding the cells (extracellular fluid). Plasma contains about 140 mmol dm^{-3}. Bone sodium is not a readily available reserve since it is locked into the crystal structure of the bone, but sodium occurring elsewhere in the body is available.

Sodium in any soluble form is absorbed from food in the intestine. Absorption is passive and low concentrations are absorbed as readily as high concentrations. Excretion is mainly via the kidneys and normal healthy kidneys have little difficulty in coping with excessive amounts of sodium provided there is sufficient water present. However, in cases of sodium deficiency there is substantial reabsorption of sodium ions in the kidney. Small amounts of sodium are also excreted in faecal material. Excretion through the skin is only significant when excessive sweating occurs.

Sodium plays an important part in many physiological functions. For example, it is involved in nerve impulse transmission, muscle contraction, transport of amino acids and carbohydrates into cells and in maintaining extra-cellular fluid osmolality and the normal balance of extracellular to intracellular water amongst other functions. Low levels of sodium in plasma (extracellular fluid) may be due to kidney failure, hormone imbalance, lung cancer and infections. It can result in loss of appetite, weakness, convulsions and coma. High levels of sodium in plasma can lead to high blood pressure, enlarged heart and kidneys. Amongst its causes are diabetes, kidney failure and dehydration.

Sodium chloride is used as an electrolyte replenisher, topical anti-inflammatory and emetic. A wide variety of other sodium salts are used as drugs and food additives (Table 30.3). Many organic drugs are used in the form of their sodium salts to improve their water solubility.

30.2.2   Potassium

The average healthy human adult contains about 3580 mmol (140 g) of potassium. Approximately 98% of this is found in the intracellular fluid whilst most of the remainder is found in the extracellular fluid. There is a negligible amount in bones.

Absorption of potassium occurs throughout the small intestine. It only occurs if the concentration of potassium in the intestine is higher than that in the blood.

Table 30.3. Some examples of the uses of sodium salts.

Use	Compound	Use	Compound
Antacid	Sodium hydrogen carbonate (sodium bicarbonate)	Metabolic acidosis	Sodium ethanoate
Purgatives	Sodium phosphate, sodium sulphate, sodium tartrate	Anticoagulant	Disodium hydrogen citrate
Sweetener	Sodium saccharin	Anticonvulsant	Sodium valproate
Flavour enhancer	Sodium 5'-inosinate, sodium 5'-ribonucleotide	Vasodilator	Sodium nitroprusside
Preservative	Sodium nitrite, sodium bisulphite (E222), sodium nitrate (E251)	Emulsifier (food)	Sodium aluminium phosphate, sodium pectate (E440a)

Excretion is mainly in the urine, about 92% of the daily intake being excreted. Excretion through the bowel is low in healthy people (about 8%). However, it can be high in humans suffering from diarrhoea and this can result in a loss of up to 30% of the potassium in the person's body. This degree of loss can result in serious medical problems. Excretion through the skin is not usually significant unless excessive sweating occurs.

Potassium plays a part in controlling the amount of water in cells. It is an essential activator for a number of enzymes and is involved in protein synthesis. Like sodium it is involved in nerve impulse transmission and preserving the acid-base balance of the body. Potassium also stimulates the normal movements of the intestinal tract. Deficiency (hypokalaemia) can result in vomiting, muscular weakness, low blood pressure, coma and respiratory failure. It can occur in drug therapy such as the use of diuretics, long-term use of corticosteroids and high doses of some penicillins. Excess potassium (hyperkalaemia) can cause muscular weakness and vomiting amongst other medical problems. Its causes include kidney failure and injuries which result in cell damage. Both hypokalaemia and hyperkalaemia can cause cardiac arrest.

Potassium compounds are used as food additives and drugs (Table 30.4).

Table 30.4. Some of the uses of potassium compounds.

Use	Compound	Use	Compound
Sedative	Potassium bromide	Potassium deficiency	Potassium chloride
Antithyroid	Potassium iodide	Antiseptic	Potassium permanganate
Buffer (food additive)	Potassium dihydrogen citrate (E332c), potassium citrate	Preservative	Potassium disulphite (E224) Potassium sorbate (E202)
Emulsifier (food additive)	Potassium polyphosphates (E450c), potassium pectate		

30.2.3   The sodium pump
Cell membranes are permeable to sodium, potassium and other metallic ions. Each of these ions has its own concentration gradient across the membrane. These gradients

have to be maintained if a cell is to remain healthy. For example, humans need to maintain the concentrations of sodium and potassium ions inside and outside their blood cells at the levels given in Table 30.5. However, the processes of dialysis (9.12) and osmosis (9.8) tend to remove these concentration gradients. The sodium pump is the name given to the process by which the cell maintains these sodium and potassium gradients across its membrane. A specific protein in the cell membrane is able, in some unknown way, to transport sodium and potassium ions across the membrane. When the concentration of sodium ions inside the cell is too high, it moves some of them out into the extracellular fluid. The protein reverses the process when the concentration of sodium ions in the cell is too low. The same protein controls the movement of potassium in a similar way. When the ions are moved against the concentration gradient the process is known as **active transport** (26.11.5). Oxygen and ATP are needed for this process.

Table 30.5. Some of the uses of potassium compounds.

Ion	Plasma concentration (mmol dm^{-3})	Erythrocyte concentration (mmol dm^{-3})
Sodium	135 to 145	10
Potassium	3.5 to 5.0	125

30.3 GROUP II

The elements that occur in group II of the periodic table are known as the alkaline earth metals (Table 30.6). Calcium and magnesium are essential to human health. Strontium is metabolised in a similar fashion to calcium but there is little evidence to suggest that the naturally occurring isotope(s) is/are essential or hazardous to human health. It is found in the same foods as calcium, namely milk, grain and fresh vegetables, and is incorporated into the bones along with calcium. However, the absorption into bones of radioactive strontium-90 is dangerous since it has a half life of 28 years. This isotope does not occur naturally but is produced in nuclear plants and explosions. Water soluble barium compounds are toxic, and fatal doses can be as low as 1 gram. However, insoluble barium sulphate is used for the **barium meal** in X-ray investigations of the gastrointestinal tract.

Table 30.6. The elements of group II.

Element	Symbol	Electronic configuration
Beryllium	Be	$1s^2\ 2s^2$
Magnesium	Mg	$1s^2\ 2s^2\ 2p^6\ 3s^2$
Calcium	Ca	$1s^2\ 2s^2\ 2p^6\ 3s^2\ 3p^6\ 3d^{10}\ 4s^2$
Strontium	Sr	$1s^2\ 2s^2\ 2p^6\ 3s^2\ 3p^6\ 3d^{10}\ 4s^2\ 4p^6\ 4d^{10}\ 4f^{14}\ 5s^2$
Barium	Ba	$1s^2\ 2s^2\ 2p^6\ 3s^2\ 3p^6\ 3d^{10}\ 4s^2\ 4p^6\ 4d^{10}\ 4f^{14}\ 5s^2\ 5d^{10}\ 5f^{14}\ 6s^2$
Radium	Ra	$1s^2\ 2s^2\ 2p^6\ 3s^2\ 3p^6\ 3d^{10}\ 4s^2\ 4p^6\ 4d^{10}\ 4f^{14}\ 5s^2\ 5d^{10}\ 5f^{14}\ 6s^2\ 6p^6\ 7s^2$

Beryllium is highly toxic. It tends to replace sodium in the bones. Since beryllium normally contains a proportion of radioactive isotopes this places a radioactive source near the site of cell production in the bone marrow.

30.3.1   Magnesium

The average human adult body contains about 1030 mmol (25 g) of magnesium. About 50% of this magnesium is found in the bones and is not readily exchangeable with the magnesium ions in the extracellular fluids. The concentration of magnesium in the extracellular fluid is low (about 1% w/v) and most of the remaining element is found in the soft tissues. If the dietary intake of magnesium falls the body compensates by lowering the amount excreted.

Magnesium is involved in many different biological processes. For example, it is a cofactor for many hormone and enzyme systems and is essential for regulating nerve function, muscle contraction and maintaining the balance of sodium, potassium and calcium across cell membranes. Deficiency in humans can lead to loss of appetite, thrombosis of major organs, convulsions, muscular tremors, epilepsy, constipation and changes in heart rhythm (arrhythmias). Excessive concentrations can decrease muscle response which can result in heart failure. However, there is also evidence to suggest that high levels of magnesium give some protection against heart disease. High concentrations of magnesium also act as a laxative. Magnesium is metabolised in a similar way to calcium. Compounds of magnesium are used mainly as magnesium supplements, antacids and laxatives. For example, magnesium ethanoate is used to supply magnesium ions in dialysis solutions, magnesium carbonate and magnesium trisillicate are antacids and magnesium hydroxide is a constituent of *Milk of Magnesia* which is used as both an antacid (low doses) and a laxative (higher doses). Magnesium compounds are often included in antacid formulations containing aluminium compounds in order to counteract the constipating effect of these compounds.

30.3.2   Calcium

The average adult human body contains about 30,000 mmol (1200 g) of calcium of which 99% is found in the skeleton and teeth largely in the form of calcium phosphate salts such as apatite $[Ca(OH)_2 \cdot 3Ca(PO_4)_2]$ embedded in a framework of organic molecules. The remaining calcium is found in nerves, muscles, including the heart, and blood. The concentration of calcium in the blood is controlled between very narrow limits by vitamin D and the hormone parathormone (which is also known as the parathyroid hormone) and calcitonin. About 50% of the calcium in the plasma occurs as Ca^{2+} ions whilst the rest is mainly bound to blood proteins and is physiologically inactive. If the concentration of calcium in the blood is high it can transfer from the blood to the bones or be deposited in the heart, kidney and other organs of the body.

Absorption of calcium occurs in the small intestine and is controlled by vitamin D and parathyroid hormone (PTH). Too little vitamin D results in poor absorption of calcium.This can result in the disease known as rickets which leads to growing children developing knock-knees, bow legs and pigeon chests among other characteristic symptoms.

Excretion occurs mainly in the urine and faeces. When the plasma calcium level is low, urinary excretion of calcium is decreased whilst absorption from the intestine and movement of calcium from bones to plasma is increased. When the plasma calcium level is high, urinary excretion is increased and more calcium moves from the plasma to the bones. The main hormones controlling these processes are PTH and vitamin D.

Calcium is involved in a large number of important biological processes including:

(1) Bone and teeth growth and maintenance: this is especially important in growing children and pregnant women. In the last few months of pregnancy the mother needs to increase her daily intake of calcium to 200 to 300 mg in order to match the rate at which it is deposited in the foetus.

(2) Blood clotting: Ca^{2+} ions play an essential part in blood clotting and any substance that lowers the concentration of calcium in the blood will reduce blood clotting. Ethylenediamine sodium and lithium salts, for example, lower the concentration of Ca^{2+} ions in blood by forming complexes with the calcium (Fig. 30.2) and so are used to prevent blood samples from clotting. These salts can also be used to treat patients because they do not damage the platelets. Sodium citrate, an anticoagulant used to help preserve the blood used in transfusions removes calcium from the blood by forming insoluble calcium citrates.

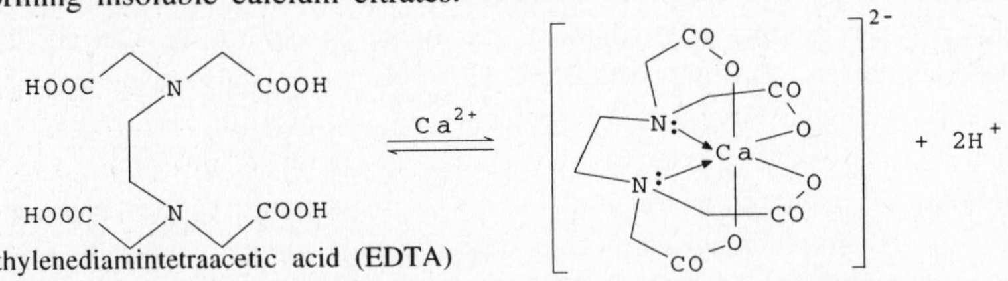

Ethylenediamintetraacetic acid (EDTA)

Fig. 30.2. The reactions of Ca^{2+} with ethylenediaminetetraacetic acid (EDTA).

(3) Control of nerves and muscles: Ca^{2+} ions are involved in nerve transmission and muscle contraction especially the rhythmic contractions of the heart. Too low a concentration of calcium ions in the blood can result in convulsions, nervous twitching and can completely stop muscle contraction. All these conditions can result in death. Too high a concentration of calcium ions can reduce nerve impulse and muscular response.

(4) Cofactors of enzymes: calcium acts as a cofactor for a wide variety of enzymes.

(5) Control of membrane permeability: calcium controls membrane permeability by binding to the proteins and lipids that form the membrane. As the concentration of calcium in the membrane increases, the membrane's permeability to other ions falls.

(6) Stimulus/response coupling: Ca^{2+} ions act as second messengers in the operation of receptors. Cells that have not been stimulated contain a low concentration of calcium. When a stimulant, such as a drug or other metabolite (the first messenger) activates a receptor on a cell membrane, calcium floods into the cell and triggers a cell response such as muscle contraction and the release of hormones, neurotransmitters and enzymes.

(7) Binding of cells to form tissue: Ca^{2+} ions appear to play a part in the binding together of cells to form tissues. In some tissues it is possible to separate the cells by shaking after removing the calcium. It has been suggested that the Ca^{2+} ions hold the cells together by bonding to negatively charged structures on the surfaces of adjacent cells. However, many tissues require additional chemical treatment to separate their

cells so Ca^{2+} ions are not the only species involved.

Calcium compounds are used in medicine to treat calcium deficiency and as antacids. Their use as antacids can induce rebound acid secretions but at normal doses this is not significant. Calcium compounds are also widely used in the food industry as buffers, thickeners, emulsifiers and calcium supplements.

30.4 THE TRANSITION METALS

The elements of the main transition series differ mainly in their inner **d** orbital configurations (Table 30.7). Consequently, as the transition elements of a series usually differ only in their inner orbital configurations they exhibit many similar general chemical and physical properties. Of particular interest is their ability to form complexes, the variable oxidation states exhibited by the majority of the elements, and the fact that their compounds have readily observable spectra.

Table 30.7. The elements of the first, second and third transition series. (Ar) represents the electronic configuration of argon, (Kr) krypton and (Xe) xenon.

Element, first series	Electronic configuration	Element, second series	Electronic configuration	Element, third series	Electronic configuration
Scandium (Sc)	$(Ar)\ 3d^1\ 4s^2$	Yttrium (Y)	$(Kr)\ 4d^1\ 5s^2$	Lanthanium (La)	$(Xe)\ 5d^1\ 6s^2$
Titanium (Ti)	$(Ar)\ 3d^2\ 4s^2$	Zirconium (Zr)	$(Kr)\ 4d^2\ 5s^2$	Hafnium (Hf)	$(Xe)\ 5d^2\ 6s^2$
Vanadium (V)	$(Ar)\ 3d^3\ 4s^2$	Niobium (Nb)	$(Kr)\ 4d^4\ 5s^1$	Tantalum (Ta)	$(Xe)\ 5d^3\ 6s^2$
Chromium (Cr)	$(Ar)\ 3d^5\ 4s^1$	Molybdenum (Mo)	$(Kr)\ 4d^5\ 5s^1$	Tungsten (W)	$(Xe)\ 5d^4\ 6s^2$
Manganese (Mn)	$(Ar)\ 3d^5\ 4s^2$	Technrtium (Tc)	$(Kr)\ 4d^6\ 5s^1$	Rhenium (Re)	$(Xe)\ 5d^5\ 6s^2$
Iron (Fe)	$(Ar)\ 3d^6\ 4s^2$	Ruthenium (Ru)	$(Kr)\ 4d^7\ 5s^1$	Osmium (Os)	$(Xe)\ 5d^6\ 6s^2$
Cobalt (Co)	$(Ar)\ 3d^7\ 4s^2$	Rhodium (Rh)	$(Kr)\ 4d^8\ 5s^1$	Iridium (Ir)	$(Xe)\ 5d^7\ 6s^2$
Nickel (Ni)	$(Ar)\ 3d^8\ 4s^2$	Palladium (Pd)	$(Kr)\ 4d^{10}$	Platinium (Pt)	$(Xe)\ 5d^9\ 6s^1$
Copper (Cu)	$(Ar)\ 3d^{10}\ 4s^1$	Silver (Ag)	$(Kr)\ 4d^{10}\ 5s^1$	Gold (Au)	$(Xe)\ 5d^{10}\ 6s^1$
Zinc (Zn)	$(Ar)\ 3d^{10}\ 4s^2$	Cadmium (Cd)	$(Kr)\ 4d^{10}\ 5s^2$	Mercury (Hg)	$(Xe)\ 5d^{10}\ 6s^2$

Many of the transition elements are essential trace elements in human metabolism. Their concentration in the body is so low that it is difficult to determine their exact role and mode of action in human metabolism. However, knowledge is increasing as analytical instruments become more sensitive. Other transition metals such as vanadium and silver occur in the body but do not appear to play an essential role in its metabolism although it is thought that vanadium may be important in lipid metabolism. Generally as one proceeds down a group in the transition series the elements and their compounds become more toxic. However, in spite of this increase in toxicity their compounds may have some medical uses. Silver nitrate, for example is used to treat warts and as a topical anti-infective in opthalmic solutions and mercury II chloride is the main constituent of *Calamine lotion* used in the treatment of pruritus.

30.4.1   Titanium

This element does not appear to be involved in human metabolism. White titanium oxide (TiO_2) is used as a protective in sun creams and as a colouring agent (E171) in certain foods and paint.

30.4.2   Vanadium

Vanadium is found in the human body but does not appear to be essential for human metabolism. It is essential for the metabolism of rats and some other animals. Deficiency in animals results in anaemia and lack of growth of bones, teeth and cartilage. Evidence suggests that excessive levels of vanadium in man may result in manic depression.

30.4.3   Chromium

Chromium is an essential trace element in man. The average adult body contains about 1.0 to 1.9 mmol (5 to 10 mg) of which 5 to 10 nmol dm^{-3} are normally found in the blood. The level of chromium is considerably lower in the bodies of older people. Absorption from food is generally poor. Most of what is absorbed is in the form of organically bound chromium. The absorption of ionic chromium is very poor.

The function of chromium in humans is obscure but it is known to be involved in the use of glucose in the synthesis of fats, the operation of insulin, the control of plasma cholesterol levels and glycogen formation. Deficiency in man leads to increased plasma-cholesterol levels, and impairs the working of insulin and the metabolism of glucose giving diabetes-like symptoms (hyperglycaemia). Chromium deficiency can also occur in alcoholism.

Chromium compounds are used mainly as supplements in tonics. Most tonics contain ionic chromium but absorption from this source is generally poor. However, it is possible to obtain organically bound chromium.

30.4.4   Manganese

The average human adult body contains between 2.2×10^{-1} and 3.6×10^{-1} mmol (12 and 20 mg) of this essential trace element which is concentrated mainly in the skeleton, kidneys, liver and heart. Dietary absorption is low, only about 4% of the manganese in food being absorbed. However, the rate of excretion is dependent on dietary intake but can be relatively high at about 4 mg daily.

Manganese appears to be involved in the control of growth, maintenance of the nervous system, and the synthesis of thyroid hormones, glycoproteins and glycogen in the liver, and the natural antiviral agent interferon. It occurs in some female hormones and is employed as a co-enzyme in active phosphate transfer, the formation of nucleic acids and bone cartilage. A deficiency of manganese has been found in humans suffering from diabetes, schizophrenia, rheumatoid arthritis, heart disease and myasthenia gravis. Manganese compounds such as the chloride, sulphate, orotate and gluconate are used to treat manganese deficiency.

30.4.5   Iron

Iron is an essential trace element. The average adult human body contains about 62.5 to 80.5 mmol (3.5 to 4.5 g) of which about 70% is found in haemoglobin and 3% in myoglobin present in muscle cells. About 1% is found in blood plasma whilst almost all the remainder occurs as the protein-iron storage complexes ferritin and haemosiderin.

Only 5 to 15% of the iron in food is absorbed, the remainder is excreted in the faeces (Fig. 30.3). Absorption occurs in the intestine mainly in the form of iron II (ferrous iron). Much of the iron III (ferric iron) in the food is converted to iron II by reducing agents in the food, such as vitamin C, prior to absorption. The iron is transported into the intestinal epithelial cells where most of the iron is incorporated into ferritin an iron-protein complex. This acts as an intracellular store for iron. Some of this iron is released into the plasma where it binds to the transfer protein transferrin. The rate of release from intestinal ferritin increases as the concentration of iron in the body falls. This leads to a subsequent increase in the rate of absorption from food to replace the ferritin in the intestinal cells. It is interesting to note that foods containing phytic acid and with a high phosphate content reduce the absorption of iron. Little iron is lost from the body as sweat, in the urine and faeces. The biological half-life of iron is about eight years. However, losses are often large during lactation and menstruation. There is often a need to increase the level of iron in the diet during menstruation. However, the DHSS does not generally recommend increasing the level of iron during pregnancy since the increased need of iron is usually equal to that which would have been lost by menstruation.

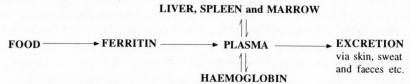

Fig. 30.3. A schematic outline of the absorption and excretion of iron in humans.

Iron is an essential element for the production of haemoglobin in animals and humans. Deficiency leads to a fall in the concentration of haemoglobin which can lead to a reduction in the number of red blood cells circulating in the blood. This condition, anaemia, results in a feeling of fatigue and apathy. It occurs in humans that suffer from gastric and duodenal ulcers and haemorroids (piles). It can also be due to excessive blood loss from wounds, deficiency in the diet and malabsorption due to either a lack of vitamin C or drug interference. For example, the antibiotic tetracycline and antacids react with iron to produce compounds that are less readily absorbed. Iron is also involved in the function of folic acid, vitamin B_6, and the production of white blood cells. Excess uptake of iron results in haemochromatosis, a general depositing of iron that leads to tissue damage. This can lead to cirrhosis of the liver and fibrosis of the pancreas resulting in diabetes mellitus and heart failure.

Iron compounds, such as iron II sulphate, gluconate, phosphate and amino-acid chelate are used as iron supplements in tonics and similar preparations. They tend to cause side effects such as nausea, vomiting, constipation, diarrhoea, blackening of teeth and faeces. Allergic reactions are common with iron sorbitol injections. Gastric irritation can be avoided if the patient takes the supplement after meals. Iron compounds are also used as food additives. Iron hydroxide and oxides, for example, are used as colourings (E172).

30.4.6   Cobalt

This essential trace element occurs in man as a constituent of vitamin B_{12} where it is in the form of cobalt II (Fig. 30.4). The average human adult body contains about 1.7×10^{-5} mmol

(1.1 mg) of cobalt, 43% of which occurs in muscle tissue, 14% in bone and most of the rest in the liver and the kidneys. Blood levels are usually small, most of the element being found in the red blood cells.

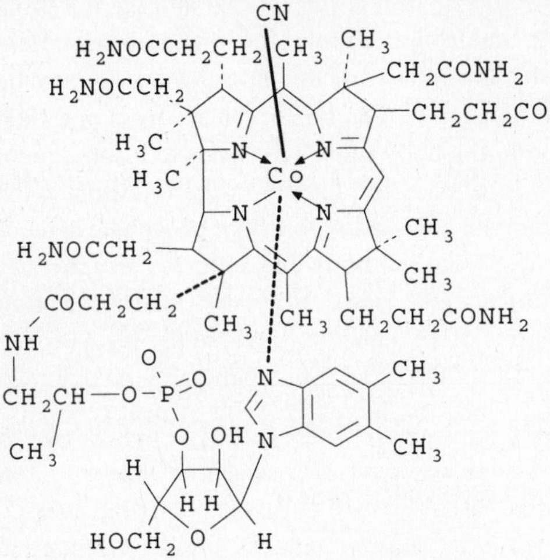

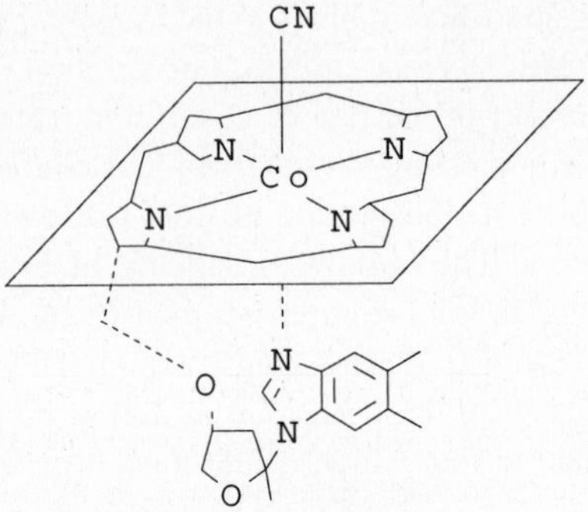

An outline of the three dimensional structure of vitamin B_{12}

Fig. 30.4. The structure of vitamin B_{12}. Co-enzyme B_{12} has an adenosine residue attached via the 5' position of the adenosine instead of the cyanide group.

Cobalt is absorbed as vitamin B_{12} in the gastrointestinal tract since humans and animals, with the exception of ruminants, cannot synthesise this vitamin. Cobalt is essential for the function of vitamin B_{12} which is involved in:

(1) The formation of myelin, the fatty sheath that insulates the nerves.
(2) The synthesis of DNA, methionine and choline. In these syntheses vitamin B_{12} is an intermediate in the transfer of methyl groups from folic acid to the receptor molecle.
(3) The detoxification of cyanide ions in food and tobacco smoke.
(4) The control of maturation of erythrocytes. Lack of vitamin B_{12} results in a failure of the body to produce sufficient erythrocytes, a condition known as megaloblastic anaemia. Too much vitamin B_{12} can cause the overproduction of erythrocytes leading to a condition known as polycythemia.

Cobalt II chloride has been used to reduce blood pressure since it dilates blood vessels. This chloride and the corresponding carbonate are used as a dietary supplement in animals whilst the nitrate has been used as a dietary supplement in both animals and humans. A deficiency of cobalt in the diet can cause a disease called pernicous anaemia which results in symptoms of weakness and fatigue.

30.4.7 Nickel

This element has been found in man but it is not known whether it is an essential trace element for human metabolism. However, deficiency diseases have been observed in animals. In animals it appears to have an antagonistic action to adrenaline, stabilises DNA and RNA in the tissue and increases the action of insulin and concentration of blood fats. High blood levels have been found in people suffering from heart attacks, serious burns, strokes, cancer of the uterus and lungs. Nickel carbonyl [$Ni(CO)_4$], produced in tobacco smoke by the reaction of nickel with carbon monoxide causes headaches, nausea, vertigo chest pains and vomiting. It also causes lung cancer in rats and is thought to be a possible cause of this disease in man.

30.4.8 Copper
This is an essential trace element for man, animals and plants. The average human adult body contains from 1.18 to 2.36 mmol (75 to 150 mg), about 50% of which is found in the muscles and bone. The liver contains about 10% but this decreases with age. There are significant amounts in the brain, heart and kidney and in these organs copper concentration increases with age. The significance of these changes is not known. Blood also contains significant amounts of copper mostly bound to the plasma albumin and ceruloplasmin the copper-binding protein. Some also occurs in red blood cells. High blood concentrations are found in women taking contraceptive pills, and those suffering from epilepsy, zinc deficiency, schizophrenia, rheumatoid arthritis and high cholesterol levels.

Copper is excreted by incorporation into the bile with subsequent secretion into the intestine. Very little is lost in the urine. Deficiency results in anaemia in humans and cattle, in retarded growth, brittle bones (osteoporosis, especially in sheep), diarrhoea and a low white cell count with a subsequent lower resistance to illness. High concentrations of copper in the body are just as damaging. Wilson's disease, which can lead to mental illness and death, is caused by high concentrations of copper in the body. This disease is treated with chelating agents that bind to the copper and so aids its excretion.

Copper acts as a co-enzyme in a wide variety of biological processes. For example, copper acts as a co-enzyme for the formation of melanin which is the dark pigment responsible for skin colour and the natural protection against the ultra-violet radiation in sunlight. Copper also acts as a co-enzyme for lipyl oxidase, the enzyme involved in the formation of the connective tissues such as collagen and elastin, cytochrome oxidase which is involved in oxygen transfer and dopamine hydroxylase which is concerned with nerve impulse transmission in the brain. It is also essential for the absorption of iron from food, bone and white cell formation.

Copper compounds are used to treat copper deficiency and other conditions. For example, copper II sulphate is used as a fungicide, to treat phosphorus poisoning and topically to treat phosphorus burns. It is believed that the action of copper bracelet used in folk medicine to treat arthritis is due to the copper forming salts with the acids in the skin. This is supported by the fact that copper salicylate is an effective anti-inflammatory for arthritis. Copper chlorophyll (E141) is a permitted green colouring agent in some jams and sugar products.

30.4.9 Zinc
Zinc is a very important trace element in humans. The average adult body contains 3.0×10^{-2} to 4.5×10^{-2} mmol (2 to 3 g) of zinc of which 63% is found in muscle and bone, 20% in the skin and about 8 to 18 μmol dm^{-3} in plasma. Significant amounts occur in the liver, kidneys, prostrate, semen, sperm, eyes and hair. There do not appear to be any specific storage sites but it complexes widely with amino acids, proteins and plasma albumin. Substances in the pancreatic juices transport the zinc into the intestinal cells and

from there into the epithelial cells (lining cells) of the blood vessels. It is stored here until required. The presence of cysteine in food helps the absorption which probably involves forming complexes with amino acids, peptides and proteins. Phytic acid in foods such as cereals, fruit and some tubers reduces the absorption by forming insoluble complexes. Deficiency can lead to anaemia, acrodermatitis enteropathica, impotence, loss of sense of taste and smell, impairment of the immune system, delayed wound healing and a general retarding of physical, sexual and mental growth, especially in children. It can also result in an increase in the absorption of the essential trace element copper and the toxic metal cadmium. Increasing the concentration of zinc reverses this situation and so zinc salts are used to treat cases of cadmium poisoning.

Excretion occurs mainly in the faeces but sweat and urine losses can be significant. It is reduced in cases of zinc deficiency but increases in cases of starvation, alcoholism, porphyria, liver and kidney disease. Thiazide diuretics also increase zinc excretion whilst ethacrynic and fruseamide diuretics cause a decrease in zinc excretion. Consequently, these drugs are often used with thiazide drugs to limit zinc losses.

Zinc has been found to play an important part in many biological systems. Some of its functions are:

(1) It is found in several liver enzymes, one of which is concerned with the conversion of ethanol to less toxic compounds. Alcoholics often have a high zinc concentration in their urine which is taken to indicate a breakdown of this enzyme.

(2) It is required for the release of insulin from the pancreas. A low zinc level has been found in many diabetics.

(3) It is involved in cell division.

(4) It plays a part in the release of vitamin A from the liver.

(5) It aids wound healing.

Some of the uses of zinc compounds are given in Table 30.8.

Table 30.8. Some pharmaceutical uses of zinc compounds.

Use	Compound (preparation)
Astringent	Zinc sulphate, zinc oxide (zinc and castor oil cream)
Supplement	Zinc sulphate, zinc gluconate, zinc orotate
Fungicide	Zinc undecenonate
Pharmaceutical aid	Zinc stearate (lubricant)

30.4.10 Molybdenum

This is an essential trace element. About 0.094 mmol (9 mg) is found in the average adult human body. Its distribution is uncertain but about 0.0037 mmol/100 g (0.35 mg/100 g) occurs in the liver and about 0.049 m mol/100 g (4.7 µg/100 g) is found in blood. It occurs in the enzyme xanthine oxidase which is concerned with iron metabolism and the production of uric acid. Molybdenum is involved in the energy transfer reactions of cells, metabolism of nucleic acids, production of sperm and the prevention of dental caries. Excessive production of uric acid is associated with gout.

About 50% of dietary molybdenum is absorbed. Deficiency in man is rare but is associated with cancer of the oesophagus, impotence in men and dental caries. Excess has been reported as causing an increase in the excretion of copper leading to a deficiency of this element and gout. Excretion occurs mainly in the urine.

30.4.11   Cadmium

This element does not occur at birth in the human body but accumulates over the years until at an age of 50+ it reaches about 0.18 to 0.27 mmol (20 to 30 mg). Over 50% of this accumulated cadmium is found in the liver and kidneys where concentrations of the order of 0.22 mmol/100 g (25 mg/100 g) can cause damage. It is detoxified by the protein metallothionein but excretion is slow.

Cadmium is very toxic to man, high levels causing anaemia by being an antagonist for copper and iron, high blood pressure and kidney damage amongst other conditions. It competes with calcium, copper, zinc and selenium in metabolic processes, probably by competing for the same active sites on proteins. Its compounds have been used as anthelmintics and cadmium sulphide is used for seborrheaic dermatitis and dandruff, but it can cause photosensitisation.

30.4.12   Mercury

Mercury does not appear to be an essential element in any living organism. The metal and its compounds are toxic to man causing brain, colon and kidney damage amongst other conditions. Two of its most toxic metabolites, ethyl and methylmercury, are formed by micro-organisms because of mercury pollution of seas, streams and rivers. These compunds have been introduced into man through the plant and fish food chains with fatal effects. Recently, some doubt has been cast on the use of mercury amalgams as dental fillings. It is claimed, somewhat speculatively, that the slight solubility of these materials may have some effect on cancer, heart disease and some menstrual and thyroid conditions.

Mercury compounds are used to prevent fungal contamination of seeds ($HgCl_2$) and as disinfectants ($HgCl_2$). Mercury II chloride is a constituent of *Calamine lotion* that is used to sooth irritated skin.

30.5 GROUP   III

The elements of group III (Table 30.9) do not appear to play a significant role in human metabolism. However, boron is an essential element in plant metabolism.

Table 30.9. The elements of group III.

Elements	Symbol	Electronic configuration
Boron	B	$1s^2\ 2s^2\ 2p^1$
Aluminium	Al	$1s^2\ 2s^2\ 2p^6\ 3s^2\ 3p^1$
Gallium	Ga	$1s^2\ 2s^2\ 2p^6\ 3s^2\ 3p^6\ 3d^{10}\ 4s^2\ 4p^1$
Indium	In	$1s^2\ 2s^2\ 2p^6\ 3s^2\ 3p^6\ 3d^{10}\ 4s^2\ 4p^6\ 4d^{10}\ 5s^2\ 5p^1$
Thallium	Th	$1s^2\ 2s^2\ 2p^6\ 3s^2\ 3p^6\ 3d^{10}\ 4s^2\ 4p^6\ 4d^{10}\ 4f^{14}\ 5s^2\ 5d^{10}\ 6s^2\ 6p^1$

Boron is absorbed from food by the body but this absorption is balanced by its excretion in the urine. It is not absorbed through whole skin but is readily absorbed through damaged skin. Boron compounds, such as boric acid and borax, have been used in dusting powders and body cavity douches because they have some fungicidal and bactericidal action but they are no longer used because of their toxicity. Fatal doses are of the order of 15-20 g for adults and 3-6 g for children.

The role of aluminium in the body is not understood but it is not thought to be an important element in normal human metabolism. However, aluminium is known to be a neurotoxin and there is evidence to suggest that aluminium can cause mental retardation. A case study, in 1990, of Canadian miners who were regularly exposed to specific doses of McIntyre powder, a form of powdered aluminium, as a protection against silicosis showed that 13% of the men exposed showed cognitive impairment compared to only 5% of a control group. High concentrations of aluminium have also been found in the brain tissues of people suffering from Alzheimer's disease. Because aluminium accumulates during the normal ageing process it is thought that it might trigger the onset of this disease in older people but no links have been established between aluminium and the disease. Aluminium can cause dermatitis, demineralisation of the bones and phosphate depletion.

Aluminium compounds are widely used as antacids to neutralise the hydrochloric acid in the gastric secretions of the small intestine. Aluminium hydroxide, aluminium glycinate and aluminium phosphate are used for this purpose but can cause constipation. Potash alum (potassium aluminium sulphate) and aluminium chloride are used as astringents to form superficial protective layers over lesions in the skin and mucous membranes. Solutions of aluminium chloride and sulphate are also used as mild antiseptics. Aluminium sulphate is added to drinking water to clarify contaminated sources whilst many aluminium compounds are used as additives in the food industry. Aluminium compounds, such as, aluminium chloride are also used as drying agents in deodorants whilst aluminium hydroxide is given orally to kidney dialysis patients to prevent excess phosphate absorption.

Practical work has shown that gallium behaves in the same way as aluminium in biological systems. Consequently, gallium-67 is used as a substitute for aluminium in physiological investigations as there are no suitable radioactive aluminium isotopes.

Thallium can be absorbed through whole skin as well as internally. It is toxic with an action similar to that of arsenic. Thallium sulphide has been used in depilatory preparations but was withdrawn after it was found to be the cause of a number of deaths.

30.6 GROUP IV

The elements of this group are listed in Table 30.10. The chemistry of carbon and its compounds is discussed in the appropriate chapters of this text. Silicon occurs in trace amounts in the body as a constituent of cartilage, arterial walls and connective tissue. It appears to be important in maintaining the strength and elasticity of these tissues. Germanium, tin and lead occur in trace amounts but with the exception of tin they do not seem to be involved in human metabolism. Excess amounts of tin and lead are highly toxic. Poisoning can result in anaemia, colic, joint and muscle pains, vomiting, convulsions and brain damage. Lead poisoning in children can cause hyperactivity and mental retardation.

Table 30.10. The elements of group IV.

Element	Symbol	Electronic configuration
Carbon	C	$1s^2\ 2s^2\ 2p^2$
Silicon	Si	$1s^2\ 2s^2\ 2p^6\ 3s^2\ 3p^2$
Germanium	Ge	$1s^2\ 2s^2\ 2p^6\ 3s^2\ 3p^6\ 3d^{10}\ 4s^2\ 4p^2$
Tin	Sn	$1s^2\ 2s^2\ 2p^6\ 3s^2\ 3p^6\ 3d^{10}\ 4s^2\ 4p^6\ 4d^{10}\ 5s^2\ 5p^2$
Lead	Pb	$1s^2\ 2s^2\ 2p^6\ 3s^2\ 3p^6\ 3d^{10}\ 4s^2\ 4p^6\ 4d^{10}\ 4f^{14}\ 5s^2\ 5d^{10}\ 6s^2\ 6p^2$

Lead is a neurotoxin displacing calcium from parts of the nervous system. Its action is probably related to the ease with which it forms complexes and the stability of those complexes. It is interesting to note that lead inhibits some enzymes involved in haemoglobin synthesis. These enzymes can also be altered in the group of diseases known collectively as porphyrias which produce symptoms similar to lead poisoning. Drugs used to treat lead poisoning usually act by forming a more stable complex that can be excreted.

30.7 GROUP V

The elements of group V are listed in Table 30.11. Nitrogen and phosphorus are essential constituents of the organic compounds found in all living matter. Both are essential to life. Arsenic, antimony and bismuth are toxic but traces of both arsenic and bismuth are found in man. Excretion is rapid and so a build-up to toxic levels does not occur with humans on a normal diet. Arsenic appears to be a trace element in animal metabolism.

Inorganic arsenic compounds have been used for medical purposes but their use is no longer recommended. Organic arsenic compounds are widely used in agriculture as insecticides and animal feed supplements. Bismuth chelate (*De-Nol*) is used as a protective agent for the gastrointestinal tract by deposition on the walls of the tract. They are often formulated with antacids to reduce gastric irritation. Bismuth carbonate is also a mild antacid in its own right but its use is not recommended as it can be neurotoxic. Bismuth subgallate is used in suppositories to treat haemorrhoids.

Table 30.11. The elements of group V.

Element	Symbol	Electronic configuration
Nitrogen	N	$1s^2\ 2s^2\ 2p^3$
Phosphorus	P	$1s^2\ 2s^2\ 2p^6\ 3s^2\ 3p^3$
Arsenic	As	$1s^2\ 2s^2\ 2p^6\ 3s^2\ 3p^6\ 3d^{10}\ 4s^2\ 4p^3$
Antimony	Sb	$1s^2\ 2s^2\ 2p^6\ 3s^2\ 3p^6\ 3d^{10}\ 4s^2\ 4p^6\ 4d^{10}\ 5s^2\ 5p^3$
Bismuth	Bi	$1s^2\ 2s^2\ 2p^6\ 3s^2\ 3p^6\ 3d^{10}\ 4s^2\ 4p^6\ 4d^{10}\ 4f^{14}\ 5s^2\ 5d^{10}\ 6s^2\ 6p^3$

30.7.1 Nitrogen

Nitrogen compounds are important constituents of living organisms (Chapters 28 and 29). Nitric oxide (NO) gas is a natural vasodilator produced by cells lining blood vessels. Investigations of this phenomenon has indicated that nitric oxide has a fundamental role in regulating blood pressure, and protecting the cardiovascular system against toxins and parasites. It has also been suggested that blocking of nitric oxide production could be useful in the treatments of both ulcerative colitis and dementia.

30.7.2 Phosphorus

Phosphorus in the form of various phosphates is a major constituent of all plants and animals. The average human adult body contains about 550 to 770 g. About 90% of this is found in the bones and teeth in the form of various calcium phosphates, a common one being calcium hydroxyapatite. The remaining phosphorus occurs largely in the muscle with about 1% in the nerve tissue.

Metabolism of phosphorus which is coupled with that of calcium and magnesium is influenced by PTH, calcitonin and vitamin D. Low body concentrations are unusual in western countries but low blood levels are observed in vitamin D resistant rickets. Excretion is mainly via the kidneys.

Phosphates have a wide variety of functions in the human body. They are required for the maintenance of the bones and teeth and with other negative ions they help to balance the positive ions in the body's tissues and fluids. The acidic salt, monosodium dihydrogen phosphate (NaH_2PO_4) and the basic salt, disodium hydrogen phosphate (Na_2HPO_4) are used to help maintain the pH of arterial blood between 7.39 and 7.41. Phosphates are involved in many metabolic processes, for example the conversion of ATP to ADP (21.7.4) is the principal source of energy for many metabolic processes.

Phosphorus compounds such as calcium and potassium phosphates, calcium, magnesium and sodium glycerophosphates are used as supplements in the hospital treatment of a number of medical conditions.

30.8 Group VI

The elements that are found in group VI are listed in Table 30.12. Oxygen and sulphur are important constituents of all living matter. The chemistry of organic compounds containing these elements is discussed in the relevant chapters of this text. Selenium was once

Table 30.12. The elements of group VI.

Elements	Symbol	Electronic configuration
Oxygen	O	$1s^2\ 2s^2\ 2p^4$
Sulphur	S	$1s^2\ 2s^2\ 2p^6\ 3s^2\ 3p^4$
Selenium	Se	$1s^2\ 2s^2\ 2p^6\ 3s^2\ 3p^6\ 3d^{10}\ 4s^2\ 4p^4$
Tellurium	Te	$1s^2\ 2s^2\ 2p^6\ 3s^2\ 3p^6\ 3d^{10}\ 4s^2\ 4p^6\ 4d^{10}\ 5s^2\ 5p^4$
Polonium	Po	$1s^2\ 2s^2\ 2p^6\ 3s^2\ 3p^6\ 3d^{10}\ 4s^2\ 4p^6\ 4d^{10}\ 4f^{14}\ 5s^2\ 5d^{10}\ 6s^2\ 6p^4$

thought to be toxic to man but since 1949 has been known to be an essential trace element. Tellurium and polonium are toxic to animals and man.

30.8.1   Sulphur

The average adult human body contains about 100 g of this element. It is found mainly in the skin, fingernails, toenails and hair. Sulphur occurs as a constituent of proteins, vitamin B_1 and other organic compounds (Fig. 30.5). It also occurs in cells as sulphate ions and substituted sulphate compounds such as heparin and chondroitin sulphate (27.12.6).

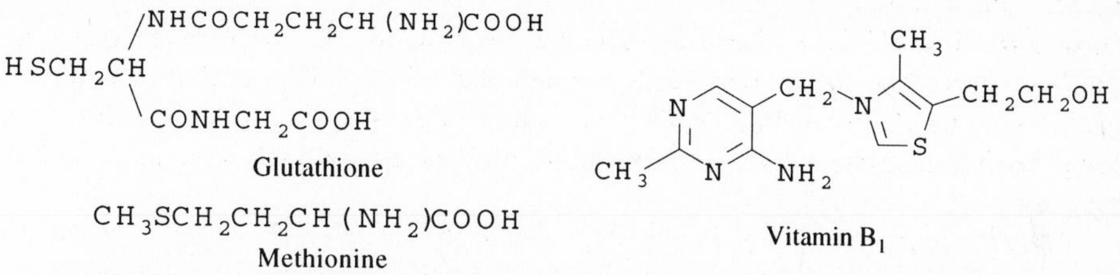

$$HSCH_2CH \begin{cases} NHCOCH_2CH_2CH(NH_2)COOH \\ \\ CONHCH_2COOH \end{cases}$$

Glutathione

$$CH_3SCH_2CH_2CH(NH_2)COOH$$
Methionine

Vitamin B_1

Fig. 30.5. Some examples of the sulphur-containing compounds found in the human body.

Sulphur is used in a variety of dusts, lotions, creams and ointments as a germicide and keratolytic. Sodium sulphite is used as a sterilising agent. The active species is sulphur dioxide which is formed by the action of acid on the sulphite.

30.8.2   Selenium

Selenium is an essential trace element in human and animal metabolism. It is absorbed into animals through the plants they eat. Animals living in regions with a low selenium concentration in the soil have a low uptake of the element and as a result can suffer from white muscle, pancreatic and liver diseases. The use of selenium animal feed supplements prevents these diseases. Conversely, animals that live in regions where the soil is rich in selenium have a high selenium uptake which can lead to a disease called 'blind staggers'. This disease causes impaired vision and muscle weakness. However, the death rate from heart disease is reported to be lower in regions of the United States where there is a high selenium concentration in the soil. Worldwide evidence suggests that a low selenium concentration in humans is associated with some forms of skin cancer.

The average human adult body contains about 0.25 mmol (20 mg) of this essential trace element. It is found in all tissues, the highest concentrations being found in the liver, kidneys and male testes. Blood levels are dependent on geographical location and range from about 1.1 to 2.8 mmol dm^{-3}. Organic selenium compounds are more easily absorbed than inorganic selenium compounds.

Selenium is involved in the maintenance of the immune defence mechanism, the liver, eyes, skin and hair. It has a synergistic action with vitamin E in the maintence of mitochondria and has a role in the production of prostaglandins and ubiquinone. Selenium also acts as an anti-inflammatory and antioxidant as part of the enzyme glutathione peroxidase. This enzyme prevents the oxidation of lipids in cell membranes and is man's main defence against peroxides. The element is also involved in counteracting the toxic affects of cadmium and mercury. It forces these metals to bind to the high molecular weight body proteins instead of the smaller more essential proteins.

30.9 GROUP VII

The elements of group VII are collectively known as the **halogens** (Table 30.13). Fluorine and iodine are essential trace elements for animals and humans. Chlorine in the form of chloride ions is an important component of the human body.

Table 30.13. The elements of group VII.

Element	Symbol	Electronic configuration
Fluorine	F	$1s^2\ 2s^2\ 2p^5$
Chlorine	Cl	$1s^2\ 2s^2\ 2p^6\ 3s^2\ 3p^5$
Bromine	Br	$1s^2\ 2s^2\ 2p^6\ 3s^2\ 3p^6\ 3d^{10}\ 4s^2\ 4p^5$
Iodine	I	$1s^2\ 2s^2\ 2p^6\ 3s^2\ 3p^6\ 3d^{10}\ 4s^2\ 4p^6\ 4d^{10}\ 5s^2\ 5p^5$
Astatine	At	$1s^2\ 2s^2\ 2p^6\ 3s^2\ 3p^6\ 3d^{10}\ 4s^2\ 4p^6\ 4d^{10}\ 4f^{14}\ 5s^2\ 5d^{10}\ 6s^2\ 6p^5$

30.9.1 Fluorine

Trace amounts are found in bones, teeth and human body tissues. Blood levels are about 3.7 to 5.0 mmol dm^{-3} (140 to 190 mg dm^{-3}) of which approximately 20% is in the form of fluoride ions. Fluorine, in the form of fluoride ions is readily absorbed from food and drinking water. Excess is excreted in the urine.

The function of fluorine in the human body is not fully understood but it is believed to be incorporated into the bone like calcium. Fluorine is believed to stabilise bone and help prevent osteoporosis in the elderly. Fluoride is also thought to protect children's teeth from dental caries. However, excess can cause fluorosis of the teeth and skeleton. Fluorosis of the teeth results in a breakdown of the enamel which initially leads to yellow-brown staining followed by pitting in severe cases. Fluorosis of the skeleton results in a condition known as osteosclerosis in which the density of some bones, such as, those of the pelvis and spine increases; calcium is deposited in ligaments, muscles and tendons and the spine stiffens. However, fluorosis is not the only cause of osteosclerosis.

30.9.2 Chlorine

Chlorine occurs in man mainly as chloride ions. The average human adult body contains about 3.23 mol (115 g) of these ions. Blood chloride levels in healthy adults are about 3.5 to 3.75 g dm^{-3} (98 to 106 mmol dm^{-3}). Body content is regulated by excretion of the excess in the urine, gastrointestinal tract and sweat. Vomiting may also cause significant losses of chloride as hydrochloric acid from the stomach.

The main functions of chloride in the human body is to provide chlorine for hydrochloric acid production in the stomach and to balance the sodium and potassium ions in body fluids. Low levels of plasma chloride are found in diabetes and fevers whilst high levels are observed in anaemia, heart and kidney diseases. Recently it has been suggested that high blood pressure is due to a high dietary chloride content and not a high sodium content as was previously thought. Increased chloride excretion is found in a number of conditions including rickets and cirrhosis of the liver whilst it decreases in cancer, pneumonia and kidney disease. However, in all the examples mentioned in this section the concentration of chloride is not thought to be diagnostically significant.

30.9.3   Iodine

Iodine is an essential trace element for animals and man. The average human adult body contains from 0.078 to 0.196 mmol (20 to 50 mg) of iodine, approximately 75% of which is found in the thyroid gland. Iodine is also found in saliva and breast milk amongst other places.

Iodine is absorbed mainly as iodide in the gastrointestinal tract and transported in the blood to the thyroid where it is stored. It is used as required to produce the hormones 3,5,3'-triiodothyronine and L-thyroxine. These hormones are stored as a complex with a protein globulin, thyroglobulin. Their release is controlled by the thyrotrophic hormones, (TSH) manufactured in the pituitary gland and TRH formed in the hypothalamus. They control the rate of growth in children and the rate at which food is metabolised. These hormones play a major role in maintenance of a healthy body.

3,5,3'-Triiodothyronine

L-Thyroxine (3,5,3',5'-tetraiodothyronine)

Iodine absorbed by the human body appears to be required solely for use in the production of the thyroid hormones. A deficiency results in the body increasing the number of cells in the thyroid in an attempt to increase hormone production. This growth of the thyroid causes a characteristic swelling on the neck known as goitre. Deficiency can also lead to myxoedema and cretinism in children (mental and physical growth retardation). It is caused by diets lacking in iodine or containing substances that reduce its uptake. Drugs, such as, p-aminosalicylic acid and resorcinol can also reduce the rate of iodine absorption by the body. Excessive absorption can also cause goitre as well as abdominal pain, vomiting, bloody diarrhoea and supression of the activity of the thyroid gland.

Potassium iodide, potassium iodate, sodium iodide and calcium iodide are used as iodine supplements to treat iodine deficiency.

30.10   STABILITY AND STORAGE

The main factors affecting the stability and storage of inorganic compounds are atmospheric moisture, oxygen and carbon dioxide in the air, heat and light. These factors can cause changes in the chemical and physical nature of the stored substance which change its stoichiometry. This means that medical and pharmaceutical preparations made with this substance will not contain the amounts calculated according to the original formula of the substance. Consequently, it is important that the substance is stored correctly and an appropriate quality assurance is carried out on any preparations made from the substance.

30.10.1 Atmospheric moisture

Atmospheric moisture usually reacts with the compound or is absorbed by it during storage. Both these processes can alter the stoichiometry of the compound.

Reaction with water vapour during storage can reduce the pharmaceutical effectiveness of a compound by converting it to an inactive compound. However, this is not always the case. For example, calcium oxide which is used as an antacid, reacts with moisture to form calcium hydroxide which is also an antacid. In this case the reaction does not change the action of the compound but it will affect formulations made with the oxide since the hydroxide has a different formula.

Compounds that absorb atmospheric moisture are said to be **hygroscopic**. Iron II sulphate (ferrous sulphate) that is used to treat iron deficiency, for example, is hygroscopic, changing its state of hydration by absorbing moisture on storage:

$$FeSO_4 \cdot H_2O + 3H_2O \longrightarrow FeSO_4 \cdot 4H_2O$$

Compounds that can absorb so much water that they form a solution are said to be **deliquescent**. For example, zinc chloride used as the sweat absorbent in some deodorants is deliquescent, as are many sodium salts.

Compounds, such as magnesium sulphate which is used as a mild laxative and in magnesium supplements, that lose water on storage are known as **efflorescent compounds**.

$$MgSO_4 \cdot 7H_2O \longrightarrow MgSO_4 \cdot 6H_2O + H_2O$$

The effects of atmospheric moisture can be minimised by storing compounds in containers with airtight lids. Deliquescent compounds should be stored in desiccators if possible. Moisture control agents such as magnesium oxide, starch and diatomaceous earth are sometimes added to preparations containing hygroscopic and deliquescent compounds. Moisture control agents are substances that will preferentially absorb atmospheric moisture but are chemically and biologically inert in the context of the use of the preparation.

30.10.2 Air

The main substances affecting the stability of inorganic compounds on storage are atmospheric oxygen and carbon dioxide. Both these substances can cause changes in the stoichiometry of a compound with its subsequent effect on formulations.

Oxygen will oxidise many inorganic compounds at room temperature, especially those containing elements with more than one oxidation state, for example;

$$Fe^{2+} \longrightarrow Fe^{3+} \qquad As^{III} \longrightarrow As^{V}$$

Inorganic compounds that are sensitive to oxygen are often formulated with antioxidants (15.9). Many sulphur IV compounds, such as, sulphites and thiosulphates, are used for this purpose. However, sulphites have a number of drug incompatibilities because of their reactivity. Hypophosphites ($H_2PO_2^-$) are also used as antioxidants.

Carbon dioxide from the atmosphere can dissolve in aqueous preparations and cause a change in pH. It can also precipitate insoluble carbonates from aqueous solution.

The effects of oxygen, carbon dioxide and atmospheric pollutants can be minimised by keeping the compound or preparation in airtight containers.

30.10.3   Heat and light

Some inorganic compounds are labile (reactive) to either heat or light. Ammonium compounds, for example, slowly release ammonia when stored in hot rooms. These compounds should be stored in lightproof containers in a cool place as appropriate. Compounds that preferentially absorb the light are used in some formulations.

30.11   WHAT YOU NEED TO KNOW

(1) The periodic table is an attempt to correlate the chemical and physical properties of the elements and their compounds. It can be used to predict the physical and chemical properties of a compound by comparison with similar compounds of other elements in the same group.

(2) The correlation between the general properties of the s block elements and their compounds within a group is good.

(3) The correlation between the general properties of the p block elements and their compounds within a group is better between the lower elements of a group than the first two elements in the group. There is often a distinct change in the nature of their properties at a specific point in the group.

(4) The correlation between the general properties of the elements and their compounds within a group of the d block elements is reasonable.

(5) Elements whose compounds often have an acidic nature are found in the top right-hand area of the table whilst elements whose compounds tend to have a basic nature are found on the left-hand side of the table.

(6) The transition elements usually exhibit multiple oxidation states and form complexes with organic compounds. Both these aspects of their chemistry are utilised in their involvement in biological processes.

(7) Essential minerals are elements other than carbon, hydrogen and oxygen that occur in the human body in quantities greater than 100 mg. They are essential components of the human metabolic processes.

(8) The principal minerals found in the human body are: sodium, potassium, magnesium, calcium, nitrogen, phosphorus, sulphur and chlorine.

(9) Trace elements are elements that are found in quantities of the order of 50 mg or less. Some trace elements are essential for human metabolism, others are not essential and not necessarily toxic.

(10) Deficiency diseases can occur when the concentration of a mineral or trace element is too low.

(11) High concentrations of minerals and trace elements often have toxic effects. High concentrations can be caused by excessive absorption or a diseased organ.

(12) Other elements that do not appear to be essential for human metabolism also occur in the body.

(13) Some elements which are not present at birth, accumulate in the body during its lifetime. These elements are often toxic at high concentrations.

(14) Metallic elements often influence the absorption of other metal ions by the body.

(15) The main factors affecting the storage of inorganics are: humidity, oxygen and carbon dioxide in the air, heat and light.

(16) Hygroscopic substances absorb water from the air.

(17) Deliquescent substances can absorb so much water that they are able in an extreme situation to form a solution.

(18) Efflorescent substances lose water to the atmosphere on storage.

(19) In general, storage problems are minimised by storing the substance in opaque, airtight containers in a cool place.

30.12   QUESTIONS

(1) Explain the meaning of the terms: (a) mineral and (b) trace element.

(2) Why is it necessary to maintain the correct balance of sodium and potassium ion concentrations between extracellular and intracellular fluids? Outline how this balance is maintained.

(3) Outline the importance of magnesium and calcium in metabolism. Describe briefly how the excretion of calcium is affected by high and low concentrations of plasma calcium. What effect do these changes of plasma calcium have on the concentration of calcium in the bones?

(4) Name five essential trace elements and list some of their uses in metabolism.

(5) Explain why high cadmium concentrations are potentially toxic to humans.

(6) List possible sources of mercury and lead pollution that could be sources of these elements in humans.

(7) Outline what you consider to be the main functions of the halogens in humans.

(8) Explain the meaning of the terms hygroscopic, effervescent and deliquescent.

(9) An antacid tablet contains magnesium carbonate, bismuth chelate, calcium carbonate and starch. Explain the function of each of these ingredients.

31

Silica and silicates

31.1 INTRODUCTION

Silica and silicates are used extensively in pharmacy as pharmaceutical aids because they have suitable physical properties and are not very reactive. Pharmaceutical aids are substances that are not usually pharmacologically active but play an essential part in the manufacture and formulation of drugs. They act as lubricants, inert carriers, whitening agents, emulsifiers, absorbents, filter aids, clarifiers, abrasives, protectives and fillers. However, other substances besides silica and silicates are used in these capacities.

Silicates have crystal lattices which consist of small cations, small anions and large, and often complex, silicate anions. All of these different types of silicate anion have structures that are based on the SiO_4 tetrahedron (Fig. 31.1). In silicate anions the oxygens of an SiO_4 unit may be either covalently bonded to the silicons of other SiO_4 in the anion or negatively charged to make up their divalency.

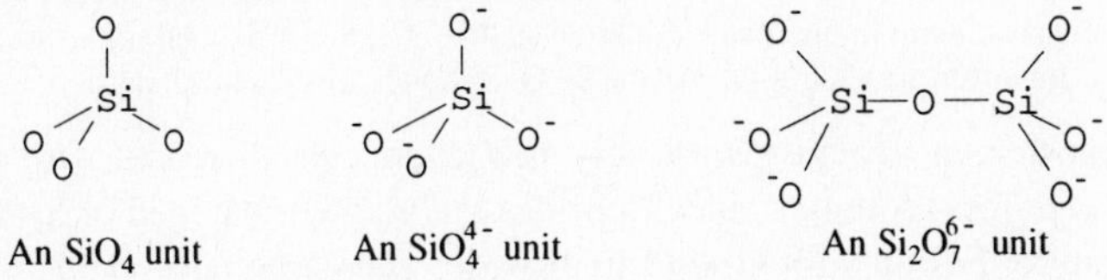

An SiO_4 unit An SiO_4^{4-} unit An $Si_2O_7^{6-}$ unit

Fig. 31.1. The structure of the SiO_4 unit that forms the basis of the structures of silicates.

Silicates have variable compositions because of isomorphous replacement of the small cations and anions in their structures by different ions of a similar size. For example, the ratio of iron to magnesium in the mineral olivine ($Mg_2Fe_2SiO_4$) can vary because iron was able to replace magnesium and vice versa in the structure of olivine when it was formed from magma containing appropriate concentrations of these elements. Similarly fluoride ions may replace hydroxide ions in some silicate structures.

A large number of silicates occur in nature. They may be conveniently classified as water soluble silicates, simple silicates with discrete silicate anions, chain silicates, layer silicates, three-dimensional lattice silicates and glasses.

31.2 SILICA

Silica (silicon dioxide, SiO_2) occurs widely in nature. Pure forms are sand, quartz and cristobalite whilst impure forms are tridymite, sandstone, flint, agate, opal, onyx and amethyst. It also occurs in bamboo, tabacco, sponges and the feathers of some birds.

Individual silica crystals usually have a diamond type of crystal lattice structure based on SiO_4 units. These SiO_4 units have a tetrahedral configuration, the silicon atom occupying the centre of a regular tetrahedron formed by the four oxygen atoms (Fig. 31.2). The oxygens are shared between adjacent silicons so that the overall stoichiometry of the crystal is SiO_2. The Si–O–Si bond angles are about 144° not 180° as shown in Fig. 31.2. The resultant structure is hard but rather more open than that of the diamond lattice.

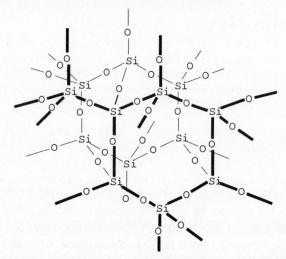

Fig. 31.2. The crystal structure of β-cristobalite, a form of silica. The silicon atoms occupy the positions occupied by the carbon atoms in an expanded diamond lattice. The Si–O–Si bond angles are not 180° as drawn but are about 144°. All the Si–O–Si bonds are the same length.

Silica is almost inert, reacting with very few chemicals. It reacts with fluorine and hydrofluoric acid to form complex silica fluorides but no other halogen or hydrogen halide at room temperature. Fusion with group I hydroxides and carbonates yields group I metal silicates.

31.2.1   Amorphous silica

Amorphous silica occurs naturally as **diatomaceous earth.** The crude diatomaceous earth is initially purified by boiling with concentrated hydrochloric acid which converts metallic impurities such as calcium, iron and sodium to their soluble chlorides. These chlorides are removed by washing the earth with water. Finally, the earth is heated to a high temperature to remove volatile and carbonaceous impurities. The product consists of a mixture of white spicules (knobbly rods), rod and star-shaped particles.

Amorphous silica is a hard chemically inert substance. Its spicules and rods form a tangled structure which give amorphous silica good absorbent properties. Amorphous silica can absorb its own mass of a liquid and still remain a powder, and up to four times its own mass of water. These good absorption characteristics are complemented by poor adsorptive properties. As a result, it will absorb a liquid but will also release that liquid relatively easily in a suitable environment. Consequently, amorphous silica is used to dispense liquids in solid form. Suitable formulations may be taken orally because the chemically inert amorphous silica passes straight through the intestinal tract. It is also used as a thickener and as an excipient for volatile oils and compounds that can be liquified.

The hardness of amorphous silica, coupled with its chemical inertness makes it useful as a filler and mild abrasive. It is used in scouring soaps, polishes and toothpastes. However, there is some evidence to suggest that it scratches the enamel coating of the teeth. Amorphous silica is also used as a filter aid and clarifying agent because of its chemical inertness, the mat-like nature of its gross structure and its absorbent/adsortive properties.

31.2.2   Synthetic amorphous silica

Two forms of synthetic amorphous silica are manufactured namely, powdered amorphous silica and hydrated amorphous silica (silica gel). Synthetic amorphous silica powder is produced by cooling silica vapour (known as silica fume). The silica vapour is produced by either vaporising silica at temperatures in excess of $2000°$, or burning silicon or compounds containing silicon in air.

$$Si \xrightarrow{\text{Vapourise}} Si_{(Vapour)} \xrightarrow{O_2} SiO_{2(Vapour)} \xrightarrow{\text{Condensation}} \text{Solid amorphous silica}$$

$$\underset{\text{Silicon disulphide}}{SiS_2} \xrightarrow{\text{Burn in air}} SiO_{2(Vapour)} + SO_{2(Gas)} \xrightarrow{\text{Condensation}} \text{Solid amorphous silica}$$

Silica gels are precipitated from solutions of soluble silicates by either acids or acid ion exchange resins. These silica gels are freed from excess moisture and are further processed to produce gels for specific purposes such as a chromatography mediums.

$$\underset{\text{A solution of sodium silicates}}{n\,Na_4SiO_4} \xrightarrow{H^+/H_2O} \underset{\text{Hydrated amorphous silica (silica gel)}}{(H_4SiO_4)_n}$$

Synthetic powdered amorphous silica is used for the same purposes as silica obtained from diatomaceous earth.

31.2.3   Silicosis

This is a lung condition resembling chronic tuberculosis. It is caused by excessive inhalation of crystalline silica dust and results in lung damage with dyspnoea (difficulty with breathing) and other complications. Amorphous silica appears to be innocuous in that it does not cause silicosis.

31.3 WATER SOLUBLE SILICATES

The alkali metal silicates are soluble in water provided the ratio of silica to metal oxide is low. Alkali metal silicates with high ratios of silica to metal oxide are insoluble in water.

Alkali metal silicates are prepared by fusion of silica with either the metal hydroxide or carbonate. The product is usually identified by its $SiO_2:M_2O$ ratio where M is an alkali metal.

$$x\,SiO_2 \ + \ 2y\,MOH \ \xrightarrow{\text{Fusion}} \ y\,(M_2O).x\,SiO_2 \ + \ y\,H_2O$$

$$x\,SiO_2 \ + \ y\,M_2CO_3 \ \xrightarrow{\text{Fusion}} \ y\,(M_2O).x\,SiO_2 \ + \ y\,CO_2$$

The soluble silicates have no direct use in pharmacy but are used as sources of silicate ions in manufacturing processes. For example, the antacid magnesium trisilicate ($2MgO·3SiO_2$) is prepared by adding the correct molar proportion of either magnesium sulphate or chloride to an aqueous solution of either ($M_2O·SiO_2$) sodium or potassium silicate.

31.4 SIMPLE SILICATES

Simple silicates have an ionic crystal lattice which contains small discrete silicate anions. Discrete anions consisting of SiO_4, Si_2O_7 and a number of anions that consist of small rings of three or more SiO_4 tetrahedra linked by sharing their oxygen atoms are known (Fig. 31.3). The sites in the silicate anion lattice are occupied by the balancing cations. For example, in the crystal structure of the mineral olivine $(MgFe)_2SiO_4^{4-}$ the oxygen atoms of the SiO_4 ions are hexagonally close-packed with the silicons occupying one-eighth of the available tetrahedral sites whilst the iron and magnesium cations occupy about one-half of the available octahedral sites.

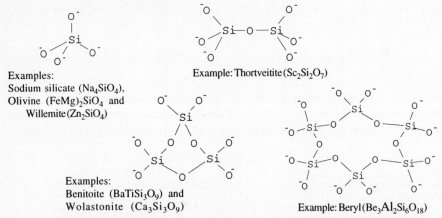

Examples:
Sodium silicate (Na_4SiO_4),
Olivine ($FeMg)_2SiO_4$ and
Willemite(Zn_2SiO_4)

Example: Thortveitite($Sc_2Si_2O_7$)

Examples:
Benitoite ($BaTiSi_3O_9$) and
Wolastonite ($Ca_3Si_3O_9$)

Example: Beryl($Be_3Al_2Si_6O_{18}$)

Fig. 31.3. The structures of some of the discrete anions found in simple silicates.

31.5 CHAIN SILICATES

Chain silicates contain single chains or several cross-linked chains of SiO_4 tetrahedra running parallel to one another, the tetrahedra being joined by shared oxygen atoms. Silicates whose structures contain single chains are known generally as **pyroxenes**

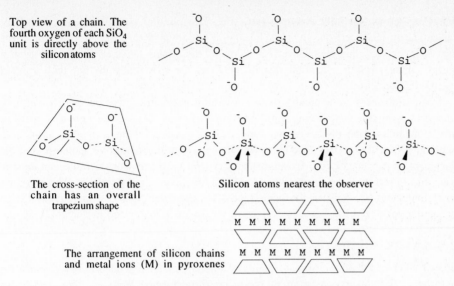

Top view of a chain. The
fourth oxygen of each SiO$_4$
unit is directly above the
silicon atoms

The cross-section of the
chain has an overall
trapezium shape

Silicon atoms nearest the observer

The arrangement of silicon chains
and metal ions (M) in pyroxenes

Fig. 31.4. The general structures of pyroxene chain silicates.

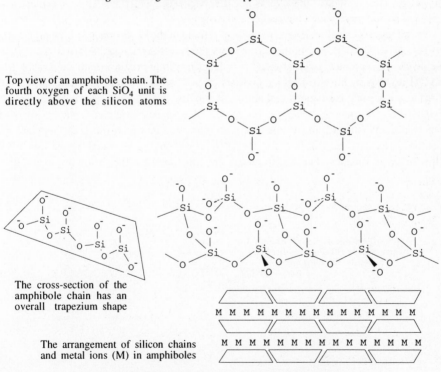

Top view of an amphibole chain. The
fourth oxygen of each SiO$_4$ unit is
directly above the silicon atoms

The cross-section of the
amphibole chain has an
overall trapezium shape

The arrangement of silicon chains
and metal ions (M) in amphiboles

Fig. 31.5. The general structures of amphibole chain silicates.

(Fig. 31.4). The chains can have a number of different configurations depending on the
relative arrangements of the SiO$_4$ units. Chain silicates with structures based on double

chains are known generally as **amphiboles** (Fig. 31.5). In both cases the chains, which are trapezoidal in cross-section, are packed parallel to one another. The cations occur in the sites between the chains and in effect 'cement' the chains together which results in a fibrous, open structure. Amphiboles differ from the pyroxenes in that they contain hydroxide ions and their chains have a bigger cross-sectional area.

Chain silicates are chemically inert and exhibit good absorptive but poor adsorptive properties. Consequently, fibrous amphibole minerals such as asbestos were used as filter aids in manufacturing processes and filter pads in gravimetric analysis. The amphibole attapulgite $Mg_5Si_8O_{20}(OH)_2 \cdot 8H_2O$, in which there is often extensive isomorphous replacement of the magnesium by aluminium, is able to absorb considerable amounts of water and so is used as an alternative to kaolin in the treatment of diarrhoea. The absorbed water molecules are held in the open spaces between the chains. It is also used as a thickening agent and has been used as a carrier for insecticides and herbicides.

31.6 LAYER SILICATES

In layer silicates the double silicate chains of amphiboles are extended in two dimensions to form a 'flat' layer consisting of repeating rings formed from six SiO_4 tetrahedra (Fig. 31.6). The 'A' face of the layer is negatively charged whilst the 'B' face is electrically neutral. The complete layer is in reality a macroanion.

The silicate macroanion layers, with the exception of the kaolin clays, are usually stacked AB-BA-AB-BA-AB-B, etc. In the kaolin clays the layers are stacked AB-AB-AB-AB, etc. The layers are cemented together at the A-A faces by either brucite which consists of Mg^{2+} and OH^- ions or hydrargillite which consists of Al^{3+} and OH^- ions. Bonding between the A-A faces is strong, the cations and anions occupying the sites between the apex oxygens of the layers. Variations occur in the cement due to isomorphous replacement of the magnesium and aluminium by other cations. The way in which the B-B faces are held together is largely responsible for the differences in the physical properties of the different layer silicates. These differences are illustrated by the following examples.

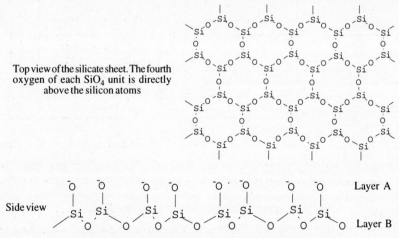

Top view of the silicate sheet. The fourth oxygen of each SiO_4 unit is directly above the silicon atoms

Side view   Layer A

                                                                                                                  Layer B

Fig. 31.6. The structure of the silicate macroanion in layer silicates.

31.6.1 Talcs

In talcs the A-A faces are cemented together by brucite whilst the B-B faces are held together by weak van der Waals' forces (Fig. 31.7). Consequently, it needs little force to slide these B layers over one another and so talcs are soft flaky materials.

Talcs are chemically inert and are insoluble in common solvents. They are good absorbents but poor adsorbents. These properties together with their soft flaky nature account for their use as dusting powders (e.g. talcum powder: these powders can act as irritants if the particles become trapped in the pores of the skin), dry lubricants (because of the ease of cleavage), fillers (their soft nature and inertness enables them to be used in soaps and cosmetics) and filter aids (lack of adsorption minimises losses of materials).

31.6.2 Micas

Micas are chemically inert substances. The A-A faces of the silicate sheets are held together by hydrargillite whilst the B-B layers are held together by electrostatic bonds involving group I and/or group II cations with the exception of lithium, beryllium and magnesium ions (Fig. 31.7). These electrostatic bonds are stronger than van der Waals' forces but not as strong as ionic bonds and so micas are still soft and flaky but are harder than talcs. Micas have no pharmaceutical use at present.

31.6.3 Bentonites

The A-A faces of the silicate sheets are bound together by hydrargillite. However, in bentonites there are one or more layers of water molecules stacked between the B-B faces. These layers are held together and to the silicate sheets by hydrogen bonding (Fig. 31.7). The water molecules in these water layers are arranged in a hexagonal pattern of similar size and shape to that of the SiO_4 tetrahedra in the silicate sheet.

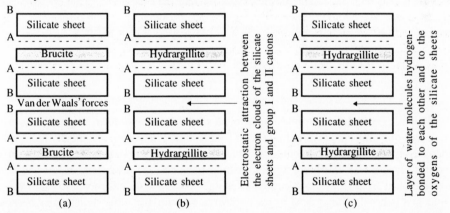

Fig. 31.7. Diagrammatic representations of the structures of (a) a talc, (b) a mica and (c) a bentonite.

Bentonites are chemically inert, soft flaky materials. The hydrogen bonding between the B-B faces of the silicate sheets makes them harder than talcs but softer than micas. Unlike the talcs and micas, bentonites have good adsorptive properties. They are good water adsorbents, the absorbed water forming new water layers between the B-B faces which increases the size of the structure, that is, the bentonite particles swell. As a result, bentonites are used as gelling, emulsifying and suspension agents. They are used to stabilise *Calamine lotion* and ointment bases. They are also used to prepare suspensions of ammoniated mercury and zinc oxide. Bentonites will also adsorb suitably sized alcohols, amines, amino acids ethers, ketones and cations. These substances form flat two-dimensional sheets between the B-B faces of the silicate sheets. They are bound to the silicate sheets by hydrogen bonds, van der Waals' and electrostatic forces. This adsorption also causes swelling of the bentonite. Bentonites are also used a clarifiers because of their adsorptive properties.

Bentonites in which some of the aluminium ions of the hydrargillite layer have been isomorphously replaced by magnesium ions exhibit ion exchange properties. The charge difference between the Al^{3+} and Mg^{2+} ions is compensated by the presence of an equivalent number of cations being incorporated into the spaces between the layers in the structure. These cations are readily exchangeable with those in a suitable solution even under extreme conditions of temperature. Ion exchange bentonites are usually supplied in the sodium form.

Bentonite is no longer used as a bulk laxative as prolonged use can result in metal ion and vitamin deficiencies because of its ion exchange and adsorptive properties. Bentonites are detergents: they are used in soaps, dental preparations and cleansing creams.

31.6.4   Kaolin clays
Kaolin clays are different from the other types of layer silicate in that the layers are stacked -AB-AB-AB-AB-, etc. The layers are held together by hydrogen bonding and van der Waals' forces between a hydrargillite layer electrostatically bonded to the A face of one silicate sheet and the B face of the next silicate sheet (Fig. 31.8).

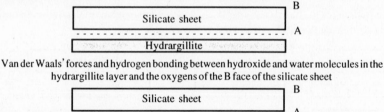

Fig. 31.8. A diagrammatic representation of the structure of a kaolin clay.

Kaolins exist as microparticles of colloidal size. They do not form large crystals. The properties of kaolins are similar to those of the bentonites. They are chemically inert, have good adsorptive properties, exhibit ion exchange and can act as detergents. Consequently, kaolin's are used as clarifying agents, dusting powders, the adsorbent in hot and cold poultices and as an intestinal adsorbent in the treatment of gastroenteritis. However, prolonged internal use can lead to trace mineral, vitamin and other deficiencies because of kaolins good adsorptive and ion exchange properties. It should be noted that the British Pharmacopoeia uses *light kaolin* for internal use and *heavy kaolin* for external use because the electrolytes cause flocculation of heavy kaolins into sticky masses.

31.7   THREE-DIMENSIONAL LATTICE SILICATES

In these silicates all the oxygens of the SiO_4 tetrahedra are shared with adjacent SiO_4 tetrahedra giving a basic structure similar to that found in silica (31.2). However, in the structures of the three-dimensional lattice silicates a proportion of the silicons are replaced by aluminium in the form of AlO_4 tetrahedra with the result that the structure gains one negative charge for every silicon replaced (Fig. 31.9). The resulting structure is very open, having many passages and chambers. These passages and chambers contain the cations that balance the negative charge of the giant silicate/aluminate anion. These **aluminosilicate** structures are classified into two types: **felspars** and **zeolites**.

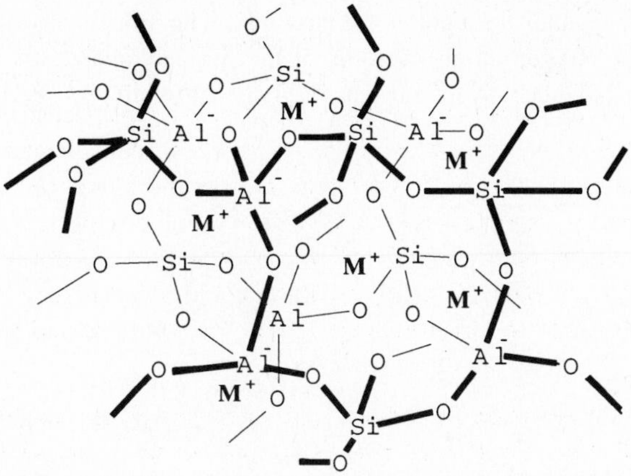

Fig. 31.9. A diagrammatic representation of the three-dimensional structure of an aluminosilicate. Note the balancing metal (M^+) cations in the cavities in the structure.

31.7.1   Felspars

Felspars have structures in which the chambers and passages in the structure do not interconnect. They contain sodium, potassium, calcium and sometimes lithium and barium ions as the balancing cations but never magnesium or iron ions. Felspars are very hard materials that make up two-thirds of all igneous rocks.

Felspars are chemically inert, do not exhibit adsorptive or ion exchange properties because the exchangeable cations are locked up in the sealed passages and chambers. However, they are used as abrasives, for example pumice in dentistry.

31.7.2   Zeolites

Zeolites have structures in which the passages and chambers are interconnected. Consequently, zeolites can act as ion exchangers because it is possible for cations to enter and leave the silicate structure provided electrical neutrality is maintained.

The interconnection of the chambers also makes it possible for the zeolites to be used as molecular filters (molecular sieves). Small molecules take longer to pass through a bed of zeolite particles than large molecules. This is because the small molecules pass into the zeolite particle structure and, once inside, need time to find a way out. Large molecules, however, are too large to enter the zeolite particles' structure and so rapidly pass through the bed around the outside of the zeolite particles.

Zeolites have strong adsorptive properties and, as a result, have some potential as carrier agents for the oral administration of gaseous drugs and drugs with small molecules. Their adsorptive properties also make zeolites useful as drying agents, the water being retained in temperatures above $100^{\circ}C$.

31.8 GLASS

Glass has a completely disordered structure which is based on the SiO_4 tetrahedron. However, the bond angles of the O–Si–O bonds in these tetrahedra can vary considerably from that expected by the sp^3 hybridisation of the silicon. These tetrahedra form a three-dimensional silicate network consisting of rings formed from five, six, seven and eight SiO_4 tetrahedra by the mutual sharing of their oxygen atoms. The rings are arranged in a completely random manner throughout the structure. The silicate structure is negatively charged because the oxygen atoms of some of the SiO_4 tetrahedra are not shared. This negative charge is balanced by cations held in the structure by electrostatic bonding (Fig. 31.10).

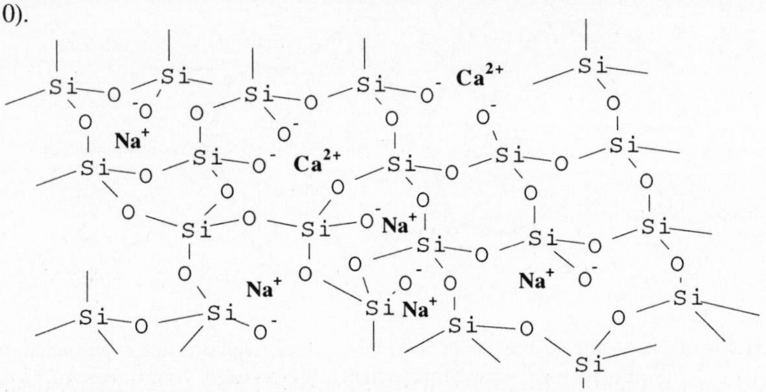

Fig. 31.10. A representation of a fragment of the structure of soda glass.

31.8.1 Soda glass

Soda glass consists of sodium and calcium silicates. It is slightly soluble in water, producing hydroxide ions. The rate of solution is slow but increases when the glass is heated.

$$Na_4SiO_4 + H_2O \longrightarrow H_3SiO_4 + 4Na^+ + 3OH^-$$

Consequently, preparations that are pH sensitive should not be stored in soda glass bottles. For example, aqueous solutions of adrenalin tartrate are slowly hydrolysed to adrenaline when stored in soda glass bottles. Adrenaline has a shorter shelf life than its tartrate. Alkalinity can be reduced by lining the bottle with a silicon film (31.9).

31.8.2 Borosilicate glass

In borosilicate glass some of the silicon has been replaced by boron and some of the calcium and sodium by aluminium. Borosilicate glass is harder and not so soluble in water as soda glass. However, it is not so easy to mould as soda glass.

31.8.3 Neutral glass

This has the composition 72-75% SiO_2, 7-10% B_2O_3, 4-6% Al_2O_3, 6-8% Na_2O, 0.5-2% K_2O and 2-4% BaO. It has a high resistance to hydrolysis and is easy to mould.

31.8.4 Special glasses

The addition of metals and metallic oxides to a glass can give the glass special properties. For example, substitution of some of the sodium by potassium will give a brown light resistant glass. The presence of praesodynium and neodynium will give a glass that absorbs

UV whilst the substitution of lead for some of the calcium gives a glass with a high refractive index. Incorporation of aluminium and boron oxides produces glass that can be used to manufacture electrodes for the potentiometric measurement of monovalent ions.

31.9 SILICONES

Silicones are organo-silicate polymers formed by the hydrolysis of compounds with the general formulae $SiClnR_{4-n}$ where R is aliphatic or aromatic. The relative molecular mass and the precise nature of the structure of the product depends on the hydrolysis conditions and the alkyl silicon chlorides used for the polymerisation. For example:

$$CH_3SiCl_3 \xrightarrow{\text{Hydrolysis}} CH_3Si(OH)_3 \longrightarrow \text{Spontaneous polymerisation}$$

Silicone polymer

```
         |           |          |        |         |
         O          CH3         O       CH3        O
         |           |          |        |         |
  —O —Si—O —Si—O —Si—O —Si—O —Si—O—
         |           |          |        |         |
        CH3          |         CH3       |        CH3
                     O                   O
        CH3          |         CH3       |        CH3
         |           |          |        |         |
  —O —Si—O —Si—O —Si—O —Si—O —Si—O—
         |           |          |        |         |
         O          CH3         O       CH3        O
         |                      |                  |
```

The physical nature of silicones ranges from oils and glasses to infusible solids. They are usually chemically inert, thermally stable and water repellent. Consequently, silicones are used as lubricants (oils and greases), implants and protective coatings.

31.10   WHAT YOU NEED TO KNOW

(1) Silica and silicates are used as pharmaceutical aids.

(2) Silica can occur in a number of crystalline forms. Their crystal lattices are based on covalently bonded SiO_4 tetrahedra. The oxygen atoms of one tetrahedron being shared with those of other tetrahedra

(3) Cristabolite, a pure naturally occurring form of silica has an extended diamond-type of crystal lattice.

(4) Amorphous silica occurs naturally as diatomaceous earth.

(5) Samples of amorphous silica have a tangled matted structure consisting of particles in the shape of rods, spicules and stars.

(6) All forms of silica are almost chemically inert.

(7) Amorphous silica is hard, and so as it is almost chemically inert, amorphous silica is used as an abrasive in soaps, polishes and toothpastes.

(8) The structures of silicates are based on covalently bonded SiO_4 tetrahedra in which none, one, two, three or four of the oxygens are shared with the oxygens of other SiO_4 tetrahedra.

(9) Simple silicates have crystal lattices which consist of silicate anions and metallic cations.

(10) The silicate anions of simple silicates have structures that consist of either short

chains of linked SiO_4 tetrahedra or cyclic structures of linked SiO_4 tetrahedra.

(11) Chain silicates have structures which contain single chains or several cross-linked chains of SiO_4 tetrahedra running parallel to one another, the tetrahedra being joined by shared oxygen atoms.

(12) These chains are stacked parallel to each other being held together by cations that occupy sites between the chains.

(13) Pyroxenes are chain silicates whose structures contain single chains of silicate tetrahedra.

(14) Amphiboles are chain silicates whose structures contain two cross-linked chains of silicate tetrahedra. These silicates also have hydroxide ions or layers of hydroxide ions associated with the silicate chains.

(15) Chain silicates are fibrous in nature.

(16) Layer silicates have macrosilicate anions with two-dimensional structures which contain 'flat' layers consisting of inteconnected rings made up from six SiO_4 tetrahedra. These macroanions have an A face which is negatively charged and a second B face which is not charged.

(17) The macroanions of layer silicates are held together by either brucite, hydrargillite, hydrogen bonded water molecules, van der Waals' forces or electrostatic attractive forces.

(18) Brucite is an inorganic substance which contains mainly Mg^{2+} and OH^- ions.

(19) Hydrargillite is an inorganic substance which contains mainly Al^{3+} and OH^- ions.

(20) Talcs are layer silicates where the silicate macro anion layers are stacked AB-BA-AB-etc. The A-A faces are held together by brucite and the B-B faces by van der Waals' forces.

(21) Talcs are soft and flaky materials, the layers readily sliding across each other.

(22) Micas are layer silicates where the silicate macro anions layers are stacked AB-BA-AB-etc. The A-A faces are held together by hydrargillite and the B-B faces by electrostatic forces involving group I and II cations.

(23) Bentonites are layer silicates where the silicate macro anion layers are stacked AB-BA-AB-etc. The A-A faces are held together by hydrargillite and the B-B faces by hydrogen bonding between layers of water molecules and the oxygens atoms in the silicate layer.

(24) Bentonites are good absorbents for water, and suitably sized amines, alcohols, amino acids, esters, ketones and cations. These species form flat, two-dimensional sheets which are bound by hydrogen bonding, electrostatic forces and van der Waals' forces between the B-B faces of the bentonite structure. This increases the size of the bentonite particles.

(25) Kaolins are layer silicates where the silicate macro anion layers are stacked AB-AB-AB-AB-etc. The layers are held together by hydrogen bonding and van de Waals' forces between a hydrargillite layer electrostatically bonded to the A face of the silicate sheet.

(26) Kaolins consist of microparticles of colloidal size.

(27) Kaolins are good absorbents for water, cations and other small molecules.

(28) Three-dimensional lattice silicates have structures that contain both SiO_4 and

AlO_4 tetrahedra in the form of giant molecular structures in which all the oxygens of the SiO_4 and AlO_4 tetrahedra are shared with adjacent tetrahedra. The presence of aluminium gives these giant structures a negative charge which is balanced by the presence of suitably sized and charged cations in the structure.

(29) Felspars are three-dimensional lattice silicates with structures in which the cations with the exception of magnesium and iron ions, are trapped in the structure.

(30) Zeolites are three-dimensional lattice silicates with structures in which the cations are not trapped in the structure because the passages and chambers in the structure are interconnected. Consequently they can act as ion exchangers.

(31) Glass has a random three-dimensional silicate lattice structure in which not all the oxygens of the SiO_4 tetrahedra are shared with adjacent tetrahedra. The structure has a negative charge which is balanced by metal cations held within the three-dimensional structure of the glass.

(32) Silicones are organo-silicate polymers which range from oils to glasses to infusible solids. They are usually chemically inert and water repellent.

31.11 QUESTIONS

(1) Describe the appearance of amorphous silica. What features makes this compound suitable for use as a filler and an abrasive?

(2) Why is the cleavage angle of pyroxenes greater than that of amphiboles when crystals of these compounds are split parallel to their chain axes?

(3) Explain, on the basis of their structures, why (a) talcs are used as lubricants in dusting powders and (b) bentonites are used as gelling agents.

(4) Distinguish between brucite and hydrargillite.

(5) Explain how a zeolite acts as (a) a molecular sieve and (b) an ion exchanger.

(6) Why is it preferable that organic salts be stored in plastic containers and not glass bottles?

32

Complex compounds and ions

32.1 INTRODUCTION

Inorganic complexes are *simply* but not *strictly accurately* defined as being chemical species which have structures in which anions and molecules, known as **ligands**, are bonded to a central metal atom or ion (Fig. 32.1). The ligands in the complexes shown in Fig. 32.1 are ammonia and 1,2-diaminoethane. However, in immunology the term ligand has a special meaning. A ligand is any molecule that binds to a macromolecule and may affect the biochemistry of that macromolecule. These ligands are usually organic in nature and range from small molecules like ATP to low molecular mass proteins.

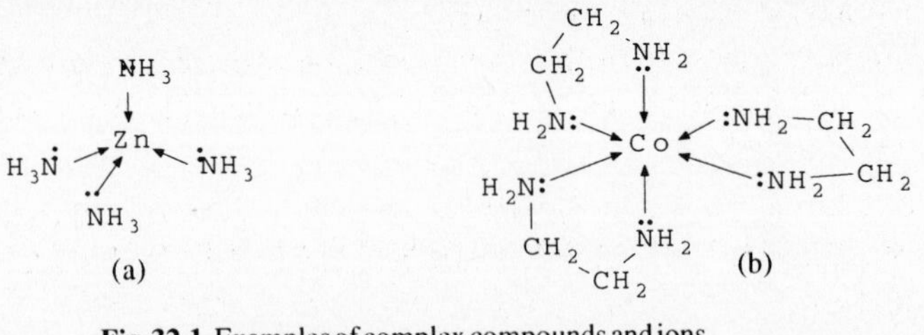

Fig. 32.1. Examples of complex compounds and ions.

Ligands are always nucleophilic in nature. They are bonded to the central metal atom or ion by either dative sigma bonds, dative pi bonds or electrostatic bonds whose nature is not fully understood. Ligands may be classified as **monodentate, bidentate, tridentate,** etc. according to the number of bonds they form with the central metal atom or ion. The 1,2-diaminoethane in the example shown in Fig. 32.1b is a bidentate ligand. The number of bonds from the ligands to the central metal atom is known as the **coordination number** of the metal. For example, the zinc atom (Fig. 32.1a) has a coordination number of 4 whilst the cobalt atom (Fig.32.1b) has a coordination number of 6. Complex compounds and ions whose structures contain ligands which form more than one link to the central metal atom are also known as **chelation compounds and ions.** For example, the complex $Co(NH_2CH_2CH_2NH_2)_3^{2+}$ shown in Fig. 32.1b is a chelated complex ion because the 1,2-diaminoethane forms two bonds with the cobalt ion. The reagent responsible for the formation of a chelation compound or ion is called a **chelating agent.** For example, 1,2-diaminoethane is the chelating agent for the formation of the $Co(NH_2CH_2CH_2NH_2)_3^{2+}$ ion whose structure is illustrated in Fig. 32.1. Chelation results in more stable complexes than those formed by monodentate ligands provided it results in structures that are five-, six- or seven-membered ring systems.

32.2   THE STABILITY OF COMPLEXES

All complex compounds and ions are regarded as being formed by thermodynamically reversible reactions. The formation of the hypothetical complex ML_n where M is a metal ion and L a ligand may be represented by the equation:

$$M \; + \; nL \; \rightleftharpoons \; ML_n \tag{32.1}$$

Therefore, the more stable the complex ML_n the more this equilibrium will lie to the right. Consequently, it is possible to use the equilibrium constant for the formation of ML_n as a measure of the stability of the complex because the equilibrium constant is also a measure of the position of the equilibrium. Consider, for example, the equilibrium constant for the equation (32.1) which is given by the expression:

$$K_c = \frac{[ML_n]}{[M][L]^n}$$

A large value for K_c will mean that the term $[ML_n]$ will be large and the value of the product $[M][L]^n$ is small. The large value for $[ML_n]$ shows that there is a high concentration of the complex present at equilibrium, that is, the complex is stable. Conversely a small value for K_c indicates the stability of the complex is low. In this case the value of the term $[ML_n]$ will be small and the value of the term $[M][L]^n$ high.

Equilibrium constants used in this way as a measure of the relative stabilities of complexes are known as **stability formation constants** (K_s) and are recorded for convenience as log K_s values (Table 32.1).

Table 32.1. Log K_s values for the complexes formed by some metals and the chelating agent, ethylenediaminetetraacetic acid. The larger the value of log K_s the more stable the complex.

Metal ion	Na^+	Mg^{2+}	Ca^{2+}	Al^{3+}	Zn^{2+}	Pb^{2+}	Cu^{2+}
$LogK_s$ (25^0C)	1.7	8.7	10.7	16.3	16.7	18.0	18.8

The reversible nature of the reactions used to form complexes has two consequences for their stability in solution:

(1) The complex could decompose and reform with the same ligands occupying different positions in space about the central metal atom. This behaviour may be of significance if the complex has stereoisomers (32.6).

(2) If a different ligand is added to the solution it may displace the original ligands

because it will compete for the metal. Some metals are labile but others such as Co^{3+} and Cr^{3+}, exchange ligands slowly

$$ML_n \overset{H_2O/L'}{\rightleftharpoons} \quad ML_{n-1} L' \text{ where L and L' are different ligands}$$

The extent of the substitution also depends on the relative stabilities of the complexes ML_n and $ML_{n-1} L'$, the more stable $ML_{n-1} L'$ the greater the degree of substitution. This type of competition is the basis of the mode of action of indicators in complexometric titrations (32.7).

Stability constants are used to explain the effects of metals on the biological activity of metabolites, drugs and enzymes. For example, the stability constants of the complexes formed by amino acids and cobalt show that the cobalt-histidine complex is considerably more stable than the complexes formed by cobalt and other amino acids (Table 32.2). It is thought that histidine neutralises the effect of cobalt on various micro-organisms and tumours by forming a stable complex which in effect takes the cobalt out of the system.

Table 32.2. The stability constants for histidine/metal complexes at $25^{0}C$

Amino acid	Alanine	Glycine	Leucine	Valine	Histidine
Log$K$	8.78	8.94	8.26	8.24	13.86

32.3 NOMENCLATURE

Systematic (Table 32.3) and trivial names are used for complexes. In general, systematic names are used for the simpler structures and trivial names for the more complex structures. The rules used for systematic nomenclature are:

(1) In salts, the name of the cation precedes the name of the anion.
(2) The order of the names in the name of a complex is: names of the ligands, name of the metal, oxidation state of the metal.
(3) Cations and neutral molecules use the English name of the metal but Latinised names with the suffix -**ate** are used for the metals of anions, for example, cuprate for complex anions containing copper and ferrate for anions containing iron.
(4) Simple ligands are either given trivial names or the common anion name but with the suffix -**o**, for example, H_2O aqua, NH_3 ammine, Cl^- chloro, CN^- cyano and OH^- hydroxo.
(5) The names of ligands of the same type are prefixed by the Greek prefixes indicating the number of that type of ligand in the complex, for example, di, tri, tetra, etc.
(6) The names of charged ligands precede those of neutral ligands, for example, chloro will precede ammine.

Table 32.3. Examples of the systematic names of complexes.

Formula	Name	Formula	Name
$[Co(H_2O)_6]Cl_3$	Hexaaquacobalt III chloride	$Na_2[CuCl_4]$	Sodium tetrachlorocuprate II
$[Cu(NH_3)_4]SO_4$	Tetraamminecopper II sulphate	$K_3[Fe(CN)_6]$	Potassium hexacyanoferrate III
$Pt(NH_3)_2Cl_2$	Dichlorodiammineplatinum II	$K_4[PtCl_6]$	Potassium hexachloroplatinate II

32.4   STRUCTURE

The structures of complexes whose ligands are covalently bonded to the central metal atom was originally explained using Pauling's hybridisation theory (2.3.4). This theory has been replaced by the ligand field and molecular orbital theories which give better accounts of the properties of complexes. However, for most purposes in the life sciences Pauling's simpler hybridisation theory is adequate and so will form the basis of the discussion of complexes in this text.

Pauling's theory is based on hybridising the empty orbitals of the central metal atom to give hybridised orbital systems with geometries that correspond to the structure of the complex. The ligands are bonded to the central metal atom by dative sigma bonds involving these hybridised orbitals, the ligand acting as the electron donor. For example, X-ray crystallography shows that the $Zn(NH_3)_4^{2+}$ ion has a tetrahedral structure. This structure may be *simply* explained by the four ammonia molecules forming four sigma bonds with the four empty hybridised sp^3 hybrid orbitals formed by the hybridisation of the empty 4s and three 4p orbitals of the zinc atom (Fig. 32.2).

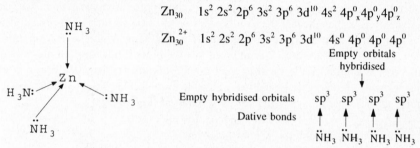

$$Zn_{30} \quad 1s^2\ 2s^2\ 2p^6\ 3s^2\ 3p^6\ 3d^{10}\ 4s^2\ 4p_x^0 4p_y^0 4p_z^0$$

$$Zn_{30}^{2+} \quad 1s^2\ 2s^2\ 2p^6\ 3s^2\ 3p^6\ 3d^{10}\ \ 4s^0\ 4p^0\ 4p^0\ 4p^0$$

Empty orbitals hybridised

Empty hybridised orbitals $\quad sp^3 \quad sp^3 \quad sp^3 \quad sp^3$

Dative bonds

$$\ddot{N}H_3\ \ddot{N}H_3\ \ddot{N}H_3\ \ddot{N}H_3$$

Fig. 32.2. The electronic structure of the $Zn(NH_3)_4^{2+}$ complex ion.

Some ligands such as carbon monoxide are weak electron donors and so form weak dative sigma bonds with a metal, and as a consequence their complexes should not be very stable. However, their complexes are stable and so it was proposed that in these cases the complex is further stabilised by the metal forming a dative pi bond as well as a sigma bond with the ligand. These proposals have now been well substantiated (Fig. 32.3).

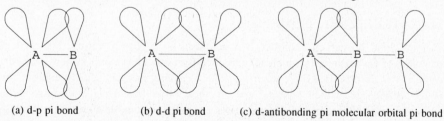

(a) d-p pi bond   (b) d-d pi bond   (c) d-antibonding pi molecular orbital pi bond

Fig. 32.3. Some of the different types of pi bonding that can occur in the structures of complexes. The single lines represent sigma bonds between the appropriate atoms

The pi bonds may either be formed between a metal orbital containing electrons and a suitable empty orbital of the ligand as in $Cr(CO)_6$ or an empty metal orbital and a suitable orbital of the ligand which contains electrons as in CrO_4^{2-} (Fig. 32.4). The former type of

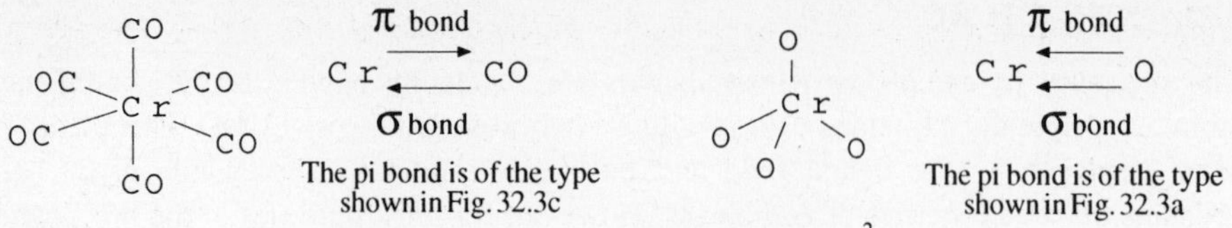

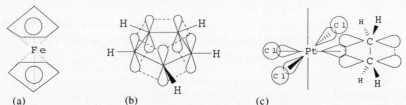

Fig. 32.4. The structures of $Cr(CO)_6$ and CrO_4^{2-}.

dative pi bonding is likely to occur when the metal atom has a low oxidation state whilst the latter is likely to be found when the metal has a high oxidation state.

The structures of a number of complexes are explained by the metal being bonded to a pi electron cloud of a ligand. For example, X-ray crystallography has shown that ferrocene, $Fe(C_5H_5)_2$, has a structure in which the iron is sandwiched between two parallel flat cyclopentadienyl rings (Fig.32.5a). The chemical properties of ferrocene indicate that these rings are aromatic in nature. This structure has been explained by sp^2 hybridisation of the carbon atoms. This gives flat rings in which all the remaining p orbitals can overlap sideways to form a pi molecular orbital (Fig.32.5b). However, these orbitals contain only five pi electrons which would not account for the aromatic nature of the rings. Consequently it is thought that the iron donates an electron to each of the rings thereby giving them a total of six pi electrons which would agree with the Huckel rule for aromaticity (12.6). Furthermore it is believed that the pi molecular orbitals are pi bonded to an empty d orbital of the Fe II ion. Pi bonding is also believed to occur in other complexes where the organic ligand has a suitable pi molecular orbital and the metal has an empty orbital which may or may not be hybridised (Fig. 32.5c).

Fig. 32.5. The structures of (a) ferrocene (b) the cyclopentadienyl ring and (c) $[PtCl_3(C_2H_4)]^-$. In (c) it is an empty dsp^2 hybrid orbital that forms a pi bond with the pi molecular orbital of the ethene.

32.5   STRUCTURAL ISOMERISM

The structural isomerism exhibited by complexes is extremely varied. This text gives examples of the main types which are classified for convenience on a system based on the original system proposed by Werner in 1920. Structural isomers usually exhibit considerable differences in their physical and chemical properties. They often have different colours, react differently with the same reagents and exhibit different biological properties.

32.5.1   Linkage isomerism

Linkage isomerism is also known as **structural** or **salt** isomerism. The ligand either exhibits structural isomerism or can be bonded by different atoms to the central metal atom. For example, the nitrito and nitro groups are isomeric and give rise to isomeric complexes in which the ligands are bonded to the central metal atom through the nitrogen atom and oxygen atoms respectively.

$[Co(NH_3)_5NO_2]^{2+}$

$$\begin{array}{c} NH_3 \\ | \\ H_3N-\overset{\displaystyle NH_3}{\underset{\displaystyle |}{\overset{\displaystyle |}{C}}}o-NH_3 \\ H_3N-\overset{\displaystyle |}{\overset{\displaystyle |}{}} \overset{\displaystyle N}{\underset{\displaystyle O}{\diagdown}}\!\!=\!\!O \\ NH_3 \end{array}$$

Nitropentamminecobalt III

$[Co(NH_3)_5ONO]^{2+}$

$$\begin{array}{c} NH_3 \\ | \\ H_3N-\overset{\displaystyle NH_3}{\underset{\displaystyle |}{\overset{\displaystyle |}{C}}}o-NH_3 \\ H_3N-\overset{\displaystyle |}{\overset{\displaystyle |}{}} O-N\!\!=\!\!O \\ NH_3 \end{array}$$

Nitritopentamminecobalt III

32.5.2   Ionisation isomerism

Ionisation isomerism is exhibited by complexes that have the same stoichiometric composition but different anions. For example:

$$[Co(NH_3)_5Br]^{2+} SO_4^{2-} \quad and \quad [Co(NH_3)_5SO_4]^+ Br^-$$
$$[Co(NH_3)_5NO_3]^{2+} SO_4^{2-} and \quad [Co(NH_3)_5SO_4]^+ NO_3^-$$
$$[Pt(NH_3)_4Cl_2]^{2+} 2Br^- \quad and \quad [Pt(NH_3)_4Br_2]^{2+} 2Cl^-$$

It is often possible to distinguish between these ionisation isomers by means of simple inorganic anion tests. For example, reaction of an aqueous solution of $[Co(NH_3)_5Br]^{2+} SO_4^{2-}$ will give a positive sulphate test and a negative bromide test whilst its isomer $[Co(NH_3)_5SO_4]^+ Br^-$ will give a negative sulphate test and a positive bromide test.

$$[Co(NH_3)_5 Br]^{2+} SO_4 + BaCl_2 \longrightarrow [Co(NH_3)_5Br]^{2+} 2Cl^- + BaSO_4 \text{ (white precipitate)}$$

$$2[Co(NH_3)_5 SO_4]^+ Br^- + AgNO_3 \longrightarrow [Co(NH_3)_5 SO_4]^+ NO_3^- + AgBr \text{ (cream precipitate)}$$

32.5.3 Coordination isomerism

Coordination isomerism usually occurs in compounds that contain both a complex cation and a complex anion. The ligands in these isomers are the same but are distributed differently between the cation and anion. Typical examples of this type of isomerism are:

$$[Cu(NH_3)_4]^{2+} [PtCl_4]^{2-} \quad and \quad [Pt(NH_3)_4]^{2+} [CuCl_4]^{2-}$$
$$[Co(NH_3)_6]^{3+} [Cr(C_2O_4)_3]^{3-} \quad and \quad [Cr(NH_3)_6]^{3+} [Co(C_2O_4)_3]^{3-}$$
$$[Pt(NH_3)_4]^{2+} [PtCl_6]^{2-} \quad and \quad [Pt(NH_3)_4Cl_2]^{2+} [PtCl_4]^{2-}$$
$$[Cr(NH_3)_6]^{3+} [Cr(SCN)_6]^{3-} \quad and \quad [Cr(NH_3)_4(SCN)_2]^+ [Cr(NH_3)_2(SCN)_4]^-$$

32.5.4   Hydration isomerism

Hydration isomerism involves water as a ligand and as water of crystallisation. The water of crystallisation occurs in the crystal lattice of the compound without being associated with a particular complex ion. It varies from isomer to isomer but the overall stoichiometric composition of the isomers is the same. For example, three chromium III chloride hydrate isomers are known:

$$[Cr(H_2O)_6]Cl_3 \text{ is a violet compound}$$
$$[Cr(H_2O)_5Cl]Cl_2 \cdot H_2O \text{ is a pale green compound}$$
$$[Cr(H_2O)_4Cl_2]Cl \cdot 2H_2O \text{ is a dark green compound}$$

It is usually possible to distinguish between these isomers by measuring the relative numbers of ions they contain. For example, the three isomers of chromium hydrate contain one, two, and three moles of chloride ions per mole of the complex respectively.

Consequently, titration of identical molar samples of each of these isomers with the same aqueous silver nitrate solution will give titration readings that are in the ratio 1:2:3 respectively.

32.6   STEREOISOMERISM

Complexes exhibit both optical and geometric isomerism. Like organic stereoisomers, the isomers of a pharmacologically active complex may be inactive, active in the same way but to a different degree, exhibit a different type of pharmaceutical activity, be toxic or give rise to side effects.

32.6.1   Optical isomerism

Optical isomerism is exhibited by complexes that are chiral in nature, that is have structures that are non-superimposable mirror images (Fig. 32.6). Like the optical isomers discussed in Chapter 5, these structures have no centre of symmetry, no plane of symmetry and no alternating axis of symmetry.

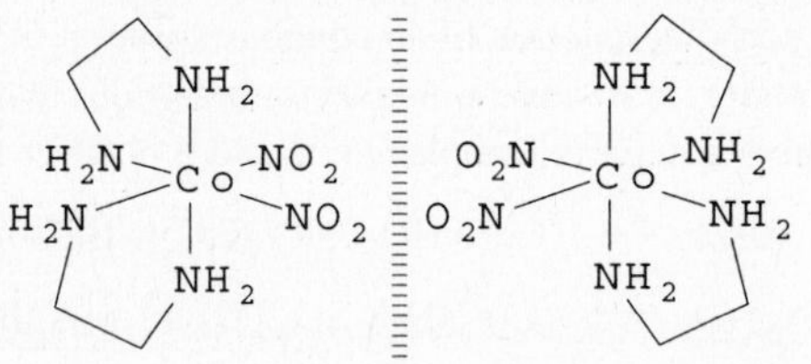

Fig. 32.6. An example of a complex that can exist as optically active isomers.

Optical isomerism is usually found in octahedral complexes with six coordination. It can occur in four coordinated tetrahedral complexes provided the ligands are all different (compare with asymmetric carbons). However, it is only found in four coordinated square planar complexes with asymmetric ligands.

The isolation of stereoisomers is often difficult as the isomeric forms rapidly interconvert. Consequently, synthesis of optically active complexes usually results in the formation of a racemate (5.4). It should be remembered that all complexes can be regarded as being formed by reversible reactions (32.2). However, it is possible to prepare the optical isomers of chromium, cobalt, platinum and some other metals of the second and third transition series as interconversion of the complexes of these metals in solution is slow.

32.6.2   Geometric isomerism

Geometric isomerism occurs in a large number of complexes with four and six coordination (Fig. 32.7). The apparently different structures for the cis and trans isomers of the cobalt complexes arise because the molecule is three-dimensional and in fact the same molecule is being viewed from a different angle. One should always be aware of this when comparing the structures of complexes with the same stoichiometric composition. Like organic geometric isomers, the geometric isomers of complexes will possess different physical, chemical and biological properties.

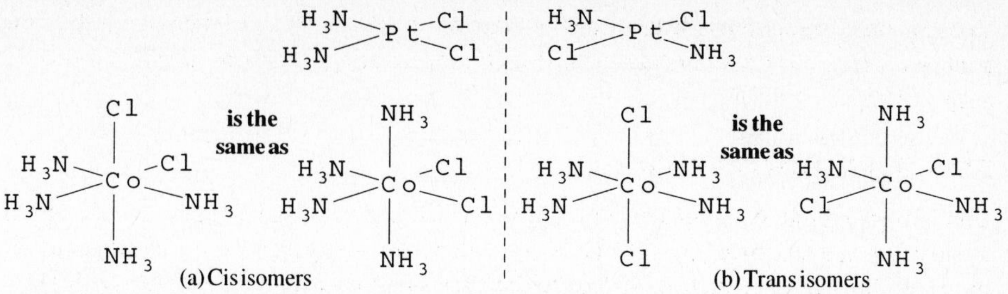

Fig. 32.7. The geometric isomers of (a) $Pt((NH_3)_2Cl_2$ and (b) $[Co(NH_3)_4Cl_2]^+$. Note the different views of the same molecule.

32.7 USE IN ANALYSIS

32.7.1 Qualitative analysis

The formation of a coloured complex is the basis of many qualitative tests for metals and organic compounds that can act as ligands (Table 32.4).

Table 32.4. Examples of the use of complexes in qualitative analysis. The structures of some of these complexes have not been fully elucidated.

Test for:	An outline of the chemistry
Esters	$RCOOR \xrightarrow{NH_2OH} RCONHOH \xrightarrow{FeCl_3}$ Purple complex
Phenols	$ArOH \xrightarrow{FeCl_3}$ Transient green, blue, violet or black complex
Cobalt	$Co^{2+} \xrightarrow{NH_4SCN} (NH_4)_2[Co(SCN)_4]$ Blue complex
Copper	$Cu^{2+} \xrightarrow{NH_3} [Cu(NH_3)_4]^{2+}$ Dark blue complex
Iron III (Fe)	$Fe^{3+} \xrightarrow{NH_4SCN} [Fe(SCN)]^{2+}$ Blood red complex
Prussian blue reaction Iron in blood films and bone marrow	$Fe^{2+} \xrightarrow[50-56^o]{HCl} Fe^{3+} \xrightarrow{K_4[Fe(CN)_6]} KFe[Fe(CN)_6]$ Prussian blue complex

Complex formation also plays a considerable part in the traditional systematic chemical schemes for identifying metals. The separation of silver I chloride from lead II chloride in a number of schemes is achieved by reacting the silver I chloride with ammonia to form the water soluble $[Ag(NH_3)_2]^+$ complex ion. The insoluble lead II chloride does not react and so can be separated from the $[Ag(NH_3)_2]^+$ ions in solution by either centrifuging or filtration.

32.7.2 Quantitative analysis

Complex formation is used as the basis of many volumetric, gravimetric and colorimetric assay methods.

(a) Volumetric analysis. Complex formation is used for a number of direct and back titrations. These titrations are known generally as complexometric titrations. The titrating agents are usually chelation agents such as ethylenediaminetetraacetic acid (EDTA). This reagent, which is normally used in the form of its disodium salt, reacts with a large number

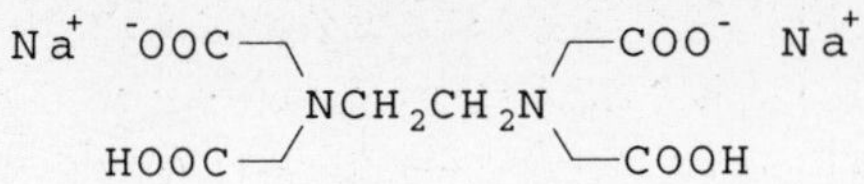

$\equiv \quad 2Na^+ \; H_2X^{2-}$

Divalent metals $M^{2+} + H_2X \rightleftharpoons MX^{2-} + 2H^+$

Trivalent metals $M^{3+} + H_2X \rightleftharpoons MX^- + 2H^+$

Tetravalent metals $M^{4+} + H_2X \rightleftharpoons MX + 2H^+$

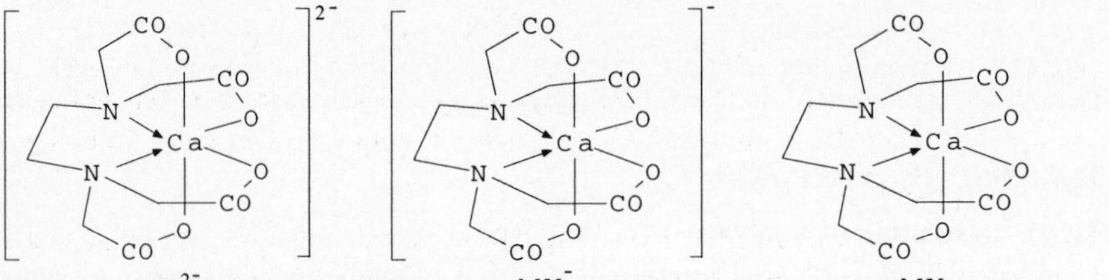

MX^{2-} MX^- MX

Fig. 32.8. EDTA usually forms octahedral complexes with 2, 3 and 4 oxidation state metals. The abbreviation H_2X^{2-} is used to represent EDTA in the equations.

of metal ions in a 1:1 ratio irrespective of the oxidation state of the metal (Fig. 32.8). The reactions are reversible and it is necessary to use a buffer to absorb the hydrogen ions and complete the reaction.

The end points of complexometric titrations can be determined by either physical methods (11.6.2) or using **metallochromic indicators**. Metallochromic indicators are dyes that react with the metals being assayed to form a dye-metal complex that is a different colour from a solution of the free dye. They are pH-sensitive and so must be used in solutions of the correct pH. Consequently, it is necessary to match the pH requirement of the indicator to that of any buffer required for the metal ion chelating agent reaction.

The use of metallochromic indicators depends on the stability of the dye-metal complex being significantly smaller than that of the titration agent-metal complex but greater than that of the metal-water complex. Consider, for example, the assay of calcium ions by titration with EDTA. This reaction of EDTA with calcium ions requires a strong ammonia-ammonium chloride buffer. The titration is carried out using the blue dye, mordant black, which has a working pH range of 7 to 11, as indicator (I). This indicator forms a red complex with calcium ions. Consequently, at the start of the titration the solution is red. The situation in the volumetric flask may be simply summarised as:

$$Ca^{2+}\text{ (total in the sample)} + I \longrightarrow Ca\text{-}I + Ca^{2+}\text{(unreacted)}$$

As the EDTA is added it reacts with the remaining metal ions until all these ions have been chelated by the EDTA. At this point the solution is still red.

$$Ca^{2+} + H_2X^{2-} + Ca\text{-}I \longrightarrow CaX^{2-} + Ca\text{-}I + 2H^+$$

Since all complexes are formed by reversible reactions, further addition of EDTA displaces the indicator I from the Ca-I complex because the Ca-EDTA complex is more stable than the Ca-I complex (equation 32.1). As the titration nears the end point the position of equilibrium moves to the right. At this stage in the titration there will be free indicator in solution and so the colour observed will change from red to purple.

$$Ca\text{-}I \rightleftharpoons Ca + I + H_2X^{2-} \rightleftharpoons CaX^{2-} + I + 2H^+ \quad (32.1)$$

The end point of the titration is reached when all the indicator in the Ca-I complex has been replaced by EDTA, that is, all the calcium has reacted. At this point the solution will be blue, having changed from red through purple to the blue of the free dye. The first appearance of this blue is the end point of the titration. It should be noted that this is different from acid-base titrations where the change in colour marks the end point.

The overall reaction for the titration is:

$$Ca^{2+} + H_2X^{2-} \xrightleftharpoons{\text{Buffer}} CaX^{2-} + 2H^+$$

It should be remembered that the buffer is necessary for the correct operation of the indicator and also to remove the hydrogen ions and move the position of equilibrium completely to the right.

(b) Gravimetric analysis. The technique of gravimetric analysis is outlined in section 11.6.5. Many metals can be assayed using this technique by using chelation agents to precipitate insoluble complexes of known stoichiometry from solutions containing ions of the metal. These complexes are washed free of mother liquor and dried to constant mass in an oven set at a suitable temperature (Figs. 32.9).

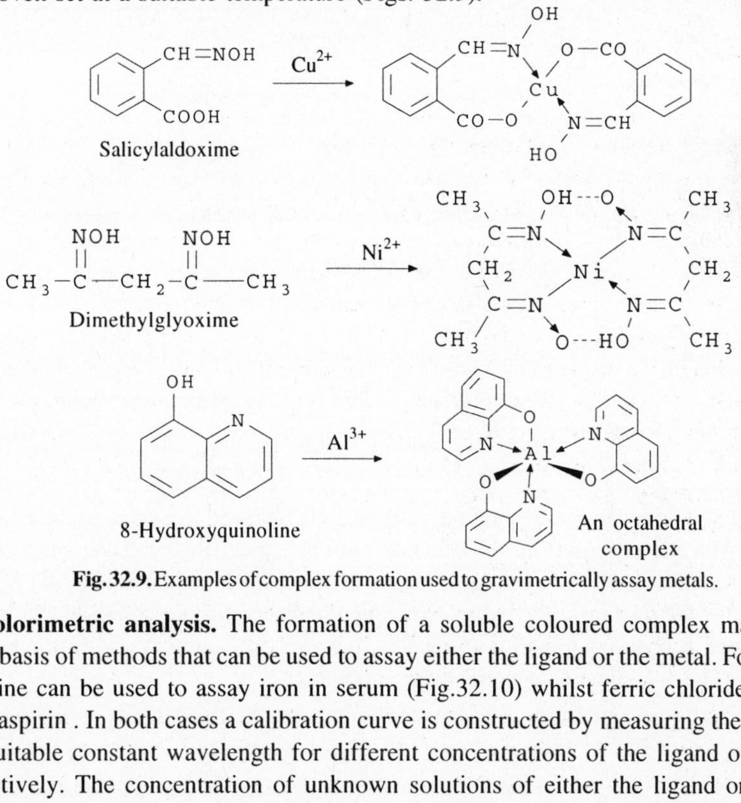
CH=NOH
COOH
Salicylaldoxime
Cu2+
CH=N
O—CO
Cu
CO—O
N=CH
OH
HO
NOH
NOH
CH3—C—CH2—C—CH3
Dimethylglyoxime
Ni2+
Ni
8-Hydroxyquinoline
Al3+
Al
An octahedral complex

Fig. 32.9. Examples of complex formation used to gravimetrically assay metals.

(c) Colorimetric analysis. The formation of a soluble coloured complex may be used as the basis of methods that can be used to assay either the ligand or the metal. For example, ferrozine can be used to assay iron in serum (Fig.32.10) whilst ferric chloride is used to assay aspirin . In both cases a calibration curve is constructed by measuring the absorption at a suitable constant wavelength for different concentrations of the ligand or the metal respectively. The concentration of unknown solutions of either the ligand or metal are obtained by measuring their absorptions at the appropriate wavelength and reading the concentration from the calibration curve.

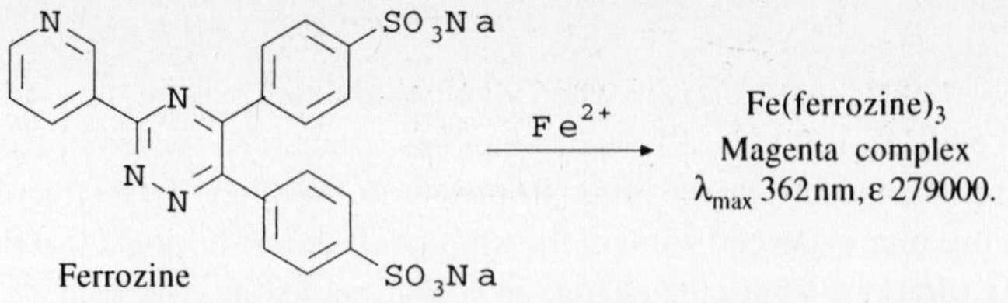

Fig. 32.10. An outline of the chemistry of the assay of iron in serum. It should be noted that cobalt II, copper I, oxalate,cyanide and nitrite interfere with the estimation of iron.

32.8 COMPLEXES IN BIOLOGICAL SYSTEMS

32.8.1 As therapeutic agents

Metals are essential for healthy life but for this to occur their concentrations must stay between concentration limits that vary from individual to individual. If the concentration of a metal fall outside this narrow concentration band a disease state may result. Chelating agents are used to reduce high concentrations of metal ions. They are often specific for certain metals (Table 32.5).

Table 32.5. Examples of the use of chelating agents to treat high metal ion concentrations.

Chelating agent	Use/comments
Dimercaprol	Used to treat gold, mercury and arsenic poisoning
Ethylenediaminetetraacetic acid (EDTA)	Used to treat lead and heavy metal poisoning in farm animals
Disodium calcium edetate	Used to treat lead and mercury poisoning. The presence of the calcium reduces the loss of calcium from the bones by complexation with the EDTA
Hydroxyurea	Used in the treatment of some cancers
D (-) penicillamine	Chelates copper. Used in the treatment of Wilson s disease
Cystenamine	Chelates copper and iron. Used as a radioprotective agent
Desferrioxamine	Parenteral chelating agent for iron and aluminium. The methane sulphonate is used to treat iron poisoning

The action of some antibiotic and antiviral drugs is believed to be due to their forming complexes with metal ions (Fig. 32.11). For example, practical evidence suggests that 8-hydroxyquinoline used to treat skin rashes chelates iron either inside the bacteria responsible for the rash or in the cytoplasmic membrane to form a complex that is toxic to the bacteria. 8-Hydroxyquinoline forms 1:1, 1:2 and 1:3 complexes with iron but it is believed that the 1:1 compound is the active complex.

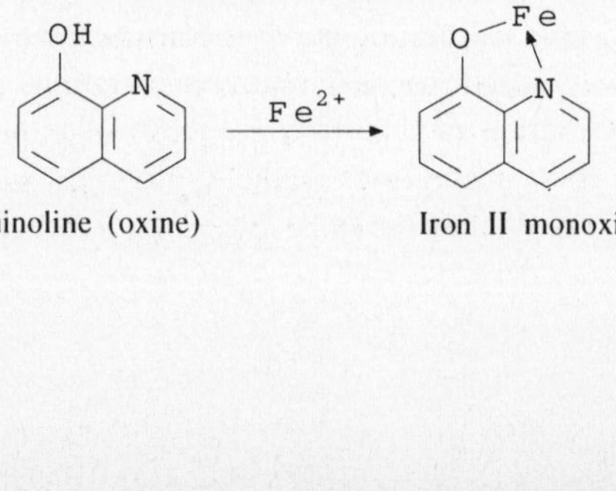

8-Hydroxyquinoline (oxine)   Iron II monoxine

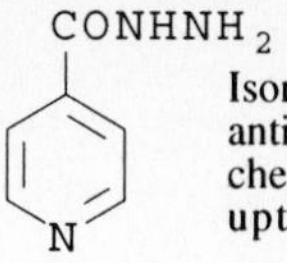

CONHNH$_2$

Isonicotinic acid hydrazide (isoniazid). An antibiotic used to treat TB. It is believed to chelate serum copper which facilitates the uptake of the drug by the bacteria.

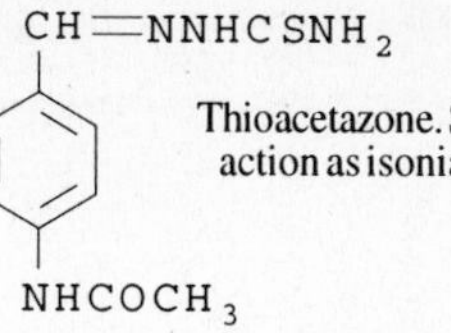

CH$=$NNHCSNH$_2$

Thioacetazone. Same action as isoniazid

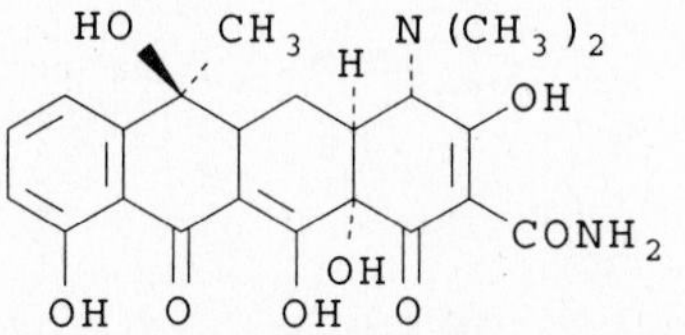

NHCOCH$_3$

HO   CH$_3$ N(CH$_3$)$_2$
          H   OH

                        CONH$_2$

OH O   OH O

NNHCSNH$_2$

$=$O

N

CH$_3$

Tetracycline, an antibiotic. It chelates calcium and magnesium. The magnesium chelate concentrates in the bacterial cell wall and prevents protein synthesis.

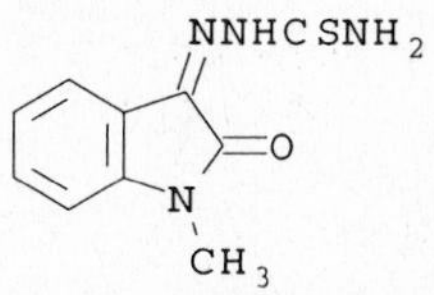

Methisazone, an antiviral. It chelates copper and is believed to interact with the nucleic acids of the virus.

Fig. 32.11. Examples of antibiotic and antiviral agents that are believed to owe their action to complex formation.

Cadmium, nickel and zinc in high concentrations and cobalt in all concentrations nullify the antibiotic effect of 8-hydroxyquinoline because these metals form more stable non-toxic complexes with the 8-hydroxyquinoline.

Cisplatin (Fig. 32.7) is used to treat cancer of the testis, ovary, squamous carcinoma of the bladder and drug-resistant choriocarcinomas.

32.8.2   Occurrence in biological systems

Complexes play an important role in many biological processes. These roles range from acting as co-enzymes to involvement in metabolite transfer and energy conversion. For example, haemoglobin which is responsible for the transfer of oxygen and carbon dioxide to relevant parts of the body and chlorophyll which is responsible for the absorption of sunlight in plants are complex ions.

(a) Haemoglobin. Haemoglobin (Hb) is a paramagnetic tetramer (28.7.4). The four monomers are conjugated proteins in which the conjugated molecule is a haem residue containing an iron II ion which is responsible for the transportation of oxygen and other small ligands. In their unoxygenated state these monomers exhibit five coordination, the four nitrogens of the phorphyrin nucleus act as four of the ligands whilst the fifth is a histidine residue. These ligands form an unusual square-based pyramid with the iron II ion occupying the body of the pyramid (Fig. 32.12a). A second histidine residue is positioned 'above' the four phorphyrin nitrogens but is too far away to coordinate with the iron. This histidine is known as the distal histidine. The ability of the iron II ion of haemoglobin to coordinate with small ligands such as water, oxygen, carbon dioxide, carbon monoxide and cyanide ions is explained by the iron having an empty d orbital. This d orbital is used for d^{2}sp^3 hybridisation which would explain the iron II ions six coordination (Fig. 32.12b).

In the lungs oxygen coordinates to the iron forming a six coordinate diamagnetic complex (Fig. 32.13). This is transported in the blood to areas where there is a high carbon dioxide content (high PCO$_2$) in the tissue. Here pressure gradients exists which result in the transfer of the oxygen in the blood (high PO$_2$) to the tissue (low PO$_2$) and carbon dioxide from the tissue (high PCO$_2$) to the blood (low PCO$_2$). The carbon dioxide is

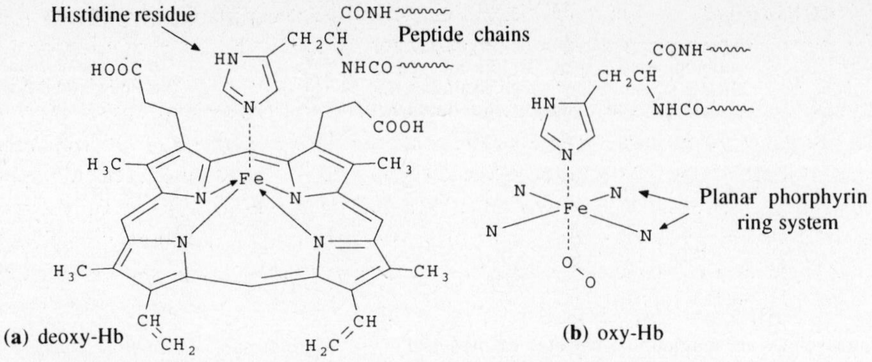

Histidine residue

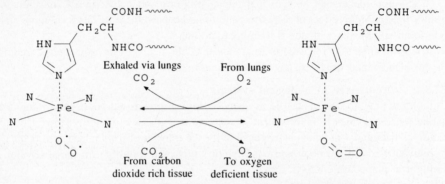

Fig 32.12. The structure of the (a) unoxygenated and (b) oxygenated haem residue in haemoglobin.

transported to the lungs where it is exchanged for oxygen. Carbon monoxide poisoning is due to the ligands forming stronger complexes than oxygen with haemoglobin (carboxyhaemoglobin) and so blocking the transport of the oxygen.

Fig. 32.13. An outline of the mode of action of haemoglobin.

(b) Chlorophyll. This is a family of compounds with very similar structures (Fig. 32.14). All forms of chlorophyll strongly absorb in the visible region but small differences in their molecular structures can have a significant effect on their absorption spectra.

Chlorophylls act as the main photoreceptor in green plants and bacteria. Experimental evidence suggests that the majority of the chlorophyll molecules act as antenna (**antenna chlorophyll**) gathering light photons. The energy of the absorbed light is transferred from these antenna chlorophyll molecules to **reaction centre chlorophyll** molecules by a process known as either **exciton** or **resonance energy** transfer. It is believed that an excited chlorophyll molecule transfers its excitation energy to a neighbouring unexcited molecule. This process can be thought of as being a form of electronic coupling between the adjacent molecules by the interaction of their molecular orbitals. The energy is transferred from one molecule to another until it reaches the photosynthetic reaction centre. The process is similar to the successive transfer of the energy of a swinging pendulum to a series of other pendulums attached to the same horizontal string. The first pendulum is swung (energy is absorbed) and in turn this sets the other pendulums swinging (energy transferred).

The chlorophyll antenna molecules are contained in a complex structure known as the **light harvesting complex** (LHC). The structures of LHCs are believed to consist of chlorophyll molecules trapped in a protein that is bound to a membrane. Experimental evidence shows that the chlorophyll molecules of the green bacterium *Prosthecochloris aestuarii* are wrapped inside a protein molecule that largely consists of a fifteen strand β pleated sheet. The structure has been likened to a string bag containing seven chlorophyll molecules. It has also been suggested that the antenna protein is a trimer consisting of three of these string bag protein subunits. In each subunit the seven chlorophyll molecules are arranged in a complex pattern which is believed to optimise the transfer of energy through the LHC.

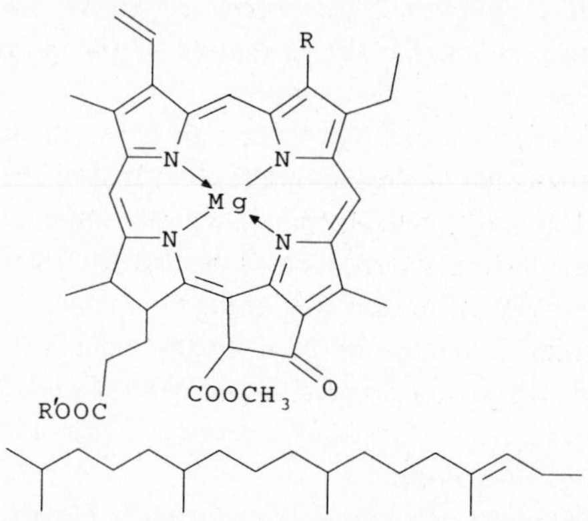

Fig. 32.14. The structures of chlorophylls a ($R=CH_3$) and b ($R=CHO$).

LHCs often contain other light absorbing compounds which absorb light in the regions of the spectrum not covered by the chlorophyll molecules. For example, the red-orange beta carotene, the red phycoerythrobilin and the blue phycocyanobilin.

32.8.3   Enzyme action

The action of enzymes that contain coordinated metal ions frequently involves a change in the oxidation state of the metal. For example, the action of the cytochromes a, b and c involve changes in the oxidation state of the iron in these complexes (Fig. 32.15).

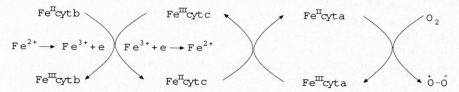

Fig. 32.15. An outline of the mode of action of the cytochromes a, b and c. The net result of the reactions is that one electron is transferred from one cytochrome (cyt) to the next until it is finally transferred to oxygen to form a superoxide ion form oxygen

32.9   WHAT YOU NEED TO KNOW

*(1)   **Complexes** may be loosely defined as being   molecules and ions which have a central metal atom to which is bonded anions and molecules known as **ligands**.*

(2) Ligands are always nucleophiles.

(3) Ligands are classified as being bidentate, tridentate, etc. depending on the number of bonds linking the ligand to the metal atom.

(4) The number of bonds to the central metal atom is known as the coordination number of the metal atom.

(5) Complexes are also referred to as coordination compounds and chelation compounds and ions.

(6) All complexes are deemed to be formed by thermodynamically reversible reactions.

(7) The stability of a complex is recorded as $^-\log K_s$ where K_s is the equilibrium constant for the reaction for the formation of the complex. The larger the value of $^-\log K_s$ the more stable the complex.

(8) The systematic names for complexes are built from the names of the ligands and the metal. In salts the name of the cation precedes that of the anion.

(9) English names are used for the metals in cations and neutral molecules but Latinised names may be used in anions. In all cases the oxidation state of the metal is shown after the name of the molecule or ion containing the metal.

(10) Trivial and systematic names are used for the ligands.

(11) Ligands are usually sigma bonded to the central metal atom. In some complexes these sigma bonds are reinforced by a dative pi bond in which either the metal or the ligand can act as the donor.

(12) The structural isomers of complexes are usually classified according to the system originally devised by Werner in 1920. A simplified version of this system is:

(a) Linkage isomerism: the ligand exhibits isomerism.

(b) Ionisation isomerism: the complexes have the same stoichiometric composition but different anions.

(c) Coordination isomerism: this is found in complexes which contain both a complex anion and a complex anion. The complexes contain the same ligands but the ligands are distributed differently between the complex ions.

(d) Hydration isomerism: these are complexes with the same stoichiometric composition which contain water which acts as either a ligand or as water of crystallisation.

(13) Complexes exhibit both geometric and optical stereoisomerism.

(14) The formation of coloured complexes is the basis of many identification and spot tests.

(15) Complex formation is the basis of complexometric titrations. These titrations must be carried out at the correct pH in order for the reaction to go to completion and the indicator to operate correctly.

(16) Complex formation is the basis of some gravimetric analysis procedures.

(17) The formation of soluble coloured complexes is the basis of some colorimetric assay methods.

(18) Complexes may be used as therapeutic agents to reduce the concentration of metal ions.

(19) The action of some drugs is believed to be due to complex formation.

(20) Complex formation is an essential part of many metabolic processes. The action can depend on a change of state of the metal ion.

32.10 QUESTIONS

(1) Specify the ligands and the coordination numbers of the metal atoms and ions in each of the following compounds: (a)tetraamminezinc II chloride, (b) cobalt hexacarbonyl, (c) cisplatin and (d) deoxy-Hb.

(2) Devise systematic names for each of the following complexes:
   (a) $[Cr(H_2O)_6]Cl_3$;   (b) $Pt(NH_3)_2Br_2$;
   (c) $[Co(NH_3)_5NO_2]\ SO_4$;   (d) $K_4[Pt(CN)_6]$.

(3) Write molecular formulae for each of the following complexes: (a) hexaaquacobalt II chloride; (b) sodium tetrabromocuprate II; (c) aquapentaaminecobalt III chloride and (d) tetraamminecopper II tetrachloroplatinate II.

(4) Draw all the possible **structural** isomers of each of the following complexes:
   (a) $[Co(NH_3)_5NO_2]\ SO_4$;   (b) $[Co(NH_3)_5Br]\ 2Cl$;
   (c) $[Cr(H_2O)_5Cl]Cl_2 \cdot H_2O$; (d) $[Cr(NH_3)_6][CrCl_6]$.

(5) Draw all the possible **stereoisomers** of each of the following compounds:
   (a) $Pt(NH_3)_2Br_2$;   (b) $[Pt(H_2NCH_2CH_2NH_2)_3]\ Cl_4$;
   (c) $[Co(H_2NCH_2CH_2NH_2)_2Cl_2]Br$; (d) $Pt(NH_3)_4Cl_2$.

(6) Outline a chemical or physical method of distinguishing between each of the members of the following pairs of complexes:
   (a) (i) $[Co(NH_3)_4(H_2O)Cl]Cl_2$ and (ii) $[Co(NH_3)_4(H_2O)Cl_2]Cl.H_2O$.
   (b) (i) $[Pt(H_2NCH_2CH_2NH_2)_3]Cl_4$ and (ii) $[Pt(H_2NCH_2CH_2NH_2)Cl_4]$.
   (c) (i) $[Cr(H_2O)_5Cl]SO_4 \cdot H_2O$ and (ii) $[Cr(H_2O)_4SO_4]Cl \cdot 2H_2O$.

(7) (a) Draw the structure of EDTA. Describe, by means of an equation, how EDTA would react with calcium. Give **two** reasons why it is necessary to carry out this reaction in the presence of a suitable buffer when it is used as the basis of an assay procedure for calcium.

   (b) 0.8721g of a laxative preparation containing a calcium salt as the active ingredient was extracted with water and the extract titrated with 0.05 M EDTA in the presence of a strong ammonia buffer using mordant black as indicator. The extract required 26.3 cm^3 of EDTA solution for complete titration. Calculate the %w/w of Ca^{2+} in the preparation. (RAM, Ca = 40.)

33

Radioactivity

33.1 INTRODUCTION

The nuclei of some naturally occurring isotopes are unstable and spontaneously decompose by emitting either particles or high energy electromagnetic radiation (Fig. 33.1). A nucleus may emit one or several different types of emission. This phenomenon is known as **radioactivity**. Isotopes which behave in this fashion are called **radionuclides** or **radioisotopes**. The radioactive decomposition of a nucleus results in the formation of a new isotope referred to as the **daughter** isotope which may or may not be stable. Unstable daughter isotopes will spontaneously decompose (**decay**) to form new isotopes. If these new isotopes are unstable they will also decay to new isotopes. This process will continue until a stable isotope is formed. The complete process is known as a **decay** or **disintegration series** (Fig. 33.1).

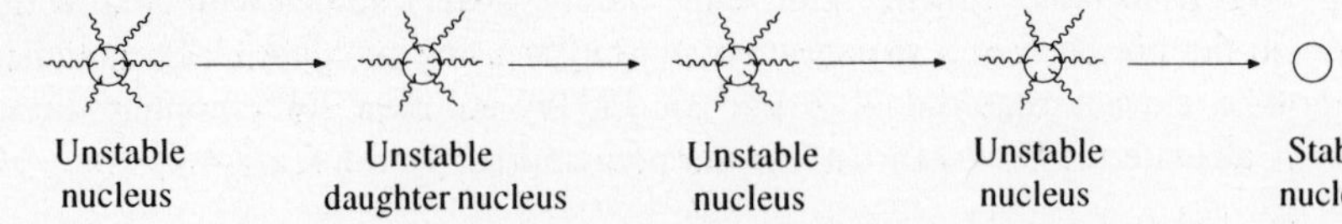

| Unstable nucleus | Unstable daughter nucleus | Unstable nucleus | Unstable nucleus | Stable nucleus |

Fig. 33.1. A schematic representation of a decay series.

Radioactive isotopes are identified by their mass numbers. The mass number is either used as a superscript in front of the symbol of the isotope or it is written after the symbol or name of the isotope. For example, the iodine isotope with a mass number of 123 would be written as either ^{123}I or iodine-123.

The names of compounds containing radioactive isotopes are prefixed by [ ] containing the symbol of the radionuclide. The position of the radionuclide in the compound is indicated by the appropriate locant. For example, the name of methionine containing radioactive sulphur-32 would be written, $[^{32}S]$-methionine and that of butanoic acid containing carbon-11 at position three, $3-[^{11}C]$-butanoic acid.

The probability of an isotope being radioactive can be related to the ratio of neutrons to protons (n/p) in its nucleus. Radionuclides are more likely to occur when n/p is greater than 1.25 for isotopes with an atomic number of less than 30, and 1.40 for those with atomic numbers greater than 30 but less than 92 It is emphasised, however, that an isotope may still be radioactive if the ratio is below the appropriate figure. The isotopes of all elements with atomic numbers greater than 92 are radioactive.

Stable isotopes, that is, isotopes that are not radioactive, are most likely to occur when a nucleus contains even numbers of both protons and neutrons. However, there are a number of stable isotopes with odd numbers of protons and neutrons. They include deuterium (hydrogen-2), lithium-6 and nitrogen-14.

Both radioactive and stable isotopes can be synthesised by bombarding a suitable isotope with high energy particles. For example, the bombardment of nitrogen-14 by helium nuclei (α particles) produces fluorine-18 which is radioactive, emitting protons to yield the stable isotope, oxygen-17.

$$^{4}_{2}He \quad + \quad ^{14}_{7}N \quad \longrightarrow \quad ^{18}_{9}F \quad \longrightarrow \quad ^{17}_{8}O \quad + \quad ^{1}_{1}p$$

The artificial conversion of one isotope into another is known as a **transmutation**. It is usually carried out in either a particle accelerator such as a cyclotron or a nuclear reactor. Transmutation is used to produce most of the radionuclides used in medicine. For example, the synthesis of technetium-99m, one of the most widely used radionuclides in medicine, involves the transmutation by neutron bombardment of molybdenum-98 to molybdenum-99 which subsequently decays to technetium-99m.

$$^{98}_{42}Mo + {}^{1}_{0}n \quad \longrightarrow \quad ^{99}_{42}Mo + \gamma$$

$$^{99}_{42}Mo \quad \longrightarrow \quad ^{99m}_{43}Tc + {}^{0}_{-1}e$$

$$^{99m}_{43}Tc \quad \longrightarrow \quad ^{99}_{43}Tc + \gamma$$

It should be noted that the **m** in the name of technetium-99m indicates that this radionuclide is metastable, that is, a high energy form of technetium-99. Technetium-99m loses its excess energy when it decays with the emission of gamma (g) radiation to technetium-99. Its use in medicine is discussed in greater detail in section 33.12.

Transmutation has also been used to produce artificial isotopes, that is, isotopes that do not occur naturally.

33.2 NUCLEAR DECAY ROUTES

This section describes the more common ways in which nuclei can decay. These nuclear decay processes result in many different types of emission, the most common of which are summarised in Table 33.1. The energy carried by these emissions is sufficient to ionise the surrounding medium, hence the use of the general term **ionising radiation** for all types of radioactive emission.

Table 33.1. Common radioactive emissions and their characteristic properties.

Emission	Symbols	Nature	Mass	Charge
Alpha	$\alpha, {}^{4}_{2}He$	Helium nucleus	4.0026	2+
Beta	$\beta, {}^{0}_{-1}e$, e	Electron	0.00055	1-
X-ray	X-ray	High energy electromagnetic radiation	none	0
Gamma	γ	High energy electromagnetic radiation	none	0
Positron	${}^{0}_{+1}e, {}^{0}_{+1}\beta$	Positively charged electron	0.00055	1+
Proton	${}^{1}_{1}p$	Proton	1.0073	1+
Neutron	n, ${}^{1}_{0}n$	Neutron	1.0087	0

Simple radionuclide decay can be summarised by the general equation:

Isotope A $\longrightarrow$ Isotope B + ionising radiation + energy
(the parent)   (the daughter)

Most of the energy liberated by the process is carried away by the ionising radiation. Its value is a characteristic property of the radiation and is recorded as part of the data for the source of that radiation. It should be noted that the daughter isotopes formed by a decay process may be in an excited state in which case they may lose energy in the form of gamma radiation. The energy of this gamma ray is the same as the energy difference between the excited and ground states of the nucleus.

33.2.1 Alpha particle emission

Alpha particles are helium nuclei and consist of two neutrons and two protons. Consequently, the loss of an alpha particle from a nucleus reduces the mass number of the nucleus by four and its atomic number by two. For example, uranium-238 decays to thorium-234 by the loss of an alpha particle.

$$ {}^{238}_{92}U \longrightarrow {}^{234}_{90}Th + {}^{4}_{2}He $$

It should be noted that nuclear equations are balanced when:

(1)   the total of the mass numbers on the   =   the total of the mass numbers on the
           right-hand side of the equation   left-hand side of the equation

and

(2)   the total of the atomic numbers on   =   the total of the atomic numbers on the
           right-hand side of the equation   left-hand side of the equation

Alpha particles have the largest charge and are the heaviest of all the types of the decay particles emitted by radionuclides. They normally have a high energy of the order of 3 to 9 MeV (mega electron volts) but are comparatively slow moving, leaving the nucleus at about 10% of the speed of light. Consequently, they do not penetrate far into the medium surrounding the source before their energy is lost by collision with the 'molecules' of that medium (Fig. 33.2).

The energy of the alpha particles emitted by a radionuclide is constant for all the alpha particles emitted by that source. Its value is a characteristic property of the radionuclide and so may be used to identify the source in much the same way as melting points are used

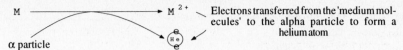

Electrons transferred from the 'medium molecules' to the alpha particle to form a
helium atom

Fig. 33.2. A schematic representation of the loss of energy by the collision of an alpha particle with a 'molecule' M of the surrounding medium.

to identify compounds in organic chemistry. However, radiation detectors can be set to detect only alpha radiation with a specific energy and so identify a particular radionuclide. The technique is called alpha-spectroscopy.

Alpha particles cannot penetrate the skin and so exposure to an external source does not pose a serious health problem to humans. However, exposure to an intense external dose of alpha radiation can cause burns. Alpha emitters will also cause a great deal of damage if they enter the body by, for example, ingestion or inhalation. This is because the particles penetrate the softer internal tissues more easily than the harder layer of skin. The ions and free radicals formed by this penetration disrupt metabolic processes occurring in the tissue which leads to tissue damage and further metabolic disruption (33.9).

The alpha emitter radium-226 was used up to the 1950s to treat some cancers. It was placed in thin-walled gold or platinum needles. These needles were placed as close to the cancer as possible to allow the alpha radiation to destroy the cancer cells.

33.2.2   Beta particle emission

Beta particles are also known as **negatrons**. They are high energy electrons which are ejected from the nucleus at velocities of 90% or more of the speed of light. Several different nuclear processes are known to lead to their production. These processes are complex but may be conveniently summarised by the equation:

$$\ce{_0^1 n} \longrightarrow \ce{_1^1 p} + \ce{_{-1}^0 e}$$

The loss of a beta particle has no effect on the mass number of the nucleus but increases its atomic number by one. Beta particles are allocated an *atomic number* of -1 in order to balance nuclear equations. Consider, for example, the beta decay of carbon-14 to nitrogen-14:

$$\ce{_6^{14}C} \longrightarrow \ce{_7^{14}N} + \ce{_{-1}^0 e}$$

The 'sums' of the atomic numbers on each side of this equation will only be equal if a beta particle has an effective atomic number of -1.

A beta-emitting radionuclide will emit beta radiation having a range of energies (Fig. 33.3). Consequently, the energies of beta particles are specified as E_{max} or E_{mean} values. These values are characteristic of the source and are recorded as part of the data of that source. They may also be used to identify a source.

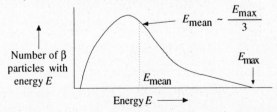

Fig. 33.3. The energy spectrum of the beta particles emitted by a simple beta-emitting source.

Beta sources may emit radiation that has more than one E_{max} value. For example, iodine-131 emits beta radiation that exhibits three E_{max} values with energies of 0.810, 0.607 and 0.337 MeV respectively.

Beta particles move at a higher velocity and are lighter than alpha particles. Consequently, beta particles penetrate further into the medium surrounding the source than alpha particles. However, the lower energy of beta particles means that the degree of ionisation they cause is considerably less than that caused by alpha particles.

Beta particles are a greater external health hazard to humans than alpha particles. High energy beta particles can penetrate the outer layer of skin and cause damage to the skin tissue. However, beta sources do not represent such a serious health hazard as alpha particles to humans unless ingested, in which case they can cause considerable damage to tissue and internal organs. However, the lower energy of beta particles means that they do less damage in this situation than alpha particles and so some beta sources are used internally in the treatment of disease and as diagnostic aids.

33.2.3   Gamma ray emission

Gamma rays are high energy electromagnetic radiations. They have short wavelengths in the region of 10^{-12} metres (1 picometre) and are emitted from the nucleus as a short burst of radiation. Gamma emission occurs when a nucleus in a high energy state decays to a lower energy state, the energy of the gamma ray corresponding to the difference in energy between the energy levels of the nucleus involved in its emission. The emission of a gamma ray does not change the mass or atomic numbers of the nucleus.

Gamma emission often accompanies alpha and beta particle emission. For example, gold-189 decays to mercury-189 by the emission of a beta particle and a gamma ray with an energy of 0.412 MeV. The gamma ray emission corresponds to a drop in the energy of the mercury-189 nucleus from a higher to a lower energy level, the two levels being separated by an energy equivalent to 0.412 MeV (Fig. 33.4).

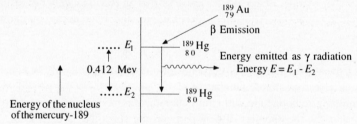

Fig. 33.4. Gamma ray emission and nuclear energy changes.

A source may emit gamma rays having different energies. This corresponds to successive energy level changes in the molecule. For example, chlorine-38 emits three beta and two gamma radiations.

$$^{38}_{17}Cl \longrightarrow \ ^{38}_{18}Ar \ + \ 3\beta \ + \ 2\gamma$$

The percentages of nuclei decaying by a particular route and energies of the emissions are:

β (MeV)   1.11 (31%)   2.71 (16%)   4.81 (53%)

γ (MeV)   1.6 (31%)   2.10 (47%)

The two gamma rays correspond to two energy transitions within the nucleus of the daughter isotope argon-38 (Fig. 33.5). To determine the order in which these gamma emissions occur one must check:

(1) The total percentage of parent and daughter nuclei that decay to a particular energy level equals the total percentage of daughter nuclei that decay from that energy level if it is an excited state of the daughter isotope. In other words, the total number of nuclei 'arriving' at an energy level must equal the total number 'departing' if that energy level is an excited state of the nucleus.

(2) The values of the energies of all the emissions fit the pattern of energy levels.

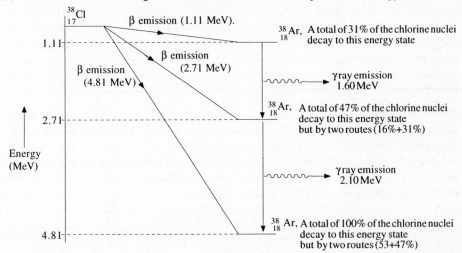

Fig. 33.5. The relationships between the energy changes that occur during the decay of chlorine-38 and argon-38 nuclei.

Gamma rays are also emitted by a process called **electron capture**. An electron from the K or L quantum shells is drawn into the nucleus where it combines with a proton to form a neutron.

$$_{-1}^{0}e \ + \ _{1}^{1}p \longrightarrow \ _{0}^{1}n \ + \ \gamma$$

The capture of the electron reduces the atomic number of the nucleus by one but does not affect its mass number. For example, argon-37 emits gamma rays by electron capture and in the process forms chlorine-37.

$$_{18}^{37}Ar \ + \ _{-1}^{0}e \longrightarrow \ _{17}^{37}Cl \ + \ \gamma$$

The energy of a gamma ray has a constant value which is a characteristic property of its source. Its value can be used to identify a source even in a mixture of sources.

Gamma rays have a high penetration power. They readily travel through the human body but can be stopped by very dense materials such as lead and concrete. High energy gamma rays can be a considerable health hazard but low energy rays are widely used as diagnostic aids and in research into the design of dosage forms (33.12) because they cause little ionisation in the surrounding medium and so little tissue damage in the human body.

33.2.4   Positron emission

Positrons are small, positively charged particles with the same mass as an electron but a charge of +1. They are sometimes referred to as positive electrons. Their formation is not understood but is believed to be due to the conversion of a proton to a neutron.

$$_{1}^{1}p \longrightarrow \ _{0}^{1}n \ + \ _{+1}^{0}e$$

Positrons are given an *apparent* atomic number of 1 in order to balance nuclear equations. They are a short-lived species, colliding with a planetary electron to form two gamma rays. These gamma rays have an energy of the order of 0.511 MeV and are emitted in almost exactly opposite directions to each other.

$$_{+1}^{0}e \ + \ _{-1}^{0}e \longrightarrow \ 2\gamma$$

Positron emitters, such as carbon-11 and oxygen-15 are used as sources for gamma imaging (33.12) because they can be readily incorporated into compounds that occur naturally within the body. These compounds can be administered to the patient and their distribution in the relevant organ followed by a positron emission tomographic device.

33.2.5   Proton and neutron emission

Radionuclides may decay by emitting either protons or neutrons at velocities of 90% or more of the speed of light. These types of emission are less common than the other types of emission described in this chapter.

Proton emission will decrease both the atomic and mass numbers of the nucleus by one. For example, fluorine-18 produced by the bombardment of nitrogen-14 with alpha particles, decays to oxygen-17 by proton emission.

$$_{9}^{18}F \longrightarrow \ _{8}^{17}O \ + \ _{1}^{1}p$$

The penetration power of protons depends on their energy. They are a serious health hazard to humans.

Neutron emissions decrease the mass number by one but have no effect on the atomic number. For example, nitrogen-17 decays to nitrogen-16 by the emission of a neutron.

$$_{7}^{17}N \longrightarrow \ _{7}^{16}N \ + \ _{0}^{1}n$$

Neutrons readily penetrate *tissue and other* materials and so all neutron sources are a serious health hazard to humans.

33.2.6   Electron capture

Electron capture usually occurs in isotopes with a high atomic number. An electron in the K ($n=1$) or L ($n=2$) shells of the atom is drawn into and captured by a proton in the nucleus. This process results in the proton being c*onverted into* a neutron and so the atomic number of the nucleus decreases by one. The process of electron capture is accompanied by a readjustment of the daughter isotope's electronic configuration to fill the gap in its K or L shell. The energy liberated by this readjustment is in the form of X-rays and/or low energy gamma rays

$$_{53}^{125}I \ + \ _{-1}^{0}e \longrightarrow \ _{52}^{125}Te \ + \ \gamma \ + \ \text{X-rays}$$

X-rays are high energy electromagnetic radiations with wavelengths in the region of 10^{-10} to 10^{-11} metres. They have a high penetrating power and can constitute a serious external health hazard for humans. However, X-rays are used as diagnostic aids but their use must be carefully controlled.

33.3 ACTIVITY

The activity (A) of a radionuclide is the rate at which it is decaying, that is, the rate at which the nuclei in a sample are decomposing. It is recorded as the number of nuclei disintegrating per second. Values are expressed in the SI unit the Becquerel (Bq) where:

$$1 \text{ Becquerel } = 1 \text{ disintegration per second (dps)}$$

Activity was formerly expressed in Curies (Ci) where:

$$1 \text{ Curie} = 3.7 \times 10^{10} \text{ disintegrations per second} = 3.7 \times 10^{10} \text{ Becquerels}$$

Activities are normally quoted in the literature in units based on the appropriate SI prefix (Table 33.2). For example, an activity of 4,000,000 Bq would be recorded as 4 MBq.

Table 33.2. The commonly used SI prefixes.

Prefix	Symbol	Factor	Prefix	Symbol	Factor
Tera	T	10^{12}	milli	m	10^{-3}
Giga	G	10^{9}	micro	μ	10^{-6}
Mega	M	10^{6}	nano	n	10^{-9}
Kilo	k	10^{3}	pico	p	10^{-12}

Activity can be measured by instruments known as radiation counters (33.8) that record the counts per minute (cpm), which is related to the number of nuclear disintegrations occurring per minute. Radiation counters are not 100% efficient and the recorded count is converted into the activity (disintegrations per second) of the source using the appropriate factor for the instrument. For example, if a radioactive source gave a count of 8000 counts per minute using a radiation counter that was 40% efficient, the actual number of disintegrations per minute is:

$$8000 \quad \text{x} \quad \frac{100}{40} \quad \text{disintegrations per minute (dpm)}$$

and the number of disintegrations / second is:

$$8000 \text{ x } \frac{100}{40} \text{ x } \frac{1}{60} = 333.3 \text{ Bq}$$

Therefore, the activity of the source is 333.3 Bq.

33.4 THE RATE OF DECAY

The rate of decay of a radionuclide is independent of its chemical or physical state. For example, the rate at which carbon-14 decays will be the same whether it occurs in diamond, graphite, calcium carbonate, carbon dioxide, ethane, toluene, glucose, cholesterol or any other compound. Changes in pressure and temperature will have no effect on the rate of decay.

The rate of decay (A) is the number of nuclei (N) disintegrating in unit time, that is:

$$A = -\frac{dN}{dt} \qquad\qquad (33.1)$$

where t is the time taken for N nuclei to decay.

Simple radioactive decay, that is, radionuclides that produce only one daughter isotope even though it may exist in several different energy states, exhibits first order kinetics (6.3). Therefore, for the decay process:

Parent radionuclide $\longrightarrow$ Daughter + Particle and/or electromagnetic radiation

The number of nuclei decaying per second is proportional to the total number of nuclei of the parent radionuclide (N), that is:

$$- \frac{dN}{dt} = \lambda N \tag{33.2}$$

where lambda (λ) is the decay constant for the process. The decay constant represents the probability of a nucleus disintegrating in a given time t. It is a characteristic property of the radionuclide. Rearrangement of equation 33.2 gives:

$$\frac{dN}{N} = -\lambda dt \tag{33.3}$$

Integration of equation 33.3 gives,

$$\log_e N = -\lambda t + c \tag{33.4}$$

where c is the constant of integration.

However, if N_0 is the number of nuclei initially present at a particular time which is arbitrary taken as $t = 0$ and N is the number present at a **lapsed time** t (time $t = t$),

$$c = \log_e N_0$$

and equation 33.4 becomes:

$$\log_e N = -\lambda t + \log_e N_0$$

that is:

$$N = N_0 e^{-\lambda t} \tag{33.5}$$

However:

$$A = - \frac{dN}{dt}$$

Substituting in equation 33.2:

$$A = \lambda N \tag{33.6}$$

that is, the activity of a radionuclide is proportional to the number of its nuclei present. Therefore as N_0 is proportional to A_0 the activity at $t = 0$ (the initial time) and N is proportional to A the activity at a time $t = t$ (the lapsed time), equation 33.5 becomes:

$$A = A_0 e^{-\lambda t} \tag{33.7}$$

It should be remembered that the time t is in fact the **lapsed time** between the two measurements of the activity of the radionuclide.

Equation 33.7 shows that the rate of radioactivity is exponential (Fig. 33.7). This equation is also the basis of experimental methods for determining the value of the decay constant of a radionuclide decaying by a simple decay process. Taking logarithms and rearranging equation 33.7 gives equation 33.8 which is of the form $y = mx + c$, the equation of a straight line.

$$\log_e A = -\lambda t + \log_e A_0 \tag{33.8}$$

Since A_0 has a constant value for a particular experiment, a plot of $\log_e A$ against t will be a straight line with a slope of $-\lambda$ (Fig. 33.7).

The mathematical treatment of decay processes where more than one type of daughter isotope is formed is more complex and is beyond the scope of this text. The reader is directed to radiochemistry textbooks for additional information.

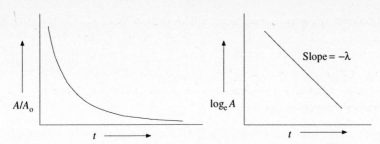

Fig. 33.7. Graphical representations of simple radioactive decay.

33.5 HALF LIFE

The half life ($t_{1/2}$) of a radionuclide is the time taken for half the nuclei in a sample of a radionuclide to disintegrate. Therefore, for $t_{1/2}$, $N = N_o/2$ and equation 33.5 becomes:

$$\frac{N_o}{2} = N_o e^{-\lambda t} \tag{33.9}$$

and so,

$$e^{-\lambda t} = \frac{1}{2}$$

hence:

$$\lambda t_{1/2} = \log_e 2$$

and so,

$$t_{1/2} = \frac{0.693}{\lambda} \tag{33.10}$$

Equation 33.10 demonstrates that the half life of a first order decay process is a constant that is independent of the mass of the radionuclide (Fig. 33.8).

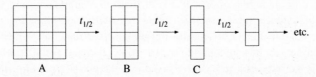

Fig. 33.8. A schematic representation of the fact that in a simple decay process the time taken for each of the masses A, B and C of a radioisotope to decay to half its value is the same no matter what its starting concentration. Each square represents one gram of a radionuclide.

Half life values are a measure of the stability of radionuclides, the larger the value the more stable the radionuclide. However, it should be noted that the half lives of different isotopes of the same element may be very different. For example, the β emitter thorium-234 has a $t_{1/2}$ of 24.1 days whilst the α emitter thorium-230 has a $t_{1/2}$ of 7.54 x 10^4 years.

Half life values are of crucial importance when selecting and calculating the quantity of a radionuclide required for medical or other work. There must be sufficient nuclei present at the completion of the work to give an accurate activity measurement. However, in medical investigations too long a half life can lead to a high level of residual activity in the patient which would constitute a serious health risk. If health risk is not a consideration, the use of a radionuclide with a $t_{1/2}$ of about 2 to 5 weeks has the advantage that changes in its concentration are small and so are less likely to affect an investigation. In practice, ten half lives are often taken as a reasonable working life for a radionuclide.

The activity (A) of a radionuclide at the end of an experiment can be calculated from;

$$A = \frac{A_o}{2^n} \tag{33.11}$$

where n is the number of half lives that are required.

33.6 ACTIVITY AND MASS

The activity of a sample of a radionuclide is proportional to the mass of the isotope. It can be calculated for a specific mass of a radioisotope using equations 33.6 and 33.10

Example 33.1
Calculate the activity of 2 g of pure iodine-125. The half life of iodine-125 is 60 days. Rearranging equation 33.10:

$$\lambda = \frac{0.693}{t_{1/2}} = \frac{0.693}{60 \times 24 \times 60 \times 60} = 0.0000001 \ s^{-1}$$

Substituting for l in equation 33.6, A = λN:

$$A = 0.0000001N \tag{33.12}$$

But 1 mole of an isotope contains 6.022×10^{23} atoms (Avogadro's number L) and so it will also contain 6.022×10^{23} nuclei. Therefore 2 g of iodine-125 will contain:

$$\frac{2}{125} \times 6.022 \times 10^{23} = 9.6 \times 10^{21} \ nuclei$$

Substituting for N in equation 33.12 the activity of 2g of iodine-125 is:

$$0.0000001 \times 9.6 \times 10^{21} = 9.6 \times 10^{14} = 960 \ TBq$$

The amount of a radionuclide required for medical and biological work is normally very small, often in the region of 10^{-12}g. It is difficult to manipulate such small quantities and so radionuclides are often supplied either in solution or diluted with **carriers** to facilitate their manipulation. Carriers are non-radioactive materials which contain non-radioactive isotopes of the radionuclide in the same chemical form as the radionuclide.

The **specific activity** of a radioactive sample is the activity per unit mass or volume of that sample. It expresses the activity of the sample relative to the total amount of both active and inactive material in the sample. Specific activity values are usually quoted as Bq per mole or gram for solids and Bq per cm^3 or dm^3 (litre) for liquids, the usual SI subunits (Table 33.2) being used as appropriate.

Specific activity is used in calculations in the same way as activity except the units are now in terms of mass or volume (Example 33.2).

Example 33.2
The specific activity of a solution of iodine-131 is 20 MBq cm^{-3} at 9am on the 10th of January. Calculate the specific activity of this solution at 9am the following day. The half life of iodine-131 is 8.4 days.

Substituting the value of $t_{1/2}$ in seconds in equation 33.10, the value for λ for iodine-131 is:

$$\frac{0.693}{8 \times 24 \times 60 \times 60} = 0.0000009 \ s^{-1}$$

At time t = 0, A_o = 20 MBq, substituting this value and that for λ in equation 33.7,

$$A = 20e^{-0.0000009 \times 24 \times 60 \times 60}$$
$$= 18.5 \ MBq \ cm^{-3}$$

It should be noted that when using equation 33.7 it is not necessary to work in seconds provided the units used for all time measurements are the same, that is, both the half life $t_{1/2}$ and the lapsed time t are measured in the same units.

33.7 SOURCES

In Britain, radioactive sources can be obtained from Amersham International, ICN and certain universities. They are supplied as either closed (sealed) or open sources. Closed sources are permanently sealed and are used mainly as external sources of radiation. Open sources are not permanently sealed. They are normally removed from their containers and added to the system being examined. Open sources are the most commonly used type of radioactive source in diagnostic medical sciences.

The label on the container of both types of radioactive source should give the following information:

(1) The type and energy of the emission. Gamma emitters are usually supplied in lead containers.
(2) The activity or specific activity of the source, which is usually given in Ci units even though these are not the legal units, together with the date on which the measurement was made. It will also include the time of measurement if the $t_{1/2}$ value is small.
(3) The half life of the radionuclide.
(4) The labels of closed sources will also show the physical nature and dimensions of the source. For example, a cylinder, disc or sphere together with the appropriate dimensions.

The labels of open sources should also show:

(1) The identity of the radionuclide responsible for the radiation and its state of chemical combination. For example, either $K^{125}I$ or iodine-125 as KI.
(2) The **radiochemical purity**. This is the percentage activity of the source that is due to the radionuclide in the form specified on the label for the source, that is:

$$\frac{\text{Radiochemical}}{\text{purity}} = \frac{\text{Activity of the radionuclide in the form specified on the label} \times 100 \ \%}{\text{Total activity of the radionuclide regardless of its chemical form}}$$

(3) The **radionuclide purity**. This is the percentage of the activity of the source that is due to the radionuclide specified on the label, that is:

$$\text{Radionuclide purity} = \frac{\text{Total activity due to the specified radionuclide} \times 100 \ \%}{\text{Total activity of the source}}$$

33.8 DETECTION AND MEASUREMENT

The methods currently used to detect and measure radiation depend on the type of emission, its energy and penetrating power. Detectors can be broadly divided into three types based on their mode of action, namely: photographic detectors, ionisation and scintillation counters. The fundamental principles underlying the operation of these general types of instrument are discussed in this section, but for detailed information concerning the mode of operation of a particular instrument you should consult specialised texts and the manufacturers' literature.

33.8.1   Photographic detectors

These take the form of badges that are worn by an individual. These badges contain photographic film that is affected by radiation. The film is developed at specific times and the degree of darkening of the negative is a measure of the total amount of radiation to which the wearer has been exposed. This type of detector is used to monitor the exposure to radiation of people such as radiographers and nurses that work with ionising radiation.

33.8.2   Ionisation counters

Ionisation counters, which include gas flow and Geiger-Muller counters, measure the ionisation produced by the radiation. The positive and negative ions formed in a suitable medium are collected by an electric field. Each ion produces a pulse of electricity and it is these pulses that are counted and recorded electronically. A summary of the different general types of ionisation counter is given in Table 33.3.

Table 33.3. General types of ionisation counter.

General type of counter	Ionising medium	Use	Notes
Gas flow (Gas ionisation counter)	A gas, e.g. a 10% mixture of methane in argon	Alpha particles	The source is placed inside the counter. They are difficult to use but can be used to measure low-level radiation sources
Geiger-Muller (GM) tube (end window type) (Gas ionisation counter)	A low pressure gaseous mixture of methane and argon	Alpha and beta particles	The source is outside the counter. GM tubes are used to count solid samples but are not suitable for low energy alpha and beta radiation
Geiger-Muller liquid counter (Gas ionisation counter)	A low pressure gaseous mixture of methane and argon	Alpha and beta particles	The source is placed inside the counter. Used for liquid samples. GM liquid counters are not suitable for low energy alpha, and beta radiation
Semiconductor detectors	A solid semiconductor	Alpha, X-ray and gamma radiation	Will distinguish between alpha, X-ray and gamma radiations of different energies

Ionisation counters are not 100% efficient. They only detect radiation emitted in the direction of the detector whereas a source will emit radiation in a random manner in three dimensions. If a full three-dimensional emission count is required the geometry of the source must be taken into account. This complex statistical procedure is not usually required in routine medical work.

Ionisation counters do not record all the radiation that is emitted by a source in the direction of the detector. Furthermore, gas ionisation counters do not record all the radiation that strikes their detectors. After a gas ionisation counter registers *one* particle or photon its electronics requires a short period of time for readjustment before they can register another particle or photon. The time taken for this to occur is known as either the **dead** or **paralysis time** of the instrument. As a result, radiation striking the detector in this time will not be counted. Some instruments have a built in automatic correction for paralysis time but in many cases the readings have to be corrected for paralysis time errors by consulting tables. However, if the instrument has a constant paralysis time equation 33.13 may be used to correct counts.

$$ N = \frac{N_o}{1 - N_o T} \qquad (33.13) $$

where T is the paralysis time in seconds, N is the corrected count rate per second and N_o is the observed count rate per second.

33.8.3   Scintillation counters

In this type of instrument the radiation raises atoms in a special medium to an excited state (Fig. 33.9). These excited atoms lose their excess energy by emitting a very short flash of light (a scintillation). These scintillations are registered by a photomultiplier tube (PM tube) which emits a pulse of electrons for each scintillation it registers. The PM tube amplifies the pulse of electrons to a level that can be counted electronically.

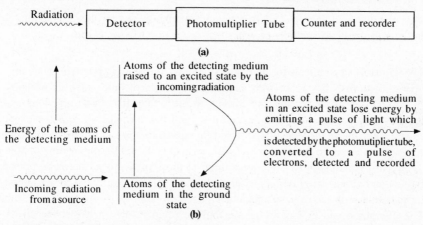

Fig. 33.9. (a) The components of a scintillation counter. (b) The mode of action of scintillation counters.

Scintillation counters may use solid or liquid scintillation media. Solid scintillators may consist of inorganic and/or organic compounds, for example, sodium iodide containing a low percentage of either thallium, anthracene or stilbene. Solid scintillation counters are used to measure X-radiation, γ and medium to high energy β radiation from external sources to the counter. Liquid scintillators are based on organic compounds such as p-terphenyl (Prr) and 1,4-diphenyloxazole (PPO) dissolved in a suitable organic solvent, such as toluene or dioxan. The sample is dissolved in the solvent containing the scintillator which is then attached to a suitable PM tube. Liquid scintillation counters are used for low energy β and γ sources.

33.8.4   Background counts

The world we live in is radioactive and so there is always a certain amount of background radiation. This radiation will register on radiation detectors but is not usually significant unless the background count is near that of the source. Lead will stop most of the background radiation and so measurements made inside a lead screen (**lead castle**) will usually reduce the background count to an insignificant level. If a lead castle is not used, adjustment for background count may have to be made. However, in very sensitive whole body counters the radiation from the lead screening is a problem as the lead is not pure.

33.9 THE BIOLOGICAL EFFECTS OF RADIATION

The passage of radiation through the cells of living tissue results in the formation of ions and free radicals. The reaction of these ions and free radicals with the compounds in the cells disrupts or alters the normal metabolic processes occurring in those cells. These changes in the metabolic processes can result in:

(1) The death of the organism or animal.
(2) A reduction of the ability of cells to divide.
(3) Abnormal cell division leading to cancers.
(4) Changes in genetic material.
(5) Increase in the rate of ageing.

The disruption of normal cell metabolism is mainly due to the interaction of the radiation with the large amounts of water that occur in all living tissue.

$$H_2O + radiation \longrightarrow H^+ + \cdot OH + e$$

The hydroxyl free radicals formed in this process are the main problem. They are very reactive and immediately react with any neighbouring molecules such as proteins and DNA to produce substances that are foreign to the tissue. It is these reactions and the compounds they produce that change and/or disrupt the normal course of the metabolic processes occurring in the tissue. Hydroxyl free radicals also react to form hydrogen, and hydrogen peroxide. Both these substances are toxic to cells and it is interesting to note that the symptoms of radiation sickness are similar to those of hydrogen peroxide poisoning. Disruption and alteration of biological processes may also occur by the direct interaction of the radiation with the organic molecules in the process.

The disruption and alterations to metabolism often results in a cascade effect, the substances formed by the initial effect of the radiation causing further disruptions and alterations to other metabolic processes. This in turn leads to further metabolic interference and so on. The initial radiation effect is magnified by the succession of secondary processes it initiates (Fig. 33.9).

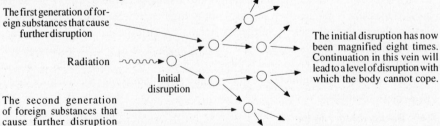

Fig. 33.9. A schematic representation of the cascade effect of radiation.

Exposure to low doses of low energy radiation over a long period of time is more likely to result in genetic mutations and cancers than low exposure to high energy radiation. Exposure to high energy radiation and high doses of low energy radiation over long periods of time will usually kill a cell or stop it reproducing. This is why high doses of low energy radiation are used to treat cancer. It is also the reason for carefully monitoring the amount of low energy radiation to which a patient is exposed in diagnostic investigations.

The biological effects of radiation depends on the type of radiation as well as its energy. Heavy particles, such as α particles and neutrons, produce dense clusters of ions and radicals along a short radiation path length whilst γ and X-rays leave ions and radicals at scattered intervals along much longer path lengths. These differences in behaviour makes it difficult to compare quantitatively the biological effect of the different types of radiation on the same type of tissue. However, it has been compared in terms of the linear energy transfer (LET) value for the radiation. This is the average energy deposited per unit path length by the radiation. Cells are more likely to be able to repair the damage caused by non-lethal doses of radiation with a low LET value than radiation with a high LET value.

33.10 EXPOSURE AND ABSORBED DOSE

External gamma and X-ray sources (and increasingly UV on unscreened skin) pose the main radiation danger to humans and other mammals since this radiation has a high penetrating power (Fig. 33.10). Alpha and beta radiations with their low penetrating powers are not usually a problem unless the recipient is exposed to a large dose of radiation. However, all sources are potentially very dangerous when they enter the body either by medical administration, breathing in contaminated air or through skin damage. It is emphasised that there is no safe level of radiation below which damage to health cannot occur, and even at so-called safe levels there is a small degree of risk to an individual.

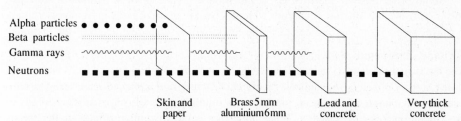

Fig. 33.10. The penetrating power of radioactive emissions in different materials. The thicknesses shown are average maximum values as penetration will also depend on the energy of the radiation.

Radiation effects are broadly classified as either **somatic** or **hereditary**. Somatic effects occur when the irradiation results in something appearing to be visibly wrong in the person irradiated, for example, the appearance of a burn or wound. Somatic effects are further subdivided into **stochastic** and **non-stochastic** effects. Stochastic effects are somatic effects with no threshold dose whose chance of occurrence is regarded as a function of dose, the greater the dose the greater the probability of an effect occurring. Cancers and genetic effects are regarded as stochastic effects. Non-stochastic effects are somatic effects whose severity is a function of the dose and in these cases there may in fact be a threshold dose. For example, impairment of fertility which is regarded as a non-stochastic effect, is not detected until a high dose is exceeded. Hereditary effects occur when irradiation of an adult leads to damage to their descendants. There is now some evidence linking irradiation of workers in the nuclear industry to the high incidence of child leukaemia in their offspring. However this evidence is not conclusive. It should be noted that not all the radiation striking a person penetrates the skin and so human exposure to radiation is measured *in terms* of both **exposure** to radiation and **absorbed dose**. Exposure in this context may be **simply** defined as *a measure of* the total amount of radiation striking the body whilst absorbed dose is the quantity of radiation that is actually absorbed by the body. A simple analogy that demonstrates the difference between exposure and absorbed dose is found in a chemistry lecture. The dose of chemistry you are exposed to in the lecture is not the same as the dose you absorb.

33.10.1   Exposure

Exposure is measured in the pre-SI units known as Roentgens (R), pronounced runtyen, where 1 R = 2.58 x 10^{-4} coulombs per kilogram (C kg^{-1}). The SI units for exposure are coulombs per kilogram (C kg^{-1}).

Exposure is defined as quantity of the ionisation produced by X- or gamma radiation as it passes through the air. Monitoring equipment functions by measuring the current produced by this ionisation.

33.10.2   Absorbed dose

The absorbed dose is a measure of the quantity of energy absorbed by the tissue. The units used are the SI unit, the gray (Gy) and the pre-SI unit, the Rad (**r**adiation **a**bsorbed **d**ose),

where: $$1 \text{ Gy} = 1 \text{ J kg}^{-1} \text{ and } 1 \text{ Rad} = 10^{-2} \text{ joules per kilogram (J kg}^{-1})$$

Hence: $$1 \text{ Gy} = 100 \text{ Rad}$$

An absorbed dose of about 10 Gy of radiation is lethal for most mammals. Although this is quite a small amount of energy (10 joules per kilogram of body mass) the disruption it causes to the biological processes in the tissue will result in the death of the mammal.

33.10.3   Dose equivalent

The biological effect of radiation depends on the type of radiation as well as its energy and dose (33.9). Consider, for example, the effect of gamma and alpha radiation with the same energy on a sample of tissue. In practice it is found that significantly different doses of these radiations are required to bring about the same disruptive effect on the tissue. Furthermore, the nature of the tissue will also affect the disruption brought about by the radiation. For example, the effect of 1 gray of beta radiation on bone will be different from the effect of the same dose of radiation on tissue. These variations make it difficult to use the Roentgen, Rad and gray to quantitatively compare the effects of different types of radiation on different types of tissue and so the dose equivalent has been devised to account for these differences.

Dose equivalent is measured in SI units known as sieverts (Sv). The pre-SI units were known as **Rems** (Roentgen equivalent man), where 1 Rem = 100 Sv. One Rem is defined as the absorbed dose of any type of radiation that will cause the same biological effect in man as 1 Roentgen. The basis of the defined value of a Rem will therefore be different for each type of radiation but a dose of 100 Rem of gamma radiation will in theory cause the same biological effect in man as a dose of 100 Rem of alpha radiation. It is emphasised that the Rem and sievert are units used to compare the biological effects of radiation in man.

The dose equivalent in sieverts is related to grays by the expression:

$$\text{Dose equivalent (Sv)} = \text{Absorbed dose (Gy)} \times Q \times N$$

where Q is an empirical **quality factor** that takes into account the nature of the radiation and the nature of the tissue and where N is an additional factor that takes into account factors not considered in the older quality factor Q. The dose equivalent is calculated in Rems from the absorbed dose using the relationship:

$$\text{Dose equivalent (Rem)} = \text{Absorbed dose (rad)} \times Q$$

Currently the International Commission for Radiation Protection (ICRP) has assigned a value of 1 to N. This means that workers are able to use the existing values for Q to calculate dose equivalents. Some typical Q values are given in Table 33.4.

Table 33.4. Typical Q values for the different types of radiation

Radiation	Q
Gamma and X-rays	1
Fast neutrons and protons	10
Alpha particles	20

The recommended dose equivalent limits for working with radioactive sources are:

(1) To prevent non-stochastic effects in all human tissues with the exception of the eye is 0.5 Sv (50 Rem) per year. A lower limit of 0.15 Sv (15 Rem) per year is recommended for the eye.

(2) To prevent stochastic effects, a limit of 50 mSv (5 Rem) per year is recommended for uniform irradiation of the whole body. However, in the case of the irradiation of particular organs, the product $W_T H_T$, where W_T is a factor that allows for the sensitivity of the organ to radiation and H_T is the annual dose equivalent of that organ should not exceed 50 mSv. Furthermore, the sum of $W_T H_T$ for all the organs irradiated should not exceed 50 mSv if more than one organ is irradiated.

The dose equivalents for the general public are one-tenth of these limits.

Dose equivalents are measured using either ionisation or scintillation counters calibrated in either Rems or sieverts. The instrument must be calibrated for both the type of radiation, its energy and the medium in which the measurement is being made.

33.11 PROTECTION

The intensity of the radiation in the atmosphere at a particular point from a source will depend on the distance from the source, the energy of the source and the type of radiation. Radiation intensity decreases with distance and as a simple rule it is usually safe to assume that doubling the distance from a source reduces the intensity of the radiation by a quarter. For example, a worker exposed to a dose rate of 10 mSv per hour at half a metre from a source will only receive a dose rate of 2.5 mSv per hour at one metre from the source.

Alpha, beta, gamma, proton and neutron emissions have different atmospheric penetrating powers. However, they can be stopped by the appropriate form of shielding. It is possible to calculate the thickness of the material required to shield a particular source but this is outside the scope of this text.

The handling and use of radioactive preparations involves a degree of risk. It should not be undertaken without the use of appropriate protective clothing, shielding, tools and procedures and carried out in a laboratory classified as being suitable for the type of radioactive work under investigation. All work should be undertaken under proper surveillance and in such a manner that the exposure of the worker is not greater than the current limit recommended by the International Commission on Radiation Protection (ICRP). Laboratories and personnel should be monitored with appropriately positioned equipment suitable for detecting the sources being handled. For example, film badge detectors worn on the chest can underestimate the exposure of a worker by a factor of three if the source is behind the worker. In this situation the worker should also wear a detector on their back. Similarly, finger detectors should be worn by workers using syringes. A procedure for dealing with spillages must be established before work with a radionuclide commences. A general protocol used by some laboratories using short half life radionuclides is to seal the laboratory and allow five half lives to pass before entering and cleaning up. This is not practical if the radionuclide has a long half life.

33.12 SOME MEDICAL USES OF RADIOISOTOPES

Medical preparations containing a radioactive source may take the form of an element, a compound, a complex or a biological preparation. The specifications of radiopharmaceuticals are given in monographs in the British Pharmacopoeia and similar publications. These monographs will usually specify:

(1)   A method of identifying the radioisotope by means of its half life, the type and energy of its emissions.
(2)   The physical characteristics of the preparation.
(3)   The radionuclide and radiochemical purity with respect to specific isotopes.
(4)   Limit tests for non-radioactive chemical impurities.
(5)   The limits of alkalinity and acidity allowed for the preparation.
(6)   The stability of the preparation and any storage conditions.
(7)   Storage conditions and the time limit for its use.
(8)   A radioactivity assay procedure.
(9)   Contraindications to use.

33.12.1   Diagnostic aids

Radionuclides are used as diagnostic aids to follow the progress of a chemical species in a metabolic process. This can give information concerning the efficiency of the operation of an organ or metabolic process in the body. Radionuclides are also used to study the physiological function of the operation of an organ by the gamma imaging technique.

Radionuclides are used to follow metabolic processes by feeding a suitable compound containing the radionuclide to the subject and following its progress through the relevant area of the body by externally monitoring the activity in the selected area or alternatively monitoring the activity of blood samples, faeces and urine. The instrument used to measure the activity is ideally set to record only the radiation that is emitted by the radionuclide. However, in practice the instrument will also record some background radiation which must be taken into account in some investigations. For example, iodine-123 which emits gamma radiation by the electron capture process, is used to check thyroid function. A measured dose of an iodine-123 preparation is administered to the patient as a solution of iodide ions and the change in activity in the patient's thyroid with time is monitored externally and recorded (Fig. 33.11). A background count is taken at the knee and a graph of percentage dose taken up by the thyroid against time drawn. The shape of this graph gives the doctor valuable information concerning the function of the patient's thyroid. Iron-59, a gamma emitter, is used in a similar manner to follow iron metabolism (Fig. 33.12).

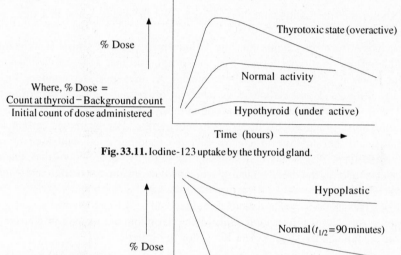

Fig. 33.11. Iodine-123 uptake by the thyroid gland.

Fig. 33.12. The use of iron-59 to follow iron metabolism.

Gamma imaging is a technique that produces an *in vivo* image of an organ or area of tissue. These images indicate the efficiency of the function of an organ. The technique of gamma imaging is carried out by introducing a suitable radionuclide into the subject and using a special detector known as a gamma camera to produce an image of the distribution of the radionuclide in the organ or tissue under examination. The radionuclide is usually administered in the form of a complex that is known to accumulate in the organ or tissue. For example, ^{99m}Tc-mercaptoacetylglycine (^{99m}Tc-MAG3) is widely used for the quantitative assessment of renal function (Fig. 33.13). The scan images, together with the plots of the variation of activity with time, show that the left kidney is functioning normally whilst the right kidney is only making a 25% contribution to total renal function. The administration of a diuretic at about 20 minutes followed by further monitoring up to 25

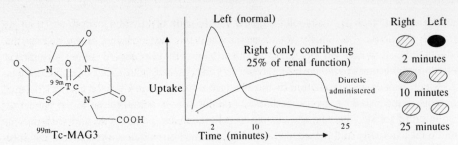

Fig. 33.13. A representation of a gamma image scan of the kidney using ^{99m}Tc-MAG3. The graphs show the rate of uptake and loss of ^{99m}Tc-MAG3 from the right and left kidneys while the diagram on the far right shows the concentration of the ^{99m}Tc-MAG3 in the kidney in the form of a shaded picture representing the type of colour variation one would obtain from the gamma image photographs at 2, 10 and 25 minutes.

minutes shows that the radionuclide has cleared from the kidneys and that there is no obstruction to urinary outflow.

A specialised form of gamma imaging is known as positron emission tomography (PET). PET differs from other forms of gamma imaging in that it shows the changes in metabolic function as they are actually occurring. It uses positron emitters such as oxygen-15, nitrogen-13 and carbon-11 as labels in drugs and naturally occurring molecules. The progress of the labelled compound in the system is followed by observing the gamma emissions produced by these positron emitters. For example, carbon-11 is incorporated into the structure of glucose and the resulting compound used to study glucose metabolism. PET scans of the brain, for example, have shown that abnormal glucose metabolism in the brain occurs in schizophrenia, manic depression and epilepsy in infants.

Radionuclides that are used as diagnostic aids should be selected on the basis of giving the maximum amount of information whilst doing the minimum amount of damage. For this reason, radionuclides should meet the following criteria:

(1) They should have a half life that is short but be stable long enough for the investigation to be completed.

(2) The radionuclide should only emit gamma radiation since particulate emissions would cause unacceptable internal damage.

(3) The daughter nuclide should not be radioactive and should be rapidly excreted.

Radionuclides are widely used as diagnostic aids. For example, compounds containing technetium-99m are used for to study the liver, spleen, bone marrow, lung and brain whilst indium-111 compounds are used for blood, liver-spleen, brain and tumour investigations.

33.12.2 Treatment of disease
Radiation therapy is used primarily to treat cancer. Cancer cells do not repair radiation damage as readily as normal cells and so may be more easily destroyed by radiation. Gamma and beta radiation are normally used for this purpose and a number of techniques are used to target the radiation at the cancer:

(1) The tissue may be bombarded by a beam of radiation from a source or generator. For example, the gamma emissions of cobalt-60 are used to treat cancers that lie deep in the body. Cobalt-60 ($t_{1/2}$, 5.3 years) emits two gamma emissions at 1.33 and

1.17 MeV and two beta particles at 0.31 MeV(99.99%) and 1.48 MeV(0.01%) per disintegration. The source is encased in a lead container with an opening at one end which is covered with an aluminium plate. This plate cuts out the beta emissions. The beam of gamma rays emitted from the covered opening is focused on the cancer. This type of therapy suffers from the disadvantage that the gamma radiation also damages the healthy tissue between the source and the cancer, with the result that the patient suffers radiation sickness. In addition, the intensity of the beam of radiation from cobalt sources declines with time and the beam is not sharp. For these reasons linear accelerators which produce a more stable output and sharper beam of radiation are frequently used instead of cobalt-60 sources.

(2) A hollow needle or capsule containing the radionuclide is implanted into the relevant area. For example, hollow needles containing strontium-90 or yttrium-90 have been implanted into the pituitary gland to treat Cushing's disease and into the breast to treat breast cancers.

(3) Radionuclides may be administered in the form of compounds that are designed to concentrate in the radiation target area. For example, the beta (1.71 MeV) emitter phosphorus-32 ($t_{1/2}$, 14.3 days) is used to treat some forms of leukaemia. It is administered as phosphate ions since these accumulate and form part of the structure of the bone. The radioactive phosphate accumulates in the bone where its radiation destroys the mutant white blood cells that are responsible for the leukaemia.

33.12.3   Radioimmunoassay (RIA)

RIA is used to study the *in vitro* interactions of antigens (immunogens) such as steroids with antibodies. The method is based on the competition of the antigen and a radioactive form of the antigen for the antibody. It involves plotting a calibration curve of concentration of the antigen against the activity of either a solution containing radioactive antigen or a radioactive antigen/antibody complex. It requires pure samples of a radioactive labelled form of the antigen, the non-radioactive form of the antigen, a suitable antibody and a method of separating the free antigen and antigen/antibody complex from mixtures of these substances. The *in vitro* interaction of drugs with the antibody can be studied using the same technique but replacing the antigen by the drug.

A constant concentration of the radioactive antigen (in excess) is incubated with a constant concentration of the antibody to produce a mixture (X) containing the antigen/antibody complex and free antigen (Fig. 33.14). A known volume of this mixture (Y) is treated with a known concentration of the non-radioactive antigen. The unlabelled antigen competes with the labelled antigen for the sites on the antibody and eventually an equilibrium is obtained. This mixture is separated into two fractions, one containing the radioactive and non-radioactive forms of the antigens and the other containing the radioactive and non-radioactive forms of the antigen/antibody complexes. The activity of both fractions is measured. The procedure is repeated using aliquots (Y) of X but different concentrations of unlabelled antigen. A calibration curve of the concentration of unlabelled antigen against the activity of either the antigen fraction or the antigen/antibody complex fraction is drawn. The concentrations of antigen in an unknown system can be found by adding the volume Y of the radioactive antigen/antibody mixture X to the system. The resulting equilibrium mixture is separated and the activity of either the antigen fraction or the antigen/antibody complex fraction measured. The concentration of antigen in the system can then be read off the relevant calibration curve.

Competitive protein binding (CPB) and immunoradiometric assay (IRA) procedures are similar to RIA. However, in CPB the antigen binds to a protein whilst in IRA it is the antibody that contains the radionuclide. Recently, similar procedures using non-radioactive labels have been developed.

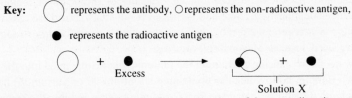

Key: ⬭ represents the antibody, ○ represents the non-radioactive antigen,

● represents the radioactive antigen

Aliquots of solution X are mixed with varying concentrations of the non-radioactive antigen ○ to form equilibrium mixtures of the type:

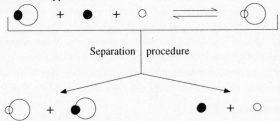

Fig. 33.14. The principle underlying of radioimmunoassay procedures.

33.12.4 Drug formulation

Radionuclides can be used to check the efficiency of the formulation of pharmaceutical preparations. The formulation is mixed with an appropriate radionuclide preparation. This mixture is used in the appropriate manner and the distribution of the radionuclide it contains followed by time lapse gamma imaging (similar to time lapse photography). The distribution of the radionuclide and the changes in this distribution will give the researcher useful information concerning the distribution of the medicaments in the pharmaceutical preparation. For example, ^{99m}Tc-DTPA is used to monitor the distribution of medicament in the nasal passages and lungs. A small volume of a solution of ^{99m}Tc-DTPA solution is mixed with a nasal spray or aerosol and then used by the subject. The time lapse images obtained by time lapse gamma imaging will give an indication of the rates of deposition and clearance of the formulation from the nasal passages (Fig. 33.15).

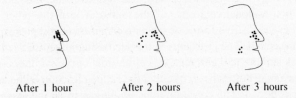

After 1 hour   After 2 hours   After 3 hours

Fig. 33.15. Typical results for the time lapse gamma imaging of a nasal spray containing a radionuclide label. The gamma camera images clearly show that the formulation is retained in the nose.

33.12.5   Anti-microbial sterilisation

Three forms of radiation are used for sterilisation: ultra-violet radiation (10.2), gamma radiation and high energy electrons. In all instances, sterilisation is carried out at about room temperature and so are useful alternatives to other 'low' temperature methods. The normal sterilising dose of radioactive emissions is 25 kGy (2.5 MRad). Since radiation effects are additive, this may be administered either in one complete dose or as separate doses totalling 25 kGy. Gamma radiation is usually produced by a cobalt-60 source but this may be replaced by caesium-137 in the future.

High energy electrons are produced by electron accelerators. Sterilisation doses are achieved more rapidly using high energy electrons than radioactive emissions.

33.13   WHAT YOU NEED TO KNOW

(1) Isotopes that spontaneously decompose with the emission of particles and high energy electromagnetic radiation are said to be radioactive and are referred to as radionuclides or radioisotopes.

(2) Nuclear decay processes may be summarised:

$$\text{Isotope A} \longrightarrow \text{Isotope B} + \text{ionising radiation} + \text{energy}$$
$$\text{(the parent)} \qquad \text{(the daughter)}$$

(3) Daughter isotopes may or may not be stable.

(4) A disintegration or decay series shows the stepwise decay of an unstable isotope via a series of unstable isotopes to a stable isotope.

$$\text{Unstable nucleus} \longrightarrow \text{Unstable nucleus} \longrightarrow \text{Unstable nucleus} \longrightarrow \text{Unstable nucleus} \longrightarrow \text{Stable nucleus}$$

(5) Isotopes with atomic numbers of 30 or below are more likely to be radioactive if the neutron to proton ratio is greater than 1:1.25 whilst isotopes with atomic numbers between 30 and 92 are likely to be radioactive if the neutron to proton ratios is greater than 1:1.40. All isotopes with atomic numbers above 92 are radioactive.

(6) Large quantities of energy are liberated when a nucleus decays. Most of this energy is carried away by the radioactive emissions.

(7) Most radioactive emissions cause ionisation of the medium through which they are passing. Consequently, they are collectively known as ionising radiations.

(8) The most common emissions are alpha, beta, gamma and X-ray emissions.

(9) The energy of the alpha, gamma and X-ray radiation emitted by a source will be constant and is characteristic of that source. It can be used to identify that source.

(10) The beta radiation emitted by a source will have a range of energies. Values are recorded in the literature as E_{max} or E_{mean} figures for each range.

(11) The E_{max} or E_{mean} energy values recorded for beta emissions can be used to identify a beta source.

(12) K electron capture is the capture of either a K or L shell electron by a nucleus with the subsequent release of X-rays during the reorganisation of the electronic configuration of the daughter isotope.

(13) Most radioactive emissions cause ionisation of some of the medium through which they pass.

(14) The penetrating power of the common radioactive emissions are generally in the order: gamma rays > X-rays > beta particles > alpha particles.

(15) The health hazard posed to humans by ionising radiation depends on the energy of the radiation as well as its tissue penetrating power.

(16) The activity (A) of a radionuclide is the rate at which it decays.

(17) The SI unit of activity is the Becquerel where 1 becquerel is 1 disintegration per second.

(18) The pre SI unit of activity is the Curie where 1 Curie is 3.7×10^{10} disintegrations per second.

(19) Instruments for measuring radioactivity count the number of emissions per unit time reaching the counter. Three-dimensional counts can only be determined by a statistical method which takes into account the geometry of the source and the random nature of nuclear disintegrations amongst other factors. Counters are not 100% efficient and so values obtained must be corrected for this inefficiency.

(20) Background counts are not usually significant unless the activity of the source is close to the value of the background count.

(21) Simple radionuclide decay processes exhibit first order kinetics:

$$A = - \frac{dN}{dt} = \lambda N$$

(22) The activity of a radionuclide sample after a time t is given by the expression:

$$A = A_o \, e^{-\lambda t}$$

(23) The half life of a radionuclide is the time taken for half the nuclei in a sample of a radionuclide to decay. It is a characteristic property of a radionuclide.

(24) The activity of a pure sample of a radionuclide is proportional to the mass of the radionuclide.

(25) Carriers are the non-radioactive substances that are used to dilute radionuclides to permit accurate handling of the very small quantities required for practical work.

(26) The specific activity of a radionuclide is the activity of a sample containing that radionuclide per unit of mass or volume of the sample.

(27)
$$\text{Radionchemical purity} = \frac{\text{Activity of the radionuclide in the chemical form specified on the label x 100 \%}}{\text{Total activity of the radionuclide regardless of its chemical form}}$$

(28)
$$\text{Radionuclide purity} = \frac{\text{Total activity due to the specified radionuclide x 100 \%}}{\text{Total activity of the source}}$$

(29) The passage of radiation through tissue results in the formation of ions and free radicals that can disrupt and/or alter the course of the metabolic processes occurring in that tissue. These disruptions and alterations can cause a cascade effect which magnifies the effect of the radiation.

(30) Exposure to low levels of low intensity radiation is more likely to result in genetic mutations.

(31) Exposure to high energy radiation or high doses of low energy radiation will usually kill a cell or stop it reproducing.

(32) Gamma and X-ray radiation are the most dangerous external hazards and alpha radiation is the most dangerous internal hazard to man.

(33) Somatic radiation effects are radiation effects which appear in the person irradiated.

(34) Hereditary effects are those that occur in the offspring of an adult who had been exposed to radiation.

(35) Stochastic effects are the somatic effects of radiation where there are no threshold dose values.

(36) Non-stochastic effects are somatic effects whose severity varies with the dose and where there may be a threshold dose value.

(37) Exposure is a measure of the total quantity of incident radiation striking an object or person. It is measured in Roentgens (R) where:

$$1\ R = 2.58 \times 10^{-4}\ C\ kg^{-1}$$

(38) Absorbed dose is the quantity of radiation that is absorbed by the object or body. It is measured in either the SI unit, the gray (Gy) or the Rad where:

$$1\ Rad = 100Gy = 10\ Jkg^{-1}$$

(39) The biological effects of radiation depend on the type of radiation, its energy and dose.

(40) Dose equivalent is used to compare quantitatively the biological effects of different types of radiation. It is recorded in either the SI unit the sievert or the Rem where 1 Rem = 100 Sv.

(41) The same biological effect will be caused by the same dose equivalents of different types of radiation.

(42) Dose equivalent is calculated from the absorbed dose by the relationships:

Dose equivalent (Sv) = absorbed dose (Gy) x Q x N

Dose equivalent (Rem) = absorbed dose (Rad) x Q

where Q is an empirical quality factor and N is an additional quality factor that currently has been given a value of 1 by the ICRP.

(43) Radiopharmaceuticals are preparations that contain a radionuclide. They may take the form of an element, a compound, a complex or a biological preparation.

(44) Radionuclides are used as diagnostic aids by either following the progress of a radioactive metabolite through a biological system or monitoring blood, faeces and urine samples.

(45) Radionuclides are used to monitor the function of organs by gamma imaging and PET scans.

(46) Radiation is used to treat a number of diseases including cancer. The radiation may be from an external or an internal source. Internal sources may be inserted into the desired area in the form of hollow needles or capsules or as part of a preparation that will target the desired area.

(47) Gamma imaging is used to check the efficiency of drug delivery systems.

(48) Radionuclides are used in radioimmuno and related assays to study the *in vitro* behaviour of antigens and drugs in biological systems.

(49) Gamma and X-ray radiation are used to sterilise some pharmaceutical products.

33.14   QUESTIONS

(1) Distinguish between alpha, beta and gamma radiation and outline the sources of these emissions and their effect on the nucleus of the parent isotope.

(2) Complete the following nuclear equations. Identify the daughter isotope formed in each case.

(a) $^{239}_{94}Pu \longrightarrow ^{4}_{2}He + ?$,

(b) $^{14}_{6}C \longrightarrow ^{0}_{-1}e + ?$,

(c) $^{60}_{27}Co \longrightarrow ^{0}_{-1}e + \gamma + ?$,

(d) $^{37}_{18}Ar \longrightarrow \gamma + _{17}Z$?

(3) Explain the meaning of the terms *activity*, *specific activity* and *carrier* as used in the context of radioactivity.

(4) A sample of technetium-99m has an activity of 20.7 MBq. Calculate the activity of the same sample after 72 hours. The half life of technetium-99m is 6 hours.

(5) The specific activity of a solution of iodine-131 is 20 MBq cm^{-3} at 9 pm on the 10th of January, 1995. Calculate the volume required to give a patient a dose of 50 MBq at 1pm on the following day. Iodine-131 has a half life of 8.4 days.

(6) List the information that should be included on the label of an open radioactive source.

(7) The label on a low energy beta source contains the following information: activity, 2.7 MBq at 2 pm on the 1st of March , 1995; $t_{1/2}$, 100 days; E_{max}, 0.066MeV. Calculate (a) the activity of the source at 2 pm on the 10th of March, 1995 and (b) the approximate value of E_{mean}.

(8) The radiation counter used to measure gamma activity in an iron-59 metabolism investigation was 15% efficient. It registered a count of 842 disintegrations per minute from a 5 cm^{3} sample of the iron solution used for the investigation. Calculate (a) the specific activity of the iron solution and (b) the mass of iron-59 in the 5 cm^{3} sample. ($t_{1/2} = 90$ minutes and Avogadro's number $= 6.022 \times 10^{23}$.)

(9) Explain the meaning of the terms: (a) somatic, (b) stochastic, (c) nonstochastic and (d) hereditary radiation effects.

(10) A mixture of K^{123}I and ^{123}I has a specific activity of 25 MBq g^{-1}. 1.267 g of this mixture were dissolved in water and the solution diluted to 100 cm^{3} .

(a)   What volume of this aqueous solution would be required to give a patient a dose of 3 MBq?

(b)   All the iodine-123 in 25 cm^{3} of the aqueous solution was extracted into chloroform. The chloroform extract was found to have a count of 0.25 MBq using a counter of 12% efficiency. Calculate the radiochemical purity of the preparation with respect to iodine-123.

Selected further reading

Analytical methods
Clarke's Isolation and Identification of Drugs, Second Edition, The Pharmaceutical Press.
D A Skoog, D M West and F J Holler, *Fundamentals of Analytical Chemistry*, Sixth Edition, Saunders College Publishing.
R V Smith, *Textbook of Biopharmaceutic Analysis*, Lea and Febiger.
W Kemp, *Qualitative Organic Analysis*, Second Edition, McGraw-Hill.
K A Conners, *A Textbook of Pharmaceutical Analysis*, John Wiley and Sons.

Biochemistry
L Stryer, *Biochemistry*, Second Edition, W H Freeman and Company.
D Voet and J G Voet, *Biochemistry*, John Wiley and Sons.
T Nogrady, *Medicinal Chemistry, a Biochemical Approach*, Oxford University Press.

Biomedical science
A H Gowenlock, *Varley's Practical Clinical Biochemistry*, Sixth Edition, Heinemann Medical Books.
J F Zilva and P R Pannall, *Clinical Chemistry in Diagnosis and Treatment*, Lloyd-Luke Medical Books Ltd.
L G Whitby, A F Smith and G J Beckett, *Lecture Notes on Clinical Chemistry*, Blackwell Scientific Publications.

Inorganic chemistry
J A Duffy, *General Inorganic Chemistry*, Longman.
W E Addison, *Structural Principles in Inorganic Chemistry*, Longman.
C A Discher, *Modern Inorganic Pharmaceutical Chemistry*, John Wiley and Sons.
K M Mackay and R A Mackay, *Introduction to Modern Inorganic Chemistry*, Blackie.
F A Cotton and G Wilkinson, *Basic Inorganic Chemistry*, John Wiley and Sons.

Natural products
D T Elmore, *Peptides and Proteins*, Cambridge University Press.
M I Gurr and J L.Harwood, *Lipid Biochemistry*, Fourth Edition, Chapman and Hall.
J J Vitale, *Vitamins*, UpJohn Scope.
G E Trease and W C Evans, *Pharmacognosy*, Thirteenth Edition, Ballière Tindall.

Nomenclature
R S Cahn, *An Introduction to Chemical Nomenclature*, Butterworth.
Handbook for Chemical Society Authors, Royal Chemical Society of Great Britain.
P Fresenius, *Organic Chemical Nomenclature*, Ellis Horwood.

Organic chemistry
J McMurry, *Organic Chemistry*, Brooks/Cole Publishing Company.
R Barker, *Organic Chemistry of Biological Compounds*, Prentice Hall.
R B Silverman, *The Organic Chemistry of Drug Design and Action*, Academic Press.
K E Suckling and C J Suckling, *Biological Chemistry*, Cambridge University Press.
M Hudlicky, *Oxidations in Organic Chemistry*, American Chemical Society.
M Hudlicky, *Reductions in Organic Chemistry*, Ellis Horwood.
A Albert, *Selective Toxicity*, Seventh Edition, Chapman and Hall.

Pharmacy
M E Aulton, *Pharmaceutics*, Churchill Livingstone.
British National Formulary, Published annually, British Medical Association and the Royal Pharmaceutical Society.
Martindale, the Extra Pharmacopoeia, Thirtieth Edition, Pharmaceutical Press.

Physical chemistry
P W Atkins, *Physical Chemistry*, Third Edition, Oxford University Press.
L Saunders, *Principles of Physical Chemistry for Biology and Pharmacy*, Second Edition, Oxford University Press.
J G Morris, *A Biologists Physical Chemistry*, Second Edition, Edward Arnold.

Physical and chemical data
Merck Index, Eleventh Edition, Merck and Co.
Heilbron and Bunbury, *Dictionary of Organic Compounds*, Eyre and Spottiswood.
Handbook of Chemistry and Physics, Sixty-second Edition, CRC Press.

Reaction mechanisms
P Sykes, *A Guidebook to Mechanism in Organic Chemistry*, Sixth Edition, Longman.
J. March, *Advanced Organic Chemistry, Reactions, Mechanisms and Structure*, Third Edition, John Wiley and Sons.
B Halliwell and J Gutteridge, *Free Radicals in Biology and Medicine*, Clarendon Press.

Stereochemistry
G Hallas, *Organic Stereochemistry*, McGraw-Hill.
H Kagen, *Organic Stereochemistry*, Edward Arnold.

Spectroscopy
D H Williams and I Fleming, *Spectroscopic Methods in Organic Chemistry*, Fourth Edition, McGraw-Hill.
A D Cross, *Introduction to Practical Infra-red Spectroscopy*, Butterworth.

Radioactivity
D Billington, G G Jayson and P J Maltby, *Radioisotopes, Introduction to Biotechniques*, Bios Scientiffic Publishers.
W J Gorey, *Radiochemical Methods*, John Wiley and Sons.

Answers to selected questions

Answers which require you to write an equation are given in terms of the names of the principal compounds formed in the reaction. This is to give you additional experience in converting the chemical names of compounds into formulae using the appropriate nomenclature sections of the text. For the answers to essay-style questions you are either referred to the appropriate section of the text or, where no answer is given, you should look up the key word(s) in the index to find the information you require. Where questions have more than one correct answer only one answer is given.

Chapter 2
(1) (a) Mg 8, (b) C 8, (c) O 8. (2) (a) zero, (b) +, (c) zero, (d) -, (e) 2-, (f) +.
(3) (a) and (b) all the carbon atoms. (4) (a) N, C and O are all sp³, (b) Pt dsp², Cl sp³, (c) C and O are sp², (d) C sp³, N sp².
(5) Draw using the appropriate angles for the hybridised atoms in each structure; sp³ 109.5°, sp² 120° and dsp² 90°.
(6) Structures with conjugated systems are:

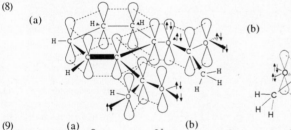

(7) (a) Fig. 2.17, p.24
     (b) Fig. 25.4, p.424
     (c) The oxygen atoms lone pairs are omited for clarity.

(8) (a)

(b)

(9) (a)

(b)

Chapter 3
(1) c, d and f. (2) (a)

(b)

(c)

(d)

(3) No.
(4) (a) Intramolecular hydrogen bonding stabilises the *o*-hydroxybenzoate ion more than intermolecular hydrogen bonding stabilises the *m*-hydroxybenzoate ion.

(4)   (b) Compounds that are intermolecularly hydrogen bonded will have higher melting points because extra energy will be needed to break these bonds.

   (c) The complex is formed because charge transfer bonding occurs between iodine and alkenes.

Chapter 4

(1)   (a) Replace one of the hydrogen atoms in Fig. 4.3 (p.45) by a methyl group. However, the energy values given on this diagram do not apply to propane.

   (b) Replace the methyl groups in Fig. 4.4 (p.46) by chlorine atoms.

(2)   (a) C_4H_9 (structure)   (b)   (structure)   The bromines are furthest apart when they are in a staggered conformation (make a model).

(3)   To check your answers convert any structures you have drawn to the one shown here by an **even number** of interchanges of two groups.

   (a)   CHO / HO—CH$_3$ / H
   (b)   COOH / CH$_3$—OH / H
   (c)   COOH / Cl—CH$_2$CH$_3$ / H
   (d)   CH$_2$—ring / CH$_3$—O—ring / H

(4)   a and b are the same; c, d and e are the same.

(5)   (a) D, (b) D, (c) D, (d) L.   (6)   (a) S, (b) S, (c) R, (d) S, (e) R, (f) S.

Chapter 5

(1)   (a)   (structures) Enantiomers   (b)   (structures) Enantiomers / Meso isomer

   (c)   (structures) Meso isomer / Enantiomers

   (2)   Centre of symmetry: a, b and e.
   Plane of symmetry: c, d and e.
   No plane or centre of symmetry: f.

(3)   Optical activity: c and h. Enantiomers c and h. Meso compounds: a, d and g. Geometric isomers, b, f and e.

(4)   (a)   COOH / H—OCH$_3$ / CH$_3$ COOH / CH$_3$O—H / CH$_3$ Enantiomers
   (b)   (structures) Enantiomers / Meso isomer
   (c)   CH$_3$ / Cl—CH=CH$_2$ / H   CH$_3$ / CH$_2$=CH—Cl / H   Enantiomers
   (d)   (structures) Geometric isomers

   (e)   CH$_3$ / Cl—C—Cl / H **A**   CH$_3$ / C—Cl / H **B**   (structures) **C**   **D**
   Enantiomeric pairs, A/B; C/D.
   Diastereoisomers (bracketed) (ACD),(BCD),(CBA) and (DAB).

(5)   The method is the same as that given in 5.6 except the diastereoisomers are the 2-butylammonium tartrate salts which may be separated by fractional crystallisation.

(6)   A is probably the trans isomer and B is the cis isomer. This is because the trans isomer *usually* has a higher melting point, a lower boiling point and a lower refractive index than the cis isomer.

(7)   (I)   (structure)   (II)   (structure)   (I)   No activity
   (II)   Possibly an anticancer agent as it has a similar shape and structure to cisplatin

Chapter 6

(1)   (a) First order, (b) third order, (c) (i) second order, (ii) pseudo first order and (d) pseudo first order.

(2)   (a) First order, (b) second order, Rate $=k[A][B]^2$ (3) 23.2 seconds. Clue: 6.3.2 and 6.5.

(4)   8.34 Hours. Clue: 6.3.2.   (5) k is $8.33 \times 10^{-5} s^{-1}$, half life is 10 minutes.

(6)   (a) $K_c = \dfrac{[H_2][I_2]}{[HI]^2}$ $K_{eq} = \dfrac{(H_2)(I_2)}{(HI)^2}$ (b) $K_c = \dfrac{[H_2]^4}{[H_2O]^4}$ $K_{eq} = \dfrac{(H_2)^4}{(H_2O)^4}$

(6) (c) $K_c = \dfrac{[\text{malate}]}{[\text{fumarate}]\,[H_2O]}$ $K_{eq} = \dfrac{(\text{malate})}{(\text{fumarate})\,(H_2O)}$

(7) (a) Yes, (b) No, not a gaseous system, (c) Yes.   (8) $K_{eq} = 4$ (no units).

(9)   Tris : TrisH$^+$ is 8.3 : 1.   (10)   Sodium ethanoate : ethanoic acid is 2.56 : 1.

(11)   +16 kJ mol^{-1}. Clue: 6.18.   (12)   (a) 0.035 kJ mol^{-1}, (b) 0.038 kJ mol^{-1}. Clue: 6.19.1.

(13)   2.31

Chapter 7

(1)   a, d and f are free radicals.   (2)   (a)   Shows the presence of canonical forms.

(3)   (a) $CH_3 - CH_3$ (b)   Move one electron.

     (b) $CH_4 + \cdot CH_2 - CH_3$ (c)   Thermodynamically reversible reaction.

     (c) $Cl - CH_2 - \overset{\cdot}{C}H_2$ (d)   Move two electrons.

     (d) $RCH = CH_2 + \cdot Cl$

(4)   (a)   See p.114, replace Fe^{2+} with Cu$^+$ and Fe^{3+} with Cu^{2+}.

     (b)   See Fig. 7.13, replace R by CH$_3$

Chapter 8

(1)   Electrophiles: c and e. Nucleophiles: a, b, d and f.

(2)   (a) Canonical forms. (b) Inductive effect. (c) Move two electrons. (d) Dative bond or reaction in this direction.

(3)   (a)   See Fig. 8.8. Y = NH$_2$, the lone pair acting as the start of the mesomeric effect transmission. (b) See Fig. 8.6. Replace the H of the CHO group by $CH_3 \rightarrow$.

     (c)

(4)   (a) $CH_3CH_2 \overset{+}{\underset{CH_3}{\diagdown}}CH + Cl^-$ (b) $CH_3CH_2CH_2\overset{+}{\underset{H}{O}}H$ (c) $CH_3\overset{OH}{\underset{\overset{+}{O}-CH_3}{\diagup}}CH$ (d)

(5)   (a) See Fig. 8.9 and Fig. 8.12. Replace ethene by propene.

     (b)

     (c)   See Fig. 8.14, first two examples.

     (6)   Section 8.6.3.

Chapter 9

(1)   See pp.138-139. The presence of groups that are able to hydrogen bond and ionise in water will increase water solubility. The absence of these groups and the presence of non-polar groups will decrease water solubility and increase lipid solubility.

(2)   (a) Good, hydrogen bonding and some ionisation of the COOH .   (b) Almost insoluble or poor, insufficient hydrogen bonding from one group. (c) Good, hydrogen bonding involving the amide and amine groups.

(3)   42.25 cm^{-3}.   (4) (a) Convert to a suitable salt, eg. tartrate or hydrochloride. (b) Convert to the sodium salt.   (5) (a) Poor, low polar group to C–H skeleton ratio. (b) Better, due to salt formation by the secondary amine group. (c) Better, due to salt formation by the phenolic hydroxyl group. (6) 18.14 kPa. (9) 0.6 mol kg^{-1}.

Chapter 10

(2)   118 mmol dm^{-3}.   (3)   For difference see p.164, Fig. 10.6. About 40.2 ng cm^{-3}.

(4)   See p.169. The main spectral changes are the strong OH peak at about 3300 cm^{-1} of the ethanol disappears as the ethanal is formed. The formation of the ethanal results in the appearance of a strong C=O peak at 1660 cm^{-1}. The formation of the ethanoic acid is accompanied by the appearence of a broad peak about 3300-2800 cm^{-1}.

(5)   A is butan-2-one and B is acetophenone.   (6)   D is benzophenone and E is adipic acid.

(7)   See p.179. EI mass spectrum of compound F has two molecular ions because it contains almost equal amounts of an isotope in its structure. a = C_2H_5, b = C_3H_7, c = C_4H_9.

Chapter 11

(1)   13.26%   (2) (a) See section 11.4.1 (b) See section 11.5 (c) Ideally the reaction should be quantitative, reproducible, consistent and involve only the drug.

(3) (a) Chemical method; see Fig. 11.1. Physical method; see 11.4.2, e.g. use its infra-red spectrum.
(b) Substance (limit test): benzyl alcohol (chromatography, 11.5.3); benzoic acid (volumetric, 11.5.2); chloride ions (Nessler tube method, 11.5.1, Table 11.1) and water (weigh to constant mass, 11.3).
(c) Use the same method as is used for aspirin in Fig. 11.8.

(4) Select functional group reactions that either involve a visible change or produce a solid product (derivative). For suitable reactions look in the appropriate chapter for the chemistry of the relevant functional group.

(5) $0.01012 \, mol \, dm^{-3}$. (6) 90.62. (7) 95.08%.

(8) (a) $1 \, cm^3$ of 0.05 M perchloric acid is equivalent to $0.02713 \, g$ of A. (b) 2.28%.

(9) (a) 91.79% (b) 95.10% (10) 97.2% w/w (11) $7.6 \, \mu g \, dm^{-3}$, see p. 202.

Chapter 12

(1) (a) 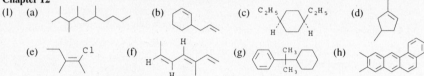 (b) (c) (d)

(e) (f) (g) (h)

(2) A is a Z isomer. (a) and (b) See Fig. 12.11, p.216. (c), (d) and (e) See p.213. Mechanisms, (c) and (d), see also Fig. 12.12. (d) This is similar to (c) and (d) but water is the nucleophile in the second step. The water loses a proton in a third step to form the final product.

(3) (a) 1,2-dibromopent-4-ene, (b) 4-chloropent-2-ene, (c) pent-1-ene.

(4)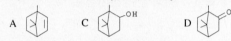

(5) Structures (b) and (d) do not contain conjugated structures.
(a) (c) (e)

(6) (a) and (d). Both obey the Huckel rule. These systems would not be very reactive. They undergo electrophilic substitution but would not readily undergo oxidation, reduction and addition.

(7) One would expect: (a) 9,10 bromine addition, (b) 9,10 diol, (c) 9,10 addition of hydrogen. Mechanism, base on the general mechanism given at the bottom of p.223.

Chapter 13

(1) (a) (b) (c) (d) (e)

(f) (g) (2) Both poor because of the large aromatic structure but improved in acid due to salt formation.

(3) (a) Base, nucleophile and oxidation. (b) Base, nucleophile, electrophilic and nucleophilic substitution and oxidisation. (c) Not very reactive. (d) Electrophilic substitution, relatively easy oxidation and reduction of the furan ring but not of the benzene ring and the furan ring may undergo a Diels-Alder reaction. (e) Similar to alkenes; oxidation, reduction and electrophilic addition.

(4) (a) See p.247, similar to the reaction of pyridine with NH_2^-.
(b)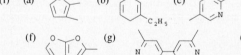

Chapter 14

(1) Fig. 14.3. Plocamene B; two identical tertiary halides and one allenyl. Drosophillin; all aryl halides. Bronopol; tertiary alkyl halide. Chlormethathiozole, primary alkyl halide. Fig. 14.4. 2-bromobutane; secondary alkyl halide. 1,2-dichloroethene; both alkenyl halides. 2,3-dichloropyridine; both are aryl halides. 3-bromoethylbenzene; aryl halide.

(2) Main products are, (a) $PhCH_2OCH_2CH_3$, (b) $PhCH=CH_2$, (c) $HOCH_2CH_2CH_2CH_2CH_3$ and $CH_2=CHCH_2CH_2CH_3$ (d) main product $CH_3CH=CHCH_2CH_3$ and minor product $CH_2=CHCH_2CH_2CH_3$

(3) (a) Allyl halide > aryl halide ~ alkenyl halide. (b) Secondary alkyl iodide > secondary alkyl chloride.
(c) Tertiary bromide (also a benzyl bromide) > secondary bromide (also an allyl bromide) > alkenyl bromide.
(4) (a) S_N2. (b) E1 or E2. (c) S_N2Ar. (d) E2.
(5) DNP formation followed by hydrolysis and TLC to show that the two isomers produce different products.
(6) In 2,4,6-trinitrochlorobenzene all three nitro groups withdraw electrons from the benzene ring weakening the
      C–Cl bond and so making it more reactive. There is no weakening of the C–Cl bond in chlorobenzene and so
      it is less reactive.

Chapter 15
(1) Primary: b; secondary: a, c, f, and g; tertiary: i; phenols: d, e and h.
(2) (a) B, G and K. (b) B , G and K. (c) B and K. (d) B, C, E, G and J.
(3) 2,4,6-trinitrophenol > 4-nitrophhenol > 4-methylphenol. See pp.279-281.
(4) (a) Secondary. (b) ⸌ Bond going away from the observer, ╱ bond coming towards the observer.
      (c) Chair and boat. (d) R. (e) (i) the ester; $CH_3COCOOR'$ where R' is the menthol residue.
      (ii)   (iii)   and

(5) (a) CH_2OH, primary alcohol and the other OH is a phenol. Antiseptic action by the phenol. (b) soluble in
      both water and dilute aqueous acid but less so in the latter. More soluble in dilute aqueous alkali due to salt
      formation. (c) See 15.11.1 and p.284. (d) See 15.10 and p.284. (e)(i) p.282, (ii) Table 15.3, (iii) Table 15.3.

Chapter 16
(1) (a) $HSCH_2CH_2SH$ (d) $(CH_3COS)_2^- Ca^{2+}$ (f)   (g)
      (b) $HSCH_2COOH$ (e)
      (c) $CH_3CH_2CH_2CH_2SH$

(2) (a) Phenyl 3,5-dinitrothiobenzoate. (b) 2,4-Dinitrophenylthiomethane. (c) Propyldithiopropane.
      (d) Cyclohexanone ethylhemimercaptal. (e) 2,2-Dimethylenedithiopropane.
(3) Most logical possibilities are: (2a) S_N2 type of mechanism (2b) S_N2Ar. (2d) See 19.9.4 and Fig. 19.7.
(4) Identification tests, see 16.6.1 and 16.6.2. Assay see p.299. Heat with NaOH or $NaOC_2H_5$ to produce sodium
      bromide from the ethyl bromide contaminent. Acidify the reaction mixture with nitric acid and add silver nitrate
      to precipitate cream-coloured silver bromide. Use the Nessler technique for the limit test (11.5.1).

Chapter 17
(1) Eugenol, CH_3O- ether. Safrole, $-O-CH_2-O-$ ethers. Ethacrynic acid, ether. Chlorpromazine,
      thioether. Methionine, CH_3S- thioether.
(2) (a)   (b)   (c)   (d) $CH_3CH_2CH_2CH_2SCH_2CH_3$
      (e)   (f) $CH_3CH_2CH_2-S-S-CH_2CH_3$ (g)
(3) (a) Sulphur is a better nucleophile than oxygen. See 17.4.3. (b) Oxygen hydrogen bonds but sulphur is not
      able to hydrogen bond. See 17.3.
(4) (a) See 17.4.2. All the possible products are likely to be obtained. (b) Diethyl sulphoxide, 17.4.1.
      (c) 4-Dimethyl methylcyclohexyl sulphonium iodide, 17.4.3. (d) ethanethiol, 17.5.
(5) (a) Oxidation of the sulphide to the sulphoxide or sulphone. (b) Reductive cleavage of the disulphide to form
      the 3,5-dimercaptopentanoic acid.

Chapter 18
(1) (a)   (b)   (c)   (d)
(2) (a) Poor in water because of the high hydrocarbon skeleton to polar group ratio. Improved in hydrochloric
      acid due to salt formation but probably the same as water in sodium hydroxide solution. Soluble in DMSO.

(b)(i) See 18.4.2. (ii) The sulphoxide is reduced to the sulphide and if the conditions are vigorous enough the benzene ring is reduced to a cyclohexane ring. (iii) The sulphoxide is oxidised to the sulphone

Chapter 19

(1) (a) [structure with CHO and C_2H_5] (b) [cyclohexanone structure with CH_3] (c) [pyridine with CHO, N, CH_3] (d) [benzene with CHO, OCH_3, OH] (e) $CH_2=CH-CHO$ (i) [benzoquinone with Ph]

(f) $CH_3COCH_2COCH_3$ (g) $PhCOCOPh$

(j) [naphthoquinone with CH_3]   (k) [anthraquinone with OH and OH]   (h) $CH_3CH_2COCH_2CH_3$

(l) $CH_3CHCH_2CSCH_3$ (with CH_3 substituent)   (m) $CH_3CHCSCH_3$ (with CH_3 substituent)

(2)(a) [benzene with CHO, CH_3, CH_3] The ring and the aldehyde group are flat, the methyls are tetrahedral. (b) Oxidation, reduction, nucleophilic addition, polymerisation and condensation. (c) See p.327, p.328 and p.329. (d) See p.328.

(3) (a) Reduction p.320, hexan-2-ol. (b) Tautomerism p.317, the corresponding enol and hydration to the hydrate p.324. (c) Nucleophilic addition p.324, the methyl hemi-acetal. (d) Reduction p.320, hexane. (e) Condensation p.326, hexa-2-one hydrazone. (f) Condensation p.331, ethyl 3-methylhept-2-enoate. (g) Condensation p.328, hexa-2-one oxime. (h) Condensation p.329, 19.11.4, N-ethylhexa-2-imine.

(4) (a) [structure: CH_3CH_2, C, H, OH, $O-C_2H_5$]   (b) [pyran ring with OH, H]   (c) [structure: CH_3, CH_3, C, OH, $O-C_3H_7$]

(5) (a) Nucleophilic addition, mechanism p322. (b) Condensation, mechanism acid catalysed nucleophilic addition followed by an acid catalysed elimination, see 15.9.1, p.288.

(6) (a) A, 3-phenylpropanal, B, 3,3-dimethylcyclohexanone C, 2-methyl-1,4-benzoquinone. (b) A will give a silver mirror with Tollens's reagent, B and C do not form siver mirrors with Tollens's reagent. C will react with benzenediazonium chloride to form a dye, A and B will not react.

(c) (i) Both A and B forms the optically active hemi-acetals. (ii) A probably forms the E hydrazone isomer but B probably forms a mixture of the E and Z hydrazones.

Chapter 20

(1) (a) [cyclohexane with COOH, NH_2] (b) [COOH] (c) [pyridine with COOH, N, COOH] (d) $CH\equiv CHCOOH$ (as $\begin{smallmatrix}COOH\\COOH\end{smallmatrix}$) (e) [cis alkene: H, COOH, CH_3, H] (f) [alkene: CH_3, COOH, H, CH_3]

(g) [benzene with OH, C, SH, S] (h) CH_3CH_2-C with O and SH (i) [benzene with CH_2-P-OH, O, OH]

(j) CH_3—[benzene]—S—OH (with O, O)   (k) H_2N—[benzene]—P—OH (with O, H)   (l) [prostaglandin-like structure: HO, H, COOH, CH_3, O, H, OH]

(2) (a) See Table 20.1. (b) Temperature, solvent, the strength of the O–H or S–H bond and the stability of the conjugate base. (c) pK_a, See 6.13.1. (d) D > C > A > B.

(3) PhCH=CHCOOH, Exhibits cis and trans geometric isomerism. (a) Oxidation to the diol PhCH(OH)CH(OH)COOH. (b) Nucleophilic displacement, ethyl cinnamate. (c) Acid, carbon dioxide and calcium cinnamate. (d) Nucleophilic displacement, cinnamoyl chloride. (e) Electrophilic substitution, 3-nitrocinnamic acid.

(4) They are all water soluble but ethanethiolic and dithioethanoic acids are unstable (p.345). All are acids, undergo nucleophilic displacement and act as electron acceptors. In addition (b) and (c) are oxidised to diacylsulphides.

(5) Propylphosphinic acid would be expected to acts as an acid, nucleophilic displacement an electron acceptor, and a reducing agent. (a) Carbon dioxide and sodium propylphosphinate. (b) Silver mirror formed. (c) Ethyl propylphosphinate.

Chapter 21

(1) (a) $CH_3CH_2COOCH_2CH_3$ (b) $CH_3CH_2\overset{O}{\underset{O}{S}}-O-CH_2CH_3$ (c) $CH_3CH_2CH_2-O-\overset{O}{\underset{OH}{P}}-OH$ (d)

(e) $CH_3CH_2CH_2-O-\overset{O}{\underset{CH_2CH_3}{P}}-OH$ (f) $CH_3CH_2-O-\overset{O}{\underset{O}{S}}-OH$ (g) (h)

(2) (a) The benzene ring; electrophilic substitution. The ester group; reduction, react with nucleophiles (p364) and the formation of carboanion under basic conditions. This carboanion could undergo nucleophilic substitution, addition and condensation. (b) Poor water solubility but good lipid solubility. (c) (i) Hydrolysis to methanol and sodium 3-phenylpropanate. (ii) Ethyl 3-phenylpropanoate and methyl ethanoate. (iii) Methyl 2-methyl-3-phenylpropanoate (iv) Methyl 2-benzyl-3-methylbut-2-enoate (v) Ethyl 3-phenylpropanoate.

(3) Similarities: nucleophilic displacement, p.365 and p.368. Differences: sulphur oxy-acid esters act as electron acceptors, phosphorus oxy-acid esters do not, phosphorus oxy-acid esters act as leaving groups in biological elimination reactions, sulphur oxy-acids do not.

(4) (a) See Fig. 21.7, p365, route a. (b) See Fig. 21.7, p365, route b.

(5) (a) A, thiocarboxylic acid ester; B, lactone; C, sulphonate. A and C will slowly decompose in both (i) and (ii) at room temperature. B is stable in (i) but will decompose in (ii). (b) A, hydrolysis to benzoic acid (identify by mp) and ethanol (identify by reacting with ethanoic acid to give ethyl ethanoate with its fruity smell). B, hydrolysis to 5-hydroxypentanoic acid. Identify the acid by TLC or measurement of one of its physical constants such as mp. (c) Hydrolysis with excess sodium hydroxide and back titration of the excess sodium hydroxide with hydrochloric acid. (d) See p.365 and p.366.

(6) Step A is a carboanion nucleophilic displacement, see Figs 21.7 and 21.10. Step B, tautomerism.

Chapter 22

(1) (a) $\underset{CH_3CHCOOH}{\overset{CN}{|}}$ (b) $NCCH_2CN$ (c) $\underset{CH_3CH_2CHCH_3}{\overset{NO_2}{|}}$ (d) (e)

(2) (a) and (c). $\underset{R \quad H}{\overset{COOH}{CH_3+CN}}$ $\underset{S \quad H}{\overset{COOH}{NC+CH_3}}$ $\underset{R \quad H}{\overset{CH_3}{CH_3CH_2+NO_2}}$ $\underset{S \quad H}{\overset{CH_3}{O_2N+CH_2CH_3}}$

(3) Reduction and as an electron acceptor. (a) 3-Nitrohexane. (b) 1-Phenyl-2-nitro-1-pentene. (c) Butyl amine. (d) Potassium butylnitrolic acid.

(4) Reduction, reaction with alcohols, as an electron acceptor and hydrolysis. (a) Sodium benzoate. (b) Benzyl amine. (c) 3-nitrobenzonitrile.

(5) (a) (A) Reduction to the aromatic amine followed by the diazo test for aromatic amines (24.9.2). (B) Hydrolysis with aqueous sodium hydroxide to benzoic acid, a white solid which can be identified by its mp. (b) Measure the boiling point. (c) Treat with aqueous sodium hydroxide. A does not react. B forms benzoic acid, see 5a.

Chapter 23

(1) (a) Carbonate. (b) Amidine. (c) Ureide. (d) Dithiocarbamic acid.

(2) (a) Strong base. (b) Weak base, nucleophile, oxidation, hydrolysis and the biuret reaction. (c) Base, hydrolysis and complex formation with transition metal ions.

(3) (a) Diethylurea. (b) Ethyl acetoacetate, tautomerism to the enol: (c) Diethylthiuram disulphide. (d) Thiobarbituric acid. (e) Methylammonium chloride, ammonium chloride and carbon dioxide. (f) Guanidine ethanoate. (h) Ethylamidine.

(4) (a) React with picric acid to from the picrate and measure its mp. (b) Titrate with hydrochloric acid (the guanidine group is a strong base). (c) Heat to form the corresponding substituted biuret and treat with copper II sulphate to form a purple complex. Compare the complex formed with that formed from a standard solution (11.5.1).

(5) B > A. The lone pair of the N=H group of B is not conjugated whereas the lone pair of the NH_2 groups of both A and B are conjugated and so are not so readily available to act as a bases.

Chapter 24

(1) (a) $H_2NCH_2(CH_2)_4CH_2NH_2$ (b) (c) $\underset{CH_3}{\overset{C_2H_5}{C_2H_5-N-C_2H_5}}$ (d)

(1) (e) (f) PhNHPh   (h) (i)

(g) $C_3H_7NHC_2H_5$

(2) Primary, a, b, i (-NH$_2$); secondary, d chain N, g, f, i (-NHCH$_3$), j; tertiary, c, d ring N, e, h.

(3) (a) 2,2-Dimethylpropylamine. (b) N-Ethyl-N-methylcyclohexylamine. (c) 2,4-Diaminopyridine.
(d) 2-Dimethylaminobutane.

(4) (a) C > A > B. (b) C > A > B. (c) C > B > A.

(5) See Fig. 24.21, replace isoprenaline sulphate by amphetamine sulphate. (a) 0.0178 mol. (b) 34.95% of $C_9H_{13}N$.

(6) Both are tertiary amines. Both act as bases and nucleophiles. The pyridine N is a strong electron acceptor which could result in the pyridine ring undergoing both nucleophilic and electrophilic substitution.

(7) (a) Piperidinium chloride. (b) N-Methylpiperidinium iodide and N,N-Dimethylpiperidinium iodide.

Chapter 25

(1) (a) $CH_3CH_2CH_2CONH_2$ (b) CH_3CONH | $CH_3CH_2CHCH_3$ (c) (d)

(e) H_2N— —SO_2NH_2

(f) H_2N— —SO_2NH—

(g) (h)

(2) (a) See p.426 (i) The lone pair of the nitrogen atom is delocalised and so is not readily available to act as a base. (ii) The N–H bond is weak because of the electron acceptor affect of the C=O group. (b) Poor. (c) The carboxamide; weak base and acid, hydrolysis, and reduction. The benzene ring; electrophilic substitution.. The methyl group; oxidation. (d) (i) propanoic acid and 2-methylanilinium chloride. (ii) N-(2-methylphenyl)propylamine.

(3) (a) Hot NaOH, see p.429, HNO$_2$, see p.430. (b) See p.428. (c) Titration with sodium hydroxide. (d) Keep dry to reduce the risk of hydrolysis by atmospheric moisture.

(4) (a) The N–H bond is weak because of the electron acceptor affects of the two C=O groups. (b) The sodium salt is more soluble in water than the parent compound but has the same pharmacological action. (c) The β lactam ring is easily hydrolysed in aqueous solution.

(5) (a) See Fig. 25.8. p.428. (b) S$_N$2 nucleophilic displacement. See p.263. The amine acts as the nucleophile displacing ⁻OCOCH$_3$ from the anhydride. (c) S$_N$1. See p.261. The nucleophile is the phthalimide anion.

Chapter 26

(1) (a) Fig. 26.3 p.437. (b) Fig. 26.7 p.442. (c) Fig. 26.9 p.444. (d) Fig. 26.8 p.443.

(2) (a) (i) The formation of acrolein p.439. (ii) Hydrolysis with NaOH and identification of the fatty acids produced by TLC. Also see Fig. 26.18. (b) Saponification with excess NaOH and back titration of the excess NaOH with hydrochloric acid.

(3) (a) Secondary alcohol. (b) Nucleophile, oxidation, source of carbocations leading to elimination, reaction with nucleophiles and rearrangements reactions.

Chapter 27

(1) (a) D (b) D (c) L (d) D.   (2) (a) β (b) α (c) β (d) α.   (3) (a) β (b) α, β (c) α.

(4) See p.467. Primary amine, primary alcohol, secondary alcohol, hemiacetal and aldehyde. Primary amine: base, nucleophile, oxidation and source of carbocations; both alcohol groups: nucleophile, oxidation and a source of carbocations; hemiacetal: hydrolysis; aldehyde: oxidation, reduction, nucleophilic addition and condensation. (a) Forms a mixture of the N-methyl, N-dimethyl and N-trimethylammonium derivatives, (p.408). (b) The ethanoate salt of the primary amine, (p.401). (c) The oxime of the aldehyde group, (Fig. 27.15, p.465). (d) Nitrogen (from the primary amine) and a variety of products, see p.4.11. (e) All the hydroxyl groups and the amino group react to form the fully saturated O-ethanoyl (p.465) and N-ethanoyl (p.409) derivative.

(5) See 27.5.2.   (6)

Glucose residue   Glalactose residue

Chapter 28

(1) S-Serine   D-Cysteine

$$HOCH_2 \overset{COOH}{\underset{H}{\rule{0pt}{1.6em}}} NH_2 \qquad H \overset{COOH}{\underset{CH_2SH}{\rule{0pt}{1.6em}}} NH_2$$

(2) See p.480.   (3) Carboxylic acid (COOH: acid, nucleophilic displacement, reduction, decarboxylation), primary amine (NH_2: base, nucleophile, oxidation and source of carbocations) and a primary alcohol (CH_2OH: nucleophile, oxidation, elimination and reaction with nucleophiles under acid conditions).

(4) (a) R and D (b) S and L (c) S and L (d) R and L.

(5) Forms: (a) the sodium salt with the liberation of carbon dioxide, (b) yellow N-2,4-dinitrophenylglycine, (c) ethyl glycinate, (d) sodium 2-benzamidoethanoate and (e) ethanolamine.   (6) See p.483.

(7)

$$\underset{\text{Alanylglycyllysylserine}}{NH_2CHCONHCH_2CONHCHCONHCHCOOH} \quad \overset{CH_3 \quad H_2N(CH_2)_3CH_2 \quad CH_2OH}{}$$

$$\underset{\text{Alanylserylarginyllysyllysylglycine}}{NH_2CHCONHCHCONHCHCONHCHCONHCH_2COOH}$$

with side chains:
$$\overset{H_2N-C=NH}{\underset{NH}{\rule{0pt}{1em}}}$$
CH_3, CH_2OH, $CH_2(CH_2)_2NH$, $CH_2NH_2(CH_2)_2$, $CH_2NH_2(CH_2)_2$

(8) See pages 485-488.   (9) (NH_2)Gly-Val-Glu-Gly(COOH)

(10) (NH_2)Gly-Gly-Ala(COOH) (NH_2)Gly-Ala-Gly(COOH) (NH_2)Ala-Gly-Gly(COOH).
Use one of the sequencing methods described on pages 493 and 494.

(11) Action is prolonged because hydrolysis of the tyrosine-D-alanine peptide link is slower. This lower rate occurs because the change in structure means that the synthetic met-enkephalin does not fit the active site of the relevant enzyme as well as the natural compound. The fact that a D configuration residue is used also reduces the rate of hydrolysis because the residues found in naturally occurring met-enkephalin have L configurations.

(13) (NH_2)Ala-Gly-Gly-Gly(COOH)

Chapte 29

(1) (a) The sugar deoxyribose and the bases adenine, guanine, cytosine and thymine (see p.505). (b) Ribose (see p.505) and uracil (see p.509) instead of thymine.

(3) Charagaff's rule (see p.505). Describe the double helix and the significance of the hydrogen bonding of the bases (see pp.506-507). AATCGAT.

(6) Use the same basic method as DNA (see pp.511-513). UACAGGAUUGCCACCGAA.

Chapter 30

(2) To keep the cell healthy. The sodium pump. See pp.521-522.

(3) Magnesium (p.523), calcium (p.524). Low plasma calcium levels result in reduced urinary excretion and an increased movement of calcium from the bones to the plasma. The reverse is true for high calcium plasma levels.

(6) Examples: mercury; ethyl and methylmercury in fish; mercury vapour in the atmosphere. Lead: waterpipes; paint and leaded petrols.

(7) Fluorine: stabilisation of bone and teeth. Chlorine: hydrochloric acid production in the stomach and as a balancing ion in body fluids. Iodine: production of thyroid hormones.

9) Magnesium carbonate: antacid and laxative; bismuth chelate: protective agent for the gastrointestinal tract; calcium carbonate, antacid and to counter the laxative effects of magnesium carbonate; starch, dissolution agent.

Chapter 31

(2) Because pyroxenes chains have a narrower cross-section.

(6) The alkaline nature of the glass would liberate the free base and form the metallic salts of the acid. This could possibility affect the absorption and mode of action of the drug.

Chapter 32

(1) (a) Ammonia, 4. (b) Carbon monoxide, 6. (c) Ammonia and chlorine, 4. (d) Haem, 5.

(2) (a) Hexaaquacobalt III chloride. (b) Dibromodiammineplatinium II.
(c) Nitropentaamminecobalt II sulphate. (d) Potassium hexacyanoplatinate II.

(3) (a) $[Co(H_2O)_6]Cl_2$ (b) $Na_2[CuBr_4]$ (c) $[CoH_2O(NH_3)_5]Cl_3$ (d) $[Cu(NH_3)_4][PtCl_4]$.

(4)   (a) $[Co(NH_3)_5SO_4]NO_2$ and $[Co(NH_3)_5ONO]SO_4$.

   (b) $[Co(NH_3)_5Cl]BrCl$, $[Co(NH_3)_4Cl_2]Br\cdot NH_3$, $[Co(NH_3)_4BrCl]Cl\cdot NH_3$ and $[Co(NH_3)_3BrCl_2]\cdot 2NH_3$.

   (c) $[Cr(H_2O)_4Cl_2]Cl\cdot 2H_2O$, $[Cr(H_2O)_6]Cl_3$ and $[Cr(H_2O)_3Cl_3]\cdot 3H_2O$.

   (d) $[Co(NH_3)_5Cl][CoCl_5(NH_3)]$ and $[Co(NH_3)_4Cl][CoCl_4(NH_3)_2]$.

(5)   (a) Geometric isomerism, see Fig. 32.7. Replace Cl with Br. (b) Optical isomerism, see Fig. 32.6. Replace the nitro groups with the third 1,2-diaminoethane ligand. (c) Geometric isomerism, see Fig.32.7. Replace the four ammonias by the two 1,2-diaminoethane ligands. (d) Geometric isomerism, see Fig. 32.7. Replace cobalt by platinum.

(6)   (a) Titrate with silver nitrate. (i) requires approximately twice as much reagent as (ii) for the complete titration of the same mass of compound.

   (b) (i) gives a positive silver nitrate test for halide ions but (ii) will not react.

   (c) (i) will give a positive sulphate ion test with barium chloride but (ii) will not react. (ii) will give a positive silver nitrate test for halide ions but (i) will not react.

(7)   (a) See Fig. 32.8, p.562. A buffer is required to move the equilibrium to the right and to allow the indicator to operate. (b) 6.03%.

Chapter 33

(1)   Alpha; reduces the mass number by 4 and the atomic number by 2. Beta; no effect on mass number but increases the atomic number by 1. Gamma; no effect on either the mass or atomic number.

(2)   (a) $^{235}_{92}U$ (b) $^{14}_{7}N$ (c) $^{60}_{28}Ni$ (d) $^{37}_{17}Cl$.

(3)   Activity; see p.577. Specific activity; see p.580. Carrier; see p.580.

(4)   5.06 kBq.   (5) 2.75 cm^{-3}.   (6) See p.581.

(7)   (a) 2.56MBq. (b) 0.022MeV.   (8) (a) 18.71 cm^{-3}, (b) 7.14 x 10^{-17}g.

(9)   See pages 585 to 586.   (10) (a) 0.95cm^3 (b) 26.3%.

Index